Gioele Zisa

The Loss of Male Sexual Desire in Ancient Mesopotamia

Medical Traditions

Edited by
Alain Touwaide

Volume 5

Gioele Zisa

The Loss of Male Sexual Desire in Ancient Mesopotamia

Nīš Libbi Therapies

With a foreword by Alessandro Lupo

DE GRUYTER

ISBN 978-3-11-127828-5
e-ISBN (PDF) 978-3-11-075726-2
e-ISBN (EPUB) 978-3-11-075733-0
ISSN 2567-6938

Library of Congress Control Number: 2021943167

Bibliographic information published by the Deutsche Nationalbibliothek
The Deutsche Nationalbibliothek lists this publication in the Deutsche Nationalbibliografie; detailed bibliographic data are available on the Internet at http://dnb.dnb.de.

This volume is text- and page-identical with the hardback published in 2021.
Cover image: Collage of illustrations in medical treatises from the 1st to the 16th century CE, from Greece and Rome, to the Arabic World and China.
Typesetting: Integra Software Services Pvt. Ltd.
Printing and binding: CPI books GmbH, Leck

www.degruyter.com

To my wife Kimia

خنک آن دم که نشینیم در ایوان من و تو
به دو نقش و به دو صورت به یکی جان من و تو

Foreword

One of the main problems in analyzing and understanding the written texts surviving from the ancient world (once we can decipher their writing and reconstruct their linguistic meaning), consists in the limited information as to the context and circumstances of their production, the needs and purposes of those who wrote them, their uses, and the characteristics of their possible reading or performance. This is also true for the written documents of recent civilizations erased by colonial subjugation, the members of which were unable to reveal to their colonizers all the complex knowledge concerning them, as happened in the New World with the forms of phonetic writing (among the Maya of South-eastern Mesoamerica) and pictography (among the Nahua and the Mixtecs of South-central Mexico) or non-strictly 'graphic' mnemonic fixation (as with the *khipu* of woven strings of the Incas in the Andean area). The numerous testimonies left by European conquerors, evangelizers, and administrators, are, for the most part, extremely lacking in content, as well as vitiated by prejudices and gross misunderstandings. Therefore, it is easy to imagine how much greater the difficulties are for the ancient Near Eastern civilizations, whose temporal distance from the present has made historical sources extremely fragmentary, so that, in order to reconstruct the context of those texts, we must rely almost exclusively on archaeological evidence, which notoriously offer very scant explicit and comprehensive information about social relations, concepts, values, forms of expression, and emotional experiences of the women and men who lived at that time.

The cuneiform textual corpus analyzed by Gioele Zisa concerning the formulas and prescriptions regarding the therapies for the treatment of the peculiar diagnostic category of the decline of sexual desire (*nīš libbi*) have constituted for about a century (i.e., since the first edition by Erich Ebeling in 1925) a challenge with no easy solution for the scholars who have undertaken their analysis. The merit and originality of this volume consists in its novel approach, which combines in a rigorous and innovative way the method of a careful philological examination of the texts with the application of Medical Anthropology's critical perspective and most up-to-date analytical tools. In this way, it presents the results of a careful re-reading of the rich literature about Mesopotamian therapeutic texts, allowing us to overcome the limitations, conjectures, reductionisms, and anachronisms that have characterized a large part of them, proposing less fanciful and more solidly argued interpretations.

Anthropology is a social science that always pays the utmost attention to the contexts in which the phenomena it deals with occur, and, as mentioned above, contextual information is precisely what is lacking in the case of the Mesopotamian therapeutic texts. However, it is exactly this awareness of having to adapt the conceptual arsenal with which the evidences are examined to the relevant historical and cultural contexts that renders this work by Gioele Zisa particularly effective. He reconstructs with meticulous attention the semantic content of the texts, brings out their possible performative functions and social uses, and outlines the complex

https://doi.org/10.1515/9783110757262-202

ways in which they could be effective. Moreover, it contributes to highlighting the fallacy of those approaches that (relying on the supposed universal applicability of the categories and perspectives of contemporary science, naively conceived as free from cultural and social conditioning) project onto the evidence of ailments that afflicted the ancient Near Eastern populations the nosographic classifications, organic reductionism, Cartesian dualism, and tendency towards reification found within present-day bio-medicine. Without pretending to replace the historical and linguistic disciplines mainly concerned with the study of the textual corpus on Mesopotamian populations' health, sicknesses, and therapies (which Gioele Zisa shows to know and master with considerable confidence), the anthropological approach has the merit of revealing the inevitable distortion that the fact of belonging to a specific historical and cultural context determines in every scholar's perspective, and thus raising our awareness as to how partial and limited the understanding it produces actually is.

I do not wish to deprive the reader of the pleasure of discovering the many innovative contributions that this book offers to the understanding of the therapeutic texts of the Mesopotamian cuneiform tradition, but I do want to highlight some that I consider particularly significant. Relying upon his sound anthropological education, Zisa does not fall into the ethnocentric error of assuming that even in Mesopotamia of the second and first millennia B.C. there was a clear conceptual distinction between the organic dimension of bodily affections and the dimension of mental processes. He therefore knows how to grasp the semantic density of terms such as *libbu*, which, in addition to indicating an internal organ such as the heart, also refers to the moral qualities and cognitive and appetitive faculties of the human being. It follows that even in the interpretation of therapeutic practices aimed at restoring sexual desire, the causes of which are often to be identified, as among the Azande studied almost a century ago by Evans-Pritchard (1937), in divine punishment or magical aggression perpetrated by human agents, it is not possible to reduce the diagnostic category of this ailment to the mere bodily dimension of "erectile dysfunction," as proposed by some scholars. On the contrary, the meticulous comparison of the texts conducted by the author demonstrates as to how this desire is configured as a «human intrinsic quality, decreed by the gods as an anthropogenic act» (ibid. 41), evidently associated with the sexual act, but that it would be simplistic and erroneous to reduce it to this alone, even considering that in some passages the ritual treatment is expressly aimed at «releasing the 'heart' (*libbu*) of the man and the woman» (ibid. 43), proving that the affection is conceived in relational terms including the two individuals involved in the desire, and is not exclusive of the male only.

How far from the individualizing paradigm of biomedicine the conceptions of sickness and therapy in ancient Mesopotamia were (very similar to those found by ethnologists in the holistic ethnomedical traditions of countless populations), is not only demonstrated by the fact that the patient to be cured was not exclusively the single male individual but included his female partner. This is also confirmed by the therapeutic practice based on the combination of the magical principles of analogy and contact, consisting in applying to the navel or to the genitals of both partners a bit of iron and magnetite powder, probably in order to produce between them on the erotic level an attraction similar to the electromagnetic one between the two mineral substances.

Moreover, the many details of the packaging and use of the "pharmaceuticals" on which Zisa focuses, examining their symbolism with anthropological sensitivity, demonstrate the fallacy of trying to rationally "explain" the use of the substances used in the potions and ointments administered to patients, looking for the possible presence of active principles capable of producing chemical reactions in the bodies, attracted by the mirage of an anachronistic capacity for systematic empirical experimentation a few millennia before Galileo. The multiplicity of substances used simultaneously in the compounds would have made it very difficult to identify the active principles that each may have had. Further evidence lies the attention to their exact quantity, mostly linked to the magical power of the number 7 (and its double 14), or in the apotropaic properties attributed to the ingredients, chosen because they were believed to be able to repel magical attacks. This "symbolic" logic obviously does not exclude the possibility that the substances could produce actual chemical reactions in the body, but it is the only one able to explain certain ritual prescriptions, such as the formulas to be recited at the time of the collection of plants, or those to be pronounced at the time of the administration of the potions and ointments. The same argument is valid for the use of such plants in the realization of amulets to be worn in order to ward off negative forces, and even more for the practice of leaving the compounds during the night on the roof of the house, exposing them to the influence of the stars, with the connected prescription to consume the remedy at dawn, so that in doing so the patient was struck by the first sun's rays. This reminds us of how deeply rooted among the ancient Mesopotamian populations the "monist" conception of the cosmos was (Augé 1982), according to which there were deep correspondences between the components of the different spheres of reality, making it necessary to take into account in detail the cyclical nature of astral influences on earthly things, and consequently regulate human actions (including primarily therapeutic acts) according to the variable qualities of temporal units (an assumption that gave rise to the so-called "zodiacal melothesia," the idea that there was a link between planets, zodiacal constellations, and body parts, the notion of which endured several millennia, until the Modern age; see Saxl 1957).

From the aforementioned, and in the light of the reasoned observations of Gioele Zisa, it is clear that the conceptions and practices about the body, its disorders, and the treatments aimed at healing them constituted an integrated whole, and cannot be reduced to a dichotomy between medicine and magic often common among present-day exponents of the "hard" sciences, which are used to contrast their own rational and empirically based "knowledge" with the irrational and baseless "beliefs" of magic: several outstanding theorists of Medical Anthropology, including Byron J.

Good (1994), have already denounced the ethnocentric prejudice and the epistemological fallacy of such a perspective, which gives biomedical knowledge a privileged ontological status (as if it were free from cultural conditionings and capable of objectively reflecting the reality of the facts), and which attributes a substantial lack in credibility, usefulness, and effectiveness to traditional medicines' holistic approach related to individual experience and to the dimension of meaning. Among its many other merits, this volume demonstrates the need to overcome similar prejudices when studying "other" medical traditions such as the Mesopotamian one, and the usefulness of examining (with equal attention and without hierarchies in terms of "truth") the body of knowledge aimed at identifying, classifying, and explaining sickness and restoring health, focusing on the close relationships existing between their different components.

Inevitably, the author dedicates considerable attention to the many implications of the use of the textual corpus in the treatment of the restoration of sexual desire. Their extraordinary formal continuity through a very long period of time demonstrates how (like so many ritual texts in the most diverse cultures) what determines their appreciation and repeated use is their ability to performatively establish the achievement of objectives, while the adherence to the context of the texts' propositional content appears much less relevant. In fact, they do not focus so much on the description of the idiosyncratic illness experienced by individuals, but they stubbornly repeat the conceptual models concerning sickness that have been codified by a long and consolidated tradition. Using the extensive theoretical production of Anthropology on the topic of linguistic acts (Austin 1962; Searle 1969; Tambiah 1968; Tambiah 1985), Zisa is able to effectively highlight the illocutionary power of healing formulas, the ability of which to produce the effects hoped for by the speaker (their "happiness" in terms of Austin 1962) depends on the different circumstances of the locution, and not on the linguistic meaning of what is uttered. Zisa convincingly demonstrates how the therapeutic formulas examined in this volume owe their complex efficacy to their formulaic style, to their metonymic and metaphoric capacity to create relationships of continuity between the patient's state and the intrinsic properties of plants, animals, and minerals employed in the "pharmaceuticals" and in the paraphernalia of the ritual action, which are intended to be 'activated' in order to heal the patient. The same is true for the ritual prescriptions regarding the most appropriate moment to pronounce them and their combination with appropriate ritual acts. Particularly noteworthy (and potentially unnoted until present) is the power that the very uttering of such words would have to directly influence reality. Sometimes this occurs through associations made by resorting to «semantic and phonetic wordplay» (ibid. 119), or by means of assonances, as in the case of the juxtaposition between the act of tying (*kīsu*, referring to tendons) and the merchant's bag (*kīsu*), or between the wind (*šāru*) and the penis (*ušāru*): it is clear that at the basis of similar phenomena lies the idea that the very sound of words reflects underlying correspondences to what they designate, and that, therefore, one believes to be able to act on reality through the manipulation of terms. This is even more evident in the illocutionary use of sounds and formulas apparently devoid of intelligible meaning, the recurrent recourse to which obviously disregards any intent to communicate a semantic content and focuses on the ability attributed to the uttered words to "act" on reality. In these cases, the rhythmic repetition of the meaningless sounds or their alternation with intelligible terms come into play, in a sort of syncopated verbal stimulation, which Zisa supposes could stimulate the projective capacities in the mind of the patient who had listened to the recitation of the formulas, according to the suggestive analytical proposal of the "symbolic efficacy" of the shamanic Kuna chant recently proposed by Carlo Severi (2004), radically reformulating the classic interpretation of 1949 by Claude Lévi-Strauss (1949b).

This last example of one of the less intuitively comprehensible aspects of the Mesopotamian therapeutic texts (on which the author advances suggestive interpretations thanks to the recourse to the most up-to-date theoretical proposals of Anthropology) demonstrates how fruitful a combination of this discipline with philology can be. With respect to formulas for arousing sexual desire, the crucial question is whether and how these could work, achieving the desired effects. The very persistence through the centuries of the use of such formulas seems to testify to what extent they were evidently considered capable of remedying the problem. Yet, the absence of appropriate conceptual tools can lead analysts to formulate fanciful hypotheses, to say the least, such as that of Robert Biggs who (convinced that the sickness consisted in a mere erectile dysfunction) supposes that mentioning certain animals, then considered emblems of sexual vigor, could produce in the patient a sexual excitement similar to that pursued by current pornography: «Surely [sic!] some of mental images provided by the incantations can be compared to the stimulation modern couples may receive from pornographic films, now often viewable in American motels and hostels» (Biggs 2002: 72). In a much more convincing way, Gioele Zisa succeeds in outlining the complex mechanisms through which the therapeutic efficacy of human acts and words unfolds, allowing us to trace also in these remote cuneiform texts the ability to solicit endogenous reactions in the organism, satisfying the need that every suffering or impaired human being feels to give meaning to his or her own condition of sickness, to identify credible causes and to configure initiatives aimed at contrasting and overcoming it. As for all societies of the past (the resources of which in terms of pharmacological, surgical and rehabilitative tools with which to effectively intervene on the bodies of the patients were incomparably more meager and inadequate than those of present-day biomedicine) the ability to concretely provide some remedy to the afflictions of its members focused above all on the capacity to establish a solid therapeutic relationship, and to provide a coherent framework of meaning within which to situate the sickness episodes and related therapies, rendering the patient and his closest social entourage active participants in the treatment, extricating them from the cognitive disorientation, the feeling of helplessness, and passive resignation in which the pathology usually throws its victims. When one is able to reconstruct with philological rigor and ethnographic precision the historical and cultural context in which certain therapeutic practices took place (much as the author of this volume has clearly succeeded in doing), it is no longer impossible to understand how words and actions such as those contained in the repertoire of *nīš libbi* therapies could be even partially effective for the populations employing them for many centuries.

Alessandro Lupo
Sapienza Università di Roma

Acknowledgments

I would like to remind all those who have helped me during the writing of this book with suggestions, criticisms, and remarks: my gratitude goes to them, any mistakes, of course, remain my responsibility. First, I would like to thank my supervisor Prof. Walther Sallaberger (LMU München) for his support and his very knowledgeable and patient guidance. From him, I learned German discipline and philological rigor: the sincerest thanks from a student almost inclined to heresy.

I thank Prof. Lorenzo Verderame (Sapienza Università di Roma), my first guide, with him I took the first steps in Assyriology and from him I learned the value of interdisciplinarity. I thank Prof. Alessandro Lupo (Sapienza Università di Roma) for teaching me to always have a critical gaze. His lessons on Medical Anthropology were for me inspirational. Acknowledgment goes also to Prof. Susanne Gödde (FU Berlin), a constant reference point for dialogue between ancient studies and the social sciences.

My sincere thanks to Dr. Érica Couto-Ferreira (Heidelberg Universität) for having shared with me her unpublished works, her ideas, and knowledge, I owe to her my passion for Mesopotamian therapies; to Dr. Claudia Lo Piccolo (Università di Palermo) who opened up to me the world of Queer Studies; to Dr. Agnès Garcia-Ventura (Universitat de Barcelona) for past and future collaborations.

Special thanks go to my colleagues and friends at the Institut für Assyriologie und Hethitologie (LMU München) who have encouraged me or spent some of their time reading and discussing drafts of my work. In particular, I would like to thank Prof. Jared Miller, Dr. Anne Löhnert and Dr. Claus Ambos, their suggestions and especially criticisms have been of great help. I also thank Dr. Paola Paoletti, Dr. Chiara Cognetti, and Dr. Nathan Morello for helping me not only scientifically but also humanly. I thank Dr. Zsombor Földi for photographing some tablets at the British Museum at my request. Thanks also to Prof. Jennifer Finn (Marquette University), who always encouraged me in Germany and now from the United States.

I would like to thank Dr. Isabel Grimm-Stadelmann (LMU München) for the collaboration with the Institut für Ethik, Geschichte und Theorie der Medizin; thanks to the staff at the British Museum for making the tablets available for research.

I was able to complete the Ph.D. thanks to the scholarship of the Doctoral Program "Distant Worlds: Munich Graduate School for Ancient Studies." I would like to thank the Director of the School Prof. Martin Hose, the postdoctoral fellows, the other doctoral fellows, and the academic coordination members. Distant Worlds has been a rich training experience. Special thanks to members of the research group "Construction of norms" with whom I shared readings, ideas, and theories. In particular, I thank two coordinators of the group: Dr. Aaron Tugendhaft (Bard College Berlin), and Dr. Paolo Visigalli (NYU Shanghai), who always meticulously read my writings and provided valuable suggestions. Thanks also to Dr. Willis Monroe (University of British Columbia) for making my English readable.

https://doi.org/10.1515/9783110757262-203

Ringrazio di cuore i miei genitori e mio fratello per essere stati sempre al mio fianco nonostante la lontananza e per aver sempre incoraggiato la mia passione per questa esotica e criptica disciplina. Non smetterò mai di ringraziarli.

Finally, a special thanks to my wife, Kimia Kamyab, because she was always close to me, sending me energy when I needed it and sacrificing, so far, her passions to build a life together. This volume is dedicated to her.

Contents

Foreword —— VII

Acknowledgments —— XIII

Introduction —— XIX

Bibliographical abbreviations —— XXI

Editorial abbreviations and sigla —— XXIII

Part I: **Interpretative analysis of the corpus**

Chapter I
***Nīš libbi* therapies and theoretical and methodological introduction —— 3**
The sources —— **3**
  Catalogues and tablet inventories —— **6**
Forschungsgeschichte of *nīš libbi* therapies and the 'reflective' approach in anthropology and gender studies —— **8**
    Forschungsgeschichte —— **8**
    Criticisms of the biomedical approach —— **12**
  Theoretical and methodological introduction —— **19**
    Illness, disease, sickness —— **19**
    Medical system as cultural system —— **21**
    The body —— **24**
    And the mind? —— **29**
    Logical thinking —— **31**
    Sexuality —— **34**
  Meaning of *nīš libbi* —— **37**
    From *libbu inaššīšu* to *nīš libbi* —— **37**
    Sexual desire or potency? —— **42**
    Sexual desire: *libbu* between physicality and psychic faculties —— **52**
  The cause of sickness —— **54**
  Symptoms —— **59**
    Nīš libbi texts —— **59**
      Fear, distress, and insomnia —— **66**
    Other Texts —— **73**
  Couples therapy —— **78**
  Who does what? —— **82**
  Authority and therapeutic efficacy —— **84**

Chapter II
***Nīš libbi* incantations — 87**
Incantation, poetry, and text function — 87
First group: animal metonymies and performer of suffering — 93
Animal metonymies — 93
The agent of suffering — 108
Unknown agent — 108
Similes concerning the action of the witch — 111
The binding — 113
Lengthening and slackening of tendons — 117
Blocking the street and the canal — 121
Second group: erotic similes with animals — 123
Invocation of animal arousal — 123
Animals tied to the bed — 124
Bestiality or figure of speech? — 127
Ascending climax relative to the mating of the animal species — 129
"I am young!" — 130
Mention of gods and operations of the therapeutic operator — 135
The formulas "*ina qibīt* DN" and "*šiptu ul yuttun šipat* DN" — 135
"I am the daughter of Ningirsu" — 142
Third group: sexuality and nature — 146
Fourth group: *historiola* and its function — 150
Fifth group: abracadabra — 153

Chapter III
Rituals and prescriptions — 157
The pharmaceutical — 157
The ingredients — 159
Plant ingredients — 160
Animal-based ingredients — 165
Potions — 173
The astral influence — 176
Ointments and amulets — 180
Libations to Ištar — 186
Etiological analysis — 187
The figurines — 188
The pig — 191
Bow ritual and battle metaphors — 195

Conclusions
Therapeutic efficacy and analogical thinking — 201
Therapeutic efficacy — **201**
Patient and recipients of therapy — **202**
Etiology, symptoms, and ingredients — **203**
Animals, metaphors, and substances — **204**
Linguistic analogy — **205**
Poeticality of incantations — **206**
Abracadabra and *historiolae* — **206**
Normativity: animality and battle metaphors — **207**
Returning to effectiveness — **208**

Part II: **Edition**

***Nīš libbi* catalogue LKA 94 — 217**

I Texts from Aššur (with duplicates from other sites) — 227
Nīš libbi A — **227**
Nīš libbi B — **257**
Nīš libbi C — **272**
Nīš libbi D — **277**
Nīš libbi E — **301**
Nīš libbi F — **333**
Nīš libbi G — **356**

II Texts from Nineveh (with duplicates from other sites) — 361
Nīš libbi H — **361**
Nīš libbi I — **368**
Nīš libbi J — **373**

III Text from Sultantepe (with duplicates from other sites) — 381
Nīš libbi K — **381**

IV Texts from Uruk (with duplicates from other sites) — 417
Nīš libbi L — **417**
Nīš libbi M — **426**

V Texts from Boghazköy — 436
Nīš libbi N — **436**
Nīš libbi O — **458**

VI Ritual fragments from Nineveh and Aššur — 465
Nīš libbi P — **465**
Nīš libbi Q — **473**
Nīš libbi R — **477**
Nīš libbi S — **480**

VII Ritual fragments from Sippar and Uruk — 485
Nīš libbi T — **485**
Nīš libbi U — **491**

VIII Ritual fragments from Boghazköy — 492
Nīš libbi V — **492**
Nīš libbi W — **496**
Nīš libbi X — **501**
Nīš libbi Y — **505**

List of mineral and botanical ingredients — 509

Index of ingredients — 523

Concordances — 531

Bibliography — 537

Index of Akkadian terms and expressions discussed — 583

Index — 585

Introduction

Nearly fifty years have passed since Robert Biggs published his edition of the šà-zi-ga incantations and rituals. šà-zi-ga in Sumerian, *nīš libbi* in Akkadian, lit. "raising of the šà/*libbu*," is the term used to indicate a group of texts intended to recover the male sexual desire. This term is used widely in ancient Mesopotamia; except for a single Old Babylonian Sumerian prescription, the material is best preserved from the Middle Babylonian period onwards. The tradition of this corpus continues in the first millennium: some texts come from the cities of Nineveh and Aššur, others from Uruk during the Neo Babylonian period and others from Sippar in the Late Babylonian and Achaemenid periods. This broad range testifies to the importance of the transmission of this material throughout Mesopotamian history.

In the fifty years since the edition by Biggs (1967), research on ancient Mesopotamian therapeutics has progressed considerably: numerous editions of medical texts, as well as monographs on specific topics, have appeared. This book, a revised version of my Ph.D. dissertation *Nīš libbi Therapies. The Loss of Male Sexual Desire in Ancient Mesopotamia* discussed at Ludwig-Maximilians-Universität Munich in February 2018 under the supervision of Prof. Walther Sallaberger, therefore aims to provide an analysis of the *nīš libbi* therapies in the light of new knowledge of ancient Near Eastern medicine.

Research on medicine and magic in Mesopotamia has been enriched in recent years by theoretical contributions from the social sciences. The draw of interdisciplinary dialogue and the need for theoretical tools have led some Assyriologists to turn to the anthropological disciplines. This book aims, too, to show how theories and methodologies of Cultural and Social Anthropology are useful for understanding the *nīš libbi* therapies.

This book, therefore, is an attempt to analyze Mesopotamian therapeutic sources both from a philological and an ethnological point of view. The choice of this dual perspective is determined by the author's background. My university education provided a grounding in Assyriology, and also in Folklore Studies and Anthropology. Fieldwork conducted during my university studies, including on magic and traditional care systems in Sicily, greatly contributed towards my understanding of the ancient Near Eastern textual sources. Inevitably, this book, though dealing with a different geographical and historical context, is influenced by questions that have been accompanying me during my years of anthropological education and research.

Studies in Mesopotamian medicine have hardly neglected the debates in social sciences, the work of Hector Avalos, Barbara Böck, M. Érica Couto-Ferreira, Mark J. Geller, Niels P. Heeßel, Ulrike Steinert, to name a few, serving as excellent examples of such cross-disciplinary research. However, this is the first attempt – to my knowledge – to make extensive use of an anthropological perspective. In recent decades, studies of Medical Anthropology and Ethno-Psychiatry have had a great impact on public debates on social, economic, and political issues, and also in other disciplines.

https://doi.org/10.1515/9783110757262-205

Medical Anthropology and Ethno-Psychiatry not only have the aim of studying the construction and production of sickness, the practices of "traditional" care systems, but have also become *critical* and *dialogical* disciplines in which the topics of health, sickness, body, and suffering are being investigated in the light of intriguing relationships with socio-cultural processes, epistemological issues, and economic and political relations. Assyriological studies should not ignore the theoretical-methodological debates in these two disciplines that have begun to rethink, thanks to the research conducted in our contemporary context, Western epistemological categories and dichotomies, such as "mind *vs.* body," "magic *vs.* religion," "medicine *vs.* pharmaceutical," "normal *vs.* pathological," "curing *vs.* healing," "rational *vs.* irrational." I refer in particular to the so-called "Critical-interpretative Medical Anthropology" (Nancy Scheper-Hughes, Margaret Lock, Allan Young) and to an Italian and French Ethno-Psychiatry, which integrates the theoretical perspectives of Michel Foucault, Franco Basaglia and Antonio Gramsci, with contributions from the field of Postcolonial Studies (Frantz Fanon, George Devereux, Andras Zempléni, Tobie Nathan, Ernesto De Martino, Roberto Beneduce, Gananath Obeyesekere).

Returning to the subject of the material under investigation, loss of sexual desire inevitably entails a crisis of some aspects of manhood. Gender issues are interconnected with those of therapy. For this reason, I have stressed the importance of contributions from Gender Studies, Queer Studies, and Post-feminist Theories in my research. In the last decades, whereas studies on the construction of manhood have appeared in many disciplines, works on this subject are still scarce in Assyriology. The purpose of this book is to contribute to the debate on manhood in the ancient Near East in addition to trying to answer the question posed by Biggs: "Do the texts work?". The results of the research are shown considering the medical anthropological debate on therapeutic efficacy.

At the end of this book, standing in front of the work now finished and the work still yet to start, I feel like E.E. Evans-Pritchard after his anthropological monograph on the Nuer population: «We feel like an explorer in the desert whose supplies have run short. He sees vast stretches of country before him and perceives how he would try to traverse them; but he must return and console himself with the hope that perhaps the little knowledge he has gained will enable another to make a more successful journey» (1940: 266). I, therefore, hope that this book provides important information on Mesopotamian care systems and that it contributes to enrich the controversial debate on the study of medicine in the ancient Near East, opening it to the anthropological disciplines as well.

Bibliographical abbreviations

AAA	*Annals of Archaeology and Anthropology*
AfO	*Archiv für Orientforschung*
AHw.	Wolfram von Soden, *Akkadisches Handwörterbuch*, 3 vols. Wiesbaden: Harrassowitz, 1965–1981
AMT	Reginald Campbell Thompson, *Assyrian Medical Texts from the Originals of the British Museum*. London: Milford, 1923
ARM	*Archives Royales de Mari*
Bab	*Babyloniaca. Études de philologie assyro-babylonienne*
BAM	*Die babylonisch-assyrische Medizin in Texten und Untersuchungen*
BID	Walter Farber, *Beschwörungsrituale an Ištar und Dumuzi. Attī Ištar ša harmaša Dumuzi*. Wiesbaden: Franz Steiner, 1977
BMS	Leonard W. King, *Babylonian Magic and Sorcery. The Prayers of the Lifting of the Hand*. London: Luzac and Co., 1896
CAD	*The Assyrian Dictionary of the Oriental Institute of the University of Chicago*
CCMAwR	*Corpus of Mesopotamian Anti-witchcraft Rituals*, 3 vols (Ancient Magic and Divination 8.1/3). Leiden, Boston: Brill (= Abusch and Schwemer 2011; Abusch and Schwemer 2016; Abusch et al. 2020)
CDA	*Concise Dictionary of Akkadian*
CSB	Culture-Bound Syndromes
DI	Dumuzi and Inana cycle
DSM	*Diagnostic and Statistical Manual of Mental Disorders*
GAG	Wolfram von Soden, *Grundriss der akkadischen Grammatik*. Rome: Biblical Institute Press, 1995
KADP	Franz Köcher, *Keilschrifttexte zur assyrisch-babylonischen Drogen- und Pflanzenkunde. Texte der Serien uru.an.na: maltakal, HAR.ra: hubullu und Ú GAR-sú*. Berlin: Akademie-Verlag, 1995
KAL	*Keilschrifttexte aus Assur literarischen Inhalts*
KAL 2	Daniel Schwemer, *Rituale und Beschwörungen gegen Schadenzauber*. Wiesbaden: Harrassowitz, 2007
KAR	Erich Ebeling, *Keilschrifttexte aus Assur religiösen Inhalts*. Leipzig: Hinrichs, 1915–1923
KBo	*Keilschrifttexte aus Boghazköi*
KUB	*Keilschrifturkunden aus Boghazköi*
LAS	*Letters from Assyrian Scholars*
LKA	Erich Ebeling, *Literarische Keilschrifttexte aus Assur*. Berlin: Akademie-Verlag, 1953
Ludlul	*Ludlul bēl nēmeqi*
MAOG	*Mitteilungen der Altorientalischen Gesellschaft*
MARI	*Mari. Annales de recherches interdisciplinaires*
MSL	*Materialen zum sumerischen Lexikon*
RlA	*Reallexicon der Assyriologie und vorderasiatischen Archäologie*
RIMB	*The Royal Inscriptions of Mesopotamia, Babylonian periods*
SEAL	*Sources of Early Akkadian Literature*
SpTU	*Spätbabylonische Texte aus Uruk*
STT	*The Sultantepe Tablets, I-II*
TCS	*Texts from Cuneiform Sources*
VAT	*Vorderasiatische Abteilung Tontafel: Siglum of Tablets in the Vorderasiatisches Museum* (Berlin)

https://doi.org/10.1515/9783110757262-206

Editorial abbreviations and sigla

ak-ka-nu	texts in italics are Akkadian
ANŠE.KUR.RA	texts in small capitals are Sumerograms
š à - z i - g a	texts in expanded character spacing are Sumerian
maš taq ti	signs with uncertain reading in Akkadian in italics lowercase separately
SIKIL BAR NUN	signs with uncertain reading in capitals lowercase separately
ša$^{!}$	emended, but certain reading (against unidentifiable or irregular sign on tablet)
ša$^{!}$(text MA)	emended, but certain reading (against identifiable sign on tablet)
ša$^{!?}$	emended, but uncertain reading (against unidentifiable or irregular sign on tablet)
ša$^{!?}$(text MA)	emended, but uncertain reading (against identifiable sign on tablet)
ša$^{?}$	uncertain reading of a single sign
(?)	uncertain reading/restoration of sign/word/phrase
x	undecipherable damaged sign
:	cuneiform division mark
{…}	erasure
[ ]	reconstructed text
[…]	break with an uncertain number of signs
[x]	indicates space available in break
⸢ ⸣	partially broken sign
{ }	sign(s) to be deleted from the text
⟨ ⟩	sign(s) to be added to the text
^{ḫepi}[[ ]]	possible restoration from an original text
…	untranscribed, untranslated sign(s), word(s) or passage(s) of text
—	indicates a missing word in a given manuscript in relation to a duplicate edited in the same score
/	end of a line if two or more lines in a given manuscript are edited on one line
+	fragments join directly
// \\	gloss
§	text section
Ass.-Mitt.	Assyro-Mittanian
col.	column
dupl.	duplicate
e.	edge
fem.	feminine
fn.	footnote
frg.	fragment; fragmentary
l., ll.	line(s)
LB	Late Babylonian
le.	left
lit.	literally
lo.	lower
masc.	masculine
MA	Middle Assyrian
MB	Middle Babylonian
Ms.	manuscript
n., No.	numeral
NA	Neo-Assyrian

https://doi.org/10.1515/9783110757262-207

NB	Neo-Babylonian
NN	*Nomen nescio*
o.; obv.	obverse
OB	Old Babylonian
or.	original
pl.	plural
r.; rev.	reverse
s.v.	sub voce
sg.	singular
u.	upper
unpubl.	unpublished

Part I: **Interpretative analysis of the corpus**

Chapter I
Nīš libbi therapies and theoretical and methodological introduction

The sources

The expression šà-zi-ga in Sumerian, *nīš libbi* in Akkadian, refers to a group of texts, rituals, and incantations, which were intended to recover one's "sexual desire." This group of texts is composed of incantations (ÉN/*šiptu*), often followed by small prescriptions or rituals (DÙ.DÙ.BI/*dudubû*), and long sequences of prescriptions.

Only one text dated to the Old Babylonian period is known. It is a Sumerian fragmentary text (UM 29-12-717) from Nippur[1] (University Museum in Philadelphia), published by Peterson (2008), containing only a prescription with several ingredients.

(lacuna)	
1′. $^{⌜\text{ĝiš}⌝}$[g]an-[na[2] . . .]	1′. [pes]tl[e . . .]
2′. ^{ĝiš}ĝišimmar[3] tur-⌜tur⌝	2′. young date palm,
3′[4]. ^{ú}ZI:ZI.LAGAB bu-ra	3′. uprooted ZI:ZI.LAGAB-rushes,
4′. kù-si$_{22}$ kù-⌜babbar⌝	4′. gold, silver,
5′. an-bar $^{\text{na4}}$nir$_7$	5′. iron, nir$_7$-stone.
6′. $^{\text{na4}}$gug $^{\text{na4}}$za-gìn	6′. carnelian, lapis lazuli
7′. ì ĝiš-ì saĝ	7′. high quality sesame oil
Reverse	
8′. ì ⌜x x ì⌝ ^{ĝiš}⌜eren⌝	8′. oil . . . cedar oil,
9′. ì-⌜bi⌝$^{?}$ 1[5] ⌜ù⌝$^{?}$-[šid$^{?}$]	9′. its oil in one, he [recites$^{?}$],[6]
10′. ⌜làl⌝ ì-[šeš$_{2/4}$/sub$^{?}$]	10′. he [anoints$^{?}$] (with) syrup.
———	
11′. šà-zi-ga [x] ⌜x⌝ [. . .] ⌜x⌝	11′. Sexual desire . . . [. . .] . . .

Unfortunately, because of its fragmentary condition, it is impossible to understand the phases of preparation of the ingredients. One can suppose that the prescription pro-

1 The text uses an archaizing script; it is perhaps a copy of an earlier manuscript from the Ur III period (see Peterson 2008: 195).

2 The restoration is proposed by Peterson 2008: 198. If this is correct, the ingredients are crushed perhaps for the creation of an ointment. This practice often appears in the later *nīš libbi* prescriptions.

3 The date palm appears once in *nīš libbi* corpus, in N prescr. 14 iii 13 together with other ingredients, and in the catalogue LKA 94 i 18. It is not a specific ingredient against the absence of sexual desire.

4 For the lines, in particular for the meaning of this type of rushes (ZI:ZI.LAGAB) and of the verb bur see Peterson's commentary (2008: 199).

5 The DIŠ sign can be interpreted as the number "1" used adverbially or as an orthography of the lexeme teš. It indicates, in some Ur III incantations from Nippur, the blending of various substances into a mixture (see Peterson 2008: 195).

6 Or "he [recites] once over$^{?}$ its oil."

https://doi.org/10.1515/9783110757262-001

vides a mixture of mineral and vegetable substances with high-quality oil (ll. 2′–7′), to be applied, together with syrup, to the patient's body as an unguent (l. 10′). One must note that many of these ingredients are used, as *materia medica*, in the later *nīš libbi* documentation, such as an-bar 'iron,' kù-si$_{22}$ 'gold,' kù-babbar 'silver,' and za-gìn 'lapis lazuli.' They are usually employed in the creation of amulets or phylacteries, or together with oil for ointment.

The *nīš libbi* documentation becomes considerably greater starting in the Middle Babylonian period. Several *nīš libbi* rituals were found in Ḫattuša (present-day Boghazköy, Turkey), the capital of the Hittite empire.[7] The tradition of this corpus of texts continues in the first millennium.

Some Neo-Assyrian texts come from the city of Nineveh,[8] in particular from the library of Ashurbanipal. Others were found in Aššur,[9] in particular in the house of the *āšipu* Kiṣir-Aššur (Library N 4).[10] Other texts were found at the Neo-Assyrian site of Sultantepe.[11] Several rituals and incantations of this corpus are from Neo-Babylonian period Uruk.[12] Two texts come from Sippar, one from the Late Babylonian period (BM

---

7 KUB 4, 48; KUB 37, 80; KUB 37, 81+AAA 3 pl. 27 No. 5; KUB 37, 82; KUB 37, 201; KBo. 36, 27. On the Babylonian magic rituals and incantations found in Ḫattuša see Schwemer 2004: 75–79; 2013: 153–158. It is difficult to affirm who wrote these Akkadian tablets. One of the criteria to determine the origins of the scribe who wrote the tablets is the script and the orthography of the manuscripts. The scripts diverge: some tablets have a 'non-Hittite' script (KUB 4, 48; KBo 36, 27), that is a script different from the New Hittite, but which cannot be assigned to a specific non-Anatolian script; one has a New Hittite script (KUB 37, 80); one has probably an Assyro-Mittanian script (KUB 37, 81 + AAA 3 No. 5); two have a Middle Assyrian script (KUB 37, 82; KUB 37, 201). Schwemer (2013: 158), regarding the presence of Babylonian rituals and incantations in Ḫattuša, supposes that foreign scholars, coming from Babylonia and Assyria, were at the Hittite court during the XIII century. They brought some tablets with them when they came to Anatolia, where they continued to write other texts. The tablets with New Hittite script show that the Hittite scribes were also taught by the foreign ones. It is possible that foreign experts were called to the Hittite court because the Babylonian therapeutic knowledge was considered prestigious and particularly efficient.

8 81-7-27, 73 (TCS 2, pl. 2); AMT 31,4 and 32,1; AMT 73, 2; AMT 65, 7; K. 2499 (TCS 2, pl. 1); K. 9415+10791 (TCS 2, pl. 2 + CT 13, 31 + Thompson 1930 pl. 17); K. 9451+11676+Sm. 818+961 (TCS 2, pl. 1); K. 8698 (TCS 2, pl. 3); K. 8907; K. 10002 (TCS 2, pl. 2); K. 11076 (TCS 2, pl. 3). K. 9451+ (fragment Sm. 818 rev. 7′–9′) includes part of an Ashurbanipal colophon similar to Hunger 1968: No. 318.

9 BAM 205; BAM 207 (BID, pl. 24); BAM 272; BAM 311 = KAR 186; BAM 318 (CCMAwR 2, pls. 53–60); BAM 319; BAM 320; BAM 369 (CCMAwR 3, pl. 1); LKA 94; LKA 95; LKA 96; LKA 97; LKA 98; LKA 99b; LKA 99c; LKA 99d (CCMAwR 3, pls. 4–5); LKA 100; LKA 101; LKA 102; LKA 103; LKA 144; K. 5901 (TCS 2, pl. 3); K. 9036 (TCS 2, pl. 1); KAR 70; KAR 236; KAR 243; WVDOG 147 (= KAL 7) No. 22.

10 Among the tablets from Aššur, LKA 96 (rev. 16′–17′, restored), LKA 100 (rev. 7–11) and BAM 369 (ll. 10′–12′), were written by the exorcist Kiṣir-Nabû (see Hunger 1968, No. 214; Abusch et al. 2020: 32, see commentary). LKA 102 (u. e. 1–2) was written by Aššur-šākin-šumi (Hunger 1968, No. 267, see commentary), while KAR 236 (rev. 24–26) by a young apprentice Issar-tarība, son of Marduk-šallim-aḫḫē (*šamallû ṣeḫru*, Hunger 1968: 238; Abusch et al. 2020: 44, see commentary). KAR 70 (rev. 35) is a copy of a Babylonian original (Hunger 1968, No. 277).

11 STT 95+295; STT 280.

12 SpTU 1, 9; SpTU 1, 10; SpTU 1, 20; SpTU 4, 135. SpTU 4, 135, found in Ue XVIII, layer II, probably belongs to Iqīšāya (see Frahm 2011: 128).

46911 = 81-8-30, 377 = TCS 2, pl. 3) and another from the Achaemenid period (before Xerxes) (81-7-1, 270+F 224, Finkel 2000: 160–161).

The *nīš libbi* tradition is documented over an extended timeframe, from the Old Babylonian to the Achaemenid period. However, most texts are dated to the Neo-Assyrian and Neo-Babylonian periods.[13]

It must be noted that many tablets from the late periods have literal duplicates from the Middle Babylonian period. This is proof of a strong tradition of these texts. Biggs (1967: 5) supposes that the traditional corpus of texts was fixed during the Kassite period (fourteenth century), but he does not exclude that the originals were from the Old Babylonian period but have not yet been recovered. According to Biggs's suggestion, the Middle Babylonian texts could be copies of older Akkadian texts from the Old Babylonian period (Biggs 1967: 6; 2003–2005: 604). This hypothesis is supported by the discovery of the Old Babylonian text, although in Sumerian, from Nippur.

Despite this long tradition, there was no standard edition of the *nīš libbi* incantations and rituals, organized into a series of numbered tablets. A text, KUB 4, 48, however, informs us that it is one/first tablet of the series "if a man's *nīš libbi*": DUB 1.KAM DIŠ LÚ ŠÀ.ZI.GA "One/first tablet of 'If a man's *nīš libbi*'" (*nīš libbi* N prescr. 25 lo. e. 5). This could refer, however, to a sequence established by the Hittite scribes at Ḫattuša.

---

13 Biggs (1967: 61) considers KUB 37 89 as a *nīš libbi* text, however it is a ritual against witchcraft (see Abusch et al. 2019: 321–322; text F.1).

The small fragment from a large library tablet found at Boghazköy, KUB 37, 7 (Chalendar 2018: 48), could be considered a *nīš libbi* text because of the precence of the *rikibtu* of the *arkabu*-bat, which is an ingredient used only in this therapeutic corpus. However, it more probable that the text deals with prescriptions against witchcraft, not necessarily referring to the loss of sexual desire. Köcher (KUB 37: ii) says that KUB 37, 2–9 belonged to the same tablet, but this remains uncertain (note in KUB 37, 8: 1′ [A].RI.A A.ZA.LU.[LU] "man's sperm") (see KUB 37, 3 Abusch and Schwemer 2016: 6–7, No. 3.2; KUB 37, 4 in ibid. 8–10, No. 3.3; KUB 37, 9, Abusch and Schwemer 2011: 67–82, No. 2.2). See Chalendar 2018: 48 fn. 205. Chalendar (2018: 34 fn. 102) suggest that also BAM 306 (esp. ll. 3′–6′) could be a *nīš libbi* text because of the mention of the human and bull's sperm: DIŠ KI.MIN A.RI.A NAM.LÚ.U$_{18}$.[LU . . .] / A.RI.A [. . .] A.RI.A GU$_4$ [. . .] / [A.R]I.A ŠAḪ [x] ÚŠ LÚ *šá ina* x x x. [. . .] / [. . .] LÚ GIG BI GU$_7$-*šú* NAG-*šú-m[a*? . . .] "If ditto: Huma[n] sperm [. . .] / sperm [of. . .], sperm of a bull [. . .] / [sper]m of a pig [. . .], blood of a man who . . .. / [. . .] that sick man, you make him eat, you make him drink and [. . .]." However, neither human nor animal sperm is used as an ingredient in this therapeutic corpus. These ingredients could be used in other therapeutic texts for other sexual dysfunctions, not necessarily related to the loss of sexual desire.

The text from Babylon VAT 13226, although it has many characteristics of *nīš libbi* texts (see the incipit *pāširu pāširu pāširu* "Releaser! Releaser! Release!" ll. 2–3; the mention of *mekku*-stone l. 13), cannot be classified as an incantation used to get the man's sexual desire. In fact, it is explicitly focused on getting a woman to have sexual intercourse with the man, acting magically on the woman's sexual desire (Zomer 2019: 236–237, 277–279).

Catalogues and tablet inventories

The *nīš libbi* texts seem to be part of the repertoire of the *āšipūtu*. This could be confirmed by the text *Manual of the Exorcist*, which lists the texts and the purview of their therapy: "'All evil' (ḪUL *kalâ*), *Maqlû*, *Šurpu*, 'to make a bad dream favourable' (MAŠ.GI$_6$ ḪUL SIG$_5$.GA) and *nīš libbi*" (l. 14, CJean 2006: 66; Geller 2018: 298). However, the *Aššur Medical Catalogue*, which according to Steinert (2018a; 2018b) is linked to *asûtu*,[14] in part II section XIX: 99–102 deals with the loss of sexual desire:[15]

> [DIŠ NA *ana* MUNUS BAR-*ti* ŠÀ-*šú* ÍL-*šú-ma*$^{?}$ *ana* MUNU]S-*šú* ŠÀ-*šú* NU ÍL-*šú*: DIŠ NA X [. . .]
> [. . .] X SAG MUŠEN DIŠ ⸢Ú⸣ ⸢ŠÀ⸣.ZI.G[A] ⸢*ana*⸣ [GÚ-*š*]*ú* GAR
> ___
> [NIGIN X DUB^{meš} DIŠ NA *ana* MUNUS BAR-*ti* ŠÀ-*šú* Í]L-⸢*šú*⸣-*ma*$^{?}$ *ana* MUNUS-*šú* ŠÀ-*šú* N[U ÍL]-*šú*[16]
> [EN$^{?}$ . . . *ana* NI]TA ZI-*tú* *š*[*ur-ši*]- ⸢*i*⸣

---

14 For a discussion of the text see Steinert (2018a; 2018b), according to whom, the *Aššur Medical Catalogue* is linked to *asû* professionals, since it was copied by a young *asû* and invokes Gula, the patron goddess of *asûtu*, in the colophon. However, as the scholar affirms, «it is very difficult to identify components or texts that pertain exclusively to *asûtu* or *āšipūtu*. It may be that both professions employed similar texts and methods for treating problems involving sexuality and libido, although it is conspicuous that Gula does not feature in this text group at all. This points to strong *āšipūtu* components in the AMC chapter SEX» (Steinert 2018a: 100).

15 Note that in the catalogue the lines 99–102 concern the absence of sexual desire, while the lines 103–108 other problems related to sex. It must be said, however, that in the section dedicated to "sex" we also find elements that are perfectly suited to the cure of the loss of sexual desire, in particular the following lines: 103. [ÉN *li-lik* IM *l*]*a i-na-áš-šá-a* ^{giš}⸢KIRI$_6$⸣^{meš} "[Incantation: 'May the wind blow], may the gardens not quake'"; 104. [1 DUB$^{?}$ ÉN *lil-lik* I]M *la i-na-áš-šá-a* ^{giš}KIRI$_6$^{meš} "[One tablet (of the section) 'Incantation: May the wi]nd [blow], may the gardens not quake'"; 106. [KA.INIM.MA ŠÀ.ZI.G]A *ù* MUNUS.GIN.NA.KÁM ŠÀ.ZI.GA.MUNUS.A.KÁM "[Wording of (the incantation) for (male) sexual desi]re and (those) to make a woman come (and for) woman's sexual desire" (Steinert 2018b: 217). Ll. 103–104 corrispond to the incipit of the *nīš libbi* incantation No. I.1: 6′.

16 The phraseology in ll. 99 and 101 of the catalogue does not appear in the *nīš libbi* corpus. See the fragmentary passage in K prescr. 1: 1–2: "[If a man does] not desire [his$^{?}$ woman] / [. . .]"; H Sympt.: 2–3: "If a man approaches his wife and [. . .] / does not desire his wife [. . .]." However, an inversed variant, where the man desires his own woman but not another woman, is attested in *nīš libbi* K.2: 23: "[If a man] desires his [woman], but do[es not desire] another woman." See also O prescr. 1: 1–2: "If a man goes to his woman, and [desires his woman, but] / he goes to another, but does not desire another woman." See the incipit of Tablet XXXIV of *Diagnostic Handbook* SA.GIG/*Sakikkû* in the catalogue l. 41, which can be restored from the fragmentary catchline of Tablet XXXIII: DIŠ ⸢NA⸣ *ana* ⸢MUNUS⸣-(*šú*) ŠÀ-*šú* ⸢ÍL⸣-*šú-ma* [*ana* MUNUS BAR$^{?}$]-*ti* ŠÀ-*šú* NU ÍL-*šú* MUNUS BI ŠÀ-[*šú* . . .] "If a man desires (his/a) woman, but does not desire [another$^{?}$] woman: this woman [. . .] his 'heart'/desire$^{?}$ [. . .]" (see Schmidtchen 2018: 141). In No. E.2: 18 the man desires neither his own woman nor other women: "[If] a man's sexual desire is taken away and (his desire) to go to (var.: he does not desire) his own woman or another woman is reduced."

[If a man desires another woman, (but)] he does not desire his (own) woman. If a man [. . .].
[. . .] . . . the head of a bird. (Instructions) to place a drug for sexual desire around his [neck].

[Total of X tablets (of the section) "If a man desires another woman], (but) he does n[ot desire] his woman."
[Including (prescriptions) . . .] to make a man have an 'elevation.'
[*Aššur Medical Catalogue* section XIX: 99–102, Steinert 2018b: 217]

ŠÀ.ZI.GA appears also in the two tablet inventories, both published by Irving L. Finkel (2018). The first one, BM 103690, written in a post-Old Babylonian or Middle Babylonian script, comes from the British Museum: 1 ŠÀ.ZI.GA *ina* ^{iti}BARA.ZAG.GAR "One (tablet for) 'Sexual desire in the month of Nisannu'" (rev. iii 5, Finkel 2018: 31). It corresponds with the incipit of *nīš libbi* N (= KUB 4, 48) prescr. 1 i 1–2: DIŠ LÚ ŠÀ.ZI.GA *ina* ^{iti}BÁRA.ZAG / TIL "If a man's sexual desire in the month of Nisannu / has finished." The inventory quotes the incipit in abbreviated form, for this reason, we could read the first sign DIŠ as *šumma* 'if' instead of 'one' (tablet). See however in *nīš libbi* N (= KUB 4, 48) prescr. 25 lo. e. 5: DUB 1.KAM DIŠ LÚ ŠÀ.ZI.GA "One/first tablet of 'If a man's sexual desire.'"

The second inventory, Ni. 2909, comes from the Istanbul Museum (second half of the second millennium): 1/DIŠ KI.MIN *ḫi-ni-iq-tam* GIG: 1/DIŠ KI.MIN *ša* ŠÀ.ZI.GA "One/if ditto, 'he suffers from strangury'; one/if ditto, for the (loss of) sexual desire" (rev. 11′, Finkel 2018: 34).

Note the possible inclusion in section V of *Diagnostic Handbook* SA.GIG/*Sakikkû* as shown by the incipit of Tablet XXXIV in the catalogue: [x (x)] DIŠ ⸢NA⸣ *ana* ⸢MUNUS⸣-(*šú*) ŠÀ-*šú* ⸢ÍL⸣-*šú-ma* "If a man desires a (var.: his) woman and" (l. 41, Schmidtchen 2018: 140). However, note that in Tablet XXXIII as catchline for Tablet XXXIV the following incipit is given: [. . . MUNUS BAR$^{?}$]-*ti* ŠÀ-*šú* NU ÍL-*šú* MUNUS BI ŠÀ-[*šú* . . .] "[. . .] (he) does not desire [another$^{?}$ woman], this woman [. . . his] 'heart'/desire$^{?}$" (see Heeßel 2000: 358). Schmidtchen (2018: 141) suggests that what is preserved in Tablet XXXIII should be considered as apodosis (see the presence of *-ma* after ÍL-*šú* in the catalogue), therefore, the complete protasis would be: "If a man desires (his/a) woman, but does not desire [another$^{?}$] woman: this woman [. . .] his 'heart'/desire$^{?}$ [. . .]."

The only commentary attributed to the *nīš libbi* corpus is SpTU 2, 39 (W 22730/2), a small fragment originating, according to its colophon, from the library of Iqīšāya, an exorcist from Uruk active in the late fourth century BCE (see Frahm 2011: 128).[17] The commentary in rev. 9 mentions a phraseology that could be attested in the *nīš libbi* texts: [. . . *šá pi um-man*]-*nu šá* ŠÀ DIŠ NA *ana* MUNUS-(*šú*) ŠÀ-*šú* ÍL-*šú-ma* (see in *nīš libbi* corpus: K.2: 23 (restored); O prescr. 1: 1–2 (restored)). According to Schmidtchen (2018: 141) the commentary could also belong to Tablet XXXIV of the *Diagnostic Handbook*.

17 The first two lines seem to quote and comment on two lines from *Udug-ḫul* (5: 1, 3), while the rest of what is preserved of the tablet is mostly devoted to the explanation of individual words.

Forschungsgeschichte of *nīš libbi* therapies and the 'reflective' approach in anthropology and gender studies

Forschungsgeschichte

Approximately fifty years have passed since the last edition of the *nīš libbi* incantations and rituals. The previous edition was completed by Robert Biggs's in 1967 and has served as the benchmark for the analysis of this documentation ever since. The study was conducted for his Ph.D. dissertation at Johns Hopkins University under the direction of Wilfred G. Lambert. It was submitted in 1962 with the title *The* ŠÀ.ZI.GA *Incantations: Sumerian and Akkadian Love Charms*. A revised version, expanded with the edition of the rituals, was published in 1967 as volume 2 in the series "Texts from Cuneiform Sources" with the title ŠÀ.ZI.GA: *Ancient Mesopotamian Potency Incantations*. The first edition of this group of texts is, however, that of Erich Ebeling in 1925 with the title *Liebeszauber im Alten Orient* in "Mitteilungen der altorientalischen Gesellschaft" 1/1.

Erich Ebeling was the first to translate the expression ŠÀ.ZI.GA/*nīš libbi* as 'sexual power' (*geschlechtliche Potenz*, Ebeling 1925: 24) in strong opposition to the traditional translation of 'courage' (*Mut*), proposed by Zimmern (1915–1916: 220). The aim of the incantations is, according to Ebeling, to acquire sexual potency. In fact, he translates the expression with *geschlechtliche Kraft* ('sexual vigor/power').

Most Assyriologists accepted Ebeling's translation. Yet it also attracted some criticisms. Benno Landsberger (1966: 263), in his philological study on the word *iṣṣūr ḫurri*, denoting a type of bird or bat,[18] which, along with deer, is often mentioned in *nīš libbi* rituals, refers to the issue of translation of the ŠÀ.ZI.GA/*nīš libbi*. He affirms that it is the word *dūtu*, not *nīš libbi*, which denotes the masculine vigor (*Manneskraft*) and that, despite this, the translation of *dūtu* as 'power' is not entirely accurate. He prefers to translate *nīš libbi* as 'sexual desire' (*sexueller Appetit*) (see also Landsberger 1967b: 161 fn. 107).

It is thus possible to recognize two different strands of interpretation in the scholarship on the *nīš libbi* texts: on the one hand a "physical" interpretation, i.e. "sexual power," that of Ebeling, on the other a "psychological" one, i.e. "sexual desire," that of Landsberger.

Reinforcing this dichotomy is a very brief analysis by James V. Kinnier Wilson in the 60s. He was the first Assyriologist to talk in terms of 'mental diseases,' and his analyses are clearly inspired by modern psychiatry and psychology (see Kinnier Wilson 1967). He writes about *nīš libbi* in his article "Hebrew and Akkadian Philological Notes": «Amongst the textbooks of primitive Babylonian "psychiatry" (a wholly unexplored field at the present time) is one given the Sumerian title of ŠA.ZI.GA. This

18 For the discussion on this topic see Chapter III § "Potions."

does not mean, with Ebeling "Potenzerhöhung," but what the modern psychiatric science would call "loss of libido" (ŠÀ = *libbu*, "desire," and on ZI.GA, "expense, loss")» (1962b: 180). He thus reads the Sumerogram ZI.GA with the Akkadian word *ṣītu* 'loss,' hence the "loss of libido." Contrary to Ebeling and Landsberger, Kinnier Wilson interprets the expression ŠÀ.ZI.GA as a mental disease, in modern psychiatry called until 1987 "inhibition of sexual desire" and today "hypoactive sexual desire." The latter is a disease officially recognized by the *Diagnostic and Statistical Manual of Mental Disorders*. The disease is characterized by a lack or absence of sexual fantasies and desire for sexual activity. To be recognized as such, according to DSM-IV, it should cause preoccupation or difficulties in interpersonal relationships, and even anguish, grief, depression, low self-esteem. Kinnier Wilson, however, contrary to Ebeling, but following Landsberger, carries the question on a 'psychological' level. The rituals and incantations thus try to eradicate a deficiency not only on a biological level but also psychological one. However, the philological interpretation of Kinnier Wilson is wrong. In fact, as Biggs demonstrated, the goal of the texts is *ana* ŠÀ.ZI.GA TUKU(-*e*)/*ana nīš libbi rašê* "to get *nīš libbi*" and that for this reason the expression cannot designate a disease.[19]

In 1967 the excellent edition of Robert Biggs was published. It provides a new edition of the texts in E. Ebeling's *Liebeszauber im alten Orient*, and includes numerous texts edited for the first time. In the introduction Biggs deals with methodological issues concerning the interpretation of ancient documents relating to sexuality:

> The interpretation of sexual phenomena in a civilization that is completely alien to our own and of which we cannot with confidence trace the survivals in contemporary societies is, of course, very difficult. The dangers of being too much influenced by our Judeo-Christian heritage (and especially by the attitudes on sexual questions which have been prevalent in Western lands since the Reformation) are even more acute in dealing with sexual matters than in dealing with ancient law, economics, religion, etc., in which progress has long been hampered by the difficulty of applying any but our own traditional patterns of thought and our own acquired prejudices. A notable difficulty here is to know to what extent certain words were intended to have a sexual connotation. [. . .] It is certain, in any case, that the attitude of ancient Mesopotamians toward sexual acts had little in common with those generally held in modern Western civilization. [Biggs 1967: 1]

Modern categories of thought and religious concerns should be kept in check to get a fair understanding of Mesopotamian sexuality. However, we shall see that Bigg's own

19 The expression ŠÀ.ZI.GA appears together with some sicknesses in a Late Babylonian Text, *Taxonomy of Uruk* (SpTU 1, 43; see Köcher 1978: 24–24; Stol 1993: 23–24; Heeßel 2010a: 30–31; Geller 2014: 3–9; Steinert 2016: 230–242). In this text, divided in four sections, diagnostic categories are listed in association with specific organs or parts of the human body. In the section, regarding the kidneys, *nīš libbi* is mentioned (l. 26). For a discussion on this text see the Chapter III § "Plant ingredients" and "The astral influence."

interpretation is not immune to the influences and limitations of modern biomedical knowledge.

He refers to Ebeling and thus translates *nīš libbi* as 'sexual power.' He argues that while *libbu inaššīšu* has the meaning 'to wish, to want, to take an interest in,' *nīš libbi* should not be construed simply as 'sexual interest, desire,' or 'libido.' For him, the patient is a man who desires a woman but is incapable of having intercourse with her. He, furthermore, observes that in several instances a man gets *nīš libbi* only after his penis has been stimulated by rubbing it with oil. He argues that *nīš libbi* is not simply the term for a man's interest in sexual relations, but for the ability to get and maintain an erection (Biggs 1967: 3).

Biggs argues, from a biomedical point of view, that the loss of potency could relate to gonorrhea. Another cause of impotence could be simply also the natural diminution of potency which affects all men sooner or later (Biggs 1967: 3). He quotes the following text (his reading and translation):[20]

> [*šumma amē*]*lu ušaršu uzaqqassu inūma šīnātīšu išattinu reḫûssu iddâ*
> [*nīš*] *libbīšu ṣabit-ma ana sinništi alāka muṭṭu šarku ginâ ina ušarīšu illak*
> [*amēlu š*]*ū mūṣâ mariṣ*
>
> [If a man]'s penis gives him sharp pains when he urinates, his 'semen' discharges,
> his [pot]ency is 'seized,' and he cannot have intercourse with a woman, pus (literally 'white blood') constantly comes out of his penis,
> th[at man] is ill with gonorrhea (literally 'discharge').
> [BAM 112 i 17′–19′, Biggs's (1967: 3) reading and translation]

Biggs returns to the argument in his article "The Babylonian Sexual Potency Texts" presented at the XLVIIth *Rencontre Assyriologique Internationale* in Helsinki. Here he clearly adopts the biomedical terminology when he argues that these texts are intended to alleviate sexual impotence, or as it is termed today, erectile dysfunction or penile erectile dysfunction. In fact, erection comes about when a man receives sensory or mental stimulation (Biggs 2002: 72). According to Biggs, then, incantations with their erotic images of sexually excited animals have the aim of causing mental stimulation, while the rituals with rubbing the penis with oil from the alabastron, often together with pulverized magnetic stones, have the aim of arousing sensory stimulation. The use of parts of the body of sexually excited, often wild, animals, in the rituals has the aim to transfer sexual power from the animals to the patient.

Gwendolyn Leick, who deals with the *nīš libbi* texts in her book *Sex and Eroticism in Mesopotamian Literature*, particularly in the chapter "Love and Magic Potency

---

20 Both Biggs (1967: 3) and Böck (2014: 147) restore line 18 with *nīš*, creating thus a link between sexual desire and 'gonorrhea' (*mūṣû*), while Geller (2005: 64, No. 4), whom I follow, restores with *ina* translating "in his innards." There are no other texts which link the absence of sexual desire with cases of gonorrhea. See also BAM 112 i 35′.

Incantations," had the same idea as Biggs. Leick (1994: 208–209) is the first scholar to explicitly assert that the expression *nīš libbi* is nothing more than a euphemism for the erection of the penis.

Returning to the problem of modern identification of the diagnostic category of *nīš libbi*, Biggs' reference, in his edition of the texts, to the biomedical aspects, was expanded on by JoAnn Scurlock and Burton R. Andersen in their book *Diagnoses in Assyrian and Babylonian Medicine*. This book aimed to provide a comprehensive review and analysis of all available ancient Mesopotamian texts with biomedical diagnosis and prognosis. After analyzing the 'sexually transmitted diseases' (Scurlock and Andersen 2005: 98–115), they focus, in the chapter "Genitourinary Tract Diseases," on sexual impotence (ibid. 111–113). This 'disease' is included in the section entitled "Sexual Dysfunctions" together with priapism, nocturnal emission, and premature ejaculation. The scholars write about impotence in Mesopotamia as follows:

> Both impotence and priapism can have multiple causes but *Schistosomiasis hematobium* is capable of causing both of them. The ancient Mesopotamian physician recognized that impotence commonly occurs with sexually transmitted diseases and advanced age, and that back or abdominal injuries could also have this affect. [. . .] What we might consider to be the psychological causes of the impotence were recognized, at least on an empirical level, by ancient physicians. [Scurlock and Andersen 2005: 112]

To understand how *nīš libbi* was interpreted by the scholars, I take into account the following sentence: *šumma amēlu nīš libbīšu* (ÍL ŠÀ-*šú*) *ekim-ma lū ana sinništīšu lū ana sinništi ahīti libbašu lā inašši* (ŠÀ-*šú* NU ÍL) (see No. E.2: 18, Ms. H STT 280 ii 62). The sentence was translated in different ways:

- "[If] a person is impotent so that he cannot get an erection (either) for his woman or for a strange woman" (Scurlock and Anderson 2005: 112);
- "[If a man]'s potency is taken away and his 'heart' does not rise for his own woman or another woman" (Biggs 1967: 27);
- "(If a man's sexual desire is taken away and) er begehrt weder seine eigene noch eine fremde Frau" (Landsberger 1966: 263).

The first logogram ÍL ŠÀ is undoubtedly read in Akkadian as *nīš libbi*, the second ŠÀ-*šú* NU ÍL shall be read *libbašu lā inašši* "to wish, desire."

The use of biomedical parameters, in particular the dichotomy "magic *vs.* medicine," in order to understand the *nīš libbi* corpus was also undertaken by René Labat as well. In the RlA entry 'Geschlechtskrankheiten,' he declared that it is important to distinguish between the therapeutic actions concerning Mesopotamian sexual and urinary diseases and those concerning 'impotence': «Wenn die Harn- und Geschlechtskrankheiten von den Akkadern im allgemeinen mit vernünftigen Mitteln geheilt werden, so fußt die Behandlung sexueller Impotenz eher auf der Magie als auf der Medizin. Dieses Leiden schreiben sie gewöhnlich dem bösen Einfluß eines Dämons oder Zauberers zu» (Labat 1957–1971: 223).

Let me now summarize the various points so far. Aside from the excellent philological analyses, such as those of Ebeling and Biggs, we also find divergent interpretations of the texts. On one hand, an interpretation that I call "psychological" and on the other a "physical" one.[21] The first is that of Landsberger who translates the expression *nīš libbi* as 'sexual desire' and of Kinnier Wilson who interprets it as 'loss of libido.' The second one comes from Ebeling as well as Scurlock and Anderson. According to these scholars, the expression has the meaning of 'sexual power.' Only in Biggs and Leick do we find an analysis of the incantations in a magical-symbolic perspective. Biggs certainly, as with Leick, does not exclude the psychological dimension of the problem in his analysis of animal images in the incantations and of the use of animal substances in the rituals.

The psychological aspect of impotence is emphasized in an article by Marten Stol "Psychosomatic Suffering in ancient Mesopotamia" (1999). Here the scholar points out the psychosomatic symptoms of sexual impotence and focuses on the "emic" causes of the sickness. A "sexual-psychological" perspective is only hinted at in the article "Freud and Mesopotamian Magic" by Markham J. Geller (1999).

Criticisms of the biomedical approach

I would like to linger on the two most divergent interpretations of *nīš libbi* therapies, those of Kinnier Wilson and Scurlock and Andersen, from a theoretical and methodological point of view. Kinnier Wilson is the first to propose a clear-cut division between physical or somatic diseases and psychological ones. A similar methodology was applied by Scurlock and Andersen using retrospective diagnosis, that is, the mechanism of applying modern biomedical disease classifications to the ancient sources.

These two approaches, despite their differences, are together equated by the use of a biomedical perspective to the analysis of ancient pathologies. This perspective is called in Medical Anthropology "biomedical reductionism." This method is criticized by Nils Heeßel (2004b: 6–7) for the fact that there is no clear criterion by which the scholars select some symptoms described in the sources instead of others, and that this approach tells us nothing about the Mesopotamian therapeutic system and about how the Babylonians understand sickness, suffering, and pain.[22]

The study of *Forschungsgeschichte* was useful to disclose the theoretical and methodological perspective used by researchers in the analysis of the *nīš libbi* corpus,

21 Wiggermann (2009–2011: 414) talks in term of reduced appetite and potency problems.

22 See also Attia and Buisson's article *Du bon usage des médecins en assyriologie* (2004).

that is the biomedical one. Biomedicine is an ethnomedicine culturally characterized by exclusive attention to the bio-psychic dimension of the individual (see Eisenberg and Kleinman 1981; Lock and Gordon 1999; Pizza 2005: 125–144). Modern biomedicine considers disease as an alteration of the psycho-physiological state of the organism, provoked by one or more organic dysfunctions. Health is therefore seen as an ideal state, the absence of disease or infirmity. Biomedicine defines, therefore, the health, on an ideal level, as a state of perfect homeostasis of an individual organism. The biomedical perspective is essentially organicist[23] and positivistic, it considers the human being as a comprehensive system of chemical, biological and physiological elements structured in different levels of complexity (chemical elements, cells, organs, apparatuses, and systems) in functional relation with each other and fighting with the external environment, in particular with pathogens. The focus of attention is the sick organ, on which one must intervene on a purely technical level, either physical or chemical.

Medical anthropologists have shown, thanks to the ethnographical research in modern Western contexts, the strong interplay between culture and processes of institutionalization in biomedical knowledge. They showed that the biomedical categories are not merely neutral descriptions of a given reality, but rather an apparatus of the cultural construction of clinical reality. Anthropological studies have revealed how the biomedical dichotomy "health *vs.* malady," along with that of "normal *vs.* abnormal or pathological" and the Cartesian one of "mind *vs.* body," are subject to a specific cultural configuration and cannot be considered universal. In other words, "health" and "malady" in the biomedical definitions are not objective conditions, which transcend the socio-cultural boundaries of a patient, but rather relative definitions, depending on historical and social contexts.

I will develop these issues in the next sections, for now, I want to emphasize that this "biomedical reductionism," which I brought out in the analysis of the *Forschungsgeschichte*, is based on a more general principle of cognitive nature, the "analogical reductionism."[24] When dealing with an unfamiliar cultural phenomenon, the researcher, as the common man, uses the interpretative parameters which, through the process of enculturation, his or her own society has provided to him. Inevitably we look at the world around us with the cognitive tools that our culture made available to us. The comparison with other cultural systems (*Weltanschauung*) shows the inefficiency of such reductionism: instead of *understanding* the others, we end up *reducing them to us*.

---

23 Social and psycho-somatic medicine are gradually emerging. The attention not only to the physiological aspects is also present in the definition of health proposed by World Health Organization: "Health is a state of complete physical, mental and social wellbeing and not merely the absence of disease or infirmity."

24 van Binsbergen and Wiggerman talk in terms of the "error of reductionism" referring to the study of Mesopotamian magic (1999: 5–6).

Much has been written on the absence of neutrality of the researcher's gaze. Especially in the humanities neutrality is a kind of "naive illusion." Unavoidably we orient our cognitive gaze according to subjective and cultural involvement. It is necessary, therefore, to adopt a "reflective" approach, which takes into account the researcher's cultural and personal background. A capacity of reflection and analysis on the projective character of the way of doing research. We need, in the words of Pierre Bourdieu (2001), *objectiver le sujet de l'objectivation*, that is analyzing the researcher who constructs his or her object of study. This construction inevitably involves the cultural level. For this reason, reflecting on the way we understand infirmity, health, therapeutic process in our Western system helps a lot. I insist on this point to avoid falling into biomedical interpretative traps, which deform the ancient sources, and which actually produce a sort of "history of biomedicine in the ancient world," leading us away from an understanding of ancient medical practices and concepts.

Anthropology has discussed the interplay between knowledge and power (see Gramsci 1975: Foucault 1969; Foucault 1971), and the consequent reduction of other cultures to one's self.[25] Similarly, Medical Anthropology and Ethno-Psychiatry have shown how ethnologists, psychiatrists, physicians, psychologists have actually used theoretical and methodological tools of biomedical and psychological knowledge for the analysis of the medical systems of other populations, reiterating a colonial process on the level of knowledge (and in fact of power). In this regard, the studies of one of the founders of the French Ethno-Psychiatry, Frantz Fanon on the biomedicalization of extra-European societies are fundamental. His studies have been applied by other ethno-psychiatrists, such as G. Devereux, T. Nathan, R. Beneduce, P. Coppo, and medical anthropologists, such as M. Lock, N. Scheper-Hughes, D. Fassin, P. Farmer.

All this presupposes an epistemological reformulation which unmasks the alleged neutrality of biomedical knowledge (see Lock and Gordon 1988; Singer 1990). As discussed by Deborah Gordon, biomedicine tends to construct for its own "identity" or "membership" based on a rhetoric of rationality and truth. It must be interpreted within a "Western tradition" based primarily on "naturalism" (= science) (1988: 20–32) and on "individualism" (ibid. 33–37). According to this worldview, nature is seen as an ahistorical, universal, rationally understandable reality, separated by culture and society, and independent of the human consciousness. Gordon argues that biomedicine relies on naturalistic epistemology as its official epistemology, using "science" as the measure of truth. Medicine's burst of development came with the displacement of the criterion of truth from tradition and rationality to "look and see" (ibid. 33). Biomedicine reiterates a series of dichotomies which underlie the theoretical and philosophical assumptions of the Western tradition: "mind *vs.* body,"

25 See Geertz 1988 and the postmodern anthropology, in particular Marcus and Fisher 1986; Clifford and Marcus 1986, see here especially Crapanzano 1986.

"rational *vs.* irrational," "material *vs.* symbolic," "natural *vs.* cultural." The "cultural practices of biomedicine"[26] are based, according to Gordon, on a precise cosmology, ontology, and epistemology. In other words, the biological reductionism, as a cultural system, goes beyond the strictly medical aspects, incorporating other notions, such as those of body, mind, person, identity, female and male, individual and community.[27] As Giovanni Pizza (2005: 130) points out, on a theoretical and ideological level, biomedicine tends to identify itself with the truth and to produce claims of the universality of its theory and ideology, concealing the sociocultural meaning of its practice. Biomedicine studies our bodies and defines their objective consistency, it is, therefore, the primary place where the construction of our own ways of perceiving and living the body takes place.

I think that these reflections developed in Critical Medical Anthropology and Ethno-Psychiatry are of great importance for the study of ancient Mesopotamian medical knowledge. It is not possible for Assyriologists not to take into account these methodological perspectives. The biomedical approach is still very strong in Assyriology. See, for example, Coleman's reply to Heeßel's criticism of employing modern diagnosis in the study of ancient sources. Coleman proposes the image of a modern medicine not subject to cultural influences and for this reason a-cultural: «The scientific knowledge of a modern physician translates across cultures in a way that is not as true of linguistics, law or other more culturally-bound disciplines. As patients from the high-rises of New York to the villages of Papua New Guinea can testify, modern medicine tends to be cross-cultural, a characteristic it shares with the other sciences. The concept of cross-cultural disciplines includes ancient-modern interactions» (2005: 44). Coleman reproduces, quoting Michael T. Taussig, «a political ideology in the guise of a science of (apparently) "real things" – biological and physical thinghood» (1980: 3). He reiterates the idea that biomedicine is a culture-free discipline (while other disciplines are culturally-bound). His argumentations are valid only in a biomedical perspective, which shall, in my opinion, understood as an ethnomedicine, which is based on philosophical assumptions of a specific ethno-epistemology.[28]

---

26 I use here the term 'cultural practice' or 'cultural production' in order to give a greater emphasis to a perspective more concerned with the real experience of the human praxis, rather than with a general and abstract concept of 'culture' or 'cultural system' (see Bourdieu 1972; 1980).

27 Foucault showed how historically the biomedical conception of the body has conditioned the power over the individual and the community, and vice versa. Biomedical knowledge and power in modernity, although separate, reinforced each other. For the philosopher in the seventeenth century the body is thought of as a machine. This was ensured by power mechanisms which characterized the disciplines: what he calls *anatomo-politics of the human body*. Towards the middle of the eighteenth century the body is thought of as the body-species, understood as support of biological processes. This created a number of regulatory controls, defined by Foucault as a *bio-politics of the population*. This initiates the era of *bio-power*, at the base of the modern capitalism (1976: 123–142).

28 For problems regarding not only the retrospective diagnostics, but also the translation of Mesopotamian pathologies with modern technical terms see Heeßel 2010b; 2016: 17–23.

The denial of any suspension of epistemological analysis to biomedicine and the attention on the social conditions of production of biomedical knowledge underpin the reflection of Allan Young. He argues that the success of the ideological practice of clinical medicine, including analysis of the medical conceptions of other populations, is due to isolation from both other ethno-concepts and other epistemologies opposed to it on the edge of rational discourse. «Through these practices – Young affirms – socialized knowledge of medicine is made to seem not so much "wrong" or "counter-productive" [. . .] as "not medicine at all" or an attempt to "politicize" medicine and science» (1982: 275). As pointed out by the historian of medicine Stengers, what characterizes the medicine in the modern sense, compared to the therapies of the ethnological or traditional societies, is not so much the doctrine or practices, but rather the claim of "rationality." *Healing proves nothing* – the researcher affirms – the charlatan is defined as one who claims as evidence the healing (Stengers 1995). It is the scientific rationality that characterizes modern medicine.

As stated above, a critical reflection of the epistemological positions of Western knowledge, intended as an ethno-epistemology (see Scheper-Hughes and Lock 1987), inevitably involves us researchers firsthand, as members of this society. As Rhodes writes: «Western biomedicine and medical anthropology are intimately connected» (1990: 159). We researchers are members of the society in which biomedicine provides the dominant form of explanation and treatment of sicknesses. This means that a critical analysis of these epistemological foundations involves a reflection on how we think about issues regarding the concepts of health and care. Therefore, it is impossible to know the sickness-event without social conditioning of some kind. The researchers, therefore, are always socially conditioned, however those who employ a critical-interpretative approach may, following Young, «claim, and what would set their accounts of sickness off from those of others, is a critical understanding of how medical facts are predetermined by the processes through which they are conventionally produced in clinics, research settings, etc. Thus, the task at hand is not simply to demystify knowledge, but to critically examine the *social condition of knowledge production*» (1982: 277). The anthropologist Ernesto De Martino uses the concept of "critical ethnocentrism," understood as an improvement of one's own cultural consciousness in comparison with "other" cultures and as a hard process of awareness of our own limits, dependent on our cultural history, in the descriptions of the "otherness."[29] Examining ourselves before and during the analysis of ancient sources is not easy, but an attempt must be made. As Taussig writes: «Medical practice inevitably produces grotesque mystifications in which we all flounder, grasping ever more pitifully for security in a man-made world which we see not as social, not as human, not as historical, but as a world of a priori objects beholden only to their own force and laws, dutifully illuminated for us by professional experts such as doctors» (1980: 5).

29 For a study on De Martino's critical ethnocentrism in English see Saunders 1993.

Or in the words of Nancy Scheper-Hughes and Margaret Lock «to admit the "as-ifness" of our ethnoepistemology is to court a Cartesian anxiety–the fear that in the absence of a sure, objective foundation for knowledge we would fall into the void, into the chaos of absolute relativism and subjectivity» (1987: 30). Analyzing critically our habitual ways of seeing sickness, health and therapeutic efficacy leads to a weakening of our epistemological certainty, but I think it is crucial for improving the quality of analysis of the sources.

The research on the *nīš libbi* diagnostic category and its therapies involves not only the therapeutic dimension (hence the dialogue with Medical Anthropology and Ethno-Psychiatry) but also that of sexuality and gender construction.[30] Even within the Gender Studies, epistemological reflections have developed. Second-wave Feminism[31] (1970s-1980s) has shown how the sexual difference "man and woman" has been culturally constructed in the frame of a male ideology that has then *naturalized* it, and had allowed acquiring different points of view, which differ from the leading ideology of clear masculine kind. Establishing woman's distinctiveness and giving value to female particularities, it has on the other side crystallized gender identities, homologating *de facto* the heterogeneity of women. This has led to the development of a concept of Woman, which remains distant from the peculiarities of each group of women, as well as from every single woman, concealing noteworthy differences between them. Such a myth of a unitary and a-historical Subject-Woman, as unrelated to single historical and cultural realities as it is, has been strongly criticized by both Afro-American and lesbian feminists. The former were, in fact, incapable of giving voice to their peculiarities, as they were on one hand excluded by the feminist movement because of their ethnic affiliation, and on the other hand by struggles for the emancipation of black people because of their gender. The latter were cut out

30 The studies on Mesopotamian manhood and male identity are limited. While in some disciplines, such as Anthropology, Sociology and Archeology, studies on masculinity have proliferated, for the ancient world and in particular for Assyriology and Archeology of the Near East there are very few. See for Archeology Alberti 2006; for Anthropology Gutmann 1997; for Sociology Connel 1995 and Kimmel 1996; for Queer Studies Halberstam 1998. For the ancient Near Eastern Studies see Winter 1996; Chapman 2004; Melville 2004; Assante 2007; Assante 2017; Asher-Greve 2008; Westenholz 2009; Svärd 2010; Suter 2012; Garcia-Ventura 2014; N'Shea 2016; N'Shea 2019; Peled 2016; Peled 2018; Cooper 2017; Westenholz and Zsolnay 2017; Helle 2018; Helle 2020; Bennett 2020. On masculinity in the ancient Near East two workshops were organized: *Mapping Ancient Near Eastern Masculinities* by Ilona Zsolnay at the Philadelphia Penn Museum in 2011 (Zsolonay 2017); *The Construction of Masculinities in Ancient Mesopotamia* by Agnes Garcia-Ventura and Lorenzo Verderame at Sapienza University of Rome in 2015 (with the author and Omar N'Shea).

31 The various trends in Gender Studies are usually presented as *waves*. For a description of these waves and their relationship with ancient Near Eastern studies see Bahrani 2001: 14–25; Svärd 2015: 8–12. For the development of Gender and Woman Studies in Assyriology and Archeology of ancient Near East see Asher-Greve 2000; Asher-Greve 2002: 33–34; Lion 2007; Bolger 2008; Pollock 2008; Vogel 2012a; Vogel 2012b; Van de Mieroop 2013; Budin and Webb 2016; Garcia-Ventura and Zisa 2017; Garcia-Ventura and Svärd 2018.

from the discourse on woman, inspired by a heterosexual perspective that excludes other sexual orientations. Such "outsider" voices have allowed for a critique of First-wave Feminism in favor of – what most interests us – an epistemological perspective, laying the foundation for that critical re-thinking, which brought in the so-called "post-Feminism" and Queer Theories.

Such thinkers find their lifeblood not only in feminist theories but also in post-colonial thought and post-structuralist philosophy.[32] Very briefly, it appears clear that the frame of our subject, the spheres of politics, ethics, socio-economics, and epistemology are extremely intertangled. The aim is to try to give voice to oppressed people for different reasons, whether ethnic, social, political, cultural, or related to their gender. A gender discourse is always, then, also a political one, which, I would argue, is related to an alterity, first constructed, and then oppressed.

Analyzing the discourse which produces such oppressed otherness means acquiring a different epistemological perspective. Following Foucault's heritage, scholars of Gender Studies have studied the interactions between knowledge and power. Acquiring a different perspective means de-constructing the concept of hegemonic knowledge and giving voice to oppressed people. Instrumental in this sense are policies of the 'scholar's location' (see Rich 1987). Under this new perspective one understands the concepts of 'marginality' of bell hooks (1996), '*mestizia*' of Gloria E. Anzaldúa (1987), 'lesbian' of Monique Wittig (1980, Engl. tr. 2013), 'eccentric subject' of Teresa de Lauretis (1990), and partly of 'cyborg' of Donna Haraway (1991). All these concepts have in common, even with minor differences, acquisition of different, "other," views, which allow for living, acting, and making research in "another" way. All these terms indicate an excessive critical position. For instance, according to bell hooks (1996: 52), the 'marginality' is the site of radical possibility, a space of resistance, offering the possibility of radical perspectives from which to see and create, to imagine alternatives, new worlds. Similarly, Wittig's 'lesbian' (1980, Engl. tr. 2013: 250) is treated as the privileged place for a reflection on the hegemonic role of heterosexuality and the cultural and social constructions of man and woman. As underlined by de Lauretis (1990: 145), Wittig's term 'lesbian,' as that of 'lesbian society'[33] is not referred to the sexual orientation to a woman; differently, they are theoretical terms of a form of feminist consciousness that can only exist historically, *hic et nunc*, as the consciousness of a *something else*.

Theories of *theoretical location* have also been developed by Donna Haraway, a biologist, philosopher and gender theorist. I cannot linger here on her thoughts on the concept of "cyborg," «a cybernetic organism, a hybrid of machine and organism, a creature of social reality as well as a creature of fiction» (1991: 149), which suggests a way out of the dualism that the theory of sexual difference imposes. As already sug-

---

32 On the relation between Feminism and Post-colonial Studies see Spivak 1988; Minh-ha 1989; Hill Collins 2000.

33 On lesbian-feminism and the concept of 'compulsory heterosexuality' see Rich 1980.

gested by other queer theorists such as Butler, de Lauretis and Preciado (2001), the concept of cyborg involves a subjectivity model that is multiple, discontinuous, articulate on several plains, not only that of gender but also those of ethnical group, social condition, age, sexual orientation, and political ideas.[34] For present purposes, her consideration of epistemology is important, especially on scientific objectivity. She strives for types of knowledge, which should be "situated," i.e. in which the specificity and peculiarity of scientific analysis are underlined (Haraway 1988: 575–599). At the base of such objectivity and rational principles is the practice of deconstruction and criticism, which does not to fall into an empty relativism. It is such 'situationality,' partiality and deconstruction that allow objective knowledge, which is held against a claimed transcendence as well as an invisible omnipotence of the scholar.[35] It is clear then that gender becomes an epistemological perspective, a *critical-analytical concept* (see Scott 1986), which allows re-analyzing some of the so-considered essential works in each specific discipline.

Theoretical and methodological introduction

Illness, disease, sickness

I would start with the words of the French Africanist anthropologist, Marc Augé, who states: «La maladie qu'elle est à la fois la plus individuelle et la plus sociale des choses» (1994: 36). For the anthropologist, the sickness should be interpreted as the elementary forms of the event (*forme élémentaire de l'événement*), of biological and social nature. Since the beginning of Anthropology, scholars have paid attention to the cultural and social aspects of the sickness, health, and healing techniques of non-European or traditional societies.[36] However, we had to wait until the 1970s, thanks to the Harvard Medical School, for a new debate and a new approach to health issues in Medical Anthropology and Cultural Psychiatry.

As a reply to the 'objectification' of malady on behalf of biomedicine, scholars of Harvard University, especially Arthur Kleinman and Byron Good, professors of

34 See Braidotti's criticisms (2002: 122–124).

35 The influence of post-structuralist and deconstructivist philosophy (Foucault, Derrida, Deleuze, Kristeva, etc.), and that of Bourdieu's epistemological reflections is clear. For the feminist epistemology see Alcoff and Potter 1993; Harding 1986; Hartsock 1983; García-Selgas 2004. On the contribution of feminist epistemology to the study of the ancient Near Eastern sources see Garcia-Ventura and Zisa 2017: 38–45.

36 See G.M. Foster's studies on the classification of etiological systems in non-Western societies (1976); Evans-Pritchard on magic among the Azande, a population of Central Africa (1937); Turner on Ndembu physician, a Bantu population of Zambia (1967); the studies of Rivers (1924), Acherknecht (1942; 1946; 1971) and Lévi-Strauss (1962a; 1962b). Although on recognizes a 'logical' to 'other' medical knowledge, they still are analyzed through the lens of biomedicine.

Medical Anthropology at Harvard Medical School, have broken down the expression 'malady' into two different dimensions: *illness* and *disease*[37] (see Fabrega 1972; Eisenberg 1977; Kleiman 1978):

- *Illness* refers to the experience of suffering and pain experienced by the patient in the first person;
- *Disease* refers to a malfunctioning of the psycho-biological processes of the individual organism.

While *disease* concerns biomedical knowledge, the *illness* represents the ideas which the family and the society have about suffering and malady. In other words, modern physicians diagnose and care for organic *disease*, while the patients experience *illness*. Kleinman, as well as the other members of the Harvard Medical School, stated the cultural character of both dimensions of malady, *illness* and *disease*. However, their analyses focused on the *illness*. The purpose of their work was in fact to mediate between the different conceptions of illness and health among patients and physicians, often in a multicultural context, to improve the effectiveness of the biomedical therapeutic process. It is then a Medical Anthropology *in* biomedicine. It is no coincidence that often the failure of therapy is caused by noncompliance, that is, the patient's refusal to follow the treatment proposed by the physician. It is brought about by a communications breakdown between the patient and physician, which originates in the contrast between different *explanatory models of illness*, between opposing cultural construction of clinical reality. On the model of the decomposition of the malady in *disease* and *illness*, regarding the recovery of health Kleinman distinguished between *curing* and *healing*. The concept of *healing* (*Heilkunst* in German) has been employed in Assyriology to indicate therapeutic systems that could not be properly considered as medical, in terms of biomedicine. Böck distinguishes, in fact, between *healing*, which addresses the wellbeing of the patient in religious, social and psychological terms, and *curing*, which refers to the «*actual* treatment and removal of illness» (2014: 192, my emphasis).

The attention to personal, family, and social perception of the malady (*illness*) from the patient and his or her family, considered as *cultural* and *social*, has effectively obscured the socio-cultural nature of biomedical diagnostic categories (*disease*). If on one hand, the anthropologists recognized the 'constructed' nature of the *disease*, on the other they saw it as the disease itself, that is a pre-social and pre-cultural, and therefore natural, reality. Not submitting the *disease* and the biomedical knowledge to critical process, they have continued to reiterate the dichotomy between "traditional and non-European medical knowledge *vs.* biomedicine" which, in turn, is

37 For previous discussion on the possible benefits obtained by the use of this theoretical perspective in the study of Mesopotamian medicine see Avalos 1995: 23–25; Heeßel 2004a; Robson 2008; Zisa 2012: 4–7. In 2016 in Berlin a conference entitled "Cultural Classification Systems: Disease, Health and Local Biologic Drugs. Interdisciplinary Approaches to the Study of Cultures in Medical Anthropology and Historical Sciences" was organized by M.J. Geller and U. Steinert.

based on the dichotomy "cultural interpretation *vs.* scientific explanation" and therefore "culture *vs.* nature" (see Young 1982; Zempléni 1985). This kind of analysis seems to me still present in Assyriology. If on one hand, it recognizes the need to use emic concepts of Mesopotamian populations in terms of diagnostic categories, etiologies, care systems, on the other hand, however, it continues to consider them *only* as cultural interpretations. On the contrary, body, emotions, diseases are considered universal psycho-physiological states, therefore natural. Culture is seen as a relevant factor in reference only to the categories. If on one hand the attempts of analysis of Mesopotamian sources using biomedical diagnosis and prognosis are criticized, on the other hand, it does not exclude the practice of explaining these sources referring to the terminology and concepts of modern medicine, psychiatry, and psychology.

This theoretical perspective has been strongly criticized by both anthropologists of *sickness*, which I will discuss shortly, and members of the so-called "Critical Medical Anthropology." Both stressed the need for an anthropology *of* biomedicine, which aims to investigate its socio-cultural and political dimension. Allan Young was one of the first to criticize the theoretical foundations of Harvard Medical School. He argues that we should not limit the epistemological issues to the interpretations of non-Western medical systems. When the epistemological question is suspended for both social sciences, including anthropology, and biomedicine, scholars merely adopt the common sense of the medical view of Western societies (Young 1982: 260). The model *disease/illness* does not take into account how social relations shape the malady, and for this reason, it is no different from the biomedical model which it wishes to overcome. Hence, according to Young, the analysis of social and political relations, which are at the basis of the malady-event, is fundamental. For this reason, he develops the concept of *sickness*. In truth, the *sickness* had been considered a third dimension of malady by the Harvard Medical School. It was defined as the *social relations of the malady*, a general concept that had not been developed. In Young's (1982: 270) reading the *sickness* refers to the historical-cultural process of the production of medical knowledge and maladies, *a process for socializing disease and illness*. The concept of *sickness* has the advantage of putting the attention on the social conditions of production of knowledge, be it a biomedical one or a Mesopotamian one. Under this perspective, I will move towards an analysis of the Mesopotamian healing. Consequently, my work cannot touch upon the dimension of *disease*, because, as said above, the biomedical clinical reality, with its vision of the body and of the pathologies, are cultural and historical products as well. It is therefore not possible to link sicknesses in medical-magic cuneiform texts to diagnostic categories of modern medicine.

Medical system as cultural system

In the previous paragraph, I showed how in the analysis of Mesopotamian medical knowledge an approach aiming to investigate its social and cultural construction is important. The Mesopotamian sources related to the sexual desire (*nīš libbi*) are incantations and rituals which have been encoded and transmitted for millennia. In other words, the sources at our disposal are highly standardized. This is confirmed by the fact that the same incantation, or the same prescription, is preserved on several manuscripts, across space and time. It is not uncommon, therefore, to find the same prescription on a text found in Ḫattuša and dated to the Middle Babylonian period, as well as on a text from Uruk from the Neo-Babylonian period. This kind of documentation inevitably influences the way we do research. Following the tripartite scheme of the malady, proposed in the previous paragraph, my research does not deal with the dimension of the *disease*, and cannot deal with the *illness*, since we did not receive any texts that refer to the experience of the patient (e.g., letters). For example, we do not find any mention of the absence of sexual desire in the correspondence between the *ummânū* (experts) at the Assyrian court (see CJean 2006: 76). Therefore, the dimension of the *sickness* will be the object of my analysis.

A useful theoretical perspective to analyze the cultural construction of the sickness-event is to consider the medical system as *a cultural system*. Kleinman elaborated the concept of "Explanatory Model." Health, malady, and those elements, which circulate in a society in terms of healing, find their expression as cultural systems:

> Such cultural systems, which I shall call *health care systems*, are, like the other cultural systems, for instance, e.g. kinship and religious system, symbolic systems built out of meanings, values, behavioural norms, and the like. The health care system articulates illness as a cultural idiom, linking beliefs about disease causation, the experience of symptoms, specific patterns of illness behaviour, decisions concerning treatments alteratives, actual therapeutic practices, and evaluations of therapeutic outcomes. Thus it establishes systematic relationship between these components [Kleinman 1978: 86].

The advantage of the "Explanatory Model" is understanding the coherence at the basis of relations between all the aspects which constitute a vision that a society has of sickness, suffering, and health. Diagnosis, etiology, prognosis, therapeutic itinerary can be understood only and exclusively if considered in relation to each other. For example, one cannot understand the therapeutic process strategies without considering the causes of the sickness-event. As we will see in Chapter III, it is not possible to understand the sense of some *nīš libbi* prescriptions, isolating them from the categorization of this diagnostic category. As we will see, some ingredients are chosen because they are considered able to act directly against the witchcraft attack, the cause of the suffering. In this perspective etiology and therapeutic practice are deeply interconnected: the second finds its *raison d'être* in the first. It is a model *of* reality and *for* the reality which is provided to individuals by society. It should be noted that Kleinman refers to explanatory models of individuals (*illness*) and not of cultures. Many criticisms have been made against the concept of the "Explanatory Model" as being too abstract and 'rationalist' (see Young 1982). On the contrary, I think that this is a model which, although it tends to simplify and does not take into account other variables of a community (such as conflict, political hegemony), has the advantage of relating all the elements which constitute the sickness-event. In other words, it is a *conceptual model* that helps the researcher understand the community's vision of health, sickness, suffering, and care.

A semiotic approach started with the interpretive turn in Anthropology (see Geertz 1973; Good and DelVecchio Good 1981: 178–180). Another member of the Harvard Medical School, Byron Good (1977) developed the concept of "semantic illness network." He carried out fieldwork for two years in Maragheh, an Azeri speaking city in the Azerbaijan province of Iran, on the 'heart distress" (*narahatiye qalb*). For Good, a "semantic illness network" is the set of illness experiences associated through networks of meaning and social interactions within a community. He makes specific reference to the thesis of Victor Turner (1967: 19–47) on "dominant ritual symbols," according to which these symbols reach their meaning not only as elements of a symbolic system but also as forces of social iteration. The dominant symbols are capable of the *unification of disparate significata* and of *polarization of meaning*. They are the manifestation of the normative principles and *significata* of a community (Turner 1967: 19–32). According to Good's ethno-semantic analysis the 'heart distress' in Iran is an element of the broader category of 'anguish' (*narahati*), associated with loss – Moharram[38] – and old age. Other recurring associations with 'heart distress' are interpersonal problems – nervous disorders – blood – insanity or troubles related to poverty (Good 1977; Good et al. 1985). This attention to the semantics allows us to analyze the paths which connect the symbolic to the emotional and the physiological, the role of language in connecting social experience and sicknesses, and the strategic use of the language of the sickness.

As we will see in Chapter II (§ "Similes concerning the action of the witch") during the analysis of incantations, the *nīš libbi* diagnostic category maintains semantic relationships with other spheres related to suffering, such as other symptoms of a witchcraft attack. The language of binding and slackening muscles clearly represents a semantic complex that transcends the circumstances of this sickness (= lack of sexual desire) to embrace the more general one of witchcraft. Such language symbolically represents the experience of, in the words of Ernesto De Martino (1959, Engl. tr. 2015), *being-acted-upon by* an external force acting on the individual, canceling his ability to act (loss of agency). This is a symbolic totality, which however goes beyond witchcraft to describe semantic associations related to a conception of suffering as immobility – fatigue – weakening – afraid – and crisis in social relations. The fundamental symbolic elements, as defined by Good, as well as the dominant ritual symbols are polysemantic, or rather they fall into different symbolic domains, placed in mutual relation. This explains the variety of elements within a "semantic illness network."

---

38 The festival which celebrates the Battle of Karbala, where the martyred Hossein, the grandson of Mohammad and son of Ali, died.

For this reason, in our case, heterogeneous elements such as weakness, panic attacks, lack of sexual desire, social and family anxiety are considered in conjunction. These symbols form a path connecting the various aspects of human life from social unrest (witchcraft), if not existential (the witch understood not as a single person, but as a generic external force), – conflict, envy, tension – to the emotional and physiological dimension of the person affected by the sickness.

The body

A critical reflection on the concepts of health and sickness must include how the body is thought of and represented. As I mentioned, the Akkadian expression *nīš libbi* literally means 'raising of *libbu*.' The term *libbu*, as we will see in the following sections, does not exactly fit into modern anatomical categorizations. It can refer to the heart, stomach, and uterus. It is also the seat of thoughts and emotions. The complexity of the term inevitably leads us to a critical reflection on the way we apply our anatomical categories to ancient sources.[39]

Linguistic terms used to represent the body necessitate understanding how the body is objectified, thought, and described within a community. It is a very widespread *cognitive* approach both in Medical and Linguistic Anthropology. The body is a physical place of inscription for cultural and social practices and thus its construction is dependent on historical and cultural context. Marcel Mauss was the first scholar who worked on the historical and cultural dimension of the body. His work has inspired a reevaluation on the body within the humanities and social sciences, and not just within biological disciplines. In his famous essay he uses the expression "body techniques," «les façons dont les hommes, société par société, d'une façon traditionnelle, savent se servir de leur corps» (1934: 5). The body, by its socio-cultural modeling, becomes the indispensable instrument of man. Man, through the process of enculturation, learns and imitates the techniques of the tool-body. Learning and imitation take us back to the historical and cultural sphere, and in this perspective, the body is understood as a product of history. Such a process of cultural absorption takes place mostly due to man's mimetic ability: the man observes others and imitates them. It is a continuous and progressive citation process. Citationality is fundamental to learning body techniques. The cultural character of these techniques, however, is subject to a process of *naturalization*. We consider the way we think and use our bodies in everyday life naturally. According to Mauss, in fact, the body technique is an *acte traditionnel efficace*, which «est senti par l'auteur comme un *acte d'ordre mécanique, physique ou physico-chimique*» (ibid. 10).

39 For a study on the Sumerian and Akkadian lexicon of the body see Couto-Ferreira 2009.

This exposure, started by Mauss, of cultural processes, which determine not only the body techniques but the idea itself which society has of the body, affected other societies. This disclosure has not been applied to our cognitive category. For a reflection on the idea which modern science, and with it Western thought, has on the body and its relationship with the mind we had to wait for phenomenology in philosophy and studies on Critical Medical Anthropology. I focus on a famous essay by two anthropologists Nancy Scheper-Hughes and Margaret Lock "The Mindful Body: A Prolegomenon to Future Work in Medical Anthropology" (1987). The scholars define a conceptual model based on the relationship between what they indicate as the "three bodies":

- Individual body, understood in the phenomenological sense of the experience of the body-self;
- Social body, as a natural symbol with which to think about nature, society, and culture;
- Politic body, referring to the regulation, surveillance, and control of bodies regarding reproduction, sexuality, sickness, work, gender.

The anthropologists critically analyzed how Western thought, from Descartes onward, thought about the body. It was placed in opposing relation with the mind. It was therefore outlined as a dichotomy "body *vs.* mind" or "matter *vs.* spirit." Descartes distinguished between *res extensa*, the physical substance, and *res cogitans*, the thinking substance. The first represents physical reality, extended, limited, and unconscious, while the second defines psychic reality, unextended, free, and conscious. This distinction has allowed biological and medical sciences to focus analytically on the physical substance. The field of biomedicine is founded epistemologically on this distinction. It is only interested in the physical dimension of the human body. The scholars reported an interesting case that showed how medical students embodied this vision of medical reality. An old woman who suffered from chronic headaches, describing her condition in front of first-year medical students, she explained that her husband was an alcoholic and sometimes beat her, over the last five years she had always remained at home to look after her mother-in-law, and that she was concerned about her son's problems at school. After the woman's explanation, a student asked the professor: "But what is the *real* cause of the head aches?" For them what was important for diagnostic purposes was information about the neurochemical changes where were understood as constituting the *true* causal explanation (Scheper-Hughes and Lock 1987: 8).

A connection between mind and body in clinical theory and practice has been rephrased by psychoanalytic psychiatry and psychosomatic medicine. However, even these tendencies within biomedicine are, according to scholars, victims of this epistemology and tend to «categorize and treat human afflictions as if they were either wholly organic or wholly psychological in origin: "it" is in the body, or "it" is in the mind» (ibid. 9–10). The concept of somatization, as used in psychosomatic medicine, does not rule out the dichotomy which was mentioned above. If it supports a unity of mind and body, it cannot escape the boundaries of this duality. It effectively treats the sickness as if it had psychological causes, the symptoms of which expand into the physical body, or it treats the sickness as if it had physically identifiable pathologies that have psychological and subjective consequences.

This duality underpins, according to their view, other conceptual oppositions of Western philosophical and scientific thought, such as "nature *vs.* culture," "individual *vs.* society," "reason *vs.* passion." Dichotomies which have crossed the boundaries of specific disciplines: the reflections of the French sociologist Durkheim on the individual-society relation; Freud's dynamic psychology according to which man is in constant struggle between natural instincts and social and moral regulation; Marx's work on the opposition between external and objective nature and the human work which can transform it (ibid. 10–11).

This critical analysis of our ethno-epistemology leads us to rethink the way we understand the dualist relationship between mind and body. Not only that, the concept of self and the person is subject to a great cultural variation as well. While we could consider as universal, in a cross-cultural perspective, intuitive perception of a bodily self, on the contrary, the idea of a self with a permanent state of consciousness, that is an independent and stable individual, and a moral and psychological, as well as legal, person is specific of our culture (see Mauss 1938; Dieterlen 1973; Geertz 1973; Augé 1980; Rosaldo 1984; Scheper-Hughes and Lock 1987: 13–16; Cohen 1994).[40]

As has been demonstrated by Ulrike Steinert (2012a: 124–125; 2012c) in Mesopotamia the person is not based on a dualistic and oppositional model "mind and body," but rather on a holistic conception. The person in Mesopotamia is conceived as pluralistic, an entity constituted by a number of components, and as a holistic, understood as a whole psycho-physical-social.

Our own scientific language is limited by the Cartesian view. As Scheper-Hughes and Lock (1987: 10) underline, we are forced, in the absence of a specific language aimed at describing the relationships between mind-body-society, to employ expressions connected by hyphens, such as bio-social, psycho-somatic, somato-social, which itself is a sign of our descriptive limitations.

One might offer a counterpoint: in Mesopotamia, the causes of suffering are not always attributed to witchcraft or a demon or divine anger. The Mesopotamians recognize that the origin of a sickness is a cause-effect relationship, determined at the

40 The anthropologist and missionary Leenhardt tells that, during his research in New Caledonia on the conception of the person in the Melanesian world, he conducted an interview with the kanak sculptor Boesoou on the changes generated by the European civilization. The anthropologist was sure that the notion of spirit was a Western importation, so he asked: «En somme, c'est la notion d'esprit que nous avons portée dans votre pensée?», but the sculptor objected: «L'esprit? Bah! Vous ne nous avez pas apporté l'esprit. Nous savions déjà l'existence de l'esprit. Nous procédions selon l'esprit. Mais ce que vous nous avez apporté, c'est le corps» (1939: 30–31).

physiological level. One might presume then that such thinking is based on biological criteria. Let us take a simple example: a Mesopotamian man drinks dirty water and knows that this is the cause of his stomachache. He does not attribute the cause of his pain to witchcraft or divine anger. He has established a causal relationship between dirty water and stomachache. Is he using biological criteria to understand his pain? No! The risk of this theoretical approach is to identify the "biological criteria" with the "criteria of biological science." Physiological processes are differently configured. The Mesopotamian man never talks in terms of toxins or pathogens which have caused biochemical and biophysical reactions in the body. This does not mean that there is no attention given to the physiological aspects of suffering or that there are no traces of forms of empirical knowledge, rather it means that the conception of the body and its functioning differs profoundly from our own.[41]

I lingered on the topic of the body and on overcoming the Cartesian dichotomy because, as we will see in the following sections and as mentioned above, Mesopotamian society had a holistic conception of self, without a clear separation between the psycho-emotional dimension and the physiological one. Obviously, these questions bring us back to the sphere of suffering and sickness. As we saw in the *Forschungsgeschichte* two general interpretative trends on the diagnostic category "absence of *nīš libbi*" have used this dichotomy. For Kinner Wilson, it was a psychological disease, found in the DSM-IV, while for Scurlock, Biggs, and Stol as erectile dysfunction, it has a physiological character (without excluding the psychological dimension). In later interpretations, the psychological dimension is still seen as something "separable" from the physiological dimension even if it is connected to it. Scholars reproduce the same model of psychosomatic medicine, as mentioned above. To understand this division of the two dimensions I take into account Biggs's hypothesis on the operation of the *nīš libbi* therapies:

> As everyone knows, erection comes about when a man receives sensory or mental stimulation. These texts – the incantations and the accompanying rituals – provide both. Surely some of the mental images provided by the incantations can be compared to the stimulation modern couples may receive from pornographic films, now often viewable in American motels and hostels (and as I learned, in Helsinki hotels as well) for a fee. [Biggs 2002: 72]

According to the scholar, the effectiveness of the therapy is based on the fact that the prescription (e.g., rubbing the penis) is directed towards the sensory stimulation of patient, while the incantations, with their images of sexually excited animals, target mental stimulation, similar to modern pornography. According to this interpretation, the therapeutic practice acts on the two levels of the Cartesian dichotomy: the phys-

41 See also Martin's (2000: 584) criticism of neurosciences which, according to the scholar, risk falling into a neuro-reductionism, for which the complexity of human behavior can be reduced to the brain's nervous function.

ical and the mental one.[42] But if the two dimensions are not intended as such by the Mesopotamian society, how is it possible that they have devised a care system based on this dichotomy? Or must we assume that, despite Mesopotamian cultural interpretation, physiology and psyche objectively and always work "as everyone knows" (where this "everyone" is the modern Western researcher)? The risk is, as the cultural historian McLaren (2007) writes, writing a history of "sexual impotence," as if the content is not subject to historical and cultural variations. As we will see, the term *libbu* embodies both dimensions, constituting an indivisible whole. This means that the lack of sexual desire is designed as a diagnostic category in physiological, psychological, and social terms.

This needs to overcome the opposition of "mind *vs.* body" for the understanding of sicknesses and suffering in Mesopotamia leads us to another current of anthropological studies on the body: the *phenomenological* approach. Previously, I discussed the cognitive one, now I will briefly mention the phenomenological. The cognitive approach provides us with theoretical and methodological tools to understand how a society thinks, orders, subjects the body, both individual and collective, establishing itself as an anthropology *on* the body; the phenomenological approach has as its starting point for analyzing the body itself. It is then an anthropology *of* the body. The body is no longer the object of analysis, but the fundamental tool of this analysis and of the cognitive process.

To understanding the phenomenological approach let us start with the concept of *embodiment* and that of *embodied subject*. This approach aims to overcome the dichotomy "mind *vs.* body" (see Csordas 1990: 34–39), to define what Scheper-Hughes and Lock call the *mindful body* (1987). As a matter of fact, Asher-Greve (1998: 8) affirms that in Mesopotamia mind *is* in the body, mind and body were inseparable, meaning and understanding were 'embodied.'[43] The concept of *embodiment* was introduced in Anthropology by Thomas Csordas (1990; 1999). His "cultural phenomenology" builds on the concept of *perception* of the philosopher Maurice Merleau-Ponty and that of *habitus* of the sociologist Pierre Bourdieu (Csordas 1990: 8–12; 2011: 139–142). According to Csordas, culture and history are *bodily phenomena*, and not only products of representations, symbols, and material conditions. The author makes a distinction between "body" and "embodiment" in Barthes's model (1984) between the "text" and "textuality": the body is seen as biological and material entity, while embodiment

42 The equation "incantation (= magic) : psychology = medical recipes (= (bio)medicine) : physical disease" is also proposed by Geller 2015: 9. According to the scholar, the incantations are essentially appeals to the psychology of the patient, while medical recipes treat physical signs of disease. For the same idea see Böck 2014: 192. In her view, medical incantations *accompany* the curing.

43 As an example, see what the old man, protagonist of the Sumerian composition *The Old Man and the Young Girl* says: "(I was) a youth, (but now) my personal god, my strength (u s u), my vitality (l a m a) / and my youthful vigor (n a m-g u r u š) have left my loins (ḫ a š$_4$) like an exhausted ass" (ll. 27–28, Alster 1975: 90–99; Leick 2015: 91–92).

is an unspecified methodological terrain, defined by the perceptual experience and forms of "presence" and agency in the world. The embodiment then is a methodological tool that helps us understand how we *are-toward-the-world*, or using Ernesto De Martino's terminology (1948) our *presence* in the world.

This approach should not replace the cognitive one, according to which the body must be investigated through cultural representations. Mary Douglas (1973), for example, studied the body as a natural source of representations which a society produces to about think nature, culture, and society; while Michel Foucault (1963; 1972; 1976) has understood the body as a product of cultural representations, intending it as a *tabula rasa* in which a society inscribes its meanings. A phenomenological approach, which has as an investigative tool the embodiment, interprets the body as a subject of knowledge and studies our perception, and our *being-toward-the-world*. Embodiment is the condition through which we construct the object structure of the social reality.

The focus on the "body" of the scholar is also at the basis of the ethno-psychiatrists' reflections, in particular by Georges Devereux. The scholar in *From Anxiety to Method in the Behavioral Sciences* (1967) criticizes the positivist model of knowledge, which neutralizes the observer/researcher position, and which is based on processes of objectification of human behavior. According to Devereux, the practice of self-analysis is of great importance for improving the quality of research in the humanities and psychological and psychiatric sciences. Its core lies in the transference–counter-transference dialectic, which allows a two-way knowledge model on the interaction subject–object–subject. The transference–counter-transference model thus becomes a general paradigm of the observer–observed relationship in the field of humanities. Focusing on himself or herself, the researcher can reach a wider knowledge, although this involves a critique of his or her theoretical-ideological background. As pointed out by Roberto Beneduce (2007: 129), in Devereux's theory the considerable influence of Claude Lévi-Strauss on the observer–observed relationship is found, according to which, during the process of objectification, the researcher becomes aware of the fact that the data of ethnographic research – in our case historical – which was obtained most objectively, is in fact integrated into his or her subjectivity. According to Devereux, the "neutrality" and the "objectivity" of the observation process turns out to be an ideological construction produced by the observer to avoid those perturbations that are generated on his own body at the moment of the encounter with the observed subject.

And the mind?

The development of Ethno-Psychiatry has inspired an epistemological debate within Psychiatry and Psychology, thanks to studying the mental otherness: from Franz Fanon's studies (1952; 1961) on the need to *decolonize the madness*, that is to disclose the processes of domination inherent in psychiatric practice when studying otherness, to recent works by Tobie Nathan (1995; 2001) aimed at establishing a new paradigm of scientific psychopathology, which takes into account for therapeutic efficacy "other" care practices. What is interesting for me is emphasizing the relationship between emic nosology and psychiatric diagnostic classifications. As I mentioned above, Kinnier-Wilson (1965; 1967) was the first in Assyriology to uphold the importance of studies on psychiatric and psychological diseases/distresses in Mesopotamia.[44] Scurlock and Andersen, as well, shared the same approach (2005: 367–385). In the chapter "Mental Illness" they read the cuneiform texts through the lenses of psychiatric diagnostic categories. Geller (2007b: 38) supports the possibility that some "diseases" come from inside (= brain?) and thus are of a psychological (and psychiatric?) order. This approach was criticized by Érica Couto-Ferreira (2020: 261–263), in fact the use of the label "psychology" is problematic because it implies the recognition of the existence of an entity "psyche," a mind that can be separated from the body, but also of the predominant role the brain plays in all processes regarding thought and feeling. As Parys (2017: 106) underlines, it would be better to talk in terms of psychic and mental symptoms, instead of mental diseases.[45]

What are the relationships between culture and mental disorders? A brief consideration should be made about the mental disorders called *culture-bound syndromes* (CBS). According to the *Diagnostic and Statistical Manual of Mental Disorders*, they combine somato-psychic disorders with specific meanings attributed to a particular cultural context. Among them, *amok* is a Malaysian diagnostic category characterized by severe psychic suffering or mental disorders which appears in a murderous rage. It mostly affects men (*pengamok*), who armed with a knife, kill anyone they encounter in their path.

The expression CBS was used for the first time by the Chinese psychiatrist Pow Meng Yap to include these mental disorders in the *Diagnostic and Statistical Manual of Mental Disorders* of American Psychiatry. He argued that these syndromes are not specific to a particular cultural context, that they do not represent distinct nosologies, but that their symptoms are determined by cultural factors (Yap 1969). According to this interpretation, a natural clinical reality exists but it is shaped so radically

44 Despite Kinnier Wilson's research, until now there are few studies on Mesopotamian insanity and mental disorders (see Ritter and Kinnier Wilson 1980; Geller 1999; Geller 2003; Reynolds and Kinnier Wilson 2004; Reynolds and Kinnier Wilson 2008; Reynolds and Kinnier Wilson 2012; Reynolds and Kinnier Wilson 2014; Stol 2009a; Yuste and Garrido 2010; Chalendar 2013; Parys 2017; Attia 2018; Attia 2019; Couto-Ferreira 2020). On the brain and marrow see Westenholz and Sigrist 2006.

45 For the cultural and social aspects of psychopathology and psychiatric care see Obeyesekere 1990. For a criticism on the use of modern psychiatric and psychological diagnostic categories on the study of other medical system see Obeyesekere's work on 'depression' in Sri Lanka (1985). On the depressive state and melancholy in Mesopotamia see Gruber 1980; Alster 1983; Stol 1993: 27–32; Barré 2001: 178–181; Kselman 2002; Maier 2009; Couto-Ferreira 2010a: 25–27; Buisson 2016; Van Buylaere 2020. Geller in the article "Freud and Mesopotamian Magic" (1999) reads Mesopotamian rituals and incantations considering Freud's theories. See also Abusch's (1985: 95) criticism to Kinnier Wilson's approach.

by culture, that its natural substrate does not clearly emerge. Obviously, psychiatry is the only science to have access to this natural dimension of mental suffering. As Hahn (1985: 166) writes, one insinuates the idea that there are *culture-free* suffering states and other *culture-bound* states, whereas the former is considered real nosological categories, that is provided with an ontological foundation.

The debate about CBS within the Ethno-Psychiatry has developed, and there have been criticisms to this concept. Ethno-psychiatrist and medical anthropologists have insisted on the fact that culture always plays a key role in shaping human suffering. The psychiatrist Inglese defines CBS as *culturally ordered syndromes*, whose intelligibility depends on the context of their phenomenology within a specific historical and cultural context. The socio-cultural dynamics as a whole give shape to the experiences of human suffering and pain (Inglese 2007: 263–264). If every "syndrome" phenomenon always has a cultural component, that means that the concept of CBS, developed by the DSM is useless (see Hahn 1985; Ciminelli 1997; Ciminelli 1998; Beneduce 2007: 205–208).

Because of the cultural dimension of sickness-event, some scholars have proposed to include among CBS some diagnostic categories of Western modernity, such as obesity or anorexia nervosa. The anthropologist Lock (1987) has gone far beyond defining the *Diagnostic Statistical Manual* III as a *culture-bound construct*. For this reason, according to the French ethno-psychiatrist Tobie Nathan (1995; 2001), scientific psychopathology should not deal with the "nosological tags" of modern psychiatry, thus defending an ideology, a doctrine, a kind of culture, but rather have as object of analysis the "therapists and techniques" of "other" therapeutic systems. According to the scholar, the observation of *real techniques* of the therapeutic actors is at the basis of the reconstruction of a *theory of these techniques*, from which to make *models of operation and of theoretical objects*.

Logical thinking

With a polemical tone Nathan (1995) affirms that "white people" distinguish between two kinds of society: those in which the thinking prevails over believing, and those in which the believing prevails over thinking. "White people" belong to the first kind. The ethno-psychiatrist emphasizes the problematic character of the concept of "belief," often used in opposition to "thinking/knowledge," implying scientific thought.[46] Robson (2008: 463) affirms the necessity to counter the dichotomy of "knowledge *vs.* belief" in the study of Mesopotamian medicine, according to which we, as modern researches or common person, "know," while they, the Mesopota-

46 For a understanding of Mesopotamian medicine as 'science' see Geller 2015: 11–18. *Pace* Böck 2014: 194.

mians, "believe."[47] For example, a statement by Böck utilizes this dichotomy: «The ancient Babylonian healing system is a fusion of views, which we would characterize as *religious and magical beliefs* and *natural concepts*» (2014: 180 my emphasis). According to the scholar, the healing strategies used by Mesopotamians are of two kinds: belief (magical and religion domain) and concept (= thinking, natural one). The religious-magical sphere is considered to be a belief, while attention to the natural dimension[48] is a (theoretical) concept.[49] The dichotomy "belief *vs.* thinking" follows a series of other oppositions such as rational/irrational, logical/prelogical, and, regarding the sphere of healing, that of magic/medicine. The continuous development of the concept of "belief" within Anthropology (see MRichter 1973: 43–52; Good 1994: 1–24; Young 1997; Severi 2004) and studies on logical thinking in non-Western society have certainly changed the previous paradigm, at least in the social sciences. Anthropologists have shown that rather than "belief," we should talk in terms of *conceptual systems*, provided with an internal logic.

Already Evans-Pritchard (1937) in his study of Zande magic, criticized the dichotomy, proposed by Lévy-Bruhul, between logical and pre-logical thinking. Evans-Pritchard demonstrated the logical coherence within Zande thought on which the witchcraft phenomenon is based. Witchcraft, as well as the therapeutic system, therefore, must be understood within a coherent cultural and logical knowledge system. Witchcraft, which is based on different premises from those of scientific thought, arrives at the conclusions, which are likewise logical and consistent within the cultural system. So magic must be understood as a knowledge system, or, according to Young (1997), as a specific rational *way of reasoning*.

In this regard Lévi-Strauss' (1962b) studies on the *pensée sauvage*, where the anthropologist demonstrated the logical classification criteria in non-Western societies, and Tambiah's works on analogical thinking (1985) are fundamental. It is impossible, using the analytical tools of modern science, to understand the logic and the consequent effectiveness of magic-therapeutic systems of other society: «It is *inappropriate* to subject these performative rites to verification, to test whether they are true or false in a referential or assertive sense or whether the act has effected a result in terms of the logic of "causation" as this is understood in science» (Tambiah 1985: 81).

47 Böck talks in terms of 'system of beliefs' (2014: 166).

48 Does 'natural' mean 'physiological' or 'referential'?

49 Though using the term 'concept,' Böck affirms that «the ancient Babylonians explained disease in metaphysical, not theoretical terms» (2014: 194–195). It seems to me that by 'theoretical' she means our Western scientific thought (see her criticism addressed to Geller, who uses the term 'science' referring to Mesopotamian medicine). This suggests that Mesopotamian thought is not theoretical because it is not like ours. It is true, there were no cuneiform theoretical texts, and the conceptual system behind Mesopotamian healing was different from ours, but affirming that it was not theoretical means reiterating the dichotomy mentioned above.

Under this perspective, a distinction between texts or practices considered as 'medical' or 'magic' is unsuitable and useless.[50] Heeßel (2004a, esp. 110–113; 2004b, esp. 7–9) underlined the necessity to overcome this dichotomy in the study of Mesopotamian therapies.[51] Studies on Mesopotamian medicine seem to be 'obsessed' with these two categories: magic and medicine. It is a frantic search, tracking down elements, text, or practices considered as 'medical' (that is 'biomedical').

Geller has always supported the need for this distinction (2007a; 2015), arguing that if one rejects the distinction between 'magic' and 'medicine' in Mesopotamia, it becomes difficult to explain the different roles adopted by these two professionals *asû* and *āšipu* (see Geller 2015: 166). So, the division between magic and medicine makes sense as the distinction between two professional operators. The problem is that, despite the reiteration of the dichotomy, our knowledge about the distinction between the actions of the *asû* and that of the *āšipu*[52] is still very scarce. The distinction between medicine and magic does not explain anything behind the action of the two professionals. As written by Biggs (1995: 1918) «the distinction is more theoretical than real», while Heeßel «assigning texts to *āšipūtu* or *asûtu* presupposes a differentiation which has to be demonstrated, not assumed» (2004a: 9). Criticisms are made by Robson as well: «It is unclear to me why anyone should try to do so [make the distinction]; it is no more likely to succeed, or produce meaningful results, than if one tried to separate Mesopotamian 'physics' from 'chemistry' or any other two post-Enlightenment scientific concepts» (2008: 476). I think the term "therapeutic" is more suitable to describe any attempt to eradicate the evil afflicting the individual or the community.

The research aim here is to understand the logical strategies on the basis of the Mesopotamian medical system.[53] I do not want to reconstruct "a history of biomedicine" starting from Mesopotamian sources, according to whose perspective "progresses" and "regresses" exist (see Scurlock 2006b). The aim of my research is not an investigation of 'medical' care techniques – in the sense of biomedicine – of loss of sexual desire, understood as an ahistorical nosological entity, nor of discerning which care practices can be attributed to magic and which cannot be.

---

50 For a discussion on Mesopotamian magic from an anthropological perspective see van Binsbergen and Wiggermann 1999.

51 See Böck 2001–2002: 230–231. On this topic see also Geller 2015: 11–18, 161–167. Böck (2014: 178) defines magic as a «motivational attitudes towards» and, quoting Czachesz, «*illusory* manipulation of the visible or invisible realities» (my emphasis).

52 On the distinction between the two professionals see Oppenheim 1964: 254; Ritter 1965; Stol 1991–1992; Haussperger 1997; Scurlock 1999; Abrahami 2003; Verderame 2004a: 15–16; Robson 2008: 467–476 esp. 475–476; Fales 2015: 21–25; Steinert 2018a: 90–116.

53 Heeßel (2004a) investigated the rationale in the *Diagnostic Handbook* SA.GIG/*Sakikkû*, see esp. p. 110. See also Rochberg 2015.

Sexuality

I would like to mention the sexual dimension briefly also, thrown into crisis by the diagnostic category "lack of sexual desire." As we will see in the next chapter, a central aspect of the male gender construction in Mesopotamia consists of sexual practices. The absence of sexual desire put a strain on some aspects of male sexuality, and consequently on the male gender.[54] In my view, the therapeutic practice, on the contrary, conveys normative ideals of sexuality and gender. The therapeutic performance, especially the recitation of incantations, organizes gender and sexuality through *citational processes* (see Butler 1990). It therefore constructs and transmits *a normative ideal* of male sexuality. Bear in mind that I use the term "ideal" to indicate aspects of the construction of normative gender, which, however, can be contradicted by the concrete practice of individuals or groups in different power relations. I also used the indefinite article "a," to emphasize the fact that normative apparatus is one of many deployed by the power to organize and to subject bodies and sexual practices in Mesopotamia. Besides, the indefinite article takes account of the fact that other regulatory models of male sexuality and gender exist. Moreover, it considers the variety of forms in which the gender apparatus is constituted in relation to other variables, such as age, social group, job. Thus, my idea is that the *nīš libbi* therapies provide elements of great importance for the understanding of a normative model of some aspects of male sexuality and the sexual relations between men and women in Mesopotamia.

We now need to clarify some terms I used to avoid confusion. The reader who is familiar with the post-structuralist philosophy and gender and queer perspectives will perhaps understand what was stated before, but it is good to clarify. As is well known, the philosopher Michel Foucault had a significant impact on gender and sexuality studies. His famous book *Histoire de la sexualité, La volonté de savoir* (1976) is one of the foundations of the post-structuralist approaches to the topic. Foucault employs an *archaeological method* to analyze how modernity thought of and constructed sexuality and how this construction created an inseparable relationship between knowledge and power. The philosopher, therefore, explains the relationship of this triad, knowledge–power–sexuality. He shows how since the end of the sixteenth-century discourses on sex have proliferated from many directions, aimed at analyzing every detail. It is a "transposition in speech" of sex. During the seventeenth and eighteenth centuries, a veritable *scientia sexualis* began to emerge. It involves medicine, psychiatry, pedagogy, and psychology, aimed at investigating the forms of peripheral sexuality and not that of the couple within the family. The discourses on peripheral sexuality have turned the latter into the "pathological" or "perverse" (when they were previously considered "unconventional" in a moral and legal per-

54 For this topic in a historical perspective see Laqueur 1990.

spective),[55] but they have put attention on the forms of "normal" sexuality as well. On the other side, governments begin to create control, administration, and management mechanisms of sexuality. Power and knowledge, therefore, are connected in a self-sustaining spiral (ibid. 26–67). Sex, in this view, retains a causal inexhaustible and polymorphous power. Sexuality is established as the truth about sex and pleasure. It is the object and field of such *scientia sexualis*. It constitutes a production and apparatus of true discourses on sex. Sex is inscribed in an ordered regime of knowing (ibid. 69–98).[56]

What was just laid out necessitates some clarification. When I refer to the term "sexuality" I do not, therefore, mean the same term employed by modern psychological and medical sciences. Talking in these terms of "sexuality," as described by

---

55 On the power of norms in differentiating 'normal' and 'pathological' see the famous volume of Canguilhem *Le normal et le pathologique* (1966). On the concept of 'norm' in Foucault and Canguilhem see Macherey 2009.

56 According to the traditional opposition "sex *vs.* gender," with the term "sex" one refers to biological and anatomical differences between male and female, as for instance genetic makeup and shape of genitals; in contrast, the term "gender" indicates the process of social and cultural construction of "man" and "woman" (i.e., their social roles). Sex is natural, what preexists us as men and women, whereas gender, being cultural, is not innate but learnt. The anthropologist Gayle Rubin, in *The Traffic in Women* (1975), introduces the concept of "sex/gender system," according to which cultural processes, starting from a biological basis, allow the assumption of a binary system that is asymmetric and hierarchical (in which what is male holds a dominion (see Bourdieu 1998)/privilege (see Ortner and Whitehead 1981) over what is female). Through an analysis of the works of Marx, Freud, Lévi-Strauss and Lacan, the anthropologist defines a "political economy of sex," through which she points out a decisive social-cultural construction in the intra-sexual relation, that have to be opposed to the biological fixity expressed by the concept/term "sex." Queer theory, but more generally the whole of Gender Studies, has tried to show how such distinction between sex and gender is inappropriate. These studies have demonstrated how such dichotomy is the result of a cultural process of categorization of human beings in a sexual dimorphism of "male/female" and "men/women," that nothing is natural, but that finds its *raison d'être* only in a hegemonic (also on a scientific level) perspective. The very act of conceptualizing the two sexes is a phenomenon which is property of the Western world, which affirms that sex is the foundation of gender and that, consequently, as "first wave" feminists affirmed, there are only two genders, and that the model of human sexuality is heterosexuality, since it is connected to reproduction. Many scholars have strongly criticized the supposed biological-essence of the concept of sex, by pointing out the social-cultural, as well as political, elements connected to its construction. It is not possible here to mention the complexity of these studies in notes. I just want to point out, for example, that studies show how the ideology of "sexual difference" also involve the biological sciences. An example is the Kessler's work (1990) on the determinations of sex and on the techniques of intervention on intersex infants. See also Butler 1990; Delphy 1991; Delphy 1996; Mathieu 1991; Hood-Williams 1996; Yanagisakosi and Collier 1997; Busoni 2006: 41–73. As Judith Butler argues: «Gender is not to culture as sex is to nature; gender is also the discursive/cultural means by which "sexed nature" or "a natural sex" is produced and established as "prediscursive," prior to culture, a politically neutral surface *on which* culture acts» (1990: 7). For a discussion on this topic and its relation with the Ancient Near Eastern Studies see Garcia-Ventura and Zisa 2017: 43–44.

Foucault, implies an ontological dimension of the sexuality or the sexual identity itself, which was unknown in Mesopotamian society. I refer rather to *the whole range of sexual practices*. They are of course subject to normative processes, but not through "dispositif de sexualité," but according to different principles that establish sexual behavior and practices as legitimate or illegitimate according to precise classification schemes. Even when the sexual practices meet the pathological sphere, as in the case of the absence of sexual desire, the texts reiterate legitimate normative models of sexual practices. No mention is made of a sexual ontology, that is as an understanding in terms of sexual nature (physical or psychological). This suggests that any attempt to explain the Mesopotamian sexuality with the parameters of modern Western sexuality is wrong because it conveys a specific modern ideology.

I discussed the relationship between power and knowledge. But what is meant by power? I refer to power in the meaning given by Foucault and later used by post-structuralist philosophy, gender and queer studies, and critical anthropology. Foucault defines power as following: «Par pouvoir, il me semble qu'il faut comprendre d'abord la multiplicité des rapports de force qui sont immanents au domaine où ils s'exercent, et sont constitutifs de leur organisation» (ibid. 122). The philosopher talks in terms of an *omniprésence du pouvoir*. It occurs at all times and in every relation: power is everywhere; not because it embraces everything, but because it comes from everywhere. Power is not an institution or a structure or a strength held by someone (no dominant–dominated relationship), on the contrary, it is the name given to a complex strategical situation in a particular society. It is exercised from innumerable points, in the interplay of non-egalitarian and mobile relations; power relations are not in a superstructural position, on the contrary, they have a direct productive role, being matrix of transformations. Power is intentional but non-subjective: it has aims and objectives, but it is not a result of individual and subjective choice. It always involves forms of resistance but to resist one needs to be inside this game of political relationships (ibid. 122–130).

Knowledge and power are two different realities with two specific roles, but they interact with each other because one supports the other and vice versa. To understand this interrelationship between technical knowledge and power, it is necessary to analyze what Foucault calls "local centers" of power-knowledge (*foyers locaux de pouvoir-savoir*) (ibid. 130). An example is given by the relationship between the therapeutic operator and the patient. Therefore, the performance of therapy constitutes a *local center of power-knowledge*, and as such, it is analyzed in this book. The therapeutic itinerary for restoring sexual desire, therefore, is to be considered in Foucault's sense as a place of power relations and organization and subjection of bodies, sexuality, and gender according to a normative apparatus of knowledge inherent in therapeutic techniques, recognized socially as a discipline (*āšipūtu*).

Meaning of *nīš libbi*

From *libbu inaššīšu* to *nīš libbi*

At this point, I must define the meaning of the Akkadian expression *nīš libbi*. It is composed of two nouns *nīšu* and *libbu*. It is the nominalization of the verbal expression *libbu inaššīšu* (*et similia*) (see Zisa 2020: 459–461). Consequently, for understanding the meaning of the nominal form we need to see it through its verbal expression. Let us start with the verb *našû* and its nominal form *nīšu*. The verb (CAD N/II 81–112; AHw. II 762–765) usually has a transitive value with the basic meaning of 'to lift, raise' and 'to carry' in very different contexts. It indicates:

- 'to lift something,' for example during a ritual or an oath ceremony, such as an image or a sacred object; 'to brandish' a weapon, a torch, a signal;
- 'to elevate a person to a high position,' like the king in relation to the gods' power; 'to raise part of the body' in medical and ritual texts; on the other hand, 'to transport, carry something.'

Therefore, because *nīšu* is the *PiRS*-form of the verb *našû* indicating a *nomen actionis*, the generic sense of the term is 'raising of something' (as accusative, see CAD N/II 294–297; AHw. II 797).

The noun is used in expressions which have both a literal and figurative meaning, here are some examples:[57]

1. *nīš qāti*
Meaning: lit. 'lifting of the hand' → 'prayer.'[58]
The verb *našû*, indeed, with the object *qātu* can be translated 'to recite a prayer with hands uplifted' (see Oppenheim 1941: 269).

2. *nīš rēši*
Meaning: lit. 'lifting of the head' → 'promotion, honor'[59] (see Oppenheim 1941: 252–256; DBaker 1976: 31–39; Gruber 1980: 607–613; Steinert 2012a: 198–199 and fn. 246).

---

57 Note the line 48 of the *Aššur Medical Catalogue* in the section "anus": DIŠ NA *ina la si-*[*ma-ni-šú* MURUB$_4$^{meš}-*šú* GU$_7$ ^{meš}-*šú* (. . .) : D]IŠ[!] NA KÚM UD.DA EN.TE.NA IM *u ša-*⌜*ra* APIN⌝ NU ÍL "If a man [has pain in his hips] prematurely. If a man cannot 'raise the seeder-plough' (due to) fever, *ṣētu*-fever, chill, flatulence, and 'wind.'" The expression APIN NU ÍL "he cannot lift/raise the seeder-plough" maybe is metaphorical (see Geller 2005a: No. 48, Scurlock 2014b: 302; Steinert 2018b: 238), since the seeder plough could metaphorically refer to the penis (see Lambert 1982; Lambert 1987a: 33; Wilcke 1987: 70, 84 fn. 1–2; Livingstone 1991: 6) and the use of the verb *našû* "to lift, to raise" as in reference of the sexual desire.

58 See the š u-í l-l á/*šuillakku* prayers.

59 See the meaning 'to pay attention, honor, exalt' of *rēša našû* (CAD N/II 108), see Nougayrol 1945–1946: 63.

3. *nīš īnī*

Meaning: lit. 'lifting of the eyes' → 'look, glance' in a benevolent sense.

It refers to the divine glance (Gula, Marduk, often thaumaturgic gods) giving protection, health, well-being.

When *īnu* is the object of *našû* it means 'to look intentionally, look for something,' and consequently 'to covet,' also in a sentimental and sexual meaning (see Stamm 1939: 125):

ana dumqi ša Gilgāmeš īnī ittaši rubūtu Ištar
The lady Ištar looked covetously on the beauty of Gilgameš.
[*Epic of Gilgameš* VI 6, George 2003: 618]

īnī tattašīšum-ma tattalkiššu
You looked at him with desire and went up to him.
[*Epic of Gilgameš* VI 67, George 2003: 622]

ša ana alti tappîšu iššû [īnīšu]
He who coveted his neighbor's wife.
[*The Šamaš Hymn* 88, Lambert 1960a: 130]

It is evident that the word *nīšu* is usually connected with parts of the body not only with a physical meaning, but also a figurative one: *nīš īni* 'glance' in a benevolent sense; *nīš rēši* 'promotion'; *nīš qāti* 'prayer.'

The expression *nīš libbi*, however, differs from the others mentioned above by the fact that it is a subject nominalization. Indeed, in the verbal expression *libbu* is the subject of the action.

When the verb *našû* has *libbu* as subject (+ person of object) (log. ŠÀ ÍL-*šú*), it means 'to wish, desire, crave' (see CAD N/II 105) in different contests. See the following examples:

ana dabābi libbašu lā inaššīšu
(If a man) does not want to talk.
[Gurney 1960: 224, l. 28, see Abusch and Schwemer 2011: 118, No. 7.2: 5]

[pā]nūšu šaknūšu libbašu našīšu k[abattašu u]blamma tiṣmuru-ma
(Šamaš' and Adad's great divinity knows the king who) has set his face, desires, has determined and plans (to march against his enemy's land).
[K. 2608+2633+3101b+3435: 2, Lambert 2007: 68, No. 5]

ana epēš Esagila / našânni libbī
I desire to rebuild the Esagila.
[Nebuchadnezzar Inscription 15 iii 18–19, Ball 1987: 101][60]

60 See for other similar expressions Lambert 2007: No 3a: 4; No 3c: 32; No. 5: 2; No. 11: 12.

ana kašād ṣibûti libbašu lā inaššīšu
He who has no interest to realize what he desires.
[KAR 26: 10, Mayer 1999: 149, see *nīš libbi* D Sympt.: 43]

In the Old Babylonian version of the *Epic of Gilgameš*, the king of Uruk is determined to fight Ḫuwawa, the elders of Uruk push him not to pursue his aim. They speak to the king as follows: *ṣeḫrēti-ma Gilgāmeš libbaka našīka* "You are young, Gilgameš, and your heart has carried you" (OB III v 191, George 2003: 202). That means that Gilgameš is carried away by the enthusiasm and the desire to combat the guardian of the Forest of Cedar.

In anti-witchcraft texts, as well as in the *nīš libbi* corpus, the recurring expression has the woman as the object of desire: *ana sinništi libbašu lā inaššīšu* "He does not desire the woman." For this reason, the expression assumes a sexual value and means "to desire sexually":

šumma amēlu ana sinništīšu iṭḫe-ma [. . .] / *ana sinništīšu libbašu lā inaššīšu* [. . .]
If a man approaches his woman and [. . .] he does not desire his woman [. . .].
[*nīš libbi* H Sympt.: 2–3, see *nīš libbi* K prescr. 1: 1; Abusch and Schwemer 2011: 118, No. 7.2: 10; ibid. 321, No. 8.6: 19; ibid. 437, No. 12.1: 88; AMT 76, 1: 6]

[*šumma amēlu nīš libbī*]*šu ṣabit-ma ana sinništīšu u ana sinništi aḫīti libbašu lā inaššīšu* [. . .]
[If a man's sexual desire] is taken and he does not desire either his woman or another woman [. . .].
[*nīš libbi* D prescr. 3: 40]

In the corpus in question, therefore, sexual value characterizes the expression, even when the object of desire is not explicit:

balu patān išattīšu-ma amē[*la šuāti libbašu i*]*naššī*
He drinks it on an empty stomach and [that] m[an will desi]re.
[*nīš libbi* N prescr. 24 le. e 7]

amēlu šū adi balṭu libbašu inaššīšu kišpī lā iṭeḫḫûšu
This man, as long as he lives, will desire (sexually). The witchcraft will not come near him.
[*nīš libbi* F prescr. 8: 28]

As with the man, so the woman feels sexual desire as well:

annâ šalāšīšu tamannū-ma (var.: *kīam iqabbi*) *zikaru u sinništu libbašunu* / *ištēniš inaššīšunūti ul inuḫḫ*[*ū*]
You recite (var.: you say) this (incantation) three times and (if) the man and the woman / desire together each other, (but) they do not find reli[ef].
[No. D.3: 35–36]

If the verbal expression, therefore, has the meaning of 'to desire,' *nīš libbi*, as its substantive, means 'desire' (CAD N/II 296; AHw. II 797 'Begierde'). In this regard, I quote a passage of an Old Babylonian letter from Mari: *nīš libbim iraššī* "He will realize the desire" (C-FJean 1950: 57, No. 23 rev. 21′). Consequently, if, as in many cases, the object is the woman, the expression means 'sexual desire.' Here an example from an Old

Babylonian physiognomic text: *šumma umṣatu ina mu[ḫḫ]i ūrīšu šakin nīš libbi ana aššatīšu* [. . .] If there is an *umṣatum*-mole in his pubic area, sexual desire to his wife [. . .]" (YOS 10 54 r. 12, Böck 2000: 296).

The expression designates a quality possessed by man, and woman as well. The *nīš libbi* is therefore an intrinsic human component which, however, can be removed. The verbs used to indicate the loss of *nīš libbi* are the following:

1. *ṣabātu* 'to seize' (CAD Ṣ 5–41; AHw. III 1066–1071 'packen, greifen, nehmen'; Couto-Ferreira 2007: 16–18)[61] in the expression "the sexual desire is seized" (*nīš libbīya ṣabit*) (D prescr. 3: 40; D Sympt.: 43; Abusch and Schwemer 2011: 30, No. 8.4: 49, 71; No. 9.1: 22; *Maqlû* I 99, Abusch 2016: 50; *Diagnostic Handbook* SA.GIG/ *Sakikkû* XIII 41, Labat 1951: 126).
2. *tabālu* 'to take away' (CAD T 11–20; AHw. III 1297 'wegnehmen, wegtragen') (see No. L.2: 20 and 22);[62]
3. *ekēmu* 'to deprive' (CAD E 64–69; AHw. I 194–195 'wegnehmen') (KAR 226 i 9, Maul and Strauß 2011: 74, No. 31);
4. *eṭēru* 'to take away' (CAD E 401–404; AHw. I 264: 'wegnehmen, retten') (No. E.2: 18; F prescr. 12: 36; N prescr. 16 iii 27 and 21 iv 11 (restored); BA M 319: 4, Farber 1977a: 236; Ashurbanipal's inscription K 2411 iv 27′, MStreck 1916: 302, also Matsushima 1988: 101).

Sexual desire, as intrinsic quality of men, can 'come to end' (*qatû* CAD Q 177–183; AHw. II 911–912 'zu Ende gehen'). A prescription in the corpus from Ḫattuša informs us that the *nīš libbi* of a man comes to end in the month of Nisannu (N prescr. 1: 1–2).

It follows that after being taken away, sexual desire can be recovered through therapeutic practice. The verb which indicates the possession and the recovery of this quality is *rašû* (TUKU) 'to acquire, get' (CAD R 193–206; AHw. II 961–962 'bekommen, erhalten, erwerben'). The patient is described as one who "does not have sexual desire" (*nīš libbi lā īši*), while at the end of the therapeutic practice one states that "he will get sexual desire" (*nīš libbi iraššī*). The ritual aim is in fact "to get sexual desire" (*ana nīš libbi rašê*). Following Biggs (1967: 4), the expression *nīš libbi*/ŠÀ.ZI.GA at the end of the prescription should be understood as an abbreviated form for *nīš libbi iraššī* "he will get sexual desire."

61 The verb *ṣabātu* is used to indicate the attack of the sickness. Such language is, in fact, used to describe the malevolent demonic character of the entity bearing the sicknesses, as well as the attack by witchcraft (see Van de Toorn 1985; Couto-Ferreira 2007; Salin 2020: 94–97). When the subject of the verb *ṣabātu* is a sickness, a demon, bad luck or even sleep, for instance, whilst the object is the patient, the word can be translated as 'to take over, overwhelm' (see CAD S 6 mng. 1). In general, the semantic root of the word means any violent action that aims for the appropriation of an object, a person, a rival city, by the use of force.

62 The verb *tabālu* is used to describe the loss of attractiveness and virility (*dūtu*), often due to witchcraft: see *Maqlû* III 9 and 12; KAR 177 r. ii 7, r. iii 3–4 (see CAD T 19 mng. 2 2′d; AHw. III 1297).

As we saw, the expression *nīš libbi* appears in this corpus as ŠÀ.ZI.GA (log. wr.), at the end of the rituals and prescriptions as a shorthand writing or an abbreviation form for *nīš libbi irašši* "he will get sexual desire." It is always written logographically in the rubric as KA.INIM.MA ŠÀ.ZI.GA "Wording of *nīš libbi* (incantation)." It identifies a category of therapeutic texts, as is confirmed by the text from Ḫattusa *nīš libbi* N (KUB 4, 48) prescr. 25 lo. e. 5: DUB 1.KAM DIŠ LÚ ŠÀ.ZI.GA "One/first tablet: 'If a man's *nīš libbi*.'" Since we have no trace of a standardization process for the corpus, this line should be understood as a sequence established by the scribes in Ḫattusa. Nevertheless, it shows that the expression in addition to indicating *strictu sensu* the sexual desire, defines the therapy, that is rituals, incantations, and prescriptions, aimed at recovering the sexual desire.

I think that the mention of *nīš libbi* in the Neo-Babylonian text *Taxonomy of Uruk* (SpTU 1, 43, Köcher 1978: 24–25; Stol 1993: 23–24; Heeßel 2010a: 30–31; Geller 2014: 3–9; Steinert 2016: 230–242) should be understood in this way. The text links a part or an organ of the body with a series of diagnostic categories. The section (ll. 25–31) dealing with the kidneys (*kalâtu*) mentions the *nīš libbi*: KI.MIN (= *ultu kalâti*) : *nīš libbi* "Ditto (= from the kidneys): sexual desire" (l. 26). It seems to me that the mention of this expression here is since it, in addition to indicating the sexual desire, represents the therapeutic practice. That is the reason for its mention in this text.[63]

The *nīš libbi* expression appears other times in the corpus:

- It is compared to flowing river water: *nīš libbīya lū mê nāri ālikūti* "May my sexual desire be constant river water!" (No. A.1: 35); See also the catalogue LKA 94 i 28. [ÉN] ÍD [ŠÀ].ZI.GA GIN.A "[Incantation]: 'River of a constant sexual desire'";
- The female partner claims to have prepared a bed for the *nīš libbi*, a clear reference to the sexual intercourse: *šiptu*: ***nīš libbi nīš libbi mayyāl nīš libbi*** "Incantation: Sexual desire! Sexual desire! The bed for the sexual desire" (No. M.1: 4);
- It is associated with brightness in an incantation: [ÉN] SU.ZI MIN ŠÀ.ZI.GA MIN "[Incantation]: "Shiver! Shiver! Sexual desire! Sexual desire!" (catalogue LKA 94 i 19, see ibid. i 26; D prescr. 4: 47–48; K prescr. 28: 77–79, also *kunuk ḫalti* series ("*ḫaltu*-seal") K. 3010+ v 24′ and 32′, Schuster-Brandis 2008: 365, text 16 A);
- When Enlil and Bēlet-ilī have created humankind, they have decreed at the same time its *nīš libbi*: x x x x ⸢šà⸣-zi-ga-b[i]? nam-e "They (= Enlil and Bēlet-ilī) have decreed his sexual desire . . ." (No. K.1: 11).

These attestations confirm what was stated above. The "sexual desire" is a human intrinsic quality, decreed by the gods as an anthropogenic act, associated, as it is obvious, with sexual intercourse (e.g., the mention of the bed).

---

63 Geller (2014: 3) translates as "impotence." Although the translation is not correct, it indicates that this expression refers as well to the therapy for defeating impotence or the absence of sexual desire.

Sexual desire or potency?

I have always translated the term in the previous pages as "sexual desire," in this section, I will show the reasons for this translation, in contrast to that of "sexual power" or "erectile capacity."

1. As we stated above, ***nīš libbi*** is the nominalization of the verbal expression ***libbu inaššīšu*** 'to want, desire.'[64]
2. If the term meant 'erectile capacity,' it would only apply to men[65]. Biggs (1967: 2) writes that only men are said to have ***nīš libbi***, while women are never addressed

---

64 There are also cases in which the verb with the subject *libbu* is used intransitively in the Gtn form, *libbašu itanašši* "his *libbu* keeps raising" (see Zisa 2020):

> [*šumma muḫḫīšu*] *ka*[*l ūmi*/*mūši it*]*teneḫpi* <u>*libbašu itanaššī*</u>*-ma mayyālu ittanasḫaršu kīma ša ana muḫḫi sinništi imqitu nīš libbi irašši qāt ar*[*da*]*t lilî*
>
> [If his crown] keeps (feeling) crushed all [day/night long], "<u>his *libbu* keeps raising</u>," and the bedding keeps turning round about him, like one who lies down upon a woman, he has sexual desire: hand of the *ardat lilî*-demoness [*Diagnostic Handbook* SA.GIG/*Sakikkû* III 4–5, see Labat 1951].

John Z. Wee (2015: 261–263) studied the commentaries of *Diagnostic Handbook* SA.GIG/*Sakikkû* in which this expression is mentioned: ŠÀ-*šú i-ta-na-ší*(SI) : Š[À]-*šú ana* BURU$_8$ *e-te-ni-la-a* "His *libbu* keeps raising" *means* "his *libbu* keeps coming up to vomit" [Comm. SA.GIG/*Sakikkû* 1–3 = STT 403, obv. 19]. In the commentary, the expression *libbašu itanašši* ("his *libbu* keeps raising") is interpreted not as the feeling of sexual desire, but as a physiological response to nauseation: "His belly keeps coming up to vomit." The commentator correctly understood that the Gtn form of the expression *libbu* + *našû* refers to nauseation and vomiting, and not to sexual desire, in fact we do not have such a form in the *nīš libbi* corpus. I disagree with Wee who argues that the commentator seems to have misunderstood a written account of sexual desire, treating it as a variation on the idiomatic expression for nauseation and vomiting. Moreover, the expression is similar to many other statements in diagnostic and therapeutic texts that describe the patient's 'belly' as 'raising' (*našu*: *libbašu ana arê itanašâ*, DPS 22: 26) or 'coming up' (*elû*: *libbašu ana pare itenellâ*, *Diagnostic Handbook* SA.GIG/*Sakikkû* III 44; BAM 578 i 27; 578 i 47) to vomit. This is confirmed by another commentary: [ŠÀ -*šú ana*] BURU$_8$ *i-ta-na-áš-šá-a* : *lib-ba-šú a-na pa-re-e* / [*i-ša*]*q-qa-a* : . . . "[His belly] *i-ta-na-áš-šá-a* (i.e., keeps raising) [to] vomit" *means* "his belly becomes raised to vomit" [Comm. SA.GIG/*Sakikkû* 7(b) = SpTU 1, 33, rev. 2′–3′]. In this statement in SA.GIG/*Sakikkû* Tablet 7 the commentator was concerned to clarify that the verb *našû* does not have its usual transitive sense ('to rise'), but an intransitive meaning ('to rise') in contexts of the belly. See also Comm. SA.GIG/*Sakikkû* 7(A) = SpTU 1, 32, rev. 5–6: . . . *i*[!]-*ta*[!]-*na-šá-a* : ÍL: *na*-⌜*šú*⌝-[u] / [Í]L : *šá-qu-u* : BURU$_8$: *ia-ár-ru* : . . . "*i*[!]-*ta*[!]- *na-šá-a*" : ÍL *means* "to rise." / ÍL *means* "to become raised." BURU$_8$ *means* "he pukes"; Comm. SA.GIG/*Sakikkû* 3(B) = BM 43854+43938, rev. 5–6: ŠÀ-*šú ana* BURU$_8$-*e* $_{a\text{-}re\text{-}e}$: *i-te-né-el-la*-⌜*a*⌝-[*ma*] / [BURU$_8$: *a-ru*]-⌜*ú* : BURU$_8$: *pa*⌝-*ru-ú* "His *libbi* keeps coming up to BURU$_8$ (subscript: 'to puke')" means "to come up" / [BURU$_8$ *means* "to puke."] BURU$_8$ *means* "to throw up" (see Wee 2015: 263 fn. 55).

65 Although the erection of the clitoris exists, it is little attested in Mesopotamian sources. For the description of woman's sexual excitement see *Moussaieff Love Song*, edited by Wasserman 2016: 133–136. Especially see the expression *ina mati appi lalêki lusuḫ*[*a*]*m* "When may I *pull out* the nose of your desire?", where *appi lalêki*, litt. "nose/tip of your desire," can euphemistically allude, according to Wasserman (2016: 134), to the clitoris.

in these incantations. It is true, that in the corpus *nīš libbi* is associated with men. However, it should be noted that it is often stated that the aim of the ritual practice is "to release the 'heart' (*libbu*) of the man and the woman" (*nīš libbi* P prescr. 4: 11). In addition, there is an explicit reference to woman's *nīš libbi*, it is a passage (from the fragment No. 9b r. 10′, Beckman and Foster 1988: 12) of the *Aššur Medical Catalogue*, edited by Steinert (2018b): [KA.INIM.MA ŠÀ.ZI.G]A *ù* MUNUS.GIN.NA.KÁM ŠÀ.ZI.GA.MUNUS.A.KÁM "[Wording of (the incantation) for (male) sexual desi]re and (those) to make a woman come (and for) woman's sexual desire" (l. 106, Steinert 2018b: 217). The expression ŠÀ.ZI.GA MUNUS.A.KÁM concerns with arousing female sexual desire. Moreover, as we saw above, in *nīš libbi* incantation No. K.1: 11 it is stated that Enlil and Bēlet-ilī have determined the *nīš libbi* for all humanity (men and women). Like the man, the woman feels sexual desire as well: *annâ šalāšīšu tamannū-ma* (var.: *kīam iqabbi*) *zikaru u sinništu libbašunu / ištēniš inaššīšunūti ul inuḫḫ*[*ū*] "You recite (var.: you say) this (incantation) three times and (if) the man and the woman / desire together each other, (but) they do not find reli[ef]" (*nīš libbi* incantation No. D.3: 35–36); *ana sinništi tibûtu* [*šurši*] "[To make] a woman have an 'elevation'" (*nīš libbi* D prescr. 7: 54).

3. A consideration on the penis and its role in incantations and rituals of the corpus needs to be made. The penis is the anatomical part which in prescriptions is often rubbed or anointed with an ointment. I should note that very few prescriptions focus the healing practices *only* on the penis. Many prescriptions, as we will see in Chapter III (§ "Ointment and amulets"), provide the realization of an oil-based ointment with magnetite and iron powder, which must be applied on both man's penis and woman's vulva:

> Its ritual: You pulverize magnetite (and) iron,
> you mix (them) with oil from the alabastron; you recite the incantation three times over it; the man's penis
> (and) the woman's vulva, you[?] anoint them and (he will get) sexual desire.
> [No. D.2: 21–23, see also No. B.1: 14–17; No. B.2: 38–40; No. E.1: 15–17; L prescr. 6: 15–18]

The aim of this therapeutic practice is the reinstatement of the sexual attraction between man and woman by means of iron and magnetite. Another prescription provides an oil-based ointment, with iron, "heals-a-thousand"-plant, sulfur, and *ru'tītu*-sulfur to be applied on the man's penis and the woman's pelvic area:

> Its ritual: Iron powder, *imḫur-līm*-plant, sulphur, *ru'tītu*-sulphur
> you pulverize together, you put (it) in oil, you recite the incantation seven times over it,
> you anoint (with it) the penis of man and the pelvic area of the woman.
> [*nīš libbi* M prescr. 1: 1–3]

When the man is the only subject of the therapeutic practice, the penis is one among many other anatomical parts to be subjected to pharmaceutical treatment.[66] The ointments are often applied not only on the penis but also on the navel, the chest, and the waist of the patient:

Its ritual: You pulverize [magneti]te (and) put (it) into oil;
He (= patient)/you (= therapeutic operator) rub(s) (with it) his penis, his chest (and) his waist, [and] he will be healthy.
[No. A.2: 47–48]

[For ditto]: You ta[ke] the blood of a ram in an unfired *pursītu*-container,
you mix [a hal]f with oil (and) you anoint your navel (and) your penis,
[and] you crush [(the other) h]alf in water, he d[rinks (it) and ditto].
[*nīš libbi* E prescr. 3: 65–67]

[. . .] you massage three times your [n]avel (and) your penis, you massage your [right?] hand (var.: which she (=witch) touched with sorceries) (and) the left one of the woman [. . .].
[*nīš libbi* D prescr. 2: 31]

Prescriptions focusing only on the penis are very few: "[If ditto]: You pound the spur of a *ballūṣītu*-bird in oil, you anoint his penis (with it) and [(he will get) sexual desire]" (L prescr. 5: 14, see also ibid. prescr. 1: 5).

However, the penis is at the primary target of the witch's attack. The penis has been sealed and shut up in clay at sunset, the sperm has been buried with a dead person:

That man's s[perm] has been buried with a dead per[son],
his penis has been sealed and shut up in a clay pit towards sunset.
[*nīš libbi* D Sympt.: 44–45, see also F prescr 5: 21]

The incantations focus on penetration as the fulfillment of the sexual act. For this reason, the erection of the penis is invoked in metaphor:

May the penis of NN, son of NN, be a stick of *martû*-wood and
hit the anus of the woman NN,
(so) he will be (never) satisfied with her charms! Incantation formula.
[No. F.5: 97–99, see No. F.3: 78]

My vulva is the vulva of a bitch! His penis is the penis of a dog,
As the vulva of a bitch took the penis of a dog, (so may I do)!
May your penis become as long as a *mašgašu*-weapon!
[No. E.1: 9–11]

May your penis which satisfies (the desire)? be compact?!
[No. F.4: 88]

May NN, son of NN, with NNfem., the daughter of NNfem.,
mate, bonk (her), mount (her),
and penetrate (her)! Incantation formula.
[No. B.2: 34–36]

66 On the concept of pharmaceutical in this book see Chapter III § "The pharmaceutical."

An incantation refers to the ability to maintain an erection during sexual intercourse. It should be noted that in this passage the *nīš libbi* and the penis are based on two different metaphorical images: in the first, it is compared to flowing river water, in the latter to a harp string. Although in direct relationship, the 'sexual desire' and the penile erection are not the same things, as demonstrated by the fact that the metaphors used in the same incantation are quite differ:

May my sexual desire be constant river water!
May my penis be a harp string,
so that it will not dangle out of her!
[No. A.1: 35–37]

The incantations, in fact, call for not only the erection, a condition considered necessary for sexual intercourse, but general well-being that embraces other anatomical parts, including the *libbu*. In an incantation one hopes for both penile erection, and that the 'heart' (*libbu*) is not tired and, in the variant, that the desire will not diminish: "May the flesh of NN, son of NN (var.: of my husband), be static, may (instead)[?] his penis be erect! / May his desire (lit. *libbu*) not abate night and day!" (No. M.1: 8–9). A fragmentary incantation mentions limbs, sperm, and perhaps penis, indicating probably a desire for the well-being not reducible to just the penis and the erection: "[. . .] . . . your limbs. / [. . .] . . . your sperm. / [. . .] . . . may your . . . be erect! [broken]" (No. G.1: 4–6, see also No. B.1: 6; P prescr. 11: 10). I have shown how the erection is often described (metaphorically or not) and the desired result in the incantations. It is mentioned primarily because in the Mesopotamian perspective penetration is considered to be fundamental to sexual practice. If the loss of sexual desire puts a strain on the sexual relationship between a man and a woman, it is obvious that the incantations describe penetration and erection. The erection is the visual manifestation of recovered sexual desire.

If the witch's actions against the patient focus on the sperm and the penis, then the therapy considers the penis as one of the anatomical parts subjected to the therapeutic treatment. However, the penis is not the organ at the heart of the healing process. As we will see, the therapeutic techniques in the corpus are more complex and embrace different spheres, not reducible to just the male sexual organ. The fact that the suffering concerns the *libbu* and not exclusively the penis is also confirmed by the fact that the *nīš libbi* is described in the *Diagnostic Handbook* SA.GIG/*Sakikkû* in tablet XIII (41–42, Labat 1951: 126), which deals with the symptoms of the torso, chest and those related to the abdomen (*libbu*).

4. The expression *nīš libbi*, literally "raising of the *libbu*," has been said to be a euphemism for "sexual power" (see Leick 1994: 208–209; Jaques 2006: 11). I disagree with the scholars who interpret *stricto sensu* the expression as a euphemism for the erection or sexual potency. The euphemism is a figure of speech which involves avoiding the use of an expression or word for several reasons,

such as moral, religious, or social, and replacing it with another, considered less coarse or strong. If we consider *nīš libbi* as a euphemism, we take for granted that some form of taboo affects "erection" or "penis." This must be demonstrated, not assumed. There needs to be concrete proof in favor of the penis or erection being considered taboo. The texts of the *nīš libbi* corpus, as well as Sumerian and Akkadian literature, have no traces of this taboo. The Mesopotamian sources are full of sexual images, whether visual (eg. plaquettes) or textual ones. It could be perhaps a taboo for "impotence," but this is not the meaning of *nīš libbi*/ŠÀ.ZI.GA. Besides, the rituals and the prescription use a technical language, which opposes the use of the euphemism.

5. Let us look at the lexical lists and other sources. We only have two sources related to "sexual desire" in lexical lists: šà-zi-[ga] = [*niš lib-bi*] (Antagal A 134, Cavigneaux et al. 1985 = MSL 17: 185); *na-šu-ú šá* ŠÀ-*bi* : zi : *n*[*a-š*]*u-ú šá* ŠÀ-*bi* : šà-zi-ga : *ni-iš lib-b*[*i* . . .] (A III/1 Comm. A 21–21a, Civil et al. 1979 = MSL 14: 324). The lexical lists inform us that, as stated above, *nīš libbi* is the nominalization of the verbal expression *libbu inaššīšu*. An Old Babylonian bilingual poem states that both *lalû*/la-la[67] and *nīš libbi*/šà-zi-ga fall under the goddess Ištar's domain:

   la-la šà-zi-ga níĝ-šu ĝál é? níĝ-ĝûn ĝá-ĝá dInana za-a-kam
   la-lu-⌜*ú*⌝ *ni-iš li-bi-im* x [. . .] x x [. . .] *bi-ši-im ra-še-e ku-*[*ma*] IŠ8.DAR
   Attractiveness, sexual desire, to have goods and property are yours, Inana/Ištar.
   [*Inana C* l. 121, ETCSL c.4.07.3, see Sjöberg 1975: 190]

   As is pointed out by Civil (1987) and Michalowski (1998), the enumeration in a text consists of a list of terms of a lexical set. This means that in an enumeration the terms are placed in a conceptual relationship. Thus, as can be seen from the text, the 'sexual desire' maintains a conceptual relationship with *lalû*/la-la 'attractiveness, charm, allure' (CAD L 49–51; AHw. I 530 'Fülle, Üppigkeit'). The latter concept is at the basis of love relations, as confirmed by No. F 5: 99 where it is hoped that the patient will not be satisfied with the female charm: *lā išabbâ lalâša* "(So) he will be (never) satisfied with her charms!" (see also K.8: 153).

6. Let us consider the following expression *ana sinništi alāka muṭṭu* (A prescr. 1: 1; A Symp.: 68; No. C.1: 9; No. E.2: 18; D Sympt: 42). See for example *nīš libbi* A Symp.: 67–68:

   [*šumma amēlu*] *lū ina šībūti lū ina ḫaṭṭi lū ina ḫimiṭ ṣēti* [*l*]*ū ina neḫēs narkabti*
   <u>*ana sinništi alāka muṭṭu*</u> *ana nīš libbi šuršīšu-ma ana sinništi alākīšu*

   [If a man], either due to the old age, or the stick, or inflammation by sun-heat, or the *neḫēs narkabti*-sickness,
   (his desire) to go to a woman is reduced, in order to make him get his sexual desire and for his going to a woman.

67 On *lalû*/la-la see Jaques 2006: 263–264; Feldman 2015.

Biggs translates this expression, frequently present in *nīš libbi* corpus and anti-witchcraft texts, with "he is not able to have intercourse with a woman."[68] According to the scholar, the phrase means that the man could not get an erection sufficient for intercourse (1967: 26–27).

The verb *muṭṭu* is the stative D of the verb *maṭû* 'to be short a given quantity, be missing, decrease in number' (CAD M/I 429–435; AHw. II 636 'gering werden, sein'). In the D-stem it has the meaning 'to reduce, diminish.' For the stative D the CAD suggests the meaning 'to have a reduced appetite' or 'capacity' (AHw. 'vermag wenig'). It is usually used in medical texts to indicate decreased appetite and thirst: *šumma amēlu akala u šikara muṭṭu* "If a man's (desire) to eat and drink beer is reduced" (Küchler 1904 pl. 10 iii 7, see ibid. 12; BAM 234: 9; ibid. 409: 28); *šumma šerru unappaq ummu iṣṣanabbassu tulâ muṭṭu* "If a baby is constipated, often has attacks of fever, (his desire) (to take) the breast is reduced" (Labat 1951: 230, l. 119). In an anti-witchcraft text, the decreased appetite and the sexual desire are linked: [*akala u*] *šikara muṭṭu ana sinništi alāka muṭṭu ana sinništi libbašu lā inaššīšu pâšu iptenette* "(The man's desire) [to eat and] drink is reduced, (his desire) to go to a woman is reduced, he does not desire a woman, he babbles" (Abusch and Schwemer 2011: 321, No. 8.6.1: 19).

These examples push us towards considering the meaning 'decrease of desire/will,' and not so much the meaning 'capacity.' Landsberger (1967b: 162 fn. 107), as opposed to Biggs, translates "er geht selten zu einem Weibe ein," interpreting with *muṭṭu* 'little, rarely.' The expression must be understood as a decrease of desire to go to a woman, and not as an inability (= impotence) to have sexual intercourse.

7. Let us consider another passage in D Sympt.: 42–45:

[*šumma amēlu*] *kašip-ma munga īši birkāšu gannā k*[*a*]*lâtūšu illaka libbašu* [. . .]
[. . .] x-*ma ana epēš ṣibûti lā inaššīšu nīš* [*libbī*]*šu ṣabit-ma ana sinništi al*[*āka muṭ*]*ṭu*
[*lib*]*bašu sinništa ḫašiḫ-ma sinništa īmur-ma libbašu itūra amēlu šū ri*[*ḫûss*]*u itti šalamt*[*i*] *šunullat*
ušaršu kanik-ma ina kullat erēb Šamši peḫi

[If a man] is bewitched and has the *mungu*-paralysis, his knees are contracted, his kid[ne]ys "go," his 'heart' [. . .] . . .
[. . .] . . . and he does not have interest to achieve (his) desire, [his] sexual [desire] has been taken and (his desire) to g[o] to a woman is [redu]ced,
his '[hea]rt' needs a woman and finds her, but his 'heart' returns: that man's s[perm] has been buried with a dead per[son],
his penis has been sealed and shut up in a clay pit towards sunset.

68 In the same way Abusch and Schwemer (2011) in anti-witchcraft texts.

Among the symptoms of a bewitched man suffering from the absence of sexual desire, it is stated: *libbašu sinništa ḫašiḫ-ma sinništa īmur-ma libbašu itūra*. How can we understand this passage?

Biggs (1967: 69) translates *libbašu sinništa ḫašiḫ* as "(if) his heart desires a woman." According to him, it describes a man who wants a woman but is not able to have sexual intercourse with her.[69] This passage could be an evidence in favor of a translation of *nīš libbi* as sexual potency. In AMT 76, 1: 6[70] we find the expression *sinnišat libbīšu ḫašiḫ-ma sinništa ippalis-ma libbašu lā inaššû*, translated by Biggs: "(If a man) desires the 'woman of his heart' and looks at the woman, but his 'heart' does not rise for him" (1967: 2). Landsberger (1967b: 161 fn. 107) translates the expression *sinništi libbīšu* with "eine Frau seines Wunsches (= Gattin, Konkubine)." The German scholar criticizes Bigg's translation, translating AMT 76, 1: 6 as following: "Wenn jemand eine Frau seines Wunsches benötigt, sie findet, aber keine Begierde spürt," which commenting: "Niemals kann *ḫašāḫu* geschlechtliche Begier sein, *naši* 'it rises' sehr bedenklich!" Therefore, the correct translation of AMT 76, 1: 6 is the following: "(If a man) needs the "favorite" (lit. of his 'heart') woman and finds her, but does not desire her."

In the quoted examples cited, therefore, the verb *ḫašāḫu* has not the meaning of 'to sexually desire,' but that of 'to need, desire' (see CAD Ḫ 134–136 'to need, desire, like'; AHw. I 332–333 'brauchen, begehren'). The expression *libbu inaššīšu* in the corpus indicates sexual desire, not the verb *ḫašāḫu*. "Needing a woman" (with the subject *libbu*) is here understood as a fundamental element of Mesopotamian manhood. The return of the 'heart' (*libbu*) perhaps indicates that his raising does not occur, since the organ returns to its original position: its movement is hindered (see Zisa 2020: 468–469).[71] The patient is described as one who has found a woman he needed, but for which does not feel desire.

8. Let us analyse the verb *tebû*/ZI.(GA). Its general meaning is 'to get up, rise, rear up' (CAD T 306–321; AHw. III 1342–1343 'aufstehen, sich erheben'; Biggs 1967: 9). When it refers to body parts it has the meaning 'to become erect, emerge' (medical and omina texts): *šumma izbu kišāssu ina papān libbīšu tebī-ma* "If the neck of a misbirth rises from his navel area (*Šumma izbu* VII 80, De Zorzi 2014: 539), or 'to pulsate, throb' (often artery, skin in medical texts) (see CAD T 318, mng. 9). However, in the *nīš libbi* corpus the word never appears in reference to parts of the body. In fact, it is used to qualify animals ready for mating. Outside the corpus, concerning the animals, the verb has the meaning 'to rear up': *šumma ṣīru ina sūqi ana pān amēli itbâm* "If a snake in the street rears up before a man" (CT 40 24

69 See also Abusch and Schwemer 2011: 103, No. 2.5.1: 8′–9′.

70 See KAR 26: 9 and the duplicate AMT 96, 7: 11, Mayer 1999: 149.

71 Other translations: Stol (1999: 58) "'heart returns'"; Thomsen (1987: 55) "Herz (d.h. Erektion?) zurückgeht"; Abusch and Schwemer (2011: 111) "'heart' falters"; Hunger (1976: 27) "seine Begierde nachäßt"; Schwemer (2010) "sein Herz kehrt macht."

K.8028:7). In the corpus in question, the reference to sexual intercourse is clear. The verb appears in the following cases:

- In the phrase "an animal which is *tebû* for mating (*ana rakābi*)":

Incantation: *Akkannu*-wild ass who is reared-up for mating (*ša ana rakābi tebû*), [wh]o has dampened [your] desire?
[No. A.2: 41; see No. L.2: 26]

If ditto: A male partridge who is reared-up for mating (*ša ana rakābi tebû*).
[*nīš libbi* N prescr. 2: 8, see also E prescr. 5: 70]

[Blood of a] male [partrid]ge, bristles of a pig reared-up for mating (*ša ana rakābi tebû*).
[*nīš libbi* F prescr. 10: 32]

The ten drugs for the sexual desire in the wool of a [ma]le lamb which for mating
is reared-up (*ša ana rakābi tebû*), you put (it) [around his neck].
[*nīš libbi* K prescr. 28: 75–76]

- An animal as *tebi* "reared-up":

If ditto: The penis (var.: testicles) of a male partridge,
the saliva of a reared-up bull,[72]
the saliva of a reared-up ram,[73] [the saliva of a reared-up buck[74]].
[*nīš libbi* N prescr. 4: 17–19]

- Imperative referring to animals or the patient as an animal:

The one (= buck) at the head of my bed, rear up, make love to me!
The one (= ram) at the feet of my bed, rear up, bleat *for* me!
[No. E.1: 7–8]

Roar on me! Roar on me! Rear up! [Rear up]!
Roar on me like a stag! Rear up lik[e a wild bull]!
[No. B.1: 1–2, see also No. F.3: 65–66]

Incantation: St]ag! Stag! Wild bull! Wild bull!
[Roa]r, stag! Rear up, wild bull!
[No. G.1: 2–3, see also ibid. 7][75]

Rear up! Rear up! Mount! [Mount!]
[No. F.2: 61]

- Precative, in connection with animals:

Together with you, may a lio[n] rear up!
Together with you, may a w[olf] rear up!

---

72 See also for the bull: N prescr. 12 iii 2.
73 See also for the ram: No. D.4: 62; F prescr. 17: 57; No. K.8: 155; N prescr. 18 iv 3.
74 See also for the buck: No. D.4: 61; N prescr. 25 lo. e. 1.
75 See the catalogue LKA 94 ii 14: *rīmī* [*t*]*i*[*bâ*] *lulīmu tibâ* "My wild bull, [rea]r [up]! Stag, rear up!".

> Together with you, may a snak[e] rear up!
> [No. B.1: 3–5]
>
> May the buck arise and repeatedly mount the goat!
> [No. E.1: 4]

As is evident from these passages from the *nīš libbi* corpus, the verb *tebû* takes as a subject the animals or the patient identified with them. The sexual context, related to mating, is clear and is confirmed by the fact that animals are qualified as *ana rakābi tebû*. But how can we understand the verb *tebû* here? Biggs (1967: 9) argues that the verb appears in the corpus with the extended meaning 'get an erection.' Although the sexual context is irrefutable, we have no evidence for this translation. As we have seen above, the verb refers to animals and therefore should be translated as 'to rear up.' As it is known, in animal mating the male rises to penetrate the female. A literal translation 'to rise' would be more legitimate. Of course, this rising is related to reproduction and consequently to the excitement.

In the *Epic of Gilgameš* the verb is used referring to Enkidu in the passage in which he lies sexually with Šamḫat: 6 *urrī u* 7 *mušâti Enkidu tebī-ma Šamḫat irḫi* "Six days and seven nights Enkidu was erect and coupled with Šamḫat" (I 194, George 2003: 548). This line would seem to support the meaning proposed by Biggs, but a clarification needs to be made. Enkidu, before sexual intercourse with Šamḫat, belonged to the wild world, he lived and behaved like animals (for example he grazed the grass with the cattle: I 122–160, George 2003: 544–546). As we will see in the next Chapter, references, more or less direct, to Enkidu are not missing in the incantations of this corpus. Enkidu makes love like animals: reared-up (*tebî*).

Mesopotamian plaquettes with scenes of *coitus a tergo* are well known (see Assante 2002: 29–36; Bahrani 2001: 51–55). An interesting case is found on a plaquette from Babylon. It is a love amulet, where on one side an embracing couple is represented, while on the other there is a continuous repetition of the sign ZI, referring to the excitement (Oberhuber 1972: 92, fig. 76–77, private collection).

The verb *tebû* appears other times in *nīš libbi* incantations, without a direct connection to the animals. In the first case, one hopes for that patient, perhaps identified with animals, not be afraid, to make love, to rise (*tibâ*), and not be afflicted: "Have sex! . . . May you do not be afraid! / Rise! May you do not be afflicted! (No. D.4: 56–57). These lines can refer both to the erect masculine position during sexual intercourse (in parallel with the verb "to copulate"), and as an exhortation for the recovery of forces and energies, which the sickness-event affected. Similarly, incantation No. F.4 encourages the patient, identified with animals (buck, ram, and wild bull, ll. 80–83), not to be afraid and hopes that his strength, his tired knees, his limbs, and his flesh can rise (*litbâ*) together with him (*ittīka*):

> Together with you, may the strength rise!
> Together with you, may your tired knees rise!

> [Together with you, may] your limbs [rise]! Together with yo[u, may your] members r[ise]!
> [Together] with you, may [your] . . . rise!
> [. . .] . . . bed [. . .].
> Do not get scared! Do not be afraid! Do not be afflict[ed] for your love-making!
> [No. F.4: 83–88]

Biggs (1967: 9) considers the terms *birku* (knee), *minâtu* (limbs), *mešrētu* (flesh), and *kululu*(?) as synonyms for the penis.[76] On the contrary, the penis is not the focus of therapeutic practice and what these lines want to emphasize is the recovery of health of his mental and physical integrity. The whole of the mentioned body parts represents the patient's physical body. The expression "together with you" is intended to create an image of recovery focused on the whole body.

I have shown two cases where the verb is used in the corpus:

1. It assumes the meaning 'to rise, rear up' refers to the mating of animals. Obviously, it is understood as 'rising' connected with sexual excitement.
2. Reference to the recovery of the forces and energies. Weakness is one of the characteristic symptoms of the lack of sexual desire. The patient *rises* from sickness.

In the corpus, the verb sometimes has *libbu* as its subject.[77] In these cases, the expression seems to be a synonym of *libbu inaššīšu*:

> If ditto: Between the man and the woman [. . .]
> they [desi]re (*libbašunu itebbū*), you put over it, thes[e . . .].
> If ditto: For the 'rising' of 'heart' (*ana libbi zikari tebî*) of the man and for [his going to] a woma[n].
> [*nīš libbi* P prescr. 5: 16–17; 8: 18]

The substantive form, *tibût libbi* 'rising of the 'heart'' (*tibûtu*: CAD T 391–393 'levy, insurrection, sexual excitement, erection'; AHw. III 1356 'Aufstehen, Erhebung'; Biggs 1967: 9–10) appears once in this corpus (*nīš libbi* K prescr. 29: 128).[78] However, the text

---

76 The reference to the erect penis is in incantation No. M.1: 8: "May the flesh of NN, son of NN (var.: of my husband), be static, may (instead)$^{?}$ his penis be erect (*lizqip*)!". The verb used is not *tebû*, but *zaqāpu* (CAD Z 51–55 'to erect'; AHw. III 1512–1513 'aufrichten'), see commentary.

77 A particular case is the following passage in a medical text where the verb is used in the Ntn form (= Gtn, AHw. III 1343): *šumma amēlu libbašu ittenetbaššum* "If a man's *libbu* continuously rise *for* him" (KK 191 + 201 + 2474 + 3230 + 3363 i 19, Küchler 1904: 2). CAD T 320 mng. 18 translates "If a man's abdomen continuously gives him pain," while Küchler (1904: 3) "Wenn einem Menschen sein Inneres sich erhebt." See Zisa 2020.

78 *Pace* Biggs 1967: 9–10, 31, who reads in STT 280 ii 61 (= No. K.5: 105) [*t*]*i*-<*bu*>/⸢ZI⸣-*ut* ŠÀ-*ka ul i-na-ḫa* "Your penis will stay erect (lit.: the risen condition of your "heart" will not get tired)." For him, the expression *tibût libbi* means "erection." However, the correct reading is [*m*]*u-ši* U_4 ŠÀ-*ka ul i-na-ḫa* "[Nig]ht and day your desire (lit. *libbu*) will not abate." See also SpTU I No. 10: 7′ (= No M.1: 105).

nīš libbi D prescr. 6: 51 and prescr. 7: 54 probably mentions the *tibûtu* of both man and woman:[79]

ana zikari tibûtu šurši	"in order to make the man have an 'elevation.'"
ana sinništi tibûtu [*šurši*]	"[in order] to make a woman have an 'elevation.'"

For these lines, Biggs (1967: 9–10) suggests the meaning 'sexual excitement' (also CAD T 293, mng. 5), for the man the penile erection, while for the woman the swelling of the clitoris. It is possible instead that here *libbu*/ŠÀ is omitted, considering *tibût libbi* as synonymous with *nīš libbi*.

In conclusion, the verb should be understood in its general meaning 'to rise' and, referring to animals, 'to rear up.' The translation 'to get an erection' is excluded.[80] As we have seen, however, it does not exclude the sexual excitement and desire, as is evident by the aforementioned plaquette and by the fact that *tibût* (*libbi*) may be understood as synonymous with *nīš libbi*.

Sexual desire: *libbu* between physicality and psychic faculties

As has been shown earlier, I translate the term *nīš libbi* with "sexual desire," rejecting the translation of "sexual potency." As I have underlined in the *Forschungsgeschichte*, two interpretative strands have characterized the study of such texts: a psychological one (sexual desire, libido) and a physiological one (penile erection). One may think at first sight that my translation of "sexual desire" is part of the psychological strand but is not.

As I have mentioned in the theoretical section above, the dualism "mind *vs.* body" is absent in the Mesopotamian anthropology (see Asher-Greve 1998; Dietrich 2010; Steinert 2012a; Steinert 2012c). It is impossible to draw a clear division between the way of feeling and thinking and human anatomy. The term *libbu* is central for the understanding of this confluence of terms. The *libbu* (CAD L 164–175;[81] AHw.

---

79 See also in the *Aššur Medical Catalogue* in the section XIX devoted to the loss of sexual desire: [EN? . . . *ana* NI]TA ZI-*tú* *š*[*ur-ši*]-⌜*i*⌝ "[Including (prescriptions) . . .] to make a man have an 'elevation'" (l. 102, Steinert 2018b: 217). See also in No. K.3: 90–91 the use of the word *tību* (ZI) 'rising': "To? 'dog's-tongue'-plant for his 'rising' [. . .] / for his 'rising': seven grains of silver, [seven grains of go]ld, in front of [. . .] x x [x]."

80 Note the text BAM 116 r. 8′ (and duplicate): *šumma amēlu ana šīnātīšu magal* ZI.ZI-*bi* "If a man repeatedly rises because his urine" (Geller 2005: 80; No. 7). It must not be translated as Biggs 1967: 9: "If a man before(?) he urinates keeps having a violent erection."

81 CAD's definition: 1. Heart, abdomen, entrails, womb; 2. Inside (or inner part) of a building, an area, a region, of a container, parts of the human body, parts of the exta, inside, pith of plants, a type of document, etc.; 3. Mind, thought, intention, courage, wish, desire, choice, preference.

I 549–551 'Leib, Inneres, Herz') can indicate the innards, abdomen, heart. However, *libbu* is also the place of feelings and thoughts (see Oppenheim 1941: 263–267; Asher-Greve 1998: 10; Karahashi 2000: 144–148; Jaques 2006: 433–445; Couto-Ferreira 2009: 251–256, 263–268; Steinert 2020; Salin 2020: 155–160; Salin forthcoming). The body in Mesopotamia is, quoting Asher-Greve, «agent of thinking, feeling, experiencing and knowing» (1998: 23). Therefore, as the anthropologist Rosaldo underlines: «Emotions are thoughts somehow 'felt' in flushes, pulses, 'movements' of our livers, minds, hearts, stomachs, skin. They are embodied thoughts, thoughts seeped with the apprehension that 'I am involved'» (1984: 143). The *libbu* is the place of union between emotion and physicality. It is therefore a kind of *psychic faculties of the human organs* or a *physicality of feelings and thoughts*. It suggests an "abdomen-centering conceptualization" of emotions, feelings, and thought (see Sharifian et al. 2008). If *libbu* is the seat of feelings, thoughts, and desires, it does not mean that they should be understood in the sense of a manifestation of a non-corporeal psyche. For this reason, I exclude the translation of *libido*, since the term refers to a conception of desire situated in the unconscious, a concept which, in addition to being from modern psychoanalysis,[82] places sexual desire at the level of a psyche separated from the body. Here, however, sexual desire is seen in its corporeality. The raising of the *libbu* should be understood, in my opinion, within the physical concreteness of the moving organ. The movement of the organ, which determines emotions[83] or pathological states (see Zisa 2020: 462–465), has been abundantly documented in Medical and Linguistic Anthropology in a transcultural perspective (see Lutz 1988; Wierzbicka 1999; Enfield 2002; Enfield and Wierzbicka 2002; Cardona 2006; Sharifian et al. 2008). The movement of the organs involves feelings, thoughts, desires, and pathologies in their physical concreteness. It is then seen as a sexual desire certainly not due to the speculations of modern sciences of the psyche, such as psychoanalysis, psychology, and psychiatry. But as a sexual desire experienced *with* and *by* the body (see Csordas 1990; 1999), in the *libbu*'s raising: *a bodily sexual desire.*

---

82 I disagree with modern psychological interpretations on ancient sources, like Geller's view: «In some cases the šaziga incantations may have been effective in being able to deal with the anxiety. This is the defence mechanism known as 'displacement', which in this case redirects the cause of the anxiety onto a witch. Externalizing the problem in the form of a witch can potentially allow a patient to control that which is beyond control, namely his own fear» (1999: 54).

83 See on the emotions in Ancient Near East, in addition to Gruber 1990, the volume edited by S. Kipfer *Visualizing Emotions in the Ancient Near East* (2017), and *The Expression of Emotions in Ancient Egypt and Mesopotamia* edited by of S. Hsu and J. Llop Raudà (2020).

The cause of sickness

In this section, I will deal with the cause of loss of sexual desire. The *nīš libbi* texts often do not mention the cause. In general, there are two possibilities: either the man is bewitched, or divine wrath has overcome him.

The text *nīš libbi* D Diagn.: 24–26 contains a ritual with etiological purpose:

> Its ritual: You mix together emmer dough and potter's clay; you make the figurines of the man and the woman; you put them one upon the other, and place them at the man's head,
> you recite [the incantation] seven times; you remove (them) and put [them near] a pig.
> If the pig approaches (them), (it means) Hand of Ištar; for the ritual procedures$^{?}$, if the pig does not approach (them), (it means that) the witchcraft seized the man.

The etiological analysis aims to understand what the cause of the sickness-event is. In the first case: if the pig approaches the figurines, it means that the cause of the suffering is the Hand of Ištar (also in X prescr. 2: 4′). The gods in Mesopotamia were the "active cause" of many malaises which afflicted humanity. The Hand of DN (*qāt*/ŠU DN) is the cause of negative forces that affect men.

What is the meaning of this expression? Heeßel (2007)[84] showed how the expression "Hand of DN" can indicate either sickness or the divine agent itself. He noted that in the therapeutic and diagnostic texts (see Labat 1951; Heeßel 2000) the expression is written differently. When referring to the "Hand of Ištar," in therapeutic texts it is written ŠU ^{d}INNIN.(NA), indicating a sickness: "Hand of the Lady/Ištar-sickness." On the contrary, in diagnostic tests it is written ŠU ^{d}*Iš-tar*/^{d}*Iš*$_{8}$*-tár*/^{d}15, indicating the divine agent of suffering (see Heeßel 2007: 122–123):

	sickness-name	divine "sender"
Hand of the Goddess/Ištar	ŠU.^{d}INNIN(.NA)	ŠU ^{d}*Iš-tar* ŠU ^{d}*Iš*$_{8}$*-tár* ŠU ^{d}15

Obviously, *nīš libbi* D Diagn.: 26 reports ŠU ^{d}*Iš-tar*, a clear reference to the divine agent. A passage from the *Diagnostic Handbook* SA.GIG/*Sakikkû* confirms that the goddess can provoke the loss of sexual desire: [*šumma* . . .] *nīš libbīšu ṣabit qāt Ištar ana ki* [. . .] "[If . . .] his sexual desire is taken: "Hand of Ištar"; to . . . [. . .]" (XIII 41, Labat 1951: 126).[85]

Since Ištar is the goddess taking away the man's sexual desire, it is she who is invoked in the incantations (No. A.2: 45; No. E.2; LKA 94 i 29, ii 8) and to whom offer-

84 See also Kinnier Wilson 1982: 349 No. 37; Avalos 1995: 135; van der Toorn 1985: 78; Heeßel 2018.

85 See other fragmentary passages from the *Diagnostic Handbook* SA.GIG/*Sakikkû* concerning the lack of sexual desire: [. . .*b*]*i-tu nīš libbi irašši qā*[*t* . . .] III 5, Labat 1951: 17; [. . .]*-ka nīš libbīšu a tab-ku qāt* [. . .] XIII 42, ibid. 126.

ings and libations of beer are dedicated (see Chapter III § "Libations to Ištar"). The incantation addresses her since she is the divine patron of love/sex and seduction. More specifically, as confirmed by an Old Babylonian bilingual text, attractiveness (*lalû*) and sexual desire fall under her domain: "Attractiveness, sexual desire, to have goods and property are yours, Inana/Ištar" (*Inana C* l. 121, ETCSL c.4.07.3, see Sjöberg 1975: 190).[86] In a love incantation one affirms that she "brought forth the sexual desire" (šà-zi-ga ba-ra-è, KAR 61: 3, Biggs 1967: 70).

The anger (*kimiltu*) of Ištar and Markuk is the cause of a set of symptoms in the sexual sphere, in relation to the lack of sexual desire in K prescr. 21: 57–59 (= K prescr. 26: 69–71) as well:

> If a man is repeatedly scared (*igdanallut*) in his bed, his 'heart' (*libbu*) is confused, in his bed his sperm comes out, over this man the wrath of Marduk and Ištar
> has come.

An inscription of the king Ashurbanipal confirms that the god Marduk can remove sexual desire:

> [As for the one who] erases my inscribed name and writes his (own) name,
> (or) effaces (and) destroys [the m]ention of the king who is assiduous towards the sanctuaries of the god Marduk (and) the goddess Zarpanītu
> by any crafty device that there is,
> may the god Marduk, king of the gods, take away his sexual desire (and) make to destroy his semen (*nīš libbīšu līṭir liḫalliq zêršu*),
> may the goddess Zarpanītu speak evil about him in the bedroom, the family head's room.
> [Ashurbanipal's inscription K 2411 iv 24′–28′, MStreck 1916: 302; Matsushima 1988: 101]

The other possible cause of loss of sexual desire, as we saw in the etiologic ritual, is witchcraft.[87] It is also possible to identify the witch's actions: for instance, the patient's sperm has been buried with a dead person, his penis has been sealed and shut up in clay towards sunset:

> [If a man] is bewitched and has the *mungu*-paralysis, his knees are contracted, his kid[ne]ys "go," his 'heart' (*libbu*) [. . .]
> [. . .] . . . and he does not have interest to achieve (his) desire, [his] sexual [desire] has been taken and (his desire) to g[o] to a woman [is redu]ced,

86 Scurlock (2014a: 109) cites the text STT 257: 3 translating: "She (= Ištar) makes (even) the ⸢impotent⸣ able to make love." The translation, in addition to being free, does not take into account the palaeographic uncertainty: *mu-ra-ʾ-i-mat* ZI.⸢GA?⸣ [. . .]. Farber (2010: 75) resolves this uncertainty by restoring: *ze-r*[*a-a-ti*] and traslating "zürnende Frauen zum Lieben bringt."

87 The cause is explicitly the witch in following texts: B prescr. 1; D Sympt; D prescr. 3–5; "[The witchcraft has] continually seized [that man. . .]" Y prescr. 1: 1′; "[If ditt]o: He has been entrusted to a ghost" F prescr. 2 (see commentary), 5–17; K prescr. 1–18; N prescr. 16–21?; No. A.4; No. E.2; No. L.2. In anti-witchcraft texts the loss of sexual desire is a symptom of the bewitched man.

> his '[hea]rt' (*libbu*)] needs a woman and finds her, but his 'heart' (*libbu*) returns: that man's s[perm] has been buried with a dead per[son],
> his penis has been sealed and shut up in a clay pit towards sunset.
> [*nīš libbi* D Sympt: 42–45, also F prescr. 5: 19–21; K prescr. 3: 24–27]

The witch attacked the man and took away his sexual desire (*nīš libbi*). To do this, she reproduced a penis (or used something associated with it), sealed, and closed it in clay in the direction of the sunset. The ritual action is easily understandable in the context of Mesopotamian magical practices. The witch often acts by burying figurines of her victims, the exorcist does the same against the witch. The figurine's burial, like its burning and destruction, is one of the most frequent magical and ritual actions used to attack a person (see Daxelmüller and Thomsen 1982; Abusch 2002: 65–78; Verderame 2013c: 302–313).[88] Here, instead of using the victim's figurine, something which reproduces his penis is used: this is an example of analogical magic. Furthermore, the action takes place in the direction of the sunset. On the other hand, some Mesopotamian therapeutic practices include ritual performance before the sun at sunrise. Together with this magical technique, the witch acts through the principle of contagion: the patient's sperm is subtracted, burying it with a dead man. The lack of sexual desire derives not only by the burial itself but from the fact that the sperm is in contact with a dead body.

The attention paid to sperm in the witch's action leads to a reflection on the instigator of the aggression. The use of sperm for magical aggression is a sign of closeness between the aggressor and the victim, probably ascribable to a close family member. It seems that the tensions, at the base of the magical attack, are situated within the couple and/or the household, a thesis supported perhaps by the presence of the woman as a ritual actor (see Chapter I § "Couple's therapy").[89] Already Abusch (2002: 79–88) had pointed out the close relationship in the Mesopotamian view between the female domain on the one hand and preparation of food and sexual activities on the other. He argues that the witch's actions are related to problems of indigestion and sexual disorders. It is often said that the victim ate or drank witchcraft (*amēlu šū kišpī šūkul u šaqi*), or that witchcraft was ingested through food or beer. Many anti-witchcraft texts describe, in fact, digestive problems. In a parallel function, this

88 See also Schwemer 2007a: 199–230.

89 *Pace* Sallaberger 2011: 28–29. He argues, against Schwemer (2007a), that the dangerous magic contacts were outside the household and therefore the best protection against witches' activities was to avoid other people's household and to restrict personal hygiene or meals to one's own household. Of course, this hypothesis is not totally excluded, but I think that the "social boundaries" regulated by witchcraft view cannot be reduced to an opposition between private and public sphere. The anti-witchcraft texts often mention, in the description of the anonymous identity of the witch, the possibility that the later (or the investigator) could be a neighbor, a family member, as well as a stranger. See for example *Maqlû* IV 80–92 (Abusch 2016: 249); LKA 115: 11–15 (Maul 1994: 502).

same magical action causes sexual dysfunction and problems, including the loss of sexual desire.

Witchcraft is a sign of social disorder (envy, social aversion, marital problems). Then, as noted by Augé (1994), one of the first goals of therapy must be to restore the order, highlighting and resolving the tensions between members of the community, considered to be the origin of the sickness. This would also explain the role of women in the therapeutic process. Anthropology provides us with many examples, which show how the family is the social seat of tension and conflict, manifested by resorting to the witch's action. The Africanist anthropologist Max Gluckman (1956: 54–80) showed how "estrangement in the family" is the basis of conflicts and tensions, not only in Africa of course. In general, the Manchester Anthropological School, principally the figures of Max Gluckman and Victor Turner,[90] has observed the ritual process as the "symbolic" space, in which to cancel the always latent tensions within a group. Turner emphasized how the ritual has the task of resolving social tensions, involving not only the individual in which the sickness is manifested but also those who are involved.

A text informs us that, in addition to divine wrath (of Marduk and Ištar) and witchcraft, other causes are considered to be the origin of the lack of sexual desire:

> [If a man], either due to the old age, or the stick, or inflammation by sun-heat, or the *neḫēs narkabti*-sickness,
> (his desire) to go to a woman is reduced, in order to make him get his sexual desire and for his going to a woman.
> [*nīš libbi* A Sympt.: 67–68]

The text states that the man does not want to go to a woman for reasons attributable neither to old age, nor to the "stick," nor to the inflammation by sun-heat, nor to the *neḫēs narkabti*-sickness, but on the contrary, as evidenced by the incantation No. A.4 following this prescription, to witchcraft. This means that the lack of desire can still be attributed to the four above-mentioned cases.

Old age (*šībūtu*/ŠU.GI.MEŠ) naturally leads to a decrease in sexual vigor (see Biggs 1967: 3; Scurlock and Andersen 2005: 112). This is confirmed, for example, in the Sumerian composition *An Old Man and a Young Girl*: "(I was) youth, (but now) my (personal) god, my strength, my vitality / and my youthful vigor left my loins like an exhausted ass" (ll. 27–28, Alster 1975: 92; Leick 2015: 91–92). Because of old age the man has lost his personal god, strength (usu), vitality (lamma), and youthful vigor

---

90 Turner (1967: 392) affirms, in fact, regarding the *ihamba* ritual of the Ndmbu community of Rhodesia (Zambia), that it seems that the Ndembu therapeutic expert's task is less that of curing an individual patient than remedying the ills of a corporate group. It means that the patient does not heal if all the tensions and aggressions in the community are not brought to light and subjected to ritual treatment.

(nam-guruš) positioned in the loins (haš$_4$). Although it does not explicitly refer to sexual desire, it is clear that old age involves its decrease.

It is not clear what stick (*ḫaṭṭu*/PA) refers to. The inflammation by sun-heat (*ḫimiṭ ṣēti*/UD.DA)[91] and *neḫēs narkabti*-sickness (*neḫēs* GIGIR)[92] can involve the loss of desire. The first is a diagnostic category, more severe than fever, whose symptoms are often located in the belly (see Stol 2007: 21–39, esp. 37–38).[93] However, the link between the two diagnostic categories and the absence of desire is not confirmed by other sources.

In conclusion the *sense of pain*, in other words, "the why of the suffering,"[94] regarding the absence of sexual desire is attributable mainly to divine wrath, especially of Ištar and Marduk, or to witchcraft, the manifestation of what the anthropologists call "social anxiety." These two causes of suffering presuppose, according to Abusch (2002: 88), two divergent attitudes from the patient and his community to the sickness-event. If the loss of sexual desire is a consequence of divine anger aroused by fault, the sickness could be experienced as a punishment, that is, as chastisement and/or divine abandonment. On the contrary, if thought to have been caused by witchcraft, it could be experienced as assault, a debilitation, and an emascula-

91 See *nīš libbi* M prescr. 6: 26: "May the sun-heat be removed, the may sun-heat regress!"

92 It is difficult to say which kind of sickness it is (CAD N/II 218–219, see Thompson 1936–1937: 340 fn. 21; Biggs 1967: 3 fn. 15; Steinert 2018b: 234). It is composed by the verb *naḫāsu* 'to go back' (CAD N/1 128; CAD N/2 218; AHw. II 713, 775a) and the noun *narkabtu* 'chariot.' See the Arabic *naḫasa* 'prick,' perhaps it referring to the vibration of the chariot. Scurlock and Andersen (2005: 23–24) translate it as "repercussion of the chariot," as negative effects of riding in a chariot, while Stol (2016: 148–149), translating it as "reversing of the warchariot," suggests a metaphorical meaning for a "slackening erection" (personal communication, in Steinert 2018b: 234). Steinert (2018b: 234) presume it is technical or figurative and not a literal meaning. The expression appears in connection with intestinal and abdomen disorders: AMT 69, 3+26, 5+BAM 55: 14′–15′; BAM 49 1′–6′ // BAM 50: 1–7; BAM 397: 3.

93 See BAM 112 ii 6–7: "If the man's penis is hot [. . .] He has been overcome by sun-heat" (Stol 2007: 35).

94 The research of the "sense," in other words the "why" of suffering is based on the possibility for suffering people to employ different systems of reference within which to configure the sickeness-event, in order to respond to the anxiety resulting from the crisis of "sense" that the experience of sickeness produces. Evans-Pritchard (1937: 25) in his study of Azande magic highlights an important issue, which is the distinction between "empirical causes" and attribution of ethical value to the suffering. In this regard he says that the Azande foreshorten the chain of events, and in a particular social situation select the cause that is socially relevant neglecting the rest. The attribution of misfortune to witchcraft does not exclude what we call its real causes but is superimposed on them and gives to social events their moral value. This distinction is evoked by Elsa Guggino (2006: 37) in her study of magic in Sicily. She distinguishes between the "how of the things," that is the misfortunes, and the "why/because of them." The "how" proposes, as in biomedicine, a consideration of the effect of a particular cause. The "why," instead, would involve various types of response according to personal and social situation, according to the particular attitude of each one of us to the world, to life, to our regime of existence. The "why" implies the expression of a moral judgement, recalls a philosophy of life, shifts the attention from the event itself to the event *with respect to the patient*, in essence to a *relationship* (ibid. 35). See also on the topic Sindzingre and Zempléni 1981; Bibeau 1982; Lupo 2012: 138–142; for Mesopotamia see Maul 2004.

tion. In both cases the absence of sexual desire compromises the manhood of the patient, but at the same time his generative capacity and consequently that of the community.[95]

Symptoms

In this section, I deal with the symptoms describing the patient (see Thomsen 1987: 55–57; Stol 1999: 57–60; Abusch 2002: 79–88). I consider it necessary to make a distinction:

1. The texts concerning the loss of sexual desire as a diagnostic category, i.e. the texts defined precisely as *nīš libbi*, the main subject of this book;
2. Other texts, in particular the anti-witchcraft ones, where the lack of sexual desire is one of many symptoms caused by the attack of the witch.

Nīš libbi texts

There are several expressions which more explicitly define the absence of sexual desire:

1. "If a man does not desire" (*libbašu lā inaššīšu*, literally "his 'heart' does not rise"):[96]
   1.a. "If a man does not desire his woman" (*ana sinništīšu libbašu lā inaššīšu*): A prescr. 20: 19; H Sympt.: 2–3; K prescr. 1: 1;

95 The crisis of the patient's manhood and vigor related to the absence of sexual desire can potentially compromise the generative power of the whole community. See on a comparative level the infantile hernia therapy performed in Pescopagano (province of Potenza, Italy) every 25th March for the feast of the Annunciation of Mary (see Di Nola 1983: 13–101), which in other Italian localities as well as in other European countries until the last century had many parallels. In this ritual practice, a bow is made from a long longitudinal cut of a branch of bramble (*Rubus*) deprived of leaves and spines (cf. *nīš libbi* E Bow ritual). A naked male infant or child under one year of age is made to "pass" through it three times with its feet forward. The function of the ritual is to prevent the male infantile hernia, more precisely the prolapse or the hernia of the scrotum (or non-scrotal forms of inguinal hernia that somehow involve the scrotum or the genital area). The success of the therapy is related to the fate of the bramble branch that is replanted in the ground after the ceremony: if it sprouts, the preventive therapy will have been successful, otherwise the ritual performance will have to be repeated. As noted by Alfonso M. Di Nola (1983: 100–101), in the emic diagnostics, the infantile hernia of the scrotum can compromise the future generative functions of the child, and it is therefore possible to trace a semantic illness network between sex–hernia–generative power. The possibility of sexual impotence is experienced as a collective critical moment, and the ritual is hence a collective action aimed at guaranteeing the future vigor of both the infant and the group.

96 Without reference to a woman in X prescr. 2: 4′.

1.b.1. "If a man desires his woman, but does not desire another woman" (*ana sinništīšu libbašu inaššīšu-ma ana sinništi aḫīti libbašu lā inaššīšu*): No. K.2: 23;[97]

1.b.2. "If a man goes to his woman and desires his woman, but he goes to another woman, but does not desire another woman" (*ana sinništīšu illak-ma ana sinništīšu libbašu lā inaššīšu ana sinništi aḫīti illak-ma ana sinništi aḫīti libbašu lā inaššīšu*): Q prescr. 1: 1–2;

1.c. "If a man desire neither his woman nor another woman (*ana sinništīšu u ana sinništi aḫīti libbašu lā inaššīšu*): D prescr. 3: 40; No. E.2: 18 (variant).

2. "The sexual desire has been seized/taken away" (*nīš libbi ṣabit/ekim/eṭir – itbala nīš libbīya*): D prescr. 3: 40; D Sympt.: 43; No. E.2: 18; F prescr. 12: 36; N prescr. 16 iii 27; 20 iv 9 (restored); 21 iv 11 (restored); No. L.2: 20 and 22.
3. "If a man's (desire) to go to a woman is reduced" (*ana sinništi alāka muṭṭu*): A prescr. 1: 1; A Symp.: 68; No. C.1: 9; No. E.2: 18; D Sympt.: 42.
4. "If a man does not have sexual desire" (*nīš libbi lā īši*): F prescr. 12: 36; catalogue LKA 94 ii 24.
5. "If a man's 'heart' needs a woman and finds her, but his 'heart' returns" (*libbašu sinništa ḫašiḫ-ma sinništa ippallas-ma libbašu itūra*): D Sympt: 42–45.
6. "If a man's sexual desire in Nisannu month has finished (*nīš libbi ina nisanni iqti*): N prescr. 1: 1–2 (from Ḫattuša).

These expressions define the main symptom described in the *nīš libbi* text category. The incantations, through evocative language and the use of metaphors, describe the symptoms: the lack of desire and erection: "[Nig]ht and day your desire (lit. *libbu*) will not abate" (No. K.5: 105, see also No. M.1: 9); "May your penis, which satisfies (the desire)$^{?}$, be compact$^{?}$! Do not [. . .] / May my crotch devour your . . . penis!" (No. B.1: 10–11); "[May his penis be a stick of *martû*-wood, may it hit the a]nus of the woman NN!" (No F.3: 78); "May the penis of NN, son of NN, be a stick of *martû*-wood and / hit the anus of the woman NN, / (so) he will be (never) satisfied with her charms!" (No. F.5: 97–99); "May your penis become as long as a *mašgašu*-weapon!" (No. E.1: 11); "May my sexual desire be constant river water! / May my penis be a harp string, / so that it will not dangle out of her!" (No. A.1: 35–37); "May [the qu]iver not be[come

97 See the incipit of Tablet XXXIV of *Diagnostic Handbook* SA.GIG/*Sakikkû* in the catalogue l. 41, which can be restored from the fragmentary catchline of Tablet XXXIII: DIŠ ⸢NA⸣ *ana* ⸢MUNUS⸣-(*šú*) ŠÀ-*šú* ⸢ÍL⸣-*šú-ma* [*ana* MUNUS BAR$^{?}$]-*ti* ŠÀ-*šú* NU ÍL-*šú* MUNUS BI ŠÀ-[*šú* . . .] "If a man desires (his/a) woman, but does not desire [another$^{?}$] woman: this woman [. . .] his 'heart'/desire$^{?}$ [. . .]" (see Schmidtchen 2018: 141). An inverted mention is mention only in the *Aššur Medical Catalogue*, section XIX dealing with the loss of sexual desire, ll. 99 and 101: 99. [DIŠ NA *ana* MUNUS BAR-*ti* ŠÀ-*šú* ÍL-*šú-ma*$^{?}$ *ana* MUNU]S-*šú* ŠÀ-*šú* NU ÍL-*šú*: DIŠ NA X [. . .] "If a man desires another woman, (but)] he does not desire his (own) woman. If a man [. . .]"; 101. [NIGIN X DUB^{meš} DIŠ NA *ana* MUNUS BAR-*ti* ŠÀ-*šú* Í]L-⸢*šú*⸣-*ma*$^{?}$ *ana* MUNUS-*šú* ŠÀ-*šú* N[U ÍL]-*šú* "[Total of X tablets (of the section) 'If a man desires another woman, (but) he does not desire his woman'" (Steinert 2018b: 217).

empt]y! May the bow not slacken! / May the batt[le of my love]-making be fought and may we lie down (together) by night!" (No. J.2: 10–11); "May the flesh of NN, son of NN (var.: of my husband), be static, may (instead)? his penis be erect! / May his desire (lit. *libbu*) not abate night and day!" (No. M.1: 8–9).

Let us now consider the texts where the cause of suffering is explicitly witchcraft: the witch buried his sperm beside a dead man. Among these, the passage *nīš libbi* F prescr. 5: 19–21 describe several symptoms associated with the loss of sexual desire:

> If a man is bewitched and his flesh feeble, has *mungu*-paralysis and his knees are contracted, his 'heart' (*libbu*)
> needs a woman and finds her, but his 'heart' (*libbu*) returns. That man's semen
> has been buried with a dead person.
> [*nīš libbi* F prescr. 5: 19–21]

The symptoms are here four, divided into two groups: 1. weak flesh (lit. "poured out"), *mungu*-paralysis,[98] contracted knees; 2. the 'heart' (*libbu*) needs a woman and finds her, but it returns. The first group is due to a state of fatigue and weakness affecting the patient. The weak flesh indicates exhaustion, while the paralysis and contracted knees refer to a state of immobility and inability to act. These symptoms are described metaphorically also in the incantations through the image of binding and the drawing of tendons, which I will discuss in Chapter II (I group). Besides, the incantations explicitly describe the fatigue and weakness of the patient, wishing for the recovery of forces and energies. The incantation No. F.4: 83–86 affirms:

> [. . .] Jump, my wild bull! Together with you, may the strength rise!
> Together with you, may your tired knees rise!
> [Together with you, may] your limbs [rise]! Together with yo[u, may your] members r[ise]!
> [Together] with you, may [your] . . . rise.
> [No. F.4: 83–86]

Weariness and weakness affect the entire body. Note that the knees are defined as "tired" (*anīḫātu*). Note that the patient's limbs (not the penis) are subjected to pharmacological treatment:

> Its ritual: You put oil from the alabastron in a boxwood container for unguents,
> [you spread] juniper three times before [Ištar], [you reci]te this [inca]ntation [over (it)] three times,
> you [an]oint his limbs (*minâtīšu*) (with the oil) and (he will get) sexual desire.
> [No. F.4: 91–93]

98 See CAD M/II 202–203 'stiffness, paralysis'; AHw. II 673 'eine Krankheit mit Krämpfen.' The *mungu*-paralysis affects mainly the knees (BAM 131 r. 9; BMS 13: 24, Ebeling 1953a; 86), arms (PBS 1/1 14: 10, Labert 1974: 274), testicles (Geller 2005: 100, No. 9: 19′). For other sources see CAD M/II 203. For the paralysis see Kinnier Wilson and Reynolds 2007: 69–72.

The second group probably refers to the fact that, although the man has found the woman he needed, his 'heart' (*libbu*) "returns." The return of the 'heart' perhaps indicates that his raising does not occur, since the organ returns to its original position: its movement is hindered (see Zisa 2020: 468–469).

The *nīš libbi* text D Sympt: 42–45 reports, in addition to those mentioned above, other symptoms:

> [If a man] is bewitched and has the *mungu*-paralysis, his knees are contracted, his kid[ne]ys "go," his 'heart' (*libbu*) [. . .]
> [. . .] . . . and he does not have interest to achieve (his) desire, [his] sexual [desire] has been taken and (his desire) to g[o] to a woman [is redu]ced,
> his '[hea]rt' (*libbu*) needs a woman and finds her, but his 'heart' (*libbu*) returns: that man's s[perm] has been buried with a dead per[son],
> his penis has been sealed and shut up in a clay pit towards sunset.
> [*nīš libbi* D Sympt.: 42–45]

The other mentioned symptoms are: "his kidneys go" (*kalâtūšu illakā*) and "he does not have interest to achieve (his) desire" (*ana epēš ṣibûti lā inaššīšu*). Here the movement of the organs determines a pathological state. It seems that the kidneys have left their position. I did not find such a symptom in the renal disease texts (see Geller 2005). It should be noted that in the text *Taxonomy of Uruk* the absence of desire is associated with the kidneys (see Zisa 2020: 469–470).

In a text describing the patient as bewitched (his sperm is buried with a dead man), we find as a symptom the weakness of the flesh (*šīrūšu tabkū*), as mentioned above. In addition, there are two symptoms related to the male genital apparatus and the sexual sphere. Here is the *nīš libbi* K prescr. 3: 24–27:

> If a man is bewitched and his flesh is weak, (and) neither when walking, nor standing
> nor he is on his bed, nor when urining,
> his sperm flows, like (that of) a woman (his) 'genital discharge' is impure,
> the sperm of this man has been buried under the earth with a dead man.
> [*nīš libbi* K prescr. 3: 24–27]

The two other described symptoms are: the patient's sperm flows[99] and like (that of) a woman (his) 'genital discharge' is impure.[100]

The tiredness (*tānīḫa*) and immobility ("contracted knees") are the symptoms also described in *nīš libbi* B prescr. 1: 18–22, where witchcraft is the cause of suffering.

---

99 See also K prescr. 21: 57–59, K prescr. 26: 69–71. These symptoms relating to the discharge of sperm have been interpreted by Biggs (1967: 3) as a clinical sign of gonorrhea. *Contra* Biggs see Chapter I fn 20. Note the opposite symptom in A prescr. 20: 19: [*šumma amēlu*] *riḫûssu lā i*[*llak*] "[If man's] sperm does not f[low]."
100 See commentary.

Other symptoms are reported: cold (*kuṣṣu*), grief (*ašuštu*), trembling chest, lack of appetite:

> If a man is constantly perturbed in his mind,
> cold (tremors) continuously afflict him and he has constantly distress,
> his knees are bound, his . . . are constantly hot,
> his body has continuously tiredness, his speech is constantly incoherent,
> he does not want to eat and drink
> [*nīš libbi* B prescr. 1: 18–22]

Thus, in addition to tiredness, other groups of symptoms are associated with loss of sexual desire, among them "the trembling chest" and "lack of appetite." "Cold" and "distress" are interesting as well. As we will see in the next section, fear and grief are often described in parallel as symptoms of the absence of *nīš libbi*.
In *nīš libbi* K prescr. 21: 57–59 (= K prescr. 26: 69–71), where the cause of suffering is the wrath of Marduk and Ištar, in addition to the fear we find two other symptoms mentioned above: "the 'heart' (*libbu*) is confused" and "the sperm comes out (in the bed)."[101] This means that beyond the cause of the malaise (witchcraft or divine wrath) the symptoms characterizing this diagnostic category are recurrent:[102]

> If a man is repeatedly scared in his bed, his 'heart' (*libbu*) is confused and in his bed
> his sperm comes out, over this man the wrath of Marduk and Ištar
> has come.
> [*nīš libbi* K prescr. 21: 57–59 = K prescr. 26: 69–71]

It is possible to affirm that the sick man's symptoms concern specifically the *libbu*: the *libbu* does not rise but returns, cold water has been poured and distress have been put on it. It is no coincidence that incantation No. D.2: 16–17 hopes for the wellbeing of the 'heart' (*libbu*): "May they grant the well-being of the 'heart' (*libbu*) to your 'heart'! May they gran[t] the well-being of the waterskin / to your waterskin!". The waterskin refers metaphorically to the penis, whose erection is desired.

Other symptoms, such as fatigue and immobility, characterize the patient's entire body. That is why the chest, limbs, the navel are often the anatomical parts subject to treatment. The incantation No. A.2: 46, in fact, looks forward to the return of health of the limbs: "May he (= Asalluḫi) make your limbs healthy through the se[duc]tion of Ištar!". One last clarification should be made: in the sequence of symptoms, those considered characterizing or expressly related to *nīš libbi*, are mentioned at the end of the sequence.

---

101 See also K prescr. 3: 24–27. Note the opposite symptom in A prescr. 20: 19.
102 See also "if a man repeatedly sees dead people in a dream" (M prescr. 5: 24).

Characterizing symptoms
If a man does not desire (*libbašu lā inaššīšu*)[103] – "If a man does not desire his woman (*ana sinništīšu libbašu lā inaššīšu*)[104] – "If a man desires his woman, but does not desire another woman" (*ana sinništīšu libbašu inaššīšu-ma ana sinništi aḫīti libbašu lā inaššīšu*)[105] – "If a man goes to his woman, and desires his woman, but he goes to another, but does not desire another woman" (*ana sinništīšu illak-ma ana sinništīšu libbašu lā inaššīšu ana sinništi aḫīti illak-ma ana sinništi aḫīti libbašu lā inaššīšu*)[106] – "If a man does not desire either his woman or another woman (*ana sinništīšu u ana sinništi aḫīti libbašu lā inaššīšu*)[107]
The sexual desire has been seized/taken away (*nīš libbi ṣabit/ekim/eṭir – itbala nīš libbīya*)[108]
If a man's (desire) to go to a woman is reduced (*ana sinništi alāka muṭṭu*)[109]
If a man does not have sexual desire (*nīš libbi lā īšî*)[110]
If a man's 'heart' needs a woman and finds her and his 'heart' returns (*libbašu sinništa ḫašiḫ-ma sinništa ippallas-ma libbašu itūra*)[111]
If a man's sexual desire has finished in Nisannu month (*nīš libbi ina nisanni iqtî*)[112]
He does not have any interest in achieving (his) desire (*ana epēš ṣibûti lā inaššīšu*)[113]

103 Without reference to a woman in X prescr. 2: 4′.
104 A prescr. 20: 19; H Sympt.: 2–3; K prescr. 1: 1–2.
105 No. K.2: 23
106 Q prescr. 1: 1–2.
107 D prescr. 3: 40; No. E.2: 18 (variant).
108 D prescr. 3: 40; D Sympt.: 43; No. E.2: 18; F prescr. 12: 36; N prescr. 16 iii 27; 20 iv 9 (restored); 21 iv 11 (restored); No. L.2: 20 and 22.
109 A prescr. 1: 1; A Sympt.: 68; No. C.1: 9; No. E.2: 18; D Sympt.: 42.
110 F prescr. 12: 36; catalogue LKA 94 ii 24.
111 D Sympt.: 42.45.
112 N prescr. 1: 1–2.
113 D Sympt.: 43.

Related symptoms	
Tiredness and immobility	Feeble flesh (lit. poured out) (*šīrūšu tabkū*)[114]
	mungu-paralysis[115]
	Contracted/bound knees (*birkāšu gannā/kasâ*)[116]
	Tiredness of the body (*šīrūšu tabkū; zumuršu tānīḫa*)[117]
Psycho-somatic	Fear and panic (*adirtu,*[118] *ḫa'attu,*[119] *gilittu*[120])
	Distress (*ašuštu*)[121]
	Insomnia (*diliptu*)[122]
	Cold (*kuṣṣu*)[123]
	Constantly perturbed in his mind (*ina ṭēm ramānīšu ittanadlaḫ*)[124]
	Repeatedly scared in his bed (*ina mayyālīšu igdanallut*)[125]
	His 'heart' is confused (*libbašu ešu*)[126]
Genital apparatus and area	Sperm's discharge (*riḫûssu illak*)[127]
	Genital discharge impure like (that) of the woman (*kīma sinništi su''ussu lā elil*)[128]
Others	The kidneys go (*kalâtū illakā*)[129]
	. . . constantly hot (*piṭrūšu ittanaṣraḫū*)[130]

114 F prescr. 5: 19.
115 F prescr. 5: 19; D Sympt.: 42.
116 F prescr. 5: 19; B prescr. 1: 20.
117 K prescr. 3: 24; B prescr. 1: 21.
118 No. 3: 23; No. 11: 4; No. 13: 4; No. 19: 18.
119 No. 21: 15.
120 Rit. 24.29: 1.
121 B prescr. 1: 19; No. 13: 4; No. 19: 19.
122 No. 3: 23.
123 B prescr. 1: 19. See No. 3: 22.
124 B prescr. 1: 18.
125 K prescr. 21: 57 = K prescr. 26: 69.
126 K prescr. 21: 57; K prescr. 26: 69.
127 K prescr. 3: 24–26; K prescr. 21: 57–59, K prescr. 26: 69–71.
128 K prescr. 3: 26.
129 D Sympt.: 42; B prescr. 1: 21.
130 B prescr. 1: 20

(continued)

Related symptoms	
	Lack of appetite (*akala u šikara muṭṭu*)[131]
	His speech is constantly incoherent (*pûšu ittanakkir*)[132]
	skull turns, his face turns (*muḫḫašu iṣâda pānašu iṣâda*)[133]

Fear, distress, and insomnia

In two lines of incantation No. A.3: 51–52 the action of the witch against humans is explicit: "Who has poured [co]ld [water] on your 'heart,' / (and) has put f[ea]r (*adirta*) upon your 'heart,' has [. . .] sleeplessness (*dilipta*)?". In these two lines, no analogy is used to express the action of the witch. The 'heart' (*libbu*) is the subject of the evil deed. Cold water is poured on it and fear (and perhaps even trembling) is put upon it. The term *adirtu*[134] (l. 52) indicates, following Stol (1999: 64), "a constant state of fear," unlike the other word for fear, *tādirtu*, which mostly occurs in the plural, indicating "single moment/s of fear." It must be noted that both *adirtu* and *tādirtu* are nouns deriving from the verb *adāru* (general mng. 'to be obscured,' CAD A/I 102–109 *adāru* A-B; AHw. I 11–12). While *tādirtu* seems to be a symptom that exists along with other various diagnostic categories, and thus not prevailing over other symptoms, *adirtu* is often present as the most important symptom, or as a diagnostic category: "if a man is afraid." Among the causes of fear, we find the wrath of the god and witchcraft.[135] In this incantation, therefore, it is precisely the witch who acts maliciously on the 'heart' (*libbu*), pouring cold water and fear on it. Fear is a symptom of great importance, which manifests witchcraft.[136] For this reason, it often appears in connection with the absence of sexual desire. The absence of desire and fear (also *galātu*) is associated

131 B prescr. 1: 22.

132 B prescr. 1: 21. See also catalogue LKA 94 ii 10.

133 Catalogue LKA 94 ii 12.

134 There are other terms in Akkadian to express states of fear, agitation, and panic. Among them the verb *paḫālu* 'to be afraid,' see Gruber 1990.

135 Other one may be the cause: confused dreams, unclear or negative oracles, other supernatural powers that threaten human existence (see Stol 1999: 63–64). On fear and panic see also Salin 2020: 140–143.

136 For other examples, not connected, however, with sexual desire see Abusch and Schwemer 2011: 152–153, No. 7.7: 19–25.

with distress (*ašāšu*/*ašuštu*),[137] understood as anxiety due to the absence of desire and the inability of sexual intercourse, and insomnia/agitation (*diliptu*):[138]

> Like shining silver, like (var.: and) reddish gold may I have no fear (*adirta ay arši*)!
> [No. E.3: 34]

> Lacking If a man's 'heart' (*libbu*) is afflict[ed an]d trembles.
> [Catalogue LKA 94 ii 11]

> Do not get scared (*ē taglut*)! Do not be afraid (*ē ta'dir*)! Do not be afflict[ed] (*ē tāšuš*) for your love-making!
> [No. F.4: 88]

> Incantation: Have sex! . . . May you do not be afraid (*ē ta'dir*)!
> Rise! May you do not be afflicted (*ē tāšuš*)!
> [No. D.4: 56–57][139]

> If a man is constantly perturbed in his own counsel,
> cold (tremors) continuously afflict him and he has constantly distress (*ašuštu irtanašši*),
> his knees are bound, his . . . are constantly hot,
> his body has continuously tiredness, his speech is constantly incoherent,
> he has no desire to eat and drink, that man is bewitched.
> [*nīš libbi* B prescr. 1: 18–22]

"Panic attacks" (*ḫa'attu*)[140] are also mentioned among the symptoms of the absence of sexual desire, as well as a state of fear (*galātu*), which the patient experiences in bed when ejaculating semen. The latter is a fear caused by the wrath of Marduk and Ištar:

> *namburbû*-ritual against panic attacks (*ana ḫa'atti*).
> [No. D.3: 37]

> If a man is repeatedly scared (*igdanallut*)[141] in his bed, his 'heart' (*libbu*) is confused and in his bed
> his sperm comes out, over this man the wrath of Marduk and Ištar has come.
> [*nīš libbi* K prescr. 21: 57–59 = K prescr. 26: 69–71]

If in the ***nīš libbi*** texts the diagnostic category to be cured is the absence of sexual desire and the cause of the malaise is witchcraft, fear will be one of the symptoms of this "sickness." If the diagnostic category is witchcraft, however, like in the formula-

137 CAD A/II 422–424 A; ibid. 479; AHw. I 79; ibid. 86. See also Salin 2020: 132–137.

138 CAD D 142; AHw. I 170.

139 See also No. B.1: 10 commentary.

140 CAD Ḫ 150–151 A; AHw. I 336; Stol 1993: 42–46.

141 See for the same symptom in combination with other more related to fear (but not sexual desire): STT 256 1–19, Abusch and Schwemer 2011: 138, No. 7.6.7.

tion "if a man is bewitched," both, absence of desire and fear, are listed as symptoms, along with many others, without attributing a dominant role to either:[142]

ummu mung[u zu]'tu sili'[tu]
šiḫḫat šīrī [. . . p]ūti irti qaqqadi dimītu ar[tanaššû]
aḫāya kimṣāya [berk]āya šēpāya ṣubbu[tā]
nīš libbīya bu[nn]ānîya? kasû
minâtīya ittanašpakā! ḫūṣ ḫīpi libbi gilit[tu]
[pi]rittu ḫurbāšu artanaššû ātanamdaru
[apt]anallaḫu itti libbīya addanabb[ubu]
[šu]nāte pardāte anaṭṭalu itti mītūti

By fever, rigid[ity, sw]eat, disease,
decay, [. . . of the fore]head, of the chest (and) of the head (and) seizures I am continuous[ly affected].
My arms, my legs, my [knee]s (and) my feet become cramped,
my sexual desire, my [conge]nial [characteristics] are bound,
My limbs are tottering, from depression, terr[or]
[pa]nic (and) fear affect me more and more, I am constantly anxious,
I am [alwa]ys fearful, I keep spe[aking] to myself,
I have terrible [dre]ams.
[LKA 154 and 155 obv. 43–49 and dupl., Abusch and Schwemer 2011: 259, No. 8.2: 52–59]

ubbiranni ukassânni uṣabbitanni urassânni
mangu lu'tu umallânni
nīš libbīya iṣbatu libbī ittīya uzannû
šer'ānīya ukanninu emūqīya unnišu
aḫīya išpuku berkīya iksû
ṣālta puḫpuḫḫâ nissata adīra
ḫatta piritta arrata
gilitta tēšâ dilipta qūla kūra
lā ṭūb libbi lā ṭūb šīri iškuna

(Who) has restricted me, tied me, grabbed me, tied me,
(who) has filled me with rigidity and decay,
(who) took away my sexual desire, it made me angry with myself,
(who) has bent my nerves, has weakened my strength,
(who) has 'weakened' my arms, tied my knees,
(who) has inflicted on me quarrel, bickering, crying, fear,
Panic, anxiety, curse,
fear, confusion, insomnia, mutism, numbness,
poor mental (and) physical health.
[*Bīt rimki* II, Abusch and Schwemer 2011: 377–378, No. 9.1: 20–28. See also Seux 1976: 388–392; Laessøe 1995: 36–47][143]

142 See also BAM 319 1–7 (dupl. BID, pls. 19–21 obv. i 1′–6′, Abusch and Schwemer 2016: 250, No. 8.29.1: 1–8. See more precisely line 4 of the text: *nīš libbīšu eṭir libbašu iltenemmen* "His sexual desire is taken, he becomes repeatedly depressed."

143 Parallel KAR 80 r. 6–10 and dupl., Abusch and Schwemer 2011: 297, No. 8.4: 49–53.

The term *adirtu* appears in the anti-witchcraft texts in combination with other symptoms and is always related to a state of fear and terror, as well as to various afflictions: *ašuštu, ḫūṣ ḫīpilibbi, gilittu, pirittu, ḫurbāšu, ḫa'attu, diliptu*. The link between these symptoms is also confirmed in *Maqlû*: *ašuštu arurtu ḫūṣ ḫīp libbi gilittu piritti u adirti yâsi taškunāni* "Distress, tremor, depression, fear, anxiety, and fear you (= witch) have inflicted on me" (V 71, Abusch 2016: 141).[144]

In the *nīš libbi* documentation, too, as we saw from the passages mentioned above, the prolonged state of fear is expressed by the word *adirtu*. Other symptoms of loss of desire are mentioned: panic attacks (*ḫa'attu*, No. D.3: 37), fright (*galātu*, No. F.4: 88; K prescr. 21: 57), distress (*ašāšu*, No. D.4: 57; No. F.4: 88, catalogue LKA 94 ii 11; *ašuštu*, B prescr. 1: 19) and insomnia/agitation (*diliptu*, A.3: 52).

The term *adirtu* in the *nīš libbi* corpus is to be understood not so much in the sense of 'bad luck, disaster,' or 'darkening,' but rather in the sense of fear, or rather an extended state of fear[145] (as indicated by Stol 1999: 64), leading to a state of apprehension and unhappiness. This hypothesis is supported by the fact that fear is provoked by witchcraft or divine wrath, which are etiologically at the basis of the loss of sexual desire. As in other kinds of documentation, from anti-witchcraft texts to wisdom literature, fear is connected to other symptoms that indicate a psycho-emotional unstable state, affected by panic, fright, fear, anxiety. The 'heart' (*libbu*) itself is the center in which the fear resides and from which it must be removed:[146]

Text:	*nīš libbi*
Diagnostic Category:	lack of sexual desire
Cause:	witchcraft or divine wrath (Marduk or Ištar)
Related Symptom:	fear (*adirtu*)
Other associated symptoms:	insomnia (*diliptu*), affliction (*ašuštu*), panic attack (*ḫattu*), fright (*gilittu*)

We have seen how fear and panic are placed in relation to states of agitation, affliction, and insomnia, which cause a state of apprehension and psycho-physical discomfort in the man. If we wanted to use a concept taken from the modern categorization of psycho-emotional disorders, we could use the concept of "melancholy" to describe the emotional state, as well as the physical one, by which the patient suffers from loss

144 See also the similar sequence in *Maqlû* VII 126–127, Abusch 2016: 185, in line 124 there is also *diliptu* 'insomnia.' The parallel *adirtu, piruttu, ḫattu* and *gilittu* is also found in the *Ludlul*, in the description of symptoms that affect the poor man (I 111–113, Oshima 2014: 84).

145 On the basis of the principle: shiny *libbu* = happiness, dark *libbu* (*adāru*) = unhappiness, suffering, fear.

146 In the incantation No. A.3: 51 -52 the *libbu* is the center in which the fear occurs or is to be found. More generally speaking, most states related to melancholy and depression consequently are described in association with the insides (Couto-Ferreira 2010b; Attia 2019; Attia 2019). See also *Ludlul* I 111–113 (Oshima 2014: 84) and a *šuilla* prayers: *liptaṭṭiru adiratu ša libbīya* "May the fear of my 'heart' be removed (litt. loosen like a knot)" (Ebeling 1953a: 120, l. 13).

of sexual desire. While there are several studies on the concept of "depression" in the area of Assyriology (see Gruber 1980; Alster 1983; Barré 2001[147]: 178–181; Kselman 2002; Maier 2009; Couto-Ferreira 2010b: 25–27; Salin 2020: 124–129; Van Buylaere 2020), there are few on "melancholy" (see Stol 1993: 27–32;[148] Maier 2009;[149] Couto-Ferreira 2010b; Buisson 2016). The definition given by Stol is as follows: «A man with a broken heart can be ill-tempered, suspicious, have a nervous breakdown, be full of apprehensions, be worried, or in panic» (1993: 31). Being aware of the theoretical-methodological difficulties of applying modern psychopathological categories to a Mesopotamian context, I use the term "melancholy," following Couto-Ferreira, «in reference to feelings, mood and behavior dominated by sadness, fear and distress in different grades and with several characteristics» (2010: 23). As with fear, "melancholy" has external causes, such as witchcraft[150] and divine wrath (Stol 1993: 29; Couto-Ferreira 2010b: 27–33).

I have previously shown the link between fear and lack of sexual desire, and, more generally speaking, the sphere of love. But what is the relationship between melancholy, lack of desire, and the domain of love?

Another term, apart from *ašuštu* seen previously, that can be traced back to the concept of melancholy, if not to that of depression, is *nissatu* 'grief, affliction' (CAD N/II A 274–275; AHw. II 725). Thanks to a pharmacological text that describes the use of specific ingredients for certain sicknesses, BAM 1 iii 35 (Attia and Buisson 2012: 29), we know that the absence of sexual desire, which is cured with the *azallû*-plant,[151] is related to symptoms of 'grief' or 'affliction' (*nissatu*):

azallâ	:	KI.MIN (= *nīš libbi*)	:	*nissata lā īši*
azallû-plant	:	for ditto (= the sexual desire)	:	he will not have any grief.

In this example, the diagnostic category to be cured is the absence of desire, while grief is a related symptom, a psycho-emotional condition from which the patient, whose sexual desire is absent, suffers. It is not surprising that the loss of sexual desire can cause melancholy. In general, love matters, and the problems arising from it, can cause states of distress (*ašuštu*) and insomnia (*diliptu*). Love as a source of worry or insomnia is confirmed by other texts, such as "love songs" (see Wasserman 2016: 45–47), here is an example from *A Dialogue Between Hammurabi and a Woman*: *râmki eli diliptim*

147 For sources in Akkadian language, the scholar mentions in the *Epic of Gilgameš* the king of Uruk's reaction to Enkidu's death (see Maier 2009), in *Enūma eliš* the one of the gods to the news of the plot of the Aspû, a passage from the *Šurpu* series (VII 36) and one from the *Šumma izbu* series (III 76).

148 Stol (1993: 31) hypothesizes a relationship between melancholy and the "black bile." Against this idea, see Geller 2007: 191; Couto-Ferreira 2010b: 21–23.

149 Maier has used the term "melancholy" to indicate the state in which Gilgameš finds himself after Enkidu's death, translating the Akkadian term *nissatu*.

150 Fear (*adirtu*, *gilittu*, etc.), insomnia (*diliptu*), and states of melancholy and depression (*nissatu*) caused by attacks of witchcraft are expressed in a passage of *Maqlû* VII 124–127, Abusch 2016: 185.

151 On the use of the plant against depression see Couto-Ferreira 2010b: 32–33.

/ *u ašuštim lā watru ina ṣērīya* "Your love nothing more than insomnia / and affliction causes me" (von Soden 1950: 170, ll. 8–9; Held 1961: 9). An Old Babylonian poem, originally from Isin, contains several incantations and rituals intended to acquire the love of a man, who is perhaps married, by a *ēntu*-priestess.[152] Here insomnia is a condition that one wishes the beloved suffer from to make sure that the woman achieves her desired goal (IB 1554: 38–40, Wasserman 2016: 261): "Be sleepless (*dilpi*) at night! / During the day may you(fem.) not sleep! / At night, may you(fem.) not sit!".

The suffering of the beloved one is certainly a recurring theme in love songs, namely his concern and distress; however, there are other examples, also taken from the Old Babylonian period, from which it is possible to deduce the connection of the melancholic state with love and sexuality in both men and women alike. It is noteworthy that the same language is employed, both in the description of the desired behavior of the beloved (man), described in the preceding text, which the incantation should activate to carry out the wishes of the woman, and the state of the sickness of love in the two poems that will be mentioned below. In the following case, it is a woman talking and crying over love:

> I always talk about you,
> I am eaten up,[153] I have convulsions,[154] I am dis[torted]$^{?}$.[155]
> Again, I craved you,
> I am afflicted continuously[156] (*ātaššuš*), I saw your [f]ace,
> you are a god! I implore you: let your he[a]rt love me$^{?}$!
> [CUSAS 10, 9: 14–18, Wasserman 2016: 90]

In the same group of Old Babylonian love compositions, another text describes the emotional condition suffered by the lover towards his beloved one. Love and sexual attraction infect the man, described as a victim of this sickness of love: his heart is broken and his mood angry:

> My mood (*kabtatī*)
> plundered the heart (*libbam*) which infects,[157]
> his sexual attraction (*dādu*) is a love that infects.
> [. . .]
> My heart [is broken], my mood is agitated.
> [CUSAS 10, 8: 4–6, 15, Wasserman 2016: 86]

---

152 Another interpretation is offered by Scurlock (1989–1990: 108) according to whom, the composition would aim at getting «economic control over an adversary». For an analysis of incantations and rituals of love and descriptions of love suffering, making use of interpretative models of modern psychology, see Geller 2002.

153 An expression typically used to describe the negative action of the sickness damaging humans.

154 *damû* CAD D 80 'to suffer from convulsions,' AHw. I 166 'taumeln.'

155 *ewû* CAD E 413–415 'to change, turn into'; AHw. I 267 'belasten mit' (*ewûm* II).

156 See Heeßel 2000: 218, XVIII 8–9.

157 *muḫattitum* (in the next line *muḫattitu*) participle of the verb *ḫuttuttum* 'to infest,' in reference to people infected with worms or lice (see George 2009: 52; CAD Ḫ 264; AHw. I 362).

In the aforementioned love songs, the lover is described as a victim of a psycho-physical state of suffering, making use of a language used in other types of documents, such as medical texts, prayers to the god or incantations of anti-witchcraft. The texts emphasize the potential pathological nature of love, especially when unrequited. I have therefore shown that a state of melancholy, characterized by fear, agitation, and insomnia is a condition of a man struck by witchcraft or divine wrath, who has been deprived of sexual desire. It is also a typical condition of a lover whose love is unrequited or who, generally speaking, suffers from the sickness of love. These texts, therefore, describe a condition of "pathologizing" of love: the lover suffering from love, as if suffering from a sickness. However, the condition of melancholy is also typical of a particular diagnostic category: 'love-sickness' (*muruṣ râmi*) (see Heeßel 2000: 264; Scurlock and Andersen 2005: 372–373;[158] Couto-Ferreira 2010b: 33–36; Wasserman 2016: 34–35; Salin 2020: 143–145). This diagnostic category is mentioned three times in *Diagnostic Handbook* (Heeßel 2000). Two times it appears in the tablet XXI (6–9), including cases of witch's attack (1–5) and pains caused by illegitimate sexual relations (10–15):

> 6–7. If he turns around restless?, the mouth constantly falls,[159] talks constantly to himself (and) laughs constantly for no reason: he suffers from "love-sickness" (*muruṣ râmi*). It is the same for man and woman.
> 8–9. If the distress (NÍG.ZI.IR/*ašuštu*) falls constantly upon him, he rotates back his throat, eats bread, drinks beer, but it is not beneficial for him, he says "O my abdomen" and he is distressed: this person is suffering from "love-sickness"; it is the same for man and woman.
> [XXII 6–9, Heeßel 2000, 251–252; see also Scurlock and Andersen 2005: 131, No. 6.83; 372–373, No. 16.23, 16.24; Couto-Ferreira 2010b: 34]

A third mention is in tablet XVIII (ll. 8–9), in the section related to fever symptoms:

> [If his body] does not have [fe]ver, he does not want to eat and drink, the crying for him is sick?, he always talks (about) impatience (and) torpidity, and is continually in anguish (*ašāšu*): this person is suffering from "love-sickness"; It is the same for man and woman.
> [XVIII 8–9, Heeßel 2000: 218]

Heeßel (2000: 264) argues, criticizing the previous interpretation of "sexual diseases" (*Geschlechtskrankheit*) proposed by von Soden (AHw. II 951), that the diagnostic category of "love-sickness" (*Liebeskrankheit*) represent quite general observations such as loss of appetite, fatigue, melancholy, corresponding to the psychological and physical stress of an (uncertain) affair of the heart.

In this section, departing from the symptoms of fear and insomnia inflicted by the witch that cause damage to humans, as mentioned in incantation No. A.3, instead we

158 They define it as "a specific form of stress."
159 For the philological commentary see Heeßel 2000: 264.

have found a series of psycho-physical disorders, such as panic, fear, grief, agitation, which characterize a patient devoid of sexual desire. Conditions that are caused by witchcraft or divine wrath, but also by negative dreams and uncertain omens. A continuous condition of melancholy and fear is often typical for those who *suffer from love*, as in the case of the Old Babylonian love songs, or those suffering from "love-sickness." When the sphere of love and sexuality is in disorder and described as a sickness by making use of literary expressions, as in the case of certain incantations and love poems, or by being defined and categorized as such, the absence of sexual desire or "love-sickness" causes, among the symptoms, a condition of melancholy, fear and constant agitation.

Other Texts

As already discussed, the lack of sexual desire is one among many other symptoms which characterize the bewitched man (see Stol 1999: 55–60; Abusch 2002: 79–88). In this case, it is not the sole purpose of the witch to take away sexual desire.[160] Many symptoms characterizing the lack of desire as a diagnostic category, are also present in the anti-witchcraft texts. But the symptoms of these texts are broader and more complex (see Schwemer 2007a: 165–179 fn. 2, 5, 6, 14, 16, 21):

– *ana sinništi alāka muṭṭu* "(the man's desire) to go to a woman is reduced":[161]

> If a man's body is afflicted with paralysis, he is constantly feverish,
> his [f]lesh is being ruined, and
> (his desire) to go to a woman is reduced, (then) f[igurines] of clay representing him have been buried (in a grave).
> [LKA 160 + BAM 140: 7′–9′ and dupl., Abusch and Schwemer 2011: 70, No. 2.2.1: 38′–40′]

> Other symptoms: paralysis; weakness; fever.

> [If a ma]n continually has vertigo, his head [. . .],
> [and] is dark, his appearance constantly [changes],
> his [appearan]ce is darkened, his ability to speak [. . .],
> he becomes more and more depressed, (his desire) to go to a woman is re[duced],
> he is constantly irritable, his ears bu[zz],
> (then) clay figurines representing [th]is [man] have been bu[ried].
> [AMT 85, 1 obv. ii 26′–31′, Abusch and Schwemer 2011: 86, No. 2.3: 42–47]

> Other symptoms: vertigo; melancholy; language problems; irritation, buzzing in the ears.

> If a man becomes increasingly depressed, [his] l[imbs are limp all the time],
> his tongue is always swollen, he bi[tes] his tongue,

160 On the relationship between fear, sexual desire and anti-witchcraft texts see Chapter I § "Fear, distress, and insomnia."

161 Also AMT 21, 2 and dupl.: 19, Abusch and Schwemer 2011: 321, No. 8.6.1: 19.

his ears buzz, his hands are numb, [his] kn[ees (and) legs]
cause him a gnawing pain, his epigastrium continually pro[trudes],
(his desire) to go to a woman is reduced, cold tremors afflict him repeatedly, he [is in turn fat and thin],
he continually salivat[es] from his mouth, [. . .],
that man was given (bewitched) bread to eat, (bewitched) beer to drink, was anoi[nted] with (bewitched) oil, [. . .].
[BAM 445 + AMT 64, 2 obv. 10–16, Abusch and Schwemer 2011: 154, No. 7.7: 47–53]

Other symptoms: melancholy; weakness; swollen tongue; buzzing in the ears; numb limbs; pain in knees and legs; epigastric problems; cold tremors.

If a man [keeps putting] his arms over h[is] head, his arms feel numb,
his feet continually produce *munû*-sores, his body feels hot [a]t sunrise,
(his desire) to go to a woman is reduced, [. . .], he eats food bu[t] it does not agree with him,
his eyes are continually staring, . . . [. . .] . . . him and
he scratches (it) constantly, his 'mouth' is constantly troubled, [his . . . is always sei]zed, they speak (lit.: "say a word") to him, but
he forgets (it), his heart is constantly troub[led].
[AMT 13, 4 + BAM 460 obv. 6–11 and dupl., Abusch and Schwemer 2011: 417, No. 10.4.1: 6–11]

Other symptoms: numb limbs; the feet produce *munu*-sores; fever; eating problems; fixed eyes; troubled mouth; memory problems.

. . . [. . .] . . . his waist . . . [. . .]
[. . .] . . . is reduced, he does n[ot w]ant to talk,
he becomes more and more *depressed*, his limbs are c[ontinuously limp],
he is continually *bloated*, he gnaws his lips, his ears b[uzz],
his hands are numb, his knees and legs cause [him] gnawing [pain],
his epigastrium continually protrudes, (his desire) to go to a woman is re[duced],
he does not desire a woman, cold tremors afflict him repeatedly,
he is in turn fat and thin, he continually saliv[ates] from his mouth,
he is often irritable, he cannot stand his bed,
(and) he is sometimes paralyzed, (then) that man is bewitch[ed];
figurines representing him have been made and bur[ied] in the lap of a dead person.
To undo the witchcraft, to save his life, to reconcil[e] him with his angry personal god.
[Tablet 65 of the canonical *ušburruda* series, Abusch and Schwemer 2011: 118, No. 7.2: 4–15]

Other symptoms: no walking; melancholy; weakness; swelling; biting lip; buzzing in the ears; numb limbs; pain in knees and legs; epigastrium problems; cold tremors; saliva from the mouth; irritation; paralysis.

[If a man]'s penis gives him sharp pains when he urinates, his 'semen' discharges,
he is affected [in] his innards, and (his desire) to go to a woman is reduced, pus (literally 'white blood') constantly comes out of his penis,
th[at man] suffers from 'discharge.'
[BAM 112 i 17′–19′, Geller 2005: 64, No. 4, see also i 34′–36′]

Other symptoms: penis' pain when urinating; semen's discharges; innards' pain; pus (literally 'white blood') from the penis. Note that the diagnostic category is not witchcraft, as in the other texts, but 'discharge.'

– *nīš libbi kasi* "sexual desire is bound":

By fever, stiffn[ess, sw]eating, sickness,
wasting away I am con[tinually affected . . . of the fo]rehead, of the chest (and) of the head (and) convulsions.
My arms, my lower legs, my [kne]es (and) my feet are cramped,
my sexual desire, my plea[sant fea]tures are bound,
my limbs keep faltering, by depression, terr[or],
[f]ear (and) fright I am more and more affected, I am constantly anxious,
I am [alw]ays fearful, I keep on talki[ng] to myself,
I have terrible [dre]ams.
[LKA 154 e 155 obv. 43–49 and dupl., Abusch and Schwemer 2011: 259, No. 8.2: 53–59][162]

Other symptoms: fever; stiffness; sweating; wasting away; convulsions; cramps; weakness; depression; terror, fear, fright, and anxiety; terrible dreams.

– *sinništa libbašu ḫašiḫ-ma sinništa ippalis-ma libbašu lā inaššīšu* "he needs a woman and he sees her, but he does not desire (her)":

[If a person's inna]rds are continually colicky, his palate continually gets dr[y],
his [arms] are continually numb, he belches, he has plenty of appetite (for food),
but when [he sees it], it does not please him; he (= his 'heart' (*libbu*)) needs a woman and he sees her, but he does not desire (her);
[his heart] is (too) depressed (for him) to speak– "hand of a ghost" is pursuing that person.
[AMT 76, 1: 6, Scurlock 2006a: 480, No. 200: 1–4][163]

Other symptoms: innards problem; dry mouth; numbness; strong appetite; melancholy.

– *nīš libbi ṣabit/ekim/eṭir* "sexual desire is seized/taken away":

(Who) has constrained me, has tied me up, has seized me, has bound me,
(who) has filled me with stiffness and decay,
(who) has seized my sexual desire, has made me angry with myself,
(who) has twisted my sinews, has weakened my strength,
(who) has 'poured out' my arms, has bound my knees,
(who) has inflicted on me quarrel, squabble, wailing, fear,
panic, anxiety, curse,
terror, confusion, sleeplessness, dumbness, numbness,
'heart' (*libbu*) (and) bodily ill health.
[*Bīt rimki* II, Abusch and Schwemer 2011: 377–378, No. 9.1: 20–28. See also Laessøe 1995: 36–47; Seux 1976: 388–392]

162 The symptoms continue until line 79.
163 The cause of absence of sexual desire is in this text the "Hand of a ghost."

Other symptoms: stiffness; decay; anger with himself; weakness; quarrel; wailing; fear; anxiety; curse; confusion; sleeplessness; dumbness; numbness; 'heart' (*libbu*) and bodily ill health

(Who) have fed me [*dīḫ*]*u*-sickness [with bread], have given me *ašû*-sickness to drink with water,
(who) have shut up [. . .] in a hole, who have performed 'cutting-of-the-throat' magic . . . ,
[who have b]eaten me, have performed 'cutting-of-the-throat' magic [against me],
(who) have caused [god, king, mag]nate (and) nobleman to be angry with me, have seized my sexual desire,
(who) have made [me] angry [with myse]lf, quarrel, fight, waili[ng],
[mis]ery, depression, confusion, convul[sions],
[panic, an]xiety, cur[se], sleeplessness,
dumb[ness, numbne]ss, 'heart' (*libbu*) (and) bodily ill health [have in]flicted on me.
[KAR 80 r. 6–10 e dupl. Abusch and Schwemer 2011: 297, No. 8.4: 46–53]

Other symptoms: *dīḫu*-sickness; *ašû*-sickness; anger of god, king, magnate and nobleman; anger with himself; quarrel; wailing; misery; depression; confusion; convulsions; panic; anxiety; curse; sleeplessness; dumbness; numbness; 'heart' (*libbu*) and bodily ill health,

Incantation: "Šamaš, these figurines are those of my warlock and witch, who constantly perform (witchcraft) against me.
You, (var. adds: judge), know them, but I do not know them, who have stung my flesh,
(who) have seized my forehead, have tied my sinews, have poured out my . . .,
(who) have immobilized my arms, have seized my sexual desire, have dried up my spittle,
(who) have poured out stiffness (and) decay over my body, have fed me bewitched bread,
(who) have given me bewitched water to drink,
(who) have bathed me with d[i]r[t]y [wash wate]r,
(who) have anointed me with an ointment containing bad herbs,
(who) have disabled my ability to speak, have slandered me, have bent my spine (var. adds: like a seal ring),
(who) have pressed my chest, have taken away the *healthy glow* of my face,
(who) (var. adds: have disfigured my features), have dulled my sense of touch,
(who) have impaired my sense of hearing, have weakened my ability to see, have dragged off my glow of health.
[KAR 80 r. 6–10 and dupl. Abusch and Schwemer 2011: 297, No. 8.4: 68–78]

Other symptoms: weakness; paralysis; drainage of saliva; weakening of the five senses.

[Incantation: "Witch, you, who has destroyed] my whole body.
[you have . . . m]y [. . .], ruined my features,
[you have . . .] me, bound me, seized me,
[you have . . .] me, filled me with stiffness and decay,
you have made me ill with [coug]h, phlegm, spittle, and mucus,
you have taken away my sexual desire, and made me angry myself,
you have weakened my strength, rendered my arms limp, my knees
you have bound, you have made my breast and shoulder hurt,
you have crushed my limbs like malt,
you have burned my body parts like fire.
[K 2467 + 80-7-19,166 obv. ii 2′–11′ and dupl. Abusch and Schwemer 2016: 158–159, No. 8.20.2: 4′–13′]

Other symptoms: immobility; decay; cough, phlegm, spittle, and mucus; anger with himself; weakness.

If a man's limbs are 'poured out' like those of a sick man,
his arms (and) his . . . are slack;
he speaks, but does not achieve (his wish),
his sexual desire has been taken away, he becomes increasingly depressed,
either when urinating or constantly, his semen
is discharged as if he had been having sex with a woman
that man is impure: god and goddess have turned away from him,
his speech does not find favor.
[BAM 319 obv. 1–8 and dupl., Abusch and Schwemer 2016: 250, No. 8.29.1: 1–8][164]

Other symptoms: weakness; melancholy; semen's discharge; removal of personal god and goddess.

Seized my mouth, made my neck tremble,
Pressed against my chest, bent my spine,
Weakened my 'heart' (*libbu*), taken away my sexual desire,
Made me turn my anger against myself, sapped my strength,
Caused my arms to fall limp, bound my knees,
Filled me with fever, stiffness, and debility,
[. . .]
Caused god, king, noble, and prince to be angry with me.
[*Maqlû* I 97–102, 109, Abusch 2016: 41–43]

Other symptoms: silence[?]; tremors; weakness; anger with himself; immobility; temperature; anger of god, king, noble, and prince.

The lack of sexual desire is thus one among many symptoms that characterize the bewitched man.[165] The symptoms vary a lot, but one can draw a "semantic illness network," using the expression of the anthropologist Good (1977), linking the absence of desire with other symptoms within the diagnostic category of the witch's attack:
- weakness, fatigue;
- immobility, paralysis, stiffness;
- decay;
- anger with himself;
- anger of god, king and noble;
- melancholic and depressive conditions;
- temperature;
- fear, panic;
- (cold) tremors;
- anxiety;

164 For a commentary of this passage in relation with the sexual impotence see Stol 1999: 59–60.
165 Or victim of the "Hand of a ghost."

- weakening of the five senses;
- buzzing in the ears;
- epigastrium problem.

Couples therapy

A question that we need to clarify is: is the loss of sexual desire only a male problem? In this section, in fact, I will show the importance of the female partner in the therapeutic process, not only as a social actor but also as a recipient of the therapy.

As I pointed out, contrary to the claims made by Biggs (1967: 2), not only men are said to have sexual desire, in fact, it is experienced by women as well. In No. K.1: 11 it is stated that Enlil and Bēlet-ilī have determined the sexual desire for all humanity, men and women. It is clear, although the sources are few, that women have sexual desire. However, in the *nīš libbi* incantations and rituals, the man is described as the suffering patient. Nevertheless, I would like to pay specific attention to the role played by women in this therapeutic process.

The woman, in fact, actively takes part in the ritual practice.[166] Many of the pharmaceuticals (potions and ointments), described in the prescriptions, are used both by the woman and the man. She drinks the potion together with her partner:

> [In order to] release the 'heart' (*libbu*) of the man and of the woman: [. . .]
> [seeds] of *sikillu*-plant and oil: You ta[ke] the green (branches) of *ēru*-tree [. . .],
> you [g]ather, equally [. . .]
> [you ta]ke and you pulverize [. . .]
> [the ma]n and the woman drink them and [. . .].
> [*nīš libbi* P prescr. 4: 11–15]

> In order to release the man and the woman and [. . .]
> you pulverize the green (part) of *nīnû*-plant [. . .]
> you do not cook/uncooked . . . [. . .]
> you throw a lump of metal into good beer and [. . .]
> they drink together and their 'hearts' will be [released].
> [*nīš libbi* B prescr. 2: 41–45]

The ointment is rubbed on anatomical parts of the man and as well on the woman's navel and her vulva. The man's right hand and the woman's left hand are massaged:

> Its ritual: You pulverize magnetite (and) iron,
> you mix (them) with oil from the alabastron; you recite the incantation three times over it; the man's penis

166 In fragmentary passages: D prescr. 1: 28; P prescr. 2: 7; P prescr. 5: 18.

(and) the woman's vulva, you? anoint them and (he will get) sexual desire.
[No. D.2: 21–23][167]

Its ritual: Iron powder, "heals-a-thousand"-plant, *ru'tītu*-sulphur
you pulverize together, you put (it) in oil, you recite the incantation seven times over it,
you anoint (with it) the penis of the man and the pelvic area of the woman.
[*nīš libbi* M prescr. 1: 1–3]

Its ritual: You pulverize magnetite, you mix (it) with oil from the alabastron,
you recite the incantation seven times over it; you apply (it) to his navel;
you pulverize iron, you mix (it) with oil from the alabastron, you recite the incantation seven times over it,
you apply (it) to the woman's navel; the man and the woman [will find relief] together.
[No. B.1: 14–17]

[. . .] you massage three times your [n]avel (and) your penis, you massage your [right?] hand (var.: which she (= witch) touched with sorceries) (and) the left one of the woman [. . .].
[*nīš libbi* D prescr. 2: 31]

The figurines of both man and woman are made and used in rituals:

[You make] the figurines of the man and the woman [. . .]
. . . them, the "desire" (lit. "rising" of the 'he[art]') of the man and the woman . . . [. . .].
[*nīš libbi* K prescr. 29: 127–128]

Its ritual: You mix together emmer dough and potter's clay; you make the figurines of the man and the woman; you put them one upon the other, and place them at the man's head,
you recite [the incantation] seven times; you remove (them) and put [them near] a pig.
[*nīš libbi* D. Diagn.: 24–25]

As we will see in Chapter III (§ "Bow ritual and battle metaphors"), in the ritual of the bow, the latter is placed over the head of the man and the woman: "You make a bow of thorn, / a tendon of the *arrabu*-mouse is [its] string, you load it [with an arrow], / you pu[t it] over heads of the man and the woman, who are lying [. . .]" (E Bow ritual: 57–59).

The woman, therefore, is an actor of great importance in the *nīš libbi* therapeutic rituals. She, together with her partner, performs the rituals. The pharmaceuticals are used by both, man and woman. Why does the woman participate in the healing ritual? If the patient, attacked by the witch or the hand of god, is the man, what is the function of the woman's presence?

As is evident from the texts, the recipient of therapeutic practice is not only the male partner but also the woman. The text P prescr. 4: 11 affirms that the aim of the therapy is: "[In order to] release the 'heart' (*libbu*) of the man and the woman [. . .]" (also B prescr. 2: 41); or, if restoration is correct in D presc. 7: 54: "[To make] a woman have an 'elevation.'" The incantation No. D.3: 35–36 confirms that the problem also

167 Also No. B.2: 38–40; No. D.3: 39; No. E.1 15–17; L prescr. 6: 15–18.

concerns the woman: "You recite (var.: you say) this (incantation) three times and (if) the man and the woman / desire together each other, (but) they do not find reli[ef]." The aim of the ritual therapy is her sexual arousal as well and that the lack of the male sexual desire directly involves the woman.

She recites many incantations from this corpus (No. B.1: 1–11; No. C.1; No. D.2; No. D.3: 32(?); No. E.1; No. E.3: 46–47(?), 48–51; No. F.3;[168] No. F.4: 80–88; No. G.1(?); No. J.3: 25–28(?); No. M.1: 4–7). The incantations mention her role in sexual intercourse, her excitement, and her role in the ritual action. The incantation No. B.1: 10–11 invokes, with typical imploring language, probably the patient's erection and the woman's ability to hold his penis, using a food metaphor: "May your penis, which satisfies (the desire)?, be compact?! Do not [. . .]! / May my crotch devour your . . . penis!". In the incantation No. F.3 the woman invites the partner to make love with her:[169]

> Make love to me! Ma[ke love to me] because I am young!
> [. . .] I am endowed with love, make love to [me]!
> Tie [. . .] of the mating/*rikibtu* of a stag! Ma[ke love to me]!
> [No. F.3: 71–73]

The invitation is made in the incantation No. E.1 as well, making use of animal metaphors:[170]

> The one at the head of my bed, rear up, make love to me!
> The one at the feet of my bed, rear up, bleat *for* me!
> My vulva is the vulva of a bitch! His penis is the penis of a dog,
> as the vulva of a bitch took the penis of a dog, (so may I do)!
> May your penis become as long as a *mašgašu*-weapon!
> I sit in a net of laughter,
> may I not miss the quarry!
> [No. E.1: 7–13]

In the incantation No. M.1, the wife[171] of the patient prepares the bed for the sexual desire as the goddesses *illo tempore* did for their lovers:

> Sexual desire! Sexual desire! The bed for the sexual desire,
> (like the one that) Ištar did for Dumuzi,
> (the one that) Nanāya did for her husband,
> (the one that) Išḫara [di]d for her lover (var.: husband), I did!
> [No. M.1: 4–7]

---

168 The one who recited the incantation is a woman, but it is not clear if she is the female partner or a female therapeutic operator.

169 Also No. C.1: 7–8.

170 See also incantation No. E.3.

171 In line 8 the variant Ms. C affirms: "M[a]y the flash of my husband be static, may (instead)? his penis be erect!"

This is one of the few passages that make explicit reference to husband and wife. In almost all cases the female partner is in fact defined as "woman" (*sinništu*). In No. A.3: 56 it states: "[Mo]unt the wife of NN, [daughter of NN!]".

The importance attributed to the woman's agency helps us to understand that it is not a single person who is cured as if he is a social atom. As the anthropologist Alessandro Lupo argues, the therapeutic action is directed at social molecules, that are believed to be affected by the sickness, which undermines their bonds and harmony, in this case, the sexual relation between man and woman (Lupo 2012: 134; see also Waldram 2000: 612). Even when it affects a single person, that is the male partner, the sickness always implies a split in the social structure, and consequently, the cure must repair these ripped relations, necessarily including the woman. Therefore, it is a "socio-therapy," which has as recipient both the man and the woman.

As mentioned above, the cause of the loss of sexual desire is often witchcraft. This leads us to suppose that the instigator of the aggression, who sought help from the witch, may not necessarily want to attack exclusively the male partner. As it is known at cross-cultural level, sexual intercourse, especially within a pair, such as the matrimonial one, is subject to envy and aversion from the community. The sexual sphere is one of the aspects of community life more susceptible to the witch's actions. It is possible to assume that the instigator of the aggression, would attack, for envy, aversion, etc., the female partner and therefore the pair as well. The fact that the sickness appears in the man, does not confirm the fact that he is the only recipient of the aggression. It is very likely, however, that the woman is the other recipient. This means that the therapy should also involve her in the ritual practice. This would explain the role of women in *nīš libbi* incantations and rituals.[172]

During the years of my research, the presence of women in the therapeutic process has been interpreted by those to whom I pointed it out as a clear example of "couples

---

172 The anthropological comparison helps us in some interpretative hypotheses. Attacking the sexual sphere may mean attacking the couple. Anthropological examples are countless. For this reason, I just take one example from my direct experience of fieldwork on magic in Sicily. In a suburban Palermo neighborhood, there was a man suffering from sexual impotence, after the medical and psychological consultation was conducted without a positive result, he decided to go to the priest. The priest, during the consultation, asked man to lower his pants. He immediately noticed the pubic hair was in a braid, fitting into the traditional Sicilian magic worldview of a magical aggression. Unable to do anything, the priest invited the man to go to a female magician (*magara*). At the magical consultation, the bride participated together the man. According to the *magara*, with the man and his wife's approval, the cousin of the new bride is the cause of such fascination. The cousin was envious of the fortune of the woman, who had found a handsome man with a steady job. The recipient of the charm was therefore not the man himself, but the couple and more specifically the female partner. The sickness appears in the man, but it cannot be reduced to him. Therapeutic itinerary must inevitably involve the bride. This example, among the thousand possible, does not not mean that it is the same in Mesopotamia, however it may lead us to wonder about the way in which sickness, witchcraft, sexuality, male and female agencies are interrelated. It allows us to look at the Mesopotamian sources from another lens.

counseling," similar to modern psychological therapy. I disagree with this interpretation because it projects a way of understanding the relationship within a modern conception of couples. The main aim of modern couples therapy is overcoming the crisis between the partners and recovering harmony. It helps couples identify and confront their conflicts. All this is completely absent in Mesopotamian sources. The absence of sexual desire does not depend on psychological distress between the partners, but, on the contrary, on an external attack, witch or divine. Of course, the witch's attack is a clear sign of social anxiety in the community or the family. Consequently, it is licit to talk in terms of "couples therapy" not in the sense of something similar to the modern psychological therapy, but for the fact that the therapy involves the female partner as therapeutic recipient.[173]

Who does what?

Another aspect of the therapeutic practice to be investigated deals with the agents of ritual performance. One must ask: "Who does what during ritual action?". As stated above, there are usually three actors: the therapeutic operator (exorcist), the patient, and the female partner. The use of Sumerian logograms, unfortunately, does not help to clarify the aspect of performativity and agency (note that the texts from the second millennium from Ḫattuša generally have the verbs written out syllabically in the second person masculine singular, that is, addressed to the *āšipu*). Of course, the exorcist is the one who prepares the pharmaceuticals described in medical prescriptions: potions, ointments, and amulets. The potions were drunk by the patient, the amulets hung on his body and the ointments massaged on the anatomical parts of man and woman. In the last case, because of the use of logograms, it is not clear who anoints the patient's body, whether himself or the exorcist.

Certainly, some incantations were recited by the woman,[174] others by the patient[175] and the exorcist.[176] The verbs employed to designate the recitation of the incantations are *manû* 'to recite' and *qabû* 'to say.' Couto-Ferreira (2015) hypothesizes that the different use of verbs *qabû* and *manû* in the ritual action can provide information on the performativity of the actors. She states that *manû* hints at the healer reciting the incantation, manifesting notions of knowledge and authority coming from the gods; while *qabû* should be understood within the general act of speech, an act of repetition.

---

173 On the ritual used to resolve the problems in the couple, for example for the partner's anger see Farber 2010.

174 No. B.1: 1–11; No. C.1; No. D.2; No. D.3: 32(?); No. E.1; No. E.3: 46–47(?), 48–51; No. F.3(?); No. F.4: 80–88; No. G.1(?); No. J.3: 25–28(?); No. M.1: 4–7.

175 No. A.1; No. A.4; No. E.2; No. E.3: 47–47; No. J.2; No. O.1

176 No. A.2; No. A.3; No. B.2; No. D.1; No. D.3; No. D.4; No. E.3: 52–53; No. F.1; No. F.2; No. K.8(?); No. K.7(?); No. H.1; No. I.1(?); No. J.1; No. K.2; No. K.3; No. K.4; No. K.5; No. K.6; No. K.9; No. L.1; No. L.2.

The verb *manû* would indicate the recitation of the incantation by the therapeutic operator, while in the same therapeutic performance the verb *qabû* the declamation by the patient.[177] The therapeutic performance, involving incantations No. E.2 (and its ritual) and No. E.3, clarifies the issue. In this therapeutic practice the actors are three: the therapeutic operator, the patient, and the female partner[178] In the text No. E.2, most verbs do not inform us about the person who performs the action. All the verbs in the first lines are written using logograms. In line 24 the verb is written syllabically in the second singular person: *ta-šár-rap* "you (= the therapeutic operator) burn." Usually, when the verb is syllabically written, the 2ps form refers to the therapeutic operator. The choice not to write this verb with a Sumerogram has, in my opinion, a precise purpose: all the earlier verbs, in logographic form, representing different phases of the ritual process, refer to the action of the therapeutic operator.

The logogram DU$_{11}$.GA (*qabû*) follows, which should be read in Akkadian with the 3ps, referring to the patient. The incantation is recited by the patient. The line 37 at the end of the incantation provides its declamation by the patient three times: *annâm šalāšīšu iqabbi* (DU$_{11}$.GA) "he declares this three times."

The following ritual is performed by the therapeutic operator and involves the creation of an amulet to hang on the patient's neck. The therapeutic operator spins the wool of ram and weaned sheep mentioned in No. E.3 and ties it around the patient's waist. He recites (*manû*, l. 45) the following incantation No. E.3(?), or rather he makes it possible for the incantation to be recited,[179] to give magical power to the *materia magica*:

19. You (= *āšipu*) set up (GIN-*an*) a *pāṭiru*-portable altar in front of Ištar of the stars, you sacrifice (BAL-*qí*) a sheep,
20. you set up (GAR-*an*) a censer with juniper, you libate (BAL-*qí*) (var. adds: premium) beer,

177 Note that the verb *manû* often (not always) appears in the corpus in relation to the indication that the incantation should be recited seven times, while on the contrary the verb *qabû* is repeated three times. See No. D.3: 35: in Ms. B and C the exorcist recites (*manû*) three times the incantation, while in Ms. D he says (*qabû*) it. In No. A.1: 39 the exorcist recites the incantation seven times in Ms. C, while three times in Ms. D. Recitation (*manû*) seven times: A prescr. 24: 72; No. B.1: 15–16; D Diagn.: 25; No. D.4: 63 (restored); No. E.1: 16 (and Ms. B rev. 24); No. E.2: 45; No. K.5: 104, No. K.9: 165; L prescr. 1: 5 (restored); L prescr. 2: 6 (restored); L prescr. 6: 17; M prescr. 1: 2; Declamation (*qabû*) three times: No. E.2: 37; No. K.3: 93. Exceptions No. D.2: 22; D prescr. 4: 48; No. F.4: 92; No. K.7: 120; L prescr. 4: 12 (restored); P prescr. 9 rev. 4; P prescr. 11 rev. 8.

178 The ritual (ll. 38–44) following the incantation No. E.2 (ll. 25–37) involves the use of the substances mentioned in the incantation. However, the mention of a ram and a weaned sheep, tied at the bed, establishes a relationship between this ritual and the incantation No. E.3. It means that both No. E.2 (with its ritual) and No. E.3 are part of the same therapeutic performance. According to Abusch et al. 2020, the incantation No. E.3 features three different speakers: the male patient (ll. 46–47); the female partner (ll. 48–51); the exorcist (ll. 52–53). It is also possible, given the parallels No. E.1 and No. F.4, that the woman also recites lines 46–47.

179 Note that the incantation features three or two different speakers, see previous footnote.

21. [you] place (GAR-*an*) (there) the shoulder, the caul fat (and) the roast meat (var. adds: you libate premium beer).
22. Two figurines of tallow, two figurines of wax, two figurines of bitumen, two figurines of gypsum, two figurines of [cl]ay,
23. two figurines of dough, two figurines of cedar wood, you make (them) (DÙ-*uš*); in an unfired *pursītu*-bowl (var.: *burzigallu*-bowl; cup)
24. you burn (*ta-šár-rap*) (them) in a fire in front of Ištar of the stars and **he** (= patient) **says** (DU_{11}.GA) the following (incantation):

25–37 = Incantation No. E.2

37. **He** (= patient) **says this three times** (*an-nam* 3-*šú* DU_{11}.GA).

38–44 = Ritual No. E.2

45. **You recite [seven]? times** ([7]-*šú* ŠID-*nu*) [this (= following)?] incantation over (them) (var.: over? the be[d]), tie (them) around his waist, and (he will get) sexual desire.

46–53 = Incantation No. E.3

To summarize:
- The therapeutic operator performs the ritual: use of logograms;
- The patient declares (*qabû*) the incantation three times (ll. 24 and 37);
- The therapeutic operator performs the medical prescription and makes it possible for the incantation No. E.3(?) to be recited (*manû*, l. 45).

From this example, we understand therefore that the therapeutic action is structured through a dialectical process between the patient (or the couple) and the healer, where the first is not just a passive object during the therapeutic itinerary of the exorcist's power, but an active agent in the healing process, albeit under the authority of the exorcist (*āšipu*).

Authority and therapeutic efficacy

The *nīš libbi* texts often mention, through specific expressions, the patient's healing. But who declares that the patient is healed and how is it determined? Responding to this question is not simple, because we are only in possession of textual sources and we ignore the social discourse around the sickness-event. However, the texts can provide us with some information.

The *nīš libbi* rituals often end with an expression indicating recovery of health by the patient: *išallim* (SILIM-*im*)[180] or *iballuṭ* (TI-*uṭ*)[181] "he will recover." More specifi-

180 No. A.2: 48; A prescr. 20: 61; A prescr. 22: 64; A prescr. 24: 74. Restoration: A prescr. 1: 5; A prescr. 2: 9; Q prescr. 1: 5.

181 B prescr. 1: 29; K precr. 3: 29; K prescr. 4: 30; N prescr. 3: 16 (Ms. B); N prescr. 4 i 22 (Ms. B); Y prescr. 1: 4′.

cally, especially in the texts from Ḫattuša (KUB 4, 48; KUB 37, 81+AAA 3 pl. 27 No. 5), after the medical prescription we read *nīš libbi iraššī* (ŠÀ.ZI.GA TUKU-*ši*)[182] "he will get sexual desire." Other formulas can ratify the patient's future well-being: "He will repeatedly have sexual intercourse" (*irtanakkab*, No. E.1: 17);[183] "Night and day desire (lit. *libbu*) will not abate" (*mūši urra libbaka ul inâḫ*, L prescr. 2: 7; see also No. M.1: 9); "That man will desire" (*amēla šuāti libbašu inašši*, N prescr. 25 le. e. 7). Other final formulas declare the well-being of the couple: "Their 'hearts' will be released" (*libbūšunu ippaššarū*, B prescr. 2: 45); "The man and woman will find relief" (*zikaru u sinništu ištēniš inuḫḫū*, No. D.3: 36 and 39, restored in No. B.1: 17).

Fragmentary text P prescr. 12 (rev. 11–12) seems to emphasize the effectiveness of therapeutic treatment: "These ritual procedures . . . [. . .] / . . . these good (results) . . . [. . .] / he has tested th[em . . .]." The text states that the exorcist "has tested" (*ultattik* from *latāku* 'to test') the ritual procedures, thus checking their effectiveness and defining them with the term *dumqūtu* 'good (results).' According to Steinert (2015: 116) in her study on the term "tested" (*latku*) in Mesopotamian medical texts, these efficacy expressions played a role in heightening the user-friendliness of the text, it could have helped the healer to select a remedy quickly for a specific treatment.

Text A prescr. 24: 72 instead provides us with information on the time of healing: "He (= patient) will repeatedly drink it for three days and he will recover on the fourth one" (*šalāšat ūmī ištanattī-ma ina erbēšu ūmi*).

An interesting case is given by the incantation No. D.3: 35–39:

> You recite (var.: you say) this (incantation) three times and (if) the man and the woman desire together each other, (but) they do not find reli[ef]:
> *namburbû*-ritual against panic attacks: in musta[rd] water [. . .]
> you mix . . . [and] keep (the medical preparation) ready,
> you apply (it) into the throat (of the man and woman) and they will find relief.

The text describes the therapeutic action: the recitation of the incantation three times, but if the man and woman desire each other, but they do not find any relief, it is necessary to perform the *namburbû*-ritual against the panic attacks. After that, the man and the woman will find relief. The text, therefore, provides for the possibility of the ineffectiveness of the medical prescription,[184] in this case, to be replaced by a *namburbû*-ritual.

These references to the patient or couple's healing are intended to confer legitimacy to the therapeutic practice. Therefore, the healing and success of therapy are

182 N prescr. 1: 7; N prescr. 2 i 11; N prescr. 3 i 16 (Ms. A); N prescr. 4 i 22 (Ms. A); N prescr. 5 i 27; N prescr. 6 i 32; N prescr. 7 ii 2; N prescr, 8 ii 16; N prescr. 13 iii 10; N prescr. 17 iv 2; Restoration No. J.2: 16; N prescr. 9 ii 26; N prescr. 10 ii 31; N prescr. 12 iii 6; N prescr. 14 iii 23; N prescr. 15 iii 26; N prescr. 16 iii 31; N prescr. 21 iv 20; V prescr. 2: 9; W prescr. 2: 3′.

183 Restoration in No. B.2: 40; L prescr. 6: 18.

184 On the Babylonian ritual failure and mistakes see Ambos 2007.

expected in the ideology of the *āšipūtu*. The affirmation of therapeutic success, therefore, strengthens the authority of the exorcist.[185] Therapeutic space is, in fact, a place of negotiating different needs and legitimacies. In this regard, the anthropologist Waldram (2000: 615) states that it is essential conceiving of efficacy not as a fixed concept anchored to a singular perspective of health, and sickness, but, rather as a fluid concept, something that is constantly shifting and being negotiated between the various role players in the sickness event. Such mentioned expressions confer therapeutic efficacy to the *āšipūtu*'s knowledge and strengthen the power of the therapeutic operator. The texts, and with them the exorcists, guarantee the future healing of the patient and the couple.

185 As it has been pointed out by many assyriologists, including Abusch, the *āšipus*, through a long process, tried to legitimize their power and authority in the field of healing, in opposition to other therapeutic operators, understood as their rivals, such as the witches, which have been subjected to a real process of demonization (Abusch 2002: 3–26).

Chapter II
Nīš libbi incantations

Incantation, poetry, and text function

This chapter will analyze the *nīš libbi* incantations. They are divided into five groups along thematic and stylistic grounds:[186]

1. Animal metonymies and agent of suffering (No. A.2; No. A.3; No. A.4; No. D.2; No. L.2);
2. Erotic animal similes (No. B.1; No. C.1; No. D.4; No. E.1: 5–13; No. E.3; No. F.3; No. F.4; No. F.5; No. G.1; No. K.7; No. K.8);
3. Sexuality and nature (No. A.1; No. B.2; No. E.1: 1–4; No. H.1; No. I.1);
4. *Historiolae* (No. K.1; No. M.1);
5. Abracadabra (No. F.1; No. J.1; No. K.3; No. K.4; No. O.1: 21–23).

The poetry of the incantations has already been stressed by Assyriologists (see Falkenstein 1931; Michalowski: 1981; Reiner 1985; Veldhuis 1990; Veldhuis 1993; Veldhuis 1999; Cooper 1996; Wasserman 2003; BFoster 2007: 92; Schwemer 2014; Geller 2015: 108–114). Niek Veldhuis, for example, in "The Poetry of Magic" (1999) analyzed the Old Babylonian incantations by tracing their poetic structure. To do so he took recourse to the semiotic theories of Yury Lotman, as well as linguistic ones of Roman Jakobson. To analyze the poetry of these texts is not to say that they are *strictly speaking* poems and were composed as such by the Mesopotamian peoples. Conversely, I shall have to track down those structural elements that contribute to a definition of the genre. These elements, as has already been emphasized, can be traced to a certain kind of "poetry."

The link between magic power and "poetry" has been analyzed in Cultural Anthropology. Anthropologist Anita Seppilli in her book *Magia e poesia* asks if it is indifferent and superfluous, concerning the magical purpose, that a formula, the narration of a myth, the ritual ceremony are poetry, or contain poetic elements, and if the fact of being "poetic" contribute to the magical aim (1962: 134). Similarly, Veldhuis: «If literary methodology clearly yields results in analyzing an incantation without literary pretensions, what do these results mean, and how are we going to use such results in a broader understanding of magic?» (1999: 38).

Before answering the questions, I shall dwell for a moment on the term "poetry" and "poeticity." To do this I refer to Jakobson's studies. He argues that at the basis of all linguistic communication, the following factors are essential (Jakobson 1960: 353):

186 For the stylistic analysis of some incantations see commentary. Some incantations were not included in these groups due to stylistic and thematic reasons or because too fragmentary: No. D.3; No. E.2; No. F.2; No. J.2; No. J.3; No. K.2; No. K.5; No. K.6; No. K.9; O.1: 15–20.

https://doi.org/10.1515/9783110757262-002

- the *sender* sends a *message* to the *recipient*;
- the message requires a *context* in order to work;
- the message is encrypted by the sender and decrypted by the recipient through a *code* shared by both;
- for the communication *contact* between the sender and recipient is required.

From each of these factors a linguistic function result. I would like to focus on two functions: the conative and the poetic one. The conative function is the orientation towards the recipient and is characterized by the use of the imperative, but also the precative. In incantations, for example, the recipient is also expressed in the third person singular, expressing a conative, hortatory or supplicatory function. As Jakobson (ibid. 355) states, the magic incantatory function is chiefly some kind of conversion of an absent or inanimate third person into an addresser of a conative message. Typical of the language of incantations, along with the conative function of the language which aims to act on the recipient, is its poetic function. It focuses on the message itself. The poetic function is not exclusive to poetry, but of all those types of linguistic communication that put the focus on the message itself. Advertising or political slogans, for example, can be characterized by the poetic function (ibid. 357). It is not useful to reduce the poetic function of language to poetry. The exorcist's magical language has, therefore, its own poetry, since there are several figures of speech such as repetition, chiasm, alliteration, climax. As we shall see, they are composed of a structure, which we might call poetic.

The language of the incantation is a language different from the ordinary one since it is built with an emphasis on the message itself. As the cross-cultural anthropological works demonstrated, the incantations are written by making use of a special language, with figures of speech, archaisms, symbolic images, literary references, especially coined terms, meaningless expressions (such as abracadabra) and redundancy (see Tambiah 1968: 181–182; Ong 1982: 36–57). Below, I will try to analyze these aspects of the poetry in incantations. In other words, as stated by Lotman, in a poem, «it is its internal organization, which transforms it (the syntagmatic level) in a structural complex. That is why, as a certain sequence of phrases in natural language can be recognized as an artistic text, one recognizes that such sentences build up a certain type of secondary structure that match the artistic level of organization» (1972, Engl. tr. 1976: 68–69). The purpose of the analysis of these incantations is to track down the elements of this structure.[187]

But I have not yet responded to the questions that anthropologist Seppilli and Assyriologist Veldhuis put to themselves. What is the function or functions of the poetic language of the texts recited within the therapeutic process?

---

187 For a general study on prayer/incantation in Anthropology see Todorov 1973; Ong 1982: 36–57; Lupo 1993; Lupo 1995.

The metaphorical language is the most obvious aspect of the poetic language used in incantations. Similes, antinomies, rhetorical metaphors are tools commonly used in this type of text.[188] According to George Lakoff and Mark Johnson (1980) the metaphor is the way we conceptualize one mental domain in terms of another; it is "mapping" a source domain to a destination domain. Simplifying drastically, we can say that what the human mind does is to set similarities between two domains, establishing analogies. Metaphor and analogy are closely interrelated. According to the anthropologist Stanley J. Tambiah (1985: 60–61), analogical thinking is a feature of the magical way of thinking of traditional societies.[189]

The theoretical conclusions of Tambiah are useful to understand not only the function of incantations but also rituals, as we shall see in the next chapter. Incantations and rituals, word and deed, are closely linked. It is impossible to separate the two aspects in therapeutic practice. If I do in this book, it is only for explanatory convenience. As Edmund Leach (1966: 497) points out, 'ritual' is a complex of words and actions, it is not the case that the words are one thing and the rite another. The uttering of the words is itself a ritual. To understand how inseparable the words recited in the incantation and the ritual with its medical prescription are, I take as an example ritual and incantation No. E.2:

> "May my body become pure like lapis lazuli!
> May [my] features (lit. head) be bright like alabaster!
> Like shining silver, like (var.: and) reddish gold may I have no fear!
> May *tarmuš*-plant, "heals-a-thousand"-plant, "heals-twenty"-plant, *ardadillu*-plant, / *usikillu*-plant, *nīnû*-plant, *bukānu*-wood (var.: "wood-of-rele[ase"]) / dispel my fascination!". He (= patient) says this three times.
>
> ---
>
> Its ritual: You string silver, gold, lapis lazuli, alabaster / *tarmuš*-plant, "heals-a-thousand"-plant, "heals-twenty"-plant, *ardadillu*-plant / *usikillu*-plant, *nīnû*-plant, *bukānu*-wood (var.: "wood-of-release") on (a cord) of flax,
> (and) put (it) around his neck. A ram at the head of his bed,
> (and) [a wea]ned [sheep] at the foot of his bed you tie.
> From the forehead of the ram and the forehead of the weaned sheep
> you pull out wool and twine (two) separate threads.
> You (= therapeutic operator) recite [this] incantation [seven] times over (them) (var.: over the be[d]), tie (them) around his waist, and (he will get) sexual desire.
> [No. E.2: 32–45]

The ritual action begins with the organization of the place of performance: a portable *pāṭiru*-altar is set up before Ištar of the stars, a sheep is sacrificed, juniper is burnt, and finally libations of beer are poured out as offerings. The figurines of

---

188 The metaphorical language is not only specific of the poetic function, but also appears in everyday language, see Lakoff and Johnson 1980.
189 On Mesopotamian analogical reasoning see Rochberg 2015.

witches are made and burnt. The patient declaims the incantation. It begins with an invocation of the goddess Ištar and is followed by the typical supplicative language, using the precative. Here are interesting similes: may the patient's body be as pure as lapis lazuli; may his features (lit. head) be as bright as alabaster; may he have no fear as bright silver and red gold. Then the invocation of some plants comes so that they can drive away the fascination: plants *vs.* fascination. The ritual consists of making an amulet containing all the above-mentioned ingredients, both mineral and plants:

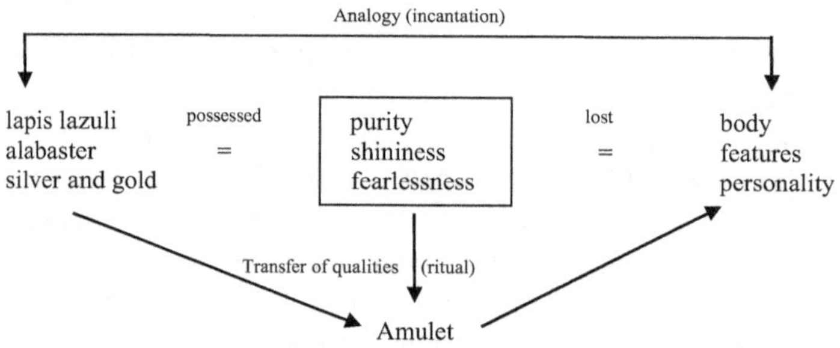

At the base of such ritual practice, there is the analogic principle: purity, shininess, and absence of fear are considered to be human qualities, characteristic of his state of mental and physical health. These qualities have been modified by the witch attack. To be restored, in the ritual those plants deemed capable of eradicating witchcraft must be used, as well as substances characterized by those qualities the patient has lost. To activate this ritual process, the incantation must be recited. Thanks to the incantation, through the invocation to the goddess, it is possible to transfer the very qualities possessed by mineral substances into the patient (lapis lazuli, alabaster, silver, and gold). It is an example of the principle Tambiah calls "analog transfer of qualities."[190]

As for this example, it is clear that it is not possible to separate the "medical" practice and the recitation of incantations. The two aspects, for therapeutic purposes, are interconnected.[191] They are related by the analog principle, which in

190 Torri (2003, esp. 6–8) uses Tambiah's perspective to study the principles of Hittite analogical magic.

191 In this regard see the criticism uttered by Tambiah (1985: 61–64) against Evans-Pritchard (1937) in his study of Azande magic and medicine. Although Evans-Pritchard admitted that the incantation was always an essential part of Zande magical rites, he emphasized that it was "medicines" which played the major part: mystical power, producing the desired end, resided in the material substance used. On the contrary, incantations were merely words of direction uttered to the "medicines" linking them to the desired ends. Regarding Mesopotamian medicine, Panayotov (2018) examines Assyrian house amulets, relating the incantations inscribed on them to the organic material, as *maštakal*-plant, the tamarisk (*bīnu*), and the "offshoot" of the date palm (*libbi gišimmari*), that was to be inserted in the

incantations often is expressed through metaphors or similes. The anthropologist has coined the expression "persuasive analogy," a term that is very reminiscent of Jakobson's conative function, aimed to modify the action of the recipient. It is structured in two parts, the first of which is to discover *similarities* between two domains, in this case, the body and the human personality on one side and minerals and metals on the other. In the rite, with the performance of the incantation as its central part, the characteristics of this interaction are transferred from one side to the other through persuasion: the qualities are transferred from elements who hold them to those who do not (see Tambiah 1968: 194; 1985: 72). To enable such a process, on the one hand, poetic language is needed to make the similarities between the two domains explicit through the spoken word (the metaphor or the simile is only one of the devices used for this purpose), and on the other hand, magic material, through which these very qualities are transferred. Some of the ingredients used in the incantation No. E.2 (lapis lazuli, alabaster, silver, and gold) are used metonymically as "processors," or allow an actual transfer by contact (= amulet) of established analogies, through the poetic function (as well as conative), in the incantation. It is "an imperative transfer," which aims to transfer specific qualities from one domain to the other (see Tambiah 1968: 193; 1985: 78).[192] This is made possible by means of the illocutionary or performative character of the rite (Tambiah 1981: 119). The performative aspect of the ritual has locutionary character-

---

holes and slots in the sides of these amulets. The scholar explains how the incantations on the amulets activated the magical power of the plant.

192 See the relation between ritual of No. E.2: 41–45 and the incantation No. E.3. These lines of the ritual No. E.2 are related because of the mention of a ram and a weaned sheep tied at the head and the foot of a bed to the incantation No. E.3: 46–47 which is maybe part of the same therapeutic performance: "[Incantation: At the hea]d of my bed a ram is tied, / [at the foot of my bed] a weaned sheep is tied. Around my waist their wool is tied." Another clear example in this corpus is the incantation No. A.1: 36–40: "May my penis be a harp string, / so that it will not dangle out of her! Incantation formula. / Its ritual: You take a harp string, tie three (var.: two) bindings (in it), / you recite the incantation seven times, you tie (it) around his right and left hands and / (he will get) sexual desire." See also No. J.2: 10–13: "May [the qu]iver not be[come empt]y! May the bow not slacken! / May the batt[le of my love]-making be fought and may we lie down (together) by night! [Incantation formula]. / [*Nī*]*š libbi* Inc[anta]tion. / [Its ritual]: You make a bow of a thorn [. . .]." In No. C.1: 7–13 it is said that the female partner is endowed with the *rikibtu* of a stag, an ingredient mentioned along with others in the ritual of the same incantation for the production of an amulet. Another example not from *nīš libbi* texts, is a love incantation in order to have sexual intercourse with a woman, VAT 13226 (Zomer 2018: 277–279). In this incantation the man, who desires the woman and wants to have sexual intercourse with her, compares himself and parts of his body to metallic and mineral substances: "Tin is of my [. . .], *pappardilû* is my tongue! / At the root of my hair is the snake-stone! / I am the *mekku* of which no equal / exists in the country!" (ll. 11–14). The ritual mentions, not by chance, the following ingredients: tin, *pappardilû*, snake-stone, *mekku*-glass, and *ittamir*-stone (ll. 17–18).

istics, which transmit information (and also regulatory models), and perlocutionary ones, creating consequences for participants:[193]

> The vast majority of ritual and magical acts combine word and deed. Hence, it is appropriate to say that they use words in a performative or illocutionary manner, just as the action (the manipulation of objects and persons) is correspondingly performative.
>
> [. . .] The rite usually consists of a close interweaving of speech (in the form of utterances and incantations) and action (consisting of the manipulation of objects). The utterance can be analyzed with respect to its "predicative" and "illocutionary" frames. In terms of predication and reference the words exploit analogical associations, comparisons, and transfers (through simile, metaphor, metonym, and so forth). The illocutionary force and power by which the deed is directed and enacted are achieved through use of words commanding, ordering, persuading. [Tambiah 1985: 80]

The anthropologist refers explicitly to John L. Austin's speech act theory[194] in *How to do Things with Words* (1962),[195] according to which assert something is to change the surrounding reality: *to say something* is *to perform an action*. According to Austin, in every linguistic enunciation it is possible to distinguish:

- Locutionary act: saying something, in a syntactically correct way and with meaning;
- Illocutionary act: the action which is sought by speaking the enunciation. It is the act, which is performed *in locutione, in saying that* (for example order, promise, warning);
- Perlocutionary act: the effects obtained with the illocutionary act, *per locutionem, by saying that.*[196]

---

193 For the use of Austin's perspective of performative speech acts in the analysis of the Limba people of Sierra Leone see Finnegan 1969. Ahern (1979) makes a comparison criticizing Tambiah's theory on rituals understood as illocutionary acts from the analysis of Chinese rituals. The scholar distinguishes between "weak" and "strong" illocutionary acts. For a critical approach to the use of Austin's theoretical perspective in the analysis of rituals see Gardner 1983.

194 Karl Bühler (1934) used yet in his theory of language functions (*Organonmodell*) the notion of linguistic act.

195 Similarly, he undertakes a critical review of the pragmatic theory of Malinowski's theory of language (1965), particularly his analysis of magical language (see Tambiah 1968: 185–186).

196 Austin's theory of language acts was criticized by John R. Searle (1969). In the article "For a Taxonomy of Illocutionary Acts" (1976) the latter distinguishes three dimensions of a linguistic act: Purpose of the act (describing, promising, ordering, claiming, etc.); Direction of fit: from the world to the words, from the words to the world; Expressed psychological states (belief, intention, desire). Based on these three dimensions, Searle distinguishes five types of illocutory acts: Assertive act: speech act committing a speaker to the truth of expressed proposition (assert, describe, conclude, swear, suggest); Directive act: speech act inducing the listener to do something (order, prohibit, ask, implore, plead, invite, challenge, etc.); Commissive act: speech act committing a speaker to a future action (promise, refuse, consent, bet); Expressive act: speech act expressing a psychological state (congratulate, thank, rejoice, apologize, regret, etc.); Declarative acts: speech act to match the content of what is expressed with the world (changing the world with linguistic proposition) (baptizing, marrying, declaring war, condemning, dismissing, etc.). The theory of linguistic acts advanced by Searle points

The studies on linguistic acts are of great importance for the comprehension of the ritual specificities, as suggested by Tambiah. They allowed the understanding of the functional peculiarities of ritual in terms of operative and operational effectiveness. They allow us to analyze the texts, in this case, rituals and incantations, inserting them in a particular context, where several needs and requests, and power relations are at stake. They allow us to study the function to which this documentation is oriented, that is, the therapeutic efficacy.

In conclusion, the poetic analysis of incantations allows us to understand the rhetorical strategies with which the "magic words" are formed, paying attention in particular on the use of metaphorical language. The latter, acting through analogies, allows an understanding of the close relationship between incantation and ritual. They act together through the principle of "persuasive analogy." Recitation and ritual practice are two aspects of the ritual performance with its locutionary and perlocutionary character. This chapter will also take into consideration the regulatory role of the therapeutic process, using not only the theoretical tools of Cultural Anthropology but also ones of Gender and Queer Studies.

First group: animal metonymies and performer of suffering

The first group of incantations includes No. A.2; No. A.3; No. A.4; No. D.2; No. L.2. They are divided into two parts. What unites them is the first section of the incantation, characterized by the following elements:

- Identification, through metonymy, of the patient with wild animals;[197]
- Rhetorical questions about the identity of the witch who has done evil;
- Description of the patient's suffering caused by the witch through similes;
- Symptoms related to fatigue and weakness expressed through metaphors of "bonding of the flesh" and "lengthening of tendons."

Animal metonymies

The animal metonymies represent one of the most obvious stylistic and conceptual means in *nīš libbi* incantations to describe, on the one hand, the state of malaise which characterizes the patient, on the other provides *normative models* to follow.[198]

---

to the purpose and effect of an enunciation/proposition. That is, the intent with which the proposition is expressed, and the extralinguistic consequence determined by the enunciation. For some criticism see the studies of the anthropologists Rosaldo 1982; Duranti 1993.

197 With the exception of No. A.4.

198 For the use of wild animal metaphors to express pain and suffering in Mesopotamian documentation see Böck 2014: 120.

The *Tierbilder* in Sumerian and Akkadian literature have been the subject of study by Assyriologists as to their stylistic component, as well as to the conceptual one (see Heimpel 1968; Marcus 1977; Black 1996; Sefati 1998: 87; MPStreck 1999: 172–176; ChWatanabe 2002).[199] In this section, I will deal with the animals used for the construction of metonymies, excluding similes and metaphors. I understand metonymy in these incantations as a figure of speech according to which a person is indicated, rather than with its own common name, in this case, for instance, "man" or "patient," but with another common or generic term to describe precise characteristics and qualities. In the examined texts, the man is defined as a wild animal, having lost his fury, strength, and sexual vigor. The following texts show this kind of metonymies: catalogue LKA 94 i 2: "My *akkannu*-wil[d as]s! My *akkannu*-wild ass! My wild bull! My wild bull!"; i 15: "My hunted *akkannu*-wild ass!"; i 16: "My stag! My stag! Horse!"; i 17: "My *akkannu*-wild ass of the mountain, who has blocked you?"; No. A.2: 41–42: "*Akkannu*-wild ass who is reared-up for mating, [wh]o has dampened [your] desire? / Impetuous horse, whose rising is a devastation, [w]ho has bound your limbs?" (see also No. L.2: 26); No. A.3: 49–50: "*Akkannu*-wild ass! *Akkannu*-wild ass! Wild bull! Wild bull! Who has slackened you (so that you are) like slackened / strings?": No. D.2: 10: "My hunted *akkannu*-wild ass! Hun[ted] onager [. . .]"; No. G.1: 2–3: "[St]ag! Stag! Wild bull! Wild bull! / [Roa]r, stag! Rear up, wild bull!"; No. K.7: 116. "Lion! Bull . . . [. . .]"; No. L.2: 19: "Rearing [onag]er, preeminent stallion, who roams the forests, who [has dampened your ardor]?".

How should these metonymies, and more generally speaking, all animal metaphors and similes be interpreted? What is the function within the therapeutic process? And to what extent can they help understand male sexuality and the subsequent gender construction in Mesopotamia of the second and first millennia?

According to scholars who have dealt with the subject, it is the immoderate animal sexual activity that plays a major role. According to Leick (1994: 210), the purpose of the animal metaphors and similes is to cause sexual images. Within the framework of magic analogy, the man derives additional strength from the invocation of coupling animals, and when some of their bodily substances are used for the rituals. The suggestion of Leick, given the lack of comparability between animal and human sexual practice, is that through the magic of analogical principles, man can get sexual vigor

199 See also the contributions concerning Mesopotamia in Mattila et al. 2019; Richardson 2019. More generally, on figurative language in Mesopotamia see MPStreck 1999 and the previous literature; ChWatanabe 2002: 14–21; Wasserman 2003. For metaphors to express malaise and body parts see the studies of Couto-Ferreiro 2013; Steinert 2013; Böck 2014: 115–119. On metaphorical language in love and sexual literature see Lambert 1987a; Westenholz 1992; Westenholz 1996; Cooper 1997: 85–97; Sefati 1998: 86–94; Haas 1999: 128–136; Besnier 2002; Wasserman 2016. On the relationship between figurative speech in medical texts related to the "love-sickness" (*muruṣ râmi*) and love literature see Couto-Ferreira 2010b: 33–36. For the animal symbolism, in addition to Heimpel 1968, see van Buren 1945: 30–42; Parayre 2000a; BCollins 2002; Corò 2005; Milano 2005; Capomacchia 2009; Couto-Ferreira 2010a; Verderame 2017a. See also the articles in the RlA for every animal.

by invoking a sexually excited animal, through the stimulation of sexual images that allow the patient to think of sharing the sexual potency of such animals. Biggs puts forth of a partly diverging perspective:

> From a study of šà.zi.ga texts, there can be no doubt that they are intended to alleviate sexual impotence, or as it is termed today, erectile dysfunction or penile erectile dysfunction. [. . .] As everyone knows, erection comes about when a man receives sensory or mental stimulation. These texts – the incantations and the accompanying rituals – provide both. Surely some of mental images provided by the incantations can be compared to the stimulation modern couples may receive from pornographic films, now often viewable in American motels and hostels. [Biggs 2002: 72]

For Biggs, the function of such images is psychological. If for the Assyriologist the diagnosis is "erectile dysfunction" and one of the causes of this sickness is psychological, these images of sexually excited animals are designed to elicit a mental stimulation, comparable to that produced by contemporary pornography (see also Cooper 1997: 92).

This idea is grounded in the principles of analogical magic according to which "similar produces similar" and therefore the mention of sexual animal activity through the figures of speech would result in the patient psychological processes such as to allow in part the acquisition of sexual vigor. If this principle seems to be clear regarding the use of substances, such as saliva and hair, of sexually excited animals in medical prescriptions, I do not think it is the only aspect regarding the incantations.

From a gender perspective, however, I would like to consider the regulatory and normative nature of these animal images. As is known, gender is a social apparatus whose purpose is to provide regulatory models to society. Sexuality is an aspect of individual and social life subject to these regulatory processes, namely socially codified rules that establish which behavior is considered correct and which deviant. This regulatory aspect of gender concerning sexuality and the body has been investigated by Foucault in his books on the history of sexuality (1963; 1969; 1971; 1976). Some exponents of Queer Theories have pointed to the regulatory power of the gender. Among others, Judith Butler (1990; 2004) makes specific reference to the works of Foucault, focusing primarily on the relationship between power/knowledge and body. The body, therefore, is a place of symbolic references established by society. The gender then it is one of the criteria by which society distinguishes, puts in order, subjects, and normalizes the body. The power, therefore, according to Foucault, is not limited to act on a pre-existing subject, but creates it, giving rise to a process of subjectification. According to the French philosopher, the body is a passive object of absorption of social norms, and the gender itself is configured as one of the regulatory mechanisms of the persons.[200]

200 Butler criticizes, therefore, this "static" view of the body, to which other scholars, in the area of philosophy and anthropology were opposed.

An innovative aspect of gender identity, as it is conceived by Butler, is the principle of the performativity of gender. In *Gender Trouble* (1990) and *Undoing Gender* (2004) Butler argues that the "subject" is a performative construct in progress, but stresses that there is no a *performer* that pre-exists to the actions, but both coexist at the same time. In this regard, in fact, the scholar implements an important distinction between "performance" (which implies the existence of a subject) and "performative" (which however does not presuppose its existence). If gender is performative, that means that «gender is not exactly what one "is" nor is it precisely what one "has"» (Butler 2004: 42), but rather a *doing*, an endless sequence of acts. Thus, the gender has no ontological value for Butler, but rather a regulatory one, and at the same time a performative one. That is why it must be investigated from a practical point of view, as well as the one of citationality. The influence of Austin and Searle's linguistic studies of the speech acts in Jacques Derrida's (1972: 365–393)[201] reading and criticism, is quite evident. To cite behavioral gender norms in daily practice, in other words, "doing gender," involves on the other hand a process of subjectification. The subject is constituted precisely through a continuous repetition of gender acts.[202] Butler, therefore, employs Louis Althusser's concept of "interpellation." Interpellation for Althusser (1970) is the process by which the ideology is addressed to the pre-ideological individual, producing it as a distinct entity. The interpellation allows the person to represent himself by incorporating an ideology. Butler employs this concept to refer to how subjects are subjected since they are involved in a series of interpellations that reiterate continuously and create a "naturalization" effect of the sex/gender system.

Returning to *nīš libbi* incantations, Gender and Queer Theories can help us understand the function of the animal metaphors. They manifest several concepts related to sexuality that one wants to be normative for a society, especially for male gender and sexuality. Wild animals in themselves convey ideas related to active, irrepressible, vigorous male sexuality (see ChWatanabe 2002: 103–106). Indomitable sexuality, that is, one which cannot be subject to control processes or limitations. The choice of wild animals is therefore functional, serving to create an idea of a strong and proud male sexuality.

The wild species, mentioned in particular in metonymies, are not subject to a process of domestication.[203] It is just such a state of wildness which represents the

---

201 See the criticism to Derrida by Bourdieu in *La domination masculine* (1998: 120). The concept of *habitus*, also focused on the practice, seems however to lead to a socio-cultural determinism which Butler, however, seems to dispute, making use of the concept of agency and of power in the Foucault's perspective. See also Butler 1999: 113–128. For a debate Butler – Bourdieu, see McNay 2004: 175–190.
202 Just to be clear: this affects all aspects of identity. There is no identity that precedes, in essence, the gender, but the identity is always performed as gender identity.
203 The Mesopotamians differentiate between domesticated and non-domesticated animals. Battini (2009) underlines that wild animals are more frequently depicted than domesticated. The most domesticated animals present in iconographic and texual sources are cattle, goat, sheep, donkey/horse, and dog. However, it is difficult to understand exactly what did "domesticated" mean to the Mesopo-

best condition in which it is possible to develop the typical active sexual characteristics of virility. The domestication, in fact, seems to be conceived as a "softening," a reduction of pride, and therefore a process of feminization. The excitement of wild animals, therefore, cannot be appeased,[204] nor should the human one.

The witch's attack resulting in the loss of sexual desire has led to a crisis of manhood: the man is diverted from *what he should be*. Incantations, with animal images, are therefore the means, within the therapeutic process, essential to the identity and reconstitution of male sexuality, according to the principles ordered by society. These images, in other words, provide role models to which the man must adapt. The description of animal sexual activity and its metaphorical use allow a normalization process of sexual behaviors that provides ideal models to follow.

But why are animals commonly employed in these political-regulatory processes?

The relationship between animality, normativity, and gender represented by metaphor has been subject of investigation by anthropologists (see Leach 1964; Wijeiewardene 1968; Tambiah 1969; Brandes 1981; Brandes 1984; Tapper 1988; SBaker 1993; Ingold 1994; Willis 1994; Descola 2005). Also, concerning Ancient Greece and Rome, there are many contributions on the subject (see Dierauer 1977; Bettini 1998; Franco 2003a; Franco 2003b, Engl. tr. 2014; Franco 2006; Franco 2008a). The animals, in the famous definition given by Claude Lévi-Strauss, are *bons à penser* (1962a: 128) and for this reason, their metaphorical use must be object of study. As highlighted by Stanley H. Brandes (1984: 207), animal metaphors play in promoting social control and reinforce the social and moral order. They constitute an important domain for talking about disapproved or undesirable attributes and remind people of human behavioral norms and physical ideals. The relationship between ideology, gender, and animal metaphors is underlined by Cristiana Franco (2008a: 73–75) in ancient Greece. According to her, animals are an important repertoire of arguments for the construction of models of "naturality" because they represent the *physis* and as such are thought to be eternally equal to themselves. For this reason, they are considered as a sure term of comparison for the variety of "human ethology." This comparison with animal ethology implies a double movement: a construction of animal species as ethical models and a reflection of man to the animal êthos thus constructed. Self-referential circularity typical of the normative model, which creates an apparatus of representations of the world – *what the world is* – to use them as proof of the objective value of the values that the model intends to impose – *what the world must be*.

---

tamians, for example, we could consider the dog as a semi-wild and ominous animal. Battini suggests that the criterion of distinction was utilitarian.

204 The indomitability is given by its wildness. It is no coincidence that the demons, often described as coming from the steppes, are compared to horses: "They are bred horses in the mountains" (*Udugḫul* V 156, 176, Geller 2016: 208). See also the Sibitti's affirmation at the beginning of the *Erra Epos* (I 46–91, Cagni 1969: 62–66).

As is evident, therefore, the use of animal behavior has the function of social control, being role models to follow. These social constructions that are founded on animality, act upon the community since they present themselves as *natural*. Animals thus provide *models of naturality* also concerning gender and sexuality. They show, based on how the male animal sexual practices *are*, what human ones *should be like* (see Lloyd 1991; Li Causi 2005; Franco 2015; Tutrone 2016).[205]

What is marked in animal images is not so much the difference, though evident, between male sexuality of wild animals and the human, but rather the similarities. It is through these similarities that men can be identified with the animals such as the stallion, the wild bull, and the stag. Through animal metaphors, the Mesopotamian man symbolically turns into an animal of the steppe, by which the untamed character of sexuality is underlined. Without these similarities, in fact, there would be no cognitive basis for linguistic and symbolic analogy.[206]

In Mesopotamian documentation, however, the wild animal-human similarities concerning sexuality are most often employed when the heart of the male sexual practice, or the sexual desire and the consequent power, are absent, having been subtracted by a witch.[207] They represent ideal models of sexuality that the sufferer would like to perform. We can then say that the animal metaphors are also effective because they aim to promote social control in terms of gender identity and organization of sexuality, as a barometer of social ideals and relationships within the community.

The reference to sexual activity, however, is not just about animal metaphors.[208] Others are in fact examples in which animal sexuality, in this case of domestic ones,

---

205 It should be noted that the relation between animality and naturality has been carefully analyzed for the Greek and Roman world. Despite the differences between the two historical and cultural contexts – just think of the absence of the concept of *physis* "nature" in the Mesopotamian languages – this association in general terms is proposed as an *ethical* analysis of animal construction for anthropological purposes.

206 On the similarity between the sexual act of the dog and the pig with the human one and their use in Akkadian compositions see Cooper 1996: 51 fn. 16; Wilcke 1985: 206.

207 For the use of animal metaphors in love literature see Wasserman 2016: 43–44. Here, «unlike flora, this semantic field focuses not on sensuality or amorous relations between the protagonists, but rather on the *force* of sexuality *per se*» (ibid. 43).

208 Also note the use of animal names to refer to sexual organs in love literature (see Wasserman 2016: 40–42 and fn. 156). The vulva is referred to by terms such as "field" or "sown field" (CUSAS 10, 10: 38 -39; ibid. 10, 13: 8$^?$; *Moussayeff Love* Song r. 9), or "canal" (CUSAS 10, 10: 36), but there are also names taken from the animal world, as *ḫuduššum*, perhaps some kind of black frog (PRAK 1 B 472 12′). Also see the animal associations with the vulva, called *rēmu* "uterus," such as "the lizard (*ṣurārû*) of [your] vulva [. . .] / the gecko (*pizalluru*) of [your] vulva [. . .] / the wild cat (*muraššû*) of [your] vulva [. . .] / the mouse (*ḫumṣīru*) of your vulva [. . .]" (K. 7924 obv. ii 10–14, Lambert 1975: 112). But, as Wasserman points out, «it must be admitted, however, that it is not clear whether the wild animals stand for the male or the female sex organ» (2016: 40). In an *irtum*-song the penis it is called metaphorically "dove": "I have thrown my coop on the young man, / that I know [I may] catch the dove; / (the coop) of my delights Nanāya$^?$ will fill for m[e]" (16056 MAH: 16–20, Wasserman 2016: 105). The snake could

is represented as a model to follow, as an educational and regulatory code of *naturality.* An example is found in Sumerian literature: the text CT 15, 28–29, published by Kramer (1973), is a poem that has as protagonists the shepherd Dumuzi and his sister Geštianna. The passage, which is of interest to us, describes the two within the sheepfold, who, during their stay there, is rich in "pure food," oil, honey, beer, and spirits. Dumuzi takes a sheep and its lamb and places them in front of the sister. The lamb mounts the mother and copulates with it. The shepherd invites his sister to look carefully and to tell him what the animals are doing. The goddess seems not to understand, she says she sees the lamb riding the crying mother. Dumuzi then explains to her that the animals are copulating: the lamb fills the vulva of the sheep with sperm. The text is incomplete. Although we are not able to fully understand the reasons for the behavior of the two gods, one could think that Dumuzi is teaching his sister, perhaps inexperienced in love matters, the meaning of the sexual act, which is why the goddess is not able to decipher the behavior of the animals:[209]

> He brought . . . there for her, a ewe and its lamb.
> The lamb has jumped on the mother,
> mounted her, copulated with her.
> The shepherd says to his sister:
> "My sister, look, what is the lamb seeking from his mother?".
> His sister replies to him:
> "He [climbed] on the back of his mother, he is making her cry out."
> "Maybe he [climbed] on her back (and) is making her cry out . . .
> Come on, what is (it): he filled her [overflow] with his sperm."
> [. . .]
> What is (it): he filled (her) vulva with ejaculated sperm."
> [CT 15 pl. 28–29: 38–46, 54, Kramer 1973: 246–248]

Another example is provided by the Akkadian poem *Ištar's Descent into the Netherworld.* Here the relationship between the animals' breeding activities, the bull and the donkey, is brought into relation to that of a young human couple. According to the poem, after the arrival of the goddess in the underworld, the sexual and reproductive cycle stops, the male animals do not mount the females and the young man does not impregnate the young female:

---

refer to the penis in an incantation to make a brothel successful: "Like a snake (= penis), going out from a hole, and birds twittering over him (= snake/penis)" (see Panayotov 2013: 293, l. 30). See also a Sumerian composition in which a man describes his retirement: "My mongoose, which used to eat strong-smelling food can no longer stretch its neck toward a jar of fine oil" (SP 17 Sec. B3: 6–7, Alster 1997: 238, including 19 SP Sec. A1: 6–7). According to Alster (1997: 436), the mongoose refers to the nose that has lost its sense of smell, while for Wasserman (2016: 40 fn. 156) interprets it as the male sexual organ.

209 Kramer (1973: 244 fn. 9) assumes that the lack of understanding of the goddess is a lie, she is concerned about a possible incestuous relationship with her brother.

After Ištar, the lady, descended into the Netherworld,
the bull does not mount the cow, the donkey does not impregnate the jenny,
the young man does not impregnate the young girl in the street,
the young man has slept in his private room,
the young woman has slept by herself.
[*Ištar's Descent into the Netherworld* ll. 76–80, Lapinkivi 2010: 19]

From the text CT 23, 10–11, we find an analogy between the recitation of the incantation, the sexual activity of the dog and the pig, and the practice of plowing. As a dog mounts the bitch, the pig the sow, as well as plowing impregnates the ground, whose seeds it receives, in the same way, the one who reads the incantation *impregnates* the sufferer's body with the latter, thus allowing the expulsion of evil:[210]

Incantation: I impregnate myself, I impregnate my body,
like a dog mounts a bitch, a pig a sow . . .
like the plow inseminates the land (and) the earth takes in its seed,
May myself take in (the incantation), may I impregnate myself! In[cantation].
[CT 23, 10–13: 26–29//4: 9–12, Cooper 1996: 50–51; Cavigneaux 1999: 267]

As can already be seen from these last examples, while wild animals metaphorically describe the active and untamed male sexual vigor and power, domestic animals instead are associated more or less explicitly with the sexual relationship between man and woman.

The patient is identified in *nīš libbi* texts for their sexual activity and potency with the following wild animals: *akkannu*-wild ass, the horse (*sisû*), the onager (*serrēmu*), the wild bull (*rīmu*), the stag (*ayyalu*) and the lion (*nēšu*).

The term *akkannu* (CAD A 274–275; AHw. I 29; Lambert 1960a: 325) indicates a wild species of donkey. In the *nīš libbi* texts, it is mentioned in the incantation No. A.2: 41: "*Akkannu*-wild ass who is reared-up for mating, [wh]o has dampened [your] desire?". In this incantation, the qualities outlined are indomitability and arousal. In No. D.2: 10 the animal is described as "hunted" (*ṭardu*). That qualification is used to indicate the status of the patient's sickness, providing an image of submission: "My hunted *akkannu*-wild ass! Hun[ted] onager [. . .]."[211] The wildness expressed by the term is also present in the *Babylonian Theodicy*: "*Akkannu*-wild ass, onager, which gets satisfaction from . . ."

---

210 On the parallels and variants of this formula see Cooper 1996; Cavigneaux 1999. One of variants is represented by an incantation of the *Maqlû* series (VII 22–24, Abusch 2016: 169–170), where we find an analogy between the recitation of incantations for therapeutic purposes and fertilization of cattle by the god Šakkan. As to the interpretation of this passage see Westenholz and Westenholz 1977: 215 fn. 31; Cooper 1996: 51; *Pace* Cavigneaux 1999: 266.

211 See il catalogo LKA 94: i 2 "Incantation: (My?) *akkannu*-ass! (My?) *akkannu*-ass! (My?) wild bull! (My?) wild bull!"; i 15 "Incantation: (My?) hunted *akkannu*-ass!"; i 17 "Incantation: (My?) *akkannu*-ass of the mountain, who has blocked you?". For the possessive adjective "my" see Hirsch 1973–1974: 66–67.

(l. 48, Lambert 1960a: 72).[212] Interestingly, a commentary mentions the animal's reproductive activities: *akk*[*annu* . . .] *ana muḫḫi* GAN : *alādu* "*Akk*[*annu*-wild ass . . .] over . . . : to generate" (ibid. 72). In the incantation No. A.3: 49–50, we find an association with the wild bull (*rīmu*): "*Akkannu*-wild ass! *Akkannu*-wild ass! Wild bull! Wild bull! Who has slackened you (so that you are) like slackened / strings?". As highlighted by Cavigneaux the animal «symbolized indomitable nature, swiftness, also sexual potency» (1999: 263).[213] The researcher claims a terminological relationship between Šakkan, the cattle god, and the *akkannu*-ass. Indeed, the god is mentioned in an Old Babylonian incantation from Meturan (H 72, Cavigneaux 1999: 258) as well as in *Maqlû*, in relation to his fertilizing capacity of herds: *kīma Šamkan irḫû būlšû* "Like Šamkan impregnates his cattle" (*Maqlû* VII 24, Abusch 2016: 170).[214]

The wild bull is another animal that is identified with the patient. The metaphorical use of the aurochs has a long tradition in Mesopotamian literature. The animal is associated with royalty, since the Sumerian times, as it is used as an epithet for some deities including Sîn and Dumuzi[215] (see van Buren 1945: 32–36; Marcus 1977; 87–88; Engel 1987: 75–76; ChWatanabe 2002: 57–64; Waetzoldt 2006–2008: 382–384; Weszeli 2006–2008: 401–403; Feldt 2007; Pfitzner 2019). It should be noted that in Akkadian documentation only wild beasts are used as metaphors for royalty since they represent «strength, force, pride, indomitability» (MPStreck 1999: 174), on the contrary, oxen seem to capture the negative value of obedience and subordination.[216] The indomitable aggressiveness value is also attributed to the animal in omens (see Weszeli 2006–2008: 402; De Zorzi 2014: 158).

The idea of strength and power, related to royalty and military campaigns,[217] symbolized by the wild bull, it is reinterpreted sexually in *nīš libbi* incantations.[218]

---

212 In these two texts, the animal is placed in relation to the *serrēmu*-onager, another species of wild ass (CAD S 318–319) For the metaphors with such wild ass in the Sumerian literature (anše-edin-na) see Heimpel 1968: 269–273. For its metaphorical use in royal Assyrian inscriptions see Marcus 1977: 90.
213 In *Epic of Gilgameš*, the funeral lament for Enkidu is described by his friend as follows: "Enkidu, [whom] your mother, a gazelle / and your father, an *akkannu*-wild donkey, [created], / whom the wild [asses] (*serrēmū*) reared with their milk, / and animals [of the wild taught] all the pastures!" (VIII 3–6, George 2003: 650); "My friend, you were a hunted mule (*kūdannu ṭardu*), an *akkannu*-ass of the mountains (*akkannu ša šadî*), a leopard of the steppe. / Enkidu, my friend, you were a driven mule, an *akkannu*-wild donkey of the mountains, a leopard of the wilderness" (VIII 50–51, George 2003: 654). See also Ponchia 2019: 200. As emphasized by M.P. Streck (1999: 104) this is a picture of strength and vigor.
214 See also the parallel AMT 67, 34: 4′ (quoted by Cavignaux 1999: 268).
215 Ištar is also defined as *rīmtu* "wild cow" (see Farber 1977a: 130, Hauptritual A iia: 42).
216 For the image of weakness and defeat conveyed from the domestic bull in the Assyrian royal inscriptions see Marcus 1977: 91. For the metaphorical use of the bull in Sumerian literature see Heimpel 1968: 79–121, of the wild bull see ibid. 133–177, for the bullock see ibid. 177–198, see also Pfitzner 2019.
217 See the relationship between battle and the sexual sphere invoked by bow metaphors in the incantations of this text corpus.
218 See also a beginning of a probable love lyric in Old Babylonian catalogue, CUSAS 10, 12: 12: "The wild bull is not standing" (Wasserman 2016: 190, for the commentary 193). In Sumerian composition

The wild bull (CAD R 359–362; AHw. II 986) is the sexual force *par excellence*. In fact, except for text No. A.3: 49 (also catalogue LKA 94 i 2), in which it is mentioned along with the *akkannu*-ass, it always appears in association with the deer, another animal characterized by sexual vigor: catalogue LKA 94 ii 14: "My wild bull, [rea]r [up]! Stag, rear up!"; No. G.1: 2–3 "[St]ag! Stag! Wild bull! Wild bull! / [Roa]r, stag! Rear up, wild bull!" (see also No. B.1: 1–2; No. F.3: 65–66). Both animals are perceived as the ones with the greatest potency. Text No. E.3.: 48–49 seems to provide a ranking of the animals relative to their sexual prowess: "[Like a ram eleven times], like a weaned ⟨sheep⟩ twelve times, like a partridge thirteen times. / [Make love to me like] a pig fourteen times, like a wild bull fifty (times), like a s[ta]g fifty (times)!". In this list the animal that is located at the lower end is the ram that copulates 11 times, then the sheep comes with its 12 times, followed by the partridge and the pig copulating 13 and 14 times, respectively. It should be noted that, except for the partridge, these are all domestic animals. Instead, the wild animals who occupy the top spot of the ranking are, in fact, the wild bull and deer which copulate 50 times. This incantation shows us how, in the imagination of Mesopotamian society, wild animals embody concepts related to sexual prowess. In text No. F.4: 83, the wild bull conveys the concepts of physical strength and sexual vigor: "[. . .] jump, my wild bull! Together with you, may the strength rise!". In an Old Babylonian love incantation the (amorous) impetuosity (*uzzum*) is compared with a wild bull: "Impetuosity, impetuosity c[omes to me] like a wild bull (*rīmāniš*)" (Wasserman 2016: 269, ll. 85–86).

Interestingly, whereas the incantations mention the bull in a metaphorical context, the medical prescriptions never include the use of animal substances taken from the aurochs. The animal from which hair or saliva is taken for medical purposes is the domestic bull (*alpu*/GU$_4$). This distinction between incantations and prescriptions is very important as it shows how the symbols can be adapted to human needs. In other words, even though the wild bull is mentioned in incantations, being of great magical-therapeutic and regulatory importance, getting substances from the animal itself is a whole different story. One can postulate that for this reason, the saliva or the hair of domestic bull lend themselves better to be therapeutic ingredients in the rituals, thanks to their greater availability.[219]

Another interesting detail about the relationship between animals and sexual images is that while the wild bull, as we have seen, is used metaphorically in incantations of this corpus, the domestic one is used to express images, clearly sexual

---

Gilgameš and the Bull of Heaven the goddess Inana addresses the Gilgameš whom she desires as a husband as am-ĝu$_{10}$: "Lord Gilgameš, my wild bull, my . . . man" (l. 7, Cavigneaux and al-Rawi 1993: 104). In *Enki and the World Order* the sexual connection is clear, as the Tigris has just been filled by Enki's sperm: "The Tigris, like having been 'delighted' inside by a great wild bull, when it gave birth . . ." (l. 258, Benito, 1969: 99). For other Sumerian sources see Pfitzner 2019: 151, 167–168).

219 Hair of leg (*šarta ša purīdi*) of a breezing bull (*alpu puḫālu*): A prescr. 2: 6–7; thigh hair of a black bull (*šārat rapalte ša alpi ṣalmi*): E prescr. 2: 63; saliva (*rupuštu*): F prescr. 12: 38–39; saliva of a reared-up bull (*rupušti alpi tebî*): N prescr. 4 i 18; phlegm of a reared-up bull (*ḫaḫḫu alpi tebî*): N prescr. 12 iii 2.

prowess, to plowing, especially in Sumerian literature. Here an example is taken from the poem *Inana and Dumuzi P*:

> ki-sikil-men$_3$ a-ba-a ur$_{11}$-ru-a-bi
> galla$_4^{la}$-ĝu$_{10}$ ki-duru$_5$ a ma-ra
> ga-ša-an-ĝen gu$_4$ a-ba-a bi$_2$-ib$_2$-gub-be$_2$
> in-nin$_9$ lugal-e ḫa-ra-an-ur$_{11}$-ru
> ^{d}dumu-zi lugal-e ḫa-ra-ur$_{11}$-ru

> I am the young girl, who will plow (for me)?
> My vulva, irrigable land, is rich in water!
> I am the mistress, who will graze the ox?
> The noble lady, may the king plow for you!
> Dumuzi, may the king plow for you!
> [*Inana and Dumuzi P* ii 26–30, Sefati 1998: 220–221]

Plowing is, therefore, a clear metaphor for sexual intercourse (Wilcke 1987: 77–78; Sefati 1998: 90–92; Couto-Ferreira 2017; Couto-Ferreira 2018a: Couto-Ferreira 2018b; Zisa 2021), and as pointed out by Ch.E. Watanabe (2002: 105): «The domestical species conveys the notions of fertility which are closely associated with the animal's role in plowing»[220] (see also Steinert 2017a: 225 and fn. 10; Zisa 2021). In the mythological work *Enki and the New World Order* the god gives vitality to men and women, the first being compared to a domestic bull (gu$_4$) with thick horns:

> Your word fills the heart (šà) of the young man of strength (usu),
> he gores in the yard like a bull with thick horns.
> Your word gives sexual attraction (ḫi-li) to the face (lit. head) of the young woman,
> they look at her in amazement in the cities.
> [*Enki and the New World Order* ll. 32–35, Benito 1969: 87]

In a Sumerian incantation that accompanies the human birth, the human reproductive practice is compared to that of bovine (UM 29-15-367: 19–20, van Dijk 1975: 55; Cunningham 1997: 70–71; Stol 2000: 60; see Krebernik 1984: 36–47, No. 6):[221] "My father, in the (pure) stables, in the pure fold, the bullock has mounted (the cow), / deposited in her womb the right seed of humanity."

The association between human fertility and that of the ox and domestic donkey is also expressed in the composition of the *Ištar's Descent into the Netherworld*, according to which the goddess's travel to the netherworld has caused a fertility crisis, with a consequent absence of sexual intercourse, in the animal world as well as in the human one:

---

220 The bull appears, always within an image of fertility in *Inanna and Dumuzi W* 19 (Sefati 1998: 261) "my calf and the bull go together."
221 See the mythological reference to the composition *A Cow of Sîn* (Veldhuis 1991).

Ištar has descended to the Netherworld and has not ascended,
from the moment she descended to the "Land of No Return,"
the bull does not mount the cow, the ass does not impregnate the jenny,
the young man does not impregnate the girl in the street,
the young man has slept in his private room,
the young woman has slept by herself.
[*Ištar's Descent into the Netherworld* ll. 85–90, Lapinkivi 2010: 19]

From the examples given here, both in Sumerian and Akkadian compositions, the wild bull and the domestic one, although employed in literary images related to sexuality, convey very divergent notions. The first is, in fact, present in *nīš libbi* incantations to normative male sexual practice, which should be active, vigorous, indomitable, not connected to the reproductive sphere. The second one repeats ideas related to fertility and reproduction and is also connected with the agricultural images, such as plowing.

The patient is also identified with the horse that represents the values of freedom, indomitability, and pride (see Weszeli 2003–2005: 479–480), so that it has been labeled "impetuous" (*ezzu*) (No. A.2: 42) and "preeminent" (*etellu*) (No. L.2: 19).

That the horse embodies the concepts of velocity and freedom, as opposed to the domestic donkey, representing the subordination to the human being, is also evident when we look at a popular saying:[222]

sisû tebû ina m[u]ḫḫi atāni parê kī ēlû
kī ša rakbu-ma ina uznīša ulaḫḫaš
u[mma m]ūru ša tullidī kī yâti lū lasim
an[a imēr]i zābil tupšikki lā tumaššalī

When a reared-up stallion was mounting a jenny-ass,
As he was mating, he whispers in her ear,
"Let the foal, which you bear, be a fast runner, like me;
Do not make it like an [ass] which bears hard labour."
[Lambert 1960a: 218, r. iv 15–18]

The horse is a symbol of military strength and the kingship (Weszeli 2003–2005: 479). The value of the horse in battle is expressed by the epithet "glorious in battle" (*na'id qabli*) in *Dispute between Ox and Horse* (Lambert 1960a: 175, text A: 24),[223] and as can be seen in the analysis of *Šumma izbu* omens (De Zorzi 2014: 158–159). In the *nīš libbi* texts,[224] its 'rising' (*tību*), a reference to the sexual vigor, is called "devastation" (*našpandu*, No. A.2: 42),[225] a term that defines the qualities of strength, vitality, and power.

222 See also in love context: "A *mare*(?) of a horse (*perdu*) / – she goes up and down" (CUSAS 10, 12: 27–28, Wasserman 2016: 191).
223 See also the *Epic of Gilgameš* VI 53.
224 See also the catalogue LKA 94 i 16: "My stag! My stag! Horse!".
225 It probably refers to a devastating flood. For the reference to the flood in love context see the

Also, within the *nīš libbi* rituals corpus, horse's urine is used as an ingredient for the acquisition of sexual desire, through the principles of contagious magic. In text E prescr. 1: 60–62, when a horse urinates on the street, the edge of the puddle of urine is mixed with beer and drunk on an empty stomach: "[When a s]tallion on the street / [ur]inates, the edge of remains/ of its uri[ne] you take, you mix (it) with beer, [he (= patient) drink]s (it) on an empty stomach."

In some Sumerian compositions, the horse is also described as an animal loved by the goddess Inana. The reference to the love affair between the goddess and the animal can be found in the composition *Dispute between Grain and Sheep*: z a-e [k ù] dI n a n a-a n-n a-g e n_7 / [a n š e]-k u r-r a k i i m-a-á ĝ "You (= wheat), as the pure goddess Inana of the sky, / love the horses" (ll. 144–145, Alster and Vanstiphout 1987: 24). Another passage refering to the relationship between the horse and the goddess is from the Sumerian hymn to Inana-Ninigalla: [a]n š e-k u r-d a k i-n ú k è-z u-n e "When you (= Inana) share the bed with the horses" (l. 61, Behrens 1998: 30).

The love affair between the goddess and the animal is mentioned in the *Epic of Gilgameš*, where the horse was one of Ištar's lovers, alike the lion and Dumuzi himself:

tarāmī-ma sisâ na'id qablī
ištuḫḫa ziqta u dirrata taltīmīšu
7 bēr lasāma taltīmīšu
dalāḫa u šatâ taltīmīšu
ana ummīšu Silili bitakkâ taltīmī

You (= Ištar) loved the horse, glorious in battle,
you have designated for it the whip, the spur, the burst,
you have designated seven leagues for the gallop,
you have designated muddy water to drink (lit. stirring up mud and drinking)
to its mother Silili you have designated constant crying.
[*Epic of Gilgameš* VI 53–57, George 2003: 620]

The identification of the patient with the horse in *nīš libbi* incantations could also allude to this very episode, since there are incantations in references to mythical time, in particular to the goddess Inana and the god Dumuzi.

Another animal mentioned in the metonymies is the lion. It represents the royal power, for this reason, it is used as an epithet of the kings, as well as the goddess Ištar and the god Ninurta (see van Buren 1945: 39–40; Heimpel 1968: 280–344; Heimpel 1987–1990a: 81–82; Cassin 1981; MPStreck 1999: 173; ChWatanabe 2000; ChWatanabe 2002: 42–56, 76–82, 89–92; Strawn 2005: 131–226).[226] It personifies the values of force,

Old Babylonian catalogue CUSAS 10, 12: 15: "He brought the flood (*iššâ abūbum*) – achieving what?" (Wasserman 2016: 190).

226 This symbolism is present also in omens, as in the *Šumma izbu* or in teratomancy at Tigunānum (see Jacobs 2010: 329–333; De Zorzi 2014: 157–158; De Zorzi 2017: 133).

aggression, and voracity of wildness. The animal, however, also embodies the sexual potency and is therefore connected with the amorous sphere. To understand this symbolism must analyze the relationship that the animal has with goddess Ištar. In the *Epic of Gilgameš* the lion is one of the lovers of the goddess, destined to a terrible end, and, as writes Ch.E. Watanabe (2002: 104) «suggests her fertility aspects, represented by lion/lioness which embodies the strong forces of sexual desire»: *tarāmī-ma nēša gāmir emūqī / tuḫtarrīšu 7 u 7 šuttāti* "You (= Ištar) have loved the lion, perfect in strength, / you have dug for him seven and seven traps" (V 51–52, George 2003: 620).

The relationship between the goddess and the lion has been extensively studied (Brown 1973: 39–52, 113–133; Wilcke 1976–1980: 82; ChWatanabe 2002: 103–105; Strawn 2005: 208–209). The goddess is, in fact, often referred to as "lioness" (*labbatu*). The reason for this identification is that the lion symbolically embodies the sexual vigor and reproductive force, aspects of human life within the domain of the goddess Ištar as confirmed by his numerous invocations in *nīš libbi* incantations:

> Incantation: Lion! Bull . . . [. . .]
> Yo[ur] lo[ve]-making . . . [. . .]
> At the command of wi[se Ištar, Nanāya],
> Ga[zba]ba (and) K[anisurra]!
> You recite this incantation three times.
> [No. K.7: 116–120][227]

From the above examples, the close symbolic relationship between the lion and the goddesses of love, amongst all Ištar,[228] becomes clear and the choice of quoting the lion always along with other wild beasts, known for their predominant sexual activity, such as stag and wild bull. In fact, an Old Babylonian love incantation the (amorous) impetuosity (*uzzum*) is compared with a lion (Wasserman 2016: 269, l. 88).

The stag is a very important animal within *nīš libbi* corpus.[229] The patient is metaphorically identified with it. It always appears along with the wild bull: No. G.1: 2–3 (also l. 7): "[St]ag! Stag! Wild bull! Wild bull! / [Roa]r, stag! Rear up, wild bull!".

The roaring of the stag, for its important role during the animal's sexual courtship, is also used in similes (No. B.1: 1–2; No. F.3: 65–66; No. G.1: 3 and 7). Other incantations refer to the animal's sexual ability:

> The ma[ting of a wild go]at six times,
> the mating of a stag seven times,
> the mating of a partridge twelve times,

227 See also No. B.1: 3 "Together with you, may a lio[n] rear up!".

228 As to the relationship between the lion and other wildlife and Nanāya, see the section dedicated to the goddess (Chapter II pp. 139–141).

229 The catalogue LKA 94 16: "My stag! My stag! Horse!"; ii 14: "My wild bull, [rear up!] Stag (LU.LIM) rear up!".

make love to me! Make love to me because I am young! [. . .]
and I am endowed? with the *rikibtu* of a stag! Make love to me!
[No. C.1: 4–8, see No. F.3: 69–73]

[Like a ram eleven times], like a weaned ⟨sheep⟩ twelve times, like a partridge thirteen times.
[Make love to me like] a pig fourteen times, like a wild bull fifty (times), like a s[ta]g fifty (times)!
[No. E.3: 48–49]

In these two ascending climaxes, related to sexual activity of the animals, the stag is mentioned, or together with wild animals such as the ibex, and the partridge (No. C.1: 4–8; No. F.3: 69–73) or with both domestic and wild animals, such as the ram, the sheep, the pig, the partridge and the aurochs (No. E.3: 48–49). While in the first climax it is stated that the animal's mating is seven times, the number in the second is fifty times (like the wild bull). This climax is of great importance because it indicates the immoderation of the deer's sexual activity, like one of the bull, both considered the vigorous wild animals *par excellance*.[230] It should be noted that in the Sumerian composition relating to the cycle of Inana and Dumuzi, both the wild bull[231] and the deer are epithets of the god. Of the god it states in *Dumuzi and Inana D* that he "copulates" (e m e-AK, literally "making tongue"[232]) well-fifty times, as the two animals:

My (= Inana's) sweet loved, lying on my heart,
one time after another, 'making tongue' one time after the other,
My brother with the beautiful eyes, did so fifty times,
like powerless I was with him,
wincing down, I was silent.
With my brother, laying (my) hands on his hips,
with my sweet loved, with him I spent the day there.
[*Dumuzi and Inana D* ll. 12–18, Sefati 1998: 152]

Substances obtained from the stag are important ingredients in *nīš libbi* prescriptions and used only in this therapeutic corpus: "*Rikibtu* of a stag, antler of a stag, penis of a stag, *takdanānu*-plant in a leather bag" (*nīš libbi* A prescr. 16: 27); "Navel of ⟨a stag⟩, antler of a sta[g],/ *rikibtu* of a stag, *aṣ*[*uṣu*]*mt*[*u*]-plant / you properly crush again and again, you sme[ar] the penis (of a stag?) with salt, / you tear off hair of the tail of a male (stag?) (and) together with a string [. . .] / [you s]pin and you insert *amašpû*-stone inside [. . .] / you tie (it) arround his waist [and (he will get) sexual desire]" (*nīš libbi* K prescr. 30: 129–234); "Make love to me! Make love to me because I am young! [. . .] / and I am endowed? with the *rikibtu* of a stag! Make love to me! [. . .] Its ritual: (You put)

230 See also *Enmerkar and the Lord of Aratta* l. 182: "Deer of the high mountains, with princely antlers."
231 Obviously, it is not exclusive to Dumuzi.
232 On this expression in the sexual sphere, referring to a love relationship see Sefati 1998: 161.

the head of a male partridge, / silver, gold, the *rikibtu* of a stag into a . . . leather bag" (No. C.1: 7–8 and 10–11).[233]

The animal thus symbolizes wildness and sexual vigor (Heimpel 1972–1975a: 420; Biggs 2002: 76). It is also used as an epithet of Inana/Ištar, whose role in *nīš libbi* corpus as already been pointed out, in *Hymn to Ištar of Nineveh of Ashurbanipal*: *gišimmaru binat Ninî* [*ayy*]*ali mātī* "Oh palm tree, daughter of Nineveh, deer of the lands" (Livingstone 1989: 18, No. 7: 1) and in the Sumerian composition *Inana and Šukaletuda*, always together with wild bulls: "My mistress is among wild bulls at the foot of the mountains, she fully owns me / Inana is among the deers on the mountain tops, she fully owns me" (ll. 11–12, Volk 1995: 117).

The agent of suffering

Unknown agent

The first group of *nīš libbi* incantations is characterized by a first section in which the recurring themes are on the one hand the patient's identification with wild animals, on the other hand the questions about the man or woman who caused the suffering, describing the evil action through similes.

In these incantations, although only explicit in one text, the cause of this malaise is the witch. The texts are not so much about finding out what might be the cause for the absence of sexual desire, but about who acted maliciously or initiated the witch's actions.

Let us take a look at some passages of this group of incantations:

> Incantation: *Akkannu*-wild ass! *Akkannu*-wild ass! Wild bull! Wild bull! Who has slackened you (so that you are) like slackened / strings? Who has block[ed] your course like (on) a ro[ad]?
> Who has poured [co]ld [water] on your 'heart,'
> (and) has put f[ea]r upon your 'heart,' has [. . .] sleeplessness?
> [No. A.3: 49–52]
>
> Who are you who has blocked up my course like on a road,
> has slackened my loincloths like taut strings?
> (who are you who), like (the string of) a merchant's leather money pouch, all my sinews
> has pulled tight and firmly bound?
> My witch and my witch! My sorceress (lit. "the superior one") and my sorceress!
> [She] has slackened me like ta[ut] strings!
> [Like] (the string of) a merchant's leather money pouch [all my sinews]
> [she has pulled tight and firmly bound]!
> [No. A.4: 75–82]
>
> My hunted *akkannu*-wild ass! Hun[ted] onager [. . .]

233 See also No. B.2, variant B r. 34; No. F.3: 73; No. H.2: 18; No. J.2: 18; N prescr. 18 iv 3; T prescr. 2: 10′.

who has blocked you like an opening of the *dilûtu*-water system
(and) who has slackened you like [taut] strings?
Who has blocked your ways like (those of) a traveler
(and) has burned your forests like (those of) Ḫumbaba?
[No. D.2: 10–14]

Rearing [onag]er, preeminent stallion, who roams the forests, who [has dampened your ardor]?
[The . . . witche]s have taken away [my] sexual desire like slack strings!
[. . . she (= witch)] practiced hate-magic against me [al]l day, bewitched me all night, bounds me like a pr[isoner].
[like sl]ackened [strings] they have taken away my sexual desire!
[No. L.2: 19–22]

As we have already seen, in incantations No. A.3 and No. D.2 the witch is not mentioned. She appears in the incantations No. A.4 and No. L.2, where the mention is always preceded by questions on the magical performer's identity. The theme of the unknown performer, who causes pain and suffering, is widespread in Mesopotamian exorcism, in particular the one dating back to the first millennium, as can be appreciated in *Maqlû* series and the group of anti-witchcraft texts.

Here are some examples from the *Maqlû* series:

Whom (= the figurines of malicious agents) you, Nuska, the judge, know, but I do not know [I 87].

Whoever you are, O witch, who has taken out clay from the river for me (= to make my own figurine) [II 183].

I do not know your city, I do not know your house, I do not know your name, I do not know your dwelling [II 208].

Whoever you are,[234] O witch, who performs the *zikurrudû*-magic against me,
If a friend or companion,
If a brother or colleague,
If 'newcomer' or citizen,
If acquaintance or stranger,
whether wizard or magician,
whether male or female,
i[f dead] or alive,
if a mistreated man or an abused woman,
if cult performer or tramp,
[if *eššebû* or] *naršindû,*
if snake charmer or *agugillu,*
or speaking a foreign language that is in the land.
[IV 80–92]

234 See also VII 45.

In anti-witchcraft texts (Abusch and Schwemer 2011):

> (The aggressors) your great deity (= Šamaš) knows, but no other god knows, only you know, but I do not know[235] [No. 8.3.1: 83].

> Šamaš, I do not kn[ow] the wiz[ard] and the witch,
> the man who destroys me, I do not know, . . . [. . .] . . . [I] d[o not know . . .],
> the man who ties me, I do not know, the woman who binds me [I do not know]
> Šamaš, my wizard and my witch [I] d[o not know . . .].
> [No 8.5: 129″–132″]

In *namburbû*-texts:

> Wizard or witch, m[an or woman],
> small or tall, a dead person or [a living person],
> father or brother or sister or [mother . . .],
> friend, colleague or [partner],
> gatekeeper or guardian of the door [. . .].
> [LKA 115: 11–15, Maul 1994: 502]

In this group of *nīš libbi* incantations, therefore, one of the recurring themes in the repertoire of anti-witchcraft texts can be found: the unknown identity of the evil performer. However, this theme is reformulated according to the needs of the *nīš libbi* therapies, with the description of specific symptoms, often expressed through similes, and by animal metaphors. The identity of the witch is only expressed through rhetorical questions, introduced by "who" (*mannu, attamannu*). Only in text No. A.4: 79 and No. L.2: 20–22 the witch[236] is mentioned.

The description of the witch and the emphasis on their unknown identity shows great tension within the community.[237] The man is the victim of an external entity, which cannot be identified. The "unknown identity" of the magical operator is an important rhetorical ploy allowing the patient to stress his complete innocence (see Abusch and Schwemer 2016: 5). As emphasized by Schwemer (2007a: 72), the use of such formulations about the unknown evil performer's identity expresses a request for help from the defenseless patient towards the god. The god is, in fact, the one that guarantees the restoration of human health and well-being, and to whom utmost devotion must be shown, and from whom protection is to be requested through the incantation. However, it should be noted that, as we are not able to grasp the social discourse surrounding the magical event by observation of social practice, it remains difficult to determine if the identity of the evil person is a real fact, evidence of social tensions characterizing

---

235 See also No. 8.4: 69.

236 On the role of women in Mesopotamian witchcraft see Sefati and Klein 2002.

237 Sallaberger (2011: 28–29 and fn. 45) also emphasizes the function of social control and normative regolamentation of Mesopotamian witchcraft.

the Mesopotamian community, or rather merely a figure of speech, which manifests devotion to the god, regarded as the deliverer of the patient (see Schwemer 2007a: 73).[238]

Similes concerning the action of the witch

In this section, I will deal with the description of symptoms in incantations and specifically those caused by the witch. As we have noted, many of these symptoms, and more generally the malaise to which the patient is subjected, are expressed through similes introduced by *kīma* 'like.' The ordinary human conceptual system, in terms of which we both think and act, is fundamentally metaphorical (Lakoff and Johnson 1980: 3). Suffering and pain, expressed through metaphors and similes, thus, represents one of the most tragic spheres of human experience. Assyriologists have stressed the importance of metaphorical language in incantations to describe, communicate, and explain pain and suffering (see Zisa 2012: 13–15; Böck 2014: 115–128). Anthropological studies[239] on "metaphors of suffering" have joined the individual and the social dimension in its historical development. The figures of speech, expressed in incantations, stress this inseparable contact between the sphere of the individual, in his physical dimension as well as the emotional one, and the socio-cultural context, in which the sufferer lives and interacts (see Pizza 2005: 103). The pain which is, however, only partly communicable, as noted by Good (1994), is contrasted with the human need to give meaning and "objectify," through the existence of "phenomenology" and "symbolism." The pain must be made thinkable and therefore imaginable for himself and others, through mythical-ritual narratives and culturally encoded metaphors, that allow one to activate the therapeutic process. In the words of Lévi-Strauss, «la cure consisterait donc à render pensable une situation donnée d'abord en termes affectifs et acceptable pour l'esprit des douleurs que le corps se refuse à tolérer» (1949b: 217).

Man needs to communicate the experience of pain, in other words, to verbalize the suffering itself. According to the argument of Elaine Scarry (1985), physical pain destroys language, but at the same time is the basis of its "construction," or rather its "rebirth" because of the relationship that is established between physical pain and imagination. Imagination, symbols, and metaphors are the key elements by which suffering can be communicated to others: «That the person in pain very typically moves through a handful of descriptive words to an "as if" construction, and an "as

---

238 On the anonymity of the witch's identity see Thomsen 1987: 21–29.

239 For the analysis of the use of metaphor in medical context see Kirmayer 1992; Kirmayer 1993a; Kirmayer 1993b; Kirmayer 2008; DelVecchio Good et al. 1992; Mattingly and Garro 2000. See also the two works by Sontag *Illness as Metaphor* (1978) and *AIDS and Its Metaphors* (1989). For the metaphor in Anthropology see Fernandez 1974; 1986.

if" construction that has a weapon on the other side, indicates the primacy of the sign in the elementary work of projection into metaphor» (ibid. 172).

Making use of metaphors to represent the pain means explaining it to oneself and others and building a "poetic" account of the suffering. In order not to fall into rhetorical and stylistic pitfalls, we should understand that these metaphors emphasize the suffering of the protagonist not only from a literary point of view, but they are also concrete realities, as they are based upon the experience of the body. The body metaphor thus completes physical and historical narratives that are set together by the experience of suffering, existence, physicality, or society. If the patient needs to "give shape and voice" to the pain and tell their family and the medical-therapeutic circle, he must employ "metaphors of suffering" which have their conceptual source in a cultural code to which both the patient and the members of the community belong, i.e. those who are more or less directly involved in the health issue and the therapeutic itinerary.[240] As Kirmayer writes:

> The metaphoric process allows all of these forms of meaning. When a patient with a life-long history of migraine headaches spontaneously remarks, "My head is made of glass," she is simultaneously revealing something about her body image, her model of migraine, and the way she wishes to be handled by the physician. She draws this metaphor from a common fund of physical experience but its nuance and full significance depend on the languages of suffering used within her family and salient in her current social context (which includes the health care system). It is not necessary that a speaker realize her statement is a formal metaphor for it to be the expression of a metaphoric relationship, whether in the speaker's own cognitive model or in the relationship of body to society. Symbolizing is the embodiment or enactment of metaphor. This is so whether the action is intentional or accidental. [Kirmayer 1992: 340]

To analyze this "poetic" time to express the suffering, one must understand how the community sees the health issue itself, the body and the emotions, and the way pain is embodied (see Csordas 1997). We must remember that such incantations do not describe the *illness* (as an anthropological concept) of a specific patient but are coded by written medical tradition over the centuries of Mesopotamian history. For this reason, the construction of *sickness*, that is the socio-cultural construction of the disorder, as the description of the symptoms, is central for the understanding of this "metaphorical language." The description of symptoms is encoded by the tradition of *āšipūtu*. As we shall see in the following sections, the similes employed in these incantations are used in a lot of other Mesopotamian documentation that deals with exorcism and therapeutics. They represent standard models with which to *give form* to human suffering. It should also be emphasized that the verbs are used in the pret-

240 It is significant that often the cause of failure of biomedical practices in the contemporary world is due to noncompliance, that is, the incommunicability between the cultural language of the patient (his or her vision of the sickness, etiology, diagnosis) and biomedical and technical language.

erite, a verbal form that expresses the malevolent action of an external aggressor, demon or witch.[241]

The binding

Incantation No. A.2: 42 describes the action of the witch through the image of binding: "Impetuous horse, whose rising is a devastation, [w]ho has bound your limbs (*mešrêtīka ukassi*)?". Also in No. L.2: 21 we read: "[. . . she (= witch)] practiced hate-magic against me [al]l day, bewitched me all night, bounds me like a pr[isoner]." The image of binding is also mentioned in the catalogue of *nīš libbi* incantations: *šiptu: irkusāma iptaṭar / šiptu: irkusānim-ma iptaṭar* "Incantation: 'They (fem.) bound, but he released!' / Incantation: 'They (fem.) bound to me, but he released!'" (LKA 94: 10–11).

It is a well-known fact from Assyriological literature that the witch (but in general the evil aggressors) often acts by binding (*kasû*) the patient's body, making any movement impossible (see Sallaberger 2007: 296; Schwemer 2007a: 85–86; Salin 2020: 104–105). Such an image is configured not as a simple rhetoric image which expresses the idea of immobility, but on the contrary, manifests a particular view of malaise; The attack of the witch, the existential condition in which the man finds himself, is a condition in which the patient is deprived of the possibility of acting according to his own will.

This line therefore must be analyzed not only in its stylistic dimension but especially for its importance in helping us understand the cultural construction of the attack of the witch. In other words, it helps us to understand how the bewitched patient's condition was experienced and seen, that is, what the psycho-physical state that the victim was forced to experience.

As I mentioned above, the image of "binding" and "bond" (*kasû*, *rakāsu*, *riksu*) is quite widespread in anti-witchcraft literature of the first millennium, so that the evil operator is defined as a "knotter" (*kāsû*, e.g. *Maqlû* IV 106, 122, Abusch 2016: 94, 97):

> They bind the young men, they kill the girls.
> [Abusch and Schwemer 2011: 165, No. 7.8.1: 18′]

> The man who ties me I do not know, the woman who binds me [I do not know],
> Šamaš, my wizard, my witch [I do] no[t know . . .].
> [Abusch and Schwemer 2011: 315, No. 8.5: 131″–132″; see also ibid. No. 9.1: 20, 24]

---

241 Falkenstein (1931: 46) had already emphasized the difference between the use of the "theme of the present" and the one "of the preterite" when describing the action of demons in Sumerian incantations of the type 'Marduk-Ea.' The "theme of the present" describes in a poetic-schematized form the demons' activity in general, their origin, residence, appearance and their actions towards animals and humans, while the "theme of preterite" reports of an attack of the demons on man in the past, for whose healing the incantation is used.

> The man's / bindings have been made (lit. bound).
> [Abusch and Schwemer 2011: 436, No. 12.1: 31–32]

> O you, who have bound me.
> [*Maqlû* III 110, Abusch 2016: 76]

> Gutee women, Elamite women, Hanigalbatee women,
> indigenous women attack me by means of bindings.
> Their bindings are six, my untyings (of bindings) are seven.
> [*Maqlû* IV 116–118, Abusch 2016: 96]

The patient invokes the god to deliver him from the "bindings" of the witch:

> You break the bad binding.
> [Abusch and Schwemer 2011: 169, No. 7.8.3: 24′, see No. 7.8.4: 6′, 8′; No. 7.8.6: 31′]

> And she, M[anz]ât, breaks the binding[s].
> [*Maqlû* IV 112, Abusch 2016: 96]

> Like the *kasû*-plant may her witchcraft bind her.
> [*Maqlû* V 31, Abusch 2016: 102][242]

> May the *kukru*-plant of the mountain brea[k] your bindings.
> [*Maqlû* VI 22, Abusch 2016: 114]

> Come here and break the strong bindings of my wizard and my witch.
> [*Maqlû* VI 28, Abusch 2016: 116]

> May Siriš, the releaser of the god and of man, loosen the bindings (of the man).
> [*Lipšur Litania*, Reiner 1956: 138, l. 116; see l. 20]

At the same time, the patient, under the supervision of the exorcist, or the therapeutic operator acts against the maker of the sickness by acting in the same way, tying together the arms and feet of the figurine that represents the wizard and/or witch. The bondage therefore also presents itself as an actual magical operation that allows the evil performer to attack his victim, but at the same time, it represents a fundamental operational tool for freeing the patient from witchcraft and binding the aggressor. It reproduces a magical practice used in both the attack, as well as the defense.[243] A thread, therefore, is not limited to the wizard and the witch, but can also be used by the patient (or the exorcist on his behalf) for defensive purposes and a counterattack.[244] The actual magic performance is

---

242 It should be noted that there is phonological similarity between the name of the plant and the verb *kasû* 'to bind.'

243 On the relationship between witchcraft and exorcism see Abusch 2002: 65–66. On the practice of bondage in exorcism as a preliminary stage for the annulment of witchcraft by means of burial and destruction see ibid. 67–78.

244 The ingredients used in the rituals are defined as capable of binding evil: "They (= ingredients)

simple: binding the arms and feet (the limbs as a whole) of the figurine representing the sorceress.[245]

> [You tie] their (= wizard and witch) arms to their shinbon[e]s.
> [Abusch and Schwemer 2011: 134, No. 7.6.4: 7, see No. 7.6.5: 11; 8.2: 30; *Maqlû* iii 97, Abusch 2016: 76]

> May my bondages be loosened, (var. adds: but) may my wizard (and) my witch be perforated!
> I unchain the [arms] (var.: limbs) of my figurine, I attach the arms (var.: limbs) of their figurines.
> May my arms be released, may their arms be tied! I shall unchain the feet of my figure, I shall bind the feet of [their figurines].
> May my feet be released, may their feet be tied! I pour fish oil over them.
> [Abusch and Schwemer 2011: 249, No. 8.1: 48″–51″; see also No 8.3.1: 10; No. 8.5: 98″]

> You make two clay figurines, two dough figurines, two [w]ax figurines, two tallow figurines, one of the man and one of the woman, and
> write their names on their [lef]t side. You turn their arms behind their backs,
> bind their feet. You put juniper in a censer before Šamaš. Before Šamaš he reads this.
> [K 22773 32–34 and dupl. (*namburbû*), Maul 1994: 448]

> I have caught you, I have bound you, I have abandoned you.
> [. . .]
> May Girra, the fire, deliver (me) of your bondage.
> [*Maqlû* IV 74, 76, Abusch 2016: 94; see ibid. IV 7]

> May the [ruler] of Eridu [untie the] bondage of the wizard and the witch,
> their evil bondage may Marduk, [the prince] among the gods, untie.
> [KAR 59: 17–18, Ebeling 1953a: 66]

It is interesting to note that in the concrete magical performance as well as in the description of malevolent action in the incantations, which I would call symptomatic and etiological, it is stated that the limbs (*mešrêtu*) are to be tied. The choice of the term is of importance as to the understanding of the state of suffering that the patient is in. The term "limbs" refers to his physical ability to move and act.[246] The limbs, as a synecdoche, however, represent the entire body of the victim, unable to act and entirely under the control of the witch:

> *ubbir[u m]inâtīya*
> *ukassû mešrêtīya ukanninu manānīya*
> [. . .]
> *minâtīki ubber mešrêtīki ukassi*
> *manānīki ukannin*

---

bind anything that causes evil" (*Udug-ḫul* 15: 216, Geller 2016: 487), as well as the gods (see Livingstone 1989: 74, No. 32 r. 17).

245 For other sources see *rakāsu* CAD R 95–96, mng. 2c2′; *riksu* CAD R 349, mng. 2c; *kasû* CAD K 252–253, mng. 3, 5c.

246 For a discussion on other terms for body and flesh see Steinert 2012a: 231–256.

> (Whoever you are, O witch,) the one who has tied my body,
> tied my limbs, twined the tendons.
> [. . .]
> I tie your body, I tie your limbs,
> I twist your tendons.
> [*Maqlû* VII 60–61, 67–68, Abusch 2016: 130]

In this text, taken from the *Maqlû* series, the limbs (*mešrêtu*) as well as the members (*miniātu*) are both associated with the action of binding and expressed by the verbs *kasû* and *ubburu*.[247]

The association of the two terms can also be found in a *nīš libbi* incantation in which the rising of the limbs, the body, and knees[248] of the patient is supplicated:

> [. . .] jump, my wild bull! Together with you, may the strength rise!
> Together with you, may your tired knees rise!
> [Together with you, may] your limbs [rise]! Together with yo[u, may your] members r[ise]!
> [Together] with you, may [your] . . . rise!
> [No. F.4: 83–86]

These examples clearly show that the practice of binding, and more precisely, of tying the limbs, is conceptualized as a magical practice to take away the victim's capacity of movement (be it the man attacked by the witch, or the latter attacked by the patient).[249] The metaphor of the limbs tied by a witch allows the description and communication of the patient's suffering. The binding which ties the patient's body, or parts of it, is an example of the "objectification" of pain. The binding is described as an external object (a rope, for example), a material manifestation of evil, that immo-

247 The wizard and/or the sorceress attack their victim also through knots (*kiṣru*, *kaṣāru*) (see Schwemer 2007a: 86). Like bondage, knots are a magical practice employed by malefic operators, as well as therapists (see Thomsen 1987: 47–49; Schwemer 2007a: 72). In *nīš libbi* incantations the image of the knot as found in witchcraft does not appear, on the contrary, it is mentioned in the requirements for therapeutic purposes. It should be emphasized that Schwemer translates both *kiṣru* (CAD K 436, mng.1; AHw. I 488–489) and *riksu* (CAD R 347–356; AHw. II 984–985; Schwemer 2007a: 86, No. 88) with 'Knoten' (knot) (see Schwemer 2007a: 80), whereas I prefer to keep a semantic difference, translating the first 'knot,' the second 'binding,' since, although the images are interconnected, they represent two different aspects of witchcraft's action.

The word *kasītu* indicates the condition of binding and being tied up: CAD K: 243–244: 'binding magic'; AHw. I 453–454 'Gebundenheit' (see Farber 1977a: 239, Hauptritual B: 13: "May my condition of tied one be untied!"; ibid. 62, l. 83; *Šurpu* IV: 70, Reiner 1958: 27; BMS 30: 11, Ebeling 1953a: 120).

248 The knees are also subject to the binding (i.e., *Ludlul* IV, Oshima 2014: 3′). In *Maqlû*, the malefic operator is defined as "knotter of the knees of the (personal) goddess" (III 50, Abusch 2016: 72).

249 It should be noted that "binding" is also a therapeutic practice of the therapeutic operator. The latter and the witch often use the same magical performances (see Abusch 2002: 7, 65–66, 93). To give an example, in the *Udug-ḫul* series (Geller 2015), the exorcist ties up the limbs (*mešrêti rakāsu*) of the patient of the ritual for therapeutic purposes (XV 133, 144, 161, 166, 173, 200; XVI 82, see also VII 109).

bilizes the victim (see Sallaberger 2007: 297). Such metaphorical construction is well known in Anthropology, as Scarry writes (1985: 172), describing the pain by using the image of a tool, here probably a rope, is equivalent to manifesting it as an object that, despite being initially thought as acting upon the body, because of its distance from the body, becomes an image which can be removed, carrying away some of the attributes of the pain. The bondage can thus be loosened, and the suffering taken away.

Tying up the limbs, as well as lengthening muscles in the following section, indicates not only the physical immobility but also a psychological perception of, using Ernesto De Martino's terminology, *being-acted-upon by* an external force (i.e., the witch), that deprives the patient of bodily control.[250] De Martino has studied the magical traditions in Basilicata, southern Italy, focusing in particular on the technique of "binding." He has detailed existential aspects of the suffering which affects the victim of this kind of magical act. "Binding" indicates a psychic condition of impediment or inhibition, a sense of domination, a being acted upon by a strong and mysterious force, that totally removes a person's autonomy as well as his capacity for decision-making and choice. Binding often also features headache, sleepiness, weakness, slackened muscles, and hypochondria, but its characteristic feature is the experience of an indomitable and ominous force (see De Martino 1959, Engl. tr. 2015: 15). Other magical practices, as well as various kinds of ailments that affect the patient due to witchcraft, should be investigated within this "binding ideology," as we shall see in the following section.[251] As De Martino (1959, Engl. tr. 2015: 27) argues, all the forms of magic are psychologically connected to the experience of domination by an obscure force lying at the basis of binding. Any manifestation of negativity bears the risk of even more serious negativity: the loss of individual presence.

Lengthening and slackening of tendons

Associated with the image of binding, is the relaxation of tendons in the *nīš libbi* texts,[252] No. A.2: 43: "Who has slackened your tendons?". The verb *ramû* 'to slacken' (see CAD R 128–133; AHw. II 953–954) refers especially to parts of the body such as the neck, arms, tendons, and has the sense of "becoming weak" and unable to perform their functions. In D-stem (CAD R 128, mng. 2) it indicates, as in the above case, the physical consequences of a witchcraft attack, which affects muscles and tendons, often compared to ropes that are loosened.

---

250 For the practice of the binding in the Greek magic see Faraone 1991.

251 For sources about the image of an external force, such as a sickness, which affects humans by means of bindings see *riksu* CAD R 349, mng. 2c. For example, the demon Alû binds (*kasû*) hands and feet of the victim (see *Udug-ḫul* VIII 4, Geller 2015: 288), as in general the demons "have bound (*iksû*) and killed the young man, caused destruction" (*Udug-ḫul* XV: 34, Geller 2015: 444).

252 CAD Š/II 308–312; AHw. III 1216; Kinnier Wilson 1962a: 60–61; Oppenheim 1962: 27–33; Heeßel 2000: 166.

Not only witches loosen tendons, but all kinds of demonic entities[253] can do so as well. An example is Lamaštu whose action is described as follows in an Old Assyrian text: "The tendons / of a lion you slacken (*turammī*) / the tendons of the youth and infant you . . ." (OA_2: 20–23, Farber 2014: 280).[254] See also a curing ritual from Ḫattuša, where it states that Lamaštu, Labāṣu, and Aḫḫazu "loosen the tendons" (KUB 29 58–59 iv 10, and dupl., Meier 1939: 206). Similarly, in *Ludlul bēl nēmeqi* the *šūlu*-spirit, the *utukku*-ghost and Lamaštu: "They have broken the muscles of (my) neck, they slackened the neck (*urammû kišādu*)" (*Ludlul bēl nēmeqi* II 61, Oshima 2014: 88). As becomes clear from these cases, it is mostly demonic entities that attack the victim by weakening their tendons. The fact that this kind of evil action is also described for the witch, leads us to the considerations already brought forth by Abusch referring to the *Maqlû* series (2002: 19–25), on the process of demonization of the witch. In this view, the witch operates as a demon and is described as such.

While in No. A.2: 43 it is the tendons that are loosened and in No. A.4: 76 the loincloths (*kannu*, "(Who) has slackened my loincloths like taut strings?"), in No. A.3: 49–50, No. A.4: 80; No. D.2: 12 it is the patient in his entirety who experiences the state of weakness and inability to move and act. This is expressed by the possessive pronoun in the second or first person (*-ka* or *-annī*). The tendons serve as a synecdoche for the entire patient's body.

The image of the slackening of the patient's tendons in the *nīš libbi* corpus is also similar: loose as taut strings (*šaddu*) or loosened ones (*ramû*). The image here is one of immobility. Unlike the previous example it is actually about threads that are not tightened, but loose or dangling, that is, incapable of performing any kind of movement or twist: No. A.3: 49–50: "Who has slackened (your body) like slackened / strings?"; No. A.4: 8: "[She] has slackened me like ta[ut] strings!"; No. D.2: 12: "Who has slackened you like [slackened] strings?"[255]

In addition to the verb *ramû* (spec. D-stem), the other verb used to describe this state of absence of agency is the verb *šadādu* 'to drag' (CAD Š/I 20–32, in part. mng 1; AHw. III 1121–1122).[256] It is not only used to describe similes with threads (*qê šaddūti*, No. A.4: 76 and 80) but also used to indicate the action of the witch in No. A.4: 77–78: "(Who are you who), like (the string of) a merchant's leather money pouch, all my sinews / has pulled tight (*ildudam*) and firmly bound?" (also ll. 81–82). As mentioned in the text, the witch, as well as stretching the tendons, binds them together: "They

253 On demons in Mesopotamia see Verderame 2011; Verderame 2013a; Verderame 2013b; Verderame 2017b; Konstantopoulos 2017.

254 For possible readings of the signs and other interpretations see Faber 2014: 215.

255 See also No. L.2: 20: "[The . . . witche]s have taken away [my] sexual desire like slack strings!"; and l. 22: "[Like sl]ackened [strings] they have taken away my sexual desire!"

256 In *Diagnostic Handbook* SA.GIG/*Sakikkû* XV 37′ (Heeßel 2000: 152): *šer'ānu libbīšu šuddudū* "Seine Bauchstränge verkrampft? sind." Heeßel translates the term with 'Strang,' following Maul (1996: 34) and Ritter (1965: 300).

(= wizard and sorceress) tied my tendons" (Abusch and Schwemer 2011: 299, No. 8.4: 70). In the *nīš libbi* text No. A.4: 77–78 and 81–82 stretched tendons and firmly bound ones are compared through a simile to the merchant's leather money pouch. The comparison is possible, not only through the image of the hanging bag, but also by means of the phonetic assonance between the bag (*kīsu*) and the verb 'to bind' (*kasû*). It is no coincidence that the term *kīsu*, following the dictionaries (CAD K A 430–432, mng. 1 'leather bag'; B 432–433, mng. 1 'bond'; AHw. II 487, II 'Geldbeutel'; III 'Bindung'), has two possible meanings: merchant's leather money pouch; binding. The metaphoric image takes place on two levels: the first associates the image of the stretching and binding of the tendons with the merchant's bag; while the second is a semantic and phonetic wordplay, between the binding (*kīsu*), to bind (*kasû*: *rakāsum-ma irkus*, ll. 78 and 82), and the bag (*kīsu*).

The binding of the tendons is used as an image for the whole patient. The tendons are the subject of the witch's action since her attack condemns the man to a state of weakness and immobility, which is expressed by the image of the binding. In other words, the image of stretching and loosening of tendons perfectly fits into the ideology according to which the action of the witch is expressed by being tied up, unable to act according to one's own will and being controlled by a powerful dark and evil external force. In fact, when the witch, or any other entities, demons or sicknesses, attack the human tendons or muscles this results in a state of fatigue and immobility for the whole body.[257]

Another aspect, which relates to the word *šer'ānu* (tendon), should be highlighted. The term can be translated as 'rope': the tendons are understood as the ropes of the body. The translation of the term with 'rope' involves two considerations:

1. The association between the tendon and the 'binding' (*riksu*);
2. Between the tendon and the bowstring and the harp string.

In the first case, I quote a comment made to tablet XIII of the *Diagnostic Handbook* SA.GIG/*Sakikkû*: *rík-su-šú ir-mu-ú* : *ri-ik-su-šú* : *šér-a-nu-šú* "His bindings have become loose: his bindings are his sinews" (Labat 1951: 110–129). The magic is based upon analogic principles, that is, the external entity that affects human activity through the binding is related to an analogous object namely the tendon. The importance of semantic connections by means of similarity of meaning has already been highlighted in the case of the binding and the merchant's leather money pouch (*kīsu*). Similarly, the term *šer'ānu* can also refer to the bowstring. As we shall see in the next chapter, the bow in the *nīš libbi* texts is one of the symbols of sexual desire, potency, and manhood. In some *nīš libbi* rituals a bow is made whose string is made of the *arrabu*-mouse's tendon or from a gazelle's left hock:

257 See also BAM 122 r. 2–10; STT 136 i 10, von Soden 1974: 341. For other sources CAD Š/II 309–312.

> You make a bow of thorn,
> a tendon of the *arrabu*-mouse is [its] string (*matnu*),[258] you load it [with an arrow],
> you pu[t it] over heads of the man and the woman, who are lying [. . .].
> [E Bow ritual: 57–59]

> The bow o[f . . .]
> [. . .] a tendon of the left hock of a gazelle is [his] string [. . .]
> [. . .] . . . a reed arrow . . . [. . .].
> [R Bow ritual: 8′–10′]

The association between the bowstring, loosening the tendons and the absence of sexual vigor is explicit in incantation No. J.2: 10 "May [the qu]iver not be[come empt] y! May the bow not slacken!". It is evident therefore that the slackening of the tendons clearly expresses the psycho-physical state experienced by the victim of the witch's attack, it allows a clear association, by means of the word 'tendon' to be made with the bow, and therefore metaphorically with sexual vigor. This idea is also supported by a passage from the *Treaty of Succession of Esarhaddon* in which the bowstring (*matnu*) metaphorically refers to the penis:

> May the *nipples*[259] of your young women,
> the *penises* (*matnu*) of your young men under your eyes the dogs and pigs
> drag here and there in the squares of Aššur.
> [*Treaty of Succession of Esarhaddon* ll. 481–483, Parpola and Watanabe 1988: 49]

Similarly, in incantation No. A.1: 36–37, we find a similar metaphor between the penis and the string *šer'ānu* of the harp (*sammû*),[260] the penis must be like a harp string that does not hang down: "May my penis be a harp string, / so that it will not dangle out of her!".

In conclusion, the loosening and stretching of the tendons neatly fit into the "binding ideology," which in turn indicates the condition experienced by the victim due to the witch's attack. The binding, as well as the loosening (the tendons serve as a synecdoche), refers to the entire body and, therefore, the physical integrity of the man, and causes a reduced ability of movement and action. The image of an external and fearsome force subtracts the power and energies from a man and remove his agency, leading the individual to a state of mental and physical exhaustion and weakness. Tiredness is, in fact, one of the most characteristic symptoms not only of the diagnostic category "lack of sexual desire," but of witch attacks in general.

---

258 This term can indicate both tendon and bowstring (CAD M 412; AHw. II 633).

259 Here *si-si*: *sissu* means 'lump' (CAD S/II 328; AHw. II 1051 'Geschwür'). Parpola and Watanabe (1988: 49 fn. 481) interpret it as a variant of *zi-zi*, that is *zīzu* (CAD Z 149 'teat'; AHw. III 1534 'Zitze').

260 The mention of the harp in love contexts is from an Old Babylonian love incantation: "Look at me and rejoice like a harp!" (Wasserman 2016: 158, l. 24).

Here another aspect should be mentioned. The term tendon allows for two possible analog connections:

- Tendon and binding;
- Tendon and bowstring: bow and sexual vigor;
- Tendon and harp string.

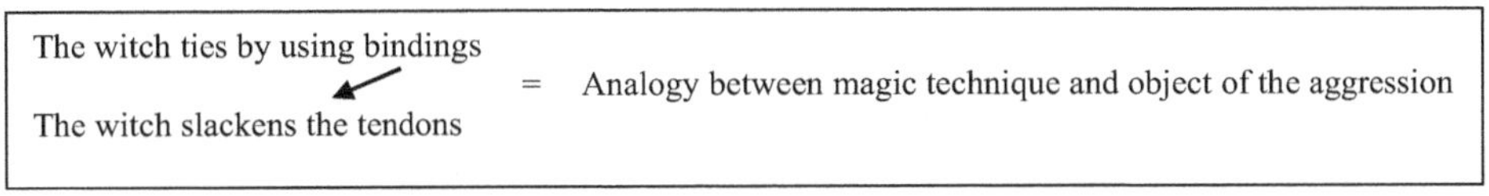

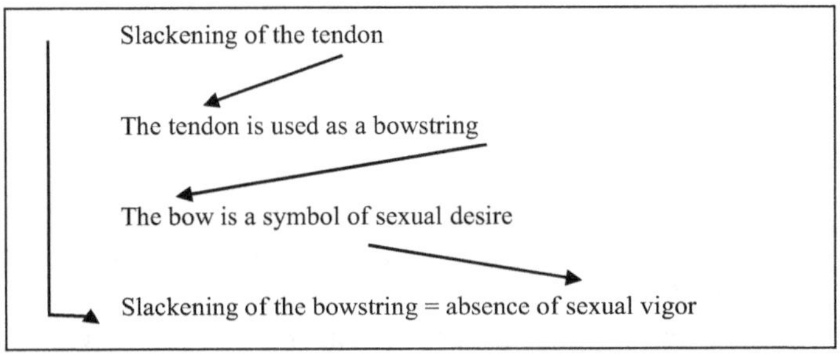

Blocking the street and the canal

Another analogy that falls within the group of incantations to express the action of the witch uses the imagery of a blocked road. It appears in three incantations of this group: No. A.3: 50: "Who has block[ed] your course like (on) a ro[ad]?"; No. A.4: 75: "Who are you who has blocked up my course like on a road";[261] No. D.2: 13: "Who has blocked your ways like (those of) a traveler."

In *Maqlû* series, the witch is described as acting on the roads, blocking trade routes, and at the same time threatening the young men and women, taking away respectively their manhood (*dūtu*) and vitality (*lamassu*), charm (*kuzbu*) and attraction (*inbu*).

izzaz ina sūqimma usaḫḫar šēpī
ina rebīti iptaras alaktu
ša eṭli damqi dūssu īkim
ša ardati damiqti inibša itbal
ina nekelmêša kuzubša ilqe
eṭla ippalisma lamassašu īki[*m*]
ardata ippalisma inibša itbal

261 See also the incipit of the catalogue LKA 94 ii 18 (with *mannu* instead of *attamannu*) (see also ibid. ii 17).

īmurannima kaššāptu illika arkīya
ina imtīša iptaras alaktu
ina ruḫêša išdīḫī iprus

Standing, in the street she (= witch) turns foot around,
in the squares she cuts off commercial traffic.
She is who robbed the fine young man of his manhood,
she is who carried off the attractiveness of the fine young woman,
with her malignant gaze she took away her charms.
She looked at the young man and robbed his vitality,
she looked at the young woman and carried off her attractiveness.
The witch has seen me and has come after me,
with her venom, she blocked commercial traffic,
with her spittle, she has cut off my trading.
[*Maqlû* III 6–15, Abusch 2016: 82–83]

As in the similes mentioned above, we find a standard description of the actions of the witch, and of malignant forces in general, such as Lamaštu, who attack human integrity: *illik ḫarrānu alaktaša iprus* "She has traveled the road and blocked the way" (*Lamaštu* I 182, Farber 2014: 89).

The blocked road is an image of hardships that affect humans and that prevents them from acting. For example, in *The Prophecy of Marduk* we find that, before the return of the god to Babylon, "lions were blocking the journey" (ii 9, Borger 1971: 8).

The blocked or obstructed road symbolizes malaise, unlike a straight and open and therefore accessible path, which represents a state of well-being and health. Often the gods are invoked to "open the way" of the suffering:

The way (of life) is open for you

(God) make straight his way, open his way, may the supplication of your servant sink into your heart.
[*A man and his God*: 54, 68, Lambert 1987b: 192, 194]

The witch attack is not always described as a blocked road, but also as a blocked irrigation canal, as in the incantation No. D.2: 11: "Who has blocked you like an opening of the *dilûtu*-water system." The latter metaphor has no parallel in the other texts in this corpus.[262] A similar passage can be found in the title of an incantation catalogue LKA 94 5: "Incantation: 'Why are you blocked like a canal?'". The association with

262 However, in medical textbooks the use of metaphors and similes with channels is not uncommon. In particular, they are used to describe the physiological processes of the female body as well as states of discomfort of a gynecological nature (see Michel 2004; Steinert 2013: 1–23; Barjamovic 2015: 74–76; Couto-Ferreira 2017; Steinert 2017b). Actually, these metaphors are not only used in the description of pregnancy and childbirth, but also to express the states of the female urogenital tract. This type of image, however, is not used for the description of male patients. See also in the same text No. D.2: 16–17: "May they (= Daughters of Anu) gran[t] the wellbeing of the waterskin (*nādu*) / to your waterskin."

the road, to describe the action of the witch, suggests an image of cancellation on behalf of the agency of the patient: the *dilûtu*-water-system is blocked and therefore has stopped working.

Second group: erotic similes with animals

The second group of incantations (No. B.1; No. C.1; No. D.4; No. E.1; No. E.3; No. F.3; No. F.4; No. F.5; No. G.1; No. K.7; No. K.8) represents a varied group as to the structure and images they evoke, which are characterized by the use of several "themes":

- Invocation of animal arousal (often through metonymy);
- Reference to a bed to which the animals are attached;
- Ascending climax, relative to the number of matings of each animal species;
- Mention of the gods to increase the power and effectiveness of the incantation.

Invocation of animal arousal

The images evoked in this "theme" are erotic, urging the patient to be sexually aroused. The invitation for arousal is described through similes (*kīma*) to wild animals, especially the stag (*ayyalu*) and the bull (*rīmu*), No. B.1: 1–2: "Roar on me (*ugga*)! Roar on me! Rear up (*tibâ*)! [Rear up]! / Roar on me like a stag! Rear up lik[e a wild bull]!" (see also No. F.3: 65–66). In another incantation the patient is identified precisely with these two animals,[263] No. G.1: 2–3: "[St]ag! Stag! Wild bull! Wild bull! / [Roa]r, stag! Rear up, wild bull!" (see also ibid. l. 7).[264]

The imperative used to express the arousal of the animals are *ugga* and *tibâ*. The latter, as we have seen in Chapter I, is from the verb *tebû* 'to rise.' From the same verb, the precative form is also used (*litbâ*). The animals are urged to rear up: the lion (*nēšu*), the serpent (*ṣerru*) and probably the wolf (*barbaru*):

> Together with you, may a lio[n] rear up!
> Together with you, may a w[olf][265] rear up!
> Together with you, may a snak[e][266] rear up!
> Together with you, may a lio[n] rear up!
> [No. B.1: 3–5]

263 Other metonymies with the lion may also be present in No. K.7: 116–117: "Lion! Bull . . . [. . .] / Yo[ur] lo[ve]-making . . . [. . .]".

264 See the catalogue LKA 94 ii 14: *rīmī* [*t*]*i*[*bâ*] *lulīmu tibâ* "My wild bull, [rea]r [up]! Stag, rear up!". See also No. F.2: 61: "Rear up! Rear up! Mount! [Mount]!"

265 See the CAD for the occurrences of *barbaru* 'wolf' after *nēšu* 'lion.' See also No. I.3: 34: "[. . .] . . . the mating of the wolf, make love to me!"

266 The snake is not mentioned elsewhere in the texts.

The other imperative employed is *ugga* from the verb *nagāgu* 'to bray' (CAD N/I 105–106; AHw. II 709).[267] The verb refers to the roaring and braying of animals (especially the stag). As we have seen, incantations employ figures of speech, similes and metonymies, borrowed from the animal world, especially the deer. In addition, during the mating period, the stag produces a hoarse and deep roaring to indicate to other males which one is the strongest and most able to mate with females.

Animals tied to the bed

The bed, as a place of love relationships,[268] is evoked in this "theme." At the bed different animals are tied, depending on the incantation to be performed. The first case, No. E.3: 46–47, mentions the ram (*puḫālu*)[269] and the weaned sheep (*immeru parsu*),[270] both attached to the bed by the ritual performer: "[At the hea]d of my bed a ram is tied, / [at the foot of my bed] a weaned sheep is tied. Around my waist their wool is tied."

To understand the incantation, we must analyze the ritual preceding this one, No. E.2. This text consists of a ritual in which pairs of figurines are made of different materials, representing the witch and the wizard, and then burned before Ištar of the stars to whom an altar made of reed has been erected, a sheep is sacrificed, and libations of beer are made. Then the pronunciation (*qabû*) of an incantation with an invocation of the goddess follows, performed by the patient. After the incantation, an amulet is made and tied around the patient's neck with the ingredients mentioned in the incantation. Subsequently, a ram is attached to the head of the bed, as well as a weaned sheep to the foot of the bed. From the front of the animal wool is plucked and spun separately. The threads then are tied around the patient's waist:

> Its ritual: You string silver, gold, lapis lazuli, alabaster / *tarmuš*-plant, "heals-a-thousand"-plant, "heals-twenty"-plant, *ardadillu*-plant / *usikillu*-plant, *nīnû*-plant, *bukānu*-wood (var.: "wood-of-release") on (a cord) of flax,
> (and) put (it) around his neck. A ram at the head of his bed,
> (and) [a wea]ned [sheep] at the foot of his bed you tie.
> From the forehead of the ram and the forehead of the weaned sheep
> you pull out wool and twine (two) separate threads.
> [No. E.2: 38–44]

---

267 See also KAR 69: r.6, Biggs 1967: 76–77.

268 On the sexual and political symbolism of the bed see Porter 2002: 523–535. For an analysis of the iconographic representations of the bed in clay sculptures see Assante 2002: 36–42.

269 The term indicates the ram or the male and uncastrated adult sheep (see CAD P 479–481; AHw. II 875 'Zuchtwidder'; Landsberger 1935–1936: 154–155; van Driel 1993: 231–233). The term also refers to a bull, a wild bull, a stallion, a male duck.

270 According to Biggs (1967: 31), it is «a young male sheep, mature enough to have produced wool». No confirmation of the animal's sex. I believe it is, for symbolic reasons, a male goat, younger than the ram. According to Hirsch (1973–1974: 65 fn. 13), it is a female.

The recitation of an incantation, maybe No. E.3, follows. The incantation No. E.3 is recited by the three actors: the male patient who mentions his bed and the animals tied (ll. 46–47); the female partner of the patient who demands to have sex repeatedly (ll. 48–51); the therapeutic operator who states the divine origin of the incantation (ll. 52–53). It is also possible, however, given the parallels in No. E.1 and No. F.4, that the woman recites the lines 46–47. The section of No. E.3 (ll. 48–49) recited by the woman provides the mention of the animals' sexual force through a climax converning the number of times a specific animal is capable of mating. Here the ram with its 11 times and the weaned sheep with it 12 times are mentioned,[271] that is, the animals that occur in ritual No. E.2 and mentioned in the first two lines of the same incantation No. E.3.

In this incantation, and its ritual, the principles of contagious magic, are used according to the following scheme:

1. The animals symbolize sexual vigor. They are tied to the patient's bed, the emblematic place of sexual intercourse;
2. The extraction of the wool and the threads made from it are tied around the patient's waist. The woolen threads are loaded with sexual vigor from the animals, since the ram and sheep, considered sexually active, were previously tied to the bed. The patient takes possession of their sexual potency, through the threads of wool. He re-acquires his sexual desire and the ability to have sexual intercourse with the woman. Similarly in ritual N prescr. 4: 20–22 hairs of the tail and the perineum of the sheep are tied around the patient's waist.

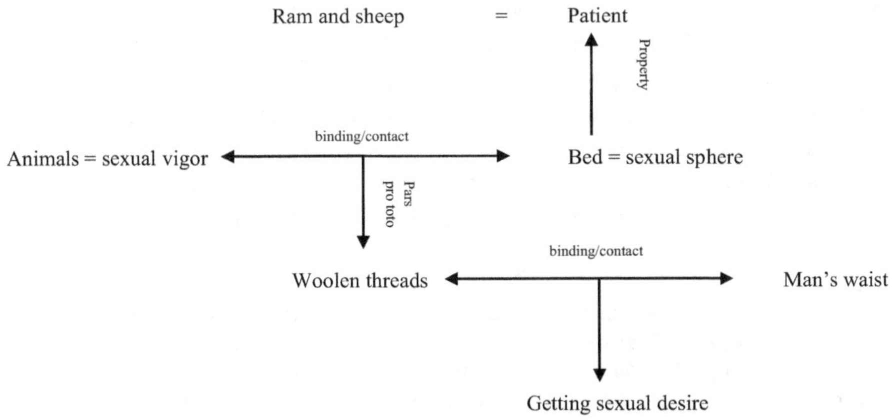

In other incantations, No. E.1 and No. F.4, we find the image of the bed connected to animals as well, this time a ram and a buck. In this case, it seems that only the

271 It is therefore possible that the mention here confirms the male sex of the sheep.

woman recites the incantation.[272] Unlike the previous case, here the image evoked in the incantation is not explicitly related to a specific ritual practice including the two animals being tied to the bed, as can be seen by the fact that the rituals that follow do not mention this practice. Thus, it is a poetic image on the model of a ritual practice, which identifies the patient with animals, based on two principles of magical practice: the redundance of ritual and the power of words.

> At the head of my bed a buck is really tied (var.: I have tied)!
> At the feet of my bed a ram is really tied (var.: I have tied)!
> The one at the head of my bed, rear up, make love to me!
> The one at the feet of my bed, rear up, bleat *for* me!
> [No. E.1: 5–8]

> [Incantation: At my head a bu]ck is tied! At my feet
> [a ram is tie]d! Buck, bleat *for* me!
> [Ram], mount me!
> [No. F.4: 80–82][273]

The one who recites the incantation in the first person is the woman who turns to the patient. The latter is identified with the animals tied to the bed: the ram (*puḫalu*) and the buck (*daššu*).[274] In No. F.4: 83 the patient is also called "my bull." The bed is qualified as "mine" in No. E.1, while No. F.4 does not include a specific reference to the woman's bed, but it is stated: "At my head, at my feet." It is possible then, that the woman is just the companion of man, who lacks sexual desire. The verbs used in imperative refer to sexual practice.

In these incantations, we face poetic images that allow for the identification of the debilitated man with animals. The patient is like an animal, ram or buck, and as such he must act, he must aspire to regain their vigor. It should be noted that while in the metonymies wild animals are used (see Chapter II § "Animal metonymies"; "Invocation of animal arousal"), here the image of the bed also includes domestic animals since, as we have seen, that image is based on a ritual practice[275] which uses domestic animals.[276] Of course, these animals are particularly sexually active, but as confirmed

272 The Mesopotamian clay sculptures often depict a bed with a woman, or a vagina (*pars pro toto*) (see Cholidis 1992 pls. 32–37). For a discussion on the topic see Assante 2002: 40–41.
273 See also the catalogue LKA 94 reverse iii 3: "Incantation: 'At my head a b[uck is tied].'"
274 The term *daššu*/MÁŠ.NITA indicates the adult uncastrated male of the goat or gazelle (see Landsberger 1935–1936: 159; D CAD: 120 'buck (said of gazelles and goats)'; AHw. I 165: 'Bock'). Given the parallelism with the ram (*puḫālu*) in this incantation and reference to ritual practice, it must be the domesticated species here. See also incantation No. E.1: 3–4 where it is placed in relation to the donkey (*imēru*). See the list *Malku* V 37: *da-aš-šu* = *gi-iz-zu* (male goat).
275 Note the use of *lû* 'real' in No. E.1: 5–6 to indicate the image of an active ritual practice.
276 It is therefore noteworthy that in prescriptions and ritual practices, substances taken from domestic animals are used, such as the breeding bull, the ram, the donkey, the dog, and the pig. Also note the pig for etiologic purposes, or the use of saliva of a reared-up ram or buck in the prescriptions of the corpus.

by the lines 48–49 of incantation No. E.3, they are by no means comparable to that of wild beasts such as the deer and the bull.

Bestiality or figure of speech?

Whether or not these incantations provide evidence of bestiality, i.e. sexual relations between humans and animals, in Mesopotamia has been a subject of discussion. There are very few testimonies of zoophilic practices in Mesopotamia.[277] One example is a *namburbû*-ritual performed to prevent future disasters to humans (and their houses), who have had sexual intercourse with a goat: "If a man approaches[278] a goat – so that misfortune may not come to a man (and his house)" (see Maul 1994: 415–420, VIII. 17; Scurlock 2002: 379). A piece of his fur is plucked, and on the roof, before Šamaš, a female goat is tied. Red goat hair is taken and placed before the female goat. A libation of beer is made. The hair is tied up in a linen cloth. With the tissue any future calamity is blocked (ll. 2–8). After placing it on the ground, the man recites his prayer to Šamaš. The man turns to the god, so that he can prevent the threat of future calamities (ll. 9–21). Finally, the linen fabric, containing the hair of the goat, is placed on the door of the tavern, a place considered to be impure[279] (ll. 22–27).

Of an entirely different nature are the reared-up animals evoked in the "theme of the bed" from the *nīš libbi* corpus, with which the patient (or the couple) is identified magically, or stylistically. Consequently, it is not the animals themselves who are invited to copulate with the woman uttering the incantation, as claimed by Leick (1994: 206). We have already underlined the metaphorical importance of using animal imagery. So, in other words, the ram, buck, bull, and other animals mentioned in this "theme" are metaphors with which the one who recites the incantation turns to the patient. The verbs used to refer to the arousal and sexual activity are in No. F.4 *ḫabābu* 'to bleat,' *rakābu* 'to mount,' and *dakāku* 'to jump';[280] in No. E.1, as well as *ḫabābu*,

277 According to Biggs (1967: 34), the excitability of goats in the presence of women is well known and we should understand here an imagined act of bestiality witnessed by the man to excite his lust. He argues that bestiality was certainly practiced in Mesopotamia and there is no evidence of a taboo against it in Mesopotamia. According to some other scholars, there are no traces of bestiality (see Hoffner 1973: 82; Budin 2015: 5), while to others, it was practiced, but without being ethically condemned (see Meissner 1925: 437; Maul 1994: 415; Scurlock 2011; Boer 2015: 67–69). See also Hirsch 1973–1974: 63 fn. 8. A tablet of the *Šumma ālu* series preserved in Berlin, contains omens related to sexual relations between humans and animals (pers. com. A. Guinan).

278 Maul (1994: 419 fn. 23) sees a sexual connotation here, translating it as "sich heranmachen."

279 The tavern is used as a place to leave these objects on which the evil has been relocated because it is a busy and heavily transited area, so anyone, possibly an impure one, can take it away (see Verderame 2017d: 85).

280 Labat (1968: 356) suggests a more erotic meaning: 'tréedigner de désir.' I think that the translation 'jump' is more feasible because it captures the concrete natural sexual action of the animal, given the image of intercourse.

râmu 'to love' and *tebû* 'to rear up.' The verbs are all in their imperative forms and are about arousal and mating for both animals (in particular *rakābu*) and humans.

An interesting case is the imperative *ḫubbibanni* from the verb *ḫabābu*, which Biggs (1967: 31) translates as "caress me," while Schwemer (2010: 121) "liebkose mich." The verb is used twice in an Old Babylonian love poem, CUSAS 10, 9 where it designates the warm thrumming of the loving heart and the murmuring of the gossiping woman:

> At daytime (when) he went away from me –
> gossiping concerning me (*ḫābibī*) was not cut off from their (the other women's) mouth.
> Those who slander me
> are not gloomy, day and night.
> Let your heart be thrumming (*liḫbub*), may it not accept the lies!
> [CUSAS 10, 9: 28–32, Wasserman 2016: 94]

The CAD distinguishes two verbs (H: 2–3, A-B): A. 'to murmur, hum, low, chirp'; B. 'to caress.' For the verb B CAD mentions a few passages, where the verb appears in an idiomatic form "*ḫabābu eli* PN," from the *Epic of Gilgameš*: "I have loved it (= ax) and caressed (*aḫabbub elšu*) as a wife" (OB II, George 2003: 174, 34, see also ibid. 555, I 284); "His sexual attraction will caress you (*iḫabbubu eli ṣērīki*)" (ibid. 548, I 186, 193). Similarly, in the oracle of the goddess Anunîtum to Zimri-Lim: *anāku elīka aḫabbab* "I will make love to you" (Dossin 1978: 32, ARM X 8, 10–11). In D-stem, the CAD quotes a passage from *Šumma ālu*, referring to the snake: "If a snake caresses (*uḫabbab*) a man" (KAR 286: 8).

More common, however, is the meaning 'to murmur,[281] hum, low, chirp,' referring to birds[282] or livestock (such as cows) (see CAD Ḫ 2, A; AHw. I 301 'murmeln, zirpen, zwitschern'). George translates the passages of *Epic of Gilgameš* mentioned above also with "caress and embrace" but he sees no reason to separate, as the CAD does, the two verbs: «Movement, as well as sound, is characteristic of lovemaking [. . .] *ḫabābu* can accompany sexual intercourse» (2003: 797). Further, it should be noted that the equation in *Malku*: *ḫa-ba-bu* = *na-šá-qu* "to kiss" (III 8). According to Groneberg (1986: 189), the verb has a figurative meaning of 'einen Beischlaf ausüben' (see also Jacobsen 1930: 70; Labat 1968: 356–357; Cooper 1977: 43 fn. 22).[283]

In my opinion, this passage should be seen in comparison with No. B.1: 1–2 and No. G.1: 2–3 where we find the imperative form of the verb *nagāgu* 'to roar.' In the latter case, the man is identified with the deer. In the text in question, as we have shown, the patient is identified with the buck. One of the features of aroused goats is very strong bleating. The incantation invites the man, identified with the buck, to get aroused, that is, to bleat. For this reason, I prefer to translate it as 'bleating,' rather than

281 On murmuring in Mesopotamia see Grayson 2000.
282 On the reference to the chirping of birds, and metaphorically the blossoming of love: "Love come about, twittering over the people/ may love twitter over me!" (CUSAS 10, 11: 5–6, Wasserman 2016: 237–238).
283 Note that in Arabic the *hbb* roots mean 'to love.'

'caressing.' The translation 'to bleat' must be interpreted within a sexual setting: the call of the animal before and during intercourse is a sign of its arousal.[284] Actually, the verb mostly has animals as its subject.[285] In only one case not an animal is the subject, rather the *dādu* 'sexual attraction, love, darling, lovemaking' (see George 2003: 797) of Enkidu, lying on the prostitute Šamḫat's body, at the moment before the sexual act: *dādūšu iḫabbubū eli ṣērīki* "His 'sexual attraction' will caress you" (I 186, 193). Here, according to George (2003: 797), it refers «both to general dalliance (the whispering of sweet nothings) and to the physical entwining of a reclining couple that is the prelude to coitus». It should be noted that Enkidu, before sex, falls within the sphere of wilderness and belongs to the animal domain. In this passage of the incantation, it is possible to see a reference to the figure of Enkidu[286] and in particular to his animal nature (see Hirsch 1973–1974: 67). In conclusion, I primarily consider the sonorous meaning of the verb, in fact, as Rendu Loisel (2016: 108) affirms the meaning of "caressing," often attributed to *habābu*, with a sexual connotation, is only an extension of the original meaning which remains sonorous.

Ascending climax relative to the mating of the animal species

Another "theme" that characterizes the second group of incantations contains an ascending climax relative to the number of times that an animal species can mate. In fact, several animals are mentioned according to a ranking scheme. Two different types of climax can be traced in this corpus. The first concerns only wild beasts:

> The ma[ting of a wild go]at six times,
> the mating of a stag seven times,
> the mating of a partridge twelve times.
> [No. C.1: 4–6, see also No. F.3: 69–70; No. K.8: 149]

---

284 According to Verderame (2017a: 403–404) the verb *ḫabābu* could indicate a repeated and rapid movement indicating both movement of the body and the sound resulting (e.g., murmuring by moving the lips or hissing of the snake's tongue or the insect's wings). In this sense, in parallel with "mount me" it could indicate a repeated movement and rubbing during the sexual act. The verb should not refer to the bleating, but to the sounds produced by the continuous bodily movement during the animal sexual intercourse.

285 Also to indicate the sound of water (CAD Ḫ 2, A mng. 1). I am not referring to the expression "*ḫabābu eli* PN."

286 It is no coincidence that to indicate his intercourse with the prostitute, instead of the verb *rakābu*, *reḫû* 'to inseminate' is used (CAD R 252, mng. 2; AHw. II 969; Groneberg 1986: 189), which is only used for men and mythological contexts: the enculturation of Enkidu was begun just from intercourse with the prostitute: 6 *urrī u* 7 *mušâti Enkidu tebî-ma Šamḫat irḫi* "Six days and seven nights Enkidu was erect and coupled with Šamḫat" (I 194, George 2003: 548).

The climax, therefore, mentions three kinds of wild beasts: wild goat,[287] stag, and partridge. Here it is the bird who is considered to be the animal with the greatest reproductive capabilities:

ibex (*turāḫu*)	6 times
stag (*ayyalu*)	7 times
partridge (*iṣṣūr ḫurri*)	12 times

The second climax, however, mentions both wild animals, such as the partridge, the wild bull and the stag, and domestic ones, such as the ram, the weaned sheep and the pig, No. E.3: 48–49: "[Like a ram eleven times], like a weaned ⟨sheep⟩ twelve times, like a partridge thirteen times. / [Make love to me like] a pig fourteen times, like a wild bull fifty (times), like a s[ta]g fifty (times)!". The evoked image, in comparison with the previous one, not only changes the animals (mentioning additional ones) but also varies the frequency of copulation and therefore their ranking:

Ram (*puḫālu*)	11 times
Weaned sheep (*immeru parsu*)	12 times
Partridge (*iṣṣūr ḫurri*)	13 times
Pig (*šaḫû*)	14 times
Wild bull (*rīmu*)	50 times
Stag (*ayyalu*)	50 times

The number of matings of the partridge is similar in both rankings, whereas it diverges in relation to the deer. In the latter ranking, the stag and the bull represent sexual vigor *par excellence*.[288]

The function of the mention of the massive sexual activity of animals is to renew the patient's sexual vigor: the patient must vigorously copulate like animals. At other times, such imagery is connected with the imperative "make love to me!" (*rāmanni*),[289] to have a sexual relationship with the woman who declares the incantation, No. C.1: 7–8: "Make love to me! Make love to me because I am young! [. . .] / and I am endowed? with the *rikibtu* of a stag! Make love to me!"

"I am young!"

In the incantation No. C.1: 7 we read: *rāmanni aššu ṣeḫrāk*[*u*] "Make love to me because I am young." Similarly, in No. F.3: 71–72: "Make love to me! Ma[ke love to me]

287 The presence of the ibex is not certain. It does not appear elsewhere in the *nīš libbi* texts.

288 For a discussion of the deer and the bull as symbols for sexuality see Chapter II § "Animal metonymies". Regarding the pig, see Chapter III § "The pig." See also the mention of the mating of the wolf (in broken context) in No. I.3: 34.

289 Restored in No. F.3: 70–72. See also No. E.1: 7; No. I.3: 34.

because I am young! / [. . .] I am endowed with love, make love to [me]!". So, the youth of the woman is a quality considered positive for purposes of sexual attraction and subsequent intercourse. The term used here is the verb *ṣeḫēru* 'to be small, young' (CAD S 120–124; AHw. III 1087–1088) from which the *ṣeḫru* adjective 'small, young' (CAD Ṣ 179–185) is derived,[290] which seems to indicate both a child and a teenager. Other terms related to it are: *batultu*/KI.SIKIL.TUR and *ardatu*/KI.SIKIL. The latter two terms indicate, respectively, 'the young teenager' and 'the older teenager' (see Finkelstein 1966; Landsberger 1968). Cooper (2002: 91) is not convinced; according to him, the second term is used in literary texts, while the first is used in the enumerations of booty for the crown, in legal texts and personal lists from the Neo-Babylonian period.[291] According to Stol, both words mean «a young teenage girl of marriageable age» (1995: 128), and since virginity was considered a feminine quality before marriage, the word denotes "the virgin"[292] (see Cooper 2002: 93).

In Mesopotamian literature, both in Sumerian and Akkadian, the association between sexual practice and youth is a recurring theme. Often, Inana is defined ki-sikil(-tur) in the poetic cycle of Dumuzi and Inana:[293]

> [The maiden (lú-ki-sikil), gleam]ing mane, perfect beauty (sig$_7$ sa$_6$-ga-àm),
> [Inan]na, gleaming mane, perfect beauty,
> [the maiden], mane of ibex (kun-sìg-dàra). . ., [d]eer (lu-lim) . . . deer,
> [Ina]nna, mane of ibex . . ., [d]eer . . . deer;
> [The m]aiden, variegated as heaps of grain, suitable to the king;
> Inana, variegated as heaps of grain, suitable for Dumuzi;
> the maiden, you are stacks of gú-nida-wheat, abundant in charm (ḫi-li šu-gi$_4$-a-mèn),
> Inana, you are stacks of gú-nida-wheat, abundant in charm.
> I am the queen, I am the queen . . . I am full of attraction (ḫi-li gur-ru-men),
> As for me, I am single, I am the maiden, I am full of attraction,
> I am the queen, the seed, generated by An, I am full of attraction.
> [*Dumuzi and Inana R* A ll. 1–12, Sefati 1998: 236]

We find the reference to sexual intercourse, using the plowing metaphors, in *Dumuzi and Inana P*:

---

290 See also Harris 2000: 184 fn. 80.

291 See Roth 1987: 38–39; Radner 1997: 148, 153–154. The only exception seems to be the myth *Enlil and Ninlil* in which the goddess is defined first ki-sikil-tur and then ki-sikil (see Cooper 2002: 91; Cooper 1980: 184; Leick 1994: 47).

292 On virginity in Mesopotamia see Cooper 2002. On the relation between sex and age see Leick 2015. On the relation between marriage and age see Roth 1987.

293 The term indicates a young woman (Römer 2001: 407). In the group of Dumuzi and Inanna (Sefati 1998): E1 obv. 17′; H obv. 24; I 7; M obv. 5; T 8, 35; V 7. In C 2 the goddess is defined as lú-tur "little one"; the goddess here reaches puberty, as is clear from ll. 39–40 that mention thelarche and pubarche: "Look: my breasts are growing bigger!/ the hair on my pubic area is growing!" (Sefati 1998: 135). In *The Dispute between Shepherd and Farmer*: "I, the maiden may marry the farmer!" (i 23, Sefati 1998: 326, see also ll. 2, 13, 87, 89).

I am the maiden, who will plow (for me)?
My vulva, wet ground, is rich in water
I am the mistress, who will feed the ox?
The noble mistress, may the king plow for you!
Dumuzi, the king will plow for you!
[*Dumuzi and Inana P* ii 26–30, Sefati 1998: 220–221]

Frymer-Kensky (2000: 92–93) defines the goddess "Lolita," "a child of privilege and leisure," "a child of luxury," "a beautiful young girl in first bloom, self-absorbed and materialistic, preening and primping and prone to daydreaming and infatuated love."[294] She is the goddess since she is young to be adorned with charm and sexual attraction (ḫi-li/*kuzbu*, see Leick 2015: 87).

Yet another goddess is given this epithet, Ninlil. In the mythological Sumerian poem *Enlil and Ninlil* (Behrens 1978) the two deities are referred to as "the young man" (ĝuruš-tur) and "the young girl" (ki-sikil-tur). The mythical story takes place in Nippur. Enlil falls in love with Ninlil, who is a virgin (ll. 30–34) and still under parental protection. Despite a prohibition by the mother,[295] the young woman goes to the river to bathe. Here, Enlil sees her, he falls in love, and with the help of his adviser Nusku reaches the canal. After an initial refusal of the girl, he takes her in the reeds, making her pregnant with their first child, Sîn. The goddess is certainly a virgin, but old enough to procreate. In this poem, both deities are called "young," which is why they are involved in a love story. A Sumerian proverb reminds us, it is up to young people to get married: "Your exuberance is something that creates a family; young people get married" (Alster 1997: 331). It is the youth of the two characters which allows procreation: «Youth was correlated to sexual potency, for males as much as females» (Leick 2015: 88). Youth is the time of love, be it voluptuous (Dumuzi and Inana) or procreative (Enlil and Ninlil).

The sexual sphere is the domain of the goddess Ištar, who finds her greatest manifestation not only in the sexual activity of animals but also in that of the young human couples. In the poem *Ištar's Descent into the Netherworld*, already mentioned above, after the arrival of the goddess to the underworld, the sexual and reproductive cycle stops, the male animals do not mount the females and the young man does not impregnate the young girl (ll. 76–80, Lapinkivi 2010: 19).

As amorous passions are identified with young age, it is no coincidence that in the description of the magic attack, the witch is represented as a threat to the vitality (*lamassu*) and virility (*dûtu*) of the young man (*eṭlu*) and the sexual appeal (*inbu*) and allure (*kuzbu*) of the young woman (*ardatu*):

---

294 On the image of Inanna as being charming see Bahrani 2000: 101–102; Westenholz 2000: 80–82.
295 Compare the Old Babylonian love incantation where the young girl is told not to listen to her parents' advice: "Do not wait for the advice of your father / overrule the advice of your mother!" (George 2009: 69, No. 11: 10–11).

ša eṭli damqi dūssu īkim
ša ardati damiqti inibša itbal
ina nekelmêša kuzubša ilqe
eṭla ippalisma lamassašu īki[*m*]
ardata ippalisma inibša itbal

She (= witch) is who robbed the fine young man of his manhood,
she is who carried off the attractiveness of the fine young woman,
with her malignant gaze she took away her charms.
She looked at the young man and robbed his vitality,
she looked at the young woman and carried off her attractiveness.
[*Maqlû* iii 8–12, Abusch 2016: 82–83]

In *Hymn to Gula of Bulluṭsa-rabi* the goddess, here called Ninigizibara, describes the qualities of her beloved, Zababa. He is young, with strength, with a robust physique, decorated with charm and sensuality:

rāmī eṭlu ša . . . [. . .]
gašru raš emūqa . . . [. . .] . . .
bēl umāši ša dannūssu lā immaḫḫar[*u*]
šīḫu šarḫu gattu mutakkip šadê . . .
ḫīpu ša kuzba za'nu u lulâ malû
mummil bāltu eṭlūtu ša ulṣa za'nu

My beloved is a young man who . . . [. . .],
powerful, with strength, . . . [. . .] . . .
master of strength, whose strength cannot be equall[ed]
noble, physically beautiful, who always "gores" mountains, . . .,
attractive, adorned with charm, full of lust,
enthusiastic, pride of youth, adorned with joy.
[*Hymn to Gula of Bulluṭsa-rabi* ll. 92–97, Lambert 1967: 122][296]

When youth on the one hand is of the epitome of attraction and sexual vigor, of both women and men,[297] old age is its opposite. Indeed, the latter is considered to be one of the possible causes of the absence of sexual desire. ***Niš libbi*** text A Sympt.: 67 mentions the old age as a possible cause of the loss of sexual desire.

Old age is considered negative for sexual activity, be it procreative or not, for women and men alike. The protagonist of the Sumerian text, also included in the teaching compendium of proverbs, entitled *The Old Man and the Young Woman*, is a white-haired man who lost his vitality and strength of youth, for which reason he asks

296 In the same text, the goddess Gula, called Ungalnibru, is defined as follows: "I am attractive among the young women" (l. 120, Lambert 1967: 122).
297 See also the love incantation: "From my cupped hands drink (f) the waters of youth (*ṣuḫurtu*)!" (VAT 13226 l. 6, Zomer 2018: 277).

the king to marry a young girl so he can regain his youthful vigor. Here is the request to the king for the old man:

> My [King], supposing that the old man (ab-ba) had taken as a wife the you[ng gi]rl (ki-sikil-tur),
> [until the rest of his days] – until they come to an end, until they will be there – the old man will get his youthful vigor (nam-ĝuruš),
> [and the young girl] will get the female maturity (nam-munus).
> [*The Old Man and the Young Woman* ll. 19–21, Alster 1975: 90–99; Leick 2015: 91–92]

Here the man's description of his senior status: "(I was usually) a warrior, but now my god, my strength, my vitality / and my youthful vigor left my abdomen like a runaway donkey" (ll. 27–28). The king agrees to the man's request and addresses the following words to the young girl: "After that, I gave him to you, he will lie on your lap like a young man (lú-tur-ra-gin$_7$)" (l. 38).

The old man, without strength and sexual stamina, can recover these qualities thanks to the relationship with a young girl. There are many references to *nīš libbi* incantations: the man's warrior activities refer to the military metaphors; the strength and force having abandoned the abdomen, the anatomic place linked to sexual desire; the exaltation of the young age of the female able to awaken the passion of love, whether it has been removed by the witch or has been lost because of old age.

Whereas for young women sexual attraction is an intrinsic quality, the old woman must claim for it. Such is the case of an elderly prostitute, who, in a bilingual proverb, defends her activities despite her age: "My vulva is good, / (but) for my people / (its use) is over for me (sum. my vulva is good, / (but) among my people / they say about me: 'it is over with you')" (Lambert 1960a: 242, iii 14–16).

It is no coincidence that in the *Šumma ālu* series tab. CIV concerning erotic omens,[298] having a relationship with an old woman is considered to be a bad omen: *šumma amēlu ana šībtim iṭhe* : *ūmišam iṣṣēl* "If a man approaches (sexually) an old woman, he will have a quarrel every day" (Pangas 1988: 212). The erotic omens must be understood not so much as a moral judgment or as a prohibition of what is expressed in the protasis, but evidence of culturally accepted or rejected sexual behavior (see ibid. 216), in this case having a relationship with an old woman is considered a sexual practice to avoid. Harris (2000: 94) reports a personal communication of Guinan, according to whom the reason for such prognosis is since the woman, being old, has lost her sexual attraction, or to the fact that the man, began a relationship with the woman, did so because she "does not have any sexual power" over him, "this is the case of a man who is not in control."

In conclusion, for the one reciting the *nīš libbi* incantation emphasizing his youth reiterates the fact that he has charm and disposes of sexual attraction, conditions

298 On the erotic omens in the *Šumma ālu* series tab. CIII-CIV see Pangas 1988; Guinan 1997; Guinan 2002.

essential for the recovery of the male sexual desire. Thanks to the sexual attraction from the woman, derived from her youth, the man can recover his vigor.

Mention of gods and operations of the therapeutic operator

The formulas "*ina qibīt* DN" and "*šiptu ul yuttun šipat* DN"

Another feature of the second group of the incantations in the corpus is the presentation of the ritual performance as wanted by the gods, or as related to them.

The most frequently used formula is "at the command of DN" (*ina qibīt* DN). Incantation No. D.4 shows such a formula in combination with another one, "the incantation is not mine, it is of DN" (*šiptu ul yuttun šipat* DN)[299]

> At to the command of Ištar, Šamaš, Ea, and Asalluḫi;
> the incantation is not mine; it is the incantation of Ea and Asalluḫi,
> the incantation of Ištar, [patro]n of love.
> [No. D.4: 58–60]

The passage quoted above contains two formulations that in Mesopotamian tradition are employed independently of one another and that always appear at the end of the incantation:[300]

1. "At the command of DN" (*in qibīt* DN);
2. "The incantation is not mine, it is of DN (*šiptu ul yuttun šipat* DN)."

What is the function of these formulas?

Falkenstein (1931) was the first to subdivide the Sumerian incantations into three types by analyzing the formulas employed: legitimatization, prophylactic, and consecration. The legitimization type (ibid. 19–35) is divided into three themes: introductive (*Einleitungsthema*); actual legitimation (*Legitimationsthema*); conclusive (*Schlußthema*). The central theme of actual legitimacy has, among others, the formula "incantation of DN" (ÉN DN) which in the Akkadian incantations is associated with the formula number 2 "the incantation is not mine."[301] Falkenstein argues that the purpose of this formula is to legitimize the exorcist. In other words, through the delivery of that formula, the exorcist acquires the legitimacy for the operation. But in the

---

299 It is not certain, if it is the woman or the exorcist who must recite the incantation.

300 According to Lenzi (2010: 160), the fact that these two formulas, usually independent of each other, appear together in this incantation, is due to the late practice of composing incantations by making use of stock phrases and recycled material from earlier time. For a discussion of these exceptions, including this incantation, see Lenzi 2010: 156–160. Biggs (1967: 39) believes that formula 2 is a mistake here.

301 It is absent in the Sumerian incantations.

eyes of whom? As is known in anthropological literature, the therapeutic operator, or more generally speaking, the one who acts for therapeutic purposes, requires legitimation. As recalled by Marcel Mauss and Henri Hubert (1902–1903: 20), the image of the magician is made up of an infinity of "they say" and the magician must only look like his portrait.

Such legitimacy is developed on two levels: the social one, that is the education undertaken by the operator to acquire the official title and be recognized as such by society; and the divine one, connected to the first, which guarantees healing power and the success of the therapeutic performance. The latter is called by Lenzi (2010: 139) "legitimacy of succession,"[302] and is based on the fact that members of the class of ritual operators trace their origins to the god Ea. Again, the two levels depend on each other: the social legitimacy is also based on the divine one. The latter is, as we have said, functional in particular to purposes of the ritual effectiveness. Ritual operators receive more power because of this divine bond. This suggests that, at the time of the therapeutic performance, the therapeutic operator, by virtue of his personal career and as someone accepted as such by society, already possesses social legitimacy. Consequently, the mention of deities does not fulfill the function of legitimizing the exorcist or guaranteeing his authority; its purpose is to be seen in terms of the therapeutic efficacy of the ritual. Those participating in the ritual performances, the patient, woman, and exorcist, are aware of the effectiveness, as it is based on the divine will. In this context, the ritual performer is already socially legitimized, described as someone who acts by divine power, and thus effective by definition.[303] This formula describes the operator as a simple executor of divine power. The incantation is effective because it does not belong to the human world, but to the divine one, of which the exorcist functions as a spokesman. Lambert (1960b: 73) already affirmed that these expressions at the end of incantations are not concerned with authorship, but with the power of the incantations, which were authoritative because they were prescribed by a higher power. So, what catches the eye is not the operator's authority, but rather the divine one, guaranteeing therapeutic success (see Lenzi 2010: 141). Rather than legitimacy, we must speak of empowerment for the therapeutic action and its effectiveness. Such a formula is intended to give greater authority to the incantation and to heighten its efficacy (see Biggs 1967: 39). Note that in the *nīš libbi* incantations recited by the patient and/or the woman, the ending formula "at the command of DN" (*in qibīt* DN) is recited

302 Lenzi (2010: 138–139) distinguishes between an *ad hoc* legitimization and a succession one. The first consists in the fact that the therapeutic operator during the ritual and the incantation defines himself as a performer legitimized by the god.

303 See Leichty's article (1988) on some medical prescriptions whose effectiveness is attributed to the fact that it has been tested and prescribed by the experts (*ummânu*). On the term *latku* 'tested' see Steinert 2015.

by the therapeutic operator (see No. B.1: 12; No. D.3: 33–34; No. E.3: 52–53; No. F.4: 89–90; No. J.3: 35; No. M.1: 8–11).

Let us return to the formula *šiptu ul yuttum* "the incantation is not mine." This expression was studied by Lenzi (2011),[304] making use in particular of theoretical formulations of Austin's "speech acts," and the "magical power of the words" of Tambiah.[305] According to this scholar, the formula «reinforces the divine status of the incantation by the human speaker disavowing, essentially, disowning it» (2010: 145). So, in incantation No. D.4: 58–60 it is stated that the incantation is of Ea and Asalluḫi and Ištar. The mention of the deities is clear: Ea and Asalluḫi as patrons of magic; Ištar as the goddess of love. While Ea and the son are commonly referred to in the formula, along with others (Damu and Ninkarrak, Ningirim) in various incantations and medical texts,[306] Ištar seems to be exclusive to the corpus in question.[307]

As mentioned above, the incantation No. D.4 ends not only with the formula "incantation is not mine," but also with the one that precedes "at the command of DN" (l. 20).[308] The ritual performer, therefore, acts according to the command of Ištar, Šamaš, Ea, and Asalluḫi, gods that, except for the sun-god, are those from which the incantation derive. This formula justifies the ritual performance as desired by the gods: the performer is just the one who executes the will of the god.[309]

The formula "*ina qibīt* DN" very commonly stands alone, both in *Maqlû* (see Biggs 1967: 38; Lenzi 2010: 146 fn. 37)[310] and in the *nīš libbi* corpus. Here the gods that appear are, on the one hand, those whose domain is related to the sphere of love, on

---

304 See the previous bibliography.

305 Lenzi (2011: 148–154), making use of the theoretical tools provided by Philosophy and Anthropology of Language, traces a relationship between the rhetorical formulation "the incantation is not mine" and legal language. Please note that the author makes a clear distinction between the "emic" (indigenous explanations) and "etic" (scholarly analysis, "outsider analysis").

306 See for example Goetze 1955: 11, ll. 31–35. For other sources see Biggs 1967: 38–39; Lenzi 2010: 145–147.

307 A *Maqlû* incantation ends with the following formula: "At the command to Ištar, Dumuzi, Nanāya, patroness of love and Kanisurra, the patroness of the witches" (V 55–56); an anti-witchcraft text: "[At the command] of Nanāya, Kanisurra, / [patron of the witch]es, and the skilfull Ištar" (*AfO* 11, pl. 11: 13–14, dupl. LKU 27 rev. 9′, Abusch and Schwemer 2016: 93, No. 7.22.1: 14–15). The command of the goddess Ištar (alone) is mentioned in an incantation of anti-witchcraft magic for a man, who becomes the victim of the attack and whose symptoms are of a sexual nature: LKA 144 obv. 4′–5′ and dupl., Abusch and Schwemer 2016: 252, No. 8.29.1: 45.

308 For the use of the formula in prayers see Mayer 1976: 303–306.

309 Lenzi (2010: 143 fn. 27) has identified, as in the previous case, an analogy with the formulations used to legitimize a war.

310 In *Maqlû*, the gods from who the command descends are the gods of the Night (I 36), Anu, Antu and Bēlet-ṣēri (I 60), Asalluḫi (I 72), Nuska (I 120, III 135), Girra (III 134, 178; IV 79), Marduk (IV 78; V 10, 111, 131; VII 169; VIII 128‴, 140‴), Ea, Šamaš and Bēlet-ilī (V 10, 111, 131; VII 169, VIII 128‴, 140‴-141‴), Enbilulu (VII 106, 113).

the other those considered patrons of magic.[311] In the second case, it is Šamaš, Ea, and Asalluḫi, as in the aforementioned example.[312] In the former it is Ištar, Nanāya, Gazbaba, Kanisurra, and Išḫara:

> At the command of Kanisurra and Išḫara, patron goddess of love.
> [No. B.1: 12]
>
> At the command of
> the wise Ištar, Nanāya, Gazbaba
> (and) Kanisurra (var. adds: lady of the [sor]ceresses). Incantation formula.
> [No. M.1: 9–11][313]
>
> A[t the command of Ištar], patron of the feminine charms, (and) Nanāya, patron of sexual attractiveness,
> T[he]y commanded (it), (and) I performed (it).
> [No. E.3: 52–53][314]

This excerpt from incantation No. E.3: 52–53 is different from others because it contains an additional formula: "They commanded (it) I performed (it)" (*šina iqbâ anāku ēpuš*). The gods have decreed the therapeutic performance, of which the ritual performer is simply the executor. Similar formulas can be found in other incantations and prayers: "[They (= Damu, Ninkarrak and Marduk) said this (= incantation)], I have repeat[ed] (*ušanni*)" (KUB 37, 43: 9′, Abusch and Schwemer 2011: 32, No. 1.1.3).[315] According to that formula, then, it is the gods who have recited the incantation (*illo tempore*), the exorcist merely repeats it, implementing it in his performative ritual. In incantation No. E.3: 52–53, however, the one who pronounces the formula enacts (*epēšu*) the command of the two goddesses, Ištar and Nanāya:[316] the incantation recitation. Here, then, as pointed out Lenzi (2010: 145),[317] the operator is presented as a messenger of the gods, who emphasizes the divinity of the incantation. The exorcist is only their spokesman, the mere executor (see Abusch 2002: 6, 56–57). Divine language becomes human discourse, and as such conveys effectiveness.

---

311 See also the mention of the gods Tutu, Šazu and Ningirima in No. D.1: 4. See the invocation of Adad in No. B.2; Enlil and Bēlet-ilī in No. K.1.

312 See No. D.3: 33–34; No. F.4: 89–90.

313 See also No. K.7: 118–119.

314 Also No. I.3: 35.

315 See for similar formulations: Abusch and Schwemer 2011: 118, No. 7.2: 34′ (= BAM 438 and dupl. ibid. 437); BAM 128 iv 22 (dupl. ibid. 124 iv 26); BAM 398, r. 20–21. See also the following passage: "They spoke to me, I have held it (*aššī*)" (STT 251: 15).

316 The formula "the incantation is not mine, it is of Ištar and Nanāya. Ištar told me this and I repeat it" is also found in an incantation for a woman whose husband is angry with her: STT 257 r. 2–16, Scurlock 2014a: 109.

317 For the image of the messenger in prayer see Mayer 1976: 62.

The gods most invoked in the *nīš libbi* corpus are Ištar, Nanāya, Gazbaba, Kanisurra, and Išḫara. Nanāya (see Westenholz 1997: 57–84; Stol 1998–2001: 146–151; Wiggermann 1998–2001: 51; Ambos 2003; Beaulieu 2003: 182–216; Schwemer 2004: 72–74) is a deity whose domain encompasses the sphere of love, seduction, and sexual attraction. Her relationship to charm and sexual attraction (ḫi-li/*kuzbu*) dates back to the Old Babylonian period. In the Sumerian hymn to Nanāya, which includes a prayer for Išbi-Erra, she is described as follows (see Hallo 1966: 243): ḫi-li-zi-da ul-šè pà-da "Forever appeared in appropriate attraction" (l. 1); nin ḫi-li-túm-ma "Mistress created in fascination" (l. 13). In the chant for Nanāya/Inana her feminine and sensual qualities are praised:

> My princess? sister, her/their . . . flour is sweet.
> on your navel let me . . .
> Nanāya . . .
> Being taken away, my sister, being taken away,
> being taken away from the entrance of the cell:[318]
> Your conversation with a man is feminine.
> Your look on a man is feminine.
> Your . . . your heart is kind,
> bowing (your) hips, they are pleasing.[319]
> [Sjöberg 1977: 17–24, No. 5, text a obv. i 10–15, see also dupl.]

In Akkadian[320] hymns,[321] the goddess is described as the patron of the sexual sphere, attraction, and seduction. Nanāya, for example in an Old Babylonian hymn, is described as follows:

318 Alster (1993: 15) translates: "Come with me, my lady, come with me, come with me from the entrance to the cell."

319 Sjöberg (1977: 24) interprets the passage as a euphemism for sexual intercourse, coitus performed while lying on ones back. This according to Wiggermann (1998–2001: 51) would link the goddess (and her priestesses) to erotic illustrations depicting women drinking beer.

320 In the inscription of Sumu-El (Frayne 1990: 133, text 2: 1–2) she is called the "mistress with the perfect charm" (nin ḫi-li-a šu-du$_7$), in that of Rīm-Sîn "mistress adorned with cham" (nin ḫi-li še-er-ka-an-di) (Frayne 1990: 275, text 3: 1–2), in that of Sîn-kāšid "mistress. adorned with charm" (nin ḫi-li-sù) (Frayne 1990: 451, text 6: 1–2). In this regard, one must look at the name of her cell in the temple in Uruk is è-ḫi-li-an-an ("House of the heavenly lust") and the personal names like Nanāya-šamhat (see Arnaud 1989: 23: 1) and Nanāya-kuzub-mātim (C-FJean 1926: 218: 31). Her dominion over the realm of seduction is also confirmed by the meaning of her names: nin-zíl-zíl/ *bēlet taknê* "mistress of affection," used in the context of beds and bedrooms; bí-zil-lá, according to Wiggermann (1998–2001: 51) "lovingly caring," but according to an unpublished commentary (BM 62741: 18, in CAD Š/I 93 *šaḫāṭu* B Lex. Sect.) it is to be understood as deriving from zi-il NUN: *qalāpu* ('to pell off') : LÁ : *šá-ḫa-ṭu ša zu-um-ri* ("to strip, said of the body").

321 The composition of *Divine Love between Nanāya and her husband Muati*, dated to the reign of Abi-ešuḫ (Lambert 1966) proposes a lot of typical phraseology of amorous texts (see a similar text: van Dijk et al. 1985: 28, No. 24). But it does not specify the relationship between Nanāya and her dominion over the realm of seduction and love. On rituals and poems related to divine love see Nissinen 2001;

[uḫ]tannamū elušša
[n]annabu mašraḫū duššupu kuzbu
[ḫūd]ī ṣīḫāti u ru'āmī tuštazna[n]
[r]āma Nanāya tazmur

[Bl]ooming upon her are
[pro]geny, the splendor, sweetness (and) sexual attraction,
[with jo]y, smile and seduction you are provide[d]
Nanāya, you have sung [the lo]ve.
[Streck and Wasserman 2012: 187, ll. 5–8]

The goddess, along with Ištar, is invoked in the Old Babylonian love poem *A Faithful Lover* (Held 1961: 6–9, i 15, 24, iv 6).[322] In an inscription of Esarhaddon, she is invoked as follows: *ana Nanāya pussumti ilāti ša kuzbu u ulṣi za'anatu lulê malâtu* "For Nanāya, the veiled one among the goddesses, she who is adorned with sexual attraction and joy and full of charm" (Borger 1956: 77, § 49 (Uruk C), l. 1; RIMB 2 [1995] 187f.). In *Maqlû* she is defined as "mistress of love" (*bēlet râmi*): "At the command of Ištar, Dumuzi, Nanāya, mistress of love / and Kanisurra, the mistress of witches" (V 55–56, Abuch 2016: 140); similarly in *Ashurbanipal's Treaty with the Babylonian Allies* (Parpola and Watanabe 1988: 68, No. 9, r. 22).

Three Old Babylonian love incantations (IB 1554: 78–98, Wasserman 2016: 268–269) refer to *uzzum* of Nanāya. The term refers to the wrath of the gods,[323] but according to the CAD U 395, it can also be translated as 'sexual arousal,'[324] while Wilcke (1985) prefers the translation of *Wildheit* ('impetuosity'), understanding it in a sexual perspective in accordance with the nature of the text. However, as the same scholar emphasizes, the word can also be read *uṣṣum* 'arrow.' Even though the arrow can be seen as an erotic symbol (think of the use of arrows and quiver in sexual metaphors in *nīš libbi* incantation texts in which the goddess Nanāya is invoked), it is preferable to translate the word as 'impetuosity,' as it is present in compositions about wild animals such as the wild bull, the lion, the wolf:

[Impetuosity], impetuosity
is constantly on his 'heart' (*libbu*).
May I give you cold [wa]ter to drink!
May I give you to drink ice and frost!
May (my) vitality (hold) you like a wolf,
may the radiance hold you like a lion!
J[u]mp, I[mpetuosity o]f Nanāya (*šeḫiṭ uzzum ša Nanāya*).
[Wasserman 2016: 268–269, ll. 78–84, see also ll. 85–98]

Lapinkivi 2004. For other compositions dedicated to the goddess, see the hymn to Nanāya of Sargon II, and that of Ashurbanipal (Livingston 1989: 13–17, No. 4–5).

322 In a Sumerian-Akkadian hymn she is called the "sacred prostitute (*ḫarimāku*) of Uruk" (see Reiner 1974a: 224, l. 3).

323 See also Wasserman 2016: 42.

324 See also Stol 1998–2001: 147.

In *nīš libbi* incantation No. O.1 the goddess is evoked:

> Enuru incantation formula. I have sex with you, oh Nanāya!
> I have sex with you, oh Nanāya! Like (that) of a ram
> it is a joyful song and like (that) of a pregnant (one) is a battle cry!
> [No. O.1: 15–17]

It is precisely in the *nīš libbi* corpus that the goddess is mentioned as the patron of the sexual sphere. In fact, she is invoked in several texts to reacquire the patient's lost sexual desire.

The goddess Nanāya often in the *nīš libbi* corpus appears along with Ištar. The couple of Ištar and Nanāya also appears in love incantations, such as Old Babylonian IB 1554:

> Lover, Lover,
> you(m.), who Ea and Enlil have "placed"
> like Ištar sits on the throne,
> like Nanāya sits in the treasury (*šutummu*)
> I encircle you.
> [IB 1554 ll. 42–46, Wasserman 2016: 262]

Even more frequently, the goddess is part of a divine feminine group invoked in *nīš libbi* texts: Ištar, Nanāya, Išḫara,[325] Gazbaba, and Kanisurra (see No. M.1: 4–7, 9–11; K.6: 118–119; K.7: 147–148). The goddess Nanāya, along with Ištar and Gazababa, is invoked in a text aimed at making a brothel prosperous: "He/she will have to say: Oh Ištar, Nanāya, Gazbaba, help it!'" (Panayotov 2013: 291, l. 10).

Gazbaba[326] and Kanisurra[327] never appear alone in the *nīš libbi* corpus, but always in relation to Nanāya (see Biggs 1967: 22; George 2000: 293, l. 23). There is no clear evidence in favor of a parent-child relationship between Nanāya and the other two

---

325 On the goddess see RIA entry (Lambert 1976–1980: 176–177) and the Prechel's monograph (1996). For the Neo-Assyrian and Neo-Babylonian sources see pp. 147–162. The dominion of the goddess, at least at this time, is certainly the loving sphere, or rather the erotic one, as confirmed by the epithet in No. B.1: 12: *ina qibīt Kanisurra Išḫara bē[l]et râmi* "At the command of Kanisurra and Išḫara, patron goddess of love." See also the incantation addressed to the goddess to eradicate witchcraft in Abusch and Schwemer 2016: 304–308 No. 8.40.

326 We do not have as much information about Gazbaba at our disposal (see Weidner 1957–1971: 153). But we know that she too must have been considered a patron of the sexual sphere, as confirmed by her very name, which is perhaps derived from the term *kazbu* "sexual attraction" and her invocation, in our group of texts, also in the incantation and ritual for the brothel. In the *Šurpu* series she is qualified as the "smiling one" (*ṣayyaḫatu*), a clear reference to the sexual sphere (Reiner 1958: 21, III 79). The smile is part of the erotic vocabulary used in Akkadian literature (see Groneberg 1999: 185–187).

327 The function of Kanisurra remains in the dark. In *Maqlû* she is called "mistress of the witches" (V 55–56, Abusch 2016: 140). For the goddess see Biggs 1967: 22; Edzard 1976–1980a: 389; Sallaberger 1993 I: 213 fn. 1008; Sallaberger 1993 II: 191; Wiggermann 1998–2001: 51.

deities. Lexical lists refer to a probable parent-child relationship between the goddess Kanisurra and Nanāya, but this is far from certain:

^{d}*Na-na-a-a*	=	[^{d}*Na-n*]*a-a*
^{d}*Bí-zíl-lá*	=	[. . .]-*e*
^{d}*Ka-ni-sur-ra*	=	*mārat* ^{d}[. . .]

[Weidner 1924–1925: 11, ll. 20–22, see Cavigneaux 1981: 82–83, 20–22; von Weiher 1988: 212, No. 108: 20–22][328]

What is clear is that the two gods fall (at least more clearly in the case of Gazbaba) in the sphere of Nanāya. A late theological text defines Gazbaba and Kanisurra as "daughters of Ezida" (*mārāt* É.ZI.DA), that is, the temple of Nabû[329] in Borsippa (79.B.1/20, l. 2, Cavigneaux 1981: 83).[330] This designation may also be found in the cultic calendar, SBH Nr. VIII v 45 (Reisner 1896: 146): [*mārāt* ÈZI]DA Ga*zbaba u Kanisurra* (see George 2000: 295). In the same text, they are given the title of "Nanāya's hairdressers" (*ṣepi*$_{x}$-(ME)-*rat* ^{d}*na-na-a* (see Edzard 1976–1980a: 389; Çağirgan 1976: 182).

"I am the daughter of Ninĝirsu"

In the parallel incantation No. F.3 and No. F5 the woman, who recites them, introduces herself[331] as the daughter of Ninĝirsu and his wife, Bau, both called "releasers" (*pāširu*):[332]

> Daughter of Ninĝirsu, the releaser I am.
> My mother is a releaser, my father a releaser.
> I, who have come, will really release!
> [No. F.5: 94–96, see also No. F.3: 74–76]

In No. F.3: 69–73 the woman, before introducing herself, turns to the patient and asks him to copulate with her, with the formulas discussed above (Chapter II § "Invocation of animal arousal"). This passage leads to a series of questions. The first one is: why does the ritual performer introduce herself as the daughter of Ninĝirsu and Bau?[333]

---

328 See also in an Assyrian ritual, the mention of the goddess Nanāya, followed three lines later by Kanisurra (Menzel 1981: T 122 Nr. 54, 3 R 66 viii 32–35). For lexical lists with the name of the goddess Nanāya see Westenholz 1997: 58–59, Stol 1998–2001: 146–147.

329 We know that in the first millennium Nabû is considered the husband of Nanāya.

330 Is it possible to complete this also in Reisner 1896: 146, No. 8 v 45 (see George 2000: 295).

331 Mayer (1976: 46–52) studied the introduction formula (*Selbstvorstellung*) in the prayers.

332 See the catalogue LKA 94 ii 19: *šiptu*: *mārat Ninĝirsu* "Incantation: 'Daughter of Ninĝirsu.'"

333 Scholars have proposed different readings of the goddess's name, depending on how the sign Ú is read. Among the others, T. Richter (2004: 118–199 fn. 526) reads Baba (ba-ba$_{6}$), Marchesi (2002) Bau (ba-ú). I keep to the "neutral" reading ^{d}ba-ú (Bau). For the goddess see Ebeling 1932a.

Ninĝirsu and Bau are otherwise never explicitly mentioned in the *nīš libbi* corpus. They do not have any direct correlation with sexual desire and its absence, as opposed to other deities who influence the sexual sphere, like Ištar or Nanāya. Other gods, Ea, Asalluḫi and Šamaš, are invoked in these texts to make the ritual more effective, as patrons of exorcism and more generally of magical practices. It is in this sense, that the mention of the divine couple must be interpreted. As is well known, the goddess Bau in the first millennium is identified with Gula, patron of the medicine.[334] Inevitably the spell calls for the divine spouses, Ninĝirsu, Ninurta, and Pabilsag. The identification between Ninurta and Ninĝirsu is a well-known fact (see MPStreck 1998–2001: 512–522). Similarly, Pabilsag, Ninisina's husband, and Zababa are also identified with both (see Lambert 1967: 109–114; MPStreck 1998–2001: 518; TRichter 2004: 214–225; Ceccarelli 2009: 39–46; Böck 2014: 14).

Returning to the incantations, the woman calls herself "daughter" of Ninĝirsu and Bau, that is Gula and her husband. The mention of goddess Gula may surprise us at first glance, as she has, in fact, never been invoked in the *nīš libbi* corpus.[335] However, as I have shown above, the gods invoked are on the one side goddesses whose domain falls under the sphere of love and sex, and on the other hand those who are competent in matters of magic and exorcism. We know that the goddess Gula, along with her son Damu, is mentioned in quite a lot of incantations, and so are Ea,

---

334 The description of the goddess Bau as a medical goddess begins during the Neo-Sumerian period, although the sources are few, and is finally firmly established in the Old Babylonian period (for the sources see Römer 1965: 245; Limet 1968: 210; 386; Sjöberg and Bergmann 1969: 32, l. 268; TRichter 2004: 516–517; Ceccarelli 2009: 36–39). Then she is identified with the goddess Ninisina (Kraus 1951: 62–75, 83–86; Römer 1969; Edzard 1998–2001: 387–388), at least in literary and religious texts (see TRichter 2004: 514–519; Ceccarelli 2009). Similarly, Ninisina is identified with the medical goddess Gula (see Kraus 1951: 64–75; Lambert 1967: 109–110; TRichter 2004: 214–225; Böck 2014: 12–13). See, for example, a bilingual Old Babylonian incantation in which the Sumerian version uses the name of Ninisina, while the Akkadian one the name of Gula and Bau (Böck 2007: 184, ll. 1 and 3). As a result, Bau, as a medical goddess, at least for the Neo-Assyrian period, is to be interpreted as a name for the goddess Gula (see Böck 2014: 13–14). An example can be found in a text that informs us about the healing power of the goddess against *di'u*-sickness: ^{d}Ba-ú tu$_6$-nam-ti-la šub-ba sag-gig-g[á-šè] : *Bau nadat šipat balāṭi ana di'*[*i*] "Bau, who recites the incantation of life against the *di'u*-sickness" (KAR 41: 5–6). The emblem of this process of identifying of Bau as the goddess of healing is the hymn KAR 109 + 343 from the Neo-Assyrian period (Aššur). Here Bau is defined as Ningirim, patron of incantations: *ina Babili nērib ilāni Ningirim* "In Babylon, the entry of the gods, (Bau) is Ningirim" (KAR 109: 12, Ebeling 1918: 50). In a text related to kidney and rectal sicknesses, KAR 73: 26 (dupl. K.2960 ii 10′) in which the incantation is directed to Gula, the identification between Bau and the goddess of medicine is obvious: "At the command of Bau, I will exalt her name among all peoples" (Böck 2014: 87). The name of Bau is also used to designate the goddess Gula in *Hymn to Gula of Bulluṭsa-rabi* dated, according to Lambert, between 1400 and 700 B.C. "The great daughter of Anu, the mother Bau, life of the people, am I (= Gula) (l. 109, Lambert 1967: 122).

335 According to Böck (2014: 147), it is possible to trace a direct link between the goddess and the absence of sexual desire through the use of the *lišān-kalbi*, a plant connected with the goddess Gula and often used in *nīš libbi* rituals. For a discussion see Chapter III fn. 369.

Asalluḫi, and Šamaš to ensure the effectiveness of the therapeutic performance. In the previous section I have shown such types of formulas within the corpus. Among them, I should also mention the formula "the incantation is not mine, it is of DN" that refers to the goddess and Damu, along with Ea and his son, as well as Ningirim (see Cunningham 1997: 16–17, 50–51; Böck 2014: 113–115):

> The incantation is not mine, it is the incantation of Ea and Asalluḫi,
> it is the incantation of Damu and Gula,
> it is incantation of Ningirim, mistress of incantations.
> They have said this and I repeat it.

So, the mention of the goddess Gula and her son is intended to make the ritual more effective. However, the formula proposed in the two *nīš libbi* incantation is different. It recalls the bond of kinship between the ritualistic performer and the deities. The woman casting the incantation is a "releaser" because she has acquired this ability by birth:[336] her parents are "releasers" and so she is. This formulation is not far from the ancient Sumerian formulas which included the mention of a divine couple, Enki/Ea and Damgalnunna/Damkina, of whom the exorcist defines himself as an emissary ("man") (see Cunningham 1997: 52). In the same formula, he introduces himself as a "messenger" of Asarluḫi/Marduk. Here is an example from the *Udug-ḫul* series:

> sum.: I am the man of Enki
> akk.: I belong to Ea
> sum.: I am the man of Damgalnunna
> akk.: I belong to Damkina
> sum.: I am herald of Asarluḫi
> akk.: I am the messenger of Marduk
> [*Udug-ḫul* III 7–8, Geller 2016: 91–92]

In more ancient texts he describes himself as a herald of the gods Enki and Damgalnunna, without mentioning Asalluḫi, with whom, however, he identifies himself (see Geller 1985: 14). The formula of incantations No. F.3 and No. F.5 despite the temporal distance, is quite close to these Sumerian expressions. The ritual efficacy is since the therapeutic operator is considered the son of a divine couple, whose domain is the sphere of healing. This is a formulation that we already find in the Sumerian texts, although reworked here with the mention of Bau and Ninĝirsu[337] instead of Ea and the consort. The ritual performer, in this way, places himself in opposition to the witch who caused the suffering, emphasizing, on the contrary, his relation to celestial

336 Of a quite different nature is the mention of the seven divine daughters of Ninĝirsu and Bau, see Falkenstein 1966: 75.

337 The deities are invoked, along with Ninurta and Gula, also in *Šurpu* VIII 28 in a divine list in order to deliver a man who has broken a taboo. In order: Ninurta, Ninĝirsu, Bau and Gula (Reiner 1958: 40).

forces, which can eradicate evil. It is his divine connection that allows therapeutic success (see Lenzi 2010: 138).

Thus, we can understand the function of mentioning the divine in this formulation. Another question, however, remains unanswered. Who is the one who recites the incantation? The woman calls herself a releaser (*pašīrat*), or more precisely the one who "came to release, and will release (the man)." We know that the social actors who take part in the therapeutic ritual, are certainly the exorcist, the patient, and his partner.[338] The epithet of "releaser" often defines the exorcist.[339] An example is given in *Maqlû*, in which the ritual operator opposes himself to the witch as being the "releaser":[340]

> Incantation: My friend is a witch, I am a releaser (*anāku pāširāk*).
> the witch is a witch, I am a releaser,
> the witch is an Elamite, I am a releaser,
> the witch is a Gutian, I am a releaser,
> the witch is a Sutian, I am a releaser,
> the witch is a Lullabitia, I am a releaser,
> the witch is a Hanigalbatian, I am a releaser,
> the witch is an *agugiltu*, I am a releaser,
> the witch is a *naršindatu*, I am a releaser,
> the witch is an enchantress of snakes, I am a releaser,
> the witch is ecstatic, I am a releaser,
> the witch is a worker of metals, I am a releaser,
> the witch is . . . of my door, I am a releaser,
> the witch is a native of my city, I am a releaser.
> [*Maqlû* IV 123–136, Abusch 2016: 126–127]

We have no information about women as exorcists in the first millennium (see Abusch 2002: 66).[341] The only attestation of an *āšiptu* (female exorcist) is in *Maqlû* III 41 (Abusch 2016: 87), where it occurs in a list of negative figures. She is to be understood more as a sort of witch rather than as a scholar and specialist for worship (see May 2018a: 151). All this makes it difficult to understand the passage, because we are

338 I do not use the term wife because the term used is *sinništu* "woman."

339 Of course, the verb *pašāru* often has the deities as a subject, who are invoked to untie the knots of evil that attack men. For the sources see CAD P 237–238, mng. 2a.

340 See also BAM 214 iii 5′–6′ (and dupl. ibid. 224 ii 16′–18′), in CAD P 238, mng. 2b.

341 We find very little information on female exorcists in the third millennium, particularly from the lexical lists (see Lecompte 2016: 35). For the Old Babylonian period see the lexical list Proto-Lú where a female doctor is mentioned (a-zu munus) and a female exorcist (munus ka-pirig) (see ibid. 40). The canonical list Lú informs us of female figures in the role of interpreters of dreams, ecstatic and necromancer (see ibid. 47, 49–50). In the Old Assyrian texts, we find traces of female interpreters of dreams (*šā'ltum*) and diviners (*bārītum*) (see Michel 2016: 198). An overview of the role of women in the love therapy in Mesopotamia see Scurlock 2014a: 106–107. For the role of women in the Hittite therapy see Beckman 1983; BCollins 2014, esp. 262–265; Vigo 2016: 329–330.

not able to understand the identity of the woman who recites the incantation. Many incantations are recited by the patient's companion, inviting him to the sexual act. Also in No. F.3 the woman makes this proposal:

> [The mating of a wild goat six times], the mating of a stag [seven times],
> [the mating of a partridge] twelve times, make [love to me]!
> Make love to me! Ma[ke love to me] because I am young!
> [. . .] I am endowed with love, Make love to [me]!
> Tie [. . .] of the *rikibtu* of a stag! Ma[ke love to me]!
> [No. F.3: 69–73]

The patient's partner plays a role of great ritual importance, as evidenced by the ritual that often follows the incantation. It seems to be her who recites the incantations and addresses the man, inviting him to coitus. However, the mention of divine origin and the role of "releaser" seem rather lead us to another interpretation. This probably refers to an exorcist, or more generally speaking, to a therapeutic operator socially legitimized to perform the ritual. In fact, both the epithet "releaser," and the mention of divine origins are prerogatives to members of the *āšipūtu*. It certainly does not refer to the female partner, but at the same time, we have no evidence of female *āšipu* at that time. The question remains open.

Third group: sexuality and nature

The peculiarity of this group of incantations is their description of natural events, such as weather, which are understood as metaphors for uninterrupted and prosperous sexuality. Here are examples from the first two lines of incantations No. A.1 and No. E.1:

> May the wind blow! May the grove quake!
> May the cloud gather! May the moisture fall!
> [No. A.1: 33–34]

> May the wind blow! Ma[y] the mountains [quake]!
> May the cloud be gathered! May the moisture fall!
> [No. E.1: 1–2]

The two pairs of lines have a parallel structure: the second section of each line is a direct consequence of the first. So, it is because of the strong blowing wind that the grove (No. A.1: 33) or the mountains (No. E.1: 1) quake and it is thanks to the gathering of clouds that the rain falls to the earth.

The same image is found in incantation No. I.1: 6′–8′, where, however, the action of the second part of each line, rather than in the above-mentioned cases, is unexpected. The reasons for these negations remain puzzling:

[May] the wind blow! May [the grove] not quake!
May [the clo]uds gather!
May [the mo]isture not f[all]!
[No. I.1: 6′–8′]

Note the use of the verb *nâšu* referring to quaking of the mountains and the grove, which recalls the semantic context of the quivering and twitching, also owner of the verb *galātu* 'to tremble' (CAD G 11–14; AHw. I 274) used to express the ejaculation. The clear reference of this image to the sexual intercourse is, perhaps, confirmed by the fact that the word 'wind' (*šāru*) is associated semantically, sharing its consonant roots, with 'penis' (*ušāru*). Maybe for this reason, in *Enūma eliš* Marduk defeats Tiamat with his wind, which is replaced in a cultic commentary by the penis: "(The king) is Marduk, who [defea]ts Tiamat with his penis" (Livigstone 1989: 94, No. 37: 18′, see Cooper 2017: 115–116; Helle 2020: 66).

According to Biggs (1967: 34), this step is intended to invoke fertility. The image of "fertility," however, bears the problem of being fairly vague in concept and open to different interpretations. I think that the purpose of such a "theme" is to integrate the human into the sexual-reproductive process of nature. To do this, human sexual activity is placed in a chain that connects the elements of nature, from atmospheric phenomena, such as wind, to the plant world represented by the grove. This chain also contains animals. In incantation No. E.1: 3–4, after the request of rain, the sexual activity of domestic animals, the donkey and buck, is invoked:

May the ass mate and mount the jenny!
May the buck arise and repeatedly[342] mount the goat!
[No. E.1: 3–4]

The arousal and the resulting mating of the animals is a clear metaphorical frame designed to evoke the absence of sexual desire and human sex life. Similarly, the mention of the grove and the rain[343] takes us back to sexuality and pleasure. The orchard (*kirû*) is often used in love literature. It is a place of charm and erotic desire, with its voluptuousness the orchard contains all the promises of sexual delights (see Lambert 1987a: 28–31; Westenholz 1992: 382; Westenholz 1995: 2482; Alster 1993: 7 and fn. 26; Leick 1994: 122–123; MPStreck 1999: 218; Besnier 2002; Paul 1997: 100; Paul 2002: 492; Couto-Ferreira 2013: 110–111; Rendu Loisel 2013; Zisa 2021). A well-known metaphor in Sumerian love literature, but also present in the Akkadian and found, for

342 Omitted in the variant.
343 The reference to the celestial Daughters of the god Anu in No. D.2: 19 should be noted. These benevolent entities are placed in association with clouds and dew (*nalšu*): "Do not release the dew of the Daughters of Anu!". The dew in turn can be associated with the rain: *asḫar kīma nalšu u imbaru lama mê tīkki* "The *asḫar*-stone (looks) like dew or fog before a rainfall" (von Weiher 1983: 126, No. 24: 28). The word *tīku* 'rain' is also used in No. L.2: 24: "May your rain keep falling for you [forev]er!".

instance, in Old Babylonian love lyrics as well as in compositions of the first millennium, is the metaphor "to go down to the garden" (*ana kirîm arādu*):[344]

You, Oh two beautiful maidens,
you are blooming!
Come down to the garden,
Come down to the garden.
[MAD 5:8: 6–9, Wasserman 2016: 242, see also l. 17]

Like a fruit of the garden come out *over* him!
[VS 17, 23: 3, Wasserman 2016: 249]

"She seeks your ripe garden of pleasures."
"The one who goes down to the garden, Oh king, the cutter of cedar."
"Oh chief gardener of the date-palm garden of delights."
"Rejoice Nanāya in the garden of Ebabbar which you love."
[KAR 158 vii 26′, 28′, 35′, 38′, Wasserman 2016: 214]

(She$^{?}$:) [Ma]ke (me) beautiful! Make me greatly flourish! Make me happy! (He$^{?}$:) Let me see the garden of the almond trees!

(She:) A word of greeting (is) the squalling of the ducks. (As) the garden – its fruit, the field – its grain, (so) did me make me grow.
[*The Moussaieff Love Song* r. 3, 5, Wasserman 2016: 133–134]

"[Nabû], my lord, put an earring on me.
Let me give you pleasure in the garden.
[. . .]
For what, for what are you adorned, my Tašmētu?
So that I may [go] to the garden with you, my Nabû.
Let me go to the garden, to the garden and [. . .].
Let me go alone to the very beautiful garden.
[TIM 9 13–14, r. 15–18, Livingstone 1989: 36–37]

As I [went down] into the garden of your (fem.) love.
[. . .]
Zarpānītum will go down to the garden.
[BM 41005 obv. II 9, 13, Lambert 1975:104]

As already pointed out, the orchard is the place of erotic pleasure, amorous attraction and lustful prelude. It is, particularly in love literature, a symbol of femininity (see Lambert 1987a). In most cases, the orchard/garden is described by female speakers. As demonstrated, plant metaphors can be linked to the female world and describe women's sexuality (see Cooper 1989; Cooper 1997; Besnier 2002: 68; Zisa 2021). It is no coincidence, that in the *nīš libbi* corpus they are completely absent. The only mention of the orchard in this group of texts is a variant of this "theme" described above. Here it is used as a meteorologic metaphor, wind and rain, which aims to describe

344 For this expression see Westenholz and Westenholz 1977: 212–113; Leick 1994: 191, 196.

male sexuality. The wind that shakes the orchard (or mountain) is a clear allusion to the male sexual desire (*nīš libbi*), that should be invoked from the atmospheric sphere to the human one, passing through the animal one. The orchard serves here as a metaphor for male sexuality. Maybe it is also possible to see a reference to sexual intercourse in which the woman (represented by the orchard) is subjected to extreme male sexual vigor.

Since the wind and rain are metaphors used to describe male sexuality, the god who presides over that domain, Adad,[345] is invoked in a *nīš libbi* incantation:

> O Adad, lock keeper of the canals of heaven (var.: of Anu), son of Anu,
> who decrees oracular decisions for all people, protector of the land,
> at your supreme command, which cannot be opposed,
> and at your authentic consent, which cannot be altered.
> [No. B.2: 30–33]

Adad in the incantation is defined as *gugal* 'lock keeper of canals,'[346] and it is no coincidence that in the *nīš libbi* incantations there are also metaphors related to canals/rivers (*nāru*):

> May the wind blow! . . . [. . .]
> May the stables fill up! . . . [. . .]
> May precisely the canals fill up! [. . .].
> [No. H.1: 6–8]

The flooding of canals is invoked indicating powerful and uncontrolled desire. See also in No. A.2: 42 "impetuous horse, whose rising is a devastation," where the term *našpandu* 'devastation' refers to a devasting flood (CAD N/II 29 mng. c). Sexual desire is compared to constantly flowing river water to indicate a continuous desire and without interruption in No. A.1: 35: "May my sexual desire be constant river water!". See also the catalogue of incipit LKA 94 i 28: "[Incantation]: 'Flow, river of sexual desire!'".[347]

Previously, I have already shown other metaphors derived from the system of canals: "Who has blocked you like an opening of the *dilûtu*-water system"

---

345 The wind that shakes the mountains, expression that can be found in the incantation, is a typical prerogative of the god Adad: *ša ina pīšu ḫuršānī inūšū* "(Adad), to whose voice the mountains tremble" (Shalm. III, Kinnier Wilson 1962a: 93, l. 6). Note an incantation against witchcraft in which the god is called upon in Abusch and Schwemer 2016: 319–322, No. 8:43. On the god see Schwemer's monograph 2001.

346 See the mention of the canal-inspector in a second millennium love text: "Ask the canal-inspector, the god and . . . me the god" (CUSAS 10, 12: 11, Wasserman 2016: 190). See the mention of Adad's lightnings in another second millennium love composition: "The lightnings of Adad passed over it, indeed, the lightning of lovemaking passed upon me" (A 7478 i 10–13, Wasserman 2016: 65).

347 See another possible reference to the canal: "[broken] I cause[ed] fear[?] [. . .] canal" (No. K.8: 144). See also in love context the reference to flood: "He brought the Flood (*iššâ abūbum*) – achieving what?" (CUSAS 10, 12: 15, Wasserman 2016: 190).

(No. D.2: 11); "Incantation: 'Why are you blocked up like a canal?'" (catalogue LKA 94 i 5); or from containers of water: "May they (= the heavenly Daughters of Anu) gran[t] the well-being of the waterskin (*nādu*) / to your waterskin" (No D.2: 16–17). It is possible that the water reference here indicates the sperm (see Couto-Ferreira 2015–2016: 52), not intended necessarily for reproductive purposes, but rather serving as a representative of the male orgasm, metonymy of consummated sexual intercourse.[348]

The use of images taken from the system of canals to indicate the human physiology is not rare. As has been analyzed by Steinert (2013) and Couto-Ferreira (2017), the female physiology is described in the incantations and rituals in terms of flowing fluids in rivers/canals. The body is thought to be formed by channels described metaphorically as canals/rivers. As mentioned by the scholar, the metaphors referring to the canal/river are mostly used to describe the female physiology (*mû* 'water' is to be understood as amniotic fluid in connection with pregnant women). This group of *nīš libbi* incantations shows us how this metaphor was also used, although very rarely, to describe the functioning of the male body. The excessive flowing of fluids from the female genitals is perceived as pathological, while the "constant river water" in a man's body is considered desirable, understood as a possible allusion to the male orgasm:

Sexual desire → Satisfaction/orgasm
↓
Sperm → Canal/river metaphor
↓
Flowing water

Fourth group: *historiola* and its function

This group is characterized by the presence of a *historiola,* a mythical story that describes an extra-human entity's exploits, carried out in ahistorical time.[349] There are two incantations of this group, containing two different *historiolae*: the preparation of the bed, on the part of female deities for their spouses and lovers (No. M.1); and the creation of mankind and the determination of the sexual desire by Enlil and Bēlet-ilī (No. K.1).

348 The relation between sperm, eros and the irrigation system in Sumerian literature was highlighted by Cooper (1989) in his study on Enki's penis, where he emphasized the reproductive value of the phallocentric actions of the god. The aspect of reproduction is, however, absent in *nīš libbi* incantations.

349 For a study on Mesopotamian *historiolae* see Sanders 2001. Marinella Ceravolo (2020) conducted a specific work on Mesopotamian *historiolae*, reflecting on the way in which the dehistoricization is carried out in them. See also TCollins 1999: 40–42.

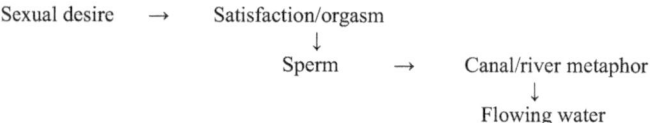

Incantation No. M.1 tells of the preparatory work *illo tempore* involved in making a bed[350] for the sexual desire, serving as a form of metonymy of the sexual act, by the goddesses Ištar, Nanāya and Išḫara for Dumuzi, her husband, and Almānu/her lover, respectively. Similarly, the narrating woman prepares a bed for her lover so that he can recover sexual desire.

It should be noted that this *nīš libbi* incantation refers to the amorous relationship in general, especially in wedlock, not only divine but also human. In the Ms. C variant, the woman who recites the incantation refers to the patient of the ritual performance as "my husband" (l. 8 *ana [ḫ]amrīya*). The role of husband and wife is almost absent in the *nīš libbi* texts corpus. Often, the female partner is referred to generically as "woman" (*sinništu*). Since the gods invoked form a bridal couple and they represent prototypes in human social life, it is no wonder that this incantation refers to the patient as "husband" and not simply "man." The bed that is being made, apart from being the place where the sexual encounter is consummated, is a symbol of wedlock.

In the incantation No. K.1 the creation of the human being[351] by the divine couple Enlil and Bēlet-ilī[352] is described. Enlil and his paredra, in the act of creation of mankind (l. 8 nam-lú-ùlu-lu nam-še$_{23}$-a lit. 'to give a name'),[353] decree (l. 11 nam-e) its *nīš libbi* and 'raise its 'heart' (*libbu*)' (l. 12 šà-bi mu-un-zi). In the other compositions, *Enki and Ninmaḫ* and *Atra-ḫasīs*, there is no mention of the determination of the sexual desire. The purpose of this *historiola* is to establish sexual desire at the origin of human creation as an essential psycho-physical characteristic of the integrity of human beings.

The presence of these *historiolae* makes us wonder about their function. Why are these episodes about the gods, *outside of historical time,* mentioned in an incantation with a therapeutic function? The fields of Anthropology and History of Religions have analyzed the function of myth in ritual contexts. Here we should consider the reflections of the Italian anthropologist Ernesto De Martino. According to him, the *historiolae* in the incantations are intended to enable a *dehistoricization* process. In *critical moments of life,* such as sickness, dehistoricization can help to overcome the momentary difficulties of a certain event, putting it in context with a similar event that has already occurred in the mystical past, *illo tempore,* outside of history.

---

350 See the catalogue LKA 94 i 6: "Incantation: 'I have set up a bed! I have now set up a [cha]ir!'"; ii 15: "Incantation: 'Se[xual] desire! Sexual desire! Bed! Bed!'".

351 On anthropogony in Sumerian and Akkadian sources see Pettinato 1971; Clifford 1994; Lambert 2013.

352 See the catalogue LKA 94 i 13: "Incantation. Enlil and Bēlet-ilī mankind." Here, instead of DIĜIR.MAḪ, we find ^{d}NIN.MAḪ, both goddesses referring to Bēlet-ilī (see Biggs 1967: 45; Krebernik 1993–1997: 504).

353 On this expression in the sense of creating mankind see *Gilgameš, Enkidu and the Underworld* l. 10, Gadotti 2014: 162.

This means that the negative event that affects a man and his community is faced and solved because they are thought of as the recurrence of an incident that has already happened and thus is likely to be overcome. The religious institute of dehistoricization removes these negative moments from the human initiative and resolves them in the *iteration of the identical,* thus carrying out the cancellation or masking of distressing history. The practice of dehistoricization of the magic-ritual institution on the one hand creates a stable and representative shared horizon and, on the other hand, provides the ideal place in which to reabsorb, overcome or block the power of negative fate (De Martino 1953–1954: 19). The principle of the *historiola* is the model of the "so-how," the *simila similibus*: just like *illo tempore* mythical characters have overcome the evil that has affected them, so can the human community. The evil that affects society is thought of as surmountable thanks to the paradigmatic model of the myth in the ritual process. The problematic present is assimilated into a paradigm in which a crisis is resolved positively (De Martino 1959, Engl. tr. 2015: 105). In incantation No. M.1 the woman behaves and acts as the female divinities. The mythical time and place are renewed in the therapeutic ritual practice. The action of the ritual actors is nothing more than the repetition of mythical actions that allowed the overcoming of the crisis and the negative outcome (see De Martino 1959, Engl. tr. 2015: 96; De Martino 1962: 44–45).

This idea is shared by the Historian of Religion David Frankfurter, who reverting to Tambiah's investigation, examined the power of words in the narrative act. He uses the expression "narrating power" referring to «a "power" intrinsic to any narrative, any story, uttered in a ritual context, and the idea that the mere recounting of certain stories situates or directs their "narrative" power into this world» (1995: 457). According to the scholar, the "narrating power" lies in the fact that the recited word confers "mythical" power to the real world. The aim is to overcome the crisis through evenemential mythic evocation. The mythic dimension comprises dramas that have been completed and tensions that have been resolved. The *historiola* involves laying out a crisis resolved, using the authoritative components of myth efficiently (Frankfurter 2016: 102).

To fulfill its function, the myth is always a foundation myth. It refers to an event that took place during a time outside of history, and which is an exemplary precedent for all situations that will arise in future human history. As described in *nīš libbi* incantation No. K.1, it was decreed by the gods at the beginning of the creation of humans and as such, it is unique to humans. It is the task of the ritual practice to recover the lost or stolen sexual desire. As it has been established in the mythical time of human creation, it helps to restore order and human balance, be it individually or for a couple. The incantation repeats and thus renews a foundational myth, that of the sexual desire. Its iteration aims at abolishing historical time, determined by the negative event of the loss of sexual desire, and at readjusting the world, according to the exemplary power of myth.

It is clear then, that the dehistoricization techniques, including the use of *historiolae* in incantations, have a defensive-protective function against the negative events of historical development. The dehistoricization, with its mythical-ritual symbolism, allows for overcoming the existential crisis of human historicity, making it possible for humans to find a solution to the negative situations they find themselves in. Witchcraft, for example, in Mesopotamia often causes the loss of desire, it thus threatens individual and collective life, a metaphor for social anxiety that characterizes small communities, and can only be eradicated through the ritual practice. The rituals provide a force to stand up against such harmful events. The magical and ritual practices give an active role to a man and his community. They grant them the opportunity to defend themselves. Human existence, thanks to mythical-ritual symbolism, is not passive, but active and helps to restore the order *hic et nunc*. This last statement allows us to point out that the mythical tale, albeit from a meta-historical context, in ritual practice leads to historical behavior. It takes place outside of history, but, on the other hand, reopens the *Einmaligkeit* of the historical decision (see De Martino 1977: 225).

Fifth group: abracadabra

These five texts (No. F.1; No. J.1; No. K.3; No. K.4; No. O.1) form the fifth group of the incantations corpus: (Pseudo-)Sumerian Abracadabra or Mumbo-Jumbo (see Lambert 1983; Veldhuis 1999: 46–48; Böck 2014: 187–190; Baragli 2020). It is a pseudo-language with magic-ritual purposes, different from day-to-day language.[354] Sumerian is apparently not the only language these texts seem to be derived from. Some are written in "Elamite" and "Hurrian" (see Finkel 1976: 58–59; Edzard and Kammenhuber 1976–1980: 509–510; van Dijk et al. 1985: 3–4; Prechel and Richter 2001; Krebernik 2018), and some in an unidentified pseudo-language. But we must also consider the possibility that they are not based on any specific language at all (see Veldhuis 1999: 47). The characteristics of the abracadabra are the repetition of sounds and the alternation of nonsense passages along with others that appear to contain an intelligible meaning. Veldhuis (1999: 48) observes that the absence of regularity on the level of syntax and grammar is compensated by other linguistic means through regular repetition of a word, so emphasizing the 'rhythm' of the text. We have to focus on the rhythm to understand the function of the abracadabra and its repetition of sounds and/or words. Let us consider four abracadabra of the corpus to emphasize the sound repetitions in the same line, as well as between the lines:[355]

354 On the relationship between ordinary language, the sacred language and ritual language see for example Tambiah 1968: 179–185.
355 Repetition in the same line in italic, repetition between lines in bold.

[ÉN x x] *aḫ* lu up *pa* di ra *aḫ* an ki *nu sum nu sum*

[. . .] x ni kab mu *bu bu* **a ḫa an til la ke$_4$**

[. . .] x *ti an* **a ḫa *an ti* la ke$_4$**

x na *ḫa an* ub *bi* **a *ḫa* *an* ti** áb *bu* uk

[No. F.1: 1-4][356]

[. . . b]u u*m ma* **ia** *ma ma* **na**

[. . . b]i ti ri **ia**

[. . .] x x mi ***na na***

[No. J.1: 4-6][357]

[ÉN É.NU].RU *ka* ab *ka* [. . .] ma na

[. . .] ⸢x x⸣ [. . .] x x [. . .] *di di* il *d*in x

lú bi ú [. . .] e te **am**

lú bi *ga* x [. . . *a*]*m* na ***am***

ki ì kaš *ga* lú [. . .] x ig *ba*

la ba il kur te [. . .] *ba*r ta

la ba il ta [n]a [. . .] x si a

la ba lu **ka ma an** [. . .] x x la ud

la ba lu **ka ma a[n** x] x x x [TU$_6$] ÉN É.NU.RU

[No. K.3: 80-88]

gìš te gìš te **da ga an ni**

da-zi **da an ni** {. . .}

gá-an-ga **da *ga an* ni** ma*ḫ ḫ*a

[No. O.1: 21-23]

The emphasis on rhythm in therapeutic processes has been subject of analysis of Medical Anthropology. Here, I would like to dwell for a moment on the studies conducted by Carlo Severi (2004). The anthropologist takes the analysis conducted by the psychoanalyst Gaetano Roi on children with autism, as well as those done by the linguist Kevin Tuite on a body of Georgian traditional songs, criticizing Lévi-Strauss' (1949b) theory of symbolic efficacy on therapeutic Kuna chants. Severi confronts lin-

356 Note the repetition of the voiced and voiceless bilabial plosive /p/ and /b/: 1. lu up pa; 2. mu bu bu; 4. ub bi; áb bu uk. See also the sound variants of a ḫa (ll. 2–4): 1. aḫ; ra aḫ; 4. na ḫa an. In addition, lines 1 and 4 share a similar sound-vocal structure: 1. aḫ/na aḫ an; 2. lu up pa di/ub bi; 3. ra aḫ/a aḫ; 4. an ki/an ti; nu sum nu sum/áb bu uk.

357 Note in lines 4 and 6 the prevalence of voiced bilabial nasal /m/ and voiced alveolar nasal /n/ and in all abracadabra of the vowels /i/ and /a/.

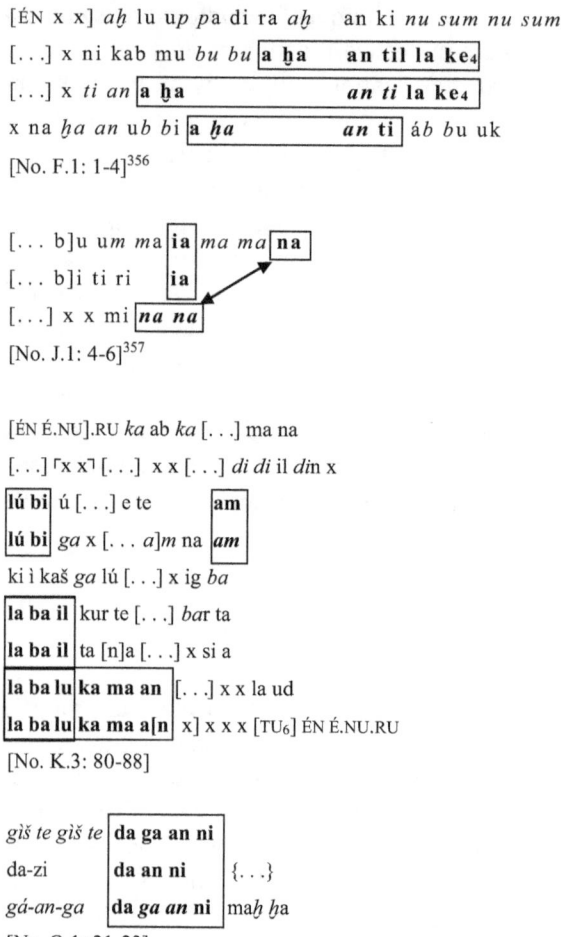

guistic symptoms of autistic children with poetry, in situations where experiences of strong emotional intensity (painful or happy) are celebrated or commemorated collectively, where affection tends to prevail over representation, often making normal communication difficult, sometimes even impossible. Poetry in non-Western societies often becomes a ritual chant, thus coming closer to sound, and at the same time closely linked to experience (see Severi 2004: 228).

This group of texts is characterized by abracadabra, which is nothing more than a meaningless combination of sounds. Yet, these "noises" are structured within an organization according to their sound, given the repetition in one line and between lines of certain syllabic sounds or groups of sounds. They are *regular sound configurations* (ibid. 229). What the patient perceives is a sequence of sounds. An organized set of sounds, without meaning, but, as we shall see below, still *meaningful*. Paraphrasing Severi, the questions to be asked are the following: why transcribe, and therefore carefully define, an abracadabra communication which does not contain any actual words? Since it still is some kind of communication, what does one want to communicate? Why is this kind of communication associated with healing practices and rituals, built on the belief in the efficacy of the words?

The *āšipu* recites formulas, codified by tradition, with a regular sound pattern, perceived by the patient, without, however, conveying any meaning to the latter. The patient knows the exorcist and thus *is acquainted with* his practices. Moreover, the use of a language unknown to the patient confers authority to the therapeutic performance (see Böck 2014: 191). Additionally, as I have mentioned above, the other characteristic of abracadabra is that it is not totally meaningless. One can observe the alternation of meaningless sounds with the ones that appear to have meaning in Sumerian (see Veldhuis 1999: 47–48). In No. F.1, at the end of line 1, the repeated formula nu sum is perhaps to be read as nu[358] sì 'to place sperm' indicating probably the evil action of the witch through the magical manipulation of a man's sperm. In lines 2–4 the expression an-ti(l-la-ke$_4$) recalls the "heaven" (an) and the "end" (til). In No. O.1: 21–23 the sign UŠ can be interpreted as gìš 'penis' or nita 'man'; TE as *ṭeḫe* 'to approach (sexually)'; da-ga-an-ni as phonetic spelling of da-ĝán 'her sleeping chamber, dwelling'; da-zi-da-an-ni 'his right side'; gá-an-ga infinitive of the verb ĝá-ĝá 'to put' or phonetic spelling for ga-na 'come on'; maḫ-ḫa infinitive form from maḫ+a 'to be superior, exalted.' Wasserman (2016: 237–238) translates the abracadabra: "Penis! Approach! Penis! Approach her bedroom! / (Approach) her right side! / Come on! (Approach) her exalted bedroom!"

What characterizes the abracadabra of this corpus, though, is the repetition of sounds, both in the same line, and between lines, and the alternation of nonsense in meaningful sequences. In almost all of these abracadabra formulas it is not pos-

358 See Proto-Ea C l. 18: nu-ú NU = *li-pi-iš-tum* (Landsberger 1951 = MSL 2: 139). For *lipištu* as 'sperm' see CAD L 199 mng. 3 'offspring'; AHw. I 554 mng. 2 'Sperma'; Cavigneaux and al-Rawi 2002: 39.

sible to trace any meaning. The rhythm of these sounds, together with the perception of meaningful segments, if any, placed in relation to the pragmatic communicative event, that is the ritual performance itself, in which the patient recognizes the exorcist's authority, offers the patient *a guided perceptual illusion* (Severi 2004: 236), a *projection* inside the matrix of belief, and therefore marks the beginning of their healing phase. It is the patient who gives meaning to what is pronounced by the exorcist. Severi (2004: 232) argues that a word ritually enunciated, even if incomprehensible, heals, because it is the patient, by means of his projection, who is the author of therapeutic efficacy.

Chapter III
Rituals and prescriptions

The pharmaceutical

In addition to incantations, the *nīš libbi* corpus contains a series of rituals and prescriptions. Some of the mentioned treatments are quite simple, consisting of a mixture of ingredients, plants, (parts of) animals and/or minerals, to make:

1. Potions;
2. Ointments and washing;
3. Amulets.[359]

Others, however, take the form of little rituals:

4. Creation of a small bow with arrows;
5. Libations of beer to the goddess Ištar;
6. Etiological analysis with a pig as well as male and female figurines.

In the case of potions, ointments, and amulets the prescription indicates the production of a real "pharmaceutical." To clarify the last term, I should address the terminological distinction between "medicine" and "pharmaceutical" which many scholars in the field of social sciences follow:

> Medicine: Substances (or objects) that, based on their inherent potency, are employed to engender transformations, such as the bodily change from ill-health to health.
> Pharmaceutical: Medicine that is based on biomedical knowledge and industrially produced. [Pool and Geissler 2005: 88]

Both Robert Pool and Wenzel Geissler, therefore, distinguish between "medicine" and "pharmaceutical": the first refers to "traditional and folk medicine," whereas the second to biomedicine. Contrary to this dichotomy, other scholars believe that we cannot make such distinction and have demonstrated the artificiality of analytical categories such as "traditional" and "folk" (see Dozon 1987; Schirripa 1996; Schirripa 2015: 22–33; Pizza 2005: 155–161; Pizza 2012). Here, however, I prefer to make use of the term "pharmaceutical" following Pino Schirripa. According to the Italian anthropologist, a pharmaceutical can be defined as a given substance, or rather a set of substances, which are perceived in a given community and in a precise historical

359 Only *nīš libbi* T prescr. 2 obv. 3′ mentions a suppository for the sexual desire. In No. A.1: 38–40 three (var.: two) bindings are made in a harp string, which is tied around the patient's right and left hands. In some prescription the therapeutic operator spins white and red wool and makes seven bindings (No. K.7: 121–124, see also M prescr. 6: 25–26).

https://doi.org/10.1515/9783110757262-003

moment as effective to fight and solve, what in that specific context is considered as "sickness." The pharmaceutical is part of that complex set of theories and practices constituting the therapeutic process. Whether it is derived from plant, animal or mineral substances, or is the product of a laboratory chemical synthesis process, and at the same time, whether it is thought that its effectiveness lies in the biochemical action of the substances used for its packaging, or that its effectiveness lies in the intrinsic spiritual qualities of the substances used, in the correctness of the ritual actions performed by those who prepare the product or in the power of the formulas accompanying its production, the pharmaceutical represents in its materiality as a concrete and visible object of the therapeutic process. Schirripa (2015: 21), therefore, places an equivalence between "traditional" pharmaceuticals, i.e. those substances found in the pharmacopeia of extra-Western populations and those of synthetic pharmaceuticals, i.e. those in use in the biomedical field.

A fundamental anthropological study on the pharmaceutical, "the material things of therapy," can be found in the book edited by Susan Reynolds Whyte, Sjaak van der Geest, and Anita Hardon (2002: 7), *The Social Lives of Medicines*, offering a social survey of drugs, understood as "things with social lives."[360] Products that not only bring about transformations of the body in the sphere of healing but also «change minds and situations and modes of understanding» (ibid. 10).[361] The power attributed to pharmaceuticals is due to their concreteness.[362] The pharmaceuticals are tangible, and used in practical ways: they can be swallowed, applied to the skin, inserted in orifices. «Practicing medicine, after all, is the art of making dis-ease concrete» (van der Geest et al. 1996: 154). The scholars, therefore, propose a biographical analysis of the pharmaceutical: «Their production and marketing, their prescription, distribution through intertwined formal and informal channels, their death through one or another form of consumption, and finally their lives after death in the form of efficacy in modifying bodies» (Whyte et al. 2002: 13–14).[363] The metaphor of "life," from production to consumption, is also fruitful for our analysis. Beginning with the production techniques within the Mesopotamian therapies the studies of Goltz (1974), Herrero (1984), Böck (2009), and Reiner (1995) are of particular interest. To be made effective, pharmaceuticals must undergo the "power of words" (see Tambiah 1985). Incantations play an important role as they convey the power of healing in medicine. The prescriptions, in fact, instruct, following the preparation of the drug, that the *āšipu* applies the incantation upon the medical substance before its consumption, according to the typical formula:

---

360 The term is taken from the book *The Social Life of the Things* (1986) by Arjun Appadurai.

361 See also the essay of Akrich (1995) who uses the expression "le médicament comme action": the drug can act not only on the individual physical level, but also on the social level of the community.

362 See also van der Geest and Whyte 1989.

363 An interesting study is the one on injection: Whyte and van der Geest 1994.

šipta sebîšu/šalāšīšu ana libbi tamannu
You recite the incantation seven/three times on it.

It is therefore clear that the drug carries in itself a series of ideological meanings and represents the concrete aspect of a particular "explanatory model of illness" (Kleinman 1978). Next to the administration of pharmaceuticals, the therapeutic practice is the recitation of incantations, offerings and libations, rituals, and a set of ceremonial performances, necessary for healing and therefore the effectiveness of the therapy. The pharmaceutical embodies this function, channeling all the elements of the therapeutic process through its materiality and substance.

The ingredients

Returning to the *nīš libbi* therapeutics, while the rituals, in particular the one which includes the manufacture of a miniature bow, are specific to this corpus, simple medical prescriptions are widespread in Mesopotamian therapeutics. The specificity of the latter therefore is to be found not so much in the preparation of therapeutic substances, that is, in the techniques employed to produce these "pharmaceuticals," but rather in the choice of ingredients.[364]

Many scholars have already pointed out the difficulty in associating specific diagnostic categories to specific ingredients, plants, animals, and minerals (see Herrero 1984: 43–47; Geller 2005: 3–6; Scurlock 2006a: 67–71). However, it is possible to trace precise indications in the medical texts in this regard. The text BAM 1, referred to as *Therapeutic Vademecum* (see Attia and Buisson 2012), for example, is composed of three columns: the first one refers to the ingredient; the second one to the diagnostic category; the third to usage techniques. The information relates to a simple use: one single ingredient used to eradicate the sickness. A study on the association between *būšānu*, *lišān-kalbi* and *ṣaṣuntu* plants and the goddess Gula by means of specific diagnostic categories was undertaken by Böck (2014: 129–163). However, in the *nīš libbi* texts they are almost always a set of ingredients to mix up the "pharmaceutical" for the patient. All of which obviously complicates matters in the analysis, since it is not possible to separately consider the isolated function of an ingredient and that of the same ingredient in a set of drugs. It is possible to identify patterns of ingredients that occur in prescriptions though.[365] A famous case is the one of the *tarmuš*, "heals-a-thousand" (*imḫur-līm*) and "heals-twenty" (*imḫur-ešrā*) plants (see Herrero 1984: 46–47) which

364 See *nīš libbi* F prescr. 14–16: 48–53, where it seems that the ingredient group can be used for the realization of a potion or an ointment or an amulet.

365 Geller (2005: 6–7) has identified patterns of ingredients in the kidney and rectal sicknesses corpus.

often appear together, not only in this corpus of prescriptions but more generally throughout Mesopotamian therapeutics.

Plant ingredients

Many plants are used in this therapeutic corpus, some more frequently than others (which are only used once or twice).[366] The text BAM 1 mentions two plants thought to be useful for regaining sexual desire.

šumutta takkal	*nīš libbi*	*umaṭṭa*
azallâ	KI.MIN	*nissata lā īši*

You eat the *šumuttu*-plant	: for the (absent of) sexual desire	: it will decrease.
The *azallû*-plant	: for ditto	: he will not have any grief.

[BAM 1 iii 34–35, Attia and Buisson 2012: 29]

The *šumuttu*-plant, however, never appears alone in the prescriptions, but always in alongside other ingredients. While the *azallû*-plant is perhaps the only ingredient in *nīš libbi* K prescr. 17: 50.[367] In *Šammu šikinšu*[368] we find other suitable plants for the sexual desire:

šammu šikinšu kīma lišān-kalbi arātūšu nepelkâ
u sām ina pān mê izzaz arki mê ina kibir nāri aṣi
amuzinna šumšu ana nīš libbi u tab[kūte]
nasāḫi damiq tasâk ina šamni tapaš[šassu]

The plant, whose appearance is similar to (that) of the "dog's-tongue"-plant, whose leaves are wide apart and which is red; it is resistant to (flooding)-water, after the water (has receded) it reappears on the riverbank,
its name is *amuzinnu*-plant, it is good for the sexual desire and to remove the fee[bleness]. You pulverize it and with it in oil you anoi[nt him (= patient)].
[*Šammu šikinšu*, text 2, § 28′: 1–4, Stadhouders 2011: 22]

The *amuzinnu*-plant appears once in the *nīš libbi* texts, in *nīš libbi* O prescr 4: 5 if we consider *anuzinnu* a scribal error. In BAM 1 it is considered a therapeutic ingredient for the weakness of the flesh. As was shown in the previous chapter, tiredness and weakness of the body are characteristic symptoms of the lack of sexual desire:

366 For studies on plants see Löw 1924 -1934; Thompson 1949. For the use of vegetal ingredients in pharmacology see Farber 1977b; Farber 1981; Herrero 1984; Powell 1993; Reiner 1995: 25–42; Stol 2003–2005a; Stol 2003–2005b; Tavernier 2008; Scurlock 2014b: 273–294; Rumor 2015.

367 [If ditto . . .] the *a[zallû]*-plant in a (bag of) leather.

368 See also *Šammu šikinšu*: [*šammu šikinšu* x x x x x x] x x *ma ṣeḫ[ḫer/rā ina? qiš]ti? aṣi? šammu š[ū]* / [. . . *šumšu ana*] *nīš* [*libbi damiq ina šikar*]*i* [*rē*]*štî* M[IN] "[The plant whose appearance . . .] . . . and is sm[all . . .], which grows? [in the fore]st? – of th[is] plant / [its name is . . .; it is good for the] sexual [desire. In bee]r [of first] quality di[tto]" (text 1, § 27′: 1–2, Stadhouders 2011: 11).

amuzinnu : *šammi šīrī tabkūte* : *ina šikari rēštî sekēru amēla raḫāṣu*
amuzinnu-plant : medicament for weak flesh : warming up in beer of best quality (and) washing the man.
[BAM 1 iii 32, Attia and Buisson 2012: 29]

In KADP 2 we find as plant for the sexual desire the "dog's-tongue"-plant (*lišān-kalbi*):[369]

369 Böck in her study on the goddess Gula (2014: 129ff.), analyzes the plants *būšānu*, *lišān-kalbi* ("dog's-tongue"), *ḫaṭṭi-rē'i* ("shepherd's staff") and *ṣaṣuntu* because, according to the scholar, they are related to the goddess. She argues that *būšānu* and *lišān-kalbi* (also *ḫaṭṭi-rē'i*) are two names for the same plant: URU.AN.NA : *maštakal* (III 9, Rumor 2017: 5). For the interpretation of the sign AŠ as DILI the scholar takes up the idea of Köcher in "Ein Text medizinischen Inhalts aus dem neubabyloischen Grab 405" (1995: 204), according to which the logogram stands for the Akkadian word *pirištu* "secret," indicating that the following name is a secret name of the plant (for the Mesopotamian *Dreckapotheke* see Böck 2008: 320–321; Rumor 2015; for the edition of 'AŠ section' of URU.AN.NA: *maštakal* III see Rumor 2017). Kinnier Wilson (2005: 48–50), however, warns against reducing all the examples of the list to this interpretation. In fact, he analyzes some cases where the second name, rather than representing a secret name, seems to be a wordplay, thus creating a link, often at the signifier level, between the first and second name. One can consider our example in the light of this interpretation, but what would be the link between the two plants as at the signifier level? We know that *būšānu* in BAM 1 iii 20–21 is defined as "Ninigizibara's dog": "The *būšānu*-plant, its name is "Ninigizibara's dog" . . . / He drinks it in first quality beer and oil" (on the goddess Ninigizida see Heimpel 1998–2001: 382ff.; according to Böck 2014: 131 fn. 8, in this passage she should be interpreted as a manifestation of the goddess Gula, on the basis of the *Hymn of Balluṭsa-rabi to Gula*, Lambert 1967: 109). In URU.AN.NA : *maštakal* (II 109–110) we find similar information: *šammu* ḪAB *šammu bu'šānu* / *šammu kalb gula šammu* MIN "The ḫ a b-plant (in Sumerian) is *būšānu* (in Akkadian) / The "dog of Gula"-plant is the same" (Böck 2014: 132 also shows another example in the medical text AMT 19, 7: 4). The link between *būšānu* and *lišān-kalbi* could then be represented in reference to the dog in the plant names. This association is thus possible in reference to a wordplay in scholastic education, as proposed by Kinner Wilson. In fact, another passage of URU.AN.NA : *maštakal* states: "The *nikiptu*-plant: 'secret name' 'dog excrements,' 'dog tongue,' 'dog bones'" (III 42, Rumor 2017: 10), while in another line: "The *būšānu*-plant: 'secret name' 'dog flea'" (III 10, Rumor 2017: 6). Conversely, Böck argues that this association refers to an identification of the plants: the two terms would be both the same plant names, so "dog tongue" is nothing more than a metonymical derivation of "Gula's dog" (2014: 140–141). Steinert (2014: 362) in her review of the Böck's volume is skeptical about these identifications. In fact, the plants *bu'šānu*, *ḫaṭṭi-rē'i*, *lišān-kalbi* and *ṣaṣuntu* appear together in different prescriptions, I shall quote only the cases in *nīš libbi* corpus: A prescr. 22: 63–64; K prescr. 19: 53–54; K prescr. 31: 138–139 (restored). Therefore, it is likely that the terms indicate similar or related plants for reasons unknown to us, rather than an identification.

According to Böck the "dog's-tongue"-plant is related to the goddess Gula. It must be noted, however, that if the plant is an important ingredient in *nīš libbi* prescriptions, the absence of sexual desire is never attributable to the goddess. In this corpus we find only a reference (No. F.3; No. F.5) to Gula as Bau. Without providing sufficient evidence, the scholar (2014: 147) connects the absence of sexual desire (*nīš libbi*) with gonorrhea, following Biggs's interpretation (1967: 3 fn. 16) of the text BAM 112 i 17′–19′ (dupl. AMT 58, 6: 2–3): "[If] the penis of the man hurts when he urinates (and) when he ejaculates / he is suffering [in] his interior ([*ina*] *libbīšu ṣabit*) and (his desire) to go to a woman is reduced (and) there is pus flowing constantly from his penis, / that man suffers from 'discharge'" (Geller 2005: 64, No. 4). Both, Biggs and Böck, restore line 18 with *nīš*, creating thus a link between

šammu ina muḫḫīšu pizalluru
rabṣu (var.: *irtabbiṣu*) : *ḫaṭṭi-rē'î lišān-kalbi šum*[*š*]*u*
ana nīš libbi damiq tasâk ina šamni tapaššassu (var.: *ana sinništu lā ālitti damiq*)

The plant upon which the gecko
lies: *ḫaṭṭi-rē'î* ('shepherd's-crook') its name is *lišān-kalbi* ('dog's-tongue').
It is good for the sexual desire (var.: for a woman who does not give birth). You pulverize it and with it in oil you anoint him.
[KADP 2 v 40–41, Stadhouders 2011: 38]

This plant also always appears within a group of drugs, only in one prescription it is used alone. In this single case, D prescr. 5: 49–50, the root of the plant is ritually extirpated with a bronze knife at dawn, after having been purified at sunset:[370] "[If ditto]: You purify "dog's-tongue"-plant when the sun sets; he st[ands] in the morning before the sun . . . [. . .] / [. . .] you remove from its root . . . with a bronze knife; you pound (it); in beer he dri[nks (it) repeatedly and (he will get) sexual desire]."

In BAM 380, a text that lists several plants and their therapeutic use, the sexual desire is included in a short list along with the "dog's-tongue"-plant, the *ḫašḫūr api*-plant, and another plant:[371]

[*lišān-kalbi*]	*šammi nīš libbi*	*ina šikari* [*tašatti*]
[*ḫašḫūr*] *api*	*šammi* KI.MIN	*ina šikari taša*[*tti*]
[x x] ⌜x x x⌝ GI	*šammi* KI.MIN	*ina karāni tašatti*

["dog's-tongue"-plant]	plant for the sexual desire	[you drink] in beer
[*ḫašḫūr*] *api*-plant	plant for ditto	you dr[ink] in beer
[. . .] . . .-plant	plant for ditto	you drink in wine

[BAM 380 r. 42–44, dupl. BAM 381 iii 37–40]

We must however not underestimate the problems of analysing the ingredients in the prescriptions. In fact, we are not able to identify some plants and as well as animal substances and minerals (see Tavernier 2008: 193); we do not know if the plants were used because of their chemical properties or their symbolic value or both; the plants "suitable" for the sexual desire are also used in other medical prescriptions for a variety of symptoms and diagnostic categories.

sexual desire and 'gonorrhea' (*mūṣû*), while Geller, whom I follow, restores with ina translating "in his innards." There are no other texts which link the absence of sexual desire with cases of gonorrhea.

370 On the temporal conditions of the ritual see Mauss and Hubert 1902–1903: 28–31. In the case of Mesopotamia see Bottéro 1987–1990: 220; Livingstone 1999.

371 See for the mention of the expression Ú ŠÀ.ZI.GA "drug for the sexual desire" in this therapeutic corpus: K prescr. 6 l- 35; K prescr. 27: 75: 10 Ú$^{!meš}$ ŠÀ.ZI$^{!}$.GA$^{!}$ "10 drugs for the sexual desire"; No. K.3: 92; N prescr. 21 iv 13; A prescr. 17: 30 7 Ú^{meš} ŠÀ.ZI.GA *ina* KAŠ NAG "he drinks in beer the seven drugs (for) the sexual desire." See also in the *Aššur Medical Catalogue*, in the section XIX devoted to the loss of sexual desire: [. . .] x SAG MUŠEN DIŠ ⌜Ú ŠÀ⌝.ZI.G[A] ⌜*ana*⌝ [GÚ-*š*]*ú* GAR "[. . .] . . . the head of a bird. (Instructions) to place a drug for sexual desi[re] around his [neck]" (l. 100, Steinert 2018b: 217).

In my opinion, the function of the drugs within a prescription can only be understood by placing them in context, the groups of ingredients in which they occur. For example, it must be noted that potions and amulets usually are composed of seven plants, indeed the number seven has a strong magical value.[372] In fact, in prescriptions often groups of seven ingredients or their double (fourteen) can be found.[373]

The cause of suffering is also an important element for understanding the choice of medical products used in the preparation of the pharmaceutical. As stated above, etiological analysis can determine the causes of the sickness. Often, the suffering is due to an attack of witchcraft.[374] Some of these plants (see list of ingredients) are employed in cases where the cause of the sickness is witchcraft. In other words, some ingredients are considered particularly effective against the witch's attack, as evidenced not only by our group of texts but also by series like *Maqlû* and the corpus of anti-witchcraft texts: *ardadillu*, *bukānu*, *ḫašû*, *imḫur-līm* ("heals-a-thousand"), *imḫur-ešrā* ("heals-twenty"), *lišān-kalbi* ("dog's-tongue"), *nuḫurtu*, *sikillu*, *tarmuš*. A clear example is given by the incantation No. E.2: 35–37: "May *tarmuš*-plant, "heals-a-thousand"-plant, "heals-twenty"-plant, *ardadillu*-plant, / *usikillu*-plant, *nīnû*-plant,[375] *bukānu*-wood (var.: "wood-of-rele[ase"]) / dispel my fascination!". These vegetal substances (together with other ingredients) are used in the prescription, following the incantation, which involves the creation of an amulet to hang around the neck, because they can dispel the cause of the patient's suffering. The anti-witchcraft power of some of these ingredients is confirmed by other texts. For example, the "heals-a-thousand"-plant (*imḫur-līm*) is defined in *Maqlû* as "the plant which releases" (VI 102, Abusch 2016: 161).[376] The same is said of the *bukānu*-wood in an *ušburruda*-text: "(I have equipped myself against you (= witches)) with the *bukānu*-wood which that undoes witchcraft" (KAL 2, 36, rev. v 40′, Abusch and Schwemer 2011: 166, No. 7.8.1: 23′), and in the same text, the "heals-twenty"-plant "which does not allow magic to come near (var. adds: the body)" (Abusch and Schwemer 2011: 166, No. 7.8.1: 24′).

372 On the mythical and magical function in relation to heaven and earth of number seven see Horowitz 1998: 208–220.

373 Exceptions are the amulets where the number of ingredients varies. Besides, in these amulets we find ingredients not often used in potions or ointments. So, stones that are mentioned only once or twice can be found there.

374 "If a man is bewitched" (*šumma amēlu kašip*), for example, is the beginning of a series of prescriptions which have as their object the absence of sexual desire and various sexual difficulties. Already Abusch and Schwemer (2011: No. 2.5; see also Abusch et al. 2020: No. 2.6) had collected these therapeutic recipes which concern, in addition to potency problems, also sexual impurity. As they emphasize, these texts were not handed down within the framework of larger *šumma amēlu kašip* collections or other collections of anti-witchcraft texts, but seem to be only a very limited common textual tradition of arranging the single units in a specific order (2011: 101).

375 In *Maqlû* it is mentioned with a metaphor: "Like *nīnû*-plant may her witchcraft give way" (V 27, Abusch 2016: 136).

376 See also Abusch and Schwemer 2011, No. 7.8.3: 17′–30′; No. 8.7: 110‴, 113‴.

Other drugs however are associated with purity[377] (the absence of which is linked to witchcraft): alabaster, lapis lazuli, gold, *bīnu* (tamarisk), *ēru*, *maštakal*, and *sassatu*. In *nīš libbi* K prescr. 3: 24–29, the absence of sexual desire is associated with other symptoms, which results in a state of impurity:

> If a man is bewitched and his flesh is weak, (and) neither when walking, nor standing nor being on his bed, nor when urinating,
> his sperm flows, like (that of) a woman (his) 'genital discharge' is impure,
> the sperm of this man has been buried under the earth with a dead man. To cure him:
> *ašqulālu*-plant, *ēdu*-plant, *sikillu*-plant, *amīlānu*-plant, seeds of *maštakal*-plant,
> seeds of *šakirû*-plant, root of *baltu*, which (grows) over a grave, you wrap up (them) in a leather bag (and) put (it) around his neck, then he will recover.
> [*nīš libbi* K prescr. 3: 24–29]

Among the ingredients used in the creation of this amulet, there are the *maštakal*-plant and *sikillu*-plant, which have a purificatory power (see Maul 1994: 6). These ingredients, therefore, are not specific to the problem of lack of sexual desire but are crucial in a vision of the sickness, where one of the causes of the sickness is witchcraft and therefore also impurity.

However, there are drugs that we find in many other medical texts for various diagnostic categories, many of which are found in the corpus concerning kidney and urinary-tract sicknesses (Geller 2005), for example, *ḫašû*, *imḫur-līm* ("heals-a-thousand"), *nuḫurtu*, *urnû*. So, can one trace a link between the prescriptions and their ingredients present in the two corpora? That is difficult to answer, but perhaps the text called *Taxonomy of Uruk*, SpTU 1, 43 (Köcher 1978: 24–25; Stol 1993: 23–24; Heeßel 2010a: 30–31; Geller 2014: 3–9; Steinert 2016: 230–242) will come in handy. The text consists of only one column and is subdivided into four parts, each of which contains a list of diagnostic categories associated with a region of the body and its organs. The four mentioned parts of the body are: *libbu* (mind/heart), *pî karši* ("mouth of the stomach"), *ḫašû* (lungs), *kalâti* (kidneys). The fourth section, which affects the kidneys, lists the following diagnostic categories:[378]

From the kidneys	contraction
ditto	(absent of) sexual desire
ditto	anal sickness
ditto	*sagallu*-sickness
ditto	infertility
ditto	twisted uterus
ditto	maintenance of "gas."

[*Taxonomy of Uruk*, SpTU 1, 43 ll. 25–31]

377 On purity in Mesopotamian religion see van der Toorn 1985; Sallaberger 2007; Sallaberger 2011; Couto-Ferreira and Garcia-Ventura 2013; Guichard and Marti 2013.

378 For the astral interpretation proposed by Geller 2014 see the section "The astral influence" in this chapter.

In this section, there are diagnostic categories associated with the area around the kidneys, mostly concerning the male and female urogenital apparatus: the "contraction" (of the bladder),[379] infertility, twisted uterus, (intestinal) gases, anal problems. In addition, we find the *sagallu*-sickness.[380] Perhaps – it remains a hypothesis – it is possible to trace a medical view that these diagnostic categories, among which the loss of sexual desire, can be eradicated by making use of the same vegetal ingredients.

As stated above, other vegetal ingredients, however, are specific to the loss of sexual desire, such as *lišān-kalbi, azallû, šumuttu,* and *ḫašḫūr api.*

Animal-based ingredients

The *nīš libbi* prescriptions are full of animal-based ingredients.[381] The most commonly used animal is the male partridge (*iṣṣūr ḫurri*).[382] In the texts from Ḫattuša, the preparation of the animal is described: its wings are plucked, it is strangled or beheaded, eviscerated, its body flattened and dried with salt and other vegetable ingredients, then pulverized for the realization of potions.[383] In many prescriptions, beyond the provenance of the texts, often the blood of the bird,[384] the head,[385] the innards[386] (which are sometimes eaten),[387] the penis,[388] or the skull[389] are used for the realization of potions or amulets. Other birds[390] used in the prescriptions included in the texts from the Hittite capital are the wren (*diqdiqqu*),[391] and the NAM.GEŠTIN-bird (only one prescription).[392]

379 In the text there is no mention of the bladder, but it is usually a problem that concerns just that organ.

380 Perhaps a thigh muscle? The sickness causes pain whilst walking and standing (CT 23.1: 1). In KAR 44 r. 9 it is associated with some forms of paralysis. The connection between paralysis, feebleness and the lack of sexual desire is clear. The thigh has a clear symbolic meaning: hair of aroused animals to be used as a medical ingredient is taken from the thigh.

381 For an introduction on animal-based *materia medica* see Chalendar 2016. Note that there is no Mesopotamian list for the therapeutic use of animals.

382 When no part is specified: A prescr. 1: 3 (potion?); innards/blood: T prescr. 3 obv. 7′ (restored, potion).

383 E prescr. 5: 70–72; N prescr. 1 i 1–7; N prescr. 2 i 8–11; N prescr. 32 i 12–16; N prescr. 24 le. e. 1–7.

384 A prescr. 21: 62 (potion); C prescr. 1: 14 (potion); E prescr. 6: 74 (potion); F prescr. 10: 31 (amulet); F prescr. 12: 36 (potion); No. F.2: 63 (?); No. K.5: 103 (?); N prescr. 25 lo. e.: 2 (restored, ointment); P prescr. 7 obv. 23 (?).

385 No. C.1: 10 (amulet).

386 A prescr. 10: 20 (amulet); K prescr. 23: 66 (amulet?) ; P prescr. 9 rev. 3 (?).

387 A prescr. 1: 4; A prescr. 18: 31; E prescr. 6: 76; F prescr. 12: 38; N prescr. 12 iii 1.

388 F prescr. 13: 42 (amulet); N prescr. 4 i 17 (var.: testicles, amulet); N prescr. 16 iii 28 (amulet).

389 N prescr. 15 iii 24 (ointment).

390 See in the *Aššur Medical Catalogue*, in the section XIX devoted to the loss of sexual desire the mention of a the head of a bird as ingredient: [. . .] x SAG MUŠEN DIŠ ⸢Ú ŠÀ⸣.ZI.G[A] ⸢*ana*⸣ [GÚ-š]*ú* GAR "[. . .] . . . the head of a bird. (Instructions) to place a drug for sexual desir[e] around his [neck]" (l. 100, Steinert 2018b: 217).

391 E prescr. 4: 68–69 (potion); N prescr. 5 i 23–27 (potion).

392 N prescr. 5 i 28–37 (potion).

Ingredients from other birds are rarely used: bood(?) of dove (*sukannīnu*);[393] wings of male eagle (*našru*);[394] spur (*ḫindūru*) of *ballūṣītu*-bird;[395] raven (*āribu*).[396]

The prescriptions of this corpus employ the hair or wool of sexually excited male animals. Except for the lion[397] and the deer, these are domesticated animals, such as bull,[398] dog,[399] pig,[400] donkey,[401] ram[402] (also lam),[403] and buck,[404]often qualified as "reared-up" ready for the mating. It is often stated that the hair of the domesticated animal must be taken from the perineum or tail after the animal has mounted the female: the hair is thus charged with the sexual potency of the animal and can be transferred, through the drug (amulet or potion), to the patient. An example is given by *nīš libbi* A prescr. 2 in which the hairs of domesticated animals are used to realize an amulet of red wool with seven bindings to be tied around the patient's waist:

> Its ritual: When a breeding bull (*alpu puḫālu*) has mounted (*išḫiṭu*) the cow
> yo[u tear off] a hair/hairs of its leg (*šarta ša purīdīšu*)
> when a ram, a donkey, a dog, a pig has mounted [. . .]
> you spin with red wool, you make seven bindings (and) [you put (them) around his waist and he will recover].
> [*nīš libbi* A prescr. 2: 6–9]

Likewise, the saliva of the sexually excited bull,[405] ram,[406] and buck[407] is an ingredient in the texts of this therapeutic corpus. Regarding wild animals, in addition to *rikibtu* as we will see, the penis, the antler, and the hair(?) of the stag are used.[408] While regard-

393 A prescr. 21: 62 (potion).
394 K prescr. 30: 130 (?) = N prescr. 8 ii 3 (eat in ball and amulet(?)).
395 No K.8: 64 (ointment); L prescr. 5: 14 (ointment); P prescr. 10 rev. 5 (?).
396 Heart of a back raven: L prescr. 1: 5 (ointment); N prescr. 25 lo. e. 1 (ointment); skull: P prescr. 7 obv. 23 (?).
397 K prescr. 4: 30 (amulet).
398 Breeding bull: A prescr. 2: 6–7 (amulet); black bull: E prescr. 2: 63 (potion).
399 A prescr. 2: 8 (amulet); something from copulating dog: N prescr. 13 iii 7 (broken, potion).
400 A prescr. 2: 8 (amulet); F prescr. 10: 31 (amulet).
401 A prescr. 2: 8 (amulet).
402 A prescr. 2: 8 (amulet); No. D.4: 62 (amulet); F prescr. 17: 57 (amulet); No K.7: 155 (no mention of hair, only of tail) (?); N prescr. 10 ii 27–28 (amulet); wool? from ram: O prescr. 6: 11 (broken, amulet); W prescr. 2: 5–6 (amulet).
403 K prescr. 27: 75 (amulet).
404 No. D.4: 61 (note also the use of the "little thing" (*mimmu ṣeḫru*) of buck's penis) (amulet).
405 F prescr. 12: 37–38 (potion); N prescr. 4 i 18 (amulet); N prescr. 12 iii 2 (potion).
406 N prescr. 4 i 19 (amulet); N prescr. 18 iv 3 (restored(?)).
407 N prescr. 4 i 119 (restored, amulet).
408 *Rikibtu*, penis and antler: A prescr. 16: 27 (amulet); navel, antler, *rikibtu*, penis(?) and hair(?): K prescr. 30: 129–132 (amulet); *rikibtu* No. B.2 Variant B r. 34; No. C.1: 8, 10 (amulet); No. F 3: 73(?); No. J.2: 18; antler: N prescr. 18 iv 3 (?).

ing domesticated ones, the throat[409] and blood of ram[410] or buck,[411] pelt of ram,[412] kidney(?)[413] or hair of the tail[414] of sheep, and pork fat[415] are also used, although to a lesser extent. Often wool is used for the creation of amulets,[416] most often red[417] but occasionally (twice only) white.[418] In many therapeutic cuneiform texts the leather bag is used for amulets as well (note the use of bag "made from a female kid that has not yet mated" in *nīš libbi* K prescr. 22: 65). As it is evident, if in the incantations the sexual qualities of wild animals, such as deer, wild bull, horse, *akkannu*-ass and *serrēmu*-onager (also wild got, partridge, lion, wolf, and snake, while the dog and other domesticated animals to a lesser extent) are exalted, on the contrary in the prescriptions hair, saliva, and other substances are taken from domesticated animals since they are more readily available.

Concerning animal urine and feces (*zû*),[419] only a prescription mentions the stallion's (*sisû*) urine: when a stallion on the street urinates, the *āšipu* takes the edge of remains of this urine and mixes it with beer, after that the patient drinks the potion on an empty stomach (*nīš libbi* E prescr. 1: 60–62). Very few times excrements are used: bat (*pīti šuttinni*),[420] dog,[421] *arkabu*-bat,[422] polychrome lizard (*ṣurāru barmu*).[423] In addition to the polychrome lizard, *šakkadirru*-lizard,[424] and copulating geckos (*pizalluru*) of the steppe[425] are used as well.[426] Urine and excrements belong to non-domesticated animals, except for the dog. Insects are rarely used: bee stinger,[427] wasp stinger,[428] head of red ant,[429] tail of scorpion.[430]

409 F prescr. 13: 42 (amulet); N prescr. 16 iii 29 (amulet); P prescr. 6 obv. 19 (amulet?).
410 E prescr. 3: 65 (ointment and potion).
411 N prescr. 25 lo. e.: 1 (ointment).
412 No. J.2: 94 (amulet).
413 K prescr. 4: 30 (restored, amulet).
414 N prescr. 4 i 20 (amulet).
415 T prescr. 1 obv. 2′ (?).
416 A prescr. 3: 10 (amulet); N prescr. 4 i 21 (from perineum, amulet).
417 A prescr. 2: 9 (amulet); No. D.4: 62 (amulet) ; F prescr. 13: 42 (amulet); K.6: 121 (amulet); M prescr. 5: 25; N prescr. 16 iii 29 (amulet).
418 K.6: 121 (amulet); M prescr. 5: 25.
419 Note that only in the incantation No. B.1 the man's urine and excrement are mentioned.
420 E Bow ritual: 74.
421 Y prescr. 1: 2′.
422 K prescr. 20: 136 (?) = N prescr. 8 ii 4 (eat in ball and amulet(?)).
423 K prescr. 31: 136–137 (?) = N prescr. 8 ii 4–5 (eat in ball and amulet(?)).
424 P prescr. 11 rev. 6 (?).
425 K prescr. 31: 136 (?) = N prescr. 8 ii 4 (eat in ball and amulet(?)); N prescr. 17 iii 32 (potion); D prescr. 6: 51.
426 Note that in a first millenium love ritual the uterus (*rēmum*) is referred to by the names of various lizards: "The lizard (*ṣurārû*) of [your] vulva [. . .] the gecko (*pizalluru*) of [your] vulva [. . .]" (K. 7924 ii 10–11, Lambert 1975: 112).
427 O prescr. 6: 10 (amulet).
428 O prescr. 6: 9 (amulet).
429 O prescr. 6: 10 (amulet).
430 No L.7: 155 (?); O prescr. 6: 9 (amulet).

In the ritual of the bow, its string is realized, as we will see, either by the tendon of the gazelle[431] or of the *arrabu*-mouse.[432] Other animal ingredients are mentioned only once: turtle;[433] wing of bat (*šuttinnu*);[434] hair of monkey;[435] meat of a mongoose (*šīr šikkî*);[436] scaly skin fish (*qulipti nūni mikî*);[437] fat of dancing(?) bear (*šaman asi muttalliki*),[438] green frog (*muṣa''irānu arqu*),[439] and its tadpole (*atmu*).[440] It is not excluded, however, that some of these animal substances, such as monkey hair, bat wing or meat of a mongoose, are actually secret names for plants.

Among the animal-based drugs, the *rikibtu*[441] of the *arkabu*-bat and the stag is frequently used and only in this therapeutic corpus. According to Civil (1984) the term *arkabu* refers to a type of bat, as already suggested by Landsberger (1934: 97), because of the frequent association in the lexical lists (e.g., see ED Birds List, OA 17: 171, l. 126; Pettinato 1981: 114, l.126) of arkab^{mušen} with su-din^{mušen} 'bat.' In *nīš libbi* corpus, as other therapeutic texts, we also find the expression U_5 ARKAB^{mušen} (*nīš libbi* P prescr. 7: 23), whose reading by the previous literature was *rikibti arkabi*.[442] As Chalendar (2018) pointed out, there is no correlation between U_5 and *rikibtu* in the lexical lists. On the contrary U_5 can be read *rikbu*. According to the scholar, *rikib arkabi* could refer to bat guano, although in other therapeutic texts, including *nīš libbi* ones, the logogram še_{10} designates the animal's excrement (*nīš libbi* K prescr. 31: 136). Chalendar's hypothesis, although suggestive, remains uncertain.

The expression *rikibti arkabi*, unlike *rikib arkabi* (U_5 ARKAB^{mušen}), seems to be peculiar to the *nīš libbi* corpus. Its meaning has been the subject of great debate among Assyriologists, who, as mentioned above, thought there was a correlation between U_5 and *rikibtu*. Some scholars considered it as a no-animal-based ingredient: Ch. Fossey (quoted in Labat 1959: 7 fn. 6) 'rut de bouquetin' (a kind of plant);

431 R Bow ritual: 9′.
432 E Bow ritual: 58. See also X prescr. 2: 10′ (?).
433 Y prescr. 1: 7′.
434 No. K.1: 13 (potion or amulet).
435 K pescr. 22: 64 (amulet).
436 X prescr. 2: 8′ (?).
437 B prescr. 1: 24 (ointment).
438 D prescr. 2: 29.
439 No. D.1: 6 (ointment).
440 F prescr. 8: 26 (ointment).
441 CAD R 344 provides the meaning of 'sexual intercourse' (see also Kwasman 1992: 42). Biggs (1967) shares this opinion, with reference to the following *nīš libbi* text: [*ri*]*kibtaka* "your [love]-making" (incantation No. C.1: 3) (also in incantation No. F.3: 68, perhaps also l. 73). For this interpretation of the term in a sexual context see A I/1 36–40: [e] [a] = *mu*-[*ú*], *ri-ḫu-tu*[*m*], *ri-kib-tum*, *ma-a-a-lum*, *ni-lum ša ra-ḫi-e*. See also the meaning of 'pollination' (CAD R 344 mng. 2, see also Chalendar 2018: 25 fn. 3).
442 Note that there are no attestations of U_5 DÀRA.MAŠ /*ayyālu* '*rikib* of deer' in cuneiform texts.

Ebeling (1925: 53) 'Brunst des Ib-Vogels'; Geller (2002: 150 No. 6), Rumor (2015: 67, 101 and 103), and Steinert (2015: 134 No. 102) consider the expression as a secret name for the *urānu*-plant (perhaps 'anise'). For others, it is an animal-based substance: Biggs (1967: 25–26) 'bat's thumb'; AHw. II 984a translates *rikibtu* in general as 'Nebenklaue, Sporn'; CDA 305 'spur of an *arg/kabi* bird'; Landsberger (1934: 97 fn. 3) an aphrodisiac substance extracted from the testicles of the bat; CAD R 344, Civil (1960: 712), and Chalendar (2018: 43–44) 'bat guano'; Civil (1984) 'bat sperm' (because logogram A is used to express *rikibtu* in the lexical lists), understood as a popular designation for guano; Finkel (2000: 151) 'bat-semen(?)'; Scurlock (2014b: 403) 'musk.'

The *rikibtu* of the *arkabu*-bat appears only in texts from Ḫattuša. It is administered orally in a potion(?):

> If ditto (= a man's sexual desire has finished in the month Nisannu): the *rikibtu* of an *arkabu*-bat, *eli*[*kulla*]-plant,
> . . . of tamarisk [. . .]
> *maštakal*-plant, date palm, *bukā*[*nu*]-wood
> and you put in water all seeds and
> let them spend the night under the stars, you pu[t] (it) on the roof,
> you let that man sit down, with *k*[*asû*]-water
> you [wash] all his body [. . .]
> the *rikibtu* of an *arkabu*-bat [. . .]
> he drinks half and [you anoint$^?$ him with the other] ha[lf]
> premium oil and . . . [of tamarisk . . .]
> on his left foot . . . [. . .]
> you pulverize and with [oil] from the a[labastron his body],
> you anoint (it) and [that] m[an will get sexual desire].
> [*nīš libbi* N prescr. 14: 11–23]

Or in the same text rubbed on the patient's body:

> [If ditto (= a man's sexual desire is taken away): The hear]t of a male raven, the blood of a reared-up buck,
> [the blood of a] male [partri]dge, the *rikibtu* of an *arkabu*-bat,
> seed of the *dadānu*-acacia you mix together, the pelvis of that man for three times
> you rub with it ⟨and⟩ he will get sexual desire. Wording of *nīš libbi* (incantation).
> [*nīš libbi* N prescr. 25 lo. e. 1–4]

These types of administration may suggest that the substance is soft or pasty. The ingredient appears also in another text, unfortunately because its fragmentary nature, we cannot be sure that it belongs to the *nīš libbi* corpus:

> [. . . ^{giš}]ÁSAL [. . .]
> [. . . *ri-ki*]*b-ti* ARKAB^{mušen} [. . .]
> [. . .] x *i-nu-uh-ha* [. . .]
> [. . .] KU RU [. . .]

> [. . .] Euphrates poplar [. . .]
> [. . . *riki*]*btu* of *arkabu*-bat [. . .]
> [. . .] . . . he will find relief [. . .]
> [. . .] . . . [. . .]
> [KUB 37, 7: 7′–10′, Chalendar 2018: 48]

In the *nīš libbi* rituals, the term *rikibtu* is related also to the stag (*ayyalu*). Ebeling (1925: 45 fn. 1) translates as 'Hirschbrunst,' indicating a kind of fungus; Landsberger (1934: 97 fn. 3) interprets *rikibtu* as an aphrodisiac substance extracted from the testicles of the stag; CAD A/I 226 *s.v. ajalu* 'an aphrodisiac obtained from the dew-claw(?) of a stag'; CAD Ḫ 246 *s.v. ḫurāṣu* 'the potency of a stag'; CDA 305 'stag's spur' (an aphrodisiac); CAD R 344 'scat(?) (of a stag)'; Finkel (2000: 162) 'stag-semen(?)'; Biggs (1967: 26) suggests the meaning of the spur of a stag, but he brings up the subject once more (2002: 75–76) claiming that the term indicates the animal's velvet; Scurlock (2014b: 403) 'musk'; Chalendar (2018: 41–43) considers it as an aphrodisiac from the stag, maybe its tears. All the identifications, however, remain uncertain.[443]

It should be remembered that the *rikibtu* of the stag is mentioned only in this category of rituals. We know that the *rikibtu* of the stag is used in the production of amulets:

> If ditto (= If a man's (desire) to go to a woman is reduced [. . .]): *rikibtu* of a stag, antler of a stag, penis of a stag, *takdanānu*-plant in a leather bag. Its ritual: You put (it) around his neck.
> [*nīš libbi* A prescr. 16: 27]

> In order to get sexual desire: Navel of a stag, antler of a sta[g],
> *rikibtu* of a stag, *aṣ*[*uṣu*]*mt*[*u*]-plant
> you properly crush again and again, you sme[ar] the penis (of a stag?) with salt,
> you tear off hair of the tail of a male (stag?) (and) together with a string [. . .]
> [you s]pin and you insert *amašpû*-stone inside [. . .]
> you tie (it) around his waist [and (he will get) sexual desire].
> [*nīš libbi* K prescr. 30: 129–234]

> Make love to me! Make love to me because I am young! [. . .]
> and I am endowed? with the *rikibtu* of a stag! Make love to me! [Incantation formula].
> Incantation: If a man's (desire) to go to a woman is reduced.
> Its ritual: (You put) the head of a male partridge,
> silver, gold, the *rikibtu* of a stag into a . . . leather bag.
> Its ritual: You recite the incantation over (it) seven times,
> [you do that] three and four [times].
> [Incantation and its ritual No. C.1: 7–13]

It has also been used in potions, where the grinding is mentioned, but not necessarily applicable to the *rikibtu* itself:

443 For a more detailed analysis of the debate on the term *rikibtu* of both bat and stag see Chalendar 2018: 24–28.

> Root of *samīdu*-plant which is in a garden-pl[ot]
> you dry and crush, *sikillu*-plant in beer . . . [. . .]
> *mūṣu*-stone, magnetite, . . .-stone,
> [innards/blood] of a male partridge in beer,
> or he drinks in wine, *dadā*[*nu*-acacia],
> seeds of *egemgiru*-plant, seeds of *ēdu*-plant [. . .]
> *rikibtu* of a stag . . . [. . .]
> you crush (and) sift, [he drinks] in strong wine [. . .].
> [*nīš libbi* T prescr. 3 obv. 7′–rev. 11′]

The term appears also in the following *nīš libbi* texts:

> Make love to me! Ma[ke love to me] because I am young!
> [. . .] I am endowed with love, make love to [me]!
> Tie [. . .] of the *rikibtu* of a stag! Ma[ke love to me]!
> [No. F.3: 71–73]

> You recite these incantations over the *rikibtu* of a stag.
> [ritual of incantation No. B.2, variant B r. 34]

> [. . .] it was created, [. . .] . . . [. . .]
> [. . .] it was created, *rikibtu* of a s[tag],
> [. . .] they were created, bet[ween]
> all [the wild animal]s of the steppe, the ani[mals . . .]
> [. . .] . . . pure . . . [. . .]
> [. . .] the mating, wing of the . . . -bird [. . .].
> [No. J.2: 17–22]

The hypothesis of considering the *rikibtu* of bat and stag as excrement seems unlikely to me. The texts of the corpus mention the excrements few times and when they do so they use the logogram ŠE_{10}, in reference to the bat and the lizard (*nīš libbi* K prescr. 31: 136–137) (see also U_5 ARKAB indicating perhaps the bat guano in *nīš libbi* P prescr. 7: 23), but never that of the stag or other animals. Therefore, the use of excrements to combat the loss of sexual desire is extremely limited in this therapeutic corpus. The prescriptions of the corpus privilege ingredients that are somehow related to sexual desire and sexual vigor, as they aim to transfer the sexual qualities from the animal to the patient: genital organs, perineum, or tail hair (often when the animal during the estrous cycle), saliva, and blood. In fact, it seems difficult to associate the excrements with a word whose *rkb* root has a sexual connotation.

Regarding the interpretation of the term as sperm, it is true that in lexical lists there is a correlation between A 'sperm'[444] and *rikibtu* and this meaning can be related

444 Proto Aa 4 : 1–6 [e] A *mu-u*$_4$, *ri-ki-i*[*b-tum*], *ṭe-ru-*[*tum*], *ni-i-lum*, [*r*]*i-ḫu-ú-tum* (Civil et al. 1979 = MSL 14: 89, ms. B); Aa I/1 36–41 [e A] *a-a-ú mu-*[*ú*], *ri-ḫu-tu*[*m*], *rikib-tum*, *ma-a-a-lum*, *ni-i-lum šá ra-hé-e*, *mi-i-lum* (Civil et al. 1979 = MSL 14: 203–204); Proto-Kagal 234–238 A$^{[mu]\text{-}u_4}$, A$^{[mi]\text{-}lum}$, A$^{[ru]\text{-}ṭi\text{-}ib\text{-}tum}$, A$^{[r]i\text{-}ki!\text{-}ib\text{-}tum}$, A$^{[ri]\text{-}ḫu\text{-}tum}$ (Landsberger et al. 1971 = MSL 13: 73).

to the *rkb* root. But why using a specific term only for the sperm of the bat and stag and not for example for the bull which is also considered a sexually potent animal? Furthermore, animal sperm is not used as an ingredient in this therapeutic corpus, and, unlike other animal-based ingredients, it is difficult to get.

The identification with the musk, secreted by the caudal gland of the musk deer (*Moschus moschiferus moschiferus*), is very fragile because this deer species is not indigenous to Mesopotamia, on the contrary to the DÀRA.MAŠ/*ayyalu* animal which seems to be. Besides, the musk deer is devoid of antlers, while cuneiform therapeutic sources often mention deer antlers as ingredients (see Chalendar 2018: 34–35). Furthermore, if we postulate that the term *ribiktu* indicates the same substance from both the deer and the bat, the latter does not produce musk, as well as velvet, as suggested by Biggs (2002: 75–76). The scholar underlines that deer velvet is produced in connection with the growth of antlers and the mating season provides a logical connection with the *rkb* root, but the identification has not sufficient proof. We can affirm the same regarding Chalendar's hypothesis to consider the *rikibtu* of the stag as its tears (it is significant that she does not give a precise identification for the bat's one).

The translation of *rikibtu* with deer dewclaw/spur and bat thumb is probably based on the semantics attributed to the *rkb* root as 'what protrudes,' but also on the role it could play during animal mating. The frequent mention of *rikibtu* together with the antler and penis of the stag fits well in this context of protuberances. However, Biggs thinks that this ingredient must be solid for the presence of a grinding operation for its preparation in the pharmaceutical (same criterion for his subsequent proposal of deer velvet). However, the grinding (*tasâk*) is only present in *nīš libbi* P presc. 7: 23 concerning U_5 ARKAB^{mušen}, which as claimed by Chalendar (2018) cannot be read as *rikibti arkabi*. Consequently, the ingredient does not need to be necessary solid, on the contrary, its solid nature could contradict the lexical attestations (A = *rikibtu*) and the methods of use of the ingredient in prescriptions. Unfortunately, due to the elusive nature of the sources, it is not possible to arrive at a definite identification.

It seems clear that the term *rikibtu* indicates a part or substance of the two animals, the stag and the *arkabu*-bat. The fact that the term contains the consonant roots of the verb *rakābu* 'to mount sexually' shows the importance of this animal-based ingredient in sexual contexts. In fact, the verb is frequently used in the *niš libbi* incantations, where it refers to the sexual activity of the animals which will hopefully be imitated by the man who has lost his sexual desire. The use of the term *rikibtu* in reference to stags and bats suggests a link between the two substances. However, morphological differences between these animals make the identification of the substance difficult. Chalendar (2018) argues that the *rikibtu* of the two animals does not refer to the same substance, but to two different ingredients both considered an aphrodisiac. This assertion cannot be considered certain, therefore we do not know if the term refers to the same substance or two different ones. I am more inclined to the first hypothesis, but without being able to identify it. What is certain is that it is a substance considered aphrodisiac, in any case, capable of acting on human sexuality, as can be seen from its etymology. The *rikibtu* is associated with the stag because the animal is considered, as we have seen in Chapter II, endowed with sexual vigor. Regarding the *rikibtu* of the *arkabu*-bat, it should be noted that also the term *arkabu* shares the consonantal root of the verb *rakābu*. The term *arkabu* is an *aPRaS*-nominalization (see GAG: 78, § 56), whereas *rikibtu* is a *PiRiSt*-form (see GAG a2b § 55). The similarity of the consonants in the bat's name and the sexual practice probably explains the reasons for the choice of the animal as an ingredient in prescriptions. It is through the name that sexual force is magically sent to the patient. This is a widespread phenomenon across many cultures: the name itself conveys these sexual qualities which man needs.

Potions

Most of the prescriptions, which are variable in length (from only one ingredient to multiple ones), include the creation of potions to be drunk in water, beer, or wine.[445] They can be subdivided into four groups according to the ingredients used: 1. plant ingredients; 2. plant and animal; 3. plants, animals, and minerals; 4. only birds.

I would like to put my attention on the last group, typical of many Middle-Babylonian *nīš libbi* prescriptions from Ḫattuša. The bird-based basic prescription makes use of the plucked, dried, pounded bird, which is drunk in high-quality beer. Of course, the information given about the preparation procedure of the potions varies from text to text. Some make mention of the blood that sometimes has to be poured out, in others they do not. In some texts, the animal must be beheaded, its wings torn off and the entrails removed. The animal is also processed along with other ingredients, such as the *dadānu*-acacia of the mountain,[446] the *nīnû*-plant, toasted wheat flour, barley, and *amānu*-salt.

The bird most used in *nīš libbi* rituals is the rock partridge (*iṣṣūr ḫurri*) (see CAD/I: 207–208 'partridge(?)'; AHw. I 390 'Steinhuhn(?)'):[447]

445 On potions see Herrero 1984: 88–92.

446 The plants growing in the mountains have a great efficacy because on the mountain's heights they are better exposed to the influence of the stars and the atmosphere is thinner (see Reiner 1995: 39).

447 The identification remains uncertain. Sommer and Ehelolf (1924: 59–61) identify the animal with the rock partridge ('Steinhuhn') based on the literal meaning of the term 'Erdlochvogel' ('bird of the hole'). Landsberger (1966: 262–264) identifies the animal with the *Tadorna ferruginea* (or *casarca*) ('Höhlenente') (*contra* Veldhuis 2004: 232 fn. 55), as well as Salonen (1973: 143–146). Janković (2004: 12–13) assumes it is the francolin, whereas Veldhuis (2004: 231–233) identifies it as the partridge. Biggs (2006: 45 fn. 57), on the contrary, considers the animal a bat. Already Reiner (1995: 87 fn. 363) had suggested this identification quoting the following text: "(Blood of) a male bat (*iṣṣūr ḫurri*) that goes about at night catching flies" (BAM 476: 10′). The identification remains uncertain. See also Minunno

If a man's sexual desire has finished in the month Nisan[nu]:
You take a male partridge,
you pluck its wings, you strangle and
flatten (it), you spread (over it) the salt,
you dry (it), together with seeds of mountain *dadānu*-acacia
you pulverize (it), he drinks in beer (it) and
this man will get sexual desire.

If ditto (= a man's sexual desire has finished in the month Nisannu): a male partridge who is reared-up for the mating
you dry, you pulverize (it), in the water,
which on the roof spent several nights, you put (it) and he drinks this and
this man will get sexual desire.

[If] ditto (= a man's sexual desire has finished in the month Nisannu): You behead a male partridge,
you put its blood in the water,
you ingests its innards and this water
you leave out during the night, when the sun rises
he drinks this and he will get sexual desire (var.: he will recover).
[*nīš libbi* N prescr. 1–3 I 1–16]

[Whe]n the partridge mates [. . .]
you [take . . . of] the male [partrid]ges in the Ayyāru [and pluck (it)],
you do not make the blood flow, you do not [. . .] their entrails, you do not . . .
you tie either two months or [three months . . .] you crush [the ta]lons, the beaks, the stomachs,
the muscles/tendons and the intestines on the stone [. . .], you soak in . . .
three handfuls of seeds of *aluzinnu*-plant, two handfuls . . . [. . .] . . . you mix, in beer in the usual wa[y]
you dilute, he drinks it on an empty stomach, and [that] m[an will desi]re.
[*nīš libbi* N prescr. 24 le.e.: 1–7]

What is certain is that in *nīš libbi* corpus, the animal is used in reference to its sexual activity. Landsberger (1934: 80 fn. 4) emphasizes the animal's symbolic role in male sexual desire and vigor. See *nīš libbi* incantations: "(Make love to me) like a partridge thirteen times!" (No. E.3: 48); "The mating of a partridge twelve times" (No. C.1: 6, see No. F.3: 70; No. K.8: 149–150). The mention of an aroused male partridge is fairly common (*iṣṣūr ḫurri zikaru ša ana rakābi tebû* "male partridge who is reared-up for the mating"). Likewise, partridges were taken during their mating period (*enūma iṣṣūr ḫurri irtanakkabu* "when the partridge mates," N prescr. 24 le.e. 1) and one of their body parts used as an ingredient was the penis. Many rituals, instead, provide the use of the blood, the heart which is eaten, or even the whole body, plucked, gutted and dried, pound, and drunk in a potion.[448] The bird is used in potions, unguents, and

2013: 127. For texts on divination in Mari see Durand 1988: 38; as to the association for the demon Asakku see *Šurpu* V-VI 42–57 and 170–172 (Reiner 1958); Lambert 1970: 114 l. 17; in the *namburbû*-rituals see Maul 1994: 240 l. 61.

448 Once the skull (UGU) is used: N prescr. 15 iii 24.

amulets, either alone or along with other ingredients. In an incantation for a prosperous brothel the following sentence probably refers to this bird species: *kīma ṣēri ša ištu ḫurri uṣṣâmma iṣṣūrū ina muḫḫīšu iḫabbubū* "Like a snake (= penis), going out from a hole, and birds twittering over him (= snake/penis)" (see Panayotov 2013: 293, l. 30).

Another bird processed in the prescriptions from Ḫattuša is the *diqdiqqu*-bird (AL.TI.RÍ.GA^{mušen}), perhaps to be identified with the wren (CAD D 159; AHw. I 173; Stol 1971: 180; Salomen 1973: 158 'wren'; Borger 2004: 465–474; Veldhuis 2004: 217–218; Fechner 2017 'Zaunkönig'; Jiménez 2017: 357–360). Veldhuis (2004: 217) suggests that the bird's name relates to the word *daqqu* 'small.'[449] The bird has a vast use in the medical sphere and in the *nīš libbi* rituals we find it only in the following prescriptions:

> In order to get sexual desire: You take and [pluck] wre[n],
> you do not make the blood come out, you dry (it), you pulverize (it), [you mix (it)] with dr[y] wheat, [he drinks (it) and ditto (= he will recover)$^?$].
> [*nīš libbi* E prescr. 4: 68–69]
>
> If ditto (= a man's sexual desire has finished in the month Nisannu): You pluck a wren,
> you eviscerate (it) and you apply *amānu*-salt (and) *nīnû*-plant (on it),
> you dry, you pulverize,
> you mix (it) with toasted grain and
> he drinks this and he (var.: that man) will get sexual desire.
> [*nīš libbi* N prescr. 5 i 23–27]

According to Fechner (2017) it is perhaps related to this corpus because of its physical appearance with its short erect tail, but we have no proofs in favor of this interpretation. In *nīš libbi* N prescr. 5 i 25 *amānu*-salt and *nīnû*-plant are applied to the bird, after it has been spun and splashed, and before it is dried and powdered. Jiménez (2017: 342) emphasized the graphic relationship between the bird and the *nīnû*-plant, which perhaps served as the reason behind this type of preparation. The bird's name can be written NI.NI-*qu*,[450] homophonous with to the name of the plant. The bird is also called *iṣṣūr samēdi* "bird of the *samēdu*-plant." The logogram of this last plant, ^{ú}KUR.ZI, is similar to that of the *nīnû*-plant (^{ú}KUR.RA.(SAR)).

---

449 The animal appears together the elephant in a Sumerian proverb SP 5.1 (Alster 1997: 121) and its Akkadian counterpart YBC 9886 (Borger 2004: 470): "An elephant spoke to himself and said: / 'Among the wild creatures of Šakkan, / There is no one who can defecate like me.' / The wren (antirigu) answered: 'And yet, I, in my own proportion, / I can defecate like you.'" The two animals, bird and elephant, are found in VAT 8807: 50–54 (Lambert 1960a: 216–218): "A wren (*diqdiqqu*), as it settled on an elephant, / said: 'Brother, did I press your side? I will make [off] at the watering-place.' / The elephant replied to the wren: / 'I do not care whether you get on – what is it to have you? – / Nor do I care whether you get off.'"
450 On the reading of *diq-diq-qu* instead of *ni-ni-qu* 'mosquito' of Lambert see Borger 2004: 462. For other Sumerian sources see Veldhuis 2004: 217.

Another bird used in the potions from the Hittite capital is the NAM.GEŠTIN-bird,[451] which appears in only one prescription:

> If ditto (= a man's sexual desire has finished in the month Nisannu): You pluck and eviscerate a NAM.GEŠTIN-bird,
> you apply *amānu*-salt (and) *nīnû*-plant,
> you pulverize (it) with barley, roasted flour,
> mountain *dadānu*-[plan]t, on an empty stomach
> [he d]rinks (it) and he will sexual desire.
> [*nīš libbi* N prescr. 6 i 28–32]

The astral influence

A *nīš libbi* prescription, F prescr. 12: 36–41, involves the construction of a water-based potion with the following ingredients: branches of the *ašāgu*-acacia, the blood of a male partridge (whereas the heart is eaten by the patient), and the saliva of a bull. The drink is left all night under the stars. At dawn, the patient stands on the branches and drinks the potion before the sun:

> If a man's sexual desire has been taken away and he does not have sexual desire:
> You throw into water branches of the *ašāgu*-acacia, you throw into water blood of a male partrid[ge] and
> he eats the innards of the male partridge,
> you take the [sa]liva of a bull, you throw (it) into water (and) let it spend the night under the stars,
> as soon as the sun rises, on the branches of *ašāgu*-acacia
> you have him(= patient) stands and he drinks (it) in front of the sun, and (he will get) sexual desire.
> [*nīš libbi* F prescr. 12: 36–41][452]

A similar prescription is also present in *nīš libbi* E prescr. 6: 73–78:

> In order to get sexual desire: [You throw] branches of *ašāgu*-acacia [in wat]er,
> you let the blood of a male partridge f[low] in river [water . . .],
> you let it spend the night in the shadows, as soon as [the sun] rises, [you have him (= patient) stand] on [the branches of *ašāgu*-acacia and]
> in front of the sun he e[ats] the innards of the male partridge.
> . . . . he drinks it . . . [. . .]
> . . . [. . .] . . . [. . .].

451 Unknown bird. Ebeling (1925) suggests 'Weinschwalbe,' reading SIM(=NAM) as *sinûntu* 'swallow.' See Salonen 1973: 233.
452 See also N prescr. 12 iii 1–6.

In this text, however, the saliva of the bull is absent as an ingredient of the potion, and the heart of the partridge, from which is the blood is extracted, is eaten before the sun. After that, the potion made of water, here specified as river water, is left under the stars during the night.

Making the ingredients spend a night under the stars (*ina kakkabī tušbât*) is a very common practice in the therapeutic tradition in Mesopotamia, and it is found in the *nīš libbi* rituals as well. See, for example, *nīš libbi* P prescr. 9 rev. 3–4: "The innards of a male partridge / you place it under the stars overnight, you re[cite] the incantation three times over it [. . .]."[453]

Commenting on this practice, Reiner (1995: 48) affirms that there are practical reasons for letting the preparation stand overnight: the ingredients must be steeped in the carrier to be properly blended. According to many scholars, the main purpose of this therapeutic practice is the maceration of the ingredients during the night (see Goltz 1974: 51 fn. 300; Schwemer 1998: 122, 131; Böck 2009: 112). However, the choice of placing the ingredients under the stars is charged with another value. Reiner (1995: 48) rightly says in this regard that the practical reason alone does not explain or justify the procedures. Indeed, the exposition of the drugs to the stars is necessary to obtain astral irradiation and, therefore, the celestial powers make the pharmaceutical efficacious.

The place where the substances would acquire astral radiation is the roof:

> If ditto (= a man's sexual desire has finished in the month Nisannu): A male partridge who is reared-up for the mating
> you dry, you pulverize (it), in the water,
> which on the roof spent several nights, you put (it) and he drinks this and
> this man will get sexual desire.
> [*nīš libbi* N prescr. 2: 8–11][454]

In the *nīš libbi* prescriptions in which such practices are formulated, the other main ingredient, in addition to the branches of the *ašāgu*-acacia, is the male partridge (particularly its blood):

> ([If ditto (= a man's sexual desire has finished in the month Nisannu): . . .])
> [you ingest] the innards of a [male] partridge,
> [you take] the phlegm of a reared-up bull, [you put (it) in the water],
> [you let] it [spend the night] under the stars on the roof,
> when the sun ri[ses, on the branches of *ašāgu*-acacia]
> he stands, in front [of the sun . . .]
> with [the water he drinks this and will get sexual desire].
> [*nīš libbi* N prescr. 12 iii 1-6]

453 See also L prescr. 4; N prescr. 14 iii 15; No O.1: 26.
454 See also N prescr. 12 iii 3; N prescr. 14 iii 15.

The water-based potion, after being exposed to the radiation, is drunk at dawn (*ištu Šamaš ittapḫa*) before the sun (*ana pān Šamši*). "Before the sun" in this context refers, not so much to the ritual representation of the god Šamaš, but rather the sun in the sky. It is clear, however, that, the two aspects cannot be separated. Parpola (1971: 182) points out that the role of the sun in magical context was that of the Supreme Judge, who doomed the wicked and let free the innocent, and whom the exorcists approached to save their patients from evil forces. Sunlight, especially that of dawn, has a purifying capacity: "For I am cleansed by the rising of the s[un]" (*Maqlû* VII 150, Abusch 2016: 187).

The influence of the stars and celestial bodies upon medical substances and upon the human body has already been analyzed by some scholars such as Reiner (1995), Casaburi (2002–2005), Scurlock (2005–2006) and most recently by Geller (2014) who analyzed the theory of melothesia in Mesopotamia.

Melothesia is the Hellenistic doctrine of the influence of the zodiac signs and planets on human anatomy. Although there are no explicit cuneiform texts on melothesia, there are some elements to be attributed to it in the Mesopotamian documentation. Let us consider in this regard a text of magic-astral character originally of the Hellenistic period from Uruk, BRM 4 20 (Biggs 1967: 5; Geller 2014: 27–39), where a specific magic practice corresponds to a zodiac sign. In the following lines the text makes mention of the sexual sphere:

KI.ÁG.ÁG NITA *ana* MUNUS	KI ^{mul}*zi-ba-nu*
KI.ÁG.ÁG MUNUS *ana* NITA	KI ^{mul}KUN.MEŠ
KI.ÁG.ÁG NITA *ana* NITA	KI ^{mul}GÍR.[TAB]
MUNUS GIN.NA	KI ^{mul}LÚ.ḪUN.GÁ
MUNUS *šu-ud-bu-bu*	KI ^{mul}LÚ.ḪUN.GÁ
MUNUS LÚ *ana* NITA *šá-nim-ma* IGI^{II} *u* IGI *la na-še-e*	KI ^{mul}MAŠ.TAB.BA GAL.[GAL]
ŠÀ.ZI.GA	KI [^{mul} . . .]

(Incantation for) 'the love of a man for a woman': the sign of the Scales.
(Incantation for) 'the love of a woman for a man':[455] the sign of the Tails.
(Incantation for) 'the love of a man for a man': the sign of Scorpion.
(Incantation for) 'going with a woman': the sign of the Hireling.[456]

(Incantation for) 'making sure that a woman has sexual intercourse':[457] the sign of the Hireling.

455 See W 23293/34 1, Weiher 1998: 243, 35 (also Geller 2014: 59).

456 In the same text, the comment on line 50: MUNUS GIN.NA // *sin-niš-tú a-na a-la-ku*. See KAR 61 (Biggs 1967: 70–71).

457 In the same text the comment on line 53: MUNUS *šu-ud-bu-bu* // MUNUS *su-un-nu-qa* MUNUS *šá e tul-la-tu-šu*$^{!}$ *mim-ma ma-la ta-sal-lu-šú i-ṭáḫ-ḫu-ka* "MUNUS *šu-ud-bu-bu* // to make an approach to a woman / a woman which you must not 'swallow' / whatever you ask of her / she will have sex with you" (Geller 2014: 36; 2005: 86–87).

(Incantation so that) 'the woman of a man not turn her eyes and face to another man': the sign of the Great Twins.

(Incantation for the) sexual desire: the sign of [. . .].
[BRM 4 20: 5–8, 17, 21, 38, Geller 2014: 28–29]

The text STT 300, dated to the seventh century from the site of Sultantepe, associates specific rituals to certain days of the moon where it is considered favorable to perform rituals.

[If] you perform *nīš libbi* incantations [on] the first day of the month [Šabāṭu], from the first day to the 30th day, it will pay off.
[If in month Ayyāru], you perform (the incantations) from the first day to the 15th day (for) having 'forehead affliction' or (for) alleviating (the problem of) the sexual desire, (or) from [the first day] to the 30th day for getting rid of Lilû-demon and seizure, it will pay off.[458]
[STT 300: 1–3, Geller 2014: 47]

In *nīš libbi* B prescr. 1 it is stated that the patient's body must be massaged the day of the new moon to make sure that he heals:

You pulverize seeds of *azallû*-plant (and) seeds of *maštakal*-plant together,
you mix (them) with *uḫūlu*-plant. He washes himself with water.
You put tamarisk into (that) water, you heat (the water) in the oven,
you rub his body (with it). On the day of the new moon,
you will repeat this (therapeutic) performance and he will recover.
[*nīš libbi* B prescr. 1: 25–29, see also X prescr. 1: 1′–3′]

The text *Taxonomy of Uruk* (SpTU 1, 43) which we have already discussed above, offers an interesting discussion on this subject. The tablet consists of a single column, divided into four parts, each of which contains a list of diagnostic categories associated with a region of the body and its organs. The four parts of the body mentioned in the text are as follows: *libbu* (mind/heart), *pî karši* ("mouth of the stomach"), *ḫašû* (lungs), *kalâti* (kidneys). The fourth section, which affects the kidneys, lists the following diagnostic categories (ll. 25–31): contraction (of the bladder); (absent of) sexual desire; anal sickness; *sagallu*-sickness; infertility; twisted uterus; maintenance of "gas." Geller (2014: 91–93) provides an astral interpretive hypothesis: he relates this text with the Babylonian astral magic, in which the signs of the zodiac play a crucial role in determining when a magical practice is to be performed. Indeed, there are associations between stones, plants, and other ingredients used in amulets and zodiac signs, which correspond to specific dates in Babylonian astral-medical texts. According to Geller, the SpTU 1, 43 text is a list of diagnostic categories associated with specific parts of the body, each influenced by a particular zodiac sign. The absence of the zodiac signs in the tablet would be attributable to the fact that the influence of specific zodiac signs

458 See also lines 9–10, 12–13, 32–33.

on particular parts of the body was well known to contemporary astrologers, and it was, therefore, unnecessary to write it down. The text is not, *strictly speaking*, a taxonomy of sicknesses, but rather a piece of, as Geller calls it, the large complex puzzle which makes up the doctrine on the influence of the zodiac signs on the human body.

The Pleiades and Venus influence the sexual sphere, and thus also the male and female sexual desire. In the text KAR 69 the incantation to win the heart of a woman is addressed to the Pleiades:

> [Incan]tion. Pleiades, Mercury,
> You are the stars of the early morning.
> [Anu] (and) Enlil created you.
> [Wi]se Nudiummud endowed you with ⟨. . .⟩.
> I am sending you to NN son of NN.
> [KAR 69: 7–11, Biggs 1967: 74]

In the same text, another incantation runs as follows: "[Incantation.] Bright [Plei]ades! Bright Pleiades! / (You) who are stationed in [the heavens]! (You) who are stationed in the heavens!" (KAR 69: 2–5, Biggs 1967: 76).

Many incantations and rituals to regain sexual desire are addressed to Ištar (see Reiner 1995: 23–24). The astral aspect of the goddess, represented by Venus, is emphasized by the expression "Ištar of the stars" (*Ištar kakkabī*) and an epithet dedicated to her in an incantation, "light of the heaven" (*nannarat šamē*, No. E.2: 25). Sacrifices and libations of beer in front of a juniper incense burner are dedicated to her, as in No. E.2: 18–24.

Ointments and amulets

Several types of medicaments are prepared to rub[459] parts of the body (penis, navel, hand, throat) or the entire body of the patient.[460] In some cases, it is not specified which part is to be anointed.[461] The ingredients listed in the prescriptions are mixed often with oil from the alabastron, in some cases with cedar or cypress oil.[462] The incantation is recited over the preparation, before being massaged onto the patient, to be made effective.

In many cases, the ointment, massaged onto the patient's body (or parts of it), can be composed of many ingredients (animal, vegetable, and mineral). In the following prescription, N prescr. 25 lo. e. 1–4, the ointment, which should be massaged three times onto the man's pelvis (*rapaštu*), is composed of mostly animal ingredients. Among them, the male raven and partridge, the latter bird very much present in

459 On ointments see Herrero 1984: 98–100.
460 *nīš libbi* N prescr. 15 iii 24–26.
461 No. D.1: 6–9; No. E.1 ritual B rev. 23–24; F prescr. 6: 26–28; F prescr. 9: 29–30; N prescr. 14 iii 11–23.
462 See index of ingredients. Only in *nīš libbi* G prescr. 7: 14 the patient's waist is repeatedly anointed only with myrtle oil.

this corpus; the reared-up buck, which appears in the theme of the ritual-incantation of animals tied to the bed (Chapter II § "Animal tied to the bed"); the *rikibtu* of an *arkabu*-bat, whose therapeutic significance has been discussed above:

> [If ditto: The hear]t of a male raven, the blood of a reared-up buck,
> [the blood of a] male [partri]dge, the *rikibtu* of an *arkabu*-bat,
> seed of the *dadānu*-acacia you mix together, the pelvis of that man for three times
> you rub with it ⟨and⟩ he will get sexual desire.
> [*nīš libbi* N prescr. 25 lo. e. 1–4]

In some cases, however, the ointment is composed only by a bird-based ingredient, such as male partridge, black raven, *ballūṣītu*-bird:[463]

> If ditto (= a man's sexual desire has finished in the month Nisannu): The skull of a [male] partridge [. . .]
> with oil from the alabastron his body
> you anoint and [that] m[an will get sexual desire].
> [*nīš libbi* N prescr. 15 iii 24–26]

> [Its ritual]: You pound [the heart of] a black raven (and) mix it with cypress oil, [you recite] the incantation [seven times over (it). You rub his penis (with it) and (he will get) sexual desire].
> [*nīš libbi* L prescr. 1: 5]

> [If ditto]: You pound the spur of a *ballūṣītu*-bird in oil, you anoint his penis (with it) and [(he will get) sexual desire].
> [*nīš libbi* L prescr. 5: 14][464]

In the following prescription with the ointment not only the man's penis and navel are massaged, but also the man's right hand and the woman's left one[465] (note that "you" is a scribal mistake):

> [. . .] . . . fungus of tanner, dust of the threshold [slab of the house . . .] . . . with fat of a danc[ing$^?$] bear [. . .],
> [. . .] you pour over it pomegranate sweet juice [and] you mix (them) [together], you take (it) at your left hand [. . .]

---

463 On the bird see commentary.

464 See *nīš libbi* P prescr. 10 rev. 5; No. K.9: 164–165.

465 A very well-known anthropological reading of the asymmetry between left and right and its symbolic components is Robert Hertz's (1909) *La prééminence de la main droite: Étude sur la polarité religieuse* (1909). See in Assyriology Sallaberger 2000: 246–248. The association between the right and the man and between the left and the woman is also confirmed by the late Assyrian version of the *Atraḫasis*, in the section dedicated to the creation of humanity. The goddess Bēlet-ilī, after preparing the clay from which human beings will be created, detaches from the latter fourteen pieces. Seven he places to the right, seven on his left, as she sits in the middle of a brick. From pieces of clay on the right she makes men, from those on left women (K 3399+3934, I iii 5–10, Lambert and Millard 1969: 60–63).

[. . .] you massage three times your [n]avel (and) your penis, you massage your [right?] hand (var.: which she (= witch) touched with sorceries) (and) the left one of the woman [. . .].
[*nīš libbi* D prescr. 2: 29–31]

In addition to the woman's hand, her pelvic area (*ūru*), as the man's penis, is massaged with the ointment:

Its ritual: Iron powder, "heals-a-thousand"-plant (*imḫur-līm*), *ru'tītu*-sulphur
you pulverize together, you put (it) in oil, you recite the incantation seven times over it,
you anoint (with it) the penis of the man and the pelvic area of the woman.
[*nīš libbi* M prescr. 1: 1–3][466]

In another prescription, perhaps for a *namburbû*-ritual against panic attacks, the mustard water-based ointment is applied to the throat of the man and woman: "In musta[rd] water [. . .] / you mix . . . [and] keep (the medical preparation) ready, / you apply (it) into the throat (of the man and woman) and they will find relief" (No. D.3: 37–39).

Once, in K prescr. 29 the ointment is rubbed not onto man and woman, but their figurines: "[You make] the figurines of the man and the woman [. . .] / . . . them, the "desire" (litt. "rising" of the *lib*[*bu*]) of the man and the woman . . . [. . .]" (K prescr. 29: 127–128).

In the incantation No. F.4 the patient's limbs are anointed only with oil from the alabastron, after that the therapeutic operator spreads juniper three times before Ištar:

Its ritual: You put oil from the alabastron in a boxwood container for unguents,
[you spread] juniper three times in front of [Ištar, you reci]te this [inca]ntation [over (it)] three times,
you [an]oint his limbs (with the oil) and (he will get) sexual desire.
[No. F.4: 91–93]

Some prescriptions require the preparation with the same ingredients of an ointment and a potion. In one case the blood of ram is divided into two parts: one part mixed with oil for the ointment to be massaged on the man's navel and penis ("your" is a scribal mistake), the other one to be mixed with water for a potion that the patient has to drink:

[For ditto (= a man's sexual desire is taken away and (his desire) to go to his own woman or another woman is reduced)?]: You ta[ke] the blood of a ram in an unfired *pursītu*-container,
you mix [a hal]f with oil (and) you anoint your navel (and) your penis,
[and] you crush [(the other) h]alf in water, he d[rinks (it) and ditto (= he will recover)?].
[*nīš libbi* E prescr. 3: 65–67]

The ritual N prescr. 14 is more complex. It contains the following ingredients: *rikibtu* of the *arkabu*-bat, tamarisk, *maštakal*-plant, date palm, *bukānu*-wood.[467] They are placed

466 See No. E.1 ritual B rev. 23–24 where the oil-based ointment, with which the patient should be anointed, is composed by similar ingredients: iron and magnetite powder, "heals-a-thousand"-plant, and sulphur.
467 The rest of the ingredients stay unknown because of the incomplete nature of the text.

in water that must spend the night under the stars. Then, the man's body is washed with *kasû*-water. The *rikibtu* of the *arkabu*-bat, probably charged by astral influence and contact with other ingredients, is divided into two parts: one part is drunk, the other one pounded and mixed with oil from the alabastron to be anointed onto the man:

> If ditto (= a man's sexual desire has finished in the month Nisannu): the *rikibtu* of an *arkabu*-bat, *eli*[*kulla*]-plant,
> . . . of tamarisk [. . .]
> *maštakal*-plant, date palm, *bukā*[*nu*]-wood
> and you put in water all seeds and
> let them spend the night under the stars, you pu[t] (it) on the roof,
> you let that man sit down, with *k*[*asû*]-water
> you [wash] all his body [. . .]
> the *rikibtu* of an *arkabu*-bat [. . .]
> he drinks half and [you anoint? him with the other] ha[lf]
> premium oil and . . . [of tamarisk . . .]
> on his left foot . . . [. . .]
> you pulverize and with [oil] from the a[labastron his body],
> you anoint (it) and [that] m[an will get sexual desire].
> [*nīš libbi* N prescr. 14 iii 11–23]

In B prescr. 1 the ingredients are pulverized and mixed and perhaps added to the water with which the patient washes himself. Then, the tamarisk is added to the water, which is heated in the oven. Once ready, the therapeutic operator massages with it the patient's body, an action that will be repeated on the day of the new moon:

> You parch the scaly skin of a fish, (peel of the) *mikû*-plant and bread.
> You pulverize seeds of *azallû*-plant (and) seeds of *maštakal*-plant together,
> you mix (them) with *uḫūlu*-plant. He washes himself with water.
> You put tamarisk into (that) water, you heat (the water) in the oven,
> you rub his body (with it). On the day of the new moon,
> you will repeat this (therapeutic) performance and he will recover.[468]
> [*nīš libbi* B prescr. 1: 24–29]

Most ointments consist of two ingredients: iron and magnetite powder. It is not to be forgotten, that these ingredients are not only to be rubbed on the male genital organ but also the female one (vulva and pelvic area):

> Its ritual: Magnetite powder (and) iron powder
> you mix with oil from the alabastron, you recite the incantation [over it] seven times;
> the man [anoints] (with it) his penis, the woman her vulva [and he (can) have repeatedly inter course].
> [No. B.2: 38–40][469]

---

468 See No. J.1: 3 where the therapeutic operator washes in juniper water the patient's hands.
469 See also No. D.2: 21–23; No. E.1: 15–17; L prescr. 6: 16–18.

The use of both ingredients is clearly of magical significance: attraction. Thompson (1936: 85) already has emphasized this function: the attraction by the magnet for the iron.[470] It is no coincidence, that in ritual No. B.1: 14-17 magnetite powder is massaged on the man's navel, whereas iron is used on the woman's navel:

> Its ritual: You pulverize magnetite, you mix (it) with oil from the alabastron,
> you recite the incantation seven times over it; you apply (it) to his navel;
> you pulverize iron, you mix (it) with oil from the alabastron, you recite the incantation seven times over it,
> you apply (it) to the woman's navel; the man and the woman [will find relief] together.
> [No. B.1: 14–17]

It is also clear that the navels indicate precisely the *libbu*, the area of the human anatomy affected by the malaise: anointing the two navels with such substances means creating a magical attraction between the man's and the woman's 'heart' (*libbu*) to regain sexual desire.[471]

In a prescription we find only magnetite powder with which the man's penis, the chest and the waist are anointed: "Its ritual: You pulverize [magneti]te (and) put (it) into oil; / He (= patient)/you (= therapeutic operator) rub(s) (with it) his penis, his chest (and) his waist, [and] he will be healthy" (No. A.2: 47–48). While in another one, an oil-based ointment with iron and magnetite powder, "heals-a-thousand"-plant (*imḫur-līm*), and sulphur is anointed onto the patient:

> Its ritual: Iron powder, magnetite powder,
> "heals-a-thousand"-plant, (and) sulphur //at the river// you put into oil,
> you recite the incantation seven times over it and anoint him.
> [No. E.1 ritual B rev. 23–24]

Regarding the amulets, the *nīš libbi* corpus makes extensive use of this medical practice. The making of amulets to put around the neck or hip indeed are among the most common medical practices in Mesopotamia (see Maul 1994: 107; Scurlock 2006a: 59–62). Amulets in *nīš libbi* corpus can be divided into two groups:

---

470 The mineral plays a role in those rituals related to the sphere of sexuality. One example is a love incantation on behalf of a woman published by Scheil 1921: 26, l. 7: The woman has magnetite on her right, while a jug made of iron on her left.

471 On the magnetic power of magnetite and iron, capable of solving marital problems see a text in Scheil 1921: 26–27, No. 17 r. ii 1′–18′ (see also Scurlock 2014a: 109).

1. A set of ingredients, often animals and stones, wrapped in sheep's wool[472] (often red[473]) or linen and hung around the man's neck or waist;[474]
2. Leather bags containing (vegetal, animal, mineral) ingredients often sprinkled with oil (see Farber 1973).[475]

It must be noted that the animal-based ingredients in amulets refer to the sexual activity of animals. Often the hair of the males (bull, buck, ram, dog, pig) who have mounted females is to be plucked. Frequently the hair is torn off the thigh, perineum,[476] or tail. A pattern of ingredients that often appears in these amulets is the navel, the penis, and the *rikibtu* of the stag.

A particular case is *nīš libbi* N prescr. 8 ii 3–15 (ll. 3–9 = K prescr. 30: 135–140) where the fourteen drugs are pulverized and mixed with *isqūqu*-flour. Then, some *kupatinnu*-pellets are made with the mix, maybe some of which are eaten on an empty stomach, while others are tied around the patient's waist:

> If ditto (= a man's sexual desire in Nisannu month): the wings of a male eagle, the wings of [. . .] /geckos of the steppe which copula[te], / excrement of an *arkabu*-bat, excrement of polychrome lizard, seeds of the . . .-tree, / seeds of the *ēru*-tree, seeds of *maštakal*-plant, / [seeds] of *azallu*-plant, seeds of *murdudû*-plant, / seeds of "dog's-tongue"-plant, seeds of *šakirû*-plant, *ṣaṣun*[*tu*-plant], /*šumuttu*-plant / [you dry] these 14 drugs together. / you pulverize, mix with the *isqūqu*-flour, and roll into [balls] / you mak[e] three *kupatinnu*-pellets, [. . .] / in the middle of the

472 The catalogue of stone amulets K. 3937 + Th 1905–4–9, 355: 15′ mentions an amulet with four stones for sexual desire (*nīš libbi*): 12 ḪUL.GIG [. . .] 4 ŠÀ.Z[I.GA] (Schuster-Brandis 2008: 209). Note also the amulet for sexual desire, which is not mentioned in *nīš libbi* corpus, with the following six stone to thread on the pelt of a ram reared-up for the mating (^{sig}ÀKA UDU.NITÁ *šá ana* U$_5$ ZI-*ú* È): KUR-*nu* DAB; *ašpû*; ŠUBA; DÚR.MI.NA; ZA.GÌN/ZA.GÌN KUR-*šu*; URUDU NITA$^?$ (BAM 419 ii 13′–iii 5; BM 56148+ v 41–42; CTMMA 2, 32 iii 12–14; K. 3010+ v 22–23, Schuster-Brandis 2008: 136, Kette 127).
473 On the role of color in Mesopotamia see Unger 1957–1971. On the colors of various types of wool see Landsberger 1967b: 155–162. The colors of the wool, red and blue, black and white, have an important role in Kiuti prayers, see Alaura and Bonechi 2012: 71–72 fn. 298–301. The red wool is associated with the sun-god Šamaš. The red color is the one of dawn (Alaura and Bonechi 2012: 72 fn. 298). The wool of this color appears also in a *namburbû*-ritual as a symbol of the god of fire (Maul 1994: 119–120). It is clearly a link to the sun-god, but «die Frage, worin die magische Kraft der Wolle bestand, kann noch nicht beantwortet werden» (Maul 1994: 120 fn. 44). On colors in Mesopotamian astrology see Verderame 2004b; on the colors in general and those of the wool in the Hittite world see Haas 1994: 894–895; 2003: 638–644.
474 A prescr. 3: 10; No. D.4: 61–63; No. E.2: 38–41; F prescr. 17; K prescr. 26–28 (?); No. K.3: 89–94; No. K.3: 103–105; K prescr. 30; K prescr. 32 (?); L prescr. 2; N prescr. 4; N prescr. 10 (?); N prescr. 16; N prescr. 23 (?); O prescr. 6; P prescr. 6.
475 A prescr. 3: 13–14; A prescr. 4–16; A prescr. 23; No. C.1: 10–13; F prescr. 4–11; F prescr. 14–16 (the ingredients here can be used also for a potion or an ointment); No. K.1: 13–15; K prescr. 3–5; K prescr. 8–24 (21$^?$); No. M.1 Ritual B; M prescr. 2–5; M prescr. 6: 23.
476 For the Akkadian term *šapru* see Kogan and Militarev 2002: 313.

kupatinnu-pellets [. . .] / all on an empty stomach you [. . .] / towards *kupatinnu*-pellets [. . .] / you tie, around your[477] waist [. . .] / [he will get] the sexual de[si]re.
[N prescr. 8 ii 3–15 (ll. 3–9 = K prescr. 30: 135–140)]

Libations to Ištar

Some of the rituals are directed to the goddess Ištar. One of them, *nīš libbi* A prescr. 24, mentions the placement of a censer with juniper in front of Ištar, libations of beer and the creation of a potion to drink in wine for three days:

Its ritual: "Heals-a-thousand"-plant, *tarmuš*-plant, "dog's-tongue"-plant, *egemgiru*-plant, *ardadillu*-plant, *kabullu*-plant, "gold-fly"-plant.
You crush (and) sift these seven plants, you set up a censer with juniper in front of Ištar,
you libate beer, you recite the incantation over it seven times (and) you let him (var.: repeatedly) drink it (= pharmaceutical) with wine. He will repeatedly drink it for three days and he will recover on the fourth one.
[*nīš libbi* A prescr. 24: 69–72]

In the same *nīš libbi* text, a ritual seems to be addressed to Ištar. However, because of the fragmentary nature of the prescription, it is not possible to say much. It is not clear whether by "his Ištar" here the text is referring to the goddess Ištar herself or a generic "his goddess":

If ditto (= If a man's (desire) to go to a woman is reduced [. . .]): [You wrap up] the . . .-stone with a sheep wo[ol tuft, you put (it) around his waist],
you dry (and) crush the 'heart' of the . . .-plant, in front of god/[the sun? . . .],
you surround in front of his Ištar (or: his goddess) with a magical circle[478] of flour [. . .] . . . [. . .],
you fumigate sulfur in fire, with oil in a leather bag.
[*nīš libbi* A prescr. 3: 10–13]

The *nīš libbi* R Ritual bow provides, besides the making of a bow, whose string is made of a left hock tendon of a gazelle, libations of *miḫḫu*-beer, juniper water for spraying and a censer once more with juniper front of Ištar:

[In order to] get sexual [desire]: In front of Ištar [. . .]
you sprinkle pure [ju]niper [water],[479] [you put] junipe[r] in a censer,
you libate [*mi*]*ḫḫu*-beer.
[*nīš libbi* R Ritual bow: 6′–8′]

477 It is probably a mistake. It should refer to the patient's waist.
478 On the magic circle in the therapeutic context see Scurlock 2006a: 57–59.
479 For the spreading of juniper in front of Ištar see also No. F.4: 91–93.

A more complex ritual aimed at Ištar of the stars is present in No. E.2, after which follows an incantation also directed to the goddess (ll. 25–37). The ritual involves the sacrifice of a sheep and the manufacturing of six pairs of figurines depicting the witch and wizard, which will be burned:

> [If] a man's sexual desire is taken away and (his desire) to go to (var.: he does not desire) his own woman or another woman is reduced,
> you set up a *pāṭiru*-portable altar in front of Ištar of the stars, you sacrifice a sheep,
> you set up a censer with juniper, you libate (var. adds: premium) beer,
> [you] place (there) the shoulder, the caul fat (and) the roast meat (var. adds: you libate premium beer).
> Two figurines of tallow, two figurines of wax, two figurines of bitumen, two figurines of gypsum, two figurines of [cl]ay,
> two figurines of dough, two figurines of cedar wood, you make (them); in an unfired *pursītu*-bowl (var.: *burzigallu*-bowl; cup)
> you burn (them) in a fire in front of Ištar of the stars and he (= patient) says the following (incantation).
> [No. E.2: 18–24]

Etiological analysis

Within the *nīš libbi* corpus, there can also be found evidence for etiological practice that allows the therapeutic operator to identify the cause of the suffering:[480]

> Its ritual: You mix together emmer dough and potter's clay; you make the figurines of the man and the woman; you put them one upon the other, and place them at the man's head,
> you recite [the incantation] seven times; you remove (them) and put [them near] a pig.
> If the pig approaches (them), (it means) Hand of Ištar; for the ritual procedures$^{?}$, if the pig does not approach (them), (it means that) the witchcraft seized the man.
> [*nīš libbi* D Diagn.: 24–26]

The etiological analysis involves the creation of two figurines, man and woman, made of emmer dough and potter's clay. They are placed upon each other and then upon the man's head. After reciting the incantation seven times they are brought close to a pig.[481] If the animal approaches them, it means that the cause of the sickness is the "Hand of Ištar" (*qāt Ištar*),[482] a clear reference to divine wrath; on the contrary, if it does not approach them, the man is bewitched (*amēlu šū kašpu iṣbassu*).

---

480 On the social aspect of etiological practices within the therapeutic procedure, see the anthropological studies of A. Zempléni at Lebou Wolof and Senegal (1969; 1985); on divination with an etiologic and diagnostic purpose and its social impact, the study of V. Turner (1975: 243–338) on the Ndembu of Zambia is a classic.

481 The association between patient's figurines and pig can be found in a reversed situation in *Maqlû* (IV 39–40, Abusch 2016: 119): the witch has made the figurines and fed it to the pigs (and dogs).

482 On the causes of loss of sexual desire see Chapter I § "Cause of sickness." For the "Hand of God" see Heeßel 2007.

The figurines

The creation of human figurines is a widespread practice in Mesopotamian magic. Often in the anti-witchcraft corpus, the figurines represent the witch and the sorcerer whose evil action must be destroyed. They, therefore, represent a substitute for human beings who are to be manipulated by means of magic power (see Abusch 2002: 65–78; Abusch and Schwemer 2011: 20–24; Verdermae 2013). Here is an example from the *nīš libbi* corpus:

> Two figurines of tallow, two figurines of wax, two figurines of bitumen, two figurines of gypsum, two figurines of [cl]ay,
> two figurines of dough, two figurines of cedar wood, you make (them); in an unfired *pursītu*-bowl (var.: *burzigallu*-bowl; cup)
> you burn (them) in a fire in front of Ištar of the stars.
> [No. E.2: 22–24]

The figures, however, do not only represent the witch and the sorcerer, but there are also indeed cases in which the patient himself is depicted. Here is another example from *nīš libbi* corpus, K prescr. 29: 127–128, where the ointment is rubbed not onto man and woman, but their figurines: "[You make] the figurines of the man and the woman [. . .] / . . . them, the "desire" (litt. "rising" of the 'he[art]') of the man and the woman . . . [. . .]."

In an anti-witchcraft text, a patient's figure is made, and a thorn is extracted from it, which is said to have been planted in the man's head by the witch and wizard. In the anti-witchcraft ritual, this thorn is then inserted into the heads of the figurines representing the malicious operators. In other words, a figurine is created that represents the figurine of the victim originally created by the witch:

> *ša muḫḫi ṣalmīšu ušaḫḫaṭ ina muḫḫi ṣalmīšunu utakkap*
> *šaman nūni ina muḫḫīšunu itabbak*
> *egubbâ ana muḫḫīšunu*
> *irammuk u kīam iqabbi*

> What (is) (= the thorn of date palm) in the skull of his (= patient) figurine he removes and sticks (it) into the skulls of their figurines (= wizard and witch)
> He pours fish oil upon them.
> On them with (water from) the holy water vessel
> he washes and says the following (incantation).
> [Abusch and Schwemer 2011: 249, No 8.1: 57″–60″][483]

---

483 See also a ritual in which a figurine, representing a patient who had been sent down to the Underworld by means of magic action, is taken away from the afterlife, while the figures of the sorcerer and the witch are buried and sent to the Underworld (Farber 1977a: 218, Hauptritual B).

The reference to the practice of the witch operating with patients' figurines, in order to act wickedly against them, is also present in an incantation from *nīš libbi* corpus, No. E.2: 31: "(I am who) against whom the witchcraft has been performed (var.: they have performed witchcraft), my figurines have been buried in the ground." So, the practice of representing the patient by means of figurines is thus not unfamiliar to Mesopotamian ritualistic magic.

Returning to *nīš libbi* D Diagn.: 24–26, in order to allow the identification of the figurines they must be fascinated. This process in this text is performed by two specific actions:

1. The male and female figurines are placed upon each other;
2. They are brought in contact with the man's head.

The first magic action probably represents the sexual act, therefore allowing the identification with the couple: the figurines represent a man and a woman performing a sexual intercourse. This identification complies with the principles of similarity in magic investigated by Mauss and Hubert: «Le semblable évoque le semblable, *similia simili bus evocantur*. [. . .] La similitude mise en jeu est, en effet, toute conventionnelle; elle n'a rien de la ressemblance d'un portrait. L'image et son objet n'ont de commun que la convention qui les associe. [. . .] L'image n'est, en somme, définie que par sa fonction, qui est de rendre présente une personne» (1902–1903: 42).

The second operation, however, ensures precise identification: *that* couple represented by the suffering man. Contact identification is another principal component of magic at a transcultural level (see Mauss and Hubert 1902–1903: 41).[484] Obviously, identification through contact occurs only at the physical presence of the person represented by the figurine, that is why it never appears in anti-witchcraft texts (see Bottéro 1987–90: 212; Verderame 2013c: 307–308).

An interesting ritual in this regard is KAR 61, whose purpose is to ensure that "a woman approaches a man" and has sex with him. The woman's statuette is made, her name is written on its left side and finally buried at the western gate of the city(?). As soon as the woman walks over the figurine the ritual is concluded.

484 On the methodological error of isolating magical techniques from the historical context with the risk of providing an ahistorical image of the magic, the words of De Martino are crucial. According to the Italian anthropologist, the reason for this apparent ahistorical character of the magic is to be sought above all in the fact that magical forces have their roots in a risk that underlies cultural life, and which concerns the very possibility of being here as a presence in human history. In this sense, the repetitions and uniformities of magical elements need to be traced to the constant of the existential risk of *being-acted-upon*. But the apparent ahistorical quality of magic could also depend on a methodological error: that is, when we isolate magical protective techniques from the concrete cultural context in which they carry out a protective function and we compare them to other, similar techniques present in other cultural contexts, in order to end up by showing a type of "magical world" that in such a fictitious isolation has never existed as a cultural fact (see De Martino 1959, Engl. tr. 2015: 111–112).

šumma KI.MIN *šumma sinništu šī lā illaku tappinna teleqqe*
ana Ea šarri ana nāri tanaddi
ṭīda kibrī nāri kilaltīn
ša ebertān u ša(text *ana*) *ebertān*
teleqqe ṣalam sinništi šuāti teppeš šumša ina naglabi šumēli tašaṭṭar
ina pān Šamaš šipta munus-sig$_5$-sig$_5$-ga
[*ana*] *muḫḫi tamannu ina bābi kamî*
bāb erēb Šamši teqebberšu x *ni ši*
ūma muṣlāla u šimītān[!?] *ana muḫḫi illak*
šipta munus-sig$_5$-sig$_5$-ga-*ma šalāšīšu*
tamannu sinništu šū illakku tarâm

If ditto. If that woman does not come, you take *tappinnu*-flour,
throw (it) into the river to King Ea;
you take clay from both river (banks),
from the far side (of the Tigris) and the far side (of the Euphrates);
you make a figurine of that woman, you write her name on its left hip;
before Šamaš, you recite the incantation "The beautiful woman"
[over] it. At the outer gate
of the West Gate you bury it . . .
During the hot part of the day[?] or during the evening[?] she will walk over it.
You recite the incantation "The beautiful woman" three times;
that woman will go to you (and) you can make love to her.
[KAR 61: 11–21, Biggs 1967: 70–71]

In this text, therefore, the identification of the statuette and the woman takes place on two levels: engraving of the name on the left hip[485] and the contiguity as the woman walks over the buried figurine. The latter guarantees the effectiveness of the ritual.

It should be emphasized that the practice of substitution is present in other forms in the *nīš libbi* rituals: the witch removes sexual desire from the man by means of an evil action that involves the use of his sperm and a realization of a copy of his penis: "That man's s[perm] has been buried with a dead per[son], / his penis has been sealed and shut up in a clay pit towards sunset" (*nīš libbi* D Symp.: 44–45).[486] It is clear that in these circumstances the sperm, according to the principles of contagious magic, represents the man himself: *pars pro toto*. However, a representation of the patient's penis was also made. The male organ, according to the principles of analogical magic, was sealed and closed by clay.

Although this appears to be an etiological medical practice specific to the *nīš libbi* corpus, it is fully embedded in Mesopotamian therapeutic strategies: the creation of

485 Mauss and Hubert: «La seule mention du nom ou même la pensée du nom, le moindre rudiment d'assimilation mentale suffit pour faire d'un substitut arbitrairement choisi, [. . .] le représentant de l'être considéré» (1902–1903: 42).
486 See also F prescr. 5: 20–21; K prescr. 3: 27.

figurines intended as a substitute for human beings to act upon magically, be they witches or patients, it is a widespread practice in Mesopotamia.

The pig

The animal used in *nīš libbi* D Diagn.: 24–26 for etiological purposes is the pig.[487] The choice of the pig for investigating the ultimate causes of suffering is due to one of the qualities attributed to the animal by the Mesopotamians: its inordinate sexual activity[488] (see Foster and Salgues 2006: 289; Weszeli 2009 -2011: 325). The impetuous sexual activity of the pig is underlined by the *nīš libbi* incantation No. E.3: 49 declaring: "[Make love to me like] a pig fourteen times." Sexual activity and excitement of the pig are also underlined by the fact that its bristles are used as an ingredient in potions or amulets in the *nīš libbi* corpus:[489]

> [If ditto (= a man is bewitched and his flesh feeble, has *mungu*-paralysis and his knees are contracted, his 'heart' (*libbu*) needs a woman and finds her, but his 'heart' (*libbu*) returns. That man's semen has been buried with a dead person): "Heals-]a-thousand"-plant, *tarmuš*-plant, iron, coral limestone, *u*[*šû*]-wood
> [blood of a] male [partrid]ge, bristles of a pig reared-up for mating,
> you wrap up (them) [in a leath]er bag, you put (it) around his neck.
> [*nīš libbi* F prescr. 10: 31–33]

Another text describes the moment when the animal ingredient, probably the hair, is to be taken: it must be torn after the pig has mounted the sow because after the sexual act the ingredient is charged with the sexual force which then can be transferred to the patient through the therapeutic practice (*nīš libbi* A prescr. 2: 6–9).

The pig is also mentioned in an Old Babylonian love incantation, along with the dog.[490] Metaphorically, it states that the two animals *lie* on the heart of the male partner to indicate his unbridled sexual desire: *ina libbīka nîl kalbum / nîl šaḫium* "In your heart lies in a dog / a pig lies" (IB 1554: 57–58, Wasserman 2016: 265). The association with the dog, in reference to its sexual activity, can be found in an incantation, in which there is an analogy between the recitation of the incantation, the sexual activity of the dog and the pig, and the practice of plowing:

487 On the pig see Parayre 2000b; Lion and Michel 2006; Weszeli 2009–2011.

488 See an Old Babylonian terracotta from Nippur showing a sow nesting four puppies while mounted by a boar (Legrain 1930 No. 309; Opificius 1961 No. 665).

489 On the medical use of the pig see Levy et al. 2006; Weszeli 2009–2011: 326–327.

490 On the association between pig and dog see Parayre 2000: 168–173. See also here the anlysis of incantation No. E.1.

šiptu: araḫḫi ramānī araḫḫi pagrī
kīma kalbu kalbata šaḫû šaḫīta irtakbu ina ṣērīšu
kīma epinnu erṣeta irḫû erṣetu imḫuru zēršu

Incantation: I impregnate myself, I impregnate my body,
like a dog mounts a bitch, a pig a sow . . .
like the plow inseminates the land (and) the earth takes in its seed.
[CT 23, 10–12: 26–28/4: 9–11, Cooper 1996: 50–51; Cavigneaux 1999: 267]

In the literary work *The Underworld Vision of an Assyrian Prince* the man is compared to a pig:

u kīma lillidu šaḫû ṣeḫru ša ina muḫḫi sinništīšu elû libbašu ittanampaḫu šāru a[n]a pīšu u arkati itteneṣṣi kabittu ušaṣriḫ-ma ū'a libbī iqabbi

Or like a sexually mature young boar, who has mounted on his mate, whose entrails are continually swollen (so that) "wind" comes out continuously from his mouth and backside,
his mood (lit. liver) complains and says "Woe, my heart!"
[*The Underworld Vision of an Assyrian Prince* rev. 30–31, Livingstone 1989: 76]

However, the comparison is not clear. Although reference is made to the pig's sexual behavior, the text rather than emphasizing sexual vigor seems to emphasize the suffering, as a metaphor of a man's agitated feeling (see BFoster 2002: 283).

In divination (see Abrahami 2006; De Zorzi 2014: 162–163, 200) the pig has an important polysemantic value.[491] The aspect that interests us most is the fact that the animal is associated with fertility, due to its unbridled sexual activity, and with the absence of fertility. Here are some examples from the *Šumma izbu* series:

šumma sinništu ūlid-ma qaqqad šaḫî šakin tālitti būli iššir bītu šū irappiš
If a woman gives birth, and (the newborn) has the head of a pig – the offspring of the herd will prosper, that home will widen.
[XXII 5, De Zorzi 2014: 394]

šumma enzu šaḫâ ūlid tālitti būli ul [iššir?]
If a goat gives birth to a pig – the offspring of the herd will not [prosper?].
[XVIII 70′, De Zorzi 2014: 777]

491 The pig plays an important role in diagnostic texts as well (see Levy et al. 2006). It appears in the first and fourth sections of the *Diagnostic Handbook* SA.GIG/*Sakikkû* (Labat 1951; Heeßel 2000). In the first section, the animal appears in the first tablet, *Enūma ana bīt marṣi āšipu illaku* ("When the *āšipu* goes to the patient's home"), which concerns the interpretation of omens perceived by the therapeutic operator on the way to the sick patient's house (see George 1991: 142–143). In the fourth part concerning cases of "epilepsy," the domesticated pig appears in the tablet 28 which in the first section deals with sicknesses that turn into other ones (Heeßel 2000: 307–317).

See also an example from the *Zaqīqu* series: [[*šumma* KI.MIN]-*ma šaḫâ imḫuršu / mārī irašši libbašu iṭâb* "[If ditto (= a man dreams)] and meets a pig / he will have children and will be satisfied" (*Zaqīqu* r. iii z+12–13, Oppenheim 1956: 319).

In divination, the pig is connected to female individuals, such as the queen, the concubine, the wife. As pointed out by De Zorzi (2014: 163), the association with women likely arises from the perception of the pig as a domestic animal:[492]

šumma sinništu šaḫâ ūlid sinništu kussâ iṣabbat
If a woman gives birth to a pig, a woman will take the throne.
[*Šumma izbu* I 8, De Zorzi 2014: 345]

[*šumma šurānu ina bīt amēli kīm*]*a šaḫî inazzuz ina māti šuāti sinništu šarrūta* [*ippuš*]
[If in the home of a man a cat] squeals [li]ke a pig, in that land a woman will [exercise] the kingship.
[*Šumma ālu* XLV 12, Freedman 2017: 43]

šumma šēn amēli šaḫû ikkal aššat amēli imât
If a pig eats the man's shoe, the man's wife will die.
[*Šumma ālu* LXXI 33, von Weiher 1988: 180, text 97][493]

We find the association with females often linked to illegitimate sexual behaviors:

šumma šaḫû ana pān amēli pīšu iptenette aššassu ittannakū
If a pig repeatedly opens its mouth in front of a man, they will fornicate repeatedly with his wife.
[*Šumma izbu* XXII 111, De Zorzi 2014: 875]

šumma šaḫû ana qereb urši amēli īrub asīrtu ana bīt bēlīšu irrub
If a pig enters the bedroom (var.: the house) of a man, a female prisoner will enter the house of her master.
[*Šumma izbu* XXII 118, De Zorzi 2014: 876]

As De Zorzi (2014: 163) notes, the pig appears in the apodosis when the theme concerns rebellion against the king by subordinated subjects, therefore, a reduction of the royal power that is interpreted as its feminization. Rebellion is a crisis of royal power, and therefore of a fundamental impact on the manhood of the king:

šumma sīsītu 2 ūlidma qaqqad šaḫî [*ru*]*bâ ummānšu ibbalakkissu*
If a mare gives birth to two (colts) and they have the head of a pig, the prince's army will rebel against him.
[*Šumma izbu* XX 20′, De Zorzi 2014: 828]

492 Another association in the *Šumma izbu* series is the one with the stranger (*ubāru*), perhaps because of the liminal character, like the dog, of the animal suspended between the inside (house) and the outside, close to the man and at the same time distant and potentially aggressive (De Zorzi 2014: 200).

493 See also l. 26 for the reference to the man's woman. For more apodosis with the pig ll. 10–12.

> *šumma sīsītu 2 ūlidma kilallān kīma šaḫî* [*rubâ*] *nišūšu išassûšu*
> If a mare gives birth to two (colts) and both are like a pig – the people [of the prince] will threaten him.
> [*Šumma izbu* XX 26′, De Zorzi 2014: 829]

The pig's relation to a decrease in sexual desire, within the therapeutic tradition, is represented by a pig-shaped amulet dating back to the eighth-seventh century. The object has been published by Arnaud (2012) and contains a small three-line inscription in the Neo-Assyrian dialect:

> *ṣalam šaḫî ṣeḫeri / mār šaḫî*
> *eṭleš lizazzī*[494]
>
> Statuette of piglet/pig's son
> "May they rise manfully!"

The animal's presentation is expressed through a typical formula purely human: "X son of Y." The verb *uzzuzzu* 'to rise' is not present in the *nīš libbi* corpus where we find the verb *tebû* instead. According to Arnaud, the different choice of verbs can be attributed to the different linguistic registers: the amulet uses a different kind of variety from the one used in the written ritualistic corpus since it is a "popular" therapeutic object. This popular use of the amulet would also justify the conjugation of the verb *uzuzzu* in plural: it would be therefore applied to a plurality of patients over time. We do not have evidence instead of the fact that it could have been one of the therapeutic objects of the *ašīpu* since in the *nīš libbi* corpus no use of zoomorphic amulets whatsoever is mentioned.

As demonstrated by the sources, the choice of the pig for etiological purposes is determined by the fact that the pig represents very unbridled sexual activity, which is absent in the patient. It is also associated with female figures. Women, in fact, play an important role in *nīš libbi* rituals and etiological practice prescribes the making of both male and female figurines. Evidently, the absence of sexual desire in humans involves a crisis of manhood, which the rituals are trying to repair, such as one that involves the creation of a small bow. Here, the pig, associated through its etiology, which determined the reason for the sickness, clearly represents this state of feminization, or the loss of a key component of Mesopotamian manhood.

---

494 *ṣa-lam*[!] *ša-ḫi* TUR[!] / A[!] ŠAḪ[!] / *eṭ-le*[!]*-eš li*[!?]*-za-zu*.

Bow ritual and battle metaphors

One of the specific ritual practices of this corpus is the creation of a small ritual bow:[495]

> If ditto (= a man's sexual desire is taken away and (his desire) to go to his own woman or another woman is reduced)?: Mountain *urnû*-plant, "heals-a-thousand"-plant, "heals-[twenty"]-plant, *ṣaṣumtu*-plant, excrement? of a bat,
> the fruit of *da*[*dānu*-acacia], *ḫašḫūr api*-plant; you pulverize together these seven plants, the blood of a partridge
> you [let flow] over it, he swallows the innards of the partridge,
> he drinks [it] with first quality beer on an empty stomach. You make a bow of thorn,
> a tendon of the *arrabu*-mouse is [its] string (*matnu*), you load it [with an arrow],
> you pu[t it] over heads of the man and the woman, who are lying [. . .].
> [*nīš libbi* E Bow ritual: 54–59]

The prescription includes a potion made with beer to be drunk on an empty stomach which includes the following ingredients: mountain mint, "heals-a-thousand"-plant, "heals-twenty"-plant, *ṣaṣumtu*, excrement(?) of a bat, fruit of *dadānu*-acacia, *ḫašḫūr api*-plant, and partridge's blood. Subsequently, a thorn (*ṣillu*, see CAD Ṣ 193; AHw. III 1101 "Dorn")[496] bow is made, while an *arrabu*-mouse's (PÉŠ.ÙR.RA)[497] tendon is used as a bowstring. The bow is placed at the head of man and woman.

---

495 For the bibliography on the bow in Mesopotamia see Stol 2015: 617–619.

496 In the rituals, however, the thorn, particularly from the date palm, has a negative connotation. It indicates an evil action of the witch, namely to prick the figure of the suffering (see Schwemer 2007a: 209–214). In a ritual, for example, the central act is represented by the removal of a thorn of the date palm from the head of the patient's figure before being inserted into the figures of the magical aggressors: "(Now) [I remove] the thorn of the date palm, which (is) in the skull of [my] figurines, I stick (it) [into the skull(s) of their figurines]" (K 3196 + 3344 r. 1 and dupl., Abusch and Schwemer 2011: 252, No. 8.3: 46″). See also Farber 2014: 150, l. 49.

497 The identification of the *arrabu*-mouse has been a topic of significant debate: CAD A/II 302–303 'dormouse(?), jerboa(?)'; AHw. I 70 'Siebenschläfer'; Landsberger (1934: 107) 'Siebenschläfer' and (1957–1958: 53) 'Springmaus'; Riemschneider (2004: 203) 'Haselmaus.' As pointed out by Heimpel (1987–1990b: 607–608), however, none of the aforementioned identifications is possible, since they are all refuted by the sources. The animal's Sumerian name péš-ùr-ra literally means "roof-mouse," indicating that the animal nests in the roof, i.e. in reed mats, palm fronds, or brushwood roofs (ibid. 607). In a poem of divine love, published by Lambert (1959: 10, l. 69), one describes a woman with the animal in front of her. Unfortunately, the text is fragmentary, but the overall theme of love is clear: "In front of her there is an *arrabu*-mouse / behind her a mouse. / He girded his clothes. / He is a shrew (*ḫulû*), son of a mouse."

The animal is an ingredient for evil action, although at the same time it is used as a magical material, as substitute of bewitched man, in order to annul the magic action: "If 'cutting-of-the-throat' magic using an *arrabu*-mouse [has been performed] against a man, and / a slaughtered (lit.: "cut") arrabu-mouse has appeared in the man's house, in [that] house / door (and) bolt are bewitched. You ta[ke] this *arrabu*-mouse, [before Sîn] / you place it. You clothe it in a pure garment, cover it with a linen cloth, [anoint it] with fine oint[ment]" (BAM 449 + 458 obv. i 8′–11′, Abusch and Schwemer

Another *nīš libbi* prescription involves the manufacturing of a bow[498] with the left hock of a gazelle (*ṣabītu*):[499]

> [In order to] get sexual [desire]: In front of Ištar [. . .]
> you sprinkle pure [ju]niper [water], [you put] junipe[r] in a censer,
> you libate [*mi*]*ḫḫu*-beer, the bow o[f . . .]
> [. . .] a tendon of the left hock of a gazelle is [his] string [. . .]
> [. . .] . . . a reed arrow . . . [. . .]
> [. . .] may [himse]lf be released [. . .]
> [. . . the sta]r of Ištar [. . .].
> Traces
> [*nīš libbi* R Bow ritual: 6′–13′]

In the *nīš libbi* incantation No. J.2 the association between combat and the sexual sphere is clear and expressed through a metaphor using the bow, which must not become loose, and the quiver which must not be empty:

> May [the qu]iver not be[come empt]y! May the bow not slacken!
> May the batt[le of my love]-making be fought and may we lie down (together) by night!
> [Incantation formula].
> [*Nī*]*š libbi* Inc[anta]tion.
> [Its ritual]: You make a bow of a thorn [. . .].
> [No. J.2: 10–13]

In this text, intercourse is compared to battle.[500] The man must always be ready to fight with his quiver and bow, the sexual act should not be interrupted. The bow (*qaštu*) in this text symbolizes the penis.

2011: 409, No. 10.3: 20′–23′; see also Schwemer 2007a: 222–225). In the therapeutic field the skin of the *arrabu*-mouse is used to make an amulet for the *bennu*-sickness (BAM 311: 55). The tendon (SA), however, is used to make a leather amulet: along with the beak of a raven against the attack of a *gallû*-demon (ND 4368, Kinnier Wilson 1957: 40 l. 24) and against "the hand of god of his city" (Labat 1951: 192, l. 39).

498 For a *namburbû*-ritual in order to hinder evil from approaching to the bow see K.9718 and LKA 113, Maul 1994: 207–208.

499 See in love context the metaphor of the red male gazelle (*armû*) used in CUSAS 10, 12: *rīmum ul izzazma armū līṣûnim* "the wild bull is not standing – let the red gazelles come out!" (ll. 12–13, Wasserman 2016: 190). Wasserman (2016: 193) suggests that in those lines the wild bull allegorically represents the husband while the red male gazelles the suitors who show up when the husband is not absent to approach the woman sexually.

500 See for association between male sexuality, sex, battle, and hunting in *nīš libbi* corpus also No. O. 1: 16–17: "Like (that) of a ram / it is a joyful song and like (that) of a pregnant (one) is a battle cry (*tanūqātu*)!". The term *tanūqātu* 'a battle cry' comes from the verb *nâqu* 'to cry, to groan,' which as suggested by Wasserman (2016: 240) could testifies to a semantic play with the quasi-homonymous *niāku* 'to have (illicit) intercourse.' See in the second millennium love literature the term *ernittu* 'cry of joy after military triumph' referring to the shout of excitement at the moment of the orgasm (see

As for the Bible, Hoffner (1966: 329) reminds us that a man's many children, as the visible proof of his sexual potency, are compared to arrows in the quiver of a strong man. The scholar underlines that in the construction of manhood in the ancient Near East, both the courage on the battlefield and the sexual sphere are fundamental elements. A man's prowess in battle and his ability to sire children are the two criteria by which his manhood was measured. For this reason, these two aspects of manhood were frequently associated with each other in ancient Near Eastern sources: those symbols relating to male military exploits refer to sexual ability as well. In the ancient Near East, the bow, as weapon,[501] is therefore a symbol of manhood, including its sexual aspects (Helle 2019). Consequently, the taking away and breaking of the enemy's bow evoked in Assyrian curse formulas emphasize his subjugation and feminization with the consequent loss of his (sexual) vigor. See, for example, the curse formula in *Treaty of Aššurnirari V with Mati'ilu, king of Arpad*: "May [Ištar, the godd]ess of men, the lady of woman, / take away their bow, bring them to shame (*bāltūšunu liškun*)" (r. v 12–13, Parpola and Watanabe 1988: 12, No. 2). This phrase is a curse formula expressing the wish that the troops lose their manhood and evoking a sort of feminization and reducing of potence of the enemies (Weidner 1932–1933: 22, fn. 40; Biggs 1967: 37 fn. 3; see Chapman 2004: 48–59).[502] Even from an iconographic point of view, the bow is a symbol of what Assante (2017: 61) calls "hypermasculinity." The king in the Neo-Assyrian period is depicted as an archer able to subdue his enemies, in fact the bow, constructed to be shot at someone, necessarily refers to a penetrator/penetrated relationship between the shooter and the target (ibid. 63). Sexual power and ability in battle are praised by the continuous use of symbolic

---

ibid. 38). See also No. E.1: 11–13: "May your penis become as long as a *mašgašu*-weapon! / I sit in a net of laughter, / may I not miss the quarry!".

501 See for the weapon as symbol of manhood the following Sumerian Ur III incantation on women in hard labour followed by a ritual carried out over the new-born child to determine its gender: "If it is a male, may it a weapon, an axe, strength of its manliness/ seize in the hand. / If it is a female, may spindle and hair-clasp be in its hand" (UM 29-15-367: 46–48, van Dijk 1975: 57; Cunningham 1997: 71; Stol 2000: 61). In *Išme-Dagan K*, the goddess Inanna is attributed the power "to turn a man into a woman and a woman into a man, to change one into the other,/ to dress young women in clothes for men on their right side, / to dress young men in clothes for women on their left side, / . . . to put spindles into the hands of [men], and to give weapons to women (ll. 21–23 and e. 1–6, Römer 1988: 32–33). For the topic see Helle 2019.

502 See also *Accession Treaty of Esarhaddon*: "[May Ištar, lady of warfare, break his bow in] the thick of battle, and have him crouch as a captive [under his enemy]" (r. 20′–21′, Parpola and Watanabe 1988: 22–23, No. 4); *Esarhaddon's Treaty with Baal, king of Tyre*: "May Astarte break your bow in the thick of battle and have you crouch at the feet of your enemy; may a foreign enemy divide your belongings" (r. iv 18′–19′, ibid. 27, No. 5); *Esarhaddon's Succession Treaty*: "May Ištar, lady of battle and war, smash your bow in the thick of ba[ttle], may she bind your arms, and have you crouch under your enemy" (ll. 453–454, ibid. 48, No. 6); *Assurbanipal's Treaty with Babylonian Allies*: "May Ištar, who resides in Arbela, goddess of battle, [break our bow in the thick of battle and] make us crouch [under the feet of] our enemy [. . .]" (r. 24′–25′, ibid. 68, No. 9).

metaphors between the two semantic domains. The bow, therefore, recalls the penis and penetration. Consequently, the depiction of the subjugated enemies, with their lowered and dropped bows, represents their lack of manhood (see Cifarelli 1998: 224; Chapman 2004: 48–59).

The bow as a symbol of manhood is also present in rituals of gender inversion operated by the goddess Ištar, where the woman becomes a man acquiring the male attributes, bow and quiver: *sinništum kī zikari šaknat ušpatam tukial*[503] *qašta* "The woman as the man is equipped with a quiver, she holds a bow" (*Ištar-Louvre* ii 6, Groneberg 1997: 26).[504]

---

503 It is for *ukial* see Gronerberg 1997: 46, 99.

504 Note that among the Hittites, the primary association of the bow is with hunting, and in particular with ritual deer hunting and with the deer-god Kuruntiya (see Cammarosano 2018: 67–72). As hunting is a male activity, the bow still ends up being a symbol of manhood, as in the Mesopotamian world. In *Ritual and prayer to Ištar of Nineveh*, the goddess is called on to destroy manhood, and thus sexual vigor, as well as the enemy's military prowess: "Take away from the (enemy) men manhood, courage, vigor and *māl*, maces, bows, arrows (and) dagger(s), / and bring them into Hatti. / For those (i.e., the enemy) place in the hand the distaff and spindle of a woman / and dress them like women. / Put the scarf' on them and /take away from them your favor" (§ 8: 56–61, BCollins 2003: 164, No. 1.65, see also: hethiter.net/: CTH 716.1 (TRit 14.02.2011)).

Another interesting Hittite text, *The First Soldiers' Oath*, describes the transformation of men into women losing their masculine attributes, bows and arrows, acquiring the female ones, distaff and mirror: "They bring a woman's garment, a distaff and a spindle / and they break an arrow (lit., reed). / You say to them as follows: 'What are these? Are they not / the dresses of a woman? We are holding them for the oath-taking. / He who transgresses these oaths and takes part in evil against the king, queen / and princes may these oath deities make (that) man (into) a woman. / May they make his troops women. / Let them dress them as women. Let them put a scarf on them. / Let them the bows, arrows, and weapons / break in their hands / and let them place the distaff and spindle in their hands (instead)'" (§ 9, BCollins 2003: 166, No. 1.66, see also Oettinger 1976: 10–11, obv. ii 42–52).

The same male and female symbols recur even in the Paškuwatti's ritual (see Hoffner 1987: 271–287; Miller 2010: 83–89; Simon 2017). In this ritual the ritual operator puts the mirror and the distaff in the man's hand. After having passed through a door, the female symbols are taken away and substituted by a bow and arrows: "I place a spindle and a distaff in the patient's [hand], / and he comes under the gates. / When he steps forward through the gates, / I take the spindle and the distaff away from him. / I give him a bow (and) [arro]w(s), / and say (to him) all the while: / 'I have just taken womanhood away from you / and given you manhood in return. / You have cast off the (sexual) behavior expected [of women]; / [you have taken] to yourself the behavior expected of men!'" (§ 4: 19–28, Hoffner 1987: 272, see also: hethiter.net/: CTH 406 (TRfr 15.02.2012); Mouton 2007: 130; Peled 2016: 311).

The bow could be a symbol of manhood in the mythological Ugaritic poem of Aqhat (for the edition, *Forschungsgeschichte* and previous bibliography see Margalit 1989). Aqhat is a young, perfect and infallible hunter who is given a bow from his father Dan'el: "Whereupon Dan'el, the Rapian, / Thereupon, the Hero, devotee of the Rainmaker; / Did bend the bow and did brace (it), / He sto[od] over Aqht (and said): / [Observe], my son, the beginning of your hunting (career), / [Behold], the beginning of your hunting; / Here! (Take it) and may you hunt well!" (Margalit 1989: 149, see also p. 150). The goddess Anat is eager to own this bow ("She covets the protruding bow, / [the arc standing erect]," Margalit 1989: 150) and asks the man to give it to her in exchange for gold, silver, and im-

In Sumerian or Akkadian love compositions the bow theme, however, is not mentioned. It seems rather that metaphor is merely used in cases of loss or attempts to regain desire and/or potency. In other words, in the literary compositions about love, in which the theme is focused more on the mutual sexual desire between men and women than the sexual act itself, the bow and quiver metaphor is not found. However, it appears with emphasis in the *nīš libbi* corpus, especially in rituals, where male vigor is lost and must be restored. The metaphor of the love act as a battle is therefore typical in this kind of text. The bow represents sexual vigor in a ritual context, as in the battle it represents the a warlike one, aspects which contribute to the construction of manhood in Mesopotamia (see Paul 2002: 493–494). This metaphor thus makes sense in those critical moments, which threaten the integrity, in this case the sexual aspect, of a man and the relation he maintains with a woman. It is no coincidence that the woman has her part in the ritualistic performance and that the bow is placed upon the heads of the couple whilst they are lying down as if imitating sexual intercourse. The manufacturing of the bow ritually restores the sexual desire, whereas the absence of the latter, due to a malevolent removal of the bow, produces a crisis of manhood and man's dignity (*bāštu*). The bow is, therefore, a symbol of manhood, and as a consequence of sexual vigor, understood as the essential component of being men in Mesopotamia. Through the metaphor and the ritual of the bow in *nīš libbi* corpus, the sexual vigor, as strong, active, powerful, appears as one of the fundamental aspects of an ideal Mesopotamian manhood.

In *nīš libbi* corpus, the bow has probably taken on additional symbolic value. As we have seen in *nīš libbi* incantations, the act of love is likened to a battle. In combat, the arrow provides a link between the archer and his target. The arrowhead is a tangible sign of the potential wound that the archer inflicts upon the victim's body. This metaphor is also present in the iconography of the Greek god Eros who falls

---

mortality. The hunter refuses ("The bow is [the weapon] of soldiers; / Now, is hunting a past time for walking-womankind?," Margalit 1989: 152), and is thus condemned to death by the goddess. Scholars disagree upon the symbolism of the bow in this poem. According to Hillers (1973) the goddess' desiring the bow has a sexual meaning (see the criticism of Hillers by Margalit 1989). An alternative view is the one of Xella (1976: 61–91, esp. 69–71) who, not denying the sexual component, or an attempt at seduction in the request of the goddess, offers a historical-religious interpretation of the myth: it is an old hunting myth later revised by the Ugaritic elite giving it new meaning and turning it into a foundation myth of the just agrarian order. Margalit (1989: 180–181) denies the interpretation of Hillers, but states that, although on the part of the goddess there is no attempt of seduction, and that there is no sexual desire towards the youngster, sexual allusion is still very present by the choice of the verb *ṣby* 'to desire': here bow and erection are charged with a sexual double meaning. The hypothesis of Margalit shows a literary device, not without importance, as it connects the two spheres: war and hunting, and sexuality. Hoffner (1987: 330), on the contrary, does not read sexuality in the episode of Anat and Aqhat. According to the scholar, the goddess seeks the bow, not to secure for herself male sexual powers, but rather to enhance her "quasi-masculine" bellicose attributes. For other references on the bow as symbol of virility in other ancient cultures see Cifarelli 1998: 228 fn. 84.

in love with someone by means of shooting his arrows of love. As we know, despite the malaise, the absence of desire, which is manifest in the man, the victim of the witch's attack is the couple, both man and woman, as can be seen in the fact that they actively participate together in the ritual. The aim of the ritual is therefore to regain mutual sexual desire or mutual attraction. It is no coincidence that the small bow is placed on the man and woman's heads while they are lying down, a clear reference to sexual intercourse. The presence of a woman as a character in the ritual bears some significance: the symbolic firing of an arrow (the arrows must actually be nocked to the bow) allows the sexual attraction to be restored through sympathetic magic, not only in the man, but also in the woman. A similar idea is to be found in the preparation of iron or magnetite-based ointment, the purpose of which is precisely the attraction of the two parts of the couple.

Conclusions

Therapeutic efficacy and analogical thinking

Therapeutic efficacy

In order to sum up this work, I think it is important to go back to the topic of therapeutic efficacy, briefly summed up by Biggs's questions as «do the texts work? Do they achieve their desired goals?» (2002: 78). Answering them is certainly not easy, not only because of the very fragmentary nature of the documentation in our possession, but also because of some epistemological issues that are inherent in our analysis, i.e. our bodies which have experienced suffering, and our vision of the world and of life.[505] Biggs' own resolution of his questions employs biomedical categories:

> While there are physical causes such as trauma to the genital area or diabetes for some erectile dysfunction, and what is now often referred to as "male menopause" in fairly recent medical parlance, it is generally thought that the principal causes are psychological. The ability of the therapist – in our case the *āšipu* – to alleviate anxiety could surely contribute to the success rate. While it is unlikely that any of the plant products in the prescriptions would be an ancient Viagra, we know too little about the plants used to judge what physiological effects they might have had. After all, an alkaloid, yohimbine hydrochloride, derived from black currant trees, has been used for over a century to treat impotence and is available by prescription under the trade name Aphrodyne. In a recent popular health magazine, mention was made of studies in Lebanese men that showed effectiveness of asafetida in treating erectile dysfunction and enhancing passion and performance, but apparently the studies have not been published in any peer-reviewed journals. [Biggs 2002: 78]

As I have tried to demonstrate, the modern biomedical and biological gaze does not provide us with an adequate response for understanding therapeutic phenomena in ancient Mesopotamia. As the anthropologist Etkin (1988) affirms, the concept of therapeutic efficacy itself is a cultural construction, and as such, it is subject to space-time variation.[506]

The interpretation of the medical-therapeutic phenomena of other cultures, ancient and modern by Western modern epistemologies is misleading (see Waldram 2000: 607–609). Likewise, the interpretative tools of psychological and psychoanalytical sciences,

505 For some aspects of the Babylonian incantations' efficacy see TCollins 1999: 51–63.

506 For a discussion on this topic in relation to Mesopotamian medical practice see Böck 2008: 304–306; Steinert 2015, esp. 113–118. Steinert studied the term "tested" (*latku*) in medical texts, making use of the concept of *skilled practice* by the anthropologist E. Hsu. According to the scholar, the efficacy of a drug depends neither on the chemical properties of the drug, the theory of the practitioner, nor the expectation of the patient, but «is the result from a skilled practice of putting practitioner-patient-plant-in-the-environment into interaction» (Hsu 2010: 16).

https://doi.org/10.1515/9783110757262-004

though stimulating, obscure the complexity of the topic in question. We find an answer to the issue of efficacy from a psychological point of view in Geller's article:

> One does not have to be particularly Freudian to grasp the point that it may be performance anxiety which caused the impotence, and in some cases the šaziga incantations may have been effective in being able to deal with the anxiety. This is the defence mechanism known as 'displacement,' which in this case redirects the cause of the anxiety onto a witch. Externalizing the problem in the form of a witch can potentially allow a patient to control that which is beyond control, namely his own fear. [Geller 1999: 54]

I want to stress that efficacy refers to a variety of multilevel processes, involving complex transformations in which physiological, psychological, emotional, social, and political dimensions combine in variable measure. Besides, the artificial analytical breakdown (biological, psychological, etc.) threatens to obfuscate the unitary nature of the healing process.

In my opinion, to understand the therapeutic efficacy of the structure inside of *nīš libbi* therapeutic itinerary, or quoting Kleinman (1978), the Mesopotamian "explanatory models of illness," it is necessary to answer the following questions: who is the patient to be cured and what are his symptoms? What is the cause of the sickness? How does therapeutic practice work?

Patient and recipients of therapy

The absence of sexual desire in the current documentation appears in the man. The patient, losing his sexual desire, does not desire the women nor wish to have sex with her. Clearly, the sickness indicates a crisis in the relationship between the man and a woman. The woman must be involved in therapeutic practice because she is directly referenced in the sickness-event. The therapy itself often has both a man and a woman as the recipients of treatment. In this perspective the ritual practice outlined in the small prescriptions which follow the incantations *logically* makes sense. They provide instructions for the creation of an oil from the alabastron-based ointment with magnetite and iron powder (or two ointments: one with iron, the other with magnetite), with which to massage the penis and the vulva (or two navels). The purpose of this practice could be analogy for the principle of magnetism. As iron and magnetite are attracted to each other, so a man and a woman will become mutually attracted for the purpose of sexual intercourse. This is the first example of how analogical principles work in the *nīš libbi* corpus: recognizing similarities between the surrounding world and the human community and/or body and activating them in ritual practice.[507]

507 Metaphorical analogy, however, is present in some modern languages such as English and Italian, where magnetism refers figuratively to sexual attraction.

The woman is also present in the ritual of the bow where the latter is placed over the heads of the lying couple. In the *nīš libbi* incantations, like in other documentation, the bow represents sexual vigor, and sexual intercourse is metaphorically described as a battle. If sex is like a battle, it is obvious that the bow is placed above the heads of both the man and the woman.

Etiology, symptoms, and ingredients

The texts describe alongside the specific symptoms of this diagnostic category, symptoms also found in other anti-witchcraft texts: including weakness and fatigue; immobility and stiffness; fear and anxiety. The reason these symptoms are found in the corpus is that they are attributable to the fact that the cause of the loss of sexual desire is an attack of a witch. If the witch is considered the agent of the sickness, the one who has taken away desire from the patient, it is obvious that the therapy must contemplate the instruments capable of annihilating the cause. For this reason, some prescriptions require the creation of potions or amulets with substances, often made from vegetables, considered capable of acting directly against the witch. Some ingredients, in fact, seem to be used because they have power against the witch or the capacity to restoring purity.

Many ingredients appear in other medical texts for various diagnostic categories, but it is good to point out that many of them are used, albeit in drug assemblies, in the text corpus of kidney and urinary-tract sicknesses: *ḫašû*, *imḫur-līm* ("heals-a-thousand"), *nuḫurtu*, *urnû*. As I have shown, at least in the late Babylonian period, the absence of sexual desire is associated, together with other sicknesses, with the kidneys, as it could be confirmed by the text the *Taxonomy of Uruk*.

Other vegetable ingredients, instead, are specific to the absence of sexual desire such as *lišān-kalbi*, *azallû*, *šumuttu* and *ḫašḫūr api*, as testified in BAM 1 and the *Šumma šikinšu* texts.

Other ingredients in amulets are used because they are endowed with characteristics that were lost by the sick man. This is the case with gold and silver, lapis lazuli, and alabaster. They represent respectively the absence of fear, purity, and brilliance, qualities which in the ritual of incantation No. E.2 are to be transmitted to the patient.

Interpreting the ingredients in the prescriptions from the perspective of modern pharmacology reduces the complexity of the choices made by Mesopotamian practitioners in favor of information about pharmaceutical production (see Etkin 1988: 303-319). That said, I do not want to attribute to these choices simply and exclusively to "symbolic" reasons. The two levels are so interconnected that it is difficult, especially in the analysis of written sources, to make a distinction between them (see Steinert 2015: 114). This interconnectedness was demonstrated by Steinert (2012b) who identified, based on a Mesopotamian gynecological text, criteria not related to the intrinsic properties of the ingredients in the choice of therapeutic remedies, and by Böck (2014: 129–163) who found a religious relationship between some plants (*būšānu*, *lišān-kalbi*, and *ṣaṣuntu*) and the goddess Gula. I tried to do the same type of analysis for the *nīš libbi* therapies.

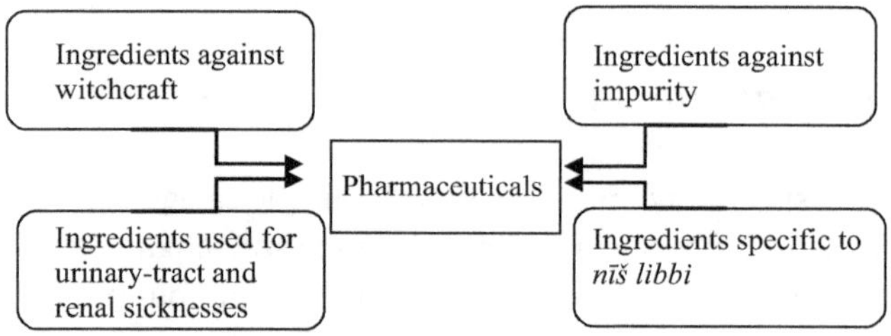

I would like to point out that I do not want to establish a cause-and-effect relationship between ingredient, function, and sickness. As is known, an ingredient may have more than one function. Besides, all ingredients cannot trace their reason for inclusion to the above scheme. My proposed solution is only a scheme of convenience which cannot reduce the complexity of the choices made in the preparation of the pharmaceuticals of the *nīš libbi* therapy. It seems that other motivations can be found in the choices of ingredients, which cannot be traced back to the scheme I have proposed. Another possibility is that the texts are passed down through the centuries, without the individual operators being aware of the reasoning behind the choice of the employed substances: but rather coming from a *vox traditionis* (see Geller 2015: 17).

Animals, metaphors, and substances

Another group of ingredients consists of animal substances: birds, especially partridges, and the saliva and hair of animals mentioned in the incantations. This is a clear example of analogical and contagious magic used together. As I have shown in Chapter II, metaphors in incantations establish similarities between an animal's sexuality and a human one. For instance, male sexuality is like that of a stag, an aurochs, a ram, and so on. At the basis of this analogy, which is expressed through metaphors and similes, there exists the idea of some relation between human and animal (and their sexuality) without which there could no analogical foundation. If the incantations establish such analogies, the ritual, through the recitation of the incantations themselves, makes them useful for therapeutic purposes. Animal substances, laden with sexual vigor through physical contact, are used in potions and amulets, allowing the patient to acquire the sexual characteristics of the animal itself.

Linguistic analogy

Through the analysis of the *nīš libbi* therapeutic practices the analogical thinking, which is at their core, emerges from the text. In the following examples, the symbolic analogy has a basis in the linguistic content, both in phonetic and graphic markers.

The first example is shown in the semantic association between the tendon and the string (*šerʾānu*). One of the most frequent descriptions of suffering is that of binding the tendons. The witch, as well as other demonic entities, strikes the man by binding his tendons. As I have shown, the term for indicating the tendons is also used to designate strings: the tendons are metaphorically understood as the ropes of the body. This allows an association between the binding and the tendon through the string as an intermediary. But there is more, the term *šerʾānu* indicates the bowstring as well. One of the ritual practices of the *nīš libbi* corpus instructs in the creation of a small bow to be placed over the head of the couple. The bow in turn is metaphorically associated with sexual vigor. The creation of the bow in the ritual practice can be explained on two levels:

1. As an analogy between the bow and the sexual vigor and between the battle and the sexual intercourse: employing a new bow one restores the patient's sexual capacity. Or symbolically: a broken bow (= devoid of sexual desire) is replaced in the ritual by a new one to be used in the battle (of love);
2. As an analogy between a bowstring and the strings of the body (= tendons). The bowstring in the ritual is free of bindings and therefore able to perform its function. Similarly, the tendons become devoid of the witch's knots.

Another example concerns the vulva and hunting practices. In the incantation No. E.1: 12–13, the woman states the following: "I sit in a net of laughter, / may I not miss the quarry!". The term for the net is here *bunzerru*. It indicates the spider's web, but as demonstrated by Civil (2006: 59) it can also refer to objects with a "weblike structure" or consequently a trap. According to Civil the word *bunzerru* derives from Sumerian be$_5$-en-zé-er, a term also used to designate the female genital organs. In this case, similarities are present on multiple spheres: animal, hunting, sexual, in total three levels which reference each other in the *nīš libbi* therapy.

Let us return to the prescriptions. Here, too, the linguistic or graphic analogy can be based on choosing some ingredients as well as a way of preparing them. The first example is given by the root *rkb* from which the verb *rakābu* 'to mount sexually' is derived. A representative ingredient of the *nīš libbi* prescriptions, given its widespread use in these texts, is *rikibtu*. It is unclear which part of the animal it refers to, but its use in this corpus is probably related to the fact that the word is a *PiRiSt*-form of verb *rakābu*. There are two animals from which the *rikibtu* is taken. The first is the stag, which incantations acclaim its reproductive abilities. The second is the bat called *arkabu*. The choice of this animal in the prescriptions could be attribute to the fact that its name also contains the above-mentioned consonantal root. The presence of the bat is then justified through linguistic reasons.

The second example deals with how the wren (*diqdiqqu*) is prepared in the Middle Babylonian prescriptions from Ḫattuša. *Amānu*-salt and *nīnû*-plant are applied to the bird after it has been spun and splashed, and before it is dried and powdered. Jiménez (2017: 342) emphasized the graphic relationship between the bird and the *nīnû*-plant: the bird's name can be written NI.NI-*qu*, homophonous with to the name of the plant. Another name of the bird is *iṣṣūr samēdi* "bird of the *samēdu*-plant," whose logogram ᵘKUR.ZI is similar to that of the *nīnû*-plant (ᵘKUR.RA.(SAR)).

Poeticality of incantations

Many of the analogies I mentioned above are expressed in incantations by means of metaphors and similes. Metaphorical language is the most evident aspect of the poeticality of incantations. I showed how a poetical analysis is important in the study of the spells. Using the studies of Jakobson (1960) on linguistics and Austin (1962) and Searle (1969) on the theory of linguistic acts, as well as ones already conducted in Assyriology, I analyzed the poeticality of such compositions, highlighting their structure, their use of figures of speech, and the importance attributed to repetition, redundancy, and sound. Such poeticality has a function in the therapeutic efficacy which should not be underestimated. It allows, by means of stylistic strategies, for the establishment of analogies, without which there would be no basis for the ritual action. For this reason, the poetry of the incantation and ritual practice cannot be separated, the former supports the latter and allows it to be effective. As I have shown with the example of incantation No. E.2, the analogies expressed in the text, between the metallic and mineral substances and the body and the personality of the patient, allow the transfer of those qualities possessed by the substances by means of which Tambiah (1968: 194) calls *persuasive analogy* to the patient during the ritual action.

In addition to this system of meaning and transference, poetry assists, thanks to its rhythmic cadence, for the transmission and memorization of the text.

Abracadabra and *historiolae*

Rhythm is the central element of abracadabra. The *nīš libbi* corpus contains five abracadabra. Their language has a clear magical-ritual purpose since it is extraneous to ordinary language. I paid attention to rhythm to understand the function of abracadabra. Following the studies conducted by Severi (2004), I have shown how these nonsense sounds are structured within an organization, through repetition in the line and between the lines of certain syllabic sounds or groups of sounds. The patient perceives *regular sound configurations*, which, though lacking in semantic meaning, are still *meaningful*. Exactly such rhythmic sounds allow the patient, who recognizes the exorcist as the therapeutic authority, a *guided perceptive illusion* (Severi 2004: 236), which, as part of a matrix of the overall system of belief, leads to healing.

Two incantations contain *historiolae*, mythical stories that describe *illo tempore* the deeds of extra-human entities. The incantation No. K.1 narrates the preparation of a bed for the sexual desire by the goddesses Ištar, Nanāya, and Išḫara for their lovers, while incantation No. M.1 describes the anthropogenic act by the divine pair Enlil and Bēlet-ilī. Following the studies of De Martino (1953–1954: 19), I found that the *historiolae* are intended to activate a process of *dehistoricization*. That is, they allow the annulment of the difficulties of the current nefarious event's progress, presenting it as if it were the repetition of a similar event already having occurred in the mythic past. This means that the negative event is thought to be the recurrence of an episode that has already occurred and therefore able to be overcome. Frankfurter (2016) argues also that the purpose is to cancel the real crisis through the mythical recalling.

Abracadabra and *historiolae* support the patient and the community's disposition towards a healing outcome. The first through an obscure form of language directed at the patient which since it was recited by the therapeutic operator, is thought to be effective. The second through the mythical-ritual story.

Normativity: animality and battle metaphors

The *Tierbilder* in Sumerian and Akkadian literature has been an object of study by Assyriologists in both stylistic and conceptual components. What I was interested to show was the normative and regulatory character of animal metaphors in the *nīš libbi* corpus. In incantations, we find two types of animal metaphors: similes with domestic animals, and metonymies with wild ones.

By means of metonymies the man is defined as a wild animal who has lost his rage, his strength, and his sexual vigor. He is identified with the *akkannu*-wild ass, the horse (*sisû*), the onager (*serrēmu*), the wild bull (*rīmu*), the stag (*ayyalu*), and the lion (*nēšu*). Borrowing from Gender and Queer Studies, I underlined the normative power of the gender apparatus expressed through such metonymies. Such images, in fact, convey the idea of an active and vigorous male sexuality. The state of wildness is the highest condition in which it is possible to develop active sexual characteristics typical of virility.

Even domestic animals are mentioned in the corpus: the dog is used for sexual similes; the pig for diagnostic and etiological purposes; the ram and the buck are tied to the bed. It should be noted, therefore, that human sexuality is represented as a mirror on to that of animals. The use of animal behavior has, in fact, the function of controlling and organizing human sexuality by providing models of alleged *naturality*. There are also some references to Enkidu and especially to his "animality."

Another metaphor domain is that of battle. This domain also emerges in ritual symbolism, specifically in the ritual of the bow. As Hoffner (1966) has shown, the bow represents male sexual vigor, if not perhaps the penis. Sexual vigor and value in battle are understood as fundamental characteristics of an ideal Mesopotamian manhood.

Returning to effectiveness

I have not yet answered the question Biggs poses about the effectiveness of these texts. In order to do this, I need to clarify the terms of the discussion. The study of the medical systems of traditional societies in an essay by Lévi-Strauss entitled "Efficacité symbolique" (1949b) had a great impact on this topic. Here he analyzed the therapeutic process conducted by a Kuna shaman, from a population of the Republic of Panama, to help a woman during a difficult delivery. The core of the therapy is a song, performed by the therapeutic operator, who tells of his "expedition" and the consequent "battle" within the woman's uterus to release her soul (*purba*) from the claws of the goddess Muu.

According to the French anthropologist, the shamanic practice is successful because it acts through *symbols*, which manifest in the patient emotions which can affect the endogenous processes of the organism. This is since the healer and the patient share the same *language*. The success of Kuna therapy is, therefore, attributable to the symbolic language, offered to the patient, so that she can overcome the suffering by *representing* it and *thinking* it (ibid. 211–218). According to the father of structuralism, shamanic care straddles modern organic medicine and psychoanalysis. It has, in fact, the aim to make *conscious* conflicts and tensions which until then had remained at an unconscious level. The author refers precisely to *abreation* in psychoanalysis to explain this phenomenon. If the anthropologist discovers similarities between shamanic and psychoanalytic practice, there are also differences: in the psychoanalysis the sick person starts an individual myth, in shamanic care it is a collective one, whose guardian is the shaman. It is the latter *who speaks for* the woman in labour. The anthropologist, however, does not merely see similarities between the two therapeutic practices, he argues in fact that they are the basis of physiological and more precisely biochemical processes (ibid. 223). Thus, according to Lévi-Strauss, the shamanic practice would induce endogenous healing processes at the neurophysiological and biochemical level.

Beginning with the article of the structuralist anthropologist, the concept of symbolic effectiveness has been the subject of debate. A series of theories followed one after the other on the functioning of symbolic effectiveness, without a consensus being reached. The research by ethno-psychiatrist Raymond Prince (1982) has shown that rituals can start endogenous healing processes through the production of endorphins with euphoric and analgesic effects. Likewise, anthropologist Gilles Bibeau (1983, It. tr. 1998: 138–139) states that this takes place through mechanisms through which the brain reacts to emotional states and establishes endocrinological communication with organs (see also Moerman 1979). This attention to the endo-crino-immune effects, though of great importance, poses some difficulties. In particular, one still cannot understand how generic and non-specific endogenous reactions can produce something specifically suited to the pathology of each patient and indeed heal a particular sickness. It is not possible to simply say that the autonomic nervous system can reach all organs: it is necessary to better understand how to cure that part of the body instead of another suffering from a particular pathological condition. We should also consider the research results of psychoneuroendocrinology, a discipline within biomedicine, which deals with the relationship between the different functions of the organism.

I do not want to go into the question of how the manipulation of symbols during the therapeutic itinerary can act organically, because the research is *in fieri* and much debated, and because the present author has not scientific knowledge to do so, and because a careful discussion of various theories would take us far from the purpose of this book. My purpose then was to trace the coherence, or rather the *logic*, at the base of *nīš libbi* therapy. To this end, I consider it essential to emphasize the cultural dimension of the sickness-event and therefore of the therapy and its effectiveness. In fact, the weakness of the theory of therapeutic efficacy, including those theories related to the production of endogenous healing processes, is that they do not consider the concrete cultural context of analysis or the conception of suffering and sickness present within a specific society. In this book, I have tried to show how there is a close relationship between the conception of sickness, sense of suffering, theories of person and self, and model of interpersonal relationships. Without understanding this interrelation, therapy and its effectiveness would not be intelligible (see Beneduce 2005: 11). Consequently, it is not possible even to distinguish between "symbolic efficacy" and "real efficacy" (as Lévi-Strauss seems to postulate) (see Le Breton 1991), on the model of a known distinction between *curing* and *healing*. The first term refers to a technical intervention onto the organic dimension of malaise (*disease*), while the second refers to the re-elaboration of suffering experienced by the patient and their community (*illness*). The problem with this distinction, which is however present in Assyriology, is that it reiterates the dichotomies "*disease vs. illness*" and "body *vs* mind," both of which were discussed in Chapter I. Moreover, it strengthens the distinction between modern Western societies and "other" societies, according to which modern biomedicine *cures disease*, while "other" medicines *heal illness* (see Waldram 2000: 604–607). This terminology obscures, on the one hand, the symbolic component present, as demonstrated by many anthropological studies (particularly in reference to the "pharmaceutical"), in modern biomedicine as well, and the ability of "other" medicines to act organically (see Lupo 2012: 131, 149). Daniel E. Moerman (1979: 60) affirms that in both the personalistic and naturalistic medical systems there is a clear symbolic metaphorical component, of course in addition to significant specific medical effects. The symbolic component of treatment, however, is very significant as well, thanks to the healing metaphors which provide the symbolic substance of general medical treatment.

The importance given to culture and meaning production in the therapeutic process was underlined by Moerman (2002: 14–15) in his study on the placebo effect. The anthropologist and physician distinguishes between three different types of patient responses during the therapy:

1. *Autonomous responses*: all those processes the organism can invoke to regain health, including the various immunological and related systems;
2. *Specific responses*: those of the body to the content of medical treatment (active substances, surgical operations, etc.);
3. *Meaning responses*: psychological and physiological effects of meaning in the treatment of sickness, meaning response attached not only to the prescription of inert medications but to active ones as well.

Meaning responses determine the physiological and biochemical effects stimulated by the cognitive, symbolic, and semantic components of experience, in other words by cultural practices. This means that the meaning that the patients attribute to the most diverse elements of the therapeutic process contributes to determining the effectiveness of therapy.

Therefore, if Biggs's answer about the efficacy of *nīš libbi* therapy does not take into account the symbolic and cultural dimension of the sickness-event, it would be unfair to Mesopotamian medicine to consider it only in empirical-rationalist terms. That is to emphasize only the dimension of the efficacy of the chemical properties of the substances used in the prescriptions (of which the operators would know), while excluding the dimension of meaning, and encapsulating it in an interpretative scheme of modern Western science (see Lupo 2009: 169–171). Geller's analysis, however, does not take into account the different levels of the therapeutic process by using the theoretical approach of Freudian psychoanalysis. Indeed, in the societies, called by the French ethno-psychiatrist Tobie Nathan (1995), "of multiple universes" (as opposed to modern society, centered on biomedicine, called "of single universe"), reducing the sickness-event to an exclusively psychological reality is a methodological mistake. In the perspective of modern psychology and psychopathology, the "problem," in this case anxiety and fear according to Geller, resides, depending on the interpretative school, within the subject, in his psyche, in his biology, and in the sediments of his individual history. In the Mesopotamian context, the interest moves on the plane of the "invisible," from the individual to the collective, in his social and religious dimension, and quoting Nathan "from what is fatal to what is repairable" (1995).

So how do I respond to Biggs's challenging question?

The recitation of incantations is a central part of *ritual performance*. In fact, it enables one, thanks to its *illocutionary* and *perlocutionary* character, to *act on the world* by changing the condition of the patient and the recipient of the therapy. The poetic language, which is different from ordinary language, contributes substantially to this purpose. It allows for the establishment of metaphorical analogies between domains such as war, animals, etc. on the one hand, and that of the body and human sexuality on the other. These metaphors guarantee the transformative processes implemented in the ritual and the "pharmaceutical," provided by the medical prescription. The poetic function of incantations is associated with the conative one (using Jakobson's terminology), or perlocutionary one (using that of Austin and Searle; see the use of precative), which allows for what Tambiah calls "persuasive analogy," that is the implementation of the analogies established by the poetic function, according to what he defines as "the power of words."[508] An emblematic case is discussed in Chapter II concerning the incantation No. E.2. Moreover, the recitation of the incantation over the "pharmaceutical" ingredient, as seen in *nīš libbi* therapy, allows for the activation of the therapeutic power within the pharmaceutical itself (Steinert 2015: 114).

Animal metaphors, like war metaphors, have a normative function as well. They reiterate ideal concepts of Mesopotamian male sexuality. Other metaphors instead serve to make communicative and representable the suffering: incantations provide cultural patterns of symptoms described through metaphors or similes codified in the tradition.

The absence of sexual desire manifests itself in the man, but he is not the only recipient of therapy: the woman is fully involved. The ritual practice has to incorporate all members of the community directly involved, for this reason the woman is an important ritual actor. Involving all those who participate in the sickness-event is intended to undo social tensions at the origin of the event, in cases where sorcery is understood as the cause of suffering.

Concerning the first cause and the patient, *strictu sensu*, the therapeutic practice acts at different levels of the sickness and seems to confirm the analysis of the ingredients used in medical prescriptions: the psycho-physiological one and one related to the sense and the cause of suffering. Some ingredients seem to act directly against the *first cause* of suffering, that is for example witchcraft. In this sense, also the diagnostic and etiological analysis, which involves the creation of the figurines of a man and a woman and the role of the pig, has a therapeutic purpose as well. As Csordas and Kleinmann (1990, It. tr. 1998: 110) claim, in some medical systems, diagnosis and etiology cannot be considered as only a pathway to therapy but as a fundamental part of the therapeutic process itself. The etiological analysis serves to understand the cause of the sickness and allows it to act against it. Other strategies are used to guarantee therapeutic efficacy: the use of *historiolae*, which serves to activate, according to De Martino, a process of dehistoricization of suffering. This strategy allows it to cancel the real malaise by relocating it to an exemplary mythical context.

Lévi-Strauss emphasized the importance of sharing language between the woman in labour and the shaman: the therapy is successful because it allows her to think and

508 For a summary of the anthropological theories on words in healing rituals see Dein 2002: 43–45.

visualise and thus overcome the suffering. This interpretative hypothesis has undergone many criticisms because the ethnographic investigations reported that the language of shamanic song is in an obscure language, which the woman cannot understand and hears only as a sequence of sounds (see Laderman 1987; Severi 2004). The use of an unknown language to the patient is not alien to Mesopotamian therapeutic practices, let us think of the abracadabra sections of the rituals. Following Severi's theory, I showed how the use of abracadabra manifests the patient's *projection* and confidence in the therapeutic operator and in his practice. The exorcist and the patient, while not sharing the same linguistic content (= *meaning*), share the same *sense horizon*. The patient knows what the exorcist is doing and so, believing in the performance, he projects. In addition, the use of a language obscure to the patient gives authority to the exorcist, just as the formulas "at the command of DN" or "the incantation is not mine, it is of DN."

The importance attributed to believing (see Lévi-Strauss 1949a: 196–199) during the therapeutic itinerary is emphasized by Moerman:

> In Western medicine, the primary device for achieving this end is the extraordinary romance medicine has with science. [. . .] . Doctors routinely argue that their work "is scientific." By this, they mean that it is somehow based on real scientific analysis or experiment; that is, that it's "true." [. . .] What is important is that doctors – healers of any sort or type – are convinced that their techniques are powerful and effective, and that there is undeniable evidence of this effectiveness. In some places, such proof comes from gods or spirits, in some places from personal experience, and in other places from the assertions of science. In so far as these convictions are somehow conveyed to patients and, in the process, convince them of their doctor's power, then they are likely (within the bounds of our physical mortality) to be effective [Moerman 2002: 43].

Any consideration of the "truth" of Mesopotamian medical practices from a biomedical perspective is inappropriate. It is not possible to establish the effectiveness of other medical practices on the basis of (scientific) "truth." [509] A therapy is effective if it provides a response to the problems posed by the sickness-event, which vary according to cultural contexts (see TSeppilli 1996: 21). It follows that, quoting Kleinman (1973: 210), efficacy is itself a cultural construct.

---

509 In this respect the famous case of Quesalid, the Kwakiutl shaman of Vancouver Island, in British Columbia, described by Lévi-Strauss (1949a), should be mentioned here. Quesalid was skeptical about the powers of the shamans, so he decided to start learning shamanic practices to debunk them. What he had learned was "tricks," including the use of a cotton ball. In fact, the shaman hides it at the corner of his mouth and at the right time, pulls out it, covered with blood, after bowing his tongue, and presents solemnly it to the patient as the pathological body extracted by sucking. However, Quesalid, as a shaman apprentice, noticed that during therapy the cotton ball technique was effective since the patients healed. Quesalid, now a great healer, has continued to exercise his profession consciously, proud of his achievements. As the anthropologist writes: «Quesalid n'est pas devenu un grand sorcier parce qu'il guérissait ses malades, il guérissait ses malades parce qu'il était devenu un grand sorcier» (ibid. 198).

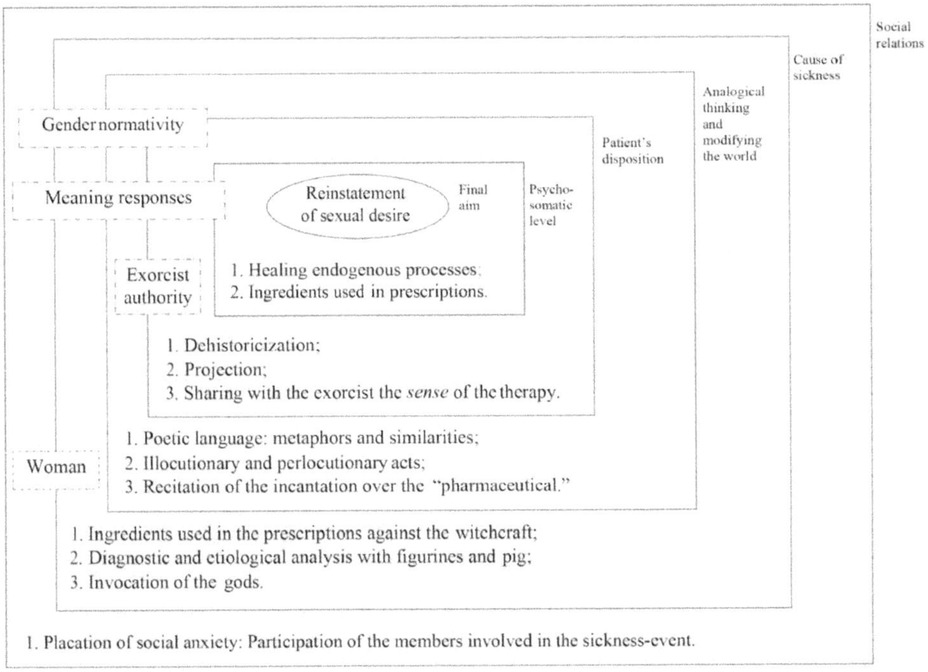
Social relations
Cause of sickness
Analogical thinking and modifying the world
Patient's disposition
Psycho-somatic level
Final aim
Gender normativity
Meaning responses
Reinstatement of sexual desire
Exorcist authority
1. Healing endogenous processes;
2. Ingredients used in prescriptions.
1. Dehistoricization;
2. Projection;
3. Sharing with the exorcist the sense of the therapy.
Woman
1. Poetic language: metaphors and similarities;
2. Illocutionary and perlocutionary acts;
3. Recitation of the incantation over the "pharmaceutical."
1. Ingredients used in the prescriptions against the witchcraft;
2. Diagnostic and etiological analysis with figurines and pig;
3. Invocation of the gods.
1. Placation of social anxiety: Participation of the members involved in the sickness-event.

Part II: **Edition**

Nīš libbi catalogue LKA 94

Museum number	Publication	Tablet	Script	Date	Provenience
Ass 13955 kb	LKA 94	Two-col. tablet	NA	8th–7th cent.	Aššur, Library N 4

Edition

Biggs 1967: 11–16

Transliteration and translation

obverse
column i

1. ÉN *i* x [x x] *mi-ni-i*	1. Incantation: "... [ ] ...".
2. ÉN *ak-k[an-n]i* MIN *ri-mi* MIN	2. Incantation: "My *akk[ann]u*-wild ass! My *akkannu*-wild ass! My wild bull! My wild bull!"
3. ÉN *pu-ḫa-li ana* U$_5$ ZI-*ú*	3. Incantation: "Ram who is reared-up for the mating."
4. ÉN GIN IM *li-nu-⟨uš⟩* KIRI$_6$	4. Incantation: "May the wind blow! May the grove quake!"
5. ÉN *am-mi-ni sek-re-ta* GIM ÍD	5. Incantation: "Why are you blocked up like a canal?"
6. ÉN *ad-di* GIŠ.NÁ *at-ta-[di* ^{giš}G]U.⸢ZA-*a*⸣	6. Incantation: "I have set up a bed! I have now set up a [cha]ir!"
7. ÉN TÚG *tu-ḫa-am-ma um-di-ṣu-u*	7. Incantation: "You . . . the clothing and I spread out (it)."
8. ÉN *am-mi-ni ar-ma-a* IGI^{II}-*ka*	8. Incantation: "Why are your eyes covered?"
9. ÉN *[a]p-šur si-ra ap-ta-šar*	9. Incantation: "[I] absolved a (reed-) shelter$^{?}$! I have absolved!"
10. ÉN *ir-ku-sa-ma ip-ta-ṭar*	10. Incantation: "They (fem.) bound, but he released!"
11. ÉN *ir-ku-sa-nim-ma ip-ta-ṭar*	Incantation: "They (fem.) bound for me, but he released!"
12. ÉN *lab-šá-ku na-ka-{ku} ḫal-pa-ku*	12. Incantation: "I am clothed with copulation! I am enveloped."

https://doi.org/10.1515/9783110757262-005

13. ÉN ^{d}BE ^{d}nin-maḫ nam-lú-ùlu-lu	13. Incantation: "Enlil and Bēlet-ilī mankind."
14. ÉN GIM dím-an-na	14. Incantation: "Like creation$^{?}$ of heaven."
15. ÉN *ak*$^{!}$*-ka-ni ṭa-ar-da*	15. Incantation: "My hunted *akkannu*-wild ass!"
16. ⸢ÉN⸣ *a-a-lì a-a-lì* ANŠE.KUR.RA	16. ⸢Incantation⸣: "My stag! My stag! Horse!"
17. [ÉN] ⸢*ak-ka*⸣*-ni šá* KUR-*i* ⸢*man-nu*⸣ *is-kir-ka*	17. [Incantation]: "My *akkannu*-wild ass of the mountain, who has blocked you?"
18. [ÉN . . .] ^{giš}GIŠIMMAR MIN	18. [Incantation: ". . .] Date palm! Date palm!"
19. [ÉN] SU.ZI MIN ŠÀ.ZI.GA MIN	19. [Incantation]: "Shiver! Shiver! Sexual desire! Sexual desire!"
20. [ÉN] *gu-ru-u*[*š* MIN] ^{giš}DÌḪ MIN	20. [Incantation:] "Have se[x! Have sex!] *Baltu*-thorn! *Baltu*-thorn!"
21. [ÉN] *e-nu-ma* [U]R.GI$_7$ *ana* ^{munus}UR.GI$_7$.⟨RA⟩	21. [Incantation]: "When a [d]og to a bitch."
22. [*ana* ŠÀ.]ZI.GA [T]UKU-*e* 3 UD 2	22. [In order to g]et sexual desire three . . . two.
23. $^{[n]a4}$*aš-pu-u*	23. *Ašpû*-stone.
24. $^{[n]a4}$KA.GI.⟨NA⟩.DAB.BA	24. Magnetite.
25. $^{[n]a4}$ZA.GÌN	25. Lapis lazuli.
26. [ÉN] SU.ZI MIN	26. [Incantation]: "Shiver! Shiver!"
27. [ÉN] ki-in-da-rab MIN	27. [Incantation]: ". . ."
28. [ÉN] ÍD [ŠÀ]-ZI.GA GIN.A	28. [Incantation]: "River of a constant sexual desire."
29. [ÉN ^{d}*iš-tar*] *be-el-tu*	29. [Incantation: "Ištar] lady."
30. [ÉN . . . g]a te-en-te	30. [Incantation: ". . .] cool down!"
One or two lines broken	

column ii

1. ÉN x [. . .]	1. Incantation: ". . . [. . .]."
2. ÉN [. . .]	2. Incantation: "[. . .]."
3. ÉN [. . .]	3. Incantation: "[. . .]."
4. ÉN [. . .]	4. Incantation: "[. . .]."
5. ^{NU} É[N . . .]	5. ^{Lacking} Incan[tation: ". . .]."
6. ^{NU} ÉN . . . [. . .]	6. ^{Lacking} Incantation: ". . . [. . .]."
7. ÉN *e-la-m*[*a-tu*$^{?}$ . . .]	7. Incantation: "Elam[ite (women$^{?}$) . . .]."

8. ÉN ^{d}*iš-tar šá* x [. . .]	8. Incantation: "Ištar of/who . . . [. . .]."
9. ^{NU} DIŠ NA KI.TUŠ-*šú it-*[*ta-na-ru-uṭ*$^{!}$]	9. ^{Lacking} If a man's seat co[ntinually sway].
10. ^{NU} DIŠ NA *ina* DU$_{11}$.DU$_{11}$-*šú* [. . .]	10. ^{Lacking} If a man in his speech [. . .].
11. ^{NU} DIŠ NA ŠÀ-*šú i-ḫa-ša-*[*aš-m*]*a*$^{!}$ UR$_{4}$	11. ^{Lacking} If a man's 'heart' is afflict[ed an]d trembles.
12. ^{NU} DIŠ NA UGU-*šú* NIGIN-[*d*]*a* IGI^{meš}-*šú* ⸢NIGIN⸣	12. ^{Lacking} If a man's skull tur[n]s, his face turns.

13. ÉN *lil-lik* I[M KU]R^{meš} *li-nu-šú*	13. Incantation: "May the wi[nd] blow! May the [moun]tains quake!"
14. ÉN *ri-mi* [Z]I.[GA] LU.LIM ZI.GA	14. Incantation: "My wild bull, [rea]r [up]! Stag, rear up!"
15. ÉN ŠÀ.ZI.[G]A MIN KI.⸢NÁ⸣ MIN	15. Incantation: "Se[xual] desire! Sexual desire! Bed! Bed!"
16. ÉN SAG x [x] x GÚ-*ia*	16. Incantation: "Head . . . [ ] . . . my neck."
17. ÉN *mu*[*š-l*]*a-li* KÙ.BABBAR MIN KÙ.SI$_{22}$	17. Incantation: "Staircase-ga[t]e of silver! Staircase-gate of gold!"
18. ÉN *man-nu šá* GIM KASKAL KUD A.RÁ DÍM	18. Incantation: "Who is the one who has made my course like a blocked road."
19. ÉN *ma-rat* ^{d}*nin-gír-su*	19. Incantation: "Daughter of Ninĝirsu."
20. ÉN x [x] x *ti-il-pan*	20. Incantation: ". . . [. . .] bow."
21. *ana* ŠÀ.ÍL TUKU-*e* UD 2 UD 2	21. In order to get sexual desire . . . two . . . two.
22. DIŠ NA ŠÀ-*šu ul-te-ni-di-*[x]	22. If a man's 'heart' . . .
23. *ana* NA ŠÀ.ZI.GA TUKU-⸢*e*⸣	23. In order to get a man's sexual desire.
24. DIŠ NA ÍL *lìb-b*[*i* N]U [TUKU]	24. If a man [does n]ot [have] sexual desi[re].
25. ÉN *ul* x *ša* [. . .]	25. Incantation: ". . . [. . .]."
26. ÉN *gur-*[*uš* . . .]	26. Incantation: "Have [sex . . .]."
27. ÉN *ram*$^{?}$ [. . .]	27. Incantation: ". . . [. . .]."
28. ÉN x [. . .]	28. Incantation: ". . . [. . .]."
29. É[N . . .]	24. Incan[tation: ". . .]."

Maybe three lines miss

reverse
column iii

1. ÉN *ar* x [. . .]	1. Incantation : ". . . [. . .]."
2. ÉN *am* x x [. . .]	2. Incantation: ". . . [. . .]."
3. ÉN *ina* SAG-*ia*$_5$ *da*$^?$-[*áš-šú ra-ki-is*]	3. Incantation: "At my head a b[uck is tied]."
4. ÉN *la-ba-an* x [. . .]	4. Incantation: "Sweet pomegranate . . . [. . .]."
5. ÉN *i-ne-en-na i* [. . .]	5. Incantation: "Now . . . [. . .]."
6. ÉN *e-piš-tú eš-še-b*[*a-a-at*]	6. Incantation: "The sorceress is an ecst[atic]!"
7. *ana* BÚR-*ti kiš-pi š*[*á ina* NIND]A^{meš} ⌜*u*⌝ [KAŠ^{meš}]	7. In order to release from the magic *from* [brea]d and [beer].
8. ÉN *ana-ku* ^{d}ŠÚ *da kan* [*n*]*i iš*	8. Incantation: "I am Marduk . . ."
9. ÉN TU$_6$^{meš} [ŠÀ.Z]I.GA	9. Incantation: Incantations for [sexual de] sire.

Commentary

This catalogue LKA 94 from the N 4 library ("Haus des Beschwörungspriesters") contains a list of incantations of the *nīš libbi* genre. At the rubric at the end of catalogue (iii 9) we read ÉN TU$_6$^{meš} ŠÀ.ZI.GA "incantations for sexual desire," however, the text contains not only incantations (some entries are preceded by NU 'lacking,' ii 5–6); but also a few incipits indicating therapeutic prescriptions: some begin with DIŠ NA "If a man," (ii 9–12, 22, 24, see some entries are preceded by NU 'lacking,' ii 9–12); other with the names of stones (i 23–25); *ana* ŠÀ.ÍL TUKU-*e* "in order to get sexual desire" (ii 21); *ana* NA ŠÀ.ZI.GA TUKU-*e* "in order to get a man's sexual desire" (ii 23); *ana* BÚR-*ti kiš-pi* "in order to release from the magic" (iii 7). It seems that LKA 94 gives an overview of a collection of this genre enumerating the incipits of text sections on multiple tablets (the tablets marked with NU were not available to the compiler). As Biggs (1967: 11) suggests, the catalogue was perhaps a complete list not of all *nīš libbi* incantations, but only of all those known from Aššur. The catalogue is divided into two sections by a double ruling, which respectively contain older compositions and more recent material added by a second compiler, in fact, there is nothing in either section to suggest that the two parts were distinguished on the basis of content (Maul 1994: 191–195; Steinert 2018a: 58).

Column i

2. See No. A.3: 49.

3. There are no incantations with this incipit, however, the mention of a ram who is reared-up for the mating is very present in the prescriptions. From the animal are taken substances used to make potions and amulets: wool of a reared-up ram: No. D.4: 61; F prescr. 17: 57; wool of tail of a reared-up ram: No. K.8: 155 (cf. A prescr. 2: 6–9; N prescr. 10: 27–28); saliva of a reared-up ram: N prescr. 4 i 19; N prescr. 18 iv 3. The animal is associated with the sexual sphere in incantations and rituals: No. E.1: 6–8; No. E.2: 41–44; No. E.3: 46–48; No. F.4: 81–82; No. K.8: 148 (broken); No. O.1: 16–17.

4. See No. A.1: 33; Cf. No. I.1: 6: "May the wind blow! May the grove not quake!", and the *Aššur Medical Catalogue*, section XX "sex": "[Incantation: 'Let the wind blow], the gardens shall not quake'" / [One tablet (of the section) 'Incantation: Let the wi]nd [blow], the gardens shall not quake'" (ll. 103–104, Steinert 2018b: 217).

5. Cf. No. D.2: 11: "Who has blocked you like an opening of the *dilûtu*-water system."

6. The bed is mentioned also in LKA 94 ii 15, and maybe in iii 3. See the preparation of a bed for the sexual desire in No. L.1 "Sexual desire! Sexual desire! I have prepared a bed for the sexual desire" (see commentary ll. 5–6 and 7). A ram and a buck are tied at the bed in No. E.1: 5–8 and No. F.4: 80–87; a ram and a weaned sheep in No. E.2: 41–42 and No. E.3: 46–47. See also No. J.3: 32; No. K.5: 98–99; No. K.8: 151; No. L.1: 2–3; No. O.1: 19, 21–23 (see commentary ll. 21–23, maybe mention of a bedroom). The only other mention of the chair is in No. B.1: 7 ". . . your chair will not be held" (see commentary). See also in this catalogue iii 9.

7. *tu-ḫa-am-ma* is unclear. It can probably be interpreted as a verbal form from the verb *ḫam/wû* (CAD Ḫ 163 'to growl'; AHw. I 338 'summen, of ghost, dog'), although D-stem not attested: "You make the clothing *sound* (so) I spread out (it)." Note that the verb corresponds to Sumerian gù dúb, whose Akkadian equivalent is *nagāgu* 'to bray' (see the imperative *ugga* in No. B.1: 1–2; No. F.3: 65–66; No. G.1: 3), *ṣarāḫu* 'to sing,' *tukku* 'alarming sound, rumor.' It is associated with the verb *šasû* 'to shout, call (out)' (for the references see CAD Ḫ 163). According to Rendu Loisel (2016: 162) the verb indicates a kind of buzzing sound like an insect. Note that the verb is used to designate the sound of thunder of the god Adad as a kind of buzzing announcing of abundance and agricultural prosperity in *Nisaba and the Grain* (i 21′, Lambert 1960a: 170). In this therapeutic corpus, the atmospheric agents are characterized by sexual values, as many metaphors from

the atmospheric realm used to describe human sexuality. The verb, therefore, could refer to the noise of clothes with a sexual value (see the verb *ḫabābu* for noise understood sexually in Chapter II § "Bestiality or figure of speech?"): the noise produced by the movement of clothes alluding to the desire for sexual intercourse. The verb *wuṣṣû* 'to spread' could refer to the act of getting undressed or undressing someone. See in the *Epic of Gilgameš* the intercourse between Enkidu and Šamḫat: *lubušīki muṣṣî-ma elīki liṣlal* "Spread your (= of Šamḫat) clothing so he (= Enkidu) may lie on you" (*Epic of Gilgameš* I 184, see also l. 191, George 2003: 548). For the phrase *lubūšī muṣṣû* see George 2003: 796.

8. See No. F.2: 59–64 "Why are your eyes covered?".

10–11. These two incipits refer to the action of the witches who bound the patient by taking away his sexual desire. The one who releases him is the god or the one who on earth takes his place, the therapeutic operator. On the binding as result of witch's action see Chapter II § "Similes concerning the action of the witch."

12. See No. D.3: 32–39 "I am clothed with copulation! I am enveloped with intercourse!". Cf. No. J.3: 25–27: "[. . .] I am washed. / [. . .] I am anointed. / [. . .] I am clothed."

13. See No. K.1: 8–17 "When Enlil and Bēlet-ilī gave mankind a name."

14. See No. K.1: 16: "Incantation: Like crea[tion? of heav]en?! Like creatio[n of heaven]!".

15. See No. D. 2: 10.

16. Cf. incantation No. G.1: 79–87 "Stag! Stag! Wild bull! Wild bull!". Here the will bull is mentioned instead of the horse.

17. Cf. No. A.2: 41. "Incantation: *Akkannu*-wild ass who is reared-up for mating, who has dampened your desire?"; No L. 2: 26.

18. As suggested by Biggs (1967: 15), it is possible to restore *gu-ru-uš* MIN, if we consider the incantation incipit in LKA 94 i 20 as a parallel: [ÉN] *gu-ru-u*[*š* MIN] ^{giš}DÌḪ MIN "[Incantation:] 'Have se[x! Have sex!] *Baltu*-thorn! *Baltu*-thorn!'". See also LKA 94 ii 26. The date palm appears in this corpus as ingredient only once: N prescr. 14 iii 13. See the date palm in amorous context in KAL 3, 75 ii 4 (Wasserman 2016: 116).

19. See LKA 94 i 26; D prescr. 4: 47–48; K prescr. 28: 77–79. See also the series *kunuk ḫalti* ("*ḫaltu*-seal") K. 3010+ v 24′–32′ (Schuster-Brandis 2008: 365, text 16 A):

[6] NA$_4$^{meš} Š[À$^?$.ZI.G]A *ina* SÍG UDU U$_5$ ZI.GA NU.NU ⸢UD⸣.D[U . . .] *ina* GÚ-*šú* GAR
broken
[ÉN SU.ZI MIN ÉN KI.IN].DA.RAB [MIN$^?$]
[2 ÉN^{meš} . . .] *ana* UGU 3 DU[R$^?$]
[*šá* ŠÀ.ZI.GA] ŠID-*nu*

[ÉN su-zi] MIN
[. . .] ⸢šà$^?$⸣-zi-ga
[. . .] x ki-ga-ke$_4$
[. . . e]n si-sá
[. . . šà$^?$-z]i-ga MIN TU$_6$ ÉN

[Six] stones for the sex[ual desi]re you insert in the spun wool of a sheep reared-up for mating [. . .] you put it around his neck.
broken
[The incantation: "Shiver! Shiver!". The incantation] ". . . [. . .]."
You recite [the two incantations . . .] over the three band[s]
[of the sexual desire].

[Incantation "Shiver!] Shiver!"
[. . .] sexual desire
[. . .] . . .
[. . .] . . .
[. . . Sexual] desire! Sexual desire!" Incantation formula.

20. See the parallel incipit in LKA 94 i 18. The *baltu*-thorn is an ingredient used several times in this therapeutic corpus (see index of ingredients).

21. This incantation is not attested. See the mention of dog and bitch in incantation No. E.1: 9–10. For the dog and its relation with the sexual sphere see the analysis of incantation No. E.1. Note that the prescriptions provide several pharmaceuticals with dog-based ingredients (see list of ingredients).

22. The meaning of 3 UD 2 is not clear. See LKA 94 ii 21 and 23.

23–25. These lines do not indicate incipits of incantation, rather ingredients used in the prescriptions in this corpus. The magnetite is mainly used for the realization of ointments, together with the iron, to be rubbed on both male and female genital organs, to activate, similarly to magnetism, the attraction between the patient and the woman. Lapis lazuli is used instead for amulets and the incantations underline its purity to instill to the patient (No. E.2: 33). The *ašpû*-stone is used for amulets as well (see list of ingredients).

26. See commentary l. 19.

27. Pseudo-Sumerian(?) incantation, whose meaning is unclear, mentioned in *nīš libbi* S prescr. 1: 9′ and K prescr. 28: 79. On kindarab-incantation see Schuster-Brandis 2008: 238, text 4: 88a′-d′. See also *kunuk ḫalti* series ("*ḫaltu*-seal") K. 3010+v 33′–36′ (Schuster-Brandis 2008: 365, text 16 A):

[ÉN ki-in-da-rab ki-i]n-da-rab
[. . .] x ki-in-da-rab
[. . .] x gar an ki-in-da-rab
[. . .] x TU_6 ÉN

28. Another possible translation "Flow, river of sexual desire! Water." A, the first word of the second half of the line, can be the logogram for 'water' as suggested by Biggs (1967: 15). Sexual desire is compared to constantly flowing river water in No. A.1: 35: "May my sexual desire be constant river water!". See for similar metaphors also No. A.2: 42; No. H.1: 7–8 (see Chapter II § "Third group: sexuality and nature").

29. See also LKA 94 ii 8. See No. D.4: 60: "(The incantation is not mine, it is) the incantation of Ištar, [patro]n of love." See the invocation of the goddess in the following incantations: No. A.2: 46; No. E.2 (esp. l. 26); No. J.3: 33; No. M.1: 5. For the relation of Ištar with sexual attraction see the analysis of incantation No. A.2. For the libations for the goddess see Chapter IIII § "Libations to Ištar." For the expression "at the command of DN" (*ina qibīt* DN) in reference to the goddess see Chapter II § "Mention of gods and operations of the therapeutic operator." Note that the "Hand of Ištar" is one of the causes of the loss of sexual desire (see Chapter I § "Cause of sickness").

i 30. See No. L.1: 2–4. See the similar fragmentary incantation K.5.

Column ii

8. See commentary ii 29.

9. For the chair/seat see commentary i 6.

10. Among the symptoms of the loss of sexual desire see B prescr. 1: 21 "his speech is constantly incoherent" (*pûšu ittanakkir*).

11. Distress (*ašāšu*/*ašuštu*) and trembles are common symptoms of the loss of sexual desire. See in particular No. D.4: 57; No. F.4: 88; B prescr. 1: 18–19. For a description of these symptoms see Chapter II § "Fear, distress, and insomnia."

12. These symptoms are not mentioned in this therapeutic corpus.

13. See No. E.1: 1: "May the wind blow! May the mountains quake!"; No. G.1: 7: "May the wind blow!".

14. This incantation is not attested. See No. A.3 "*Akkannu*-wild ass! *Akkannu*-wild ass! Wild bull! Wild bull"; No. B.1 "Roar on me! Roar on me! Rear up! Rear up!"; No. F.3 "Roar on me! Roar on me! Rear up! Rear up!"; No. G.1. "Stag! Stag! Wild bull! Wild bull!". Note that the term LU.LIM/*lulīmu* 'stag' is not attested in this corpus.

15. See commentary i 9.

18. See No. A.4: 73 and 75. Cf. No. D.2: 13: "Who has blocked your ways as you are a traveler."

19. See No. F.5: 93–99 "Daughter of Ninĝirsu, the releaser I am." See also No. F.3: 74.

20. On the bow in *nīš libbi* ritual see Chapter III § "Bow ritual and battle metaphors."

21. The logograms ŠÀ.ÍL should be inverted (see ibid. ii 24: ÍL *lìb-bi*; No. E.2: 18: ÍL ŠÀ-*šú*). It is more probable that the sign DIŠ is here to be read *ana* rather than *šumma*, see the formula used in the corpus *ana* ŠÀ.ZI.GA TUKU "in order to get sexual desire." The meaning of UD 2 UD 2 is not clear to me, see LKA 94 i 22.

22. Biggs (1967: 13) restores and translates *ul-te-ni-di-i*[*l*?] "constantly gets blocked up(?)." The Štn-stem of the verb *edēlu* 'to lock' is, however, not attested in the dictionaries.

24. See the symptom "(If a man) does not have sexual desire" in F prescr. 12: 36.

Column iii

3. See No. F.4: 80–93 (Ms. A, F) "At my head a buck is tied!". See also No. E.1: 5 "At the head of my bed is really tied (var.: I have tied) a buck!". See also No. E.3: 46–47.

4. *labbānu* 'sweet pomegranate' is uncertain here, other possabilities: *labânu* 'neck(tendons)'; *la-ba-an a*[*p-pi* . . .] 'stroking of the nose.'

6. See *Maqlû* IV 133 (Abusch 2016: 127).

7. The restoration is suggested by Biggs 1967: 18.

9. It could be amended TU_6 ÉN^{meš} "incantation formulas."

I Texts from Aššur (with duplicates from other sites)

Nīš libbi A

List of manuscripts

Manuscript	Museum number	Publication	Tablet	Script	Date	Provenience	Incantations and prescriptions
A	–	LKA 95: 1–rev. 29	Single-col. tablet	NA	8th–7th cent.	Aššur, Library N 4	Prescr. 1–10; 11–19; No. A.1; No. A.2; No. A.3
B	VAT 13758	LKA 96: 1–rev. 15	Frg. of a single-col. tablet	NA	8th–7th cent.	Aššur, Library N 4	Prescr. 1–7; 20–24; Sympt.
C	–	LKA 101 rev. 12–19	Single-col. tablet	NA	8th–7th cent.	Aššur, Library N 4	No. A.1
D	SU 52/139 + 161+170+ 250+ 250A+323	STT 280 i 44, 51–55; iv 37–41	Two-col. tablet	NA	8th–7th cent.	Sultatepe	Prescr. 7; 9–12; 16; No. A.1
E	W. 22307/ 4+68	SpTU 1, 9: 28′–29′	Frg. of a single-col.? tablet	NB/LB	4th–3rd cent.	Uruk U 18	Prescr. 18
F	K 5991	AMT 88, 3: 1–18	Frg. of two-col. tablet	NA	7th cent.	Nineveh, 'Ashurbanipal's Library'	Sympt.; Prescr. 24; No. A.4
G	VAT 13731	LKA 100 rev. 1–6	Frg. of a single-col. tablet	NA	8th–7th cent.	Aššur, Library N 4	Sympt.; Prescr. 24
H	Sp II 976 = BM 35394	CCMAwR 3, pls. 2–3 rev. 6′–10′	Three-col. tablet	LB	4th–3rd cent.	Babylon?	No. A.3
I	BM – Sm 1514	AMT 66, 1: 6–10	Small frg. of a single-col. tablet	NA	8th–7th cent.	Nineveh, Ashurbanipal's Library'	Prescr. 22–23

https://doi.org/10.1515/9783110757262-006

Editions

Thompson 1930–1931: 18 (trans. of Ms. F)
Biggs 1967: 17, 19–21, 35–36, 52, 61–64, 66–68 (Ms. A, B, C, D, F, G)
Buccellati 1976: 67–68 (trans. of No. A.3)
Hunger 1976: 26–27, No. 9 (Ms. E)
Sefati and Klein 2002: 577 with fn. 53 (trans. of Ms. F)
Schwemer 2010: 115–120 (trans. of Ms. D)
Abusch and Schwemer 2011: 106, No. 2.5.4 (Ms. D, E)
Abusch et al. 2020: 31–34, No. 4.1 (Ms. B, F, G)

Structure of the text

Text *nīš libbi* A is based on Ms. A. It begins with nineteen prescriptions (Ms. A 1–rev. 5) with duplicates from Ms. B (prescr. 1–17), D (prescr. 7, 10–12, 16) and E (prescr. 18). The text, following Ms. A, continues with the incantation No. A.1 "May the wind blow! May the grove quake! (with duplicates Ms. C and D), No. A.2 "*Akkannu*-wild ass who is reared-up for mating, who has dampened your desire?", and No. A.3 "*Akkannu*-wild ass! *Akkannu*-wild ass! Wild bull! Wild bull!" (with duplicate Ms. H). The text continues, following Ms. B (rev. 2–15), with five prescriptions (prescr. 20–24) and the symptom description (with duplicates: Ms. I prescr. 22–23; F Sympt.; G Sympt.). Finally, following Ms. F, the text includes the incantation No. A.4 "Who are you who have blocked up my course like a road."

A – B – D	Prescr. 1–19
A – C – D	No. A.1
A	No. A.2
A – H	No. A.3
B – I	Prescr. 20–23
B – F – G	Sympt.
B – F – G	Prescr. 24
F	No. A.4

I. Prescriptions

1.	ll. 1–5	(Ms. A 1–2 // B 1–5)
2.	6–9	(Ms. A 5–8 // B 6–8)
3.	10–13	(Ms. A 9–12 // B 9–12)
4.	14	(Ms. A 13–14 / B 13)
5.	15	(Ms. A 15 // B 14)
6.	16	(Ms. A 16 // B 15)

7. 17 (Ms. A 17 // B 16 // D i 44)
8. 18 (Ms. A 18)
9. 19 (Ms. A 19 // D i 52)
10. 20 (Ms. A 20–21 // D i 53)
11. 21 (Ms. D i 54)
12. 22 (Ms. A 22 // D i 55)
13. 23 (Ms. A 23)
14. 24 (Ms. A 24)
15. 25–26 (Ms. A 25–26)
16. 27 (Ms. A 27–28 // D i 51)
17. 28–30 (Ms. A rev. 1–3)
18. 31 (Ms. A rev. 4 // E 28′–29′)
19. 32 (Ms. A rev. 5)

II. Incantation and its ritual No. A.1 "May the wind blow! May the grove quake!" (Biggs 1967 No. 15), ll. 33–39 (Ms. A rev. 6–11 // C rev. 12–19 // D iv 37–41)

III. Incantation and its ritual No. A.2 "*Akkannu*-wild ass who is reared-up for mating, who has dampened your desire?" (Biggs 1967 No. 1), ll. 41–48 (Ms. A rev. 12–19)

IV. Incantation and its ritual No. A.3 "*Akkannu*-wild ass! *Akkannu*-wild ass! Wild bull! Wild bull!" (Biggs 1967 No. 3), ll. 49–58 (Ms. A rev. 20–29 // H rev. v 6′–10′)

V. Prescriptions
20. ll. 59–61 (Ms. B rev. 2–4)
21. 62 (Ms. B rev. 5)
22. 63–64 (Ms. B rev. 6–7 // I obv. 6–8)
23. 65–66 (Ms. B rev. 8–9 // I obv. 9–10)
Symptoms ll. 67–68 (Ms. B rev. 10–11 // F 1–3 // G rev. 1–3)
Prescription
24 69–74 (Ms. B rev. 13–15 // F 4–10 // G rev. 2–6)

VI. Incantation No. A.4 "Who are you who have blocked up my course like a road" (Biggs 1967 No. 4), ll. 75–82 (Ms. F obv. 11–18)

Summary of the sections of manuscripts not included in the transliteration:

- Ms C = LKA 101 obv.! 7–rev.! 11 = No. D.2 (Ms. A)

- Ms. D = STT 280
  i 1–7 = K prescr. 1–2 (Ms. A)
  i 8–17 = D Sympt.; prescr. 4 (Ms. E)
  i 18–21 = F prescr. 10–11 (Ms. E)

i 22–51 = K prescr. 3–18 (Ms. A) (note i 44 = A prescr. 7 (Ms. D))
i 56–ii 9 = K prescr. 19–22 (Ms. A)
ii 10–21 = No. M.1 (Ms. B)
ii 22–35 = K prescr. 23–28 (Ms. A)
ii 36–50 = No. K.3 (Ms. A)
ii 51–53 = No. K.4 (Ms. A)
ii 54–61 = No. K.5 (Ms. A)
ii 62–iii 23 = No. E.2 (Ms. H)
iii 24–33 = No. K.6 (Ms. A)
iii 34–42 = No. K.7 (Ms. A)
iii 43–iv 7 = K prescr. 29–31 (Ms. A)
iv 8–23 = No. K.8 (Ms. A)
iv 24–31 = No. K.9 (Ms. A)
iv 32–36 = K prescr. 32 (Ms. A)

– Ms. E = SpTU 1, 9
1′–4′ = Fragmentary lines
5′–7′ = D Diagn. (Ms. C)
8′–12′ = D prescr. 1–2 (Ms. C)
13′–16′ = No. D.3 (Ms. C)
17′–18′ = D prescr. 3 (Ms. C)
19′–22′ = D Sympt. (Ms. C)
23′–27′ = D prescr. 4–5 (Ms. C)

– Ms. G = LKA 100 obv. = Not preserved

– Ms. H = Sp II 976 = BM 35394 (CCMAwR 3, pls. 2–3)
i 1′ = Undecipherable
i 2′–8′ = D Diagn. (Ms. G)
i 9′–21′ = No. E.1 (Ms. D)
ii 1′–4′ = Fragmentary lines
ii 5′–22′ = No. E.2 (Ms. D)
iii 1′–4′ = Undecipherable
rev. iv = Not preserved
rev. v. 1′–5′ = Fragmentary lines
rev. vi 1′–3′ = Undecipherable

– Ms. I = AMT 66, 1
1–5 = Q prescr. 1
6–10 = A prescr. 22–23 (Ms. I) = Q prescr. 2–3
11 = Q prescr. 4

Transliteration

I. A obv. 1–rev. 5 // B 1–16 // D i 44, 51–55 // E 28′–29′ = Prescriptions 1–18

1.	A o. 1	traces
	B o. 1	DIŠ NA *ana* MUNUS *a-la-k*[*a*] *m*[*u*$^{!}$*-uṭ-ṭu* . . .]
2.	A o. 2	traces
	B o. 2	PIŠ$_{10}$.^{d}ÍD *ta-mar-raq ina* ^{giš}[. . .]
3.	A o. 3	traces
	B o. 3	x x *a tu* BURU$_5$.ḪABRUD.DA^{mušen} [NITÁ . . .]
4.	A o. 2	traces
	B o. 4	ŠÀ BURU$_5$.ḪABRUD.DA^{mušen} *i-al-lu*[*t* . . .]
5.	B o. 5	GIM ^{d}UTU È N[AG-*ma* SILIM-*im*]
6.	A o. 5	DÙ.DÙ.BI *e-nu-ma* GU$_4$ [UGU] GU$_4^{!}$.ÁB *iš-*[*ḫi-ṭu*]
	B o. 6	DÙ.DÙ.BI *e-nu-ma* GU$_4$*p*[*u-ḫ*]*a-lu i*[*na* UGU]
7.	A o. 6	SÍG ⸢*ša*⸣ *pu*$^{?}$*-ri*$^{?}$(text Ú)*-di-šu t*[*a-na-saḫ*$^{?}$]
8.	A o. 7	*e-nu-ma* ⸢UDU.NITÁ⸣ ANŠE UR.GI$_7$ ⸢ŠAḪ⸣ *iš-ḫi-ṭu* [. . .]
	B o. 7	*e-nu-ma* UDU.NITÁ ANŠE UR.GI$_7$ Š[AḪ . . .]
9.	A o. 8	*ina* SÍG.ḪÉ.MED NU.NU 7 ZÚ.KEŠDA ⸢ZÚ⸣.KEŠDA [*ina* MÚRU-*šú* GAR-*an* SILIM-*i*]
	B o. 8	[NU].NU 7 ZÚ.KEŠDA [*ina* MÚRU-*šú* GAR-*an* SILIM-*i*]
10.	A o. 9	DIŠ KI.MIN NA$_4$ x *ni* x *ḫu ina* ^{síg}[ÀKA NIGIN-*mi ina* MÚRU-*šú* GAR-*an*]
	B o. 9	broken
11.	A o. 10	ŠÀ Ú$^{?}$ x x x ḪÁD.A GAZ SIM *ana* IGI ⸢^{d}⸣[. . .]
	B o. 10	broken
12.	A o. 11	*ana* IGI ^{d}⸢15-*šu*⸣ ZÍD.SUR.RA NIGIN-*mi* [. . .] *ta* [. . .]
	B o. 11	[ZÍD.SUR.]RA NIGIN-[*mi* . . .]
13.	A o. 12	PIŠ$_{10}$.ÍD *ina* IZI SAR-*šu ina* Ì *ina* ⸢KUŠ⸣
	B o. 12	[ ] IZI SAR[-*šu*]
14.	A o. 13–14	^{ú}AŠ.TÁL.TÁL ^{ú}*ṣa-ṣu-un-tú* NUMUN ^{giš}ḪAB ⸢^{ú}⸣[ ] / *ina* Ì(text KAŠ) *ina* KUŠ
	B o. 13	[^{ú}]*ṣa-ṣu-un-tú* NUMUN ^{gi}[^{š}ḪAB]
15.	A o. 15	^{ú}*an-ki-nu-tu*$^{?}$ ^{ú}EME.UR.GI$_7$ ⟨KU⟩ *ina* Ì *ina* KUŠ
	B o. 14	[^{ú}]EME.UR.GI$_7$ [ ]

16.	A o. 16	^{ú}IGI.NIŠ IM.SAḪAR.NA$_4$.KUR.RA ^{na4}*su-u ina* Ì *ina* KUŠ
	B o. 15	[IM.SAḪAR.]NA$_4$.KUR.RA ^{na4}*su-*[*u*]
17.	A o. 17	— — ^{ú}SUMUN.DAR ^{ú}ÁB.DUḪ ^{ú}A.ZAL-*ú* ^{ú}DILI *ina* Ì *ina* KUŠ
	B o. 16	broken
	D i 44	DIŠ ⸢KI.MIN⸣ ^{ú}SUMUN.DAR ^{ú}ÁB.DUḪ ^{ú}[ ]⸢^{ú}⸣DILI — — *ina* KUŠ
18.	A o. 18	Ú.GIŠ$^{!}$.ḪAŠḪUR$^{!}$ Ú.NAM.TI.LA ^{ú}A.ZAL-*ú* Ú.[KU$_6$ *ina* Ì] *ina* KUŠ
19.	A o. 19	— — ^{ú}GIŠ$^{!}$.ḪAŠḪUR$^{!}$.^{giš}GI ^{ú}*ṣa-ṣu-un-tú* ^{ú}A.ZAL-⸢*ú*⸣ ^{ú}[ ] *ina* KUŠ
	D i 52	[DIŠ KI].MIN GIŠ. ḪAŠḪUR-^{giš}GI ^{ú}*ṣa-ṣu-*⸢*un*⸣*-tú* ^{ú}A.ZAL.L[Á] x *ina* KUŠ
20.	A o. 20–21	— — ^{ú}SUMUN.⸢DAR⸣$^{?}$ ŠÀ BURU$_5$.ḪA[BRUD NI]TÁ ^{ú}GAG.KU ⸢^{ú}⸣MI.PÀR / *ina* ⸢Ì⸣ *ina* KUŠ
	D i 53	[DIŠ KI].MIN — ÚŠ ⸢BURU$_5$⸣.ḪABRUD.DA^{mušen} N[I]TÁ ^{ú}KU.GAG ^{ú}MI.[PÀR — — *ina* K]UŠ
21.	D i 54	[DIŠ KI].MIN ^{na4}AD.BAR ⸢^{ú}⸣[x ^{ú}B]ÚR ^{ú}*tar-m*[*uš ina* KU]Š
22.	A o. 22	— — ^{šim}SES ^{šim}BULUḪ KU.KU ⟨^{na4}⟩KUR-*nu* ⸢DAB⸣ *ina* Ì *ina* KUŠ
	D i 55	[DIŠ KI.MIN KU.K]U ^{na4}KUR-*nu* ⸢DAB⸣ *ina* Ì *ina* [x]
23.	A o. 23	^{ú}IGI.NIŠ ^{ú}*nu-ṣa-bu* ^{ú}*ti-ia-tú ina* KUŠ
24.	A o. 24	^{ú}ÁB.DUḪ ^{ú}A.ZAL-*u* ^{ú}SUMUN.⸢DAḪ⸣ [$^{ú?}$ . . . *ina* Ì *ina*] KUŠ
25.	A o. 25	KÙ.BABBAR A.BÁR KÙ.SI$_{22}$ *ši*$^{?}$ x x x x *man nu* AN.BAR KÙ.BABBAR *ni* [. . .] x
26.	A o. 26	*ina* KUŠ DÙ.DÙ.BI [. . .] x *ina* GÚ-*šú* GAR-*an*
27.	A o. 27–28	DIŠ KI.MIN ⸢*ri-kib-*⟨*ti*⟩ *a-a-li*⸣ SI *a-a-li* GÌŠ *a-a-li* / ^{ú}*tak-da-na-nu ina* KUŠ DÙ.DÙ.BI *ina* GÚ-*šu* GAR-*an*
	D i 51	DIŠ KI.MIN *ri-kib*$^{!}$*-ti a-a-lì* SI *a-a-lì* [GÌŠ *a-a-li* — *ina*] KUŠ — — — —
28.	A r. 1	^{ú}*ka-zal-lu* SU[ḪUŠ ^{ú}]A.ZAL-*e* SUḪUŠ ^{ú}*an-*[*ki-nu-ti*]
29.	A r. 2	SUḪUŠ ^{ú}ḪAR.ḪAR SUḪUŠ [^{ú}]*e*(text [Z]U)*-di* SUḪUŠ ^{ú}x [. . .]
30.	A r. 3	SUḪUŠ ^{ú}NÍG.GIDRU 7 Ú^{meš} ŠÀ.ZI.GA *ina* KAŠ NAG
31.	A r. 4	— — ŠÀ BURU$_5$.ḪABRUD^{mušen} NITÁ *ina* MUN *be ba la i*(text Ú)*-al-lu*[*t*] — — — — — — —
	E 28′–29′	[DIŠ KI.MIN Š]À BURU$_5$.ḪABRUD.DA^{mušen} NITÁ *ina* MU[N N]Á*-al* NU *pa-tan i-al-lut* x [/ ^{ú}*ṣa*]*-ṣu-ut-tú* ^{ú}GU[R$_5$.U]Š *ina* KAŠ N[AG.MEŠ-*ma* ŠÀ.ZI.GA]
32.	A r. 5	⸢ÉN⸣ SU.ZI MIN ŠÀ.ZI.GA MIN *ina* UGU SIKIL BAR NUN Š[ID-*nu*]

II. A rev. 6–11 // C rev. 12–19 // D iv 37–41 = Incantation and its ritual No. A.1

33.	A r. 6	ÉN *li-lik* IM *li-⸢nu-uš⸣* ^{giš}KIRI$_6$
	C r. 12	ÉN *lil-lik* IM *li-nu-uš* ^{giš}KIRI$_6$
	D iv 37	ÉN ⸢*lil*⸣-[*lik* I]M ⸢*li-nu-uš* ^{giš}⸣KIRI$_6$
34.	A r. 6–7	*liš-tak-ṣir* ⸢IM.DIRI^{meš}-*ma*⸣ / *ti*$^{!}$(text AN)-*ki* *li-tuk*
	C r. 13	⸢*liš-tak*⸣-*ṣir* *er-pe-tum$_4$-ma* *ti*$^{!}$-*ku*$^{!}$ ⸢*lit-tuk*⸣
	D iv 37–38	*liš-ta*[*k-ṣ*]*ir* *ur-pa-tum$_4$* / ⟨*ti*⟩-⸢*ku*⸣$^{!}$ *lit-tuk*
35.	A r. 7	*niš* *lìb-bi-ia* *lu* A^{meš} ÍD *a*-⟨*li*⟩-⸢*ku-u-ti*⸣
	C r. 14	⸢*ni-iš*⸣ *lìb-bi-ia* *lu* A⸢$^{meš?!}$⸣ ÍD GIN^{meš}
	D iv 38	⸢*ni*⸣-[*iš lìb*]-*bi*-⸢*ia*⸣(text DU) *lu* A^{meš} ÍD GIN^{meš}-⸢*te*⸣
36.	A r. 8	*ú-šá-ri* *lu-u šèr-an sa-mi*-⸢*e*⸣
	C r. 15	⸢*i*⸣-*šá-ri* *lu* SA-*an sa-am-mi-e*
	D iv 38–39	*i-šá-ri* / [ ] ^{giš}ZÀ.M[Í]
37.	A r. 8	*la* *ú*-⸢*ra-da*⸣ / *ul-tú* *muḫ-ḫi-šá* TU$_6$ ÉN
	C r. 16	*la* *ur-ru-da* *ul-tu* *muḫ-ḫi-šá* TU$_6$ ⸢ÉN⸣
	D iv 39	[*l*]*a-a* *ur-ra*-⟨*da*⟩ *ul-tú* ⸢*muḫ*⸣-*ḫi-šá*(text ŠÚ) ÉN
38.	A r. 10–11	KA.INIM.MA ŠÀ.ZI.GA DÙ.DÙ.BI SA *sa*-⸢*me-e* TI-*qé*⸣ / ⸢3⸣ ZÚ.KEŠDA ZÚ.KEŠDA
	C r. 17	— — DÙ.DÙ.BI SA ZÀ.MÍ TI-*qé* 3 ZÚ.KEŠDA ⸢KEŠDA⸣
	D iv 40	[— — ^{g}]^{iš}ZÀ.⸢MÍ TI⸣-*qé* 2 ZÚ.KEŠDA KEŠDA
39.	A r. 11	— — — *ina* ŠU^{II} ZAG *u* KAB ⸢KEŠDA⸣-*ma*
	C r. 18	ÉN ⸢7⸣-*šú* ⸢ŠID-*nu*⸣ *ina* ŠU$^{III!}$(text LU) X 15 *u* 150 ⸢KEŠDA⸣-[*ma*]
	D iv 40–41	[É]N 3-*šú* ŠID / [ ] *u* 150 KEŠDA-*ma*
40.	C r. 19	X X X
	D iv 41	ŠÀ.ZI.GA

III. A rev. 12–19 = Incantation and its ritual No. A.2

41.	A r. 12	ÉN *a-kan-nu šá a-na* U$_5$-*bi ti-bu-u* [*man*]-*nu ú-ni-iḫ*-[*ka*]
42.	A r. 13	ANŠE.⟨KUR⟩.RA *ez-zu šá* ZI-*šu na-aš*-⸢*pan*⸣-[*du ma*]-*nu* ⸢*meš-re-ti-ka ú-ka-si*⸣
43.	A r. 14	*man-nu* SA^{meš}-*ka ú-ra-me a-me-lu-tú* [*ú-k*]*a*$^{?}$-*an-ni-ka* X X
44.	A r. 15	^{d}15-*ka is-ḫur-ka* ^{d}*asal-lú-ḫi* [E]N *a-ši-pu*-[*t*]*i*
45.	A r. 16	*ina šam-me šá* KUR-*e* Ú^{meš} *šá naq-bi* [*li*]-*pa*-⸢*šir*⸣-*ka*-⸢*ma*⸣
46.	A r. 17	*li-na-ḫi-iš*(text ZU) *meš-re-tú-ka ina r*[*u*]-⸢*a*⸣-*mu šá* ^{d}15 [ÉN]
47.	A r. 18	KA.INIM.MA ŠÀ.ZI.GA DÙ.DÙ.B[I ^{na4}KUR]-*nu* DIB SÚD *ina* Ì ŠUB

48.	A r. 19	GÌŠ-*šu* GABA-⟨*su*⟩ MÚRU-*šú* EŠ.MEŠ-[*ma*] SILIM-*im*

IV. A rev. 20–29 // H rev. v 6′–10′ = Incantation A.3

49.	A r. 20	ÉN *a-kan-nu* MIN *ri-mu* MIN *man-nu ú-*[*ram-me-k*]*a ki-ma qi-i*
	H r. v 6′–7′	⸢ÉN *a*⸣-*kan-ni a-kan-*⸢*ni*⸣ [ ] / ⸢*man*⸣-*nu* ⸢*ú-ram-me-ka*⸣ [ ]
50.	A r. 21	⸢*ra*⸣-*mu-ti man-nu ki-ma ḫu-l*[*i a-lak-t*]*a-ka ip-ru-u*[*s*]
	H r. v 7′–8′	[ ] / *man-nu ki-ma* x *a-*⸢*lak-ta*$^{?}$⸣-[*ka*]
51.	A r. 22	*man-nu it-bu-uk ana* ŠÀ-*ka* [A^{meš} *ka*]-⸢*ṣu-ti*⸣
	H r. v 9′	[*man-n*]*u* ⸢*it-bu-uk ana* ŠÀ⸣-[*ka* A^{meš}]
52.	A r. 23	*ana* UGU ŠÀ-*ka iš-kun a-d*[*ir-t*]*a* [*di-l*]*ip-ta i-*[. . .]
	H r. v 10′	traces
53.	A r. 24	3 x *ma-ti* ^{d}*na-na-a i*[*na* ŠÀ]-*k*[*a* . . .]
54.	A r. 25	*li-it-bu-ku ina* ŠÀ-*ka ri-*[*šá-t*]*ú* [. . .]
55.	A r. 26	*šu-lu-ṣu šá* NENNI A NENNI [x x (x)] ⸢*lu*⸣-*ma* [. . .]
56.	A r. 27	[*ri*]-⸢*ka*⸣-*ab áš-ti* NENNI [DUMU.MUNUS NENNI TU$_6$ ÉN]
57.	A r. 28	[DÙ.DÙ.BI] x *me*$^{?}$ TI-*q*[*í*$^{?}$ x] x *ina* ŠÀ x [. . .]
58.	A r. 29	[. . .] x x [x] ŠUB$^{?}$ Ì EŠ *ana* [. . .]

V. B rev. 2–15 // F 1–10 // G rev. 1–6 // I 6–10= Prescriptions 19–23

59.	B r. 2	[DIŠ NA] ⸢*ri-ḫu*⸣-*us-su la* [*il-lak*] *ana* MUNUS-*šú* ŠÀ-*šú* NU Í[L-*šú*]
60.	B r. 3	[. . .] x SUḪUŠ ^{ú}EME.UR.GI$_7$ SUḪUŠ(text UZU) Ú(text KAL).KUR.RA KI KAŠ ḪE.ḪE-*ma* NAG-*šú*
61.	B r. 4	[EGIR]-*šú* ^{giš}GEŠTIN NAG-*ma i-šal-lim*
62.	B r. 5	[DIŠ KI.MIN ÚŠ] TU.KUR$_4$^{mušen} ÚŠ MUŠEN *ḫur-ri* NITA TÉŠ.BI ḪE.ḪE-*ma* NAG-*šú* EGIR-*šú* ^{giš}GESTIN NAG-*ma* KI.MIN
63.	B r. 6	[DIŠ KI.MIN ^{ú}AŠ.TÁL.]TÁL ^{ú}MUNZER NUMUN ^{ú}ḪAB ^{ú}EME.UR.GI$_7$
	I o. 6–7	DIŠ KI.MIN ^{ú}AŠ.TÁL.TÁL ^{ú}M[UNZER] / ^{ú}EME.UR.GI$_7$
64.	B r. 7	[*ina* KAŠ] NAG-*šú ù* EGIR-*šú* ^{giš}GEŠTIN NAG-*ma* SILIM-*im*
	I o. 7–8	*ina* KAŠ [ ] / EGIR-*šú* GEŠTIN NA[G-*ma*]
65.	B r. 8	[DIŠ KI.MI]N ^{ú}*an-ki-n*[*u-t*]*i* ^{ú}EME.UR.GI$_7$ ^{na4}KA.GI.NA.DAB.[BA]
	I o. 9	DIŠ KI.MIN ^{ú}*an-ki-nu-te* ⸢^{ú}⸣[ ]
66.	B r. 9	*ina* Ì.GIŠ ŠÉŠ-*su ina* KUŠ DÙ.DÙ *ina* GÚ-*šú* GAR-*ma* SILIM-[*im*]
	I o. 10	*ina* Ì EŠ-*su ina* KUŠ [ ]

67.	B r. 10	[DIŠ NA] *lu-ú ina* ŠU.GI-*t*[*um l*]*u-ú ina* ^{giš}GIDRU *lu-ú* ⟨*ina*⟩ *ḫi-miṭ* UD.[DA]
	F o. 1–2	[ ] *lu-u ina* ⸢ŠU⸣.GI-*tum lu-ú ina* ^{giš}GIDRU *lu-u ina ḫi-miṭ* UD.DA / [*lu*]-*ú ina né-ḫe-es* ^{giš}GIGIR
68.	B r. 11	[ ] *a-la-ka muṭ-*⸢*ṭu*⸣ *a-na* ŠÀ⸣.ZI.GA *šur-*⸢*ši*⸣*-šú-ma* [ ]
	F o. 2–3	*a-na* MUNUS *a-la-ka mu-uṭ-ṭú* / ⸢*a*⸣*-na* ŠÀ.ZI.GA *šur-ši-šu-ma ana* MUNUS DU-*šu*
	G r. 1	— — — — *ana* ŠÀ.ZI.GA *š*[*ur-ši-šú-ma*]
69.	B r. 12	[DÙ.DÙ.B]I ^{ú}I[GI]-*lim* ^{ú}*tar-muš*$_8$ ^{ú}⸢EME⸣.[UR.GI$_7$]
	F o. 4	DÙ.DÙ.BI ^{ú}IGI-*lim* ^{ú}*tar-muš*$_8$ ^{ú}EME.UR.GI$_7$ ^{ú}NÍG.GÁN.GÁN
	G r. 2–3	DÙ.BÙ.BI ^{ú}IG[I-*lim*] / ^{ú}NÍG.GÁN.GÁN
70.	B r. 13	[^{ú}]⸢*ar-da*⸣*-dil-la* ^{ú}*ka-bu-*[*ul-la*]
	F o. 5	^{ú}*ar-da-dil-lu*$_4$ ^{ú}*ka-bul-lu* NIM.KÙ.SI$_{22}$
	G r. 3–4	^{ú}[ ] / NIM.KÙ.SI$_{22}$
71.	B r. 14	[x] ⸢Ú⸣.ḪI.A *an-nu-ti* GAZ SIM *ina* [ ]
	F o. 6	7 Ú.ḪI.A *an-nu-tì* GAZ SIM *ana* IGI ^{d}15 NÍG.NA ^{šim}LI GAR-*an*
	G r. 4–5	7 Ú.ḪI.A *an-*[*nu-ti*] / NÍG.NA ^{šim}[LI] GAR-*an*
72.	B r. 15	KAŠ BAL-*qí* ÉN 7-*šú ana lìb-bi* ŠID-*nu* [ ]
	F o. 7–8	KAŠ BAL-*qí* ÉN 7-*šú ana* ŠÀ ŠID-*nu ina* GEŠTIN NAG-*šú* / U$_4$.3.KAM NAG.MEŠ-*ma ina* U$_4$.4.KAM SILIM-*im*
	G r. 5–6	K[AŠ 7-*š*]*ú ana l*[*ìb-bi*] / *ina* GEŠTIN NAG.MEŠ-*š*[*u*$^?$] — — *ina* [4] ⸢*u*$_4$⸣*-me i-*[*šal-lim*]
73.	F o. 9	ÉN *at-ta-man-nu ša* GIM *ḫar-ra-ni ip-ru-su a-lak-ti*
74.	F o. 10	*a-na muḫ-ḫi* ŠID-*nu*

VI. F 11–18 = Incantation No. A.4

75.	F o. 11	ÉN *at-ta-man-nu ša* GIM KASKAL *ip-ru-su a-lak-ti*
76.	F o. 12	GIM *qé-e šad-du-ti ú-ram-mu-u kan-ni-ia*
77.	F o. 13	GIM ^{kuš}NÍG.NA$_4$ *šá* ^{lú}DAM.GÀR *gab-bi* SA^{meš}*-ia*
78.	F o. 14	*il-du-dam-ma ra-ka-su-um-ma ir-ku-us*
79.	F o. 15	*kaš-šap-ti u* MIN ⸢*e*⸣*-le-ni-ti u* ⸢MIN⸣
80.	F o. 16	⸢*ú*⸣*-ra-man-ni ki-i* GU^{meš} *šad-d*[*u-ti*]
81.	F o. 17	[*ki-i k*]*i-*⸢*si*⸣ *šá* ^{lú}⸢DAM.GÀR⸣ [*gab-bi* SA^{meš}*-ia*]
82.	F o. 18	[*il-du-dam-ma ra-ka-su-um-ma ir-ku-us*]

Transcription

I. A 1–rev. 5 // B 1–16 // D i 44, 51–55 // E 28′–29′ = Prescriptions 1–18

1. 1. *šumma amēlu ana sinništi alāk[a] m[uṭṭû . . .]*
2. *kibrīta tamarraq ina [. . .]*
3. *. . . iṣṣūr ḫurri [zikari . . .]*
4. *libbi iṣṣūr ḫurri iʾallu[t . . .]*
5. *kīma Šamšu uṣṣû iš[attī-ma išallim]*

2. 6. *dudubû: enūma alpu puḫālu ina [muḫḫi] litti išḫiṭu*
7. *šārta ša purīdīšu t[anassaḫ]*
8. *enūma immeru imēru kalbu šaḫû išḫiṭu [. . .]*
9. *ina tabribi taṭammi sebet riksī tarakkas [ina qablīšu tašakkan išallim]*

3. 10. *šumma* KI.MIN (= *amēlu ana sinništi alāka muṭṭu [. . .]*) *. . . ina it[qi talammi qablīšu tašakkan]*
11. *libbi . . . tubbal taḫaššal tanappi ana pāni [. . .]*
12. *ana pān Ištarīšu zisurrâ talammi [. . .] ta [. . .]*

13. *kibrīta ina išāti tuqattaršu ina šamni ina maški*

4. 14. *ardadillu ṣaṣuntu zēr ḫûrati [. . .] ina šamni ina maški*

5. 15. *ankinūtu lišān-kalbi ina šamni ina maški*

6. 16. *imḫur-ešrā gabû sû ina šamni ina maški*

7. 17. *šumuttu kamantu azallû ēdu ina šamni ina maški*

8. 18. *ḫašḫūru šammi balāṭi azallû šim[ru*(or *urânu) ina šamni] ina maški*

9. 19. *ḫašḫūr api ṣaṣuntu azallû [. . .] ina maški*

10. 20. *šumuttu libbi iṣṣūr ḫurri zikari kušru lipāru ina šamni ina maški*

11. 21. *[šumma* KI].MIN (= *amēlu ana sinništi alāka muṭṭu [. . .]*) *atbaru [. . . u]rnû tarm[uš ina ma]ški*

12. 22. *murru buluḫḫu sīkti šadâni ṣābiti ina šamni ina maški*

13. 23. *imḫur-ešrā nuṣābu tīyatu ina maški*

14. 24. *kamantu azallû šumuttu [. . . ina šamni ina] maški*

15. 25. *kaspu abāru ḫurāṣu ši . . . man nu parzillu kaspu . . . [. . .] . . .*
26. *ina maški dudubû [. . .] . . . ina kišādīšu tašakkan*

16. 27. *šumma* KI.MIN (= *amēlu ana sinništi alāka muṭṭu [. . .]*) *rikibti ayyali qaran ayyali ušar ayyali takdanānu ina maški dudubû ina kišādīšu tašakkan*

17. 28. *kazallu šu[ruš] azallî šuruš an[kinuti]*

Translation

I. A 1–rev. 5 // B 1–16 // D i 44, 51–55 // E 28′–29′ = Prescriptions 1–18

1. 1. If a man's (desire) to g[o] to a woman is re[duced . . .]
2. you pulverize sulfur in [. . .]
3. . . . [male] partridge [. . .]
4. he eat[s] the innards of the partridge [. . .],
5. when the sun rises, he d[rinks (it) and he will recover].

2. 6. Its ritual: When a breeding bull has mounted the cow
7. yo[u tear off] (some) hairs of its leg,
8. when a ram, a donkey, a dog, a pig has mounted [. . .]
9. you spin with red wool, you make seven bindings (and) [you put (them) around his waist and he will recover].

3. 10. If ditto (= If a man's (desire) to go to a woman is reduced [. . .]): [You wrap up] the . . .-stone with a flee[ce, you put (it) around his waist],
11. you dry (and) crush the 'heart' of the . . .-plant, in front of [. . .],
12. you surround in front of his goddess with a magical circle of flour [. . .]. . . [. . .],
13. you fumigate sulfur in fire, with oil in a leather bag.

4. 14. *ardadillu*-plant, *ṣaṣuntu*-plant, seeds of *ḫûratu*-plant, [. . .]-plant with oil in a leather bag.

5. 15. *ankinūtu*-plant, "dog's-tongue"-plant with oil in a leather bag.

6. 16. "heals-twenty"-plant, alum, *sû*-stone with oil in a leather bag.

7. 17. *šumuttu*-plant, *kamantu*-plant, *azallû*-plant, *ēdu*-plant with oil in a leather bag.

8. 18. Leaves[?] of apple-tree, "plant-of-life," *azallû*-plant, *šimru*-plant (or *urânu*-plant) with oil in a leather bag.

9. 19. *ḫašḫūr api*-plant, *ṣaṣuntu*-plant, *azallû*-plant, [. . .]-plant in a leather bag.

10. 20. *šumuttu*-plant, innards of a [mal]e partri[dge], *kušru*-plant, *lipāru*-fruit/plant with oil in a leather bag.

11. 21. [If dit]to (= If a man's (desire) to go to a woman is reduced [. . .]): Basalt, [. . .]-plant, [*u*]*rnû*-plant, *tarm*[*uš*-plant in a leather] bag.

12. 22. *murru*-gum, *baluḫḫu*-plant, magnetite powder with oil in a leather bag.

13. 23. "heals-twenty"-plant, *nuṣābu*-plant, *tīyatu*-plant in a leather bag.

14. 24. *kamantu*-plant, *azallû*-plant, *šumuttu*-plant [. . . with oil in] a leather bag.

15. 25. Silver, lead, gold . . . iron, silver. . . [. . .] . . .
26. in a leather bag. Its ritual: [. . .] . . . you put (it) around his neck.

16. 27. If ditto (= If a man's (desire) to go to a woman is reduced [. . .]): *Rikibtu* of a stag, antler of a stag, penis of a stag, *takdanānu*-plant in a leather bag. Its ritual: You put (it) around his neck.

17. 28. *kazallu*-plant, ro[ot of] *azallû*-plant, root of *an*[*kinutu*]-plant,

29\. *šuruš ḫašî šuruš ēdi šuruš . . .* [. . .]
30\. *šuruš ḫaṭṭi-rē'i sebet šammī nīš libbi ina šikari išatti*

18\. 31. *libbi iṣṣūr ḫurri zikari ina ṭābti* [*tuš*]*nāl lā patān i'allut . . .* [*. . . ṣa*]*ṣuttu ša*[*rma*]*du ina šikari iš*[*attī-ma nīš libbi*]

19\. 32. *šipta šalummatu* MIN *nīš libbi* MIN *ina muḫḫi . . .* [*tamannu*]

II. A rev. 6–11 // C rev. 12–19 // D iv 37–41 = Incantation and its ritual No. A.1

33\. *šiptu: lillik šāru linūš kirû*
34\. *lištakṣir erpetu-ma tīku littuk*
35\. *nīš libbīya lū mê nāri ālikūti*
36\. *ušarī lū šer'ān sammî*
37\. *lā urrada ultu muḫḫīša tê šipti*

38\. (var. adds: *šipat nīš libbi*) *dudubû: šer'ān sammî teleqqe šalaš* (var.: *šina*) *riksī tarakkas*

39\. *šipta sebîšu* (var.: *šalāšīšu*) *tamannu ina qāti imni u šumēli tarakkas-ma*

40\. *nīš libbi*

III. A rev. 12–19 = Incantation and its ritual No. A.2

41\. *šiptu: akkannu ša ana rakābi tebû* [*man*]*nu unīḫ*[*ka*]

42\. *sisû ezzu ša tībūšu našpan*[*du m*]*annu mešrêtīka ukassi*
43\. *mannu šer'ānīka uramme amēlūtu* [*uk*]*anni-ka . . .*
44\. *ištarka isḫurka Asalluḫi* [*b*]*ēl āšipū*[*t*]*i*
45\. *ina šammī ša šadî šammī ša naqbi* [*li*]*pašširka-ma*

46\. *linaḫḫiš mešrêtīka ina r*[*u*]*'āmu ša Ištar* [*šiptu*]

47\. *šipat nīš libbi dudub*[*û: šadâ*]*na ṣābita tasâk ina šamni tanaddi*

48\. *ušaršu irassu qabalšu ta/iptaššaš-*[*ma*] *išallim*

29. root of *ḫašû*-plant, root [of] *ēdu*-plant, root of . . .-plant [. . .],
30. root of "shepherd's-crook"-plant. He drinks in beer the seven drugs (for) the sexual desire.

18. 31. [You pic]kle in salt the innards of a male partridge (and) he eats (them) on an empty stomach, . . . [. . . *ṣa*]*ṣuttu*-plant (and) *ša*[*rma*]*du*-plant, he drin[ks] (them) in the beer [and (he will get) sexual desire].

19. 32. [You recite] over . . . the incantation: "Shiver! Shiver! Sexual desire! Sexual desire!"

II. A rev. 6–11 // C rev. 12–19 // D iv 37–41 = Incantation and its ritual No. A.1

33. Incantation: May the wind blow! May the grove quake!
34. May the cloud gather! May the moisture fall!
35. May my sexual desire be constant river water!
36. May my penis be a harp string,
37. so that it will not dangle out of her! Incantation formula.

38. (var. adds: Wording of *nīš libbi* (incantation)) Its ritual: You take a harp string, tie three (var.: two) bindings (in it),
39. you recite the incantation seven (var.: three) times, you tie (it) around his right and left hands and
40. (he will get) sexual desire.

III. A rev. 12–19 = Incantation and its ritual No. A.2

41. Incantation: *Akkannu*-wild ass who is reared-up for mating, [wh]o has dampened [your] desire?
42. Impetuous horse, whose rising is a devastation, [w]ho has bound your limbs?
43. Who has slackened your tendons? The mankind . . . your . . .
44. Your goddess has turned to you. [May] Asalluḫi, p[at]ron of exorcism knowled[ge],
45. release you by means of the plants of the mountain (and) the plants of the spring, and
46. may he make your limbs healthy through the se[duc]tion of Ištar! [Incantation].

47. Wording of *nīš libbi* (incantation). Its ritual: You pulverize [magneti]te (and) put (it) into oil;
48. He (= patient)/you (= therapeutic operator) rub(s) (with it) his penis, his chest (and) his waist, [and] he will be healthy.

IV. A rev. 20–29 // H rev. v 6′–10′ = Incantation and its ritual No. A.3

49. *šiptu: akkannu akkannu rīmu rīmu mannu urammēka kīma qî*
50. *ramûti mannu kīma ḫūl[i] alaktaka ipru[s]*

51. *mannu itbuk ana libbīka [mê ka]ṣûti*
52. *ana muḫḫi libbīka iškun ad[irt]a [dil]ipta i-[. . .]*
53. *šalaš . . . māti Nanāya i[na libbī]k[a . . .]*
54. *litbukū ina libbīka rī[šāt]u [. . .]*
55. *šūluṣu ša annanna mār annanna [. . .] . . . [. . .].*
56. *[rik]ab ašti annanna [mārat annanna tê šipti]*

---

57. *[dudubû:] . . . me teleqq[i . . .] . . . ina libbi . . . [. . .]*
58. *[. . .] . . . [. . .] tanaddi šamna tapaššaš ana [. . .]*

V. B rev. 2–15 // F 1–10 // G rev. 1–6 // I 6–10 = Prescriptions 19–23

20 59. *[šumma amēlu] riḫûssu lā i[llak] ana sinništīšu libbašu lā ina[ššīšu]*
60. *[. . .] . . . šuruš lišān-kalbi šuruš nīnî itti šikari taballal-ma išattīšu*

61. *[arkī]šu karāna išattī-ma išallim*

---

21. 62. *[šumma* KI.MIN (= *amēlu riḫûssu lā illak ana sinništīšu libbašu lā inaššīšu*) *dām] sukannīni dām iṣṣūr ḫurri zikari ištēniš taballal-ma išattīšu arkīšu karāna išattī-ma* KI.MIN (= *išallim*)

---

22. 63. *šumma* KI.MIN (= *amēlu riḫûssu lā illak ana sinništīšu libbašu lā inaššīšu*) *ardadilla supāla zēr būšāni lišān-kalbi*
64. *ina šikari išattīšu u arkīšu karāna išattī-ma išallim*

---

23. 65. *šumma* KI.MIN (= *amēlu riḫûssu lā illak ana sinništīšu libbašu lā inaššīšu*) *ankinūte lišān-kalbi šadâna ṣābi[ta]*
66. *ina šamni tapaššassu ina maški tašappu ina kašādīšu tašakkan-ma išall[im]*

Symp. 67. *[šumma amēlu] lū ina šībūti lū ina ḫaṭṭi lū ina ḫimiṭ ṣēti [l]ū ina neḫēs narkabti*

68. *ana sinništi alāka muṭṭu ana nīš libbi šuršīšu-ma ana sinništi alākīšu*

---

24. 69. *dudubû: imḫur-līm tarmuš lišān-kalbi egemgiru*

70. *ardadillu kabullu zumbi-ḫurāṣi*
71. *sebet šammī annûti taḫaššal tanappi ana* (var.: *ina*) *pān Ištar nignak burāši tašakkan*

IV. A rev. 20–29 // H rev. v 6′–10′ = Incantation and its ritual No. A.3

49–50. Incantation: *Akkannu*-wild ass! *Akkannu*-wild ass! Wild bull! Wild bull! Who has slackened you (so that you are) like slackened / strings? Who has block[ed] your course like (on) a ro[ad]?

51. Who has poured [co]ld [water] on your 'heart,'

52. (and) has put f[ea]r upon your 'heart,' has [. . .] sleeplessness?

53. Three . . . Nanāya i[n] your ['heart' . . .]!

54. May they(m.) pour out j[o]y into your 'heart!' [. . .]

55. To bring joy of NN, son of NN [. . .] . . . [. . .].

56. [Mo]unt the wife of NN, [daughter of NN! Incantation formula].

57. [Its ritual]: . . . you tak[e . . .] . . . inside? . . . [. . .]

58. [. . .] . . . [. . .] you throw, you anoint (with)? oil [. . .].

V. B rev. 2–15 // F 1–10 // G rev. 1–6 // I 6–10 = Prescriptions 19–23

20 59. [If man's] sperm does not f[low] (and) does not desi[re] his woman:

60. You mix [. . .] root of "dog's-tongue"-plant, root of *nīnû*-plant with beer and he drinks it,

61. [after] that he drinks wine and he will recover.

21 62. [If ditto (= If man's sperm does not flow and does not desire his woman)]: You mix together [blood] of a dove (and) blood of a male partridge and he drinks (it) and ditto (= he will recover).

22 63. If ditto (= If man's sperm does not flow and does not desire his woman): *ardadillu*-plant, *supālu*-plant, seeds of *būšānu*-plant, "dog's-tongue"-plant

64. he drinks in the beer and after that he drinks wine and will be heal.

23 65. If ditto (= If man's sperm does not flow and does not desire his woman): *ankinūtu*-plant, "dog's-tongue"-plant, magneti[te]

66. you anoint with oil, you wrap it in a leather bag, you put it around his neck and he will be hea[l].

Symp. 67. [If a man], either due to the old age, or the *ḫaṭṭu*-sickness, or inflammation by sun-heat, or the *neḫēs narkabti*-sickness,

68. (his desire) to go to a woman is reduced, in order to make him get his sexual desire and for his going to a woman.

24. 69. Its ritual: "Heals-a-thousand"-plant, *tarmuš*-plant, "dog's-tongue"-plant, *egemgiru*-plant,

70. *ardadillu*-plant, *kabullu*-plant, "gold-fly"-plant.

71. You crush (and) sift these seven plants, you set up a censer with juniper in front of Ištar,

72. *šikara tanaqqi šipta sebîšu ana libbi tamannu ina karāni tašaqqīšu* (var.: *taštanaqqīšu) šalāšat ūmī ištanattī-ma ina erbēšu ūmi išallim*

73. *šipta attamannu ša kīma ḫarrāni iprusu alaktī*
74. *ana muḫḫi tamannu*

VI. F 11–18 = Incantation No. A.4

75. *šiptu: attamannu ša kīma ḫarrāni iprusu alaktī*
76. *kīma qê šaddūti urammû kannīya*
77. *kīma kīsi ša tamkāri gabbi šer'ānīya*
78. *ildudam-ma rakāsum-ma irkus*
79. *kaššaptī u kaššaptī elēnītī u elēnītī*
80. [*u*]*rammânni kī qê šadd*[*ūti*]
81. [*k*]*ī kīsi ša tamkāri* [*gabbi šer'ānīya*]
82. [*ildudam-ma rakāsum-ma irkus*]

72. you libate beer, you recite the incantation over it seven times (and) you let him (var.: repeatedly) drink it (= pharmaceutical) with wine. He will repeatedly drink it for three days and he will recover on the fourth one.

73–74. You recite over it the incantation "Who are you who has blocked up my course like on a road."

VI. F 11–18 = Incantation No. A.4

75. Incantation: Who are you who has blocked up my course like on a road,
76. has slackened my loincloths like taut strings?
77. (who are you who), like (the string of) a merchant's leather money pouch, all my sinews
78. has pulled tight and firmly bound?
79. My witch and my witch! My sorceress (lit. "the superior one") and my sorceress!
80. [She] has slackened me like ta[ut] strings!
81. [Like] (the string of) a merchant's leather money pouch [all my sinews]
82. [she has pulled tight and firmly bound]!

Commentary

6. For the reading of DÙ.DÙ.BI as *dudubû* see Maul 2009.

11. ŠÀ Ú$^{?}$ x: Biggs (1967) reads DIŠ KI.MIN Ú$^{?}$ *tu*$^{?}$.

17. A prescr. 7 = K prescr. 12: 46.

18. Ú.GIŠ.ḪAŠḪUR could refer to the leaves of the plant.

20. In Ms. A the sign ^{ú}GAG.KU must be inverted as in Ms. D: ^{ú}KU.GAG/*kušru*.

25 *ši*$^{?}$ x x x x *man nu*: Biggs (1967) reads the last two unintelligible signs *bu u*.

27. *ri-kib-ti a-a-li*: the *emendatio* in Ms. A *ri-kib-⟨ti⟩* is based on the Ms. D. According to Chalendar (2018: 39–41), the manuscripts are not parallel and Ms. A does not mention the ingredient *rikibti ayyali*, but *rikbi ayyali*. The scholar translates the expression as "'cornichon' de cerf," that is the initial phase of the annual regrowth of the deer antlers, which is accompanied by new ramifications. She argues also that the *emandatio ri-kib-ti*(text SI) DÀRA.MAŠ in *nīš libbi* K prescr. 30: 130 is not correct, preferring the reading *rikbi qaran ayyali* (*ri-kib* SI DÀRA.MAŠ), which she translates also as "'cornichon' de cerf." Chalendar's hypothesis, however, does not consider the fact that in the *nīš libbi* corpus, as in other Mesopotamian medical texts, the ingredient *rikbi ayyali* or *rikbi qaran ayyali*, to my knowledge, never appears. For this reason, the *emendatio* is more appropriate.

30. See for other mention of the expression also Ú ŠÀ.ZI.GA "drug for the sexual desire": K prescr. 6: 35 K prescr. 26: 75: 10 Ú$^{!meš}$ ŠÀ.ZI$^{!}$.GA$^{!}$ "10 drugs for the sexual desire"; No. K.3: 92; N prescr. 21 iv 13. See also in the *Aššur Medical Catalogue*, in the section XIX devoted to the loss of sexual desire: [. . .] x SAG MUŠEN DIŠ ⸢Ú⸣ ⸢ŠÀ⸣.ZI.G[A] ⸢*ana*⸣ [GÚ-*š*]*ú* GAR "[. . .] . . . the head of a bird. (Instructions) to place a drug for sexual desir[e] around his [neck]" (l. 100, Steinert 2018b: 217). See for specific plants for sexual desire's problem BAM 380 r. 42–44 (dupl. BAM 381 iii 37–40) (see Chapter III § "Plant ingredients").

32. According to Biggs (1967) after UGU the sign DIŠ follows, I do not see it in the copy. SIKIL BAR NUN is unclear to me.
For the incantation "SU.ZI MIN" see the catalogue LKA 94 i 19 and 26; D prescr. 4: 47–48; K prescr. 28: 77–79. See also the series *kunuk ḫalti* ("*ḫaltu*-seal") K. 3010+ v 25′ and 28′: [ÉN SU.ZI MIN ÉN KI.IN].DA.RAB [MIN$^{?}$] "[Incantation: 'Shiver! Shiver!'. Incantantion] '. . . [. . .]'" and [ÉN SU.ZI] MIN "[Incantation: 'Shiver!] Shiver!'" (Schuster-Brandis 2008: 365, text 16 A).

33. See the catalogue LKA 94 i 4. Cf. No. I.1: 6: "May the wind blow! May the grove not quake!", and the *Aššur Medical Catalogue*, section XX "sex": "[Incantation: 'Let the wind blow], the gardens shall not quake'" / [One tablet (of the section) 'Incantation: Let the wi]nd [blow], the gardens shall not quake'" (ll. 103–104, Steinert 2018b: 217).

41. See No. L: 26.
Nougayrol (1968: 94) proposed here for the verb *nâḫu* the meaning of 'finding sexual satisfaction, joy,' as well as in No. B.1: 17 (restored) and No. D.3: 39. However, the context is different: in this passage, it is the witch who acts, whereas in the other incantations the verb is used at the end of the text to indicate the future recovery of the couple. See also Labat 1968: 357.

42. *sisû ezzu ša tībūšu našpandu* "Impetuous horse, whose rising is a devastation": term *našpandu* refers to a devasting flood (CAD N/II 29 mng. c). Sexual desire is compared to constantly flowing river water to indicate a continuous desire and without interruption (No. A.1: 35), while in No. H.1: 8 the flooding of canals is invoked, again indicating powerful and uncontrolled desire. See also the catalogue of incipit: [ÉN] ÍD [ŠÀ].ZI.GA GIN.A "[Incantation]: Flow, river of sexual desire!" (LKA 94 i 28). See also in love context the reference to flood: "He brought the Flood (*iššâ abūbum*) – achieving what?" (CUSAS 10, 12: 15, Wasserman 2016: 190).
Reading ⸢*meš-re-ti-ka ú-ka-si*⸣ follows Biggs 1967.

43. The second part of this line is obscure and its relation with the first part of the following line is uncertain. Biggs (1967) excludes the restoration [*ú-m*] *a-an-ni-ka*.

46. *li-na-ḫi-iš*: for the *emendatio* of the sign ZU with IŠ see Biggs 1967. It is a precative D-stem of the verb *naḫāšu* (CAD N/I: 133–134 'to prosper, thrive, be in good health'; AHw. II 713–714 'füllig sein, werden'). Biggs (1967) translates "make it attractive," but here I prefer a more generic translation "to be in good health."

48. EŠ.MEŠ: it is not clear who rubs the patient's body, the logogram could be read in fact with the third (= patient) or second (= therapeutic operator) singular person.

49. See the catalogue LKA 94 i 2 and parallel No. A.4: 75–76.

51–54. I translate the term *libbu* as 'heart,' as in the other texts of the corpus, to emphasize – as it occurs in modern Western languages – that it is here understood as the seat of feelings, emotions, and desire.

51. *mê kaṣûti* "cold water": Biggs's (1967) suggestion. It is probable, given the presence of the verb *tabāku* 'to pour,' that it could be water, but it is the quality of the latter that remains doubtful. The use of cold water is indeed positive connotations; it is used, for example, for libations in ritual contexts. In witchcraft contexts, the patient usually drinks magically modified water given by the witch (think of the so-called "water of *zikurudû*," or bewitched water, *mê kaššāpūti*, see for example Abusch and Schwemer 2011: 299, No. 8.4: 73). See, for instance, in *Maqlû*: "May the water that you have extracted be yours" (V 8, Abusch 2016: 134, see also IV 49). The evil operators can make funeral offerings for the patient, as if he were a dead man, pouring water (*Maqlû* IV 44–48, Abusch 2016: 119–120). In a text (Abusch and Schwemer 2011: 181, No. 7.8.1: 13′–19′) they take water from the sea and pour it on the road spreading silence and death. In the same text (No. 7.8.2: 19″) we read that "their arms are full of putrid water for washing." Biggs (1967: 20), however, argues that the restoration and the reading are appropriate, since he interprets *libbu* as penis, and therefore the cold water would have the function of reducing the capacity of penile erection.

52. Buccellati (1976: 68) restores with *išpuk* from *šapāku* 'heap up,' however, the restoration is impossible since the only visible sign is *i-*, whereas *iš-pu-uk* is expectend. In addition, the verb in reference to the 'heart' (*libbu*) is unfitting. Here the sleeplessness (*diliptu*) is a sign of the witch's action who has taken away the patient's sexual desire. For both sleep and insomnia in second millennium love literature see Wasserman 2016: 45–47.

53. The first part of the line is unclear. Biggs (1967) reads 3 *a-me-la-ti*, however, instead of *a-me*, the copy has only an illegible sign. The verb of the sentence cannot be *tabāku* 'to pour' in line 54 (*litbukū*), since the -*ū* designates a plural masculine subject, not feminine.

56. The sentence "[Mo]unt the wife of NN" seems to allude to the wife of someone other than the patient. This, however, is not reflected in other texts. In general, the term 'wife' (*aštu*) is under-used in the corpus, unlike the term 'woman' (*sinništu*).

60. At the beginning of the line Biggs (1967) doubtfully reads MUL; maybe based on Q prescr. 1: 3, one could restore [^{ú}*ka-bu-u*]*l-lu*.

62. On TUM$_{12}$.GUR$_4$^{mušen}/*sukannīnu* 'dove' see Landsberger 1964–1966: 267–268; Veldhuis 2004: 292. Note that in an *irtum*-song the penis is called metaphorically "dove": "I have thrown my coop on the young man, / that I know [I may] catch the dove; / (the coop) of my delights Nanāya$^?$ will fill for m[e]" (16056

MAH: 16–20, Wasserman 2016: 105). See also the fragment K 6082 + 81-7-27, 241: 7–8, Lambert 1975: 118.

63–66 Prescr. 22–23 = Q prescr. 2–3.

63. The term *supālu* (CAD S 390–391; AHw. II 1059–1060) refers to two different plants: a variety of juniper (^{giš}ZA.BA.LUM, ^{ú}NIGIN^{sar}) and the *supālu*-plant (^{ú}MUNZER). Here, as in Q prescr. 2: 6, the plant is mentioned (while the juniper only once in F prescr. 5: 22). For the reading ^{ú}MUNZER of the signs ^{ú}KI.^{d}NANNA see Civil 1966: 122–123.

67. In Ms. B the unintelligible sign is not MEŠ, like in the duplicate Ms. F. Only in Ms. F we find MEŠ.

75. See the catalog LKA 94 ii 18: ÉN *man-nu šá* GIM KASKAL KUD A.RÁ DÍM.

72. Both Ms. B (LKA 96) and G (LKA 100) come from Library N 4 in Aššur (see May 2018b) and were written by the exorcist Kiṣir-Nabû (see Hunger 1968, No. 214; Abusch et al. 2020: 32):

B r. 16′.	[*ina*] ZAG *ú-ìl-ti* ^{I}*aš-šur*-LUGAL-[*a-ni* x x x x x]
B r. 17′.	[GIM SUMUN]-*šú šà-ṭir ba-rì tup-pi* ^{I}*ki-ṣ*[*ir-*. . . x x x x]
G r. 7.	*ina* ZAG *ú-ìl-ti* ^{I}*aš-šur*-LUGAL-*a-ni šá-ṭir* [*bari*]
G r. 8.	*ú-ìl-tì* ^{I}*ki-ṣir-*^{d}AG *šá* ^{d}PA *tuk-*[*la*]*t-s*[*u*]
G r. 9.	DUMU ^{Id}AŠ.ME-DÙ ^{lú}MAŠ.MAŠ É AN.ŠÁR
G r. 10.	DUMU ^{Id}PA-*bi-su-nu* ^{lú}MAŠ.MAŠ É AN.ŠÁR
G r. 11.	DUMU ^{Id}BA.Ú-*šúm-íb*-[*n*]*i* ^{lú}ZABAR.DAB.BA *é-šár-ra*

"Written [(and) collated] in accordance with an *u'iltu*-tablet of Aššur-šarrāni. An *u'iltu*-tablet of Kiṣir-Nabû, who puts h[is] tru[st] in Nabû, son of Šamaš-ibni, exorcist of the Aššur temple, son of Nabû-bēssunu, exorcist of the Aššur temple, son of Bābu-šumu-ibni, *zabardabbû*-official of Ešarra."

76. *K/gannu*: Biggs (1967: 21) supposes that the term should be read *gannu* and understood as a part of the body of animals (see also CAD G 41; AHw. I 280) since the patient is described metaphorically as an *akkannu*-wild ass. However, as suggested by Abusch et al. 2020: 34, the reading *kannu* 'band, bandage (of a dress), (a wrestler's) loincloth' (CAD K 156–157) is more probable. The loosening of the loincloth, in fact, refers metaphorically to the loosening of the loins, one of the symptoms of the loss of sexual desire.

Incantations: stylistic and functional analysis

III. A rev. 12–19 = Incantation and its ritual No. A.2

This text is divided into two parts: an incantation and a small prescription. The latter, introduced by the formula *dudubû* ('its ritual,' l. 47) refers to ointments with magnetite and oil to be rubbed upon the patient's penis, chest, and waist. The incantation is divided into two sections each of three lines:

- The first section (ll. 41–43) describes the patient and their condition of suffering, comparing him to sexually active animals whose sexual desire has been taken away, and at the same time speculates on the identity of the instigator of this malaise, i.e. who might have caused such suffering;
- The second (ll. 44–46), on the other hand, introduced by the news that the goddess has come back to contact the patient, invokes Asalluḫi to ensure that the man is healed.

The first section includes three lines, of which the first two share the same structure, which is only partially taken up in the third one. I divide the lines into two parts: the first is the identification of the patient with the wild beast and its sexual quality; the second one poses rhetorical question who (*mannu*) caused the loss of sexual desire.

41.	*Akkannu*-wild ass (*akkannu*	who is reared-up for mating *ša ana rakābi tebû*)	who has dampened your desire? (*mannu unīḫka*)
42.	Impetuous horse (*sisû ezzu*	whose rising is a devastation *ša tībūšu našpandu*)	who has bound your limbs? (*mannu mešrêtika ukassi*)

The two lines begin with the name of an animal, the *akkannu*-wild ass and the impetuous horse (*sisû ezzu*), and their sexual characteristics (introduced by the relative *ša*). In the first line, it is stated that the wild ass is "reared-up for the mating" (*ša ana rakābi tebû*), while in the second one such rising (*tību*) is qualified as "a devastation" (*našpandu*).

The sexual virility of the man, compared to the vigor of wild animals, however, fails, as it has been taken away by someone maliciously, probably a witch. The incantation speculates on who might have caused this crisis. The second half of each of the first two lines as well as the first half of the third line[510] show the damage caused by the removal of sexual desire: "Who has dampened your desire?" (l. 41); "who has bound your limbs?" (l. 42); "who has slackened your tendons?" (l. 43).

510 The second part of the line is not clear.

Akkannu-wild ass	reared-up for the mating	who has dampened your desire?
Impetous horse	whose rising is a devastation	who has bound your limbs?
who has slackened your tendons?		Mankind … your …

The second part of the incantation (ll. 44–46) has, unlike the first, a positive tone, thanks to the invocation of the god Asalluḫi, called "patron of exorcism knowledge,"[511] using the precative. The first line maintains, compared to the previous three, a binary structure (l. 15), which is absent in the following two lines:

44. Your goddess has returned to you   Asalluḫi, patron of the exorcistic knowledge

The first part seems to guarantee the beginning of a change of the patient's situation: the return of the personal goddess (*ištarka*) to the man.[512] It appears that only the intervention of the god Asalluḫi ensures the restoration of relations between the patient and his goddess. As we know, the witch may cause the removal of the patient's personal god and goddess, thus taking away their protection (see Abusch 2002: 29–57).[513] This text is the only one in the corpus that mentions a personal goddess and her removal.[514] Since the primary cause of suffering is to be attributed to witchcraft, it is she who has caused the removal of the goddess from the patient. The removal of the personal goddess does not appear to be a typical condition of a man whose sexual desire has been taken away since it does not appear in any other texts. The resulting relationship between witchcraft and the wrath of the goddess (and the god) is confirmed by other texts in which it is clear that the cause of the rupture of relations between a man and his god is due to magic: "Witchcraft was practiced against him, he was cursed before the god and the goddess" (BAM 315: 8–9// Bu. 91-5-9, 214: 10′, Abusch 2002: 41–42).[515] In the corpus analyzed here, the idea of the patient as an innocent victim emerges, that is, someone without guilt, who was struck by witchcraft, which is able to take away the personal god or goddess or cause their anger and

511 See for other epithets of Asalluḫi related to the therapeutic range e.g., the series *Udug-hul* XI 102–104, Geller 2016: 362.

512 On the personal god see Oppenheim 1964: 198–206; Jacobsen 1976: 155–160; Klein 1982: 295–306; van der Toorn 1996: 66–93; Abusch 1998: 378–383; Abusch 1999; Abusch 2002: 48–57; Steinert 2012a: 395–404.

513 See *Maqlû* I 6; III 49–50, 112 (Abusch 2016).

514 In A prescr. 3 l. 12 ^{d}15-*šú* (*Ištaršu*) is mentioned, but here it seems to refer to the goddess Ištar.

515 See SpBTU 2, No. 22, i 39′–40′ (Abusch 2002: 43–44). On the development over the millennia of witchcraft as a cause of suffering see Abusch 2002: 45–47. On the relationship between witchcraft and the wrath of the gods (the hand of the god) see Stol 1999: 59–61.

the end of their protection over them (see Abusch 2002: 48). Only the invocation to the god, in this case Asalluḫi, can undo the magic power of the witch's action, allowing the reconciliation of the deity with the man and bring him back to the initial state of wellbeing. Stylistically, this line seems to contain a *hysteron proteron,* a figure of speech, which consists of saying first what happened last: the return of the goddess after the intervention of the god.

Interestingly, in this incantation there is no mention of the divine couple, the personal "god and goddess" (*ilu* and *ištaru*), but only of the goddess. The personal goddess never appears alone in the texts, but always together with the personal god. Why is she mentioned alone in this incantation? Why did seem the witchcraft cause the removal and consequently the anger of the personal goddess only? What is does the goddess's function? Answering these questions is very difficult, but we can follow the observations offered by Zimmern (1927: 574–577) which have been taken up by Oppenheim (1964: 201–206), who has analyzed what he calls Mesopotamian "psychology," studying the four external manifestations of the Mesopotamian ego: *ilu, ištaru, lamassu,* and *šēdu* (see also Mayer 1976: 472). According to these scholars, in understanding the function of *ištaru,* one needs to analyze the term *šīmtu* "fate, destiny" (CAD Š/II 11–20; AHw. III 1238–1239). The word *šīmtu,* in the first millennium, seems to become a synonym for *ištaru,* as it appears in texts where we would expect to find the latter, often in connection with *ilu* "personal god":

ša ilšu isbusu tusaḫḫar kišāssu
ša zēnat šīmtašu tusallam ittīšu

Whose god became angry, his (= personal god's) neck you (= Nabû) will turn back again,
whose fate is in anger, you will make peace with him.
[*šuilla to Nabû* ll. 9–10, Mayer 1976: 470]

[lib]bi ilīka libbi šīmtīka qāt ilīka qāt Ištarīka lippa[ṭir]
May the '[hea]rt' of your god, the 'heart' of your destiny, the hand of your god, the hand of your goddess be relea[sed].
[BBR No. 61 r. 7, par. 62 r. 5, CAD Š/II :16]

ilānī ša na[q]bi lippirdū
šīmātu Ištarāt māti lū hadâ

May the gods of the aby[ss] shine.
May the destinies, goddesses of the land, rejoice.
[*Ašurnaṣirpal Hymn to Ištar*, KAR 107: 48–49, Zimmern 1927: 575]

Ištarāt šamāmī šīmat naqab erṣeti lini[ḫḫāka]
May the goddesses of the heavens, the destinies of abyss of the earth calm you!
[*Hymn of Tukulti-Ninurta to Aššur*, KAR 128 r. 32, Zimmern 1927: 575]

These examples show us how *šīmtu* and *ištaru* are closely related, whereas the latter is a sort of personification of the first, owned by the individual. Oppenheim (1964: 202) defines *šīmtu* as «a disposition originating from an agency endowed with power to act and to dispose, such as the deity, the king, or any individual may do, acting under specific conditions and for specific purposes». In other words, *ištaru* is the personification of what the man has been intended to be, the entity which guarantees the man his realization as an individual. But why does the *nīš libbi* incantation mention the removal and return of the personal goddess? One answer might lie in the fact that the absence of sexual desire has led to a severe crisis of manhood. Active sexual practice is one of the most important aspects of the Mesopotamian construction of manhood. The man, to be a man, must be sexually active. The personal goddess is therefore also the manifestation of one of the provisions integral to being a *man* in the sexual sphere. If the witch has taken away his sexual desire that means the victim loses his personal goddess and is thus deprived of the necessary conditions to fulfill his agency (one could also say: his fate) of being a sexually active man, leading to a crisis that only an exorcism ritual can solve. The relationship between destiny and sexual desire is confirmed by the Sumerian *nīš libbi* texts No. K.1: 11: x x x x ⌜šà⌝-zi-ga-b[i]$^{?}$ nam-e "They (= Enlil and Bēlet-ilī) have decreed his sexual desire. . ."; and M prescr. 5: 27: [. . . e]gir nam-tar-ra-zu al-gub ". . . a]fter$^{?}$ it, your destiny is assigned."

Let us return to the second section of the incantation. The second part of line 44 "Asalluḫi, patron of exorcism knowledge,"[516] initiates the invocation of the god (ll. 45–46):

> by means of the plants of the mountain (and) the plants of the spring (may) release you, and may he make your limbs healthy through the se[duc]tion of Ištar!

The two verbs, expressed in precative, alongside the object, and adverbial of means are placed in chiastic position[517]:

ADVERBIAL OF MEANS	VERB AND OBJECT
by means of the plants of the mountain (and) the plants of the spring (*ina šammī ša šadî šammī ša naqbi*)	may he release you (*lipašširka*)

VERB AND OBJECT	ADVERBIAL OF MEANS
may he make your limbs healthy (*linaḫḫiš mešrêtīka*)	through the seduction of Ištar! (*ina ruʾāmu ša Ištar*)

The *āšipu* invokes the god so that he can deliver the man from the attack of the witch, by means of mountain and spring plants. The mountain and the spring are emblem-

516 For similar bynames see *Udug-hul* XI 18 (Geller 2016: 344); *Maqlû* I 62, 72; II 171; VI 57 (Abusch 2016).
517 Note also that the second section of the incantation begins and ends with the word *Ištaru*/Ištar.

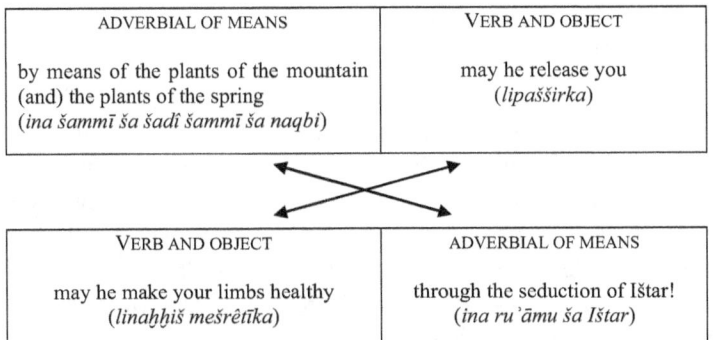

atic places and sources of purity (see Sallaberger 2007; 2011). It is possible that here we face a mythological reference to the source of the Apsû and the Cedar Forest. The mention of both, mountain and spring, and access to water, is present in a *Kiutu* prayer:[518]

> The water from the pure source, which comes from Eridu,[519]
> (which) escapes from the mountain of the pure source, from the Cedar Mountain,
> (with this water) you wash your hands, make your hands shining!
> [*Kiutu* prayer ll. 28–30, Cooper 1972: 74–75]

The ingredients that the god Asalluḫi often uses to destroy the evil action of demons, their ability to cause suffering, explicitly come from Apsû and the mountains, as in *Udug-hul*:

> The sulphur, which was created in the Apsû,
> bright salt and *qarnānu uḫḫulu*-plant (salicornia), brought from the mountains,
> the *azupīru*-plant, well-suited to the garden,
> and powder of antler of the stag, well-suited to the mountain,
> (all) cleanse the patient, seven of these cleanse the patient,
> they bind whatever causes evil.
> [*Udug-hul* XIII-XV 211–216, Geller 2016: 486–487]

The mountain is referred to as a place from which one can acquire ingredients intended to heal magic attacks in *Maqlû*:

> [Incantation]: Šamaš [has ri]sen, I reach the mountains$^?$,
> [. . .] I reach the mountains$^?$,
> I [wait for yo]u, my lord, Šamaš,
> [my hands], O Šamaš, hold up the plant of deliverance,
> [. . . may I] send you to the daughters of Šamaš, my releaser.
> [*Maqlû* VIII 24–28, Abusch 2016: 141]

The god Asalluḫi is invoked to heal the patient's flesh "by the seduction of Ištar" (*ina ruʾāmi ša Ištar*). The term *ruʾāmu* (CAD R 392 'charm, seductiveness'; AHw. II 991 'Liebreiz'), derived from the verb *râmu* 'to love,' is used in poetic language (see GAG § 55 k: 75).[520] In another *nīš libbi* incantation, the goddess Ištar is defined as "mistress of seduction": *i*[*na qibīt Ištar*] *bēlet ruʾāmi Nanāya bēlet kuzbi* "At [the command of Ištar], patron of the feminine charms, (and) Nanāya, patron of sexual attractiveness" (No. E.3: 52). This kind of seduction is therefore an attribute of the goddesses Ištar or

---

518 For other references see Cooper 1972: 80–81.

519 See also *Maqlû* VII 115, Abusch 2016: 184.

520 See also its use in texts concerning the hierogamy, in the expression *bīt ruʾāmi* "house of seduction/love," for example, between Bēl and Bēltiya and between Marduk and Ṣarpānītu (see Matsushima 1988: 108).

Nanāya, and should be interpreted in a sensual quality, typical of women. As for the name of the goddess, we find in an Old Babylonian *Hymn to Ištar*:

šāt mēleṣim ru'āmam labšat
za'nat inbī mēqiam u kuzbam
Ištar mēliṣim ru'āmam labšat
za'nat inbī mēqiam u kuzbam

She (= the goddess) of joy is clothed in seduction,
She is adorned with fruits (= erotic enjoyment), cosmetics and sexual attraction,
Ištar of joy is clothed in seduction,
She is adorned with fruits, cosmetics, and sexual attraction.
[OB *Hymn to Ištar* ll. 5–8, Thureau-Dangin 1925: 170; SEAL No. 7495]

Equally, the goddess Nanāya in an Old Babylonian hymn is described as "provided with seduction":

[uḫ]tannamū elušša
[n]annabu mašraḫu duššupu kuzbu
[ḫūd]i ṣīḫātim u ru'āmi tuštazna[n]
[r]āmam Nanāya tazmur

[Bl]ooming upon her are
[pro]geny, splendor, sweetness (and) sexual attraction,
[with jo]y, smile and seduction you are provide[d]
Nanāya, you have sung [the lo]ve.
[Streck and Wasserman 2012: 187, ll. 5–8]

In an Old Babylonian love dialogue, a woman betrayed by her lover who abandoned her for another woman, curses her rival, hoping that the goddess Ištar can remove the seduction from her rival:

mutakkiltaki lilqē
ru'āmki tēkiatīki ḫulliqī

May your supporter (the goddess) take away
your seduction, put an end to your complaint!
[i 29–30, Held 1961: 6; see Held 1962: 37; Wasserman 2016: 176]

The second section of the incantation, therefore, helps us to understand some aspects of *nīš libbi* therapeutics. The god is invoked so that he can, on the one hand, deliver the patient from the witch attack using medical ingredients, using plants from the mountain and the water spring, a clear reference to the practice of prescriptions that accompany incantations, and on the other hand, heal the patient's limbs by using Ištar's seduction. The result is that the man with his restored body functions feels a renewed sexual desire, thanks to the alluring and attractive action of the goddess.

IV. A rev. 20–29 = Incantation and its ritual No. A.3

Text No. A.3 consists of an incantation (ll. 49–56) and a prescription (ll. 57–58). The incantation is divided into two parts of four lines each:

- Identification of the patient with wild beasts (*akkannu* and *rīmu*, l. 49), and rhetorical questions about the unknown performer of the evil deed (ll. 49–52). This section shares the same structure with incantation No. D.2;
- Invocation of Nanāya and other deities(?) (ll. 53–55) and exhortation of sexual activity (l. 56).

The structure of the first part of the incantation begins with two metonymies, which are repeated twice, followed by three questions introduced with the pronoun "who" (*mannu*), the last of which is composed of two lines (using three verbs). While the first two questions, which form a parallel pair, contain similes introduced by *kīma* 'like,' the third one makes explicit reference to psychophysical states (fear and insomnia).

Stylistically note the chiastic word order in ll. 49–50; the anticipation of the predicate over complement and direct object in ll. 49 and 51; chiastic structure between ll. 51–52; and in l. 52 the deletion of *mannu* in the last sentence, resulting in a longer sentence at the end with a noticeable climactic effect (see Buccellati 1976: 68). Below the chiastic structure is diagramed:

Akkannu-wild ass! *Akkannu*-wild ass! Wild bull! Wild bull!
akkannu akkannu rīmu rīmu

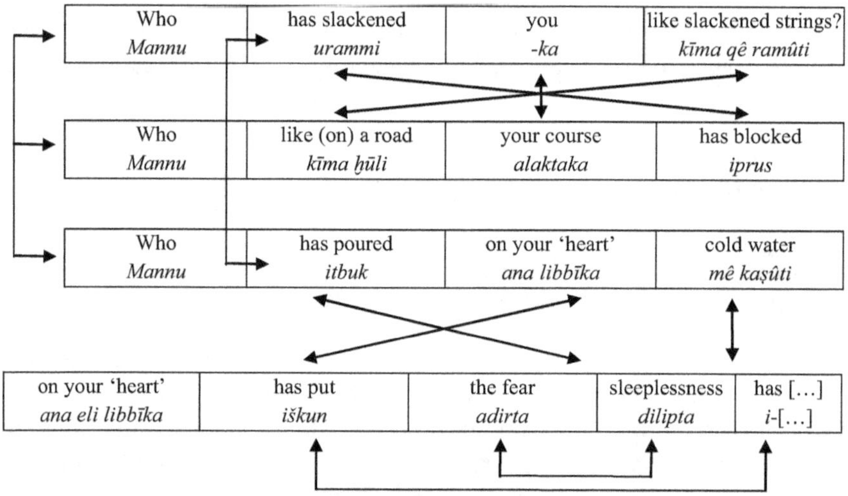

The second section of the incantation includes the invocation to the gods, in particular Nanāya. It appeals to the divinities(?) so that they can "pour joys (*rīštu*) on the 'heart.'" It is interesting here that the verb *tabaku* 'to pour' is used; in fact, the expression “pouring on ‘heart’ (*libbu*)” is used above (l. 22) to indicate the negative action of the witch, probably the pouring of cold water. The aim is to bring joy to the patient. It is to the patient which the last line (l. 56) is addressed by using the imperative: “mount the wife!”.

VI. F 9–18 = Incantation No. A.4

Incantation No. A.4, unlike the other incantations, explicitly mentions the witch as an evil performer causing the loss of human sexual desire. The incantation can be divided into two separate parallel sections, starting at the line which mentions the performer of the witchcraft:

- Rhetorical questions about the identity of the cause of evil (ll. 75–78);
- Mention of the witch and description (based on the model of the questions) of her action (ll. 79–82).

As can be seen from the diagram, the witch is identified in l. 79. The description of the witch’s deed follows by using the third person singular, building up a parallel structure to the previous lines. This use of “she” is in opposition to the first line of the incantation, which uses the pronoun *attamannu* (“whoever you are”). The second section of the incantation (l. 80), therefore, starts with a preterite verb according to the person who creates (along with the object complements) a chiasmic structure making use of the three verbs of line 82.

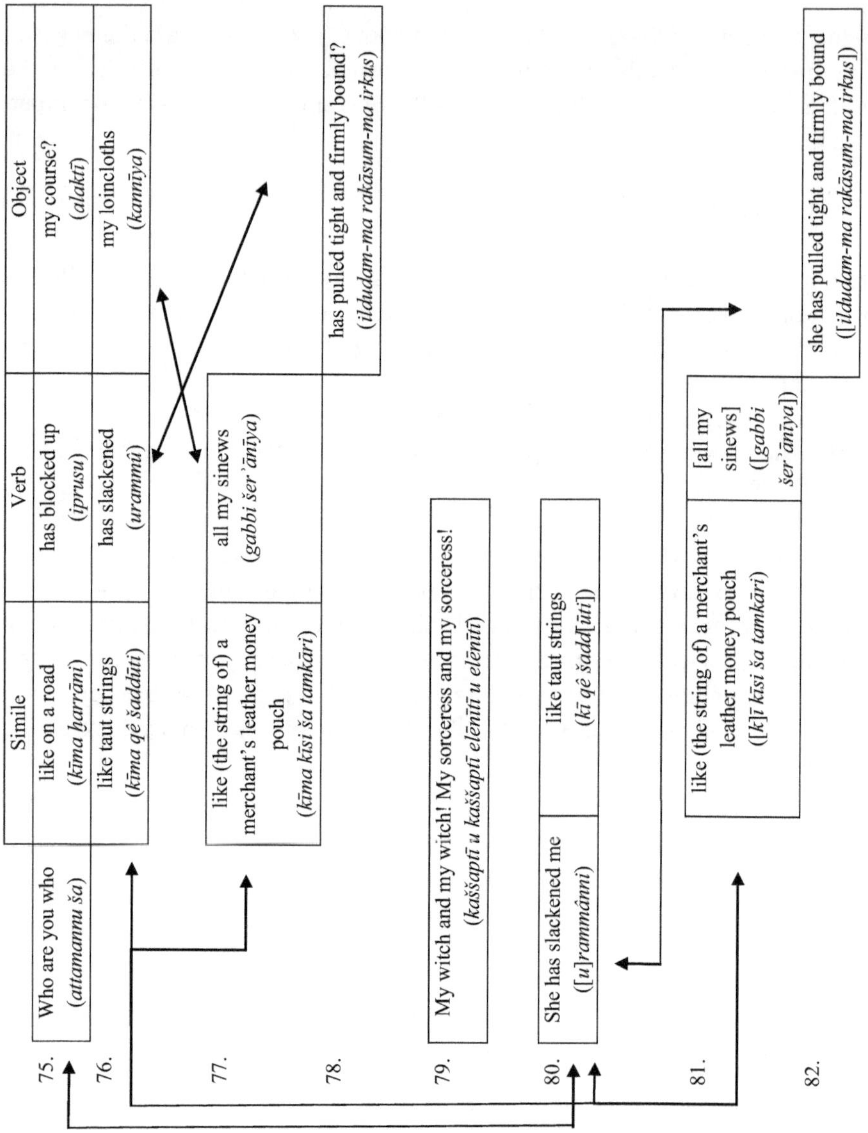
Simile
Verb
Object
75.
Who are you who (attamannu ša)
like on a road (kīma ḫarrāni)
has blocked up (iprusu)
my course? (alaktī)
76.
like taut strings (kīma qê šaddūti)
has slackened (urammû)
my loincloths (kannīya)
77.
like (the string of) a merchant's leather money pouch (kīma kīsi ša tamkāri)
all my sinews (gabbi šer'ānīya)
78.
has pulled tight and firmly bound? (ildudam-ma rakāsum-ma irkus)
79.
My witch and my witch! My sorceress and my sorceress! (kaššaptī u kaššaptī elēnītī u elēnītī)
80.
She has slackened me ([u]rammânni)
like taut strings (kī qê šadd[ūti])
81.
like (the string of) a merchant's leather money pouch ([k]ī kīsi ša tamkāri)
[all my sinews] ([gabbi šer'ānīya])
82.
she has pulled tight and firmly bound ([ildudam-ma rakāsum-ma irkus])

Nīš libbi B

List of manuscripts

Manuscript	Museum number	Publication	Tablet	Script	Date	Provenience	Incantations and prescriptions
A	VAT 13610	LKA 102: 1–rev. 21	Single-col. tablet	NA	8th–7th cent.	Aššur, Library N 4	No. B.1; Prescr. 1; No. B.2; Prescr. 2
B	VAT 8916	KAR 70 rev. 31–34	Single-col. tablet	NA	8th–7th cent.	Aššur, Library N 4	No. B.2
C	A 483	BAM 369 = CCMAwR 3, pl. 1 rev. 2′–9′	Fragment of a single-col. tablet	NA	8th–7th cent.	Aššur, Library N 4	Prescr. 1

Editions

Ebeling 1925: 34–35, 41–44, 64 (Ms. A, B)
Biggs 1967: 22–23, 41–45 (Ms. A, B)
Schwemer 2010: 120–122 (trans. of Ms. B)
Abusch et al. 2020: 28–30, No. 3.16 (Ms A, C)

Structure of the text

Text *nīš libbi* B is based on Ms. A. It begins with the incantation No. B.1 "Roar on me! Roar on me! Rear up! Rear up!". A prescription (with duplicate Ms. C) and the incantation No. B.2 "O Adad, locker keeper of the canals of heaven" (with duplicate Ms. B) follow. The text ends with another prescription (Ms. A).

A	No. B.1
A – C	Prescr. 1
A – B	No. B.2
A	Prescr. 2

I. Incantation and its ritual No. B.1 "Roar on me! Roar on me! Rear up! Rear up!" (Biggs 1967 No. 6), ll. 1–17 (Ms. A 1–17)

II. Prescription 1, ll. 18–29 (Ms. A 18–rev. 5 // C rev. 2′–9′)

III. Incantation and its rituals No. B.2 “O Adad, locker keeper of the canals of heaven” (Biggs 1967 No. 23), ll. 30–40 (Ms. A rev. 6–16 // B rev. 31–34)

IV. Prescription 2, ll. 41–45 (Ms. A rev. 17–21)

Summary of the sections of manuscripts not included in the transliteration:

- Ms. A = LKA 102
  u. e. 1–2 = colophon of Aššur-šākin-šumi, see commentary l. 45

- Ms. B = KAR 70
  1–5 = No. F.1 (Ms. A)
  6–10 = D Diagn. (Ms. F)
  11–44 = F prescr. 5–7, 10–17 (Ms. A)
  45–rev. 9 = No. F.4 (Ms. A)
  rev. 10–24 = No. E.1 (Ms. B)
  rev. 25–30 = No. F.5 (Ms. A)
  rev. 35 = Fragmentary colophon, see commentary l. 40

- Ms. C = BAM 369
  1–12 = No. E.1 (Ms. C)
  rev. 10′–12′ = Fragmentary colophon, see commentary l. 29

Transliteration

I. A 1–17 = Incantation and its ritual No. B.1

1. ÉN *ug-ga ug-ga ti-ba* [*ti-ba*]
2. *ug*$^{!}$*-ga* GIM *a-a-lì ti-ba* GI[M *ri-mi*]
3. *it-ti-ka lit-ba-a ni-e-š*[*ú*$^{?}$]
4. *it-ti-ka lit-ba-a ba*[*r*$^{!}$*-ba-ru*]
5. ⸢*it*⸣*-ti-ka lit-ba-a* MU[Š . . .]
6. *p*[*u-ḫ*]*ur* SA^{meš} ŠID-*ka ni-il-k*[*a* x] x [x]
7. x ⸢*ka*⸣$^{?}$ *a ku-us-su-ka la ik-kal-l*[*a*]
8. ⸢*su*⸣ [x] x *da ki-ma ze-e-k*[*a*]
9. ⸢*uk*⸣ x *ki-ma ši-na-ti-*[*ka*]
10. *li-ma-aṣ-ra ú-šar-ka mu-ni-iḫ e* x [. . .]
11. GÌŠ-*ka ku-ut-mi-ni-tu-*⸢*ma*⸣ *li-ku-la pu-ri-di-iá*
12. *ina qí-bit* ^{d}*ka-ni-sur-ra* ^{d}*iš-ḫa-ra be-*[*l*]*et ra-me* ÉN

---

13. KA.INIM.MA ŠÀ.ZI.GA

---

14. DÙ.DÙ.BI ^{na4}KA.GI.NA.DAB.BA SÚD *ina* Ì BUR ḪE.ḪE
15. ÉN 7-*šú ana* UGU ŠID-*nu* LI.DUR-*su* TAG-*at*

16. {KU.KU} AN.BAR SÚD *ina* Ì BUR ḪE.ḪE ÉN 7-*šú* *ana* ŠÀ ŠID
17. LI.DUR MUNUS TAG-*at* NITA *u* MUNUS TÉŠ.BI [*i-nu-uḫ-ḫu*]

II. A obv. 18–rev. 5 // C rev. 2′–9′ = Prescription 1

18.	A o. 18	DIŠ NA *ina ṭe-em ra-ma-ni-šu it-*[*ta-na-ad-laḫ*]
	C. r. 2′	[*ra-m*]*a-*⸢*ni-šu it*⸣*-ta-*⸢*na-ad*⸣*-laḫ*
19.	A o. 19	SED ŠUB.ŠUB-*su a-šu-uš-tú* TUKU.TUKU-⸢*ši*⸣?
	C. r. 2′–3′	SED ⸢ŠUB.ŠUB⸣-*s*[*u*] / [ ] ⸢TUKU⸣.TUKU-*ši*
20.	A o. 20	*bir-ka-šú ka-sa-a* DU$_8^{meš}$-*šú it-ta-na-a*[*s-ra-ḫu*]
	C. r. 3′–4′	*bir-ka-šú ka-sa-a* DU$_8^{me}$[^{š}-*šú*] / [*it*]-⸢*ta-na-as*⸣-[*r*]*a-ḫu*
21.	A o. 21	SU-*šú ta-ni-ḫa* TUKU.TUKU-*ši* KA?-*šú* KÚR.KÚR
	C. r. 4′–5′	SU-*šú ta-ni-ḫa* TUKU.TUKU-*ši p*[*u*?-*šú*] / [ ]
22.	A lo. e. 1	NINDA *u* KAŠ LAL NA BI [ ]
	C r. 5′	NINDA *u* KAŠ.⸢SAG⸣ LAL NA BI *ka-šip*
23.	A lo. e. 2	EN INIM-*šú kiš-pi* [*is-ḫu-ur-šu*]?
	C r. 5′	EN INIM-*šú kiš-pu* [*is-ḫu-ur-šu*]?
24.	A lo. e. 3	BAR KU$_6$ *mi-ki-i* ⸢*ù*⸣ [ ]
	C r. 6′	[*mi-ki*]-⸢*i ù* NINDA *tur*⸣-*ár*
25.	A r. 1	NUMUN ^{ú}A.ZAL.LÁ NUMUN ^{ú}IN$_6$.ÚŠ 1-*niš* SÚD
	C r. 6′–7′	NUMUN ^{ú}A.ZAL.LÁ [NUMU]N ⸢IN⸣.NU.[ÚŠ] / [1-*ni*]*š* ⸢SÚD⸣
26.	A r. 2	KI ^{ú}NAGA ḪE.ḪE *ina* A^{meš} *i-ra-muk*
	C r. 7′	⸢KI ^{ú}NAGA⸣ [ ] *ina* A^{meš} *i-*⸢*ra*⸣*-*[*mu*]*k*
27.	A r. 3	^{giš}*bi-nu ana* A^{meš} ŠUB *ina* NINDU ÚŠ-*er*
	C r. 8′	[^{giš}]⸢ŠINIG⸣ *ana* A^{meš} Š[UB]-⸢*di*⸣ *ina* NINDU Ú[Š-*er*]
28.	A r. 4	SU-*šú tu-maš-šá-aʾ* 〈*ina*〉 U$_4$.NÁ.À[M]
	C r. 9′	[SU-*š*]*ú tu-maš-*⸢*šá-aʾ i-na* U$_4$⸣.[N]Á.ÀM
29.	A r. 5	*an-nam* DÙ.DÙ-*uš-ma* TI-*uṭ*
	C r. 9′	*an-nam* DÙ.DÙ-⸢*uš*⸣-[*ma*] T[I-*uṭ*]

III. A rev. 6–16 // B rev. 31–33 = Incantation and its rituals No. B.2

30.	A r. 6	ÉN ^{d}IŠKUR GÚ.GAL AN-*e* DUMU ^{d}*a-nim*
	B r. 31	ÉN ^{d}IŠKUR GÚ.GAL ^{d}*a-nim* DUMU ^{d}*a-nim*
31.	A r. 7	TAR-*is* EŠ.BAR *šá kiš-šat* UN^{meš} *la-mas-si* KUR
	B r. 31–32	TAR-*is* EŠ.BAR / *šá kiš-šat* UN^{meš} ^{d}LAMA *ma-a-ti*
32.	A r. 8	*ina* DU$_{11}$.GA-*ka ṣer-ti šá* NU KÚR-*rù*
	B r. 32	*ina qí-bi-ti-ka* / — — — —
33.	A r. 9	*ù an-ni-ka ki-nim šá* NU BAL-*u*
	B	— — — — — —
34.	A r. 10	NENNI A NENNI *ana* NENNI-*ti* DUMU.MUNUS NENNI-*ti*
	B r. 33	NENNI A NENNI — — — —
35.	A r. 11	*li-im*(text E)-*gu-ug lim-ḫaṣ li-ir-kab*
	B r. 33	— *lim-ḫa-aṣ li-ir-kab*
36.	A r. 12	*ù li-še-rib* TU$_{6}$ ÉN
	B r. 33	*ù li-še-rib* TU$_{6}$ ÉN
37.	A r. 13	KA.INIM.MA ŠÀ.ZI.GA
38.	A r. 14	DÙ.DÙ.BI KU.KU ^{na4}KUR-⸢*nu* DIB.BA KU.KU AN.BAR⸣
39.	A r. 15	*ina* Ì BUR ḪE.ḪE ÉN 7-*šú* [. . .] ⸢ŠID-*nu*⸣
40.	A r. 16	NITA GÌŠ-*šú* MUNUS GAL$_{4}$.LA-*šá* [. . .]

Variant B r. 34 ÉN^{meš} *an-na-a-tu ana* UGU *ri-kib-tú a-a-lì* ŠID-*nu-ma* ŠÀ.ZI.GA

IV. A rev. 17–21 = Presciption 2

41.	A r. 17	*ana* NITA *u* MUNUS *šup-šu-ri-im-ma* [. . .]
42.	A r. 18	Ú.KUR.RA SIG$_{7}$-*su ta-sàk* [. . .]
43.	A r. 19	NU ŠEG$_{6}$.GÁ *ga šu ú ma la* x [. . .]
44.	A r. 20	*ku-ub-tú ana* ŠÀ KAŠ.DU$_{10}$.GA ŠUB-*ma* [. . .]
45.	A r. 21	TÉŠ.BI NAG-*ma* ŠÀ-*šú-nu ip*-[*pa-aš-ša-ru*]

Transcription

I. A 1–17 = Incantation and its ritual No. B.1

1. *šiptu: ugga ugga tibâ [tibâ]*
2. *ugga kīma ayyali tibâ kī[ma rīmi]*
3. *ittīka litbâ nēš[u]*
4. *ittīka litbâ ba[rbaru]*
5. *ittīka litbâ ṣer[ru . . .]*
6. *p[uḫ]ur gīdī minâtīka nīlk[a . . .] . . . [. . .]*
7. . . . *kussûka lā ikkall[a]*
8. . . . [. . .] . . . *kīma zêk[a]*
9. . . . *kīma šīnātī[ka]*
10. *limmaṣra ušarka munīḫ ē* . . . [. . .]
11. *ušarka* . . . *līkulā purīdīya*
12. *ina qibīt Kanisurra Išḫara bē[l]et râmi šiptu*

---

13. *šipat nīš libbi*

---

14. *dudubû: šadâna ṣābita tasâk ina šaman pūri taballal*
15. *šipta sebîšu ana muḫḫi tamannu abbunassu talappat*
16. *parzilli tasâk ina šaman pūri taballal šipta sebîšu ana libbi tamannu*

17. *abbunat sinništi talappat zikaru u sinništu ištēniš [inuḫḫū]*

II. A obv. 18–rev. 5 // C rev. 2′–9′ = Prescription 1

18. *šumma amēlu ina ṭēm ramānīšu ittanadlaḫ*
19. *kuṣṣu imtanaqqussu ašuštu irtanašši*
20. *birkāšu kasâ piṭrūšu ittanaṣ[r]aḫū*
21. *zumuršu tānēḫa irtanašši pûšu ittanakkir*
22. *akala u šikara* (var. adds: *rēštâ*) *muṭṭu amēlu šū kašip*
23. *bēl amātīšu kišpī [isḫuršu]*[?]
24. *qulipti nūni mikî u akalu turrar*
25. *zēr azallî zēr maštakal ištēniš tasâk*
26. *itti uḫūli taballal ina mê irammuk*
27. *bīna ana mê tanaddi ina tinūri tesekker*
28. *zumuršu tumašša' ina bubbuli*
29. *annâ tēteneppuš-ma iballuṭ*

Translation

I. A obv. 1–17 = Incantation and its ritual No. B.1

1. Incantation: Roar on me! Roar on me! Rear up! [Rear up]!
2. Roar on me like a stag! Rear up lik[e a wild bull]!
3. Together with you, may a lio[n] rear up!
4. Together with you, may a w[olf] rear up!
5. Together with you, may a snak[e] rear up!
6. A[l]l the muscles of your limbs, your sperm . . .
7. . . . your chair will not be held.
8. . . . like you[r] excrements.
9. . . . like y[our] urine.
10. May your penis, which satisfies (the desire)$^?$, be compact$^?$! Do not [. . .]!
11. May my crotch devour your . . . penis!
12. At the command of Kanisurra and Išḫara, patron goddess of love. Incantation.

---

13. Wording of *nīš libbi* (incantation).

---

14. Its ritual: You pulverize magnetite, you mix (it) with oil from the alabastron,
15. you recite the incantation seven times over it; you apply (it) to his navel;
16. you pulverize iron, you mix (it) with oil from the alabastron, you recite the incantation seven times over it,
17. you apply (it) to the woman's navel; the man and the woman [will find relief] together.

II. A 18–rev. 5 // C rev. 2′–9′ = Prescription 1

18. If a man is constantly perturbed in his own counsel,
19. cold (tremors) continuously afflict him and he has constantly distress,
20. his knees are bound, his . . . are constantly hot,
21. his body has continuously tiredness, his speech is constantly incoherent,
22. he has no desire to eat and drink, that man is bewitched.
23. The man's litigant [has employed]$^?$ witchcraft [against him].
24. You parch the scaly skin of a fish, (peel of the) *mikû*-plant and bread.
25. You pulverize seeds of *azallû*-plant (and) seeds of *maštakal*-plant together,
26. you mix (them) with *uḫūlu*-plant. He washes himself with water.
27. You put tamarisk into (that) water, you heat (the water) in the oven,
28. you rub his body (with it). On the day of the new moon
29. you will repeat this (therapeutic) performance and he will recover.

III. A rev. 6–16 // B rev. 31–33 = Incantation and its rituals No. B.2

30. *šiptu: Adad gugal šamê* (var.: *Anim*) *mār Ani*
31. *pāris purussêša kiššat nišī lamassi māti*
32. *ina qibītīka ṣīrti ša lā nakru*
33. *u annīka kīni ša lā enû*
34. *ananna mār annanna ana annannīti mārat annannīti*
35. *limgug limḫaṣ lirkab*
36. *u lišērib tê šipti*

---

37. *šipat nīš libbi*

---

38. *dudubû: sīkti šadâni ṣābiti sīkti parzilli*
39. *ina šaman pūri taballal šipta sebîšu [ana libbi] tamannu*
40. *zikaru ušarašu sinništu biṣṣūraša [iptaššaš-ma irtanakkab]*

Variant B r. 34. *šiptī annātu ana muḫḫi rikibti ayyali tamannū-ma libbi*

IV. A rev. 17–21 = Presciption 2

41. *ana zikara u sinništa šupšurim-ma* [. . .]
42. *nīnû arqūssu tasâk* [. . .]
43. *lā tabaššal* . . . [. . .]
44. *kubtu ana libbi šikari ṭābi tanamdī-ma* [. . .]
45. *ištēniš išattû-ma libbūšunu ip[paššarū]*

III. A rev. 6–16 // B rev. 31–33 = Incantation and its rituals No. B.2

30. Incantation: O Adad, lock keeper of the canals of heaven (var.: of Anu), son of Anu,
31. who decrees oracular decisions for all people, protector of the land,
32. at your supreme command, which cannot be opposed,
33. and at your authentic consent, which cannot be altered,
34. May NN, son of NN, with NNfem., the daughter of NNfem.,
35. mate, bonk (her), mount (her),
36. and penetrate (her)! Incantation formula.

---

37. Wording of *nīš libbi* (incantation).

---

38. Its ritual: Magnetite powder (and) iron powder
39. you mix with oil from the alabastron, you recite the incantation [over it] seven times;
40. the man [anoints] (with it) his penis, the woman her vulva [and he (can) repeatedly have intercourse].

Variant B r. 34. You recite these incantations over the *rikibtu* of a stag and (he will get) sexual desire.

IV. A rev. 17–21 = Prescription 2

41. In order to release the man and the woman and [. . .]:
42. You pulverize the green (part) of *nīnû*-plant [. . .]
43. you do not cook/uncooked . . . [. . .]
44. you throw a lump of metal into good beer and [. . .]
45. they drink together and their 'hearts' will be [released].

Commentary

1–12. Incantation No. B.1 features two speakers: the female partner (ll. 1–11) and the therapeutic operator (l. 12).

1. *ugga*: imperative from *nagāgu* 'to bray' (CAD N/I 105–106; AHw. II 709) + ventive. The verb refers to the roaring and braying of animals (especially the donkey).

3–5. The expression *ittīka litbâ* is not easily translatable. Here, following Biggs (1967: 22), I translate it as "together with you." This expression aims at establishing a dialogue between the animal sex world and the human one. The incantation invites the man to get aroused along with the animal world, to which the man himself, metaphorically, belongs. The formula also appears in incantation No. F.4: 83–86, Biggs (1967: 32) translates "for you," while Nougayrol (1968: 94) proposes the translation of *ittīka* with "regarding you" (see Heimpel 1970: 191). Here too, we cannot exclude the translation "together with you/with you," to indicate that the man both in his psycho-physical integrity and in his anatomical parts, through the figure of speech of *pars pro toto*, rises and regains strength. Note that the snake (here in l. 5) could refer to the penis in an incantation to make a brothel successful: "Like a snake (= penis), going out from a hole, and birds twittering over him (= snake/penis)" (Panayotov 2013: 293, l. 30). It should be noted that in the *Šumma izbu* series the serpent symbolizes, like the wild bull, aggression (see De Zorzi 2014/I: 163–164). The lion and the wolf represent power and strength. On the lion and wolf motif in Old Babylonian incantations see Mertens-Wagshal 2018.

7. *ik-kal-la*: Biggs (1967: 22) translates "will not hurt you" from the verb *kullu*. The form can only be present in the N-stem. The dictionaries, however, do not report sources for this form. The only other mention of the seat in the corpus is in the catalogue LKA 94 i 6: *šiptu addi erša attadi* [*kus*]*sâ* "Incantation: 'I have set up a bed! I have now set up a [cha]ir!'". Here the chair, like the bed, has a positive value and alludes to the sexual sphere.

10. CAD N/I 149 restores *ē ta*[*rši*$^?$], but doubts persist. Possible restorations: *ē taglut* ("do not get scared"); *ē taʾdir* ("do not be afraid"); *ē tāšuš* ("do not be afflicted"). See incantation No. F.4: 88.
limmaṣra ušarka munī ("may your penis which satisfies (the desire)$^?$ be compact$^?$!"): *munīḫ* is a D-stem participle of the verb *nâḫu* (CAD N/I 143–150, G-stem 'to relent, became peaceful, abate,' D-stem 'to appease, pacify, dampen a desire, satisfy'; AHw. II 716–717, G-stem 'ausruhen, zur Ruhe kommen, sich beruhigen,' D-stem 'beruhigen'). In this line, therefore, I

prefer to translate "to satisfy" (see CAD N/I mng. 7). The verb *nâḫu* appears several times in the *nīš libbi* corpus with different meanings:

- G-stem in No. D.3: 36 and 39 (and restored in No. B.1: 17) *ištēniš inuḫḫū* "(the man and the woman) will find relief together." Here the verb appears at the end of the ritual. It indicates the future healing of the couple. In this case, it has the sense of 'calm down' for healing (see CAD 'to find relief' from a sickness). Biggs (1967) translates: "they will find satisfaction together"; Nougayrol (1968: 94) suggests the meaning of 'finding sexual satisfaction, rejoicing.'
- D-stem in No. A.2: 41 (and No. L.2: 26): *akkannu ša ana rakābi tebû* [*man*] *nu unīḫ*[*ka*] "*akkannu*-wild ass who is reared-up for the mating, [wh]o has dampened [your] desire?". The subject is the witch who acts against the man, subtracting his sexual desire. Here I have chosen the meaning of 'to dampen the desire' (see CAD 'to dampen a desire').

The verb *li-ma-aṣ-ra* is difficult to understand. It is a precative N, probably from the verb *maṣāru* (CAD M/I 329–330 'to move in a circle, make a detour, linger'; AHw. II 619–620 'etwas umschreiten(?)'). The N-stem is used in *Enūma eliš* (I 129; II 5; III 19, 77): *im-ma-aṣ-ru-nim-ma*, often translated with "they form a circle" (see Lambert 2013: 471). AHw. proposes, for both cases in the N-stem, albeit with reserve, the meaning of 'kompakt sein.' Here I follow AHw.'s suggestion.

11. The writing *ku-ut-mi-ni-tu-*⌜*ma*⌝ is obscure. Biggs (1967: 23) suggests the reading *tuš-tam-mi ṣal-tu-ma*, which however does not make sense in this context. Biggs' translation (1967: 22) of the line is different: "Let your . . . penis hurt my crotch!". The scholar interprets the penis as the subject (here logographically written GÌŠ) and the woman's crotch as object of a verbal form from *kullu* 'to hurt' (see AHw. I 503, mng. 8c). My solution is different: 1. The verb cannot be *kullu*, but on the contrary *akālu* ('to eat'); 2. The verb, *līkulā*, is a feminine plural precative, and therefore can have the crotch as a subject (*purīdu*, fem. pl.); 3. The word for penis is written logographically, and therefore can be read *ušarka*, as an accusative; 4. The image of the female genitals who take (here they eat) the penis appears also in incantation No. E.1: 10 with the verb *ṣabātu*: *kīma ūri kalbati iṣbatu ušar kalbi* "As the vulva of a bitch took the penis of the dog!".

16. The presence of KU.KU/*sīktu* is here perhaps a mistake. It does not appear in line 14 of the same text. Moreover, it is evident that the presence of the verb *sâku* 'to pulverize' implies the need to realize the powder. In the same prescriptions, as Biggs stresses (1967: 23), we find either KU.KU AN.BAR "iron powder" or AN.BAR SÚD "you pulverize iron."

17. For the restoration see No. D.3: 39.

18–22. This prescription does not mention either the loss of sexual desire or general sexual problems.

18. *ina ṭēm ramānīšu* lit. “in the mind of himself.” On *ṭēmu* see Steinert 2012a: 385–404; on *ramānu* ibid. 257–270.
Ms. C rev. 1′ is the final line of a prescription: [. . .] x ⸢GAR-*an-ma*⸣ Š[À]?.Z[I]?.G[A]?

20. *piṭru*: the term indicates an unidentified body part (see CAD P 449–550, mng. 2).

23. For the restoration see Abusch and Schwemer 2011: 222, No. 7.10.1: 187‴, 190‴, 205‴.

24. For the use of the peel of the *mikû*-plant in a medical prescription see BAR *mi-ki-i* in K 4023+ i 3′ (AMT 102, 1). For the use of bread with the verb *urruru* see Abusch and Schwemer 2011, No. 2.3.1: 114⁗.

29. Ms C (BAM 369) rev. 10′–12′ has a fragmentary colophon of Kiṣir-Nabû from Library N 4 in Aššur (see esp. rev. 11′ [*ki*]-*ṣir*-[d]AG).

30. See *nīš libbi* L prescr. 6: 16.

35. *limgug* “may mate (her)”: precative from the verb *magāgu* (CAD M/I 28, esp. mng. 1b, ‘to become stiff, taut’; AHw. II 574 ‘(weg)spreizen’). The verb referring to animals means ‘to cover, mate’ (ibid. mng 4 ‘bespringen’).
limḫaṣ “may he bonk (her)”: precative from the verb *maḫāṣu* ‘to beat.’ Biggs 1967: “come into contact with”; Schwemer 2010: “zustoßen.” Note that in many modern Western languages the verb ‘to bonk’ has a sexual connotation.

38. Note that the incantation No. E.1 Ms. A: 15–17 has the same procedure concerning an ointment with magnetite and iron powders and oil from the alabastron to be rubbed on the male penis and the female vulva.

40. Ms. B (KAR 70) rev. 35 has a fragmentary colophon from Library N 4 in Aššur: “. . . [. . .] . . . exemplar from Babylon, written and collated” (Hunger 1968, No. 277).

42. Ú.KUR.RA SIG$_7$-*su*: “green (part) of *nīnû*-plant” or “green (leaves).” SIG$_7$/*arqūtu* means the green and fresh part of the plant. The term is used with verbs for the preparation of plants for pharmaceutical purposes. One of these verbs is *sâku* ‘to pound’ (see AMT 34, 1: 33; BAM 264: 19; Labat 1951: 222: 43; AMT 91, 5: 6).

43. *ga šu*: Ebeling (1925) reads *šizbu*(GA)-*šú*? “its milk,” but it is uncertain (note in the copy *šu*). The rest of the line is unclear.

44. *ku-ub-tú*: one can read as well ZÍD *ár-tú* (see Biggs 1967).

45. In Ms. A u. e. 1–2 a colophon of Aššur-šākin-šumi follows: "Hand of Aššur-šākin-šumi, son of [. . .], *šangû*-priest of [. . .]" (see Hunger 1968, No. 267). For the activities of Aššur-šākin-šumi see Capraro 1998: 2; Maul 2010: 216–217.

Incantations: stylistic and functional analysis

I. A obv. 1–17 = Incantation and its ritual No. B.1

Text No. B.1 consists of an incantation (ll. 1–12) and a prescription (ll. 14–17), separated by the rubric "*nīš libbi* incantation." The incantation was recited by the female partner, as confirmed by the possessive adjectives: *-ya* "my" in line 11, referring to the woman's crotch; *-ka* "you" which refers to the patient's penis. The woman's presence is also confirmed by the prescription of two oil units from the alabastron-based ointments: one with magnetite powder to be applied on the man's navel; the other one with iron applied on the woman's navel.[521] An incantation is recited seven times for both applications.

The incantation can be divided into six smaller sections. The first four sections have a binary structure. The second one presents an anaphora with the repetition of *ittīka litbâ* "together with you, may it rear up":

1.	Roar on me! Roar on me! (*ugga ugga*)	Rear up! Rear up! (*tibâ tibâ*)
2.	Roar on me like a stag! (*ugga kīma ayyali*)	Rear up like a bull! (*tibâ kīma rīmi*)

3-5.	Together with you may rear up (*ittīka litbâ*)	the lion (*nēšu*) the wolf (*barbaru*) the snake (*ṣerru*)

6.	A[l]l the muscles of your limbs, your sperm (*p[uḫ]ur gīdī minâtīka nīlk[a]*)	…
7.	…, your seat (… *kussûka*)	will not be held … (*lā ikkalla*)

8.	…	like you[r] excrements (*kīma zêka*)
9.	…	like yo[ur] urine (*kīma šīnātīka*)

521 On sexual magnetism see Chapter III § "Ointments and amulets."

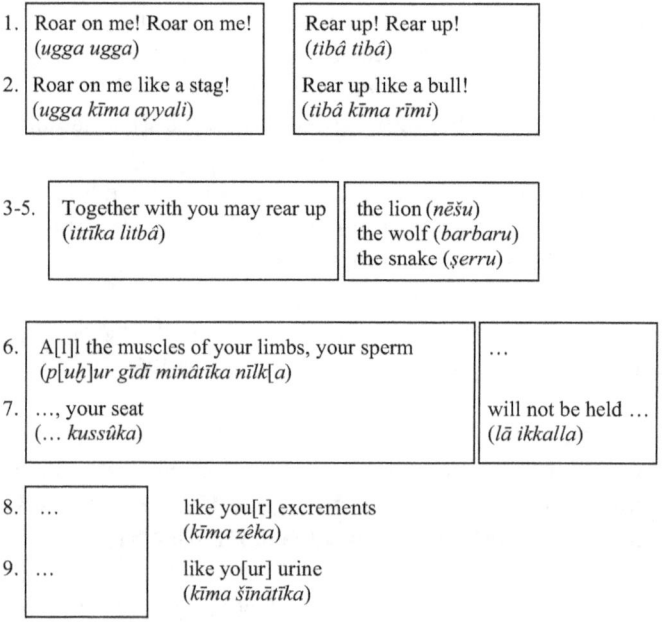

The fifth section provides a chiastic structure, using the verb-subject and the direct object/other complements:

10.	may your penis which satisfies (the desire)? be compact? (*limmaṣra ušarka munīḫ*) V-S	… ?
11.	your penis … *ušarka ku-ut-mi-ni-tu-ma* O	may my crotch devour *līkula purīdīya* V-S

The last section, made up of a single line (l. 12), is composed of the usual formula of an invocation of divine power to make the incantation effective: "At the command to Kanisurra and Išḫara, pa[tr]oness of sex."

The first part of the incantation provides the use of imperatives that motivate the patient, identified with the deer and the wild bull, to sexual arousal (ll. 1–2). The precative forms are related to wild predatory beasts: the lion, the wolf, and maybe the snake (ll. 3–5). Here no form of human identification can be found. The arousal of the animals is in relation to that of humans (*ittīka* "together with you").

The second section, unfortunately incomplete, describes the human body. It mentions parts of the patient's body (muscles, sperm), for which, however, an interpretation is hard to find (ll. 6–7). Expelled bodily substances, i.e. urine and excrement, are used in similes we do not understand fully (ll. 8–9). Finally, the penis and the woman's crotch (*purīdu*) are subject to precative expressions related to the sexual act (ll. 10–11): "May your penis, that satisfies, be compact?/May my crotch devour your penis. . .".

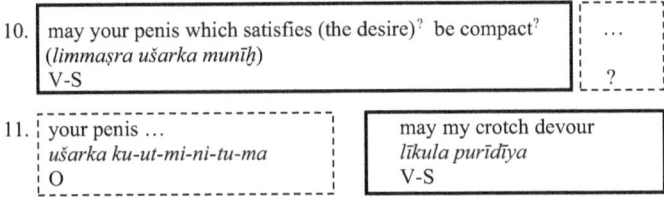

Nīš libbi C

List of manuscripts

Manuscript	Museum number	Publication	Tablet	Script	Date	Provenience
A	–	LKA 103	Frg. of a single-col. tablet	NA	8th–7th cent.	Aššur, Library N 4

Editions

Ebeling 1925: 44
Biggs 1967: 26–27, 64–65
Chalendar 2018: 46

Structure of the text

I. Incantation and its ritual No. C.1 "Your lovemaking" (Biggs 1967 No. 9), ll. 1–13

II. Prescriptions
   1. ll. 14–16
   2. 17–18
   3. l.e. i 1–3
   4. l.e. ii 1–3

Transliteration

I. A 1–13 = Incantation and its ritual No. C.1

1. [. . .] x [. . .]
2. [. . .] *an* [. . .]
3. [. . . *ri*]-*kib-ta-ka* x [. . .]
4. *r*[*a-am* DÀR]A$^{?}$ 6-*šú*
5. *ra-am a-a-lì* 7-*šú*
6. *ra-am* BURU$_5$.ḪABRUD.DA^{mušen} 12-*šú*
7. *ra-man-ni ra-man-ni áš-šu ṣe-eḫ-ra-k*[*u* . . .]
8. *ù ri-kib-tú* DÀRA.MAŠ GAR-*ku ra-man-ni* [TU$_6$ ÉN]

9. KA.INIM.MA DIŠ NA *ana* MUNUS GIN-*ka* LÁ

10. DÙ.DÙ.BI SAG.DU BURU$_5$.ḪABRUD.DA^{mušen} NÍTA

11. ^{na4}KÙ.BABBAR ^{na4}KÙ.SI$_{22}$ *ri-kib-te a-a-lì ina* KUŠ {x}
12. DÙ.DÙ.BI ÉN 7-*šú ana* UGU ŠID-*nu* [*an-na-a*]
13. ⸢*ša*⸣-*la-áš u e*[*r-be-šú* DÙ].

II. A 14–18, l.e. i 1–3, l.e. ii 1–3 = Prescriptions 1–4

1. 14. ⟨DIŠ⟩ KI.MIN ÚŠ BURU$_5$.ḪABRUD.DA^{mušen} [NITÁ]
15. [GE]ŠTIN.ŠUR.RA K[AL] ^{ú}[. . .]
16. [*ina* UG]U ŠID-*nu* NAG-*šú* [. . .]

---

2. 17. [DIŠ KI.MIN . . .] ŠE ^{ú}x x [. . ..]
18. [. . . ^{ú}*a*]*m-ḫa-ra* x [. . .]
Rest destroyed

3. left edge i
1. [DIŠ KI.MIN . . . ^{ú}ḪAR.Ḫ]AR 1-*niš* SÚD
2. [. . . N]AG-*ma*
3. [. . . ŠÀ.ZI].⸢GA⸣

4. left edge ii
1. DIŠ KI.[MIN] ^{ú}[. . .]
2. ^{ú}LÚ.⸢U$_{19}$⸣.[LU . . .]
3. x x x x [. . .]

Transcription

I. A 1–13 = Incantation and its ritual No. C.1

1. [. . .] . . . [. . .]
2. [. . .] . . . [. . .]
3. [. . . *ri*]*kibtaka* . . . [. . .]
4. *r*[*âm turā*]*ḫi šeššīšu*
5. *râm ayyali sebîšu*
6. *râm iṣṣūr ḫurri šinšerīšu*
7. *rāmanni rāmanni aššu ṣeḫrāk*[*u* . . .]
8. *u rikibti ayyali šaknāku rāmanni* [*tê šipti*]

---

9. *šiptu*: *šumma amēlu ana sinništi alāka muṭṭu*

---

10. *dudubû*: *qaqqad iṣṣūr ḫurri zikari*
11. *kaspu ḫurāṣu rikibti ayyali ina maški*
12. *dudubû*: *šipta sebîšu ana muḫḫi tamannu* [*annâ*]
13. *šalaš u e*[*rbešu teppuš*]

II. A 14–18, l.e. i 1–3, l.e. ii 1–3 = Prescriptions 1–4

1.
   14. *šumma* KI.MIN *dām iṣṣūr ḫurri* [*zikari*]
   15. [*ka*]*rānu ṣaḫtu a*[*qru*] [. . .]
   16. [*ina muḫ*]*ḫi tamannu išattīšu* [. . .]

2.
   17. [*šumma* KI.MIN . . .] *še'u* . . . [. . .]
   18. [. . . *a*]*mḫara* . . . [. . .]

3. left edge i
   1. [*šumma* KI.MIN . . . *ḫa*]*šâ ištēniš tasâk*
   2. [. . . *i*]*šattī-ma*
   3. [. . .] *nī*[*š libbi*]

4. left edge ii
   1. *šumma* KI.[MIN] [. . .]
   2. *amīlā*[*nu* . . .]
   3. . . . [. . .]

Translation

I. A 1–13 = Incantation and its ritual No. C.1

1–2. Traces

3. [. . .] your [love]-making . . . [. . .].
4. The ma[ting of a wild go]at six times,
5. the mating of a stag seven times,
6. the mating of a partridge twelve times,
7. make love to me! Make love to me because I am young! [. . .]
8. and I am endowed? with the *rikibtu* of a stag! Make love to me! [Incantation formula].

---

9. Wording of (the incantation): If a man's (desire) to go to a woman is reduced.

---

10. Its ritual: (You put) the head of a male partridge,
11. silver bead, gold bead, the *rikibtu* of a stag into a leather bag.
12. Its ritual: You recite the incantation over (it) seven times,
13. [you do that] three and four [times].

II. A 14–18, l.e. i 1–3, l.e. ii 1–3 = Prescriptions 1–4

1. 14. If ditto: Blood of a [male] partridge,
15. pre[cious grape [jui]ce, [. . .]-plant [. . .].
16. you recite (the incantation) [ove]r (it) and he drinks it [. . .].

---

2. 17. [If ditto: . . .] wheat, . . .-plant [. . .]
18. [. . . *a*]*mḫara*-[plant] . . . [. . .].

3. left edge i
1. [If ditto: . . . *ḫa*]*šû*-[plant] you pulverize together
2. [. . . he] drinks and
3. [. . .] sexua[l desire].

4. left edge ii
1. If di[tto]: [. . .]-plant
2. *amīlā*[*nu*]-plant [. . .]
3. Traces

Commentary

3–8. See the parallel No. F.3: 68–71. In line 7 space too small to restore *ra-man-ni*.

4–6. Cf. also incantations No. E.3: 48–49; No. K.8: 149–150.

12–13. It is possible to restore also ⸢4⸣-[*šú te-pu-uš*]. For the restorations see Chalendar 2018: 46 and fn. 187. However, the prescription to repeat the ritual four times is surprising. In fact, this indication never appears in the corpus.

17. Biggs's (1967) reading ᵘ̉A.ZAL.L[Á . . .] remains uncertain.

Nīš libbi D

List of manuscripts

Manuscript	Museum number	Publication	Tablet	Script	Date	Provenience	Incantations and prescriptions
A	–	LKA 101: 7–rev. 11	Single-col. tablet	NA	8th–7th cent.	Aššur, Library N 4	No. D.2
B	K. 2499	TCS 2, pl. 1: obv.! 1′–rev.! 11	Single-col. tablet	NA	8th–7th cent.	Nineveh	No. D.1; No. D.2; No. D.3; Prescr. 6–7
C	W. 22307/4+68	SpTU 1, 9: 5′–27′	Frg. of a single-col.? tablet	NB/LB	4th–3rd cent.	Uruk, U 18	Diagn.; Prescr. 1–2; No. D.3; Prescr. 3; Sympt.; Prescr. 4–5
D	VAT 13721	LKA 97 ii 5?–26	Large frg of a two-col. tablet	NA	8th–7th cent.	Aššur, Library N 4	Prescr. 2; No. D.3; No. D.4
E	SU 52/139+161 + 170+250 + 250A +323	STT 280 i 8–17	Two-col. tablet	NA	8th–7th cent.	Sultantepe	Sympt.; Prescr. 4
F	VAT 8916	KAR 70: 6–10	Single-col. tablet	NA	8th–7th cent.	Aššur, Library N 4	Diagn.
G	Sp II 976 = BM 35394	CCMAwR 3, pls. 2–3 i 2′–8′	Three-col. tablet	LB	4th–3rd cent.	Babylon?	Diagn.

Editions

Biggs 1967: 17–19, 38–40, 42–43, 63, 65, 66–68 (Ms. A, B, D, E)
Hunger 1976: 26–27, No. 9 (Ms. C)
Scurlock and Andersen 2005: 257 (Ms. C)
Schwemer 2010: 115–120 (trans. of Ms. E)
Abusch and Schwemer 2011: 106, No. 2.5.4 (Ms. C, E)

Structure of the text

Text *nīš libbi* D begins with the incantations No. D.1. (Ms. B) and No. D.2 "My hunted *akkannu*-wild ass! Hunted onager!" (Ms. A, B). The diagnostic ritual (Ms. C, F, G) and two prescriptions (prescr. 1–2, Ms. C, D) follow. It continues with the incantation No. D.3 "I am clothed with (Ms. B, C, D), followed by five prescriptions (prescr. 3–7, Ms. C, B, E) and the symptom description (Ms. C, E). The text ends with the incantation No. D.4 "Have sex!" (Ms. D).

B	No. D.1
A – B	No. D.2
C – F – G	Diagn.
C	Prescr. 1
C – D	Prescr. 2
B – C – D	No. D.3
C	Prescr. 3
C – E	Sympt.
C – E	Prescr. 4
C	Prescr. 5
B	Prescr. 6–7
D	No. D.3

I. Incantation and its ritual No. D.1 (Biggs 1967 No. 24), ll. 1–9 (Ms. B obv.! 1′–9′)

II. Incantation and its ritual No. D.2 "My hunted *akkannu*-wild ass! Hunted onager!" (Biggs 1967 No. 2), ll. 10–23 (Ms. A 7–rev. 11 // B obv.! 7–rev.! 1)

III. Diagnostic ritual ll. 24–26 (Ms. C 5′–7′ // F 6–10 // G i 2′–8′)
Prescriptions
1. 27–28 (Ms. C 8′–9′)
2. 29–31 (Ms. C 10′–12′ // D ii 5?–9)

IV. Incantation and its ritual No. D.3 "I am clothed with copulation!" (Biggs 1967 No. 21), ll. 32–39 (Ms. B rev. 1–6 // C 13′–16′ // D ii 10–17)

V. Prescriptions
3. 40–41 (Ms. C 17′–18′)
Symptoms ll. 42–45 (Ms. C 19′–22′ // E i 8–14)
Prescriptions
4. 46–48 (Ms. C 23′–25′ // E i 14–17)
5. 49–50 (Ms. C 26′–27′)
6. 51–53 (Ms. B rev. 7–9)
7. 54–55 (Ms. B rev. 10–11)

VI. Incantation and its ritual No. D.4 "Have sex!" (Biggs 1967 No. 19), ll. 56–64 (Ms. D ii 18–26)

Summary of the sections of manuscripts not included in the transliteration:

- Ms. A = LKA 101
  rev. 12–19 = No. A.1 (Ms. C)

- Ms. C = SpTU 1, 9
  1′–4′ = Fragmentary lines
  28′–29′ = A Prescr. 18 (Ms. E)

- Ms. D = LKA 97
  i = Fragmentary lines

- Ms. E = STT 280
  i 1–7 = K prescr. 1–2 (Ms. A)
  i 18–21 = F prescr. 10–11 (Ms. E)
  i 22–51 = K prescr. 3–18 (Ms. A) (note i 44 = A prescr. 7 (Ms. D))
  i 52–55 = A prescr. 9–12, 16 (Ms. D)
  i 56–ii 9 = K prescr. 19–22 (Ms. A)
  ii 10–21 = No. M.1 (Ms. B)
  ii 22–35 = K prescr. 23–28 (Ms. A)
  ii 36–50 = No. K.3 (Ms. A)
  ii 51–53 = No. K.4 (Ms. A)
  ii 54–61 = No. K.5 (Ms. A)
  ii 62–iii 23 = No. E.2 (Ms. H)
  iii 24–33 = No. K.6 (Ms. A)
  iii 34–42 = No. K.7 (Ms. A)
  iii 43–iv 7 = K prescr. 29–31 (Ms. A)
  iv 8–23 = No. K.8 (Ms. A)
  iv 24–31 = No. K.9 (Ms. A)
  iv 32–36 = K prescr. 32 (Ms. A)
  iv 37–41 = No. A.1 (Ms. D)

- Ms. F = KAR 70
  1–5 = No. F.1 (Ms. A)
  11–44 = F prescr. 5–7, 10–17 (Ms. A)
  45–rev. 9 = No. F.4 (Ms. A)
  rev. 10–24 = No. E.1 (Ms. B)
  rev. 25–30 = No. F.5 (Ms. A)

rev. 31–33 = No. B.2 (Ms. B)
rev. 35 = Fragmentary colophon, see commentary ***nīš libbi*** B: 40

– Ms. G = Sp II 976 = BM 35394 (CCMAwR 3, pls. 2–3)
i 1′ = Undecipherable
i 9′–21′ = No. E.1 (Ms. D)
ii 1′–4′ = Fragmentary lines
ii 5′–22′ = No. E.2 (Ms. D)
iii 1′–4′ = Undecipherable
rev. iv = Not preserved
rev. v. 1′–5′ = Fragmentary lines
rev. v 6′–10′ = No. A.3 (Ms. H)
rev. vi 1′–3′ = Undecipherable

Transliteration

I. B obv.$^{!}$ 1′–9′ = Incantation and its ritual No. D.1

1.	B 1′	x ⸢UZU⸣ [. . .]
2.	B 2′	x *ud* x [. . .]
3.	B 3′	⸢TU$_6$⸣.TU$_6$ ⸢ABZU⸣ [. . .]
4.	B 4′	^{d}*tu-tu* ^{d}*šà-zu* ⸢^{d}*nin-gìrima* EN$^{?}$⸣ [ÉN TU$_6$ ÉN]
5.	B 5′	KA.INIM.MA ŠÀ.ZI.GA
6.	B 6′	DÙ.DÙ.BI BIL.ZA.ZA SIG$_7$ ḪÁD.A GAZ *ina si-ik-ti* ⸢*ú*⸣[. . .]
7.	B 7′	1-*niš ina* Ì+GIŠ BUR ḪE.ḪE TA *me-e ni* x [x x *ina*]
8.	B 8′	Ì+GIŠ EŠ-*su* A ^{giš}*bi-ni* 7-*šú ana* IGI-*šú* 7-*šú ana* EGI[IR-*šú*]
9.	B 9′	*i-šal-lu-ma* ŠÀ.Z[I.GA]

II. A obv. 7–rev. 11 // B obv.$^{!}$ 7′–rev.$^{!}$ 1 = Incantation and its ritual No. D.2

10.	A o. 7	ÉN *ak-kan-ni*	*ṭar-du*	*sèr-re-mu*		*ṭar*$^{!?}$-*d*[*u*$^{?}$ . . .]
	B 10′	ÉN *a-kan-ni*	*ṭa-ar-du*	*sìr-ri*$^{!}$(text ḪU)-*mu*		*ṭ*[*a*$^{?}$-*ar*$^{?}$-*du*$^{?}$ . . .]
11.	A o. 8	*man-nu* ⸢*ik*⸣-*ri-ik*!(text IN)-[*k*]*a*	*ki-i*	⸢*pi*$^{!}$(text MA)-*ì*		*di*$^{?}$-*lu*$^{?}$-*ti*$^{?}$⸣
	B 11′	*man-nu ik-ri-ik-ka*	*ki*	*pi-ì*		*d*[*i-lu-ti*]
12.	A o. 9	unreadable				
	B 12′	*ú-ram-mì-ka* ⸢*ki-i qé-e*⸣ [*ra-mu-ti*]				
13.	A r. 1	*man-nu ki-i* DUMU	*al-la*-⸢*ki*	KASKAL⸣^{meš}-*ka ip-ru-us*		
	B 13′	*man-nu ki-i* DUMU	*al-l*[*a*	KASKAL^{m}]^{eš}-[*ka*		]

14.	A r. 2	*ki-i* DUMU ^{d}*ḫum*$^{!}$(text GU)-*ba-ba qí-šá-ti-ka ú-ṣar-rip*
15.	A r. 3	*a-lik-ma ana* DUMU.MUNUS ^{d}*a-nim šá* AN-*e lil-li-ku s*[*u*$^{?}$-*pu*$^{?}$-*ka*$^{?}$]
16.	A r. 4	⸢SIG$_5$ *lìb*-$^{!?}$*bi*$^{!?}$ *lid*$^{!}$⸣-*da-a* ⟨*ana*⟩ *lìb-bi-ka* : SIG$_5$ *na+a-di lid-da*-[*a*]
17.	A r. 5	*ana na+a-di-ka*
18.	A r. 6	*a-mur en-dam-ma*$^{!}$ *la ta-kal-la*
19.	A r. 7	*ù na-al-ši* DUMU.MUNUS ^{d}*a-nim la tu-maš-šar*
20.	A r. 8	TU$_6$ ÉN
21	A r. 9	DÙ.DÙ.BI ^{na4}KA.GI.NA.DAB.BA AN.BAR SÚD
22.	A r. 10	*ina* ⸢Ì⸣+GIŠ BUR ḪE.ḪE ÉN 3-*šú ana* ŠÀ ŠID-*nu šá* NITA GÌŠ-⸢*šú*⸣
23.	A r. 11	*šá* MUNUS GAL$_4$.LA-*šá* EŠ.MEŠ-*ma*$^{!}$ ŠÀ.ZI.GA

III. C 5′–12′ // D ii 5–9 // F 6–10 // G i 2′–8′ = Diagnostic ritual and prescriptions

24.	C 5′	[–] IM KI.GAR 1-*niš* ḪE.ḪE NU NITA *u* MUNUS x D[U$^{?!}$-*uš*]
	F o. 6–7	DÙ.DÙ.BI NÍG.SILA$_{11}$.GÁ ZÍZ.AN.NA *u* IM KI.GAR 1-*niš* ḪE.ḪE NU NITA *u* MUNUS DÙ-[*u*]*š* / *ana* UGU *a-ḫa-meš* ŠUB-*di-šu-nu-ti ina* SAG.DU LÚ GAR-*an-ma*
	G o. i 2′–5′	[ ] ⸢ZÍZ⸣.AN.NA *u* IM KI.GAR 1-*niš* Ḫ[E.ḪE /] *ana* UGU ⸢*a*⸣-*ḫa-meš* {*ú*} x / ŠUB -⸢*di-šú-nu-ti*⸣ / [GAR-*an*]-⸢*ma*⸣
25.	C 6′	[ÉN 7-*šú* Š]ID-*nu tu-nak-ka-ram-ma a-n*[*a*]
	F o. 7–8	[ÉN] / 7-*šú* ŠID-*nu tu-nak-ka-ram-ma ana* ŠAḪ *tu-q*[*ar-rab-šu*]
	G o. i 5′–6′	⸢ÉN⸣ 7-*šú* ŠID-*nu tu*-⸢*nak*⸣-*kara-ma* / [*t*]*u-qar-ra*[*b-š*]*u*
26.	C 7′	[^{d}]15 *ana* GARZA BE-*ma* ŠÁḪ *la iq-te-ru*-[*ub*]
	F o. 9–10	BE-*ma* ŠAḪ *iq-te-ru-ub* ŠU ^{d}EŠDAR *ana* GAR[ZA] / ŠAḪ *la iq-ru-ub* NA BI *kiš-pu* DAB-[*su*]
	G o. i 6′–8′	BE-*ma* ŠAḪ *iq-te-ru-ub* / [ ] GARZA BE-*ma* ⸢ŠAḪ *la*⸣ *iq-te-ru-ub* / ⸢NA BI *kiš-pu* DAB⸣

27.	C 8′	[. . .] x *iš ki-lal-le-e-šú-nu* TI-*qé* [. . .]
28.	C 9′	[. . . *r*]*am*$^{?}$-*šú-nu-ti* NITA *u* MUNUS x x [. . .]

29.	C 10′	[. . .] x *kam-ma šá* ^{lú}AŠGAB SAḪAR I.D[IB$^{!}$ É$^{?}$ x]-*ti ina* Ì AZ *mut-tal-li*-[*ki*$^{?}$ . . .]
	D ii 5$^{?}$–6	$^{[ḫe-pí]}$ / x ⸢Ì⸣ AZ DU.DU $^{[ḫe-pí]}$

30.	C 11′	[. . .] A ^{giš}NU.ÚR.MA KU$_7$.KU$_7$ *ana* ŠÀ ŠUB-[*ma* TÉŠ.BI$^?$] ḪE.ḪE *ina* GÙB-*ka* TI-*qé* [. . .]
	D ii 6–7	$^{\text{[ḫe-pí]}}$ / *er*$^?$-*ši* ŠUB-*ma* $^{\text{[ḫe-pí]}}$
31.	C 12′	[LI.DUR]-*ka* ⸢GÌŠ⸣-*ka* 3-*šú* TAG ŠU-*ka* [*šá* ZAG$^?$] ŠU MUNUS *šá* GÙB T[AG-*ma* . . .]
	D ii 8–9	[L]I.DUR-*ka* ⟨*u*⟩ GÌŠ-*ka* 3-*šú* TAG.TAG-*at* / [Š]U-*ka šá up-šá-še*$^!$-*e* TAG ŠU MUNUS *šá* GÙB TAG-*ma* [. . .]

IV. B rev. 1–6 // C 13′–16′ // D ii 10–17 = Incantation and its ritual No. D.3

32.	B r. 1	ÉN *lab-šá-ku na-a-ku ḫal-pa-*[ ]
	C 13′	[*n*]*a-a-ka ḫal-pa-ku gur-ša*
	D ii 10	ÉN *lab-šá-ku na-a-ku ḫal-pa-ak gur-u*[*š*]
33.	B r. 2	*ina qí-bit pi-i te-li-tu*$_4$ ^{d}*iš-tar*
	D ii 11	*ina qí-bit pi-i te-li-te* ^{d}⸢15⸣
	C 13′	*ina qí-b*[*it pi*]-*i te-li-te* ^{d}[ ]
34.	B r. 2	^{d}[ ]
	C 13′–14′	[/]
	D ii 12	^{d}UTU ^{d}*é-a* ^{d}⸢*asal*⸣-*lú-ḫi* T[U$_6$ ÉN]
35.	B r. 3	— 3-*šú* ŠID-*nu-ma*(text GIŠ) NITA *u* MUNUS TÉŠ.BI [ ]
	C 14′	[*an-n*]*a-a* 3-*šú* ŠID-*nu-ma* NITA *u* MUNUS ŠÀ-*šú-nu*
	D ii 13	— 3-*šú* UR$_5$.GIM DU$_{11}$.GA-⸢*ma*⸣$^?$ NITA ⸢*u*⸣ MUNUS ŠÀ-*š*[*ú-nu*]
36.	B r. 4	— *i-na-áš-ši-šú-nu-ti-ma ul i-*[*nu-uḫ-ḫu*]
	C 14′	— ÍL-*šú-nu-ti ul i-nu-*[*uḫ-ḫu*]
	D ii 14	TÉŠ.BI ⟨*i*⟩-*na-ši-šú-nu-ti* [x] *i-nu-uḫ-ḫ*[*u*]
37.	B r. 5	NAM.BÚR.BI ŠÈ *ḫa-a-a-ti ina* A(text MIN) ^{ú}GAZI^{sa}[^{r} . . .]
	C 15′	[NAM.BÚR].BI ŠÈ *ḫa-a-a-at-ti ina* A GÁZI^{sar} [. . .]
	D ii 15	NAM.BÚR.BI ŠÈ$^?$ *ḫ*[*a*]-⸢*a*⸣-[*a-ti* {*q*]*í*} *ina* A GAZI^{s}[^{ar} . . .]
38.	B r. 5–6	[ ] / SAG-*ka ú-kal*
	C 16′	[SA]G-*ka ú-kal*
	D ii16	x x x *ti* ḪE.ḪE-[*m*]*a*$^?$ SAG-*ka ú-ka*[*l*]
39.	B r. 6	ZI — — — TAG-*ma* TÉŠ.BI [ ]
	C 16′	ZI — — — TAG-*ma* 1-*niš uš ta aḫ* [ ]
	D ii 17	ZI NITA *u* MUNUS TAG-*ma* TÉŠ.BI *i-nu-uḫ-ḫu*

V. C 17′–27′ // E i 8–17 // B rev. 7–11 = Presciptions 3–7 and symptoms

40.	C 17′	[DIŠ NA ŠÀ.ZI.GA]-⸢*šú*⸣ DAB-⸢*ma*⸣ *ana* MUNUS-*šú u ana* MUNUS BAR-*ti* ŠÀ-*šú là* ÍL-*šú kit-pu-lu* [. . .]
41.	C 18′	[. . .] *a* x *an* ^{ú}IN.NU.UŠ *ina* NINDA ^{še}IN.NU.ḪA GU$_7$-*ma* ŠÀ.ZI.[GA]
42.	C 19′	[DIŠ NA] *ka-šip-ma mun-ga* TUK-*ši* *bir-ka-šú* *ga-an-na* *k*[*a*]-*la-tu-šú* DU-*ak*$^?$ [ ]
	E i 8–9	[TUK]U-⸢*ši*⸣ *b*[*ir-k*]*a-šú* / [ ] ⸢DU-*ak*⸣ ŠÀ-*šú* [x] *ub tú*$^?$
43.	C 20′	[. . .] x-*ma ana e-peš* Á.ÁŠ NU ÍL-*šú ni-i*[*š* ŠÀ-*šú*] DAB-*ma ana* MUNUS DU-*k*[*a* LAL]
	E i 10–11	[. . .] NU ÌL-*šú* ÌL [Š]À-*šú* / [L A]L
44.	C 21′	[*ḫa-ši*]*ḫ-ma* MUNUS IGI-*ma* ŠÀ-*šú* GUR-*ra* NA BI [*ri-ḫu-s*]*u* KI ADD[A]
	E i 11–13	[Š]À-*šú* MUNUS *ḫa*-[*ši*]*ḫ-ma* / [GU]R-*ra* NA BI *ri*-[*ḫu-us-s*]*u* / [ ] *šu-nu-lat*
45.	C 22′	[GÌŠ]-⸢*šú*⸣ [*k*]*a-nik-ma* *ina* KI.GAR ^{d}UTU.ŠÚ. ⸢A⸣ [ ]
	E i 13–14	GÌŠ-*šú* ⸢*ka*⸣-[*nik-ma*] / [ ] ÚŠ-*ḫi*
46.	C 23′	⸢*ana*⸣ [BÚR ^{ú}]*tar-muš* ^{ú}ḪAR.ḪAR — ^{ú}*ak-tam* ^{ú}EME.UR.⸢GI$_7$⸣ *túb*-[*bal* TÉŠ.BI SÚD]
	E i 14–16	*ana* BÚR ^{ú}[/ Ú].KUR.KUR ^{ú}*ak-tam* ^{ú}[SÚ]D
47.	C 24′	[*šum-ma ana*] ^{giš}[GEŠ]TIN.ŠUR.RA *šum-ma ana* KAŠ ŠUB-*ma* ÉN SU.[ZI SU.ZI ŠÀ.ZI.GA]
	E i 16–17	*lu* *ina* KAŠ.SAG *lu* *ina* GEŠTIN.ŠUR ŠU[B-*ma*] / [ ]
48.	C 25′	[ŠÀ.Z]I.GA GUR – *ana* ŠÀ ŠID-*nu-ma* NAG.MEŠ-*ma* [ŠA.ZI.GA]
	E i 17	– – 3-*šú ana* [*lìb-b*]*i* ŠID-*nu*-⸢*ma*⸣ NA[G.MEŠ-*ma*]
49.	C 26′	[DIŠ KI.MIN] ^{ú}EME.UR.GI$_7$ GIM ^{d}UTU.ŠÚ.A *tu-qad-daš ina še-rì ana* IGI ^{d}UTU G[UB$^?$-*az*$^?$ . . .]
50.	C 27′	[. . .] x x *ina* GÍR ZABAR TA *šur-ši-šú* ZI-*aḫ-šú* SÚD *ina* KAŠ N[AG.MEŠ-*ma* ŠA.ZI.GA]

51.	B r. 7.	*ana* NITA ZI-*ta*$_5$ *šur-ši-i* MUŠ.DÍM.GURUN.[NA]
52.	B r. 8.	*rit-ku-ba-ti ina* ⸢IGI⸣ [. . .]
53.	B r. 9.	*ina* SAG.DU-*šú* GAR-*m*[*a* ŠÀ.ZI.GA]
54.	B r. 10	*ana* MUNUS ⸢ZI-*ta*$_5$⸣ [*šur-ši-i* . . .]
55.	B r. 11.	x x [. . .]

Rest destroyed

VI. D ii 18–26 = Incantation and its ritual No. D.4

56.	D ii 18	ÉN *gu-ru-uš ka-na-a sar e ta-aʾ-dir*
57.	D ii 19	*ti-ba-a e ta-šu-uš*
58.	D ii 20	*ina qí-bit* ^{d}15 ^{d}UTU ^{d}*é-a u* ^{d}*asal-lú-ḫi*
59.	D ii 21	ÉN *ul ya-ut-tu-un* ÉN ^{d}60 *u* ^{d}*asal-lú-ḫi*
60.	D ii 22	ÉN ^{d}15 *b*[*e-l*]*et ra-a-mi* TU$_6$ ÉN
61.	D ii 23	[DÙ.DÙ].BI [SÍ]K$^{?}$ MÁŠ.[NÍTA] ZI-*i* NÍG.TUR *šá* GÌŠ-*šú*
62.	D ii 24	[. . . SÍG$^{?}$ UD]U.NÍTA ZI-*i* SÍG.ḪÉ.ME.DA x [. . .]
63.	D ii 25	[*ina* MÚRU-*šú* KEŠDA A] ⸢KÙ$^{?}$⸣ DUB-*ak* ÉN 7-[*šú* ŠID-*nu*]
64.	D ii 26	traces

Rest destroyed

Transcription

I. B 1′–9′ = Incantation and its ritual No. D.1

1. . . . *šīru* [. . .]
2. . . . [. . .]
3. *šipāt Apsî* [. . .]
4. *Tutu Šazu Ningirima bēlet*? [*šipti tê šipti*]

---

5. *šipat nīš libbi*

---

6. *dudubû*: *muṣa''irāna arqa tabbal taḫaššal ina sīkti* [. . .]
7. *ištēniš ina šaman pūri taballal ištēniš išti mê* . . . [. . . *ina*]
8. *šamni tapaššassu mê bīni sebîšu ana maḫrīšu sebîšu ana ar*[*kīšu*]

9. *išallu-ma n*[*īš*] *libbi*

II. A 7–rev. 11 // B obv.! 7–rev.! 1 = Incantation and its ritual No. D.2

10. *šiptu*: *akkannī ṭardu serrēmu ṭard*[*u* . . .]
11. *mannu ikrikka kī pî dilûti*
12. *urammīka kī qê* [*ramûti*]
13. *mannu kī mār allāki ḫarrānātīka iprus*
14. *kī mār Ḫumbaba qišātīka uṣarrip*
15. *alik-ma ana mārāt Ani ša šamê lillikū s*[*uppûka*]
16. *dumuq libbi liddâ ana libbīka* : *dumuq nādi lidd*[*â*]
17. *ana nādīka*
18. *amur endam-ma lā takalla*
19. *u nalši mārāt Ani lā tumaššar*
20. *tê šipti*

---

21. *dudubû*: *šadâna ṣābita parzilla tasâk*
22. *ina šaman pūri taballal šipta šalāšīšu ana libbi tamannu ša zikari ušaršu*

23. *ša sinništi bișșūrša taptanaššaš-ma nīš libbi*

III. C 5′–12′ // D ii 5–9 // F 6–10 // G i 2′–8′ = Diagnostic ritual and prescriptions

Diagn. 24. *dudubû*: *līš kunāši u ṭīd kullati ištēniš taballal ṣalmū zikari u sinništi tepp*[*u*]*š ana muḫḫi aḫāmeš tanaddīšunūti ina qaqqad amēli tašakkan-ma*

25. [*šipta*] *sebîšu tamannu tunakkaram-ma ana šaḫî tuqarra*[*bš*]*u*

Translation

I. B 1′–9′ = Incantation and its ritual No. D.1

1. . . . flesh [. . .]
2. . . . [. . .]
3. Incantations of the Apsû [. . .]
4. Tutu, Šazu (and) Ningirima, mistress? [of incantation. Incantation formula].
5. Wording of *nīš libbi* (incantation).
6. Its ritual: You dry and crush a green frog, in the powder of the [. . .]-plant
7. you mix (it) together with oil from the alabastron, together with water . . . [. . . with]
8. oil you anoint (the patient). Tamarisk water seven times in front of him, seven times be[hind him]
9. he smears(?) (it) and (he will get) sexual de[sire].

II. A 7–rev. 11 // B obv.! 7–rev.! 1 = Incantation and its ritual No. D.2

10. Incantation: My hunted *akkannu*-wild ass! Hun[ted] onager [. . .]
11. who has blocked you like an opening of the *dilûtu*-water system
12. (and) who has slackened you like [slackened] strings?
13. Who has blocked your ways like (those of) a traveller
14. (and) has burned your forests like (those of) of Ḫumbaba?
15. Go and may [your] sup[plications] proceed to the heavenly Daughters of Anu!
16–17. May they grant the well-being of the 'heart' to your 'heart'!
May they gran[t] the well-being of the waterskin / to your waterskin!
18. Look, come close, and do not hold back (= do not tarry),
19. and do not release the dew of the Daughters of Anu!
20. Incantation formula.
21. Its ritual: You pulverize magnetite (and) iron,
22. you mix (them) with oil from the alabastron; you recite the incantation three times over it; the man's penis
23. (and) the woman's vulva, you? anoint them and (he will get) sexual desire.

III. C 5′–12′ // D ii 5–9 // F 6–10 // G i 2′–8′ = Diagnostic ritual and prescriptions

Diagn. 24. Its ritual: You mix together emmer dough and potter's clay; you make the figurines of the man and the woman; you put them one upon the other, and place them at the man's head,
25. you recite [the incantation] seven times; you remove (them) and put [them near] a pig.

26. *šumma šaḫû iqterub qāt Ištar ana parṣi šumma šaḫû lā iqterub* (var.: *iqrub*) *amēla šā kišpu iṣbas*[*su*]

---

1. 27. [. . .] . . . *kilallēšunu teleqqe* [. . .]
28. [. . . *r*]*āmšunūti zikaru u sinništu* . . . [. . .]

---

2. 29. [. . .] . . . *kamma ša aškāpi eper askup*[*pi bīti* . . .] . . . *šaman asi muttalli*[*ki* . . .]

30. [. . .] *mê nurmî matqi ana libbi tanaddī-*[*ma ištēniš*] *taballal ina šumēlīka teleqqe* . . .]
31. [. . . *ab*]*unnatka ušarka šalāšīšu talappat* (var.: *tulappat*) *qātka* [*ša imitti*[?]] (var.: *ša upšaššê ilputu*) *qāt sinništi ša šumēli talappat* [. . .]

IV. B rev. 1–6 // C 13′–16′ // D ii 10–17 = Incantation and its ritual No. D.3

32. *šiptu: labšāku nâka ḫalpāk gurša*
33. *ina qibīt pî telīti Ištar*
34. *Šamaš Ea Asalluḫi* [*tê šipti*]

---

35. *annâ šalāšīšu tamannū-ma* (var.: *kīam iqabbi*) *zikaru u sinništu libbašunu*

36. *ištēniš inaššīšunūti ul inuḫḫ*[*ū*]
37. *namburbû ana ḫa'atti: ina mê ka*[*sî* . . .]
38. . . . *taballal-*[*m*]*a rēška ukāl*
39. *napišti* (*zikari u sinništi*) *talappat-ma ištēniš inuḫḫū*

V. C 17′–27′ // E i 8–17 and B rev. 7–11 = Presciptions 3–7 and Symtomatology

3 40. [*šumma amēlu nīš libbī*]*šu ṣabit-ma ana sinništīšu u ana sinništi aḫīti libbašu lā inaššīšu kitpulu* [. . .]
41. [. . .] . . . *maštakal ina akal ennēni ikkal-ma nī*[*š*] *libbi*

Sympt. 42. [*šumma amēlu*] *kašip-ma munga īši birkāšu gannā k*[*a*]*lâtūšu illaka libbašu* [. . .] . . .
43. [. . .] . . . *ana epēš ṣibûti lā inaššīšu nīš* [*libbī*]*šu ṣabit-ma ana sinništi al*[*āka muṭ*]*ṭu*
44. [*lib*]*bašu sinništa ḫašiḫ-ma sinništa īmur-ma libbašu itūra amēlu šū ri*[*ḫûss*]*u itti šalamt*[*i*] *šunullat*
45. *ušaršu kanik-ma ina kullat erēb Šamši peḫi*

---

46. [*ana pašāri*] *tarmuš ḫašâ atā'iša aktam lišān-kalbi tub*[*bal ištēniš tasâk*]

26. If the pig approaches (them), (it means) Hand of Ištar; for the ritual procedures?, if the pig does not approach (them), (it means that) the witchcraft seized the man.

1. 27. [. . .] . . . you take both [. . .]
28. [. . . them]self, the man and the woman. . . [. . .].

2. 29. [. . .] . . . fungus of tanner, dust of the threshold [slab of the house . . .] . . . with fat of a danc[ing?] bear [. . .],
30. [. . .] you pour pomegranate sweet juice over it [and] you mix (them) [together], you take (it) at your left hand [. . .]
31. [. . .] you massage three times your [n]avel (and) your penis, you massage your [right?] hand (var.: which she (= witch) touched with sorceries) (and) the left one of the woman [. . .].

IV. B rev. 1–6 // C 13′–16′ // D ii 10–17 = Incantation and its ritual No. D.3

32. Incantation: I am clothed with copulation! I am enveloped with interco[urse]!
33. At the command of wise Ištar,
34. Šamaš, Ea, Asalluḫi. [Incantation formula].
35. You recite (var.: you say) this (incantation) three times and (if) the man and the woman
36. desire together each other, (but) they do not find reli[ef]:
37. *namburbû*-ritual against panic attacks: in musta[rd] water [. . .]
38. you mix . . . [and] keep (the medical preparation) ready,
39. you apply (it) into the throat (of the man and woman) and they will find relief.

V. C 17′–27′ // E i 8–17 and B rev. 7–11 = Presciptions 3–7 and Symtomatology

3 40. [If a man's sexual desire] is taken away and he does not desire either his woman or another woman . . . [. . .]
41. [. . .] . . . he eats *maštakal*-plant in *ennēnu*-barley bread and (he will get) sexual de[si]re.

Sympt. 42. [If a man] is bewitched and has the *mungu*-paralysis, his knees are contracted, his kid[ne]ys "go," his 'heart' [. . .] . . .
43. [. . .] . . . and he does not have interest to achieve (his) desire, [his] sexual [desire] has been taken and (his desire) to g[o] to a woman is [redu]ced,
44. his '[hea]rt' needs a woman and finds her, but his 'heart' returns: that man's s[perm] has been buried with a dead per[son],
45. his penis has been sealed and shut up in a clay pit towards sunset.
46. [In order to release him]: You dr[y (and) crush together] *tarmuš*-plant, *ḫašû*-plant, *atā'išu*-plant, *aktam*-plant, "dog's-tongue"-plant.

4 47. [*šumma ana*] *karāni ṣaḫti šumma ana šikari tanaddī-ma šipta šalum*[*matu šalummatu nīš libbi*]

48. [*nī*]*š* [*libbi*] *itūr šalāšīšu ana libbi tamannū-ma ištanattī-ma* [*nīš libbi*]

5 49. [*šumma* KI.MIN (= Sympt.: 42–45)]: *lišān-kalbi kīma erēb Šamši tuqaddaš ina šēri ana maḫar Šamši iz*[*zâz*? . . .]

50. [. . .] . . . *ina paṭar siparri ištu šuršīšu tanassaḫšu tasâk ina šikari išta*[*nattī-ma nīš libbi*]

6 51. *ana zikari tibûta šurši pizallur*[*tī*]

52. *ritkubāti ina maḫar* [. . .]

53. *ina qaqqadīšu tašakkan-m*[*a nīš libbi*]

7 54. *ana sinništi tibûta* [*šurši* . . .]

55. . . . [. . .]

VI. D ii 18–26 = Incantation and its ritual No. D.4

56. *šiptu*: *guruš* . . . *ē ta'dir*

57. *tibâ ē tāšuš*

58. *ina qibīt Ištar Šamaš Ea u Asalluḫi*

59. *šiptu ul yattun šipat Ea u Asalluḫi*

60. *šipat Ištar b*[*ēl*]*et râmi tê šipti*

61. [*dudu*]*bû*: [*šāra*]*t daš*[*ši*] *tebî mimma ṣeḫerta ša ušarīšu*

62. [. . . *šīpāt im*]*meri tebî nabāsa* . . . [. . .]

63. [*ina qablītīšu tarakkas mê*] *ellēti tašappak šipta sebî*[*šu tamannu*]

64. Traces

4 47. You throw (it) [either in] grape juice or in beer and the incantation "Shi[ver! Shiver! Sexual desire!]

48. [Sex]ual desire returns!" you recite over it three times and he drinks (it) repeatedly and [(he will get) sexual desire].

---

5 49. [If ditto (= Sympt.: 42–45)]: You purify "dog's-tongue"-plant when the sun sets; he st[ands] in the morning before the sun . . . [. . .]

50. [. . .] you remove from its root . . . with a bronze knife; you pound (it); in beer he dri[nks (it) repeatedly and (he will get) sexual desire].

---

6 51. To make a man have an 'elevation': Gecko[s]

52. which copulate in front of [. . .].

53. you put (them) on his head and [(he will get) sexual desire].

---

7 54. [To make] a woman have an 'elevation':

55. . . . [. . .]

VI. D ii 18–26 = Incantation and its ritual No. D.4

56. Incantation: Have sex! . . . May you not be afraid!

57. Rise! May you do not be afflicted!

58. At the command of Ištar, Šamaš, Ea and Asalluḫi;

59. the incantation is not mine; it is the incantation of Ea and Asalluḫi,

60. the incantation of Ištar, [patro]n of love. Incantation formula.

---

61. Its [ritual]: [Hair] of a reared-up buck, the "something little" of its penis,

62. [. . . wool of a] reared-up [r]am, red wool, . . . [. . .]

63. [you bind (them) around his waist], you libate pure [water], [you recite] the incantation seven [times].

64. Traces

Commentary

1–4. According to Biggs (1967) this is the final part of a Sumerian incantation. In my opinion, it is a fragmentary Akkadian incantation. Note that Ningirima's epithet is often written logographically EN ÉN (see e.g., KAL 2, 7 ii 8′, Abusch et al. 2020: 191, No. 8.49: $17^{\text{vii } 44}$ Ms. E).

4. On the god Tutu see e.g. TRichter 2014–2016. He is identified with Marduk (Sommerfeld 1982: 37) or Nabû. Šazu is a name for Marduk, see Krebernik 2009–2011.

6. *muṣa''irānu arqu* "green frog": For the use of the frog in therapeutic texts see Bácskay 2018. See F prescr. 8: 26 where the tadpole (*atmu*) of a green frog is an ingredient for the realization of an ointment. Note that it is a 'secret name' for the *kukru*-plant and for the drug *baqqu*-gnat (URU.AN.NA: *maštakal* III 43–43a, Rumor 2017: 10). In a second millennium love incantation from Kiš a kind of dark frog called *ḫuduššu* refers to the female sex organs (PRAK 1 B 472 i 12′, Wasserman 2016: 151).

11. *dilûtu* is a water-system tool of a well or a canal. AHw. I 170 translates 'Schöpfarbeit, -vorrichtung,' CAD D 142, B 'hoisting device for drawing water from a well.' See Læssøe 1953: 14; van Laere 1980: 36. Biggs sees it as a kind of small canal (1967: 18). The verb *karāku* 'to block' is typical of contexts related to irrigation systems (CAD K: 199; AHw. I 446).

12. For the restoration *ramûti* 'slackened' see No. A.3: 50; No. L.2: 20. Another possible restoration: *šaddūti* 'taut,' see No. A.4: 76.

16–17. *nādu* 'waterskin' possibly is a metaphor to indicate the penis, on the model of the previous lines that mention the *dilûtu*-water system. Another word for waterskin is *indūru* which appears in following love context: "Waterskin (and) food ration *give* [*me*. . .]" (KAR 158 ii 53′, Wasserman 2016: 208).

24–26. The first four lines of Ms. C are poorly preserved: 1′. [. . .] x [. . .] / 2′. [. . .] *ti ti* [. . .] / 3′. [. . .] *lit* [. . .] / 4′. [. . .] x É$^{?}$ [. . .].

29. Ms. D (ii 3–5) has a poorly preserved prescription before: ^{ú}[. . .]/ 10 Ú$^{!}$.ḪI$^{!}$.A$^{!}$ ŠÀ.Z[I.GA] / *ina* [S]ÍK ⸢UDU$^{!}$.NITÁ$^{!}$⸣ ZI-*i* $^{[\text{\textit{ḫe-pí}}]}$.
kam-ma: It is maybe a king of fungus. When it is followed by the word *aškāpu* it indicates a particular type called "of the tanner" used for therapeutic purposes for rectal problems (BAM 1 iii 3), as a suppository (BAM 104: 16), for eyes (BAM 165 ii 12′) for the ears (AMT 86, 2: 11). AHw. I 433 *s.v. kammu* II 'Gerbstoffe' (tannery); According to CAD K 125 a type of fungus; Landsberger, pointing out that the determinative used is ú for plants and that both *kibšu* and *šuḫtu* are synonymous with *kammu* (see URU.AN.NA: *maštakal* II 364–365),

translates it as "Schimmelpilz" (mold) (Landsberger 1967b: 172 fn. 135; AHw. I 472 *s.v. kišbu*).
mut-tal-li-[ki]: Hunger's (1976: 27) idea of the restoration in Ms. C based on DU.DU in Ms. D.
SAḪAR I.D[IB! . . .]-*ti ina* Ì AZ *mut-tal-li-[ki]*: for the reading see Farber 1979: 301. For *šaman asi muttalliki* "fat of a dancing? bear" see AMT 101, 3: 3 (see CAD M/II 306 *s.v. muttalliku* mng. 2a). Hunger (1976: 27) reads Ì AZ as NI.AS translating the line: "[. . .] . . . des Färbers . . . [. . .] in einem beweg[lichen? . . .]."

30. See also the mention of the fruit *labbānu* 'sweet pomegranate' in the catalogue LKA 94 iii 4. The pomegranate sweet juice is used in several prescriptions of this corpus for the realization of potions. See the love incantation in order to make a woman come and make love to her: "The beautiful woman has brought forth love. / Inana, who loves apples and pomegranates, / has brought forth sexual desire," and its ritual: "Either to an apple or to a pomegranate / you recite the incantation three times. You give (the fruit) to the woman (and) have her suck the juices. / That woman will come to you: You can make love to her" (KAR 61: 1–3, 8–10, Biggs 1967: 70). See also KAR 69: 4–5 (Biggs 1967: 74).

31. Ms. D ii 9 has the D-stem of the verb *lapātu* (TAG.TAG) (*contra* Ms. C 12′ G-stem). The D-Stem of the verb is also used in therapeutic texts to indicate 'to rub, massage' parts of the body (CAD L 91–92 mng 4 i).
In Ms. D the reading *up-šá-še*!*-e* TAG is suggested by Farber (1979: 301), who translates "(mit deiner Hand), die die Zaubereien 'berührt' hat (berührst du die linke Hand der Frau)." Note, however, that in Ms. C the space is too small to restore *šá up-šá-še-e* TAG. For this reason, I restore *šá* ZAG "right (hand)," since the previous line mentions the left one. It seems that it is the patient who performs the ritual (his penis is mentioned), although with the second singular person one usually addresses the therapeutic operator.

32. See the catalogue LKA 94 i 12. Note that No. D.3 features two speakers: the female partner (l. 32) and the therapeutic operator (ll. 33–34).

36. The verb *nâḫu* is translated by Biggs (1967: 40) with "to find satisfaction": "You recite thus three times; (If) the man's and the woman's hearts both want, (but) they (still) [cannot] find satisfaction." Biggs's idea is that, although the man and woman desire each other, they are still incapable of having sexual intercourse. Hunger (1976: 27), on the other hand, translates the line 14 of Ms. B: "[di]es rezitirst du dreimal, und Mann und Frau werden in sexuelle Erregung geraten und nicht müde werden." In the following sentence – NITA *u* MUNUS ŠÀ-*šú-nu* ÍL-*šú-nu-ti ul i-nu-[uḫ-ḫu]* – the subject of the two verbs is the 'heart' (*libbu*) of woman and man. When *libbu* is the subject of verb *nâḫu* it is possible to translate with "to find relief" in reference to states of being

(see CAD N/I 143–150; AHw. II 716–717). See discussion on the verb *nâḫu* in the commentary *nīš libbi* B: 10

37. ŠÈ *ḫa-a-a-ti*: Biggs (1967: 40) and Hunger (1976: 27) read ZÍD *ḫa a a ti* ". . . flour$^{?}$." The latter word could be *ḫa'attu*/*ḫa(y)yattu* (CAD Ḫ 1 'panic'; AHw. I 309 'krankhafte Schreckhaftigkeit'). It indicates a state of fear considered pathological (on the fear as a symptom of loss of sexual desire see Chapter II § "Fear, distress, and insomnia"). Previous scholars have considered the term as a qualification of ZÍD, that is, an unknown kind of flour. Their idea is that this kind of flour should be put into mustard water. Another problem is that they restore in all three texts the verb ŠUB. My proposal is as follows: instead of reading ZÍD, we could read ŠÈ/*ana* 'for, against' (although ŠÈ for *ana* is mainly used in MB texts) and delete the verbal restoration.

38. For the *rēška ukāl* formula see CAD K 517 *s.v. kullu* mng. 5 f c′.

39. See incantation and its ritual No. B.1: 17: *abbunat sinništi talappat zikaru u sinništu ištēniš* [*inuḫḫū*] "You apply (it) to the woman's navel; the man and the woman [will find relief] together."
ZI: According to Biggs (1967) it refers to the sexual organs. On the contrary, it indicates the throat (*napištu*).

40. *kitpulu*: CAD K 466–467 'entwined'; AHw. I 494 'umeinander gewickelt,' said of snakes, birds, lizards in divinatory literature such as *Šumma ālu*.

44. See F prescr. 5 for similar a symptom description.

47. For the incantation "SU.ZI MIN" see the catalogue LKA 94 i 19, 26; A prescr. 19: 32; K prescr. 28: 77–79. See also *kunuk ḫalti* series ("*ḫaltu*-seal") K. 3010+ v 25′ and 28′ (Schuster-Brandis 2008: 365, text 16 A).

48. GUR: it could refer to the return (Akk. *târu*) of the sexual desire.

49. *tu-qad-daš*: The verb *quddušu* indicates a ritual purification (CAD Q: 46–47; AHw. II 891). See incantation No. L.2: 22 *kukrū quddušūtu* "the holy *kukru*-plants."
G[UB$^{?}$-*az*$^{?}$] suggested by Abusch and Schwemer 2011.

51. *Aššur Medical Catalogue* section XIX devoted to the loss of sexual desire: [EN$^{?}$ . . . *ana* NI]TA ZI-*tú* *š*[*ur-ši*]-⌈*i*⌉ "[Including (prescriptions) . . .] to make a man have an 'elevation'" (l. 102, Steinert 2018b: 217).

51–52: The copulating geckos are ingredients also in the following prescriptions: K prescr. 31: 136 = N prescr. 8 ii 4; N prescr. 17 iii 32.

54. See the explicit reference to woman's sexual desire in a passage of the *Aššur Medical Catalogue*, in the section XIX devoted to the loss of sexual desire: [KA.

INIM.MA ŠÀ.ZI.G]A *ù* MUNUS.GIN.NA.KÁM ŠÀ.ZI.GA.MUNUS.A.KÁM "[Wording of the (incantation) for (male) sexual desi]re and (those) to make a woman come (and for) woman's sexual desire" (l. 106, Steinert 2018b: 217). The expression ŠÀ.ZI.GA MUNUS.A.KÁM concerns with arousing female sexual desire.

56. See the catalogue LKA 94 i 20: [ÉN] *gu-ru-u*[*š* MIN] ^{giš}DÌḪ MIN "[Incantation:] Have se[x! Have sex!] *Baltu*-thorn! *Baltu*-thorn!"; and also ibid. ii 26 ÉN *gur*-[*uš* . . .] "Incantation: Have se[x! . . .]." For the imperative of the verb *garāšu* see No. J.3: 7′.
ka-na-a sar is unclear.

57. *tibâ* "rise": This has been translated in other incantations as "rear up," in reference to animal metaphors. Animals here, however, are absent, so it is clear that the sexual reference of the verb alludes to sexual excitement.

61. NÍG.TUR/*ṣeḫertu* "'something little' (of the penis)" is not clear. Biggs (1967: 39) thinks that it could be "pre-coital seminal secretions"; Nougayrol (1962: 94) hypotheses a euphemism for the foreskin.

Incantations: stylistic and functional analysis

II. A 7–rev. 11 // B obv.! 7–rev.! 1 = Incantation and its ritual No. D.2

The text contains an incantation (ll. 10–20) and a prescription (ll. 21–23). The latter provides instructions to produce an ointment, consisting of magnetite and iron powder together with oil from the alabastron, on which the incantation is recited three times and with which the man's penis and the woman's vulva are to be anointed (it is not clear who anoints, whether the exorcist or the patients themselves).[522]

The incantation, recited by the woman (l. 10 "*my* hunted *akkannu*-wild ass") should be divided into two parts, each consisting of five lines:

- The first section (ll. 10–14) identifies the patient with a wild animal (*akkannu*-wild ass and *serrēmu*-onager), qualified as prey (*ṭardu* 'hunted'), and through the usual rhetorical figures about the anonymity of the malevolent performer the patient's symptoms are described by making use of similes;
- The second section (ll. 15–19) consists of an invocation of the Daughters of Anu and recommendations for the exorcist or the patient (use of imperatives).

522 It should be noted that the use of magnetite and iron implies that in addition to the man, the woman must be a social performer of therapeutic practice. This is because, as seen in Chapter III § "Ointments and amulets," magnetite and iron are used for their ability of attraction.

The first section is composed of an initial line in which the patient is identified by metonymy with two wild animals and by two pairs of lines introduced by the interrogative pronoun "who" (*mannu*). The two couples are in antithetical syntactic order, with the verbs[523] and the similes with the antipodes and the direct object (in the first pair the personal pronoun) in the middle:

Who *mannu*	has blocked *ikrik*	you *-ka*	like an opening of the *dilûtu*-water system *kī pî dilute*
	has slackened *urammi*	you *-ka*	like slackened strings *kī qê ramûti*

Who *mannu*	like (those of) a traveller *kī mār allāki*	your ways *ḫarrānātīka*	has blocked *iprus*
	like (those of) Ḫumbaba *kī mār Ḫumbaba*	your forests *qišātīka*	has burned *uṣarrip*

The second section is also divisible into two parts of two lines each (excluding the variant line). Each section is introduced by an imperative, probably addressed to the patient: "go" (l. 15, *alik*) and "look" (l. 18, *amur*). The first imperative is followed by two precative phrases, the second by other mandatory forms, positive and negative. Incantation makes use of imperatives, although it is impossible to say to whom they are addressed, whether to the therapeutic operator or the patient himself (see Biggs 1967: 19): "go" (l. 15, *alik*); "look, approach, and do not go back" (l. 18, *amur endam-ma lā takallâ*). This is clearly an exhortation of positive character. A similar formulation can be found in the inscriptions of Esarḫaddon (Borger 1956: 43 and note, § 27, l. 61): *alik lā kalâta* "go without going back." In this text, the king turning to the gods, receives from the latter, as an oracle, the call to action: they will be at his side and his enemies will die.[524]

The heavenly Daughters of the god Anu are invoked[525] (*mārāt Ani ša šamê*). Unfortunately, no mythological context clarifies the nature of these entities (see Farber

523 It should be noted that the verbs of the first line of the first pair, and of the second line of the second pair are in D-stem, whereas those of the two neighboring lines (the second of the first pair and the first of the second pair) stand in G-stem.

524 A parallel is given by a therapeutic text whose purpose is to eradicate fever. In this case the expression, which acquires a negative tone, refers to fever, which is forced to leave the body of man: *ṭardāta tattalak lā takallâ* "You are driven out, go, do not come back!" (KUB 29, 58 ii 32–33, also l. 14, Meier 1939: 202–204). In general, the verb *kalû* with negation and placed in relation to other verbs, often in the imperative, means 'to do immediately something' (see CAD K 102, mng. 5b).

525 The expression is translated "heavenly Daughters of Anu" instead of "Daughters of Anu-of-the-sky" in order not to confuse these benign entities with the dangerous demon Lamaštu, also called "daughter of Anu" (see Biggs 1967: 18; Farber 1990: 301 fn. 12). Actually, in the text we find the Sumerogram DUMU.MUNUS in singular, but as noted by Biggs, it is not uncommon to refer to this group

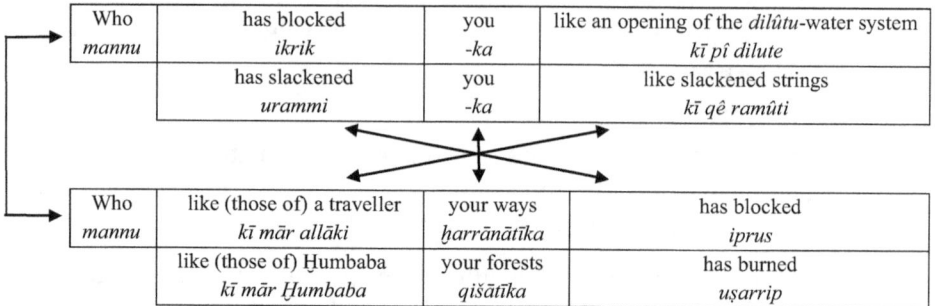

1990: 302), whose number is often seven (maybe seven plus seven)[526] or two/three (see Goetze 1955: 8; Landsberger and Jacobsen 1955: 15–16; Farber 1990: 306). AMT 10, 1 iii 18–19 states: "Two are the Daughters of Anu, between whom a mud wall is built / one sister has never seen the other" (Landsberger and Jacobsen 1955: 16).[527] They are invoked in different incantations against various sicknesses and malaises, from the Old Babylonian period up to the compositions of the first millennium.[528] In particular, they are mentioned in a formula used in certain incantations (from the Old Babylonian period): *mannan lušpur ana mārāt Anim ša šamê* "Whom should I send to the heavenly Daughters of Anu" (see Farber 1990).

Amongst the various diagnostic categories against which they are evoked, there is witchcraft:

> Incantation. Two they are, the heavenly Daughters of Anu,
> three are they, the heavenly Daughters of Anu.
> Holding the rope, they descend to me from heaven.
> (I ask them): "For what have you arisen, whither do you go?"
> "We have come to search the sorcerer and the witch of NN, son of NN;
> to take their sticks,
> to collect their waste,
> to illuminate the brazier at night have we come."
> [*Maqlû* III 31–38, Abusch 2016: 85–87]

Similarly, they can be the authors of witchcraft:

> Incantation. I am the pure River and Sacred Light (*namru*)
> my sorcerers are the Sages (*apkallū*) of the Apsû,
> my sorceresses are the heavenly Daughters of Anu.[529]
> [*Maqlû* III 61–63, Abusch 2016: 90]

---

of benevolent entities with the singular (see Biggs 1967: 18). The expression is also used as an epithet for other supernatural entities, such as the Lamaštu, but also gods like Ištar (KAR 144: 16) or Nin-Karrak (KAR 16: 4).

526 For the number seven in reference to divine entities (Sebettu) see e.g. Verderame 2017b; 2017c.

527 The same formula can be found in lines 25–28, referring to the eyes, described as sisters, separated by a mountain and a wall of mud (see Landsberger 1958: 57–58).

528 For the sources see Goetze 1955; Landsberger and Jacobsen 1955; Biggs 1967: 18–19; Farber 1990; CAD M/I 334 (sse also Botterò 1987–1990: 229). See also STT 73: 61–62 (ibid. 71–72, Reiner 1960: 33): "Incantation 'Constellation of the Cart'. [The C]art of heaven, its yoke is Ninurta / the rod (of the yoke) is Mar[duk], its axis are the pure heavenly Daughters of Anu."

529 See also ibid. VI 86–87. Abusch (2002: 202–203) argues that the witches, the heavenly Daughters of Anu, constitute a threat to *namru* (the sun-god Šamaš), since they seem to represent clouds who can cover over the sun, even prevent it from rising in the morning in the eastern mountains (see ibid. 209–210).

A text informs us about the origin of these entities, who were created by Anu in primordial time alongside Heaven and Earth and the oath:

> [Incantation. We] are the heavenly Daughters of Anu,
> creation of Ayabba, the extended Sea.
> Anu our father created us and
> Heaven and Earth were created together with us,
> [and] the oath was created with us.
> [King 1975: 126, No. 61: 5–9, dupl. LKA 153: 6–9][530]

Their creation is enacted in primordial times by Anu, in the Ayabba, the vast sea.[531] The waters of this ocean are used, according to different incantations, by the heavenly Daughters of Anu, often along with those of the Euphrates and Tigris,[532] and sprayed during treatment against sicknesses:

> May I send someone to the Daughters of Anu.
> May they take for me their pitchers of [*ḫulālu*]-stone (and)
> their mugs of lapis laz[uli].
> May they take the ocean waters (*ayabba*), the vast sea, (and)
> water from the Tigris and Euphra[tes].
> May they spray it and eliminate the *sikkatu*, *miqtu*, and *a*[*šû*]-sicknesses.
> [BAM 543 iv 30′–35′, dupl. STT 136 iii 37–42, Farber 1990: 317][533]

The connection between these figures and the ocean waters becomes clear. They are created in the Ayabba and use its water for therapeutic purposes. In incantation No. D.2: 19, however, there is no mention of ocean waters, but of dew: "Do not release the dew of Anu's daughters." The term *nalšu* describes the dew (see CAD N/I 202–203 'dew'; AHw. II 724–725 '(Nacht-)Tau'; Landsberger 1934: 160–161; Biggs 1967: 19). It does not refer to a specific type of rain, although it is connected to it (see Landsberger 1934: 160; Biggs 1967: 19).[534] In fact, it is used with the verb *zanānu* 'to rain': "[After] the sickness has rained (*iznunu*) like dew" (KAR 375 rev. iv 25–26, CAD N/I 202);[535] alternatively the verb used is *nalāšu* 'to wet with dew,' but also in the sense of 'to rain':[536] "(The demon) rains on earth [like de]w" (CT 17, 27: 3–4, Landsberger 1934: 161; CAD N/1 199).[537] The fact that the dew comes from heaven is confirmed by other

530 See Landsberger 1958: 57 fn. 6.
531 On Ayabba see Horowitz 1998: 303–304.
532 Other sources mention the pure water of the Sea (*mê tâmtim ellūtim*) (see Farber 1990).
533 See also Msk 731030, Arnaud 1987: 345, l. 11; CT 23, 2: 6–7 and dupl. AMT 31, 2 : 7, Farber 1990: 311; BAM 510 iii 1–2, Farber 1990: 312; Hunger 1976, No. 44: 72.
534 See the following love composition: "[The Daugh]ters of Anu the sky/[day]? Lights have purified, / the sk]y of Anu" (CUSAS 10, 11: 2–4, Wasserman 2016: 236).
535 See also CT 17, 33: 36–37, dupl. STT 179 r. 53–54.
536 See CT 18, 24 K.4219 r. ii 3 (dupl. LTBA 2 2: 310): *na-la-šu* = *za-na-nu* (CAD N/I: 199)
537 See LKA 70 ii 24, Ebeling 1931: 52.

texts, which would also explain its relationship with the Daughters of Anu. It is the god Adad[538] who governs the domain of the atmospheric beings, as well as the dew:

> May he (= Adad) make the rain fall at day,
> may he at night furtively
> make the dew rain down,
> so that the field can, furtively, support the grain.
> [*Atraḫasis* II ii 16–19, Lambert and Millard 1969: 74]

In another text, a prayer to Marduk,[539] it is the latter who causes the dew to fall from the "breast of heaven (*ina ṣerret šamāmi*)[540]": "(Marduk) [the one who makes] the dew fall from the breast of heaven" (Lambert 1959–1960: 61: 9). Here the dew, as in the previous text, plays a positive role in terms of abundance and prosperity for the fields (l. 10–12).

The absence of dew, as well as rain, is considered a curse:

> Th[us] may the r[ain and the d]ew on your fields
> and [you]r pastures not come, instead of [d]ew,
> may coal r[ain] on your land.
> [*Vassal Treaty of Esahaddon* vii 531–533, Wiseman 1958: 69]

There are no clear astrological omens regarding dew,[541] and contrary to what Biggs says (1967: 19), it does not seem to be considered as an unfortunate phenomenon.[542] The scholar quotes a passage from *Maqlû*, in this regard, where the witches are considered to be the cause of this phenomenon:

---

538 See CT 24, 40: 50 [^{d}*Ša*]-*la* : MIN (= ^{d}*Ša-la*) *šá nam-še* "Šāla of the dew"; CT 25, 10: 38 ^{d}*Ša-la* : MIN (= ^{d}*Ša-la*) *nišē u na-al-ši* "Šāla of the people of the dew." On the goddess and her relationship with Adad see Schwemer 2006–2008: 565–567.

539 Wheras in a *Hymn to Šamaš*, it depends on the sun god: "Dew, fog, snow [. . .] without Šamaš they are not given" (KBo 1, 12: 5–8, Ebeling 1954: 213).

540 For the expression "breast of the heavens" see Borger 1964: 55; referring to our incantation, Biggs writes: «Behind this concept probably lies the observation that a cow's udder, when full, often drips. It is possible that *uššuru/muššuru* is the technical term for the physiological process of letting down milk» (1967: 19).

541 Possibile mention of the dew in *Enūma Anu Enlil*: "If Adad shouts in the middle of the sky and the middle of the sky(?) [. . .], in that year (there will be) rain, (variant): no rain(?)"; where A.AN can indicate indeed raining/the rain (*zanānu*/*zunnu*), but the dew (*nalšu*) as well (XLV 42′, Gehlken 2012: 62; see also XLVI 45′).

542 It must not be interpreted that the use of dew in the similes for the description of sicknesses or demons is evidence for considering the dew as an unfavorable sign (as in CT 23, 10: 18; Lambert 1970: 43, l.30; *Ludlul* III 81, Oshima 2014: 98; LKA 70 ii 24; *Lamaštu* II 17, Farber 2014: 98). Consider for example the positive similarity in VAT 17347, Old Babylonian *Love Dialogue between Nanāya and Muati*: "Like dew charms (of love) (*irimmu*) rain down (on him)" (rev.$^{?}$ 11′, Wasserman 2016: 125).

> Incantation. Whoever you are, O witch, who like the wind of South has accumulated (*ikkimu*) for fifteen days,
> nine days of fog, dew for one year.
> She has formed a cloud against me and has put it over me.
> [*Maqlû* V 76–78, Abusch 2016: 141–142]

The passage in question outlines the action of the witch that causes human suffering. She has accumulated fog and dew, like the southern wind, and formed a cloud against the man. It is in this case, that dew or other atmospheric phenomena are not so much a negative sign per se, but rather metaphors for witchcraft drawn from meteorological phenomena that can be found all over the series, for example in line 81 we even find: "I disperse your witchcraft w[hich you] have acc[umulated against me day and night]." It makes use of the same verb *nakāmu*, used in line 76 to express the evil doings of the witch, by using meteorological metaphors.

Returning to incantation No. D.2, the heavenly Daughters of Anu are invoked so that they may grant the "wellbeing of the 'heart' (*libbu*) to the 'heart' and that of the 'waterskin' to 'waterskin'" of the patient. The patient or the exorcist is advised: "Not to release the dew of Anu's daughters." One possible interpretation of the dew not being released is that it should be metaphorically collected for therapeutic purposes. As we have seen above, it is invoked elsewhere to refer to abundance and well-being, and its absence is seen as a curse. To not drop it means to not lose it because it is a manifestation of the aid of benevolent entities.

Nīš libbi E

List of manuscripts

Manuscript	Museum number	Publication	Tablet	Script	Date	Provenience	Incantatios and prescriptions
A	VAT 8233	KAR 236: 1–rev. 23	Single-col. tablet	NA	8th–7th cent.	Aššur, Library N 4	No. E.1; No. E.2; No. E.3
B	VAT 8916	KAR 70: rev. 10–24	Single-col. tablet	NA	8th–7th cent.	Aššur Library N 4	No. E.1
C	A 483	BAM 369 = CCMAwR 3, pl. 1: 1–12	Fragment of a single-col. tablet	NA	8th–7th cent.	Aššur, Library N 4	No. E.1
D	Sp II 976 = BM 35394	CCMAwR 3, pls. 2–3 i 9′–ii 22′	Three-col. tablet	LB	4th–3rd cent.	Babylon?	No. E.1; No. E.2
E	VAT 8265	KAR 243 obv.! 1–rev.! 12′	Fragment of a single-col. tablet	NA	8th–7th cent.	Aššur, Library N 4	No. E.1; No. E.2
F	VAT 13643bis	LKA 99b: 1–11	Fragment of a single-col. tablet	NA	8th–7th cent.	Aššur, Library N 4	No. E.2
G	VAT 10697+ 10830	LKA 99d = CCMAwR 3, pls. 4–5 rev. iii 1′–16′, iv 4′–30′	Fragment of a two-col. tablet	MA	13th–11th cent.	Aššur, Library N 4	No. E.2; No. E.3; Bow ritual; Prescr. 4–6
H	SU 52/ 139+161+ 170+250 + 250A +323	STT 280 ii 62–iii 23	Two-col. tablet	NA	8th–7th cent.	Sultantepe	No. E.2
I	K. 11076	TCS 2, pl. 3 rev. 1–8	Fragment	NA	7th cent.	Nineveh, Ashurbanipal's Library'	No. E.2
J	VAT 10090	BAM 272: 1′–23′	Frg. of a single-col. tablet	NA	8th–7th cent.	Aššur, Library N 4?	Bow ritual; Prescr. 1–5
K	K 916	AMT 73, 2: 3–8	Small frg. of a single-col. tablet	NA	8th–7th cent.	Nineveh, Ashurbanipal's Library'	Bow ritual

Editions

Ebeling 1925: 32, 36, 56 (Ms. A, F, K)
Thompson 1934: 93–94 (Ms. K)
Biggs 1967: 27–35, 52–54, 63–64 (Ms. A, B, E, F, G, H, I, J, K)
Seux 1976: 400–401 (trans. of No. E.2: 25–37)
BFoster 2005: 676–677 (trans. of No. E.2: 25–37)
Schwemer 2010: 119 (trans. of Ms. H)
Scurlock 2014b: 548–550 (Ms. J)
Abusch et al. 2020: 35–44, No. 4.2 (Ms. A, B, D, E, F, G, H, I)

Structure of the text

Text *nīš libbi* E is based on Ms. A. It is constituted by the following incantations: No. E.1 "May the wind blow! May the mountains quake!" (with duplicates Ms. B, C, D, E); No. E.2 "Light of the heavens, wise Ištar" (with duplicates D, E, F, G, H, I); No. E.3 "At my bed a ram is tied" (with duplicate G rev. iv 26′–30′). The text continues, following Ms. G and J, with the bow ritual (Ms. G, J, K) and six prescriptions (prescr. 1–6, Ms. G, J).

A – B – C – D	No. E.1
A – D – E – G – H -I	No. E.2
A – G	No. E.3
G – J – K	Bow ritual
J	Prescr. 1–3
G – J	Prescr. 4–6

I. Incantation No. E.1 "May the wind blow! May the mountains quake!" (Biggs 1967 No. 14), ll. 1–17 (Ms. A 1–17 // B rev. 10–24 // C 1–12 // D i 9′–21′ // E obv.! 1–14)

II Ritual and incantation No. E.2 "Light of the heavens, wise Ištar" (Biggs 1967 No. 11), ll. 18–45 (A 18–rev. 15 // D ii 5′–22′ // E obv!. 15′–rev! 12′ // F 1–11 // G rev. iv 4′–25′ // H ii 62–iii 22 // I rev. 1–8)

III. Incantation No. E.3 "At my bed a ram is tied" (Biggs 1967 No. 12), ll. 46–53 (Ms. A rev. 16–23 // G rev. iv 26′–30′)

IV. Bow ritual ll. 54–59 (Ms. G rev. iii 1′–5′ // J 1′–6′ // K 3–8)
Prescription
1. 60–62 (Ms. J 7′–9′)
2. 63–64 (Ms. J 10′–11′)
3. 65–67 (Ms. J 12′–14′)

4. 68–69 (Ms. G rev. iii 6′–7′ // J 15′–16′)
5. 70–72 (Ms. G rev. iii 8′–10′ // J 17′–19′)
6. 73–78 (Ms. G rev. iii 11′–16′ // J 20′–23′)

Summary of the sections of manuscripts not included in the transliteration:

- Ms. A = KAR 236
  rev. 24–26 = a colophon of Issar-tarība, see commentary l. 53

- Ms. B = KAR 70
  1–5 = No. F.1 (Ms. A)
  6–10 = D Diagn. (Ms. F)
  11–44 = F prescr. 5–7, 10–17 (Ms. A)
  45–rev. 9 = No. F.4 (Ms. A)
  rev. 25–30 = No. F.5 (Ms. A)
  rev. 31–33 = No. B.2 (Ms. B)
  rev. 35 = Fragmentary colophon, see commentary *nīš libbi* B: 40

- Ms. C = BAM 369 (CCMAwR 3, pl. 1)
  rev. 2′–9′ = B prescr. 1 (Ms. C)
  rev. 10′–12′ = Fragmentary colophon, see commentary *nīš libbi* B: 29

- Ms. D = Sp II 976 = BM 35394 (CCMAwR 3, pls. 2–3)
  i 1′ = Undecipherable
  i 2′–8′ = D Diagn. (Ms. G)
  ii 1′–4′ = Fragmentary lines
  ii 5′–22′ = No. E.2 (Ms. D)
  iii 1′–4′ = Undecipherable
  rev. iv = Not preserved
  rev. v. 1′–5′ = Fragmentary
  rev. v 6′–10′ = No. A.3 (Ms. H)
  rev. vi 1′–3′ = Undecipherable

- Ms. G = LKA 99d = CCMAwR 3, pls. 4–5
  i 1′–12′ = Fragmentary
  ii 1′–19′ = Fragmentary
  rev. iii 18′–27′ = Fragmentary prescription
  rev. iii 28′–29′ = Fragmentary prescription
  rev. iv 1′–3′ = Fragmentary incantation

- Ms. H = STT 280
  i 1–7 = K prescr. 1–2 (Ms. A)
  i 8–17 = D Sympt.; Prescr. 4 (Ms. E)

i 18–21 = F prescr. 10–11 (Ms. E)
i 22–51 = K prescr. 3–18 (Ms. A) (note i 44 = A prescr. 7 (Ms. D))
i 52–55 = A prescr. 9–12, 16 (Ms. D)
i 56–ii 9 = K prescr. 19–22 (Ms. A)
ii 10–21 = No. M.1 (Ms. B)
ii 22–35 = K prescr. 23–28 (Ms. A)
ii 36–50 = No. K.3 (Ms. A)
ii 51–53 = No. K.4 (Ms. A)
ii 54–61 = No. K.5 (Ms. A)
iii 24–33 = No. K.6 (Ms. A)
iii 34–42 = No. K.7 (Ms. A)
iii 43–iv 7 = K prescr. 29–31 (Ms. A)
iv 8–23 = No. K.8 (Ms. A)
iv 24–31 = No. K.9 (Ms. A)
iv 32–36 = K prescr. 32 (Ms. A)
iv 37–41 = No. A.1 (Ms. D)

Transliteration

I. A 1–17 // B rev. 10–24 // C 1–12 // D i 9′–21′ // E obv.$^{!}$ 1–14 = Incantation No. E.1

1.	A o. 1	[IM] KUR-*e* *l*[*i-nu-šu*]
	B r. 10	ÉN *li-lik* IM KUR^{meš} *l*[*i-nu-šu*]
	C o. 1	[ ]^{meš} ⸢*li-nu*⸣-[*šu*]
	D o. i 9′	[*l*]*i-lik* IM KUR^{meš} [ ]

2.	A o. 2	[*ur-pa*]-*tum*$_4$-*ma ti-ku* [ ]
	B r. 10–11	*lik-ta-ṣir* / *ur-pa-tùm-ma ti-ku lit*-⸢*tuk*⸣
	C o. 2	[*ur-pa-tu*]*m*$_4$-*ma ti-ku* [ ]
	D o. i 10′	[*lik-ta*]-*ṣir ur-pa-tùm-ma* ⸢*ti-ku lit*⸣-[*tuk*]

3.	A o. 3	[*lim*]-*gu-ug* ANŠE-*ma* ÈME *li*-[*ir-kab*]
	B r. 11–12	*lim-gu-ug* ANŠE-[*ma*] ÈM[E] / ⸢*li*⸣-*ir-kab*$^{!}$
	C o. 3	[È]ME *li-ir*-[*kab*]
	D o. i 11′	[*lim-gu*]-⸢*ug*⸣ ANŠE-*ma* ÈME ⸢*li-ir-kab*⸣

4.	A o. 4	[*li*]*t-bi da-áš-šú li-ir-tak-ka-bu ú-ni-qé* x
	B r. 12	*lit-bi da-áš-šú li-ir*-[*kab*] ⸢*ú*⸣-*ni-qé* x
	C o. 4	[ ] ⸢*li-ir-tak*⸣-*ka-ba* ⸢*ú-ni*⸣-*q*[*é-ti*]
	D o. i 12′	[*d*]*a-áš-šú li-ir-tak-ka*-⟨*bu*⟩ *u*⸢*ni-qé*⸣-*ti*(text TE)
	E o. 1	[ ] ⸢x x⸣ [ ]

5.	A o. 5	ina SAG GIŠ.NÁ-ia lu ra-ki-is da-áš-⸢šú⸣
	B r. 13	ina SAG GIŠ.⸢NÁ⸣-ia lu-ú ú-ra-ki-is da-á[š-šú]
	C o. 5	[ ] lu ra-⸢ki⸣-is da-áš-[šú]
	D o. i 13′	[SA]G GIŠ.⸢NÁ-ia$_5$⸣ lu ra-ki-⸢is da⸣-[áš-šú]
	E o. 2	[GIŠ.NÁ-i]a lu ra-ki-is da-áš-[šú]
6.	A o. 6	ina še-pit GIŠ.NÁ-ia lu ra-ki-is pu-ḫa-lu$_4$
	B r. 14	ina še-pit GIŠ.NÁ-ia lu-ú ú-ra-ki-is pu-ḫa-l[u]
	C o. 6	[ ] lu ra-ki-is pu-ḫa-l[u$_4$]
	D o. i 14′	[GI]Š.⸢NÁ⸣-ia$_5$ lu ra-ki-is p[u-ḫa-lu]
	E o. 3	[GIŠ.N]Á-ia lu ra-ki-is ^{udu}pu-ḫ[a-lu$_4$]
7.	A o. 7	šá — SAG GIŠ.NÁ-ia ti-bá-a ra$^!$-man-ni
	B r. 15	šá ina SAG GIŠ.NÁ-ia ti-ba-a ra(text IŠ)-man-n[i]
	C o. 7	[ ] ti-ba-a ra-ma-[an-ni]
	D o. i 15′	[SA]G GIŠ.NÁ-ia$_5$ ti-ba-a ra-man-[ni]
	E o. 4	[GIŠ.N]Á-ia ti-ba-a ra-man-ni
8.	A o. 8	šá še-pit GIŠ.NÁ-ia ti-bá-a ḫu-ub-⟨bi⟩-ba-an-ni
	B r. 16	šá še-pit GIŠ.NÁ-ia — ḫu-ub-bi-ba-an-n[i]
	C o. 8	[ ] ḫu-ub-bi-ba-an-n[i]
	D o. i 16′	[še-pi]t GIŠ.NÁ-⸢ia$_5$ ti⸣-ba-a ḫu-⸢ub⸣-bi-ban-n[i]
	E o. 5	[GIŠ.]⸢NÁ⸣-ia ti-ba-a ḫu-bi-ba-an-n[i]
9.	A o. 9	ú-ru-[ú]-a ⸢ú⸣-ru MUNUS.UR ú-šar-šú ú-šar UR.GI$_7$
	B r. 17	ú-ru$^!$-ú-a ú-ru kal-ba-ti GÌŠ-⸢šú⸣ ú-šar U[R.GI$_7$]
	C o. 9	[MU]NUS.UR ⸢GÌŠ-šú⸣ ú-šar U[R.GI$_7$]
	D o. i 17′	[ú-ru]-ú-a ú-⸢ru MUNUS⸣.UR GÌŠ-šú ú-šar UR.GI$_7$
	E o. 6	[ú-ru-ú]-a ú-ru [MUNUS].UR ú-šar-šú ú-šar UR.GI$_7$
10.	A o. 10	GIM ú-ru MUNUS.UR iṣ-ba-⸢tú⸣ ú-šar UR.GI$_7$
	B r. 18	GIM ú-ru MUNUS.UR iṣ-ba-tu ú-⸢šar kal$^{?!}$⸣-[bi$^{?!}$]
	C o. 10	[ ]
	D o. i 18′	[x ú]-⸢ru⸣ MUNUS.UR iṣ-ba-tu GÌŠ UR.GI$_7$
	E o. 7	[x] ⸢ú⸣-ru MUNUS.UR iṣ-ba-tu ú-šar UR.GI$_7$
11.	A o. 11	[GÌŠ]-ka li-ri-ka ma-la maš-ga-šú
	B r. 19	GÌŠ-ka li-ri-ka ma-la maš-ga-[šú]
	C o. 10	[GÌŠ]-ka l[i-ri-ka ma-l]a maš-ga-š[i$^?$]
	D o. i 19′	[l]i-⸢ri-ka⸣ ma-la maš-ga-ši(text RI)
	E o. 8	ú-šar-ka li-ri-ka ma-la maš-ga-ši

12.	A o. 12	*[áš-b]a-ka* [x] *bu-un-zer-ri* *šá ṣi-ḫa-a-te*
	B r. 20	*áš-ba-ku ina bu-un*(text E$^{?!}$)*-zer-ri šá ṣi-ḫa-a-*[*ti*]
	C o. 11	[*bu-un-zer*]*-ri* [x *ṣi-ḫa*]-⸢*a*⸣-[*ti*]
	D o. i 20′	[*bu-un-ze*]*r-ri* *šá ṣi-ḫa-a-*[*ti*]
	E o. 9	[*bu*]-⸢*un-zer*⸣*-ri* *šá ṣi-ḫa-a-*⸢*ti*⸣

13.	A o. 13	[*bu-ʾ*]*u-ra a-a aḫ-ṭi* TU$_6$ ÉN
	B r. 21	*bu-ʾu-ú-ra a-a aḫ-ṭi* TU$_6$ ÉN
	C o. 11	⸢*bu*⸣*-ʾu-ra a-a* ⸢*aḫ-ṭi* TU$_6$$^?$⸣ [x]
	D o. i 21′	[*aḫ*]-⸢*ṭi* TU$_6$⸣ [x]
	E o. 10	*bu-ʾu-ra a-a aḫ-ṭi* TU$_6$ ÉN

14.	A o. 14	[KA].INIM.MA ŠÀ.ZI.GA
	B r. 22	KA.INIM.MA ŠÀ.ZI.GA
	C o. 12	[ ] ŠÀ.ZI.G[A]
	E o. 11	KA.INIM.MA ŠÀ.ZI.GA

15.	A o. 15	[K]U$^?$.KU ^{na4}K[A$^!$.GI].NA.DAB.BA KU.KU A[N.BAR]
	C o. 13	traces
	E o. 12	DÙ.DÙ.BI KU.KU ^{na4}KA.GI.NA.DAB.BA KU.[K]U AN.B[AR$^!$]

16.	A o. 16	[x] ⸢Ì⸣.GIŠ BUR ŠUB-[*di*] ÉN 7-*šú ana* ŠÀ [NIT]A
	C o. 13	traces
	E o. 13	[N]A$_4$$^?$ x x [ ] ÉN 7-*šú ana* ŠÀ ŠID-*nu* [x]

17.	A o. 17	[GÌ]Š-*šú* MUNUS GAL$_4$.⸢LA⸣-*šá* ⸢EŠ⸣$^{?}$.[MEŠ] ⸢x x *ir-ta*⸣-[*na-ka-ab*]
	E o. 14	[ ] ⸢x⸣ [GAL$_4$.L]A-*šá* EŠ.MEŠ-*ma ir-ta-*[*na-ka-ab*]

Variant B of the lines 15–17

B r. 22 DÙ.DÙ.BI KU.KU AN.BAR KU.KU ^{na4}KA.GI.NA.DAB.BA(text AB)!
B r. 23 ^{ú}IGI-*lim* {⸢MIN⸣} PIŠ$_{10}$.^{d}ÍD : //*a-na* ^{d}ÍD// *ina lìb-bi* Ì.GIŠ ŠUB-*di*
B r. 24 ÉN 7-*šú a-na lìb-bi* ŠID-*nu-ma* ŠÉŠ-*su*

II. A 18–rev. 15 // D ii 5′–22′ // E obv$^!$. 15′–rev.$^!$ 12′ // F 1–11 // G rev.$^!$ iv 4′–25′ // H ii 62–iii 23 // I rev. 1–8 = Ritual and incantation No. E.2

18.	A o. 18	[x] NA ÍL ŠÀ-*šú* KAR-*ma lu* ⸢*ana*⸣ MUNUS-[*šú* x] *ana* MUNUS BAR ⸢DU-*ka* LAL⸣
	D o. ii 5′–6′	DIŠ NA ⸢ÍL⸣ ŠÀ-*šú* ⸢KAR⸣-*ma* ⸢*lu*⸣ *ana* M[UNUS-*šú*] *lu ana* MUNUS [BAR]-*ti* DU-[*ka* x]

E o. 15′ [ ] ⸢*ana*⸣ M[UNUS]

G r. iv 4′ [BAR-*t*]*e* DU LAL

H ii 62 [Í]L ŠÀ ⸢KAR⸣ — *ana* ⸢MUNUS⸣-*šú lu* — MUNUS BAR-*ti* ŠÀ-*šú* NU ⸢ÍL⸣

19\. A o. 19 ⸢*ana*⸣ IGI 15 MUL^{meš} ^{gi}DU$_8$ GIN-*an* ^{udu}SISKUR(text GAZ) BAL-[*qí*]

D o. ii 6′–7′ [^{d}]⸢*iš-tar*⸣ M[UL^{meš}] / ^{gi}DU$_8$ GIN-*an* ^{ud}[^{u}SISKUR B]AL-*q*[*í*]

G r. iv 5′ [ ]

H ii 63 [^{d}*i*]*š-tar* MUL^{meš} — — ^{udu}SISKUR DÙ-*uš*

20\. A o. 20 NÍG.NA ^{šim}LI GAR-*an* KAŠ.SAG [ ]

D o. ii 8′ NÍG.N[A] ^{šim}LI GAR-*an* [B]AL-*qí-m*[*a*]

G r. iv 5′ [^{ši}]^{m}LI GAR-*an* \ KAŠ BAL-*qí*

H ii 63–iii 1 NÍG.NA ^{šim} L[I] / [ ]

21\. A o. 21 ^{uzu}ZAG ^{uzu}ME.ḪÉ ^{uzu}KA.NE [ ] — —

D o. ii 9′–10′ ^{u}[^{zu}Z]AG ^{uzu}ME.ḪÉ ⸢^{uzu}KA⸣.N[E GA]R-*an* / ⸢KAŠ⸣.SAG BAL-*qí*

G r. iv 6′ [GA]R-*an* — —

H iii 2 [ ] — —

22\. A o. 22 2 NU Ì.UDU ⸢2⸣ NU DUḪ.LÀL 2 NU *kup-ri* 2 NU IM.BABBAR [ ]

D o. ii 10′–11′ 2 N[U Ì.UD]U [x N]U [DUḪ.LÀ]L / 2 NU *kup-ri* 2 NU ⸢IM.BABBAR 2⸣ NU [I]M

G r. iv 7′–8′ [ ] ⸢Ì.UDU 2 NU DUḪ.LÀL⸣ [*ku*]*p-ri* [ ] IM.BABBAR / [I]M

H iii 3–4 [ ] / [2 N]U I[M]

23\. A o. 23 2 NU NÍG.SILA$_{11}$.GÁ 2 NU ^{giš}ERIN DÙ-*uš* *ina* — ^{dug}BUR.ZI NU AL.[ŠEG$_6$.GÁ]

D o. ii 12′–13′ 2 NU NÍG.SILA$_{11}$.GÁ 2 ⸢NU ^{giš}ERIN⸣ D[Ù-*u*]*š* / *ina* — ^{dug}BUR.ZI.⸢GAL NU AL.ŠEG$_6$.GÁ⸣

G r. iv 8′–9′ 2 NU ⸢*li*⸣-*še* 2 NU *e-re-ni* [DÙ]-*uš*/ [*ina* Š]À$^{?}$ ⸢*ka*⸣-*si* NU ⸢ŠEG$_6$⸣.GÁ

H iii 4–5 [/] — ⸢^{dug}⸣[ ]

I r. 1 [DÙ-*u*]*š* *ina* — ^{dug}BUR.ZI.GAL NU [ ]

24.	A o. 24	ina IZI ana IGI 15 MUL^{meš} ta-šár-rap-ma kam DU$_{11}$-GA
	D o. ii 13′–14′	i[na IZ]I / ana IG[I] ^{d}iš$_{8}$-tár ⸢MUL^{meš} ta⸣-šár-rap-ma kam D[U$_{11}$-GA]
	G r. iv 9′	ina ⸢IZI⸣ ana IGI ^{d}[15] — ta-šár-rap — —
	H iii 5–6	[/ [t]a-šár-r[ap]
	I r. 2	[MUL^{m}]^{eš} ta-šár-rap-ma kam [ ]

25.	A o. 25	ÉN na-na-rat AN-e te-[l]i⸣-tu ^{d}⸢iš⸣-[tar]
	D o. ii 15′	ÉN na-na-rat ⸢AN-e te-li-tu ^{d}⸣[ ]
	F 1	[A]N-⸢e⸣ AN.ZÍB ^{d}iš-tar
	G r. iv 10′	[na-na]-rat AN-e ta-li-tu ^{d}iš-tar
	H iii 7	ÉN ^{d}na-[na-rat]
	I r. 3	[AN]-e te-li-tum ⸢^{d}⸣[ ]

26.	A o. 26	be-let DINGIR^{meš} šá an-[na-š]á a[n-nu]
	D o. ii 16′	be-let DINGIR^{meš} ⸢šá⸣ an-na-⸢šá⸣ [ ]
	F 2	[be-l]et DINGIR^{meš} šá an-na-šá an-nu
	G r. iv 10′–11′	^{d}be-⸢let⸣ DINGIR^{meš} / [ ] an-nu
	H iii 8	be-let DINGIR^{m}[^{eš}]
	I r. 4	[š]á an-na-šá a[n-nu]

27.	A o. 27	mu-tal-la-at DINGIR^{meš} šá q[í-bit-sa]
	D o. ii 17′	⸢mu⸣-tal-la-⸢at DINGIR⸣^{meš} šá qí-⸢bit-sa⸣ ṣ[i-rat]
	F 3	[mu]-⸢tal⸣-la-at DINGIR^{meš} šá qí-bit-sa ṣi-rat
	G r. iv 11′	⸢mu⸣-tal-la-at ⸢ DINGIR^{meš} ša qí-bit$^{?}$-sa ṣi-rat⸣
	H iii 8	[mu-tal-l]a-at DIN[GIR^{meš}]
	I r. 5	[DINGIR^{m}]^{eš} šá qí-bit-sa [ ]

28.	A o. 28	⸢be⸣-[let] AN-e u KI-tim muma-ʾi-[rat]
	D o. ii 18′	[be]-⸢let⸣ AN-⸢e⸣ [x K]I-⸢tim mu⸣-ma-ʾi-rat ⸢ka⸣-l[a]
	F 4	[b]e-let AN u KI mu-maʾ-i-rat DÙ URU.URU
	G r. iv 12′	[m]u-ma-⸢ʾi⸣-rat DÙ.A.BI URU.URU^{meš}
	H iii 9	⸢be-let⸣ A[N mu-ma]-⸢ʾi⸣-rat ⸢ka⸣-la [ ]
	I r. 6	[mu-ma]-ʾi-rat ka-la [ ]

29.	A o. 29	⸢^{d}iš-tar ina ni⸣-ip-⸢ḫi-ki kit-mu⸣-s[u]
	D o. ii 19′	⸢^{d}⸣iš-t[ar ni]-⸢ip-ḫi-ki⸣ kit-mu-su ŠU.NI[GIN]
	E 5	^{d}iš-tar ina KUR(text MU)-ki kit-mu-su ŠU.NIGIN EN^{meš}-e
	F r. iv 13′	[ni-ip-ḫi-]-ki kit-⸢mu-su be-lu be-le-e⸣
	H iii 10	⸢^{d}iš-tar ina KUR-ki kit⸣-mu-su nap-ḫar be[-le-e]
	I r. 7	[kit]-mu-su ŠU.NI[GIN]

30.	A obv 30	⸢*ana-ku*	NENNI A NENNI⸣	*a[k-ta-mis*	]
	D o. ii 20′	*ana-k[u*		*a]k-ta-mis*	*m[a-ḫar-ki]*
	F 6	*ana-ku*	NENNI A NENNI	*ak-ta-mis*	IGI-*ki*
	G r. iv 14′	[	] NENNI	*ak-ta-mi-*⸢*is*⸣	*ma-a-*⸢*ḫar-ki*⸣
	H iii 11	*ana-ku*	NENNI A NENNI	*ak-ta-mis*	I[GI-*ki*]
	I r. 8	[		*ak-ta-m]i-is*	*m[a-ḫar-ki]*

31.	A r. 1	*ša kiš-pi*	*ep*$^{!}$(text LU)-*šu*	*ina* KI	*šu-nu-lu*	N[U^{meš}-*ia*]
	D o. ii 21′	*š[á*				]
	F 7	*šá kiš-pi*	*e-pu-šú-ni*	*ina* KI	*šú-nu-lu*	NU^{meš}-*ia*
	G r. iv 15′	[	D]Ù-⸢*ni*⸣	— —	*šu-nu-lu*	NU^{meš}-*ia*
	H iii 12	*šá kiš-pu*	*ep-šú-ni-ni*	*ina* KI	*šu-nu-lu*	NU[^{meš}-*ia*]
	I r. 9	[			*šu-nu]-*⸢*lu*⸣	[ ]

32.	A r. 2	GIM	$^{\text{na4}}$ZA.GÌN	*lu-bi-ib*	*zu-um-[ri]*
	D o. ii 22′	G[IM			]
	F 8	GIM	$^{\text{na4}}$ZA.GÌN	*líb-bi-ib*	*zu-u[m-ri]*
	G r. iv 16′	[		*lu-bi-i]b*	SU-*ri*
	H iii 13	GIM	$^{\text{na4}}$ZA.GÌN	*lu-bi-ib*	*z[u-um-ri]*

33.	A r. 3	GIM	$^{\text{na4}}$GIŠ.NU$_{11}$.GAL	*lu* ZALÁG^{meš}	SAG^{meš}-[*ia*]
	E r. 1′	[x	$^{\text{na4}}$GIŠ.N]U$_{11}$.⸢GAL⸣	[	]
	F 9	[GI]M	$^{\text{na4}}$GIŠ.NU$_{11}$.GAL	— ZALÁG^{meš}	SA[G^{meš}-*ia*]
	G r. iv 16′–17′	GIM ⸢	$^{\text{na4}}$GIŠ. NU$_{11}$.GAL⸣ /	[	]
	H iii 14	GIM	$^{\text{na4}}$GIŠ.⸢NU$_{11}$⸣.GAL	— ZALÁG^{meš}	SA[G^{meš}-*ia*]

34.	A r. 4	GIM	KÙ.BABBAR	*eb-be*	GIM KÙ.SI$_{22}$	*ru-še-e*	*a-dir-ta*
		a-[a	]				
	E r. 2′	[	KÙ.BABBA]R	⸢*eb*⸣*-be*	GIM KÙ.S[I$_{22}$		
			]				
	F 10	[GI]M	⸢KÙ.BABBAR⸣	*eb-be*	*u* KÙ.SI$_{22}$	*ru-uš-*⸢*ši*⸣*-[i*	*a-dir-t]a*
		⸢*a-a*⸣	*ar-ši*				
	G r. iv 17′	[			K]Ù.SI$_{22}$	*ru-še-e*	*a-dir*(text ḪU)*-ta*
		ia	*ar-ši*				
	H iii 15	—	—	—	[GI]M KÙ.SI$_{22}$	⸢*ḫuš-ši*⸣*-e*	*a-dir-t[ú*
			]				

35.	A r. 5	^{ú}*tar-muš*	^{ú}IGI-*lim*	^{ú}IGI.NIŠ	^{ú}AŠ.TÁL.[TÁL]
	E r. 3′	[^{ú}*ta]r-*⸢*muš*⸣	^{ú}IGI-*lim*	^{ú}I[GI.NIŠ]	⸢^{ú}⸣AŠ.TÁL.⸢TÁL⸣
	F 11	⸢^{ú}⸣[ ] ⸢	^{ú}IGI-*lim*	^{ú}⸣IGI.[NIŠ	]
	G r. iv 18′	[	]	⸢^{ú}⸣IGI.NIŠ	^{ú}AŠ.TÁL.TÁL
	H iii 16	^{ú}*tar-muš*	⸢^{ú}⸣[IG]I-*lim*	^{ú}⸢IGI⸣.NIŠ	⸢^{ú}⸣AŠ.[TÁL.TÁL]

36.	A r. 6	^{ú}SIKIL Ú.KUR.RA ^{giš}GAN.⸢U$_5$⸣
	E r. 3′–4′	⸢^{ú}⸣[ ] / ⸢Ú⸣.KUR.RA ^{giš}GAN.U$_5$
	G r. iv 18′–19′	⸢^{ú}SIKIL⸣ / [^{giš}GAN].⸢U$_5$⸣
	H iii 17	— ⸢Ú⸣.KUR.RA ⸢GIŠ⸣ B[ÚR]
37.	A r. 7	⸢*liṭ-ru*⸣*-du ru-ḫe-e-a an-nam* 3-*šú* DU$_{11}$-GA-[*ma*]
	E r. 4′–5′	*liṭ-r*[*u*]*-du ru-ḫe-e-*[*a*] / *an-nam* 3-*šú* DU$_{11}$-GA-*ma*
	F r. iv 19′	*li-iṭ-ru-du ru-ḫe-e-ia* — — —
	H iii 17	*li-iṭ-*⸢*ru-du ru*⸣-[*ḫe-e-ia*] — — —
38.	A r. 8	DÙ.DÙ.BI KÙ.BABBAR KÙ.SI$_{22}$ ^{na4}ZA.GÌN ^{na4}GIŠ.NU$_{11}$.[GAL]
	E r. 6′	DÙ.DÙ.BI KÙ.BABBAR KÙ.SI$_{22}$ ^{na4}[ ] ^{na4}⸢GIŠ⸣.NU$_{11}$.⟨GAL⟩
	G r. iv 20′	— [^{na4}GI]Š.NU$_{11}$.GAL
	H iii 18	— KÙ.BABBAR KÙ.SI$_{22}$ ⸢^{na4}ZA⸣.GÌ[N ^{na4}GI]Š.NU$_{11}$.GAL
39.	A r. 9	^{ú}*tar-muš* ^{ú}IGI-*lim* ^{ú}IGI.NIŠ ^{ú}AŠ.TÁL.[TÁL]
	E r. 6′–7′	^{ú}*tar-muš*$_8$ ^{ú}IGI-*lim* / ^{giš}IGI.NIŠ ^{ú}AŠ.TÁL.TÁL
	G r. iv 20′–21′	^{ú}*tar-muš* ^{ú}IGI-*lim* / [ ]
	H iii 18–19	⸢^{ú}*tar-muš* ^{ú}⸣[ ] / ⸢^{ú}IGI.NIŠ ^{ú}⸣AŠ.TÁL.[TÁL]
40.	A r. 10	^{ú}SIKIL Ú.KUR.RA ^{giš}GAN.U$_5$ *ina* ^{túg}GADA È-⸢*ak*⸣
	E r. 7′–8′	⸢^{ú}⸣[SI]KIL Ú.KUR.RA ^{giš}GAN.U$_5$ / *ina* ^{túg}GADA È-*ak*
	G r. iv 21′	[— — — —]
	H iii 19	⸢^{ú}⸣SIKIL ⸢Ú.KUR.RA⸣ GIŠ [BÚR]
41.	A r. 11	*ina* GÚ-⸢*šú*⸣ [GAR-*a*]*n* ^{udu}*pu-ḫa-la ina* SAG GIŠ.NÁ-*šú* —
	E r. 8′	*ina* GÚ-*šú* GAR-*an* ^{udu}*pu-ḫa-lu ina* SAG GIŠ.NÁ-*šú* —
	G r. iv 21′–22′	[*i+n*]*a* GÚ-*šú* GAR-*an* ^{udu}*pu-ḫa-la* / [ ] —
	H iii 20	*ina* ⸢ GÚ⸣-[*šú* ^{udu}*pu-ḫa-l*]*a ina* SAG G[IŠ.N]Á-*šú* ⸢KEŠDA⸣
42.	A r. 12	[*š*]*e-pit* GIŠ.NÁ-*šú tara-kás*
	E r. 9′	[x K]U$_5$-*su ina še-pit* GIŠ.NÁ-*šú tara-kás*
	G r. iv 22′	[ ] GIŠ.⸢NÁ⸣-*šú* KA.⸢KEŠDA⸣
	H iii 20	⸢UDU KU$_5$⸣-*s*[*u*]
43.	A r. 13	[*p*]*u-ha-la* — — *u* SAG.KI UDU KU$_5$-*si*
	E r. 9′–10′	*ina* SAG.KI *pu-ḫa-la* / — — [x] ⸢SAG.KI⸣ UDU KU$_5$-*si*
	G r. iv 23′	[ ] / — — [ ] ⸢UDU KU$_5$-*si*⸣
	H iii 21	— ⸢SAG.KI! *pu*⸣-[*ḫa-la*] ⸢SÍG⸣^{m}[^{eš} Z]I-⸢*aḫ*⸣ — SAG.KI U[DU]
44.	A r. 14	[*ta-n*]*a-saḫ-ma* DUR^{meš} *a-ḫe-na-a* NU.N[U]
	E r. 10′–11′	SÍG^{meš} *ta-na-saḫ-ma* / [DUR^{me}]^{š} *a-ḫe-na-a* NU.NU

G r. iv 24′ [-*t*]*a-te* NU.⸢NU⸣
H iii 22 $^{he\text{-}pí}$ [x x]-*te* NU.N[U]-*ma*

45. A r. 15 [ÉN *an-ni-ta*$_5^?$ 7$^?$]-⸢*šú*⸣ *ana* UGU ŠID-*nu ina* MÚRU-*šú* KEŠDA-*ma* ŠÀ.ZI.GA
E r. 11′–12′ ÉN ⸢*ina* UGU$^?$⸣ GIŠ.N[Á 7$^?$]-*šú* / [— — MÚRU]-⸢*šú*⸣ KEŠDA-*m*[*a*]
G r. iv 25′ [ ] ŠÀ.ZI.GA
H iii 22 — — — — — — — — *ana* [MÚRU-*šú* KEŠDA-*ma* ŠÀ.ZI.GA]

III. A rev. 16–23 // G rev. iv 26′–30′ = Incantation No. E.3

46. A r. 16 [ÉN *ina* SA]G GIŠ.NÁ.MU KEŠDA ^{udu}*pu-ḫa-lu*
G r. iv 26′ [GIŠ.N]Á-*ia* KEŠDA ^{udu}*pu-ḫa-lu*

47. A r. 17 [*ina še-pit* GIŠ.NÁ.M]U KEŠDA UDU KU$_5$-*su ina* MÚRU^{meš}-*ia*$_5$ SÍG^{meš}- *šú-nu rak-sa*
G r. iv 27′–28′ [KU$_5$-*s*]*u i+na* M[ÚRU^{meš}]-*ia* / [ ]

48. A r. 18 [GIM *pu-ḫa-li* 11-*šú*] GIM KU$_5$-*si* 12-*šú* GIM BURU$_5$.ḪABRUD.DA^{mušen} ⸢13⸣-*šú*
G r. iv 28′–29′ [KU$_5$-*s*]*i* 12-*šú* / [ ]

49. A r. 19 [*ra-man-ni* GIM] ŠAḪ 14-*šú* GIM AM 50 GIM [DÀ]RA.MAŠ 50
G r. iv 29′–30′ [1]4-*šú* / [ ]

50. A r. 20 [. . .]-⸢*ta*⸣ *at-tú-ka* DÙ *da-ád-me*
G r. iv 30′ [. . . *da-á*]*d-m*[*e*]

51. A r. 21 [. . .]-*ta at-tú-ka* DÙ *ḫur-šá-a-ni*
52. A r. 22 *i*[*na qí-bit* ^{d}15] *be-let ru-a-me* ^{d}*na-na-a be-let* ḪI.LI
53. A r. 23 *š*[*i-n*]*a iq-ba-a ana-ku* DÙ-*uš* TU$_6$ ÉN

IV. G rev. iii 1′–17′ // J 1′–23′ // K 3–8 = Bow ritual and prescriptions 1–6

54. K 3–4 DIŠ KI.MIN ^{ú}*úr-na-a šá* KUR-*e* ^{ú}IGI-*lim* ^{ú}IGI.[NIŠ] / ^{ú}*ṣa-ṣu-um-tú pi-ti* SU.TIN^{mušen}

55.	G r. iii 1′	G[URUN] ^{giš}KI[ŠI$_{16}$.ḪAB]
	J 1′–2′	traces / [ÚŠ BURU$_{5}$.ḪABRUD^{mušen}] NITÁ
	K 4–5	GURUN ^{giš}KI[ŠI$_{16}$.ḪAB] / GIŠ.ḪAŠḪUR.GIŠ.GI 7 Ú^{meš} ŠEŠ-*ti* 1-*niš* SÚD ÚŠ MUŠEN *ḫur-ri* —
56.	G r. iii 2′	*ana* ŠÀ ⸢*tu*⸣-*maš-šar* ŠÀ [ ]
	J 2′–3′	*ana* [/] *i-ʾa*-[*lut*]
	K 5–6	*ana* ŠÀ *tu*-[*maš-šar*] / ŠÀ BURU$_{5}$.ḪABRUD.DA^{mušen} *i-al-lut*
57.	G r. iii 3′	*ina* KAŠ.SAG NU *pa-tan* [NAG-*šú* ^{giš}BAN *šá* ^{giš}DÁLA DÙ-*uš*]
	J 3′–4′	[/ NAG]-*šú* ^{giš}BAN *šá* ^{giš}MI [DÙ-*u*]*š*
	K 6–7	*ina* KAŠ.SAG NU *pa-tan* NAG-[*šú*] / [^{giš}]BAN *šá* ^{giš}DÁLA DÙ-*uš*
58.	G r. iii 4′	SA ⟨PÉŠ⟩.ÙR.RA *ma-ta-an*$^{!}$-[*ša*]
	J 5′	[x PÉŠ.ÙR.R]A *ma-ta-an-ša* G[I SA$_{5}$-*ši*]
	K 7–8	SA PÉŠ.ÙR.RA *ma-ta-an*-[*ša*] / [x] SA$_{5}$-*ši*
59.	G rev iii 5′	*ina* SAG NITÁ *u* MUNUS *šá* *ṣa-a*[*l-lu* . . .]
	J 6′	[NIT]Á *u* MUNUS *šá* *ṣa-lu* GAR-[*a*]*n* [. . .]
	K 8	*ina* SAG NITÁ *u* MUNUS *šá* *ṣa-al-lu* GAR-[*an* . . .]
60.	J 7′	[UD-*ma* ANŠ]E.KUR.RA NITÁ KÀŠ^{meš}-*šú* *ina* KASKAL-*ni*
61.	J 8′	[*iš-t*]*i-nu si-ḫi-ir mi*-⸢*du*⸣-*u*ʾ-⸢*ri*⸣
62.	J 9′	[KÀŠ^{m}]^{eš}-*šú* TI-*qé* *ina* KAŠ ḪE.⟨ḪE⟩ NU *pa-tan* [N]AG-[*ma* KI.MIN]
63.	J 10′	[*ana* KI.MIN] *šá-rat ra-pal-te šá* GU$_{4}$.NITÁ GE$_{6}$ *ta* x x
64.	J 11′	[. . . *t*]*u-bal* SÚD *lu ina* KAŠ *lu ina* GEŠTIN.ŠUR NU *pa*-⸢*tan*⸣ [NAG-*ma* KI.MIN]
65.	J 12′	[*ana* KI.MIN] ÚŠ UDU.MÁŠ *ina* DUG.⸢BUR⸣.ZI NU AL.ŠEG$_{6}$.GÁ *ta-ma*[*ḫ-ḫar*]
66.	J 13′	[*mi-iš-l*]*a ina* Ì.GIŠ ḪE.ḪE ⸢LI⸣.DUR-*ka* GÌŠ-*ka* ŠÉŠ-*áš*$^{?}$
67.	J 14′	[*ù m*]*i-iš-la-ma ina* A^{meš} ⸢GAZ⸣ NA[G-*ma* KI.MIN]
68.	G rev iii 6′	*ana* ŠÀ.ZI.GA ⸢TUKU⸣ AL.TI.RÍ.G[A^{mušen}]
	J 15′	[ ] AL.TI.RÍ.G[A^{muš}]^{en} DAB-*bat* *t*[*a-ba-qa-an*]
69.	G rev iii 7′	ÚŠ^{meš} NU *tu-še-ṣa-a* *tu-bal* [ ]
	J 16′	[*t*]*u-še*-⸢*ṣa*⸣-*a* *tu-bal* SÚD KI ^{zi}ŠE.SA.[A ḪE.ḪE NAG-*ma* KI.MIN]

70.	G rev iii 8′	*ana* ŠÀ.⸢ZI.GA⸣ TUKU BURU$_5$.ḪABRUD$^!$.DA NITÁ *šá* [*ana* U$_5$ ZI DAB-*bat*]
	J 17′	[ ] ⸢BURU$_5$.ḪABRUD⸣^{mušen} NITÁ *šá ana* U$_5$ [ ]
71.	G rev iii 9′	⸢*kap-pí*⸣ *ta-ba-qa-an* ÚŠ^{meš} NU *t*[*u-še-ṣa-a*]
	J 17′–18′	[/ [*ta-ba-q*]*a*-[*a*]*n* ÚŠ^{meš} [ ]
72.	G rev iii 10′	*tu*$^!$-*bal* SÚD *ina* KAŠ.SAG ⸢NU⸣ *pa*-[*tan* NAG-*ma* KI.MIN]
	J 19′	[*tu-bal* SÚD] *ina* KAŠ.SAG NU *p*[*a-tan*]
73.	G rev iii 11′	*ana* ŠÀ.ZI.GA TUKU 7 PA^{meš} $^{[gi]š}$KIŠI$_{16}$ *ina* A[^{meš} ŠUB-*di*]
	J 20′	[P]A^{meš} ^{giš}K[IŠI$_{16}$]
74.	G rev iii 12′	ÚŠ MUŠEN *ḫur-ri* NITÁ *ana lìb-b*[*i* A] ÍD *tu-m*[*aš-šar*$^?$ . . .]
	J 20′–21′	[ ] / traces
75.	G rev iii 13′	*i-na* GI$_6$ *tuš-bat* TA ^{d}[UTU] ⸢È⸣ *ina* UGU [PA^{meš} ^{giš}KIŠI$_{16}$ GUB-*su-ma*]
	J 22′	[*it-tap*]-⸢*ḫa*$^?$⸣ *ina* UG[U]
76.	G rev iii 14′	*ana* ⸢IGI ^{d}UTU⸣ ŠÀ BURU$_5$.⸢ḪABRUD.DA⸣ NITÁ *i*-[*al-lu-ut*]
	J 23′	traces
77.	G rev iii 15′	*a* x x x *nu* NAG$^?$-*šú* x x x [. . .]
78.	G rev iii 16′	*an* x [. . .] x x [. . .]

Transcription

I. A 1–17 // B rev. 10–24 // C 1–12 // D i 9′–21′ // E obv.[!] 1–14 = Incantation No. E.1

1. *šiptu: lillik šāru šadû linū[šū]*
2. *liktaṣṣir urpatum-ma tīku littuk*
3. *limgug imēru-ma atāna lirkab*
4. *litbi daššu lirtakkaba unīqēti*
5. *ina rēš eršīya lū rakis* (var.: *urakkis*) *daššu*
6. *ina šēpīt eršīya lū rakis* (var.: *urakkis*) *puḫālu*
7. *ša* (var.: *ina*) *rēš eršīya tibâ rāmanni*
8. *ša šēpīt eršīya tibâ ḫubbibanni*
9. *ūrūya ūrū kalbati ušaršu ušar kalbi*
10. *kīma ūrū kalbati iṣbatū ušar kalbi*
11. *ušarka līrika mala mašgaši*
12. *ašbāku ina bunzerri ša ṣīḫāte*
13. *bu''ura ay aḫṭi tê šipti*

---

14. *šipat nīš libbi*

---

15. *dudubû: sīkti šadâni ṣābiti sīkti parzil[li]*
16. *[ana] šaman pūri tanad[di] šipta sebîšu ana libbi tamannu [zikar]u*

17. *[uša]ršu sinništu biṣṣūrša iptaššaš-ma irta[nakkab]*

B r. 22 *dudubû: sīkti parzilli sīkti šadâni ṣābiti*
B r. 23 *imḫur-līm kibrīta: //ana nāri// ina libbi šamni tanaddi*
B r. 24 *šipta sebîšu ana libbi tamannū-ma tapaššassu*

II. A 18–rev. 15 // D ii 5′–22′ // E rev[!]. 1–12 // F 1–11 // G rev. iv 4′–25′ // H ii 62–iii 23 // I rev. 1–8 = Ritual and incantation No. E.2

18. *[šumma] amēlu nīš libbīšu eṭir-ma lū ana sinništīšu lū ana sinništi aḫīti alāka muṭṭu* (var.: *libbašu lā inašši*)
19. *ana maḫar Ištar-kakkabī pāṭira tukān nīqa tanaqq[i]* (var.: *teppuš*)
20. *nignak burāši tašakkan šikara* (var. adds: *rēštâ*) *tanaqqi*
21. *imitta ḫimṣa šumê [ta]šakkan* (var. adds: *šikara rēštâ tanaqqi*)

Translation

I. A 1–17 // B rev. 10–24 // C 1–12 // D i 9′–21′ // E obv.! 1–14 = Incantation No. E.1

1. Incantation: May the wind blow! Ma[y] the mountains [quake]!
2. May the cloud be gathered! May the moisture fall!
3. May the ass mate and mount the jenny!
4. May the buck arise and repeatedly[543] mount the goat!
5. At the head of my bed a buck is really tied (var.: I have tied)!
6. At the feet of my bed a ram is really tied (var.: I have tied)!
7. The one at the head of my bed, rear up, make love to me!
8. The one at the feet of my bed, rear up, bleat *for* me!
9. My vulva is the vulva of a bitch! His penis is the penis of a dog,
10. as the vulva of a bitch took the penis of a dog, (so may I do)!
11. May your penis become as long as a *mašgašu*-weapon!
12. I sit in a net of laughter,
13. may I not miss the quarry! Incantation formula.

---

14. Wording of *nīš libbi* (incantation).

---

15. Its ritual: Magnetite powder (and) iro[n] powder
16. you put [into] oil from the alabastron, you recit[e] the incantation over (it) seven times. [The ma]n
17. anoints (with it) his [pen]is, the woman her vulva, then he will repeate[dly have intercourse].

B r. 22 Its ritual: Iron powder, magnetite powder,
B r. 23 "heals-a-thousand"-plant, (and) sulphur //at the river// you put into oil,
B r. 24 you recite the incantation seven times over it and anoint him.

II. A 18–rev. 15 // D ii 5′–22′ // E rev!. 1–12 // F 1–11 // G rev. iv 4′–25′ // H ii 62–iii 23 // I rev. 1–8 = Ritual and incantation No. E.2

18. [If] a man's sexual desire is taken away and (his desire) to go to (var.: he does not desire) his own woman or another woman is reduced,
19. you set up a *pāṭiru*-portable altar in front of Ištar of the stars, you sacrifice a sheep,
20. you set up a censer with juniper, you libate (var. adds: premium) beer,
21. [you] place (there) the shoulder, the caul fat (and) the roast meat (var. adds: you libate premium beer).

---

543 Omitted in the variant.

22. *šina ṣalmī lipî šina ṣalmī iškūri šina ṣalmī kupri šina ṣalmī gaṣṣi šina ṣalmī ṭī]di*

23. *šina ṣalmī līši šina ṣalmī erēni teppuš ina pursīti* (var.: *ina burzigalli*; [*ina lib*]*bi*? *kāsi*) *lā ṣaripti*
24. *ina išāti ana maḫar Ištar-kakkabī tašarrap-ma kâm iqabbi*

---

25. *šiptu: nannarat šamê telītu Ištar*
26. *bēlet ilī ša annaša annu*
27. *muttallat ilī ša qibīssa ṣīrat*
28. *bēlet šamê u erṣeti muma'irat kala ālanī*
29. *Ištar ina nipḫīki kitmusū napḫar* (var.: *bēlū*) *bēlē*
30. *anāku annanna mār annanna aktamis maḫarki*
31. *ša kišpī epšū* (var.: *ēpušūni*; *epšūninni*) *ina erṣeti šunūlū ṣalmūya*

32. *kīma uqnê lūbib* (var.: *libbib*) *zumrī*
33. *kīma ašnugalli lū namrā rēšā*[*ya*]
34. *kīma kaspi ebbe kīma* (var.: *u*) *ḫurāṣi r/ḫuššê adirta ay arši*
35. *tarmuš imḫur-līm imḫur-ešrā ardadillu*
36. *usikillu nīnû bukānu* (var.: *iṣ pi*[*šri*])
37. *liṭrudū ruḫêya annâ šalāšīšu iqabbī-ma*

---

38. *dudubû: kaspa ḫurāṣa uqnâ ašnugalla*
39. *tarmuš imḫur-līm imḫur-ešrā ardadilla*
40. *usikilla nīnâ bukāna* (var.: *iṣ* [*pišri*]) *ina kitê tašakkak*
41. *ina kišādīšu tašakkan puḫāla ina rēš eršīšu* (var. adds: *tarakkas*)
42. *immeru parsu ina šēpit eršīšu tarakkas*
43. *ina pūt puḫāli* (var.: *šipā*[*ti tan*]*assaḫ*) *u pūt immeri parsi*
44. *šīpātī tanassaḫ-ma ṭurrī aḫennâ taṭammi* (var. adds: *-ma*)
45. *šipta* [*annīta sebî*]*šu ana muḫḫi* (var.: *ina muḫḫi may*[*yāli*]) *tamannu ina* (var.: *ana*) *qablīšu tarakkas-ma nīš libbi*

III. A rev. 16–23 // G rev. iv 26′–30′ = Incantation No. E.3

46. [*šiptu: ina rē*]*š eršīya rakis puḫālu*
47. [*ina šēpīt eršī*]*ya rakis immeru parsu ina qablītīya šīpātūšunu raksā*
48. [*kīma puḫāli ištešerīšu*] *kīma* ⟨*immeri*⟩ *parsi šinšerīšu kīma iṣṣūr ḫurri šalaššerīšu*

49. [*rāmanni kīma*] *šaḫî erbēšerīšu kīma rīmi ḫanšā kīma* [*ay*]*yāli ḫanšā*

50. [. . .] . . . *attūka kal dadmē*
51. [. . .] . . . *attūka kal ḫuršānī*
52. *i*[*na qibīt Ištar*] *bēlet ru'āmi Nanāya bēlet kuzbi*

22. Two figurines of tallow, two figurines of wax, two figurines of bitumen, two figurines of gypsum, two figurines of [cl]ay,
23. two figurines of dough, two figurines of cedar wood, you make (them); in an unfired *pursītu*-bowl (var.: *burzigallu*-bowl; cup)
24. you burn (them) in a fire in front of Ištar of the stars and he (= patient) says the following:

---

25. Incantation: "Light of the heavens, wise Ištar,
26. mistress of the gods, whose "yes" is "yes."
27. Most noble among the gods, whose command is supreme,
28. Mistress of heaven and earth, ruler of all the cities.
29. Ištar, at your rising, all the lords (var.: the lords of the lords) kneel down.
30. I, NN, son of NN, have knelt down before you.
31. (I am who) against whom the witchcraft has been performed (var.: they have performed witchcraft), my figurines have been buried in the ground.
32. May my body become pure like lapis lazuli!
33. May [my] features (lit. head) be bright like alabaster!
34. Like shining silver, like (var.: and) reddish gold may I have no fear!

35–37. May *tarmuš*-plant, "heals-a-thousand"-plant, "heals-twenty"-plant, *ardadillu*-plant, / *usikillu*-plant, *nīnû*-plant, *bukānu*-wood (var.: "wood-of-rele[ase"]) / dispel my fascination!". He (= patient) says this three times.

38–40. Its ritual: You string silver, gold, lapis lazuli, alabaster / *tarmuš*-plant, "heals-a-thousand"-plant, "heals-twenty"-plant, *ardadillu*-plant, / *usikillu*-plant, *nīnû*-plant, *bukānu*-wood (var.: "wood-of-release") on (a cord) of flax,

41. (and) put (it) around his neck. A ram at the head of his bed,
42. (and) [a wea]ned [sheep] at the foot of his bed you tie.
43. From the forehead of the ram and the forehead of the weaned sheep
44. you pull out wool and twine (two) separate threads.
45. You recite [this] incantation [seven] times over (them) (var.: over the be[d]), tie (them) around his waist, and (he will get) sexual desire.

III. A rev. 16–23 // G rev. iv 26′–30′ = Incantation No. E.3

46. [Incantation: At the hea]d of my bed a ram is tied,
47. [at the foot of my bed] a weaned sheep is tied. Around my waist their wool is tied.
48. [Like a ram eleven times], like a weaned ⟨sheep⟩ twelve times, like a partridge thirteen times.
49. [Make love to me like] a pig fourteen times, like a wild bull fifty (times), like a s[ta]g fifty (times)!
50. [. . .] . . ., yours are all the dwellings,
51. [. . .] . . ., yours are the mountain ranges.
52. A[t the command of Ištar], patron of the feminine charms, (and) Nanāya, patron of sexual attractiveness,

53. *š[in]a iqbâ anāku ēpuš tê šipti*

IV. G rev. iii 1′–17′ // J 1′–23′ // K 3–8 = Bow ritual and prescriptions 1–6

Bow ritual 54. *šumma* KI.MIN (= *amēlu nīš libbīšu eṭir-ma lū ana sinništīšu lū ana sinništi aḫīti alāka muṭṭu*)$^?$ *urnâ ša šadê imḫur-līm imḫur-[ešrā] ṣaṣumta pīti šuttinni*

55. *inib da[dāni] ḫašḫūr api sebet šammī annûti ištēniš tasâk dām iṣṣūr ḫurri*

56. *ana libbi tu[maššar] libbi iṣṣūr hurri iallut*
57. *ina šikari rēštî lā patān išattī[šu] qašta ša ṣillî teppuš*

58. *šerʾān arrabi matanša qanâ tamallâši*
59. *ina rēš zikari u sinništi ša ṣallū tašakk[an . . .]*

1. 60. *[inūma s]isû zikaru šīnātišu ina ḫarrāni*
61. *[išt]īnu siḫir midduʾri*
62. *[šīnā]tišu teleqqe ina šikari taballal lā patān [iš]attī-m[a* KI.MIN (= *išallim*)$^?$]

2. 63. *[ana* KI.MIN (= *amēlu nīš libbīšu eṭir-ma lū ana sinništīšu lū ana sinništi aḫīti alāka muṭṭu*)$^?$] *šārat rapalte ša alpi ṣalmi . . .*

64. *[. . . t]ubbal tasâk lū ina šikari lū ina karāni ṣaḫti lā patān [išattī-ma* KI.MIN (= *išallim*)$^?$]

3. 65. *[ana* KI.MIN (= *amēlu nīš libbīšu eṭir-ma lū ana sinništīšu lū ana sinništi aḫīti alāka muṭṭu*)$^?$] *dām immeri ina pursīti lā ṣaripti tama[ḫḫar]*

66. *[mišl]a ina šamni taballal abunnatka ušarka tapaššaš*
67. *[u m]išla-ma ina mê taḫaššal iša[ttī-ma* KI.MIN (= *išallim*)$^?$]

4. 68. *ana nīš libbi rašî: diqdiqq[a] taṣabbat t[abaqqan]*
69. *dāmī lā tušeṣṣa tubbal tasâk itti qalīt[i taballal išattī-ma* KI.MIN (= *išallim*)$^?$]

5. 70. *ana nīš libbi rašî: iṣṣūr ḫurri zikara ša ana rakābi [tebû taṣabbat]*

71. *kappī tabaqqan dāmī lā t[ušeṣṣâ]*
72. *tubbal tasâk ina šikari rēštî lā pa[tān išattī-ma* KI.MIN (= *išallim*)$^?$]

53. T[he]y commanded (it), (and) I performed (it). Incantation formula.

IV. G rev. iii 1′–17′ // J 1′–23′ // K 3–8 = Bow ritual and prescriptions 1–6

Bow ritual

54. If ditto (= a man's sexual desire is taken away and (his desire) to go to his own woman or another woman is reduced)$^{?}$: Mountain *urnû*-plant, "heals-a-thousand"-plant, "heals-[twenty"]-plant, *ṣaṣumtu*-plant, excrement$^{?}$ of a bat,
55. the fruit of *da*[*dānu*-acacia], *ḫašḫūr api*-plant; you pulverize together these seven plants, the blood of a partridge
56. you [let flow] over it, he swallows the innards of the partridge,
57. he drinks [it] with first quality beer on an empty stomach. You make a bow of thorn,
58. a tendon of the *arrabu*-mouse is [its] string, you load it [with an arrow],
59. you pu[t it] over heads of the man and the woman, who are lying [. . .].

1.

60. [When a s]tallion on the street
61. [ur]inates, the edge of remains
62. of its uri[ne] you take, you mix (it) with beer, [he (= patient) drink]s (it) on an empty stomach and [ditto (= he will recover)$^{?}$].

2.

63. [For ditto (= a man's sexual desire is taken away and (his desire) to go to his own woman or another woman is reduced)$^{?}$]: The thigh hair of a black bull . . .
64. [. . . y]ou dried up, you pulverize (it) neither in beer nor in grape juice, [he drinks (it)] on an empty stomach [and ditto (= he will recover)$^{?}$].

3.

65. [For ditto (= a man's sexual desire is taken away and (his desire) to go to his own woman or another woman is reduced)$^{?}$]: You ta[ke] the blood of a ram in an unfired *pursītu*-container,
66. you mix [a hal]f with oil (and) you anoint your navel (and) your penis,
67. [and] you crush [(the other) h]alf in water, he d[rinks (it) and ditto (= he will recover)$^{?}$].

4.

68. In order to get sexual desire: You take and [pluck] a wre[n],
69. you do not make the blood come out, you dry (it), you pulverize (it), [you mix (it)] with dr[y] wheat, [he drinks (it) and ditto (= he will recover)$^{?}$].

5.

70. In order to get sexual desire: [You take] a male partridge [reared-up] for the mating,
71. you pluck the wings, you do not [make] the blood [come out],
72. you dry (it), pulverize (it), [he drinks (it)] with the first quality beer on [empty] stoma[ch and ditto (= he will recover)$^{?}$].

---

6. 73. *ana nīš libbi rašî sebet arī ašāgi ina m[ê tanaddi]*

74. *dām iṣṣūr ḫurri zikari ana libbi mê nāri tum[aššar . . .]*
75. *ina mūši tušbât ištu [Šamšu] uṣṣû* (var.: *ittapḫa*) *ina muḫḫi [arī ašāgi tušzāssu-ma]*

76. *ana pān Šamši libbi iṣṣūr ḫurri zikari i'[allut]*
77. . . . . *išattīšu* . . . [. . .]
78. . . . . [. . .] . . . [. . .]

6. 73. In order to get sexual desire: [You throw] branches of *ašāgu*-acacia [in wat]er,
74. you let the blood of a male partridge f[low] in river [water . . .],
75. you let it spend the night in the shadows, as soon as [the sun] rises (var.: has risen) [you have him (= patient) stand] on [the branches of *ašāgu*-acacia and]
76. in front of the sun he e[ats] the innards of the male partridge.
77. . . . . he drinks it . . . [. . .]
78. . . . . [. . .] . . . [. . .].

Commentary

1. See catalogue LKA 94 ii 13: "Incantation: 'May the wi[nd] blow! May the [moun]tains quake!'". Cf. No. G.1: 7; No. A.1: 33; No. I.1: 6; catalogue LKA 94 i 4; H.1: 6; *Aššur Medical Catalogue*, section XX "sex" ll. 103–104 (Steinert 2018b: 217). See also the following passage of The Kurba'il Statue of Shalmaneser III: "(Adad) at whose voice the mountains shake" (l. 6, Kinnier Wilson 1962a: 93).

4. Ms. D has *ú-ni-qé* TE, which we can emend as acc. pl. *ú-ni-qé-ti*. The proposal of CAD R 87 to read *ú-ni-qé-*⌜*ti*⌝ must be rejected for epigraphical reasons. Another possibility is to consider TE the logogram for the verb *ṭeḫû* 'to come near, approach' (CAD Ṭ 71–82, AHw. III 1384–1385 'ganz nah herankommen, sich nähern'), which has also a sexual connotation in reference to both humans and animals (see for the references CAD Ṭ mng. 1 a.1´e´ and 1.b.2´). Ebeling (1925), Schwemer (2010), and CAD D 120b read in the other manuscripts the last sign as ⌜EDIN⌝, but, as Biggs (1967: 34) argues, *daššu* represents an old category of male goats (buck) and therefore not a wild animal (see Landsberger 1960 = MSL 8/1: 59).
lirtakkab-unīqi in Ms. A is maybe a sandhi form.

5–8. See catalogue LKA 94 iii 3; No. F.4: 80–93. Cf. No. E.3: 46–47.

12. For the form *ašbāka* see GAG § 75 c. 6: 122.
Laughs (*ṣīḫtu*) refer to sexual attraction and pleasure, also *ṣiāḫu* 'to laugh' describes amorous activity (see Groneberg 1999: 185–187; Wasserman 2016: 54; for verb *ṣḥq* in Hebrew see Paul 2002: 498).

15. See the same ritual No. B.2: 38–40. See also No. D.2: 21–23; L prescr. 2 15–18. The ritual in Ms. C obv. 15–16 is very fragmentary. See also in Ms. C rev. 1′ the final line of a prescription: [. . .] x ⌜GAR-*an-ma*⌝ Š[À]?.Z[I]?.G[A]?.

21. The repetition of *šikara rēštâ tanaqqi* in Ms. D is probably due to a scribal mistake.

26. In Ms. F the divine determinative before Ištar's epithet *bēlet ilī* "mistress of the gods" testifies an identification of Ištar with the goddess Bēlet-ilī (see Abusch et al. 2020: 43).

27. For *muttallu* see Landsberger 1954: 132–133.

29. For the emendation in Ms. F of MU-*ki* to *ina* KUR-*ki* "at your rising" see Biggs 2002: 75 fn. 48.

32–35. Note that in the second millennium love compositions both love and body parts (in particular genitalia) are compared with mineral and metal substances, emphasizing light and brilliance: "Your love is an obsidian-blade, your lovemaking is golden"; "Your genitals (*ribītum*) are lapis lazuli of the mountain" (catalogue KAR 158 vii 43′–44′ and 49′, Wasserman 2016: 212). In rites related to the death of Dumuzi the sperm of the god is identified with gold (*ḫurāṣu riḫûssu* "gold is his sperm"): CT 15, 44 r. 4′ (restored), Livingstone 1989: 95, No. 37; LKA 72 rev. 15, ibid. 98, No. 38; KAR 307: 12, ibid. 99, No. 39, here also "silver is his skull."

32. Manuscripts A and G have *lu-bi-ib*, with Krecher (1970: 354) a form for *lībib* "may it become pure."

33–34. Light equals sexual desire while darkness equals its absence. On this opposition in second millennium love songs see Wasserman 2016: 47–48.

34. See for a similar *šuilla* passage: *kīma parûti nūrī limmir idirtu ay āršī* "As alabaster let my light shine, I will not be afraid" (Ebeling 1953a: 80, l. 69). Ebeling translates the term *adirtu* with 'Trübnis' (AHw. I 13 'Verfinsterung'; CAD A/I 126 'misfortune, calamity, mourning'). CAD distinguishes two nouns for *adirtu*, A and B, resulting from two distinct *adāru* verbs (A/I: 103): A. 'to be worried, be restless'; B. 'to get scared, be scared.' AHw. (I 11–12) distinguishes between: A. 'finster sein.' B. 'fürchten.' Given the importance attributed to fear, as a symptom in the magical context and in particular in its connection with the absence of sexual desire, and distress (*ašāšu*/*ašuštu*) here I prefer the meaning of 'to be afraid,' and not the one of darkening or bad luck. See on this topic Chapter II § "Fear, distress, and insomnia."

40. The plant names *bukānu* and *iṣ pišri* are interchangeable. See Abusch and Schwemer 2011 No. 7.10.1: 97‴; 7.8.1: 23′; Abusch and Schwemer 2016, No. 3.4.1: 17 (cf. also No. 8.18: 17′). GIŠ BÚR/*iṣ pišri*/*gišburru*, lit. 'wood-of-release,' served also as a designation of the exorcist's ceremonial mace (*gamlu*). See its Sumerogram variant ^{giš}ŠÍTA 'weapon.' For *bukānu* = ^{ú}GAN.NA = ^{ú}GAN.U$_5$ = *iṣ pišri* = ^{giš}ŠÍTA/ ^{giš}ŠITA see Schwemer 2007b: 114; 2009: 66, commentary on BM 40568 rev. 7.

Biggs (1967:30) reads *ina* ÉŠ GADA. Abusch et al. 2020: 39 read *ina* KU GADA. We expect DUR (*ina ṭurri kitê*), but as argued by Abusch et al. 2020 a secondary use of DÚR as a logogram for *ṭurru* (DUR) is without parallel. I prefer, although uncertain, the reading *ina* ^{túg}GADA "on (a cord of) flax."

In Ms. F there is not enough space in the break for restoring the phrase *ina kitê tašakkak*.

45. Ms. H does not mention the recitation of the incantation over the two threads. In Ms. E the reading ⌜*ina* UGU?⌝ GIŠ.N[Á] "over the be[d]" is uncertain. A rubric follows in Ms. H iii 23: 30 MU.MEŠ.N]I?.

46–53. As argued by Abusch et al. 2020, the incantation features three different speakers: the male patient who describes his treatment (ll. 46–47); the female partner of the patient is speaking, demanding to have sex repeatedly (ll. 48–51); the therapeutic operator states the divine origin of the incantation (ll. 52–53). It is also possible, given the parallels No. E.1 and No. F.4 that the woman recites the lines 46–47.
Note that the fragmentary Ms. D (Sp II 976 = BM 35394, CCMAwR 3, pls. 2–3) obv. ii 1′–4 is maybe a duplicate of the final part of the incantation No. E.3, in fact in line 4′ we read [d]*na-na-[a] be-let* Ḫ[I.LI]. It could be, alternatively, a duplicate of the final part of incantation No. M.1.

48–49. See No. C.1 3–6; No. K.8: 149–150.

50–51: Biggs (1967: 30–31) interprets the repeated sign sequence *ta at tú ka* as unclear fragmentary verbal forms with *kal dadmē* and *kal ḫuršānī* as their subject. Here I follow the suggestion of Abusch et al. 2020.

53. In Ms. A rev. 24–26 a colophon follows (see Hunger 1968, No. 238; Abusch et al. 2020: 44):

[LIBIR.R]A.BI.GIM AB.SAR BA.AN.È
[BUD id1]5?-*ta*-SU lúŠÁMAN.LÁ TUR
[DUMU idAMAR.UTU-SILIM-PAP?].MEŠ lúA.BA É *aš-šur*!

kīma [labir]īšu šaṭir bari
[tuppi Iss]ar?-tarība šamallî ṣeḫri
[mār Marduk-šallim-aḫḫ]ē? tupšar bīt Aššur

Written according to its [original], collated.
[Tablet of Iss]ar-tarība, the young apprentice,
[son of Marduk-šallim-aḫḫ]ē, the scribe of the Aššur temple.

54. This prescription in Ms. I is preceded by the final part of another prescription, ll. 1–2: TÉŠ.BI SÚD *lu ina* KAŠ *lu ina* Ameš *šá ina ú-⌜ri⌝ [bu-ut-tu]* / NU *pa-tan* NAG [. . .] "You pulverize together, neither with beer nor with water which on the roof [has spent several nights] / he drinks on an empty stomach [. . .]."

57. In Ms. I *ṣillu* (gišMI) is a mistake for *ṣillû* (gišDÁLA).

61. *siḫru* (CAD S 239–240; AHw. II 1034) means 'edge,' while *minduḫru* 'remains' (CAD M/II 86; AHw. II 655).

63. The broken space is too small to restore *ana* ŠÀ.ZI.GA TUKU-*e* (the same in l. 65).

65. UDU.MÁŠ: mistake for NITÁ.

66. We have likely a mistake when the text says "your navel" and "your penis." For *mišla . . . mišla* see *mišla išattī-ma miš[la tapaššassu]* "He drinks half and [you anoint? him with the other] ha[lf]" (N prescr. 14: 19).

68. See N prescr. 5.

73–78. Similarities with F prescr. 12; N prescr. 12.

78. Ms. F contains another broken prescription: ii 17 KI.MIN x [. . .] x x [. . .].

Incantations: stylistic and functional analysis

I. A obv. 1–17 // B rev. 10–24 // C 1–12 // D i 9′–21′ // E obv.! 1–14 = Incantation No. E.1

The text is composed of an incantation and a ritual. The incantation contains four "themes": 1. meteorological, ll. 1–4; 2. animals tied to the bed, ll. 5–8; 3. canine metaphor, ll. 9–10; 4. warfare, ll. 11–13.

The first section is composed of four lines, all with precative verbs.[544] They are divisible into two lines each: in the first one, the blowing of the wind and the shaking of the mountain are invoked, as well as the gathering of the clouds and the falling rain; in the remaining two the mating of the donkey and the goat. All lines, except for the fourth and final section, are characterized by a chiastic structure made up of V-N-N-V:

544 Note the repetition of the sound *li* of the precative at the beginning and at the end of the four lines.

Verb	Noun/Subject	N./Subject (1-2) N./Object (3)	Verb
May blow (*lillik*)	the wind (*šāru*)	the mountains (*šadû*)	ma[y quake] (*l*[*inūšū*])
May be gathered (*liktaṣṣir*)	the cloud (*urpatu*)	the rain (*tīku*)	may fall (*littuk*)[545]
May mate (*limgug*)	the ass (*imēru*)	the jenny (*atāna*)	may mount (*lirkab*)

May rear up (*litbi*)	the buck (*daššu*)	may mount/ repeatedly (*lir*[*kab*]/*lirtakkab*)[546]	the female kids (*unīqēti*)

The second section is related to the first one only through the mention of the buck (*daššu*). If the first section describes the arrival of the rain and the reproduction of domestic animals, the second leads to the human dimension by linking the buck to it, and the ritual event *hic et nunc*. We can say that the first section has the function of a frame, capable of putting in relation the sexual sphere of animals, in their natural dimension, with humans, and to contextualize the ritualistic practice. This section has a symmetrical structure:

At the head of my bed (*ina rēš eršīya*)	is really tied/I have tied (*lū rakis*/*urakkis*)	a buck (*daššu*)
At the feet of my bed (*ina šēpīt eršīya*)	is really tied/I have tied (*lū rakis*/*urakkis*)	a ram (*puḫālu*)

The one at the head of my bed (*ša rēš eršīya*)	rear up! (*tibâ*)	Make love to me! (*rāmanni*)
The one at the feet of my bed (*ša šēpīt eršīya*)	rear up! (*tibâ*)	Bleat *for* me! (*ḫubbibanni*)

The third section (ll. 9–10) is not similar to any other part in the corpus in question. It is a canine metaphor: the woman who recites the incantation defines her vulva (*ūru*)[547] as of a bitch and the man's penis of a dog:

545 Note the alliteration between the verbal form *littuk* and the noun *tīku*.

546 Note in lines 3–4 the use of the verb *rabāku* 'to mount.' The two lines in fact describe the mating of the animals. At the same time, the donkey mates (*magāgu*) and buck arises (*tebû*).

547 The term *ūru* (CAD U: 265–266: 'genitalia'; AHw. III 1435 'Blöße, (weibliche) Scham') refers to the genitals of both sexes, although in most cases it is used to designate the vulva (see Holma 1911: 100–101; Biggs 1967: 34–35; Cooper 2002: 106. For other terms indicating the sexual organs see Kogan and Militarev 2002; Couto-Ferreira 2009: 243–250.

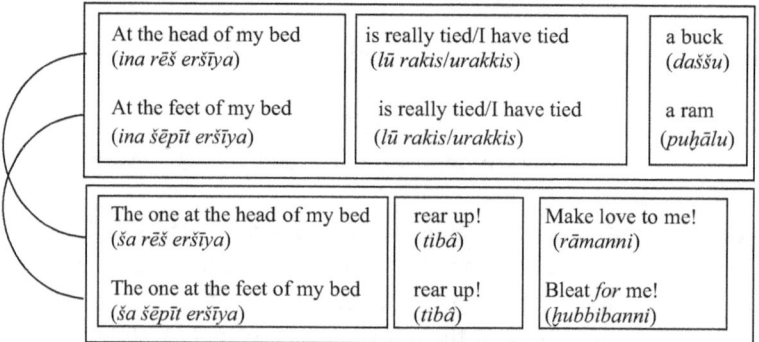

My vulva is the vulva of a bitch! His penis is the penis of a dog,
as the vulva of a bitch took the penis of a dog, (so may I do)!

The dog is an animal that is imbued with enormous symbolic value.[548] Such symbolic richness is perhaps attributable to the strong proximity of the animal to the human community.[549] In this *nīš libbi* incantation, it is precisely the sexual dimension of the cultural construction of the dog which must be underlined here. The dog is never used in metonymy. Moreover, unlike wild beasts, the dog is used in a comparison of nature primarily female: the woman is the one who recites the incantation.

The hair of the dog, which has mounted the bitch, is used in *nīš libbi* prescriptions since it is charged by sexual power: "When a ram, a donkey, a dog, a pig has mounted [. . .]" (A prescr. 2: 8).[550] In an Old Babylonian love incantation the (amorous) impetuosity (*uzzum*) is compared with a dog: "(The impetuosity) jum[ps on me] ag[ain and again] like a dog (*kalbāniš*)" (Wasserman 2016: 269, l. 87).

The dog is often seen in relation to the pig,[551] in regard to its sexual activity. This association can be found in an incantation, already cited above, in which there is an

548 On the dog in general see Heimpel 1972–1975b; Finet 1993; Villard 2000; Selz 2019: 38–39. The dog embodies both positive and negative conceptual associations (see George 1999: 298). See on the rabid dogs Yuhong 2001; on black dogs Sibbing Plantholt 2017; on the canine qualities of the demon Lamaštu see Hirvonen 2019: 320–324. The dog metaphor is used to describe the subordination and loyalty, particularly in the context of royalty (see Villard 2000: 246–249; Liverani 2011: 24–26), but also "loathsomeness" and "cowardice." On the image of the dog in divination, in particular in the *Šumma izbu* series as a symbol of confrontation, pestilence and death, and its relationship-opposition to the lion see De Zorzi 2014: 160–162. On sexual behavior of the dog in the above series see ibid. 910–911. On the dog as a symbol of the goddess Gula see Fuhr 1977; Groneberg 2007; Böck 2014: 38–44; Tsouparopoulou 2020. On the role of the dog in incantations see Groneberg 2007.

549 Such proximity between dog and man was underlined by the father of anthropological structuralism, Claude Lévi-Strauss, in particular in his study of names given to animals in modern France (1962a: 126). The anthropologist hypothesized that the choice of the names given to animals varies according to the position (metonymic/metaphoric, subject/object) that a given species occupies in relation to human society. The "metonimity" is the category of the dog. This category is defined as the category of animals "close" to men, such as domestic ones, which by their proximity to the human community are considered part of it. "Distant" animals (birds and horses in the anthropologist's essay) – such as those at the base of metonymys in *nīš libbi* incantations – instead, represent a different case: the humans entertain with them a "metaphorical" relationship. An example is given in the Greek world where the dog is at the top of the animal hierarchy, given the qualities attributed to it, like fidelity, intelligence, and devotion; when it comes to human society, however, it is placed near the bottom because it feeds on scraps and is subject to the master (see Franco 2003a; 2003b, Engl. tr. 2014; 2008b).

550 See also N presc. 7: 13: "I[f ditto] . . . of a dog that co[pulates? . . .]." See the use of dog excrements in Y presc. 1: 2′.

551 This "metonymic" animal is also associated with the woman, albeit less than the dog, for example in the *Šumma izbu* omens. On the figure of pig in the text corpus in question see Chapter III § "The pig." On the relationship between dog and pig in Mesopotamia see Parayre 2000b: 168–173;

analogy between the recitation of the incantation, the sexual activity of the dog and the pig, and the practice of plowing:

> Incantation. I impregnate myself, I impregnate my body,
> like a dog mounts a bitch, a pig a sow . . .
> like the plow inseminates the land (and) the earth takes in its seed.
> [CT 23, 10–12: 26–28/4: 9–11, Cooper 1996: 50–51; Cavigneaux 1999: 267]

In an Old Babylonian love incantation, the two animals are metaphorically described as lying in the 'heart' of the male partner to indicate his unbridled sexual desire: *ina libbīka nîl kalbum / nîl šaḫium* "In your 'heart' lies in a dog / a pig lies" (IB 1554: 57–58, Wasserman 2016: 265). In the same tablet the pair dog–pig is present in an incantation *Place Your Mind with my Mind* where the woman tries to act magically upon the partner, to make him feel emotionally attached to her:[552] *rabiṣ kalbum rabiṣ šaḫium / attā ritabbiṣ ina ḫallīya* "The dog is lying, the boar is lying, / you lie forever in between my thighs" (IB 1554: 21–22, Wasserman 2016: 257). To understand the passage, as already pointed out by Wasserman (2016: 259), we can compare it to a love ritual from the first millennium with Marduk as the protagonist addressing his wife: *ana [biṣ]ṣūrīki ša taklāte kalbī ušerreb bāba arakkas* "To your [vu]lva, in which you trust, I will make my dog enter (and) will lock the door" (Lambert 1975: 123, l. 1).[553] In this passage, the dog seems to refers metaphorically to the penis and the image of the door seems to allude to the sexual practice of the dogs. During the mating, its penis' bottom portion widens into a big turgid bulge, called the bulb of the penis, which is retained by the vulva, and prevents the escape of the penis during mating before the ejaculation has occurred. In the incantation analyzed here, the verb used to describe the penetration is *ṣabātu* 'to grab' (l. 10), indicating an iron grip.

In iconography, the two animals also are present together, when it comes to depicting, according to some archaeologists, erotic contexts (see Eichmann 1997).[554] A Babylonian terracotta depicts a musician (male or female)[555] probably with a lute in his/her hands, with a pig on his/her left and a dog on the right (Legrain 1936, pl. 17 n. 94).[556] Another terracotta from Nippur, depicts a man with a belt, holding a lute

Villard 2000: 244–245; De Zorzi 2014: 200–201; De Zorzi 2017: 136–137. The association between the two animals is also present in the Greek world, see Franco 2008b.

552 See line 35 of the incantation perhaps uttered by the man who feels once again attraction to the woman: "May I swell like a dog!".

553 Also Lambert 1975: 104, l. 7. For a comment on the passage see Groneberg 2007: 91–92.

554 According to Wilcke (1985: 206) and Cooper (1996: 51 fn. 16) – opinions that I find too general – the recurrence of the dog and the pig together and their metaphorical use for the description of the human coitus is due to the similarities between the animal activities to the human ones.

555 The previous interpretation interpreted the human as a shepherd musician remembered by animals of the steppe. Parayre (2000b: 169 and fn. 144) interprets the human figure as a male/female prostitute, but this hypothesis is unlikely.

556 The same model in McCown et al. 1967, pl. 138 No. 1.

in the right hand, and a small stick in the left. We can see behind him a pig, while in front of him a dog. Between the man and the dog, there is an object composed of a grooved foot and a thick plate, perhaps to be understood as a chair or a table (Opificius 1961: 159, pl. 17 n. 579; McCown et al. 1967, pl. 138 No. 5; Eichmann 1997).

The metaphoric mention of the dog in incantation No. E.1 occurs therefore for several reasons:

1. It is a metonymic animal, quite close to man, which strengthens the analogical power of the similes (introduced by *kīma* 'like');
2. It refers to femininity, to the extent, that the incantation is to be recited by the female partner;
3. It is considered an animal with sexual vigor, along with the pig, and symbolizes sexual desire (note the love poems).

The last section of the incantation consists of three lines. In the first one, the woman turns to the patient so that the penis might expand as much as the *mašgašu*-weapon:[557] *ušarka līrika mala mašgaši* "May your penis become as long as the *mašgašu*-weapon!" (l. 11) The other two lines form a pair in which the sexual sphere, as well as the animal and of hunting spheres, are intertwined: *ašbāku ina bunzerri ša ṣīḫāte* / *buʾʾura ay aḫṭi*[558] "I sit in a net of laughter, / may I not miss the quarry!" (ll. 12–13). The term *bunzerru* (CAD B 322 'web (of the spider), blind'; AHw. I 138 'Rohrschirm, Netz') indicates the web, as we can see in the following proverb:

> [*et*]*tūtu ana zumbi iḫtadal bunzerru*
> [*ṣu*]*rārû e*[*li*] *bunzerri*
> [*i*]*ttašiš ana* [*e*]*ttūti*
>
> [The spi]der spun a web for a fly,
> [a li]zard o[n] the web
> [was] caught for the spider!
> [Lambart 1960a: 220, ll. 23–25]

As emphasized by Civil (2006: 59–60), the noun has a semantic extension, in which, in addition, to designate the web, it also indicates objects with a "weblike structure":

557 For *mašgašu* CAD M/I 364: 'a tool'; AHw. II 625 'eine Streitkeule?.' The noun is a *maPRaS*-form of the verb *šagāšu* 'to kill' (see GAG § 56b, p. 78): «'Instrument für die Handlung *šag/kāšu*,' also 'Totschläger'» (Edzard 1975–1980b: 579). Here, given the context, it is to be understood as a weapon, perhaps a club, and not a pestle (see Poebel 1933–1934: 256 and fn. 23; Biggs 1967: 35). For a bibliography on the club in Mesopotamia see Stol 2015: 616. References to the world of the armory are present in other texts that have intercourse as their theme. In a catalogue of Old Babylonian texts, mostly on love matters, CUSAS 10, 12: 18 (Wasserman 2016: 190, No. 18), the opening words are "I carry a knife and a whetstone" and must have a sexual connotation.

558 Note the chiastic position of the two verbs, at the beginning and at the end of the lines.

^{gi}kid-níg-nigin-na	*nalmû*
^{gi}kid-á-ùr-ra	MIN
^{gi}kid-á-ùr-ra	*kīt bunzerri*

[ḪAR-ra = *ḫubullu* VIII 303–305, Civil 2006: 59]

^{kuš}kin-tur	*bunzerru*

[ḪAR-ra = *ḫubullu* XI 144, Civil 2006: 5; cf. Landsberger 1959 = MSL 7: 130]

The last line refers to an object, *kīt būnzerri*, shaped like a net, manufactured with leather strips, indicating a "reed mats web." *Nalmû*, not attested in dictionaries, is derived from the N-stem of the verb *lawû* 'to surround,' that is, 'to be surrounded.' Civil therefore does not exclude that the latter term means 'trap.'[559]

The meaning widens until it acquires a nuance of hunting. The images of hunting are, in fact, explicit in the incantation in question: the woman hopes to not miss her prey (l. 13 *bu''ura ay aḫṭi*). This would connect the second part of the section with the first line and its reference to the *mašgašu*-weapon. The context explicates the relationship between the web and the trap (Schwemer 2010: 122 fn. 259). Hence a woman is metaphorically associated with the spider, which waits for prey to fall into the trap of its web. I have shown the semantic relation between the sphere of animals (spider) and the one of hunting and war (*mašgašu*) contained in the word *bunzerru*. Civil (2006: 55–58) analyzed the etymological relationship between the term *bunzerru* and the one used for vulva *biṣṣūru*. According to him, the word *bunzerru* comes from the Sumerian word be$_5$-en-zé-er (ibid. 58–59), which apart from a cobweb, can also designate the female genital organs both in lexical lists and in Sumerian literary texts:

gal$_4$-la
gal$_4$-la
siki-gal$_4$-la
be$_5$-en-zé-er
[Proto-Lu list: 379–382, Civil 1969 = MSL 12: 46]

The Emar version of the Lu list about the female pelvic area provides us with the following terms:

^{gal}gal$_4$	*ūru*
^{gal}gal$_4$	*biṣṣūru*
siki-gal$_4$-la	*su-uḫ-šu*[560] : *iz-bu*

[Arnaud 1987: 191, n. 602: 368′–370′]

559 For the possible meaning of "enclosure" see Civil 2006: 56 and 60. On the use of animal metaphors to describe traps and enemies see Milano 2005.

560 The term is a hapax; Civil (2005: 55 fn. 3) suggests that it might derive from the verb *saḫāšu* 'to catch in a net' (CAD S 54b; AHw. II 1008) and thus confirm the semantic association between "pelvis" and "web/net."

The hypothesis of Civil (2006: 57) is that the Sumerian term be$_5$-en-zé-er is a loan from Semitic 'clitoris,' attested in Akkadian *biṣṣūru* and Arabic *baẓr* or *bunẓur*. It thus indicates both the female pelvic area and the spiderweb: «Unless one assumes two separate, but homophonous or quasi-homophonous words, an assumption with no etymological basis at present, the only solution is to postulate a semantic extension from "pubic hair" to "spider web"» (ibid. 59). The relationship between the words is summarized as follows (ibid. 60):

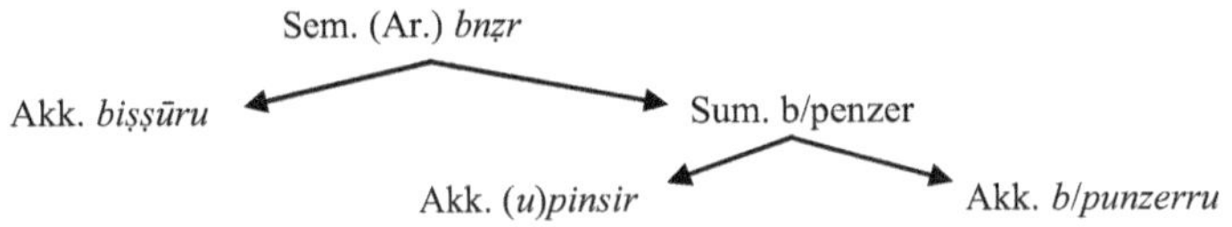

Thanks to the study of Civil, we can understand the different semantic references in the two lines of the incantation:

- Animal sphere: the woman is called a spider waiting in its web (*bunzerru*);
- Hunting Sphere: the web is a trap (semantic extension of *bunzerru*) for prey;
- Sexual sphere: the vulva is represented as a web/net (*bunzerru*/*biṣṣūru*) for joy and the man is a victim beckoned by lust.[561]

561 For the representations of the female pelvic area in the Old Babylonians terracotta see e.g. Legrain 1930 pls. 1–5.

Nīš libbi F

List of manuscripts

Manu-script	Museum number	Publication	Tablet	Script	Date	Proveni-ence	Incantations and prescriptions
A	VAT 8916	KAR 70 obv. 1–rev. 9; rev. 25–30	Single-col. tablet	NA	8th–7th cent.	Aššur, Library N 4	No. F.1; Prescr. 6–7, 10–13; No. F.4; No. F.5
B	81-7-27, 73	TCS 2, pl. 2 obv. 1′–rev. 7′	Small frg. of a two-col.	NA	7th cent.	Aššur, Library N 4	Prescr. 1–5, 8; No. F.2
C	A 2715	LKA 101 rev. 12–19	Large frg. of a single-col. tablet	NA	8th–7th cent.	Aššur, Library N 4	Prescr. 5, 8–9
D	K. 9451+ 11676+ Sm. 818 +961[562]	TCS 2, pl. 1 obv. i 2′–6′	Frg. of a two-col. tablet	NA	7th cent.	Nineveh Ashurba-nipal's Library	Prescr. 10–11
E	SU 52/139+ 161+170+ 250+250A +323	STT 280 i 18–21	Two-col. tablet	NA	8th–7th cent.	Sultantepe	Prescr. 10–11
F	BM 46911 (= 81-8-30, 377)	TCS 2, pl. 3 obv. 1–rev. 15	Fragment	NB/LB	6th–3rd cent.	Sippar	No. F.3; No. F. 4

Editions

Ebeling 1925: 28–29, 30–31, 34 (Ms. A)
Biggs 1967: 24–26, 31–32, 40–41, 46–47, 50, 53, 65–69 (Ms. A, B, D–F)
Farber 1987: 273–274 (trans. of Ms. A)
Scurlock and Andersen 2005: 257 (Ms. B, C)
Schwemer 2010: 115–122 (trans. of Ms. A, E)
Abusch and Schwemer 2011: 101, 103–104, 107 No. 2.5 (Ms. B, C, D, E)

562 Ms. B: joins of K. 9451+K. 11676+Sm. 818+961 recognized by Biggs (1967: 45, pl. 1) (see for the photo CDLI No. P398122).

Structure of the text

Text *nīš libbi* F is based on Ms. A starting with the abracadabra No. F.1 and a diagnostic ritual (= *nīš libbi* D: 24–26). Seventeen prescriptions follow (Ms. A, B, C, D, E). The text continues with Ms. B containing the incantation No. F.2 "Why are your eyes covered?", and then with Ms. F with the incantation No. F.3 "Roar on me! Roar on me! Rear up! Rear up!". Both Ms. A and F provide the incantation No. F.4 "At my head a buck is tied!". The rest of *nīš libbi* F follows Ms. A with three incantations: = No. E.1 "May the wind blow! May the mountains quake!"; No. F.5 "The daughter of Ninĝirsu, the releaser I am"; = No. B.2 "O Adad, locker keeper of the canals of heaven."

A	No. F.1
A (+ dupl.)	= *nīš libbi* D Diagn.: 24–26
B	Prescr. 1–4
A – B – C	Prescr. 5
A	Prescr. 6–7
B – C	Prescr. 8
C	Prescr. 9
A – D – E	Prescr. 10–11
A	Prescr. 12–17
B	No. F.2
F	No. F.3
A – F	No. F.4
A (+ dupl.)	= No. E.1
A	No. F.5
A (+ dupl.)	= No. B.2

I. Incantation No. F.1 (abracadabra) (Biggs 1967 No. 27), ll. 1–5 (Ms. A obv. 1–5)

II. ll. 6–10 (Ms. A obv. 6–10) = *nīš libbi* D Diagn.: 24–26

III. Prescriptions
1. 11–12 (Ms. B obv. 1′–2′)
2. 13 (Ms. B obv. 3′)
3. 14–16 (Ms. B obv. 4′–6′)
4. 17–18 (Ms. B obv. 7′–8′)
5. 19–23 (Ms. A obv. 11–14 // B obv. 9′–15′ // C obv. 7′–13′)
6. 24 (Ms. A obv. 15)
7. 25 (Ms. A obv. 16)
8. 26–28 (Ms. B obv. 16′–18′ // C obv. 14′–16′)
9. 29–30 (Ms. C obv. 17′–18′)
10. 31–33 (Ms. A obv. 17–19 // D obv. i 2′–4′ // E i 18–20)

11. 34–35 (Ms. A obv. 20–21 // D obv. i 5′–6′ // E i 21)
12. 36–41 (Ms. A obv. 22–27)
13. 42–47 (Ms. A obv. 28–33)

IV. Incantation No. F.2 “Why are your eyes covered?” (Biggs 1967 No. 35), ll. 59–64 (Ms. B rev. 2′–7′)

V. Incantation No. F.3 “Roar on me! Roar on me! Rear up! Rear up!” (Biggs 1967 No. 8), ll. 65–79 (Ms. F obv. 1–15)

VI. Incantation No. F.4 “At my head a buck is tied!” (Biggs 1967 No. 13), ll. 80–93 (Ms. A obv. 45–rev. 9 // F rev. 1–15)

VII. Ms. A rev. 10–24 = Incantation No. E.1 “May the wind blow! May the mountains quake!” (Biggs 1967 No. 14)

VIII. Incantation No. F.5 “Daughter of Ninĝirsu, the releaser I am” (Biggs 1967 No. 22), ll. 93–99 (Ms. A rev. 25–30)

IX. Ms. A rev. 31–34 = Incantation and its rituals No. B.2 “O Adad, locker keeper of the canals of heaven” (Biggs 1967 No. 23)

Summary of the sections of manuscripts not included in the transliteration:

- Ms. A = KAR 70
  6–10 = D Diagn. (Ms. F)
  rev. 10–24 = No. E.1 (Ms. B)
  rev. 31–34 = No. B.2 (Ms. B)
  rev. 35 = Fragmentary colophon, see commentary *nīš libbi* B: 40

- Ms C = BAM 205
  obv. 1′–8′ = Abusch and Schwemer 2011, No. 2.5: 26–30
  obv. 19′–rev. 2′ = K prescr 21 (Ms. G)
  rev. 3′–5′ = K prescr. 26 (Ms. G)
  rev. 6′–8′ = K prescr. 22 (Ms. G)
  9′–14′ = Farber 1977a, Hauptritual B (with duplicates)
  15′–20′ (// BAM 320 rev. 15′–20′)= Instruction for a therapeutic ritual against sexual impurity

- Ms D = K. 9451+11676+Sm. 818+961
  Obv. i 1′ = Fragmentary prescriptions (see commentary l. 31)
  obv. i 7′–16′ = No. K.1 (Ms. B)

obv. ii = Fragmentary
rev. iv 1′–6′ = No. K.2 (Ms. B)
rev. iv 7′–9′ = Part of an Ashurbanipal colophon similar to Hunger 1968, No. 318

– Ms E = STT 280
i 1–7 = K prescr. 1–2 (Ms. A)
i 8–17 = D Sympt.; prescr. 4 (Ms. E)
i 22–51 = K prescr. 3–18 (Ms. A) (note i 44 = A prescr. 7 (Ms. D))
i 52–55 = A prescr. 9–12, 16 (Ms. D)
i 56–ii 9 = K prescr. 19–22 (Ms. A)
ii 10–21 = No. M.1 (Ms. B)
ii 22–35 = K prescr. 23–28 (Ms. A)
ii 36–50 = No. K.3 (Ms. A)
ii 51–53 = No. K.4 (Ms. A)
ii 54–61 = No. K.5 (Ms. A)
ii 62–iii 23 = No. E.2 (Ms. H)
iii 24–33 = No. K.6 (Ms. A)
iii 34–42 = No. K.7 (Ms. A)
iii 43–iv 7 = K prescr. 29–31 (Ms. A)
iv 8–23 = No. K.8 (Ms. A)
iv 24–31 = No. K.9 (Ms. A)
iv 32–36 = K prescr. 32 (Ms. A)
iv 37–41 = No. A.1 (Ms. D)

Transliteration

I. A obv. 1–5 = Incantation No. F.1

1.	A o. 1	[ÉN x x] aḫ lu up pa di ra aḫ an ki nu sum nu sum
2.	A o. 2	[. . .] x ni kab mu bu bu a ḫa an til la ke_4
3.	A o. 3	[. . .] x ti an a ḫa an ti la ke_4
4.	A o. 4	x na ḫa an ub bi a ḫa an ti áb bu uk
5.	A o. 5	KA.INIM.MA *maš taq ti* [Š]À.ZI.GA

II. ll. 6–10 (Ms. A) = *nīš libbi* D Diagn.: 24–26

III. A obv. 11–44 // B obv. 1′–17′ // C obv. 7′–18′ // D 2′–6′ // E i 18–21 = Prescriptions 1–17

11.	B o. 1′	[. . .]^{ú}x x [. . .]
12.	B o. 2′	[1-*niš* S]ÚD NU *pa-ta*[*n* NAG-*ma* TI]
13.	B o. 3′	[DIŠ KI.MI]N *ana* GIDIM *pa-qid* ÚḪ(text ÙḪ).^{d}ÍD x [. . .]
14.	B o. 4′	[DIŠ] KI.MIN ^{ú}*tu-lal* ^{ú}*aš-ta-til-la* ^{ú}*k*[*a-bul-la*$^{?}$]
15.	B o. 5′	^{ú}*a-ṣu-ṣu-um-tú* ^{ú}SIKIL SUḪUŠ ^{ú}x [. . .]
16.	B o. 6′	7 Ú. ḪI.A *an-nu-ti* 1-*niš* SÚD NU *pa-tan ina* KAŠ NA[G-*ma* TI]
17.	B o. 7′	DIŠ KI.MIN ^{ú}A.ZAL.LÁ-*a ina* KAŠ SILA$_{11}$[-*aš* . . .]
18.	B o. 8′	*ina* KUŠ *ina* MÚRU-*šú* KE[ŠDA-*ma* TI]
19.	A o. 11	DIŠ NA *ka-šip-ma* — — *mu-un-ga i-šu* — *bir-ka-šú* *ga-*[*an-na*]
	B o. 9′–10′	DIŠ NA *ka-šip-ma* UZU^{meš}-*šú tab-ku mun-ga* TU[KU-*šú*] / *ù bir-ka-a-šú* *ga-an-na* ŠÀ-*šú*
	C o. 7′–8′	DIŠ NA *ka-šip-ma* UZ[U^{meš}-*šú ta*]*b-ku mun-ga* T[UKU-*šú*]/*u bir-ka-šú* *ga-*[*an-n*]*a* ŠÀ-*šú*
20.	A o. 12	⸢MUNUS *ḫa-šiḫ-ma* MUNUS IGI-*ma*⸣ ŠÀ-*šú i-tu-ra* {*šá*} N[A]
	B o. 10′–12′	MUNUS *ḫa-ši*[*ḫ-ma*] / MUNUS IGI.BAR-*ma* ŠÀ-*šú* GUR-[*ra*] / NA — *ri- ḫu-su*
	C o. 8′–9′	MUNUS *ḫa-šiḫ-*[*ma*] / MUNUS IGI.BAR-*ma* ŠÀ-[*šú* GU]R-*ár*$^{!}$ NA BI *ri- ḫu-*[*su*]
21.	A o. 13	KI [x] ⸢*šu-nu-lat ana* TI-*šú* ^{ú}NU.LUḪ.ḪA⸣ – ^{ú}Ḫ[AR.ḪAR]
	B o. 12′–13′	KI ADDA *šu-*[*nu-lat*] / *ana* TI-*šú* Ú.KUR.RA ^{ú}ḪAR.ḪAR ^{ú}[ ]
	C o. 10′–11′	KI AD$_{6}$ *šu-nu-*[*lat*] *ana* TI-*šú* Ú.KUR.RA/^{ú}ḪAR.ḪAR ^{ú}IGI-*lim*
22.	A o.	– – – – –
	B o. 14′–15′	^{ú}NU.LUḪ.ḪA ^{ú}NIGIN^{sar} ^{ú}*tu-lal* ^{ú}A[Š.TÁL.TÁL] / ⸢1⸣-*niš* SÚD
	C o. 11′–13′	^{ú}NU.LUḪ.ḪA / ^{ú}*su-pa-lam* ^{ú}*tu-lal* ^{ú}AŠ.TÁL.TÁL / 1-*niš* SÚD
23.	A o. 14	*ina* GE[ŠTIN. Š]UR – – NAG.MEŠ-*ma* [ ]
	B o. 15′	*ina* GEŠTIN. ŠUR.RA NU *pa-tan* NAG.[MEŠ]
	C o. 13′	*ina* [G]EŠTIN.ŠUR NU *pa-tan* NAG.MEŠ
24.	A o. 15	DIŠ K[I.MIN] $^{\text{na4}}$*mu-ṣa* $^{\text{na4}}$KA A.AB.BA AN.[BAR SÚ]D-*ma ina* [x NAG.MEŠ]
25.	A o. 16	⸢^{ú}⸣GAN.U$_{5}$ ^{ú}*tar*$^{!}$-*muš ina* KUŠ DÙ.DÙ-*pí ina* GÚ-*šú* G[AR-*an*]
26.	B o. 16′–17′	[^{ú}]KUR.KUR ⸢*at*⸣-*ma* BIL.ZA.ZA / [x]
	C o. 14′	DIŠ KI.MIN ^{ú}KUR.KUR *at-ma* BIL.ZA.ZA SIG$_{7}$

27.	B o. 17′	[^{giš}ER]EN EŠ.M[EŠ-*su*]
	C o. 14′–15′	GIŠ.ḪAŠḪUR.^{giš}GI / *ina* Ì.GIŠ ^{giš}EREN EŠ.MEŠ-*su*
28.	B o. 18′	Rest broken
	C o. 15′–16′	NA BI EN TI.LA / ⸢ŠÀ⸣-*šú* ÍL-*šú kiš-pi* NU TE-*šú*

29.	C o. 17′	DIŠ KI.MIN ^{šim}GÚR.GÚR ^{giš}LI$^{!}$ MUN *a-ma-nim*
30.	C o. 18′	⸢^{ú}⸣KUR.KUR *ina* Ì.GIŠ ŠUR.MÌN EŠ.MEŠ-*s*[*u*]

31.	A o. 17	DIŠ KI.MIN ^{ú}IGI-*lim* ^{ú}⸢*tar*⸣-*muš* AN.BAR KA *tam-tim* ^{giš}ESI
	D 2′	[^{ú}I]GI-*lim* ^{ú}*tar-muš* AN.BAR KA A.AB.BA ^{giš}E[SI]
	E i 18	[^{ú}*tar*]-*muš* AN.BAR KA A.AB.BA ^{giš}ESI
32.	A o. 18	ÚŠ [BU]RU$_5$.ḪABRUD.DA$^{mušen!}$ NITA$^{!}$ *zap*$^{!}$-*pí* ŠAḪ *šá ana* U$_5$ ZI-*ú*
	D 3′	[x BURU$_5$.ḪABRU]D.DA^{mušen} NITÁ *zap-pi* ŠAḪ *šá ana* U$_5$ ZI-*u*
	E i 19–20	[x BURU$_5$.ḪABRUD.DA]^{mušen} NITÁ *zap-pi* ŠAḪ *šá ana* U$_5$ / [ZI]-⸢*ú*⸣
33.	A o. 19	*ina* KUŠ DÙ.DÙ-*pí* *ina* GÚ-*šú* GAR-*an*
	D 4′	[x KU]Š DÙ.DÙ *ina* GÚ-*šú* GAR-*an*
	E i 20	⸢*ina* KUŠ⸣ DÙ.DÙ *ina* GÚ-*šú* ⸢GAR⸣-*an*

34.	A o. 20	DIŠ KI.MIN ^{ú}IN.NU.UŠ ^{ú}E[ME.U]R.GI$_7$ Ú.K[U$_6$] ^{ú}AŠ.TÁL.TÁL
	D 5′	[ ] ⸢^{ú}⸣IN$_6$.ÚŠ ^{ú}EME.UR.GI$_7$ Ú.KU$_6$ ^{ú}AŠ.TÁL.TÁL
	E i 21	[ ] ⸢^{ú}⸣IN$_6$.ÚŠ ^{ú}EME.UR.GI$_7$ Ú.KU$_6$ ^{ú}AŠ.TÁL.TÁL
35.	A o. 21	*ina* KUŠ DÙ.DÙ-*pí* *ina* [G]Ú-*šú* GAR-*an*
	D 6′	[x KU]Š DÙ.DÙ *ina* GÚ-*šú* GAR-*an*
	E i 21	*ina* KUŠ – – – –

36.	A o. 22	DIŠ NA *ni-iš* ŠÀ-*šú e-ṭir*-⸢*ma*⸣ *ni-iš lìb-bi* NU TUKU-*ši* PA^{meš}
37.	A o. 23	^{giš}KIŠI$_{16}$ *ina* A^{meš} ŠUB-*di* ÚŠ BURU$_5$.ḪABRUD^{ru}.[D]A NITA *ana* A^{meš} ŠUB-*ma*
38.	A o. 24	ŠÀ BURU$_5$.ḪABRUD⸢^{ru}⸣.DA NITÁ *i-al-lu-ut* [*ru-p*]*u-uš-ti*
39.	A o. 25	GU$_4$ TI-*qé ana* A^{meš} *ta-nam-di ina* UL *tuš-bat*
40.	A o. 26	*iš-tu* ^{d}UTU *it-tap-ḫa ina* UGU PA^{meš} ^{giš}KIŠI$_{16}$
41.	A o. 27	GUB-*su-ma ana* IGI ^{d}UTU NAG-*ma* ŠÀ.ZI.GA

42.	A o. 28	DIŠ KI.MIN [GÌŠ BURU$_5$.ḪABRUD^{ru}].DA^{mušen} NITÁ ^{uzu}*nap-šat* UDU.NITÁ *ina* ^{síg}ḪÉ.ME.DA
43.	A o. 29	NIGIN-[*ma ina* MÚRU-*šú* KEŠ]DA$^{?}$ ŠÀ.ZI.GA *ina* Ì.GIŠ ŠÉŠ-[*su* . . .]
44.	A o. 30	[. . . Š]À.ZI.[GA . . .]
45.	A o. 31	x [. . .] x TÉŠ.BI ZI-*ú*(text GA) [. . .] *an* [. . .]
46.	A o. 32	[. . .] TU$_6$ ^{d}EN.KI [. . .] SI.SÁ
47.	A o. 33	[. . . Š]À.ZI.GA MÚRU [. . .] x ⸢MIN⸣$^{?}$

48.	A o. 34	[DIŠ KI.MIN] ⸢^{ú}⸣EME.UR.GI$_7$ [. . . ^{ú}IGI]-*lim* ^{na4}PA
49.	A o. 35	[. . . *ina* KA]Š NAG-⟨*šú*⟩ *ina* Ì$^{?}$.[GIŠ ŠÉŠ-*su ina* KUŠ] *ina* GÚ-*šú* GAR-*an*
50.	A o. 36	[DIŠ KI.MIN] ^{ú}x x x ^{ú}IN$_6$.ÚŠ SUḪUŠ ^{ú}ŠAKIR ^{ú}GAB.LAM
51.	A o. 37	[*ina* KAŠ NAG-*šú ina* Ì.GIŠ] ŠÉŠ-*su ina* KUŠ *ina* GÚ-*šú* GAR-*an*
52.	A o. 38	[DIŠ KI.MIN . . .] x *ba ina* KAŠ NAG-*šú ina* Ì.GIŠ ŠÉŠ-*su*
53.	A o. 39	[*ina* KUŠ DÙ.DÙ-*p*]*í*$^{?}$ *ina* GÚ-*šú* GAR-*an*
54.	A o. 40	[DIŠ KI.MIN ^{na4}AMAŠ.PA].⸢È⸣ ^{na4}ZÚ ^{na4}ZA.GÌN
55.	A o. 41	[^{na4}PA ^{na4}KUR-*nu*] DAB.BA ^{na4}*a-ba-aš-mu*
56.	A o. 42	[^{na4}ALGAMEŠ ^{na4}]*ri*$^{?}$-*ḫu* ^{na4}*saḫ-ḫu-u* ^{na4}*bil-li*
57.	A o. 43	[SÍG UDU.NIT]Á$^{?}$ ZI-*i* TI-*qé* NU.NU NA$_4^{meš}$ È-*kak*
58.	A o. 44	[*ina* GÚ-*šú* GAR-*an*]-*ma* ŠÀ.ZI.GA

(Ms. A continues with section VI)

IV. B rev. 2′–7′ = Incantation No. F.2

59.	B r. 2′	[ÉN *am-mi-ni*] *ar-ma* IGI^{II}-*ka* [*am-mi-ni* . . .]
60.	B r. 3′	[*i-b*]*a-áš-ši ina* ŠÀ-*ka šá* MUNUS DU-*k*[*u*$^{?}$ . . .]
61.	B r. 4′	*ti-bi ti-bi* ⸢GU$_4$.UD⸣ [GU$_4$.UD]
62.	B r. 5′	*ina* ^{na4}*me-ek-ki* x [. . .]
63.	B r. 6′	[*i*]*na* ÚŠ BURU$_5$.ḪABRUD.DA[^{mušen} NITA . . .]
64.	B r. 7′	[. . .] *lip-pa-šir* [. . . TU$_6$ ÉN]

V. F obv. 1–15 = Incantation No. F.3

65.	F o. 1	[ÉN *ug-ga ug-ga*] [ZI]-⸢*a*⸣ [ZI-*a*]
66.	F o. 2	[*ug-ga* GIM *a-a-li*] ZI-*a* GIM [*ri-mi*]
67.	F o. 3	[. . .] *lu ú-da-a-ni* [. . .]
68.	F o. 4	[. . .] *ri-kib-ta-ka* [. . .]
69.	F o. 5	[*ra-am* DÀRA$^{?}$ 6-*šú*] *ra-am a-a-*⸢*li*⸣ [7-*šú*]
70.	F o. 6	[*ra-am* BURU$_5$.ḪABRUD.DA^{mušen}] 12-*šú ra-*[*man-ni*]
71.	F o. 7	*ra-man-ni*$^{?}$ *áš-šú ṣe-eḫ-re-ku r*[*a-man-ni*]
72.	F o. 8	[. . .] *šak-na-ku ra-mu ra-*[*man-ni*]
73.	F o. 9	[. . .] *šá ri-kib-ti* DÀRA.MAŠ KEŠDA-*ma r*[*a-man-ni*]
74.	F o. 10	[DUMU.MUNUS ^{d}*nin-gír-su*] *pa-ši-ru* [*ana-ku*]
75.	F o. 11	[*um-mi pa-ši-rat* A]D-*ú-a pa-*[*ši-ir*]
76.	F o. 12	[*ana-ku šá al-li-ka pa-ša-á*]*r*$^{?}$-*um-ma a-*⸢*pa*⸣[-*áš-šar*]
77.	F o. 13	[. . .]-⸢*ḫat tu*$_4$⸣ *šá* NENNI [A NENNI]
78.	F o. 14	[GÌŠ-*šú lu-u* ^{giš}GIDRU *mar-te-em-ma*$^{?}$ *li-duk* KÁ *š*]*u-bur-ri šá* MUNUS NEN[NI-*ti*]
79.	F o. 15	[. . .] ⸢*tu*$_4$⸣ [. . .]

VI. A obv. 45–rev. 9 // F rev. 1–15 = Incantation No. F.4

80.	A o. 45	[ÉN ina SAG-ia da-á]š-šú ra-ki-is : ina še-pi-ti-ia
	F r. 1	[še-pi-ti-i]a$^{!}$
81.	A o. 46	[$^{\text{udu?}}$pu-ḫa-lu ra-ki-i]s$^{!}$(text A) da-áš$^{!}$(text PA)-šú ḫu-ub-bi-ban-ni
	F r. 2–3	[^{ud}]$^{\text{u?}}$pu-[ḫa-lu] / [ḫ]u-ub-⸢bi⸣-ba[n-ni]
82.	A o. 47	[pu-ḫa-lu] rit-ka-ban-ni
	F r. 3–4	[ ] /rit-ka-ban-[ni]
83.	A o. 48	[. . . du]-ku-uk ri-i-mi it-ti-ka lit-ba-a
	F r. 4–5	[. . .] / du-ku-uk ri-i-[mi]
84.	A o. 49	[e-mu-q]a-an it-ti-ka lit-ba-a bir-ka-ka a-ni-ḫa-tu
	F r. 6–7	e-mu-qa-[an] / bir-ka-ka a-ni-ḫa-⸢a⸣-[tu]
85.	A r. 1	[it-ti-ka lit-ba]-⸢a⸣ mi-⸢na-tu⸣-ka it-ti-k[a l]it-[ba-a] meš-re-⸢tu⸣-[ka]
	F r. 7–9	[ ] / mi-na-tu-[ka] / meš-re-⸢tu⸣-[ka]
86.	A r. 2	[it-t]i-ka ⸢lit⸣-ba-a ⸢ku⸣-[l]u$^{?}$-l[i$^{?}$-ka$^{?}$]
	F r. 9–10	[ ] / ⸢ku⸣-lu-[li-ka]
87.	A r. 3	[. . .]-⸢ka-ma⸣$^{?}$ ma-a-a-al-⸢ta⸣-[x x]
	F r. 10–11	[. . .] / [m]a-a-a-a[l-ta-x x]
88.	A r. 4	⸢e tag⸣-lu-ut ⟨e⟩ ⸢ta⸣-a’-dir ra-am-ka ⸢e⸣ ta-šú-[uš]
	F r. 11–12	[ ] / [r]a-am-k[a]
89.	A r. 5	ina qí-bit {ina} te-e te-⸢li⸣-ti ⸢^{d}⸣iš-ta[r]
	F r. 12–13	[ ] / [^{d}]15
90.	A r. 6	^{d}é-a ^{d}UTU ù ^{d}a[sal-lú-ḫi TU$_{6}$] ÉN
	F r. 13	^{d}IDIM [ ]
91.	A r. 7	DÙ.DÙ.BI Ì BUR ina ^{giš}DÍLIM$^{!}$.⸢Ì⸣$^{!}$.[ŠÉ]Š$^{!?}$ ⸢^{giš}⸣TAŠKARIN TI-qé [. . .] ⸢x⸣ [x]
	F r. 13–14	[ ] / [^{gi}]^{š}TAŠKARIN T[I-qé]
92.	A r. 8	^{šim}LI 3-šú ana IGI ⸢^{d}⸣[15 MAR$^{?}$ É]N an-ni-tú 3-šú [ana UGU ŠI]D-nu
	F r. 14–15	[/ an-ni]i$^{?}$-⸢tú$^{?}$⸣]
93.	A r. 9	mi-na-ti-šú [Š]ÉŠ-ma ŠÀ.ZI.GA

VII. Ms. A rev. 10–24 = Incantation No. E.1

VIII. A rev. 25–30 = Incantation No. F.5

94.	A r. 25	ÉN DUMU.MUNUS ^{d}*nin-gír-su pa-ši-ru*(text RI) *ana-ku*
95.	A r. 26	*um-mi pa-ši-rat a-bu-ú-a pa-ši-ir*
96.	A r. 27	*ana-ku šá al-li-ka pa-šá-ru-um-ma a-pa-áš-šar*
97.	A r. 28	*šá* NENNI A NENNI GÌŠ-*šú lu-u* ^{giš}GIDRU *mar-te-em-ma*
98.	A r. 29	*li-duk* KÁ *šu-bur-ri šá an-na-ni-tu-ú-a*
99.	A r. 30	*la i-šab-ba-a la-la-a-šá* TU_6 ÉN

IX. Ms. A rev. 31–34 = Incantation and its rituals No. B.2

Transcription

I. A o. 1–5 = Incantation No. F.1

1–4. Abracadabra
5. *šipat . . . nīš libbi*

II. ll. 6–10 (Ms. A) = *nīš libbi* D Diagn.: 24–26

III. A obv. 11–44 // B 1′–17′ // C obv. 7′–18′ // D 2′–6′ // E i 18–21 = Prescriptions 1–17

1. 11. [. . .] . . . [. . .]
   12. [*ištēniš tas*]*âk lā patā*[*n išattī-ma iballaṭ*]

2. 13. [*šumma* KI.MI]N *ana eṭemmi paqid ru'tītu* . . . [. . .]

3. 14. [*šumma*] KI.MIN *tullal aštatilla k*[*abulla*]
   15. *aṣuṣumtu usikillu šuruš* . . . [. . .]
   16. *sebet šammī annûti ištēniš tasâk lā patān ina šikari išat*[*tī-ma iballaṭ*]

4. 17. *šumma* KI.MIN *azallâ ina šikari tal*[*âš* . . .]
   18. *ina maški ina qablīšu tarak*[*kas-ma iballaṭ*]

5. 19. *šumma amēlu kašip-ma šīrūšu tabkū munga īšu u birkāšu gannā libbašu*

   20. *sinništa ḫašiḫ-ma sinništa ippalis-ma libbašu itūra amēlu šū riḫûssu*
   21. *itti mīti šunūlat ana bulluṭīšu nīnâ ḫašâ imḫur-līm*

   22. *nuḫurta supāla tullal ardadilla ištēniš tasâk*

   23. *ina karāni ṣaḫti lā patān ištanatti*

6. 24. *šumma* K[I.MIN (= *amēlu kašip-ma šīrūšu tabkū munga īšu u birkāšu gannā libbašu sinništa ḫašiḫ-ma sinništa ippalis-ma libbašu itūra amēlu šū riḫûssu itti mīti šunūlat*)]: *mūṣa imbû tâmti parzil*[*lu tasâk*]*-ma ina* [. . . *ištanatti*]

7. 25. *bukāna tarmuš ina maški tašappi ina kišādīšu tašak*[*kan*]

8. 26. *šumma* KI.MIN (= *amēlu kašip-ma šīrūšu tabkū munga īšu u birkāšu gannā libbašu sinništa ḫašiḫ-ma sinništa ippalis-ma libbašu itūra amēlu šū riḫûssu itti mīti šunūlat*): *atā'išu atma muṣa''irāni arqi*

   27. *ḫašḫūr api ina šaman erēni taptanaššassu*
   28. *amēlu šū adi balṭu libbašu inaššīšu kišpī lā iṭeḫḫûšu*

Translation

I. A obv. 1–5 = Incantation No. F.1

1–4. Abracadabra
5. Wording of (the incantation) . . . (for the loss of) sexual desire.

II. ll. 6–10 (Ms. A) = *nīš libbi* D Diagn.: 24–26

III. A obv. 11–44 // B 1′–17′ // C obv. 7′–18′ // D 2′–6′ // E i 18–21 = Prescriptions 1–17

1. 11. [. . .] . . .-plant [. . .]
   12. [you pul]verize (them) [together], [he drinks (them)] on an emp[ty] stomach [and he will recover].
2. 13. [If ditt]o: He has been entrusted to a ghost, sulphur . . . [. . .].
3. 14. [If] ditto: *tullal*-plant, *aštatillu*-plant, *k*[*abullu*]-plant,
   15. *aṣuṣumtu*-plant, *sikillu*-plant, root of the . . .plant [. . .];
   16. you pulverize together these seven plants, he drink[s] (them) with beer on an empty stomach [and he will recover].
4. 17. If ditto: You knea[d] *azallû*-plant with beer [. . .]
   18. you pu[t] (it) in a leather bag around his waist [and he will recover].
5. 19. If a man is bewitched and his flesh feeble, has *mungu*-paralysis and his knees are contracted, his 'heart' (*libbu*)
   20. needs a woman and finds her, but his 'heart' (*libbu*) returns. That man's semen
   21. has been buried with a dead person. To cure him: *nīnû*-plant, *ḫašû*-plant, "heals-a-thousand"-plant,
   22. *nuḫurtu*-plant, *supālu*-juniper, *tullal*-plant, *ardadillu*-plant, you pulverize (them) together,
   23. he drinks it repeatedly in grape juice on an empty stomach
6. 24. If d[itto (= a man is bewitched and his flesh feeble, has *mungu*-paralysis and his knees are contracted, his 'heart' needs a woman and finds her, but his 'heart' returns. That man's semen has been buried with a dead person)]: [you pulverize] *mūṣu*-stone, coral limestone, iro[n] and [he drinks (it)] in [. . .].
7. 25. You wrap up *bukānu*-plant (and) *tarmuš*-plant in a leather (bag), you pu[t] (it) around his neck.
8. 26. If ditto (= a man is bewitched and his flesh feeble, has *mungu*-paralysis and his knees are contracted, his 'heart' needs a woman and finds her, but his 'heart' returns. That man's semen has been buried with a dead person): *atāʾišu*-plant, tadpole of a green frog,
   27. *ḫašḫūr api*-plant in cedar oil, you rub him repeatedly (with it),
   28. this man, as long as he lives, will desire (sexually). The witchcraft will not come near him.

9. 29. *šumma* KI.MIN (= *amēlu kašip-ma šīrūšu tabkū munga īšu u birkāšu gannā libbašu sinništa ḫašiḫ-ma sinništa ippalis-ma libbašu itūra amēlu šū riḫûssu itti mīti šunūlat*): *kukuru burāšu ṭābat amānim*
30. *atā'išu ina šaman šurmēni taptanaššass*[*u*]

10. 31. *šumma* KI.MIN (= *amēlu kašip-ma šīrūšu tabkū munga īšu u birkāšu gannā libbašu sinništa ḫašiḫ-ma sinništa ippalis-ma libbašu itūra amēlu šū riḫûssu itti mīti šunūlat*): *imḫur-līm tarmuš parzilla imbû tâmti ušâ*

32. *dām* [*iṣ*]*ṣur ḫurri zikari zappī šaḫî ša ana rakābi tebû*
33. *ina maški tašappi ina kišādīšu tašakkan*

11. 34. *šumma* KI.MIN (= *amēlu kašip-ma šīrūšu tabkū munga īšu u birkāšu gannā libbašu sinništa ḫašiḫ-ma sinništa ippalis-ma libbašu itūra amēlu šū riḫûssu itti mīti šunūlat*): *maštakal lišān-kalbi šimra* (or *urâna*) *ardadillu*

35. *ina maški tašappi ina kišādīšu tašakkan*

12. 36. *šumma amēlu nīš libbīšu eṭir-ma nīš libbi lā īši arī*
37. *ašāgi ina mê tanaddi dām iṣṣūr ḫur*[*r*]*i zikari ana mê tanaddī-ma*

38. *libbi iṣṣūr ḫurri zikari i'allut* [*rup*]*ušti*
39. *alpi teleqqe ana mê tanaddi ina kakkabī tušbât*

40. *ištu Šamšu ittapḫa ina muḫḫi arī ašāgi*
41. *tušzāssu-ma ana pān Šamši išattī-ma nīš libbi*

13. 42. *šumma* KI.MIN (= *amēlu nīš libbīšu eṭir-ma nīš libbi lā īši*): [*ušar iṣṣūr ḫur*]*ri zikari napšat immeri ina nabāsi*
43. *talammī-*[*ma ina qablīšu tarak*]*kas nīš libbi ina šamni tapaššas*[*su . . .*]

44. [. . .] *nīš lib*[*bi . . .*]
45. . . . [. . .] . . . *ištēniš itbû* [. . .] . . . [. . .]
46. [. . .] *tî Ea* [. . .] *išaru*
47. [. . .] *nīš* [*lib*]*bi qablu* [. . .] . . . MIN

14. 48. [*šumma* KI.MIN (= *amēlu nīš libbīšu eṭir-ma nīš libbi lā īši*):] *lišān-kalbi* [. . . *imḫur*]-*līm ayyarta*

49. [. . . *ina šika*]*ri išattīšu ina ša*[*mni tapaššassu ina maški*] *ina kišādīšu tašakkan*

15. 50. [*šumma* KI.MIN (= *amēlu nīš libbīšu eṭir-ma nīš libbi lā īši*):] . . . *maštakal šuruš šakirî* GAB.LAM
51. [*ina šikari išattīšu ina šamni*] *tapaššassu ina maški ina kišādīšu tašakkan*

9. 29. If ditto (= a man is bewitched and his flesh feeble, has *mungu*-paralysis and his knees are contracted, his 'heart' needs a woman and finds her, but his 'heart' returns. That man's semen has been buried with a dead person): *kukru*-plant, *burāšu*-juniper, *amānu*-salt,

30. *atāʾišu*-plant in cypress oil, you rub hi[m] repeatedly (with it).

10. 31. [If ditto (= a man is bewitched and his flesh feeble, has *mungu*-paralysis and his knees are contracted, his 'heart' needs a woman and finds her, but his 'heart' returns. That man's semen has been buried with a dead person): "Heals-a-thousand"-plant, *tarmuš*-plant, iron, coral limestone, *u*[*šû*]-wood

32. [blood of a] male [partrid]ge, bristle of a pig reared-up for mating,

33. you wrap up (them) [in a leath]er bag, you put (it) around his neck.

11. 34. [If ditto (= a man is bewitched and his flesh feeble, has *mungu*-paralysis and his knees are contracted, his 'heart' needs a woman and finds her, but his 'heart' returns. That man's semen has been buried with a dead person)]: *maštakal*-plant, "dog's-tongue"-plant, *šimru*-plant (or *urânu*-plant), *ardadillu*-plant

35. you wrap up (them) [in a leath]er bag, you put (it) around his [ne]ck.

12. 36. If a man's sexual desire has been taken away and he does not have sexual desire:

37. You throw into water branches of the *ašāgu*-acacia, you throw into water blood of a male partrid[ge] and

38. he eats the innards of the male partridge,

39. you take the [sa]liva of a bull, you throw (it) into water (and) let it spend the night under the stars,

40. as soon as the sun has risen, on the branches of *ašāgu*-acacia

41. you have him (= patient) stand and he drinks (it) in front of the sun, and (he will get) sexual desire.

13. 42. If ditto (= a man's sexual desire has been taken away and he does not have sexual desire): [The penis of a] male [partrid]ge, the throat of a sheep with red wool,

43. you wrap up [and you ti]e (them) [around his waist], sexual desire, you anoint [it] with oil [. . .]

44. [. . .] sexual desi[re . . .]

45. . . . [. . .] . . . together rise [. . .] . . . [. . .]

46. [. . .] incantation of Ea [. . .] right$^{?}$

47. [. . .] sexual [desi]re, waist [. . .] . . . ditto.

14. 48. [If ditto (= a man's sexual desire has been taken away and he does not have sexual desire)]: "Dog's-tongue"-plant, [. . ., "heals]-a-thousand"-[plant], *ayyartu*-shell,

49. [. . .] he drinks it [in be]er, (or$^{?}$) [you anoint it] with o[il], (or$^{?}$) you put (it) [in a leather bag] around his neck.

15. 50. [If ditto (= a man's sexual desire has been taken away and he does not have sexual desire)]: . . .-plant, *maštakal*-plant, root of *šakirû*-plant, GAB.LAM-plant,

51. [he drinks (them) in beer], (or$^{?}$) you anoint it [with oil], (or$^{?}$) you put (it) [in a leather bag] around his neck.

16. 52. [*šumma* KI.MIN (= *amēlu nīš libbīšu eṭir-ma nīš libbi lā īši*): . . .] . . . *ina šikari išattīšu ina šamni tapaššassu*

53. [*ina maški tašappi*] . . . *ina kišādīšu tašakkan*

17. 54. [*šumma* KI.MIN (= *amēlu nīš libbīšu eṭir-ma nīš libbi lā īši*): *amašp*]*û ṣurru uqnû*

55. [*ayyartu šadânu*] *ṣābitu abašmû*
56. [*algamišu*] *riḫu saḫḫû billi*
57. [*šīpāt immer*]*i tēbî teleqqe taṭammi abnī tašakkak*
58. [*ina kišādīšu tašakkan*]*-ma nīš libbi*

IV. B rev. 2′–7′ = Incantation No. F.2

59. [*šiptu: ammīni*] *armā īnāka* [*ammīni* . . .]
60. [*ib*]*ašši ina libbīka ša sinništu* DU-*k*[*u* . . .]
61. *tibi tibi šaḫiṭ* [*šaḫiṭ*]
62. *ina mekki* . . . [. . .]
63. [*i*]*na dām iṣṣūr ḫurri* [*zikari* . . .]
64. [. . .] *lippašir* [. . . *tê šipti*]

V. F obv. 1–15 = Incantation No. F.3

65. [*šiptu: ugga ugga*] [*tib*]*â* [*tibâ*]
66. [*ugga kīma ayyali*] *tibâ kīma* [*rīmi*]
67. [. . .] . . . [. . .]
68. [. . .] *rikibtāka* [. . .]
69. [*râm turāḫi šeššīšu*] *râm ayyali* [*sebîšu*]
70. [*râm iṣṣūr ḫurri*] *šinšerīšu rā*[*manni*]
71. *rāmanni aššu ṣeḫrēku r*[*āmanni*]
72. [. . .] *šaknāku râmu rā*[*manni*]
73. [. . .] *ša rikibti ayyali rukus-ma r*[*āmanni*]
74. [*mārat Ningirsu*] *pāširu* [*anāku*]
75. [*ummī pāširat a*]*būya pā*[*šir*]
76. [*anāku ša allika pašā*]*rum-ma apa*[*ššar*]
77. [. . .] . . . *ša annanna* [*mār annanna*]
78. [*ušaršu lū ḫaṭṭi martê-ma lidūk bāb š*]*uburri ša sinništi anna*[*nnīti*]
79. [. . .] . . . [. . .]

VI. A obv. 45–rev. 9 // F rev. 1–15 = Incantation No. F.4

80. [*šiptu: ina rēšīya da*]*ššu rakis : ina šēpītīya*
81. [*puḫālu raki*]*s daššu ḫubbibanni*
82. [*puḫālu*] *ritkabanni*

16. 52. [If ditto (= a man's sexual desire has been taken away and he does not have sexual desire)]: . . . he drinks (it) in beer, (or$^?$) you anoint it [with oil],
53. (or$^?$) [you wrap up (it) in a leather (bag)], you put (it) around his neck.

17. 54. [If ditto (= a man's sexual desire has been taken away and he does not have sexual desire): *amašp*]*û*-stone, *ṣurru*-black stone, lapis lazuli,
55. [*ayyartu*-shell, magne]tite, *abašmû*-stone,
56. [*algamišu*-stone], *riḫu*-[stone], *saḫḫû*-stone, *billu*-stone;
57. you take [wool of a] reared-up [ra]m, you spin (it), you align the stones (over it),
58. [you put (it) around his neck] and (he will get) sexual desire.

IV. B rev. 2′–7′ = Incantation No. F.2

59. [Incantation: Why] are your eyes covered? [Why . . .]?
60. [(It) i]s in your 'heart,' which the woman . . . [. . .].
61. Rear up! Rear up! Mount! [Mount!]
62. With the *mekku*-stone . . . [. . .],
63. [w]ith the blood of a [male] partridge [. . .],
64. [. . .] may he be released! [. . . Incantation formula].

V. F obv. 1–15 = Incantation No. F.3

65. [Incantation: Roar on me! Roar on me]! [Rear up! Rear up]!
66. [Roar on me like a stag]! Rear up like [a wild bull]!
67. [. . .] . . . [. . .]
68. [. . .] your love-making [. . .].
69. [The mating of a wild goat six times], the mating of a stag [seven times],
70. [the mating of a partridge] twelve times, make [love to me]!
71. Make love to me! Ma[ke love to me] because I am young!
72. [. . .] I am endowed with love, Make love to [me]!
73. Tie [. . .] of the *rikibtu* of a stag! Ma[ke love to me]!
74. [Daughter of Ninĝirsu], the releaser [I am].
75. [My mother is a releaser, my fa]ther is a rel[easer].
76. [I, who have come], will really release!
77. [. . .] the . . . of NN, [son of NN],
78. [may his penis be a stick of *martû*-wood, may it hit the a]nus of the woman NN!
79. [. . .] . . . [. . .].

VI. A obv. 45–rev. 9 // F rev. 1–15 = Incantation No. F.4

80. [Incantation: At my head a bu]ck is tied! At my feet
81. [a ram is tie]d! Buck, bleat *for* me!
82. [Ram], mount me!

83. [. . .] *dukuk rīmī ittīka litbâ*
84. *emūqān ittīka litbâ birkāka anīḫātu*
85. *[ittīka litbâ] minâtūka ittīk[a li]t[bâ] mešrêtū[ka]*

86. *[itt]īka litbâ kulūl[īka?]*
87. [. . .] . . . *mayyalta* [. . .]
88. *ē taglut ē ta'dir râmka ē tāš[uš]*
89. *ina qibīt tê telīti Ištar*
90. *Ea Šamaš u A[salluḫi tê] šipti*

---

91. *dudubû: šaman pūri ina napša[št]i taskarinni teleqqe* [. . .] . . .
92. *burāšu šalāšīšu ana pān [Ištar tazarru šip]ta annītu šalāšīšu [ana muḫḫi tam]annu*

93. *minâtīšu [tap]aššaš-ma nīš libbi*

VII. Ms. A rev. 10–24 = Incantation No. E.1

VIII. A rev. 25–30 = Incantation No. F.5

94. *šiptu: mārat Ninĝirsu pāširu anāku*
95. *ummī pāširat abūya pāšir*
96. *anāku ša allika pašārum-ma apaššar*
97. *ša annanna mār annanna ušaršu lū ḫaṭṭi martê-ma*
98. *lidūk bāb šuburri ša annannītūa*
99. *lā išabbâ lalâša tê šipti*

IX. Ms. A rev. 31–34 = Incantation and its rituals No. B.2

83. [. . .] jump, my wild bull! Together with you, may the strength rise!
84. Together with you, may your tired knees rise!
85. [Together with you, may] your limbs [rise]! Together with yo[u, may your] members r[ise]!
86. [Together] with you, may [your] . . . rise!
87. [. . .] . . . bed [. . .].
88. Do not get scared! Do not be afraid! Do not be afflict[ed] for your love-making!
89. At the command of the incantation of wise Ištar,
90. Ea, Šamaš e As[alluḫi]. Incantantion [formula].

---

91. Its ritual: You put oil from the alabastron in a boxwood container for unguents,
92. [you spread] juniper three times in front of [Ištar, you reci]te this [inca]ntation [over (it)] three times,
93. you [an]oint his limbs (with the oil) and (he will get) sexual desire.

VII. Ms. A rev. 10–24 = Incantation No. E.1

VIII. A rev. 25–30 = Incantation No. F.5

94. Incantation: Daughter of Ninĝirsu, the releaser I am.
95. My mother is a releaser, my father is a releaser.
96. I, who have come, will really release!
97. May the penis of NN, son of NN, be a stick of *martû*-wood and
98. hit the anus of the woman NN,
99. (so) he will be (never) satisfied with her charms! Incantation formula.

IX. Ms. A rev. 31–34 = Incantation and its rituals No. B.2

Commentary

1–4. Note that a blank space divides the lines into two columns, where many sound elements of the second column sound generally Sumerian. In line 1 an-ki "heaven and earth"; nu sum might be read as nu sì "to place sperm";[563] in lines 2–4 the expression an-ti(-la-ke$_4$) recalls the "heaven" (an) and the "end" (til). In line 4 the last three signs *âb bu uk* can be read *lit-bu-uk* da *tabāku* 'to pour': "May he pour!" (see Ebeling 1925: 28; Biggs 1967: 47).

5. Ebeling (1925) interpreted the line as "incantation for an old woman" reading *pár-šum-ti*. Another possible reading suggested by Biggs (1967: 47) is *mas-taq-ti*. For this word see Labat 1951: 64 fn. 117 who suggests the meaning of "physical deficiency." Maybe the very fragmentary line Sp II 976 = BM 35394 (CCMAwR 3, pls. 2–3) obv. i 1′ is a duplicate of Ms. A (KAR 70) obv. 5, since both the texts are in the following lines duplicates (see *nīš libbi* D Diagn.: 24–26).

13. The verb *paqādu* has in this context a negative connotation, in fact "to have been entrusted to a ghost" is a sign of the malevolent witch's attack: the witch entrusts her victim to the ghost. See for similar expression in the anti-witchcraft texts: No. 8.1: 9; 8.3.1: 2; 8.6.1: 23 (Abusch and Schwemer 2011); No. 7.11.2: 7, 17; 8.25: 127, 150, 172, 190; 11.4: 16 (Abusch and Schwemer 2016); No. 3.15: 30″; 6.2: 10′; 7.35.2: 13‴ (Abusch et al. 2020). See also a diagnosis in *Diagnostic Handbook* SA.GIG/*Sakikkû* XVI 10 (Heeßel 2000: 172; Scurlock 2014b: 151), and *Aššur Medical Catalogue* section "pregnancy" l. 109 (Steinert 2018b: 217 and 269).

14. *aštatillu* is a variant of *ardadillu*.

19. Similar passages in *nīš libbi* D Sympt.: 42–45.

22. The term *supālu* (CAD S 390–391; AHw. II 1059–1060) refers to two different plants: a variety of juniper (^{ú}NIGIN^{sar}) and the *supālu*-plant (^{ú}MUNZER). Here, the juniper is mentioned (while the *supālu*-plant in A prescr. 22: 63; Q prescr. 2: 6).

23. The logograms GEŠTIN.ŠUR are read *karānu ṣaḫtu*, but according to CAD Ṣ 63–64, esp. 64, the reading **ṣiḫtu* is not excluded. In ḪAR-ra = *ḫubullu*, in MB literary texts, in the medical texts from Boghazköy and in AMT 40, 4: 9 (dupl. ibid. 54, 3: 15) we find the determinative GIŠ. It is, therefore, possible that the term indicates the grape must as well as the liquid decanted by the must, producing a liquid containing a high percentage of tannins, suitable only for medical use.

---

563 See Chapter II § "Fifth group: abracadabra."

27. For *ḫašḫūru* 'apple' in amorous context see the following love incantation: "The beautiful woman has brought forth love. / Inana, who loves apples and pomegranates, / has brought forth sexual desire," and its ritual: "Either to an apple or to a pomegranate / you recite the incantation three times. You give (the fruit) to the woman (and) have her suck the juices. / That woman will come to you: you can make love to her" (KAR 61: 1–3, 8–10, Biggs 1967: 70). See also KAR 69: 4–5 (Biggs 1967: 74); A 7478 i 8 (Wasserman 2016: 65); 16056 MAH i 4 (ibid. 104). Note that in the Sumerian myth of *Enki and Ninḫursaĝa* apples are among the fruits which Enki presents to Uttu before having sex with her (l. 177, Attinger 1984: 22). For apples in love literature see Paul 2002: 490; in Biblical love lyrics see Paul 1997: 100.

31. Ms. D contains the final part of another prescription: 1′. [. . .] ⌜GEŠTIN.ŠUR.RA NAG⌝ x x x x "[. . .] he drinks grape juice . . .".

42–43. A similar prescription is *nīš libbi* N prescr. 16 iii 27–31.

43. The sense of *nīš libbi* here is not clear to me. Perhaps *ana* is omitted.

52. Biggs (1967) restores [DIŠ KI.MIN UŠ^{meš} x MUŠ]EN NITÁ, but it is uncertain.

54–58. Note that the *riḫu*-stone is not attested, the reading is therefore uncertain. An amulet with the same ten stones is present in K. 3010+ v 17–21, with the only exception of *ašpû* instead of *amašpû* (Schuster-Brandis 2008: 365):

[^{n}]^{a4}*aš-pú-u* ⌜^{na4}⌝ZÚ GE$_6$ ^{na4}ZA.GÌN ^{na4}PA
[^{n}]^{a4}KUR-*nu* [D]A[B ^{n}]^{a4}*àb-aš-mu* ^{na4}ALGAMEŠ
[^{n}]^{a4}x$^{?}$ [x ^{na4}*s*]*aḫ-ḫu-u* ^{na4}*bil-li*
10 NA$_4$^{meš} Š[À.ZI.G]A *ina* SÍG UDU U$_5$ ZI.GA
NU.NU ⌜UD⌝.D[U . . .] *ina* GÚ-*šú* GAR-*an*

59. The incantation is in the catalogue LKA 94 i 8.

60. *ša* MUNUS DU-*ku* cannot be read in Akkadian as *ša sinništi alāku* because in this expression it is always found in the corpus together with the preposition *ana*.

62. See in a love incantation to have sexual intercourse with a woman where the man declares: "I am the *mekku* of which no equal / exists in the country!" (VAT 13226: 13–14, Zomer 2018: 277, note its use in the ritual in l. 18).

65–66. For the restoration of the lines see No. B.1: 1–2; No. G l: 7.

69–71: See No. C.1: 4–6; No. K.8: 149–150; cf. No. E.3: 48–49.

71–73. See No. C.1: 7–8. Chalendar (2018: 42) restores on the basis of *nīš libbi* C: *áš-šú ṣe-eḫ-re-ku* ⸢x⸣ [. . . *rikibti*$^{?}$ *ayyali*$^{?}$] *šak-na-ku ra-mu ra-*[*man-ni* . . .] *šá ri-kib-ti* DÀRA.MAŠ KEŠDA-*ma r*[*a-man-ni*]. She considers the passage a proof for considering *rikibti ayyali* as an aphrodisiac from the stag. The restoration is uncertain.

74–78. This section of the incantation No. F.3 has as parallel No. F.5. Here the incantations are presented separately because they are not duplicates. For *pāširu* 'releaser' see a love incantation to have sexual intercourse with a woman, where the man declares: "Releaser! Releaser! Releaser!" (VAT 13226: 2–3, Zomer 2018: 277).

80–90. Incantation No. F.4 features two speakers: the female partner (ll. 80–88) and the therapeutic operator (ll. 89–90).

80. The restoration is based on the incantation No. E.1: 5. See also the incipit of an incantation, although partially preserved, in the catalogue LKA 94 iii 3: ÉN *ina* SAG-*ia*$_5$ x [. . .] "Incantation: In my head . . . [. . .]."

83. *rīmī* "my wild bull" could also be read *rîmi* "my beloved."

86. The restoration of *kulūlu* is here uncertain. It is either a headgear or a kind of band around the head, used for both people and divine statues (see CAD K 527–528 'headdress, headband'; AHw. I 505 'Kranz, Schleier$^{?}$'). See in *Maqlû*: *ina birīt kalbī lisūrū kulūlūša / ina birīt kulūlūša lisūrū kalbū* "May her (= witch's) headbands circle between the dogs, / may the dogs circle between its headbands" (V 44–45, Abusch 2016: 139). According to Biggs (1967: 32) it should be part of the body, but we have no evidence supporting this thesis.

89. The formula *ina qibīt ina tê* does not appear elsewhere. The formula *ina tê* "with the incantation" is used to indicate the healing effect of divinity through the incantation. For the sources see CAD T 443–444. It does not appear therefore with the *ina qibīt*-formula. Either we consider only *ina* as a scribal error, translating "at the command of the incantation of the wise Ištar," or *te-e* has to be omitted, leaving the classical formula at the end of the incantation "according to the wise Ištar." Here I follow the first option.

91. See for *taskarinnu* 'boxwood' in amorous context: "Encircle(f.) me between the boxwood trees, as the shepherd encircles the flock" (MAD 5, 8: 21–21, Wasserman 2016: 243).

94–99. 74–78. This incantation has as parallel No. F.3: 74–78.

95. For *abūya* "my father" see GAG § 65 i.

97. *martû* (CAD M/I: 300 'stick, pole(?), (a tree and its wood)'; AHw. II 614–615 'ein Hartriegelstrauch, Hartholz'; Biggs 1967: 41) is probably a tree from whose strong wood it is possible to make sticks (for example, for shepherds, see Lambert 1960a: 160 r. 14). For the realization of a bat in a war ritual (*ḫutpalû*) with this wood, see Elat 1982: 11, l. 3. See *Maqlû*: "(Bearer) of a stick of *martû*-wood to hit the sorcerer" (I 66, Abusch 2016: 36). According to Jiménez (2017: 217–223) *martû* and *ēru* are two names of the same tree.

98–99. Biggs's translation (1967: 41) is the following: "May it hit the anus of (my rival) NN (and injure her) / so that he cannot satisfy himself with her charms! Incantation formula." He suggests, however, other possible translations: "May the penis of NN son of NN be a stick of *martû*-wood, may it hit the anus of the woman NN whose desire is not satisfied." See in *Nergal and Ereškigal* the following passage: *ul ašbâ lalâšu ittalkanni* "(Erra, lover, my lust) I was not sated with his charms when he went away!" (l. 290, Ponchia and Luukko 2013: 19, see also l. 292).

Incantations: stylistic and functional analysis

V. A obv. 45–rev. 9 // F rev. 1–15 = Incantation No. F.4

Text No. F.4 contains an incantation and a short prescription. The prescription (ll. 91–93) prescribes the preparation of an oil from the alabastron-based ointment to be rubbed upon the patient's limbs, after having recited the incantation on the medical preparation three times and having scattered juniper before the goddess Ištar.

The incantation consists of four sections. The first (ll. 80–83) provides the image of the bed; the second (ll. 83–87) uses precative forms aimed at recovering the patient's forces; in the third (l. 88) the patient is asked not to be afraid, in the fourth (ll. 89–90), the final section, we find the formula of efficacy by means of evoking divine names.

The first section is therefore defined by the image of the bed[564] to which animals are tied, namely the buck (*daššu*) and the ram (*puḫalu*). A synecdoche is used to refer to the bed, the animals are in fact tied "to my head, to my feet" (*ina rēšīya*, *ina šēpītīya*). It is the patient's female partner who recites the incantation, and it is to her bed (marked with *-ya*) that the animals are tied. The animals are magically identified with the man. The man is also given the name "my bull" (*rīmī*, l. 83). Clearly, the rhetorical figure of the ascending climax concerning the sexual activity of animals is associated with the principle of stylistic identification, and through the possessive adjective, between man and animals. We have already seen that the wild bull along with the deer are considered

564 For the bed in the rituals see Verderame 2018: 793–794.

to be the beasts with the greatest potency ("the mating of the wild bull 50 times"). It is with the bull that the man is stylistically (-*ī*) identified:

Stylistic and magic identification between man and animals

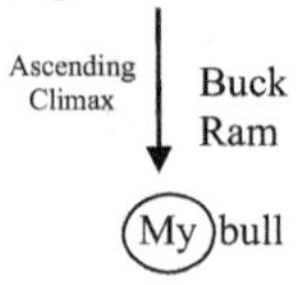

The identification between the patient and the animals is clear from the way in which the woman turns to the animals. She calls the buck, the ram, and the bull to perform sexual acts. "Bleat *for* me," "mount me," "jump" are more or less explicit imperatives, related to the sphere of mating animals. Here there is no question of zoophile practices, so the text uses stylistic figures with magical therapeutic value.

The structure of the first section of the incantation contains two groups of parallel lines and a third one, which, due to its differentiation, allows for continuation to the next section.

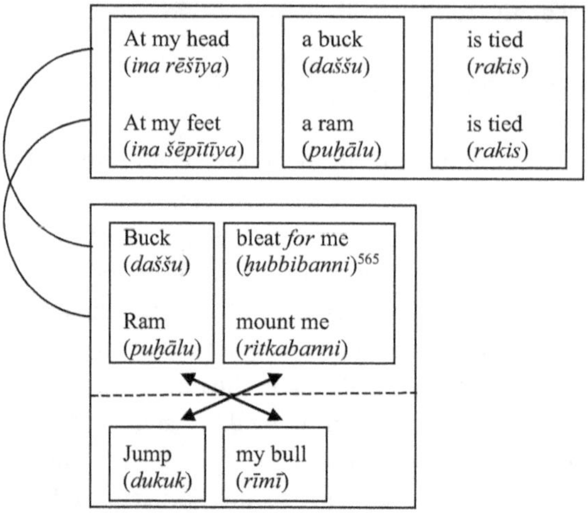

Section two (ll. 83–87) deserves special attention for its poetic structure. It contains an anaphora of the repetition[566] of *ittīka litbâ* "along with you may they rise" at the beginning of the sentences:

565 Note the homeoteleuton between the two lines.
566 On the importance of repetition and redundancy for the definition of "poetic texts" see Lotman 1972 (Engl. transl. 1976); in ritual contexts see Leach 1966: 403–408, esp. 408; for an analysis of redundancy in the rituals from the information theory see Tambiah 1981: 130–132. For an analysis of repetition in Mesopotamian literature see for example Vanstiphout 1992 with previous literature.

Together with may rise (*ittīka litbâ*)	the strength (*emūqān*) your tired knees (*birkāka anīḫātu*) your limbs (*minâtūka*) your members (*mešrêtūka*) [your] . . . (*kulūl*[*īka*?])

Here, the desire of the man to restore his health, his mental and physical integrity is emphasized. The entirety of the mentioned body parts represents the patient's physical integrity. They constitute a hyperbole (see Hirsch 1973–1974: 65). The expression "along with you" here has, therefore, the goal to create an image of recovery of the forces in which the whole body, with all its individual parts, participates. It is clear then, that the term "knees" (*birku*, l. 84) here is not, as in other texts, to be understood as a euphemism for penis (*Pace*: Paul 2002: 494; Schwemer 2010: 121 fn. 250), but rather in its literal meaning: the weary knees may recover!

Nīš libbi G

List of manuscripts

Manuscript	Museum number	Publication	Tablet	Script	Date	Provenience
A	–	LKA 99c: 2–10	Frg. of a two-col.? tablet	NA	8th–7th cent.	Aššur Library N 4

Editions

Biggs 1967: 23–24

Structure of the text

I. Incantation No. G.1 "Stag! Stag! Wild bull! Wild bull!" (Biggs 1967 No. 7), ll. 2–10

Transliteration

I. A 2–10 = Incantation No. G.1

A 2 [ÉN DÀRA].MAŠ KI.MIN ^{gu4}AM KI.MIN
A 3 [*ug-g*]*á*! DÀRA!.MAŠ ZI!-*a*!(text PI?) ^{gu4}AM
A 4 [x] *zi-ni mi-na-tu*$_4$*-ka*
A 5 [x] *zi*!*-ni ni-il-ka*
A 6 [x x] x x x *tú-ka lu zaq-*⸢*pat* $^{he⸣-pí}$
A 7 [*ug*]*-gá* GIM DÀRA!.MAŠ ZI-*a* GIM ^{gu4}AM
A 8 x *zi-ne* Á^{meš}*-ka*
A 9 [T]U$_6$ ÉN É.NU.RU

A 10 KA.INIM.MA ÉN ŠÀ.ZI.GA

Transcription

I. A 2–10 = Incantation No. G.1

2. [*šiptu: ayy*]*alu ayyalu rīmu rīmu*
3. [*ugg*]*a ayyalu tibâ rīmu*
4. [. . .] . . . *minâtūka*
5. [. . .] . . . [. . .] *nīlka*
6. [. . .] . . . *lū zaqpat* *ḫepi*
7. [*ugg*]*a kīma ayyali tibâ kīma rīmi*
8. . . . *emūqīka*
9. [*t*]*û šipti* É.NU.RU

---

10. *šipat nīš libbi*

Translation

I. A 2–10 = Incantation No. G.1

2. Incantation: St]ag! Stag! Wild bull! Wild bull!
3. [Roa]r, stag! Rear up, wild bull!
4. [. . .] . . . your limbs.
5. [. . .] . . . [. . .] your sperm.
6. [. . .] . . . may your . . . be erect! broken
7. [Roa]r like a stag! Rear up like a wild bull!
8. . . . your strength.
9. Enuru incantation [for]mula.

---

10. Wording of *nīš libbi* incantation.

Commentary

1–2. See parallels B.1: 1–2; No. F.2: 61; No. F.3: 65–66. See also catalogue LKA 94 ii 14.

3. The sign ZI is written with an extra wedge.

4–5. See No. B.1: 6 for the mention of limbs and sperm.

7. See commentary ll. 1–2.

8. The imperative *zibil* from *zabālu* 'to bring' is possible, although meaningless here (see Biggs: 1967: 24). Nougayrol (1968: 94) suggests the restoration [*liz*]-*ze-ni* from *zenû* 'to be angry,' but without sense in this context.

10. The incantation is not followed by a ritual, but only by a very fragmentary colophon in ll. 11–13.

II Texts from Nineveh (with duplicates from other sites)

Nīš libbi H

List of manuscripts

Manuscript	Museum number	Publication	Tablet	Script	Date	Provenience	Incantations and prescriptions
A	K. 8790	AMT 65, 7 obv. 1–11	Small frg. of a single-col. tablet	NA	8th–7th cent.	Nineveh Ashurbanipal's Library	Prescr. 1–3; Sympt.; No. H.1
B	VAT 13703	WVDOG 147 (= KAL 7) No. 22 obv. 1–rev. 7	Fragment	NA	8th–7th cent.	Aššur	No. H.1

Editions

Thompson 1934: 82–83, 131–132 (Ms. A)
Biggs 1967: 36, 51–52 (Ms. A)
Meinhold 2017: 58–59 (Ms. B)

Structure of the text

Text *nīš libbi* H is based on Ms. A containing three prescriptions and the symptom description. Both Ms. A and B provide the incantation and its ritual No. H.1 "May the wind blow!".

A Prescr. 1–3; Sympt.
A – B No. H.1

I. Prescription (Ms. A obv. 1)
1. 1
Symptoms (Ms. A obv. 2–3) ll. 2–3
Prescriptions
2. 4 (Ms. A obv. 4)
3. 5 (Ms. A obv. 5)

https://doi.org/10.1515/9783110757262-007

II. Incantation and its ritual No. H.1 "May the wind blow!" (Biggs 1967 No. 16), ll. 7–18 (A obv. 6–11 // B obv. 1–rev. 7)

Transliteration

I. A obv. 1–5 = Prescriptions 1–3 and symptoms

1.	A o. 1	[PI]Š$_{10}$.^{d}ÍD SAḪAR SILA LIMMÚ.BA ^{ú}IGI-*lim* x [. . .]
2.	A o. 2	DIŠ NA *ana* MUNUS-*šú iṭ-ḫe-ma* [. . .]
3.	A o. 3	*a-na* MUNUS-*šú* ŠÀ-*šú* NU ÍL[-*šú* . . .]
4.	A o. 4	DIŠ KI.MIN ^{ú}*ur-ṭú* ^{ú}*a-r*[*a-ri-a-nu* . . .]
5.	A o. 5	DIŠ KI.MIN SUḪUŠ Ú.NAM.TI.LA ^{ú}*a-r*[*a-an-tu*$^{?}$ . . .]

I. A obv. 6–11 // B obv. 1–rev. 7 = Incantation and its ritual No. H.1

6	A o. 6	ÉN *lil-lik* IM *la* x [. . .]
	B o. 1	ÉN *lil-li*[*k* . . .]
7	A o. 7	[*li*]*m-la-a ú-re-e-tú la* [. . .]
	B o. 2	⸢*lim*⸣-*la-a ú-*[*re-e-tú* . . .]
8	A o. 8	[*lim*]-*la-a* ÍD^{meš}-*ma* *l*[*a* . . .]
	B o. 3	⸢*lim-la*⸣-*a* ÍD⸢^{meš}⸣-[*ma* . . .]
9	A o. 9	[*lim-g*]*u-ug* NENNI A NENNI x [. . .]
	B o. 4	*lim-gu-*⸢*ug*⸣ NEN[NI . . .]
10	A o. 10	[x *p*]*i-til-ti* [. . .]
	B o. 5	GIM ⸢*pi-til-ti*⸣ [. . .]
11	A o. 11	[. . . D]UG x [. . .]
	B o. 6	traces
12.	B r. 1	⸢*ra*⸣-x (x) [. . .]
13.	B r. 2	*ina na-*⸢*a*⸣$^{?}$-x [. . .]
14.	B r. 3	*ina* ÚŠ GÌŠ x [. . .]
15.	B r. 4	*at-ta l*[*a*$^{?}$ . . .]
16.	B r. 5	DÙ.DÙ.BI IGI [. . .]
17.	B r. 6	*u ú sar a* [. . .]
18.	B r. 7	traces

Transcription

I. A obv. 1–5 = Prescriptions 1–3 and symptoms

1.	1. [*ki*]*brītu eper sūq erbetti imḫur-līm* . . . [. . .]
Sympt.	2. *šumma amēlu ana sinništīšu iṭḫema* [. . .]
	3. *ana sinništīšu libbašu lā inaššī*[*šu* . . .]
2.	4. *šumma* KI.MIN (= Sympt.: 2–3) *urṭû ar*[*ariānu* . . .]
3.	5. *šumma* KI.MIN (= Sympt.: 2–3) *šuruš šammi balāṭi ar*[*antu*$^{?}$ . . .]

II. A obv. 6–11 // B obv. 1–rev. 7 = Incantation and its ritual No. H.1

6. *šiptu: lillik šāru* . . . [. . .]
7. *limlâ urētu* . . . [. . .]
8. *limlâ nārātu-ma* . . . [. . .]
9. *limgug annanna mār annanna* . . . [. . .]
10. *kīma pitilti* [. . .]
11. [. . .] . . . [. . .]
12. . . . [. . .]
13. *ina* . . . [. . .]
14. *ina dām išāri* . . . [. . .]
15. *atta* . . . [. . .]

---

16. *dudubû*: . . . [. . .]
17. . . . [. . .]
18. Traces

Translation

II. A obv. 1–5 = Prescriptions 1–3 and symptoms

1.	1. [Sul]phur, dust from a crossroads, "heals-a-thousand"-plant . . . [. . .].
Sympt.	2. If a man approaches his woman and [. . .] 3. does not desire his woman [. . .].
2.	4. If ditto (= Sympt.: 2–3): *urṭû*-plant, *ar*[*ariānu*]-plant [. . .].
3.	5. If ditto (= Sympt.: 2–3): Root of "plant-of-life," *a*[*rantu*$^{?}$]-plant [. . .].

II. A obv. 6–11 // B obv. 1–rev. 7 = Incantation and its ritual No. H.1

6. Incantation: May the wind blow! . . . [. . .]
7. May the stables fill up! . . . [. . .]
8. May precisely the canals fill up! [. . .]
9. May NN, son of NN, mate! . . . [. . .]
10. Like the cord [. . .]
11. [. . .] . . . [. . .]
12. . . . [. . .]
13. in/with . . . [. . .]
14. with blood of the penis . . . [. . .]
15. you do n[ot . . .]
16. Its ritual: . . . [. . .]
17. . . . [. . .]
18. Traces

Commentary

1. Road crossings have a strong magical value (see Mauss and Hubert 1902–1903: 29). For example, in an anti-witchcraft text the sufferer affected by a curse tells Šamaš that his magical attackers made figurines of him that were buried at the crossroads: "Ditto (= they have made figurines of me and) buried them at the crossroads, ditto and [bu]ried them under the dog Kū[bu]" (Abusch and Schwemer 2011: 276, No. 8.3.1: 50). The connection between the crossroads and impurity is underlined by Maul 1994: 105. In *Šumma ālu* (tab. CIV, CT 39, 45: 29, see Guinan 1997: 474) we find the crossroad as a place of sexual relations between men and women: "If a man seizes a woman at a crossroad and has sex with her, that man will not prosper. If a man frequents a prostitute at a crossroads, either the hand of the god or the hand of the king will catch him" (CT 39 45: 29–30, Guinan 2014: 112–114.). The dirt collected from crossroads is a magic ingredient in the ritual "for a prosperous brothel." It is mixed with the dust from many other places, with river water and cypress oil in a first ritual, and with beer or water in a second one, to prepare an ointment to be anointed on the door and the roof of the man's house (Panayotov 2013: 291, l. 3 and 37). The crossroads is a place of passage, frequented by many people. Its use in this ritual is justified by the fact that through magical sympathy it ensures that the brothel will be as frequented as the crossroad. This interpretation is also useful for understanding the function of the dust of the crossroads in texts for sexual desire. It symbolizes the continuous action of people passing by: sexual desire must always be active and never wane. The dirt of the crossroads is used to eradicate the attack of the witch (Schwemer 2007b: 57 fn. 109). In BAM 237 i 9, a direct incantation against *naḫšātu,* an abnormal flow of blood during pregnancy, a protruding shard taken from a crossroads is used to remove the forces that plague the woman (Finkel 1980: 50, l. 9′). It cannot be excluded that "dust of crossroad" is a 'secret name' for a plant. In this regard note that the 'secret name' of the *supālu*-plant is "dust of the crossroad from burnt ashes" (URU.AN.NA: *maštakal* III 21, Rumor 2017: 7).

6. See catalogue LKA 94 i 4, ii 13; No. A.1: 33; No. E.1: 1; No. G.1: 7; No. I.1: 6; *Aššur Medical Catalogue*, section XX "sex" ll. 103–104 (Steinert 2018b: 217).

7. *urētu*: Thompson's proposal (1934) is to read *šam-re-e-tú*. It should be a nominalization of the adjective *šamru* 'violent, impetuous,' referring to floods or waves. Although it fits the context describing abundance of water considered favorable (the river floods the fields, making them fertile and productive, see also l. 8), the reading *urētu* is more appropriate. In fact, the stables filling up refer in the same way to the abundance connected to the reproductive sphere.

8. The translation "may *precisely* the canals fill up!" is given by the use of -*ma* with emphasis value.

9. See for the precative *limgug* "may he mate" No. B.2: 35; No. E.1 l 3; K.6: 109.

16. IGI perhaps is to be understood as the beginning of the name of a plant (without determinative), such as *imḫur-līm*.

17. u ú sar: the three possible interpretations are:
   - *u* [ú]*(w)arqu* "and the (w)*arqu*-plant." The plant is not mentioned in other texts of the corpus;
   - *u šamma tuqattar* "and the plant you fumigate." It is very uncertain: SAR for *qatāru* D-stem 'to fumigate';
   - *ú-šar status costructus* for *ušāru*. As Meinhold (2017: 59) stressed, however, the term "penis" is present often in the corpus with the possessive suffix -*šu*. The first hypothesis is the most likely.

Nīš libbi I

List of manuscripts

Manuscript	Museum number	Publication	Tablet	Script	Date	Provenience
A	K. 8698	TCS 2, pl. 3 6′–11′	Fragment	NB (see the signs *ta*, *aṣ*, *kí*)	8th–7th cent.	Nineveh

Edition

Biggs 1967: 37

Structure of the text

I. Incantation No. I.1 "May the wind blow! May the grove not quake!" (Biggs 1967 No. 17), ll. 6–11

Transliteration

I. A 6′–11′= Incantation No. I.1

6′. [ÉN *l*]*il-lik* IM *a-a i-nu-u*[*š* ${}^{\text{giš}}$KIRI$_6$]
7′. [IM].DIRI^{meš} *lik-ta-aṣ-ṣi-r*[*a*]
8′. [*t*]*i-ik-ku*(text KI) *a-a i*[*t-tuk*]
9′. [x] *ta*$^?$ *mu šá al-du* x [. . .]
10′. [. . . *š*]*u ip-pa-lu* [. . .]
11′. [. . .] *i* x [. . .]

Transcription

I. A 6′–11′= Incantation No. I.1

6′. [*šiptu*: *l*]*illik šāru ay inū*[*š kirû*]
7′. [*er*]*pētu liktaṣṣir*[*ā*]
8′. [*t*]*ikku ay i*[*ttuk*]
9′. [. . .] . . . *ša aldu* . . . [. . .]
10′. [. . . *š*]*u ippalu* [
11′. [. . .] . . . [. . .]

Translation

I. A 6′–11′= Incantation No. I.1

6′. [Incantation: May] the wind blow! May [the grove] not quake!
7′. May [the clo]uds gather!
8′. May [the mo]isture not f[all]!
9′. [. . .] . . . he who was born [. . .]
10′. [. . .] . . . answers [. . .]
11′. [. . .] . . . [. . .]

Commentary

See the parallel Mo. H.1: 6 and the catalogue LKA 94 i 4. Note that the incantation is mentioned in the *Aššur Medical Catalogue*, in the section XX devoted to sexual problems (while section XIX to the loss of sexual desire).

[ÉN *li-lik* IM *l*]*a i-na-áš-šá-a* $^{\text{giš}⌜}$KIRI$_{6}^{⌝\text{meš}}$
[1 DUB$^{?}$ ÉN *lil-lik* I]M *la i-na-áš-šá-a* ^{giš}KIRI$_{6}^{\text{meš}}$

[Incantation: 'Let the wind blow], the gardens shall not quake.'
[One tablet (of the section) 'Incantation: Let the wi]nd [blow], the gardens shall not quake.'
[*Aššur Medical Catalogue*, section XX, ll. 103–104, Steinert 2018b: 217]

It is, therefore, possible that the incantation does not refer to the loss of sexual desire, but to other problems concerning sex. However, it should be noted that line 106 of the catalogue in the section XX explicitly mentions therapies for the recovery of sexual desire, testifying how fluid is the medical knowledge related to sexual problems (after all, a series known as ŠÀ.ZI.GA did not exist): [KA.INIM.MA ŠÀ.ZI.G]A *ù* MUNUS.GIN.NA.KÁM ŠÀ.ZI.GA.MUNUS.A.KÁM "[Wording of (the incantation) for (male) sexual desi]re and (those) to make a woman come (and for) woman's sexual desire" (l. 106, Steinert 2018b: 217).

NĪŠ LIBBI J

List of manuscripts

Manuscript	Museum number	Publication	Tablet	Script	Date	Provenience	Incantations and prescriptions
A	K.10002	TCS 2, pl. 2 i 1′–ii 5′	Frg. of a two-col. tablet	NA	8th–7th cent.	Nineveh	No. J.1; No. J.2
B	K. 9415 +10791	TCS 2, pl. 2 + CT 13, 31 + Thompson 1930 pl. 17 obv. 1′–rev. 14′	Fragment	NA	8th–7th cent.	Nineveh	No. J.2; No. J.3

Edition

Biggs 1967: 37–38, 47–48 (Ms. A, B)
Chalendar 2018: 47 (Ms. B)

Structure of the text

Text *nīš libbi* J, following Ms. A, starts with the end of a fragmentary incantation and its ritual, followed by an abracadabra (No. J.1). The incantation and its ritual No. J.2 "May the quiver not become empty!" follow in both Ms. A and B. The text ends with the incantation and its ritual No. J.3 "I am washed" present in Ms. B.

A	No. J.1
A – B	No. J.2
B	No. J.3

I. Fragmentary incantation and its ritual (ll. 1–3), and incantation No. J.1 (abracadabra) (Biggs 1967 No. 28), ll. 4–7 (Ms. A i 1′–7′)

II. Incantation and its ritual No. J.2 "May the quiver not become empty!" (Biggs 1967 No. 18), ll. 8–24 (Ms. A ii 1′–5′ // B obv. 1′–16′)

III. Incantation and its ritual No. J.3 "I am washed" (Biggs 1967 No. 20), ll. 25–37 (Ms B rev 2′–14′)

Transliteration

I. A i 1′–7′ = Incantation fragment and incantation No. J.1

1.	A i 1′	[. . . *l*]*ik-šu-ud* T[U$_6$ ÉN]
2.	A i 2′	[KA.INIM.MA ŠÀ].ZI.GA
3.	A i 3′	[DÙ.DÙ.BI$^?$] *ina* A^{meš} ^{šim}LI ŠU^{II}-*šú* LUḪ
4.	A i 4′	[. . . b]u um ma ia ma ma na
5.	A i 5′	[. . . b]i ti ri ia
6.	A i 6′	[. . .] x x mi na na
7.	A i 7′	[. . .] x x

II. A ii 1′–5′ // B obv. 1′–16′ = Incantation and its rituals No. J.2

8.	A ii 1′	x [. . .]
9.	A ii 2′	*s*[*a* . . .]
	B o. 1′	[. . .] x [. . .]
10.	A ii 3′	*a-a* ⸢*i*⸣-[*ri-qa*]
	B o. 2′	[ ] x [*iš*]-⸢*pa-tu*$_4$ *a-a ir-ma*⸣-*a* ^{giš}BAN
11.	A ii 4′	*ta-ḫ*[*a-az*]
	B o. 3′	[*r*]*a-me-ia li-in-ni-pu-uš-ma mu-ši-ta*$_5$ *i ni-iṣ-lal* [TU$_6$ ÉN]
12.	A ii 5′	KA.[INIM.MA]
	B o. 4′	[KA.INI]M.MA ŠÀ.ZI.GA
13.	B o. 5′	[DÙ.DÙ.BI ^{giš}]⸢BAN *šá* ^{giš}DÁ⸣LA D[Ù-*uš* . . .]
14.	B o. 6′	[. . .] x *la* [. . .]
15.	B o. 7′	[. . .] x *ut* [. . .]
16.	B o. 8′	[. . . ŠÀ-*šú* Í]L-*ma* ŠÀ.[ZI.GA TUKU-*ši*]
17.	B o. 9′	[. . .] *ib-ba-ni* [. . .] ⸢x⸣ [x]
18.	B o. 10′	[. . .] *ib-ba-ni ri-kib-ti a*-[*a-li*]
19.	B o. 11′	[. . .] *ib-ba-nu-ú ina bi*-[*rit*]
20.	B o. 12′	[NÍG.KI].⸢A⸣$^?$ *šá* EDIN DÙ-*šú-nu* NÍG.K[I . . .]
21.	B o. 13′	[. . .] x *nu el-lu l*[*a* . . .]
22.	B o. 14′	[. . . *r*]*am* PA MUŠEN x [. . .]
23.	B o. 15′	[. . .] x *ub-šú sa-a* [. . .]
24.	B o. 16′	[. . .] ⸢TU$_6^?$.ÉN⸣ [. . .]

III. B rev. 2′–14′ = Incantation and its ritual No. J.3

25.	B r. 2′	[. . . *r*]*a-am-ka-ku*

26.	B r. 3′	[. . . *p*]*a-áš-šá-ku*
27.	B r. 4′	[. . . *l*]*ab-šá-ku*
28.	B r. 5′	[. . .] x *nu-ul-lu-šá pa-na-tu-u-a*
29.	B r. 6′	[. . . D]Ù?-*ma* NENNI A NENNI
30.	B r. 7′	[. . . *gu-ru*]-*uš u gu-ru-uš*
31.	B r. 8′	[. . .] x-*šú pu-ḫur* SA^{meš}-*ka*
32.	B r. 9′	[. . .] *mu* ^{giš}NÁ-*šá* GAR-*at*
33.	B r. 10′	[. . .] x-*šá it-bu-ka* AN.ZÍB ^{d}*iš-tar*
34.	B r. 11′	[. . .] ÁD.DA *ra-am* UR.BAR.RA *ra-man-ni*
35.	B r. 12′	[*ina qí-bit iq-bu-ú* AN.]ZÍB ^{d}*iš-tar*
36.	B r. 13′	[DÙ.DÙ.BI . . .] x ŠÉŠ-*su*
37.	B r. 14′	[DIŠ KI.MIN . . . *ana* A]^{meš} ŠUB-*di*

Rest destroyed

Transcription

I. A i 1′–7′= Incantation fragment and incantation No. J.1

1. [. . . l]*ikšud t*[*ê šipti*]

2. [*šipat*] *nīš* [*libbi*]

3. [*dudubû*]: *ina mê burāši qātēšu temesse*

4–7. Abracadabra

II. A ii 1′–5′ // B obv. 1′–16′ = Incantation and its rituals No. J.2

8. Traces
9. . . . [. . .] . . . [. . .]
10. *ay i*[*rīq*]*a* [*iš*]*patu ay irmâ qaštu*
11. *tāḫ*[*āz r*]*âmīya linnipuš-ma mušīta i niṣlal* [*tê šipti*]

12. *ši*[*p*]*at* [*nī*]*š libbi*

13. [*dudubû*:] *qašta ša ṣilli te*[*ppuš* . . .]
14. [. . .] . . .
15. [. . .] . . . [. . .]
16. [. . . *libbašu ina*]*ššī-ma* [*nīš*] *libbi* [*iraššī*]

17. [. . .] *ibbani* [. . .] . . . [. . .]
18. [. . .] *ibbani rikibti a*[*yyāli*]
19. [. . .] *ibbanû ina bi*[*rīt*]
20. [*nammaš*]*šû ša ṣēri kalûšunu nam*[*maššû* . . .]
21. [. . .] . . . *ellu* . . . [. . .]
22. [. . . *r*]*âm kappi iṣṣūri* . . . [. . .]
23. [. . .] . . . [. . .]
24. [. . .] *tê šipti* [. . .]

III. B rev. 2′–14′ = Incantation and its ritual No. J.3

25. [. . . *r*]*amkāku*
26. [. . . *p*]*aššāku*
27. [. . . *l*]*abšāku*
28. [. . .] . . . *nullušā pānātūya*
29. [. . .] . . . *annanna mār annanna*
30. [. . . *gur*]*uš u guruš*
31. [. . .] . . . *puḫur šerʾānīka*
32. [. . .] *mu eršuša šakintu*
33. [. . .] . . . *itbûka telītu Ištar*
34. [. . .] . . . *râm barbari rāmanni*

Translation

I. A i 1′–7′= Incantation fragment and incantation No. J.1

1. [May] he achieve [. . .]! In[cantation formula].
2. [Wording of] *nīš* [*libbi* (incantation)].
3. [Its ritual:] You wash in juniper water his hands.

4–7. Abracadabra

II. A ii 1′–5′ // B obv. 1′–16′ = Incantation and its rituals No. J.2

8. Traces
9. . . . [. . .] . . . [. . .]
10. May [the qu]iver not be[come empt]y! May the bow not slacken!
11. May the batt[le of my love]-making be fought and may we lie down (together) by night! [Incantation formula].
12. Wor[din]g of [*nī*]*š libbi* (incantation).
13. [Its ritual]: You make a bow of a thorn [. . .]
14. [. . .] . . .
15. [. . .] . . . [. . .]
16. [. . . he will desi]re and he [will get] the [sexual] desire.
17. [. . .] it was created, [. . .] . . . [. . .]
18. [. . .] it was created, *rikibtu* of a s[tag],
19. [. . .] they were created, bet[ween]
20. all [the wild animal]s of the steppe, the ani[mals . . .]
21. [. . .] . . . pure . . . [. . .]
22. [. . .] the mating, wing of the . . . -bird [. . .],
23. [. . .] . . . [. . .]
24. [. . .] Incantation formula [. . .].

III. B rev 2′–14′ = Incantation and its ritual No. J.3

25. [. . .] I am washed.
26. [. . .] I am anointed.
27. [. . .] I am clothed.
28. [. . .] are bedewed in front of me$^?$.
29. [. . .] . . . NN, son of NN,
30. [. . . have] sex and have sex!
31. [. . .] . . . all your muscles
32. [. . .] . . . her bed is placed,
33. They have risen for you [. . .], wise Ištar,
34. [. . .] . . . mating of a wolf, make love to me!

35. [*ina qibīt iqbû*] *telīti Ištar*
36. [*dudubû*: . . .] . . . *tapaššassu*
37. [*šumma* KI.MIN . . . *ana m*]*ê tanaddi*
Rest destroyed

35. [At the command of w]ise Ištar.

36. [Its ritual:] . . . you anoint him.

37. [If ditto: . . .] you put [. . . into water]

Rest destroyed

Commentary

20. Restoration suggested by Chalendar (2018: 47 and fn. 194): [NÍG.KI].⌜A⌝? *šá* EDIN DÙ-*šú-nu* NÍG.K[I.GAR? . . .] "Tous [les animaux sauvages] de la steppe, [tous] les ani[maux rampants . . .]."

23. As Chalendar (2018: 47 and fn. 195) suggests, the reading *ár* of sign UB allows to restore *lu uš*]*aršu šer'ān* [*sammê*] "May his penis be a harp string," as in the incantation No. A.1: 36.

25–35. Incantation No. J.3 features two speakers: the female partner (ll. 25–34) and the therapeutic operator (l. 24).

III Text from Sultantepe (with duplicates from other sites)

Nīš libbi K

List of manuscripts

Manu-script	Museum number	Publication	Tablet	Script	Date	Proveni-ence	Incantations and prescriptions
A	SU 52/139+ 161 + 170+ 250 + 250A +323	STT 280 i 1–7, 22–51, 56–ii 9, 22–61, iii 24–iv 36	Two-col. tablet	NA	8th–7th cent.	Sultan-tepe	Prescr. 1–28; No. K.3; No. K.4; No. K.5; No. K.6; No. K.7; Prescr. 28–31; No. K.8; No. K.9; Prescr. 31
B	K. 9451+ 11676+ Sm. 818 +961[567]	TCS 2, pl. 1 obv. i 7′–16′, rev. iv 1′–6′	Frg. of a two-col. tablet	NA	7th cent.	Nineveh, Ashurba-nipal's Library	No. K.1; No. K.2
C	BM 68033	CCMAwR, 1, pl. 18 rev.[?] 9′–13′	Fragment	NB/LB	6th–3rd cent.	unknown proveni-ence	Prescr. 3, 22
D	VAT 13616	LKA 144 rev. 21–33	Sin-gle-col. tablet	NA	8th–7th cent.	Aššur, Library N 4	Prescr. 3–4, 22
E	VAT 13917	BAM 207 BID, pl. 24 obv. 10′–rev. 3	Frg. of a small sin-gle-col. tablet	NA	8th–7th cent.	Aššur	Prescr. 3–4
F	VAT 13893 +13982	BAM 320 obv. 5′–6′, 13′–16′, rev. 5′–14′	Small sin-gle-col. tablet	NA	8th–7th cent.	Aššur, Library N 4	Prescr. 3–4, 21–22, 26
G	A 2715	BAM 205 obv. 19′–rev. 8′	Large frg of a sin-gle-col. tablet	NA	8th–7th cent.	Aššur, Library N 4	Prescr. 21–22, 26

567 Ms. B: the joins of K. 9451+K. 11676+Sm. 818+961 recognized by Biggs (1967: 45, pl. 1) (see for the photo CDLI No. P398122).

https://doi.org/10.1515/9783110757262-008

(continued)

Manu-script	Museum number	Publication	Tablet	Script	Date	Proveni-ence	Incantations and prescriptions
H	A 522	BAM 318 = CCMAwR 2, pls. 53–60 rev. iii 16–18	Large frg. of a two-col. tablet	NA	8th–7th cent.	Aššur, Library N 4	Prescr. 22
I	VAT 14111	BAM 319 rev. 3′–4′	Frg. of a sin-gle-col. tablet	NA	8th–7th cent.	Aššur, Library N 4	Prescr. 22
J	BM 54650	BID, pls. 19–21 rev. iv 12′–16′	Frg. of a two-col. tablet	NB/LB	6th–3rd cent	unknown prove-nance	Prescr. 22
K	SU 1951, 93 +SU unnum-bered	STT 95+295 obv. i 16–22	Two-col. tablet	NA	8th–7th cent.	Sultan-tepe	Prescr. 21
L	VAT 8914	BAM 311 = KAR 186 obv. 10′–13′	Sin-gle-col. tablet	NA	8th–7th cent.	Aššur Library N 4	Prescr. 21
M	K. 8907 obv. 12′–16′	Stadhoud-ers forth-coming	Frg. of a sin-gle-col. tablet	NB	7th cent.	Nineveh 'Ashurba-nipal's Library'	Prescr. 21
N	Bo 4894	KUB 4, 48 ii 3–9	Two-col. tablet	'Not Hittite[568]	13th cent.	Ḫattuša	Prescr. 30
O	K. 2417	AMT 31, 4; 32, 1 rev. iv 4–5	Frg. of a two-col. tablet	NA	7th cent.	Nineveh 'Ashurba-nipal's Library'	Prescr. 26

Edition

Ebeling 1925: 28–31, 50–51 (Ms. G, L)
Biggs 1967: 21–22, 27, 45–49, 65–68 (Ms. A, B, L)
Farber 1977a: 234–235, 260 (Ms. E–H)

568 Note the use of *bá*, see Schwemer 2013: 154.

Scurlock and Andersen 2005: 256–257 (Ms. C–I).
Schwemer 2010: 116–120 (trans. of Ms. A)
Abusch and Schwemer 2011: 101–114, No. 2.5 (Ms. A, B, C–G)
Scurlock 2014b: 653 (Ms. K)
Chalendar 2018: 45–46 (Ms. A)
Abusch et al. 2020: 21–27, No. 3.15 (Ms. O)
Stadhouders forthcoming (Ms. K, M)[569]

Structure of the text

Text *nīš libbi* K is based on Ms. A starting with two prescriptions, the description of the symptoms (= *nīš libbi* D Sympt.: 42–48), and other two prescriptions (= *nīš libbi* F prescr. 10–11).[570] The text *nīš libbi* K continues with Ms. B which provides two incantations: No. K.1 "When Enlil and Bēlet-ilī gave mankind a name"; and the fragmentary No. K.2. It returns to Ms. A containing several prescriptions: sixteen with duplicates Ms. D, E, F and G (prescr. 3–18); five corresponding to *nīš libbi* A prescr. 9–12 and 16; four with duplicates Ms. D, F, G, H, I, J, L, M. The text, always following Ms. A, provides an incantation (= No. M.1), five prescriptions with duplicates Ms. F, G and O (prescr. 23–28), six incantations (No. K.3, No. K.4; No. K.5; = No. E.2; No. K.6; No. K.7), three prescriptions with duplicate Ms. N (prescr. 29–31), two incantations (No. K.8; No. K.9), another prescription (prescr. 32), and an incantation (= No. A.1).

A	Prescr. 1–2
A (+ dupl.)	= *nīš libbi* D Sympt.
A – B (+dupl.)	= *nīš libbi* F prescr. 10–11
B	No. K.1
B	No. K.2
A – C–F	Prescr. 3–18
A(+ dupl.)	= *nīš libbi* A prescr. 9–12, 16
A – C – D – F–M	Prescr. 19–22
A (+dupl.)	= No. M.1
A – F – G – O	Prescr. 23–28
A	No. K.3
A	No. K.4
A	No. K.5
A (+ dupl.)	= No. E.2

569 I would like to thank Dr. Henry Stadhouders for making his unpublished edition of text STT 95+295 (+duplicates, including K. 8907) available to me.

570 In this two prescriptions Ms. A (STT 280 i 18–21) and Ms. B (K. 9451+11676+Sm. 818+961 obv. i 2′–6′) are duplicates together with KAR 70 obv. 17–21 (= Ms. A in *nīš libbi* F).

A	No. K.6
A	No. K.7
A – N	Prescr. 29–31
A	No. K.8
A	No. K.9
A	Prescr. 32
A (+ dupl.)	= No. A.1

I. Prescriptions
   1. ll. 1–6 (Ms. A i 1–6)
   2. 7 (Ms. A i 7)

II. Symptoms (Ms. A i 8–17) = *nīš libbi* D Sympt.: 42–48

III. Two prescriptions (Ms. A i 18–21// Ms. B obv. i 2′–6′) = *nīš libbi* F prescr. 10–11: 31–35

IV. Incantation and its ritual No. K.1 "When Enlil and Bēlet-ilī gave mankind a name" (Biggs 1967 No. 26), ll. 8–17 (Ms. B obv. i 7′–16′)

V. Incantation fragment and ritual No. K.2 (Biggs 1967 No. 34), ll. 18–23 (Ms. B rev. iv 1′–6′)

VI. Prescriptions
   3. 24–29 (Ms. A i 22–27 // C rev.$^{?}$ 9′–11′ // D rev. 23–32// E obv. 10′–lo. e. // F obv. 13′–15′)
   4. 30 (Ms. A i 28 // D rev. 33// E rev. 2–3// F obv. 16′)
   5. 31 (Ms. A i 29)
   6. 32–35 (Ms. A i 30–33)
   7. 36–39 (Ms. A i 34–37)
   8. 40–41 (Ms. A i 38–39)
   9. 42 (Ms. A i 40)
   10. 43–44 (Ms. A i 41–42)
   11. 45 (Ms. A i 43)
   12. 46 (Ms. A i 44)
   13. 47 (Ms. A i 45)
   14. 48 (Ms. A i 46)
   15. 49 (Ms. A i 47)
   16. 50 (Ms. A i 48)
   17. 51 (Ms. A i 49)
   18. 52 (Ms. A i 50)

VII. Five prescriptions (Ms. A i 52–55) = ***nīš libbi*** A prescr. 9–12, 16: 19–22, 27

VIII. Prescriptions
- 19. 53–54 (Ms. A i 56–57)
- 20. 55–56 (Ms. A i 58–59)
- 21. 57–63 (Ms. A ii 1–7 // F rev. 5′–11′ // G obv. 19′–27′ // K obv. i 16–22 // L obv. 10′–13′ // M obv. 12′–16′)
- 22. 64–65 (Ms. A ii 8–9 // C rev.? 12′–13′ // D rev. 21–22 // F obv. 5′–6′ // G obv. 31′–33′ // H rev. iii 16–18 // I rev. 3′–4′ // J rev. iv 12′–16′)

IX. Ms. A ii 10–21 = *nīš libbi* incantation No. M.1 "Sexual desire! Sexual desire! The bed for the sexual desire" ll. 4–15

X. Prescriptions
- 23. 66 (Ms. A ii 22)
- 24. 67 (Ms. A ii 23)
- 25. 68 (Ms. A ii 24)
- 26. 69–72 (Ms. A ii 25–38 // F rev. 12′–14′ // G rev. 3′–5′ // O rev. iv 4–5)
- 27. 73–76 (Ms. A ii 29–32)
- 28. 77–79 (Ms. A ii 33–35)

XI. Incantation and its ritual No. K.3 (abracadabra) (Biggs 1967 No. 29), ll. 80–94 (Ms. A ii 36–50)

XII. Fragmentary incantation No. K.4 (Biggs 1967 No. 30), ll. 95–97 (Ms. A ii 51–53)

XIII. Fragmentary incantation and its ritual No. K.5 (Biggs 1967 No. 31), ll. 98–105 (Ms. A ii 54–61)

XIV. Ms. A ii 62–iii 23 = *nīš libbi* incantation No. E.2 "Light of the heavens, wise Ištar" ll. 18–45

XV. Fragmentary incantation No. K.6 (Biggs 1967 No. 32), ll. 106–115. (Ms. A iii 24–33)

XVI. Incantation and its ritual No. K.7 "Lion! Bull!" (Biggs 1967 No. 10), ll. 116–134 (Ms. A iii 34–42)

XVII. Prescriptions
- 29. 125–128 (Ms. A iii 43–46)
- 30. 129–134 (Ms. A iii 47–52)
- 31. 135–141 (Ms. A iv 1–7 // N ii 3–9)

XVIII. Fragmentary incantation and its ritual No. K.8 (Biggs 1967 No. 5), ll. 142–157 (Ms. A iv 8–23)

XIX. Fragmentary incantation and its ritual No. K.9 (Biggs 1967 No. 33), ll. 158–165 (Ms. A iv 24–31)

XX. Prescription 32 ll. 166–170 (Ms. A iv 32–36)

XXI. Ms. A iv 37–41 = *nīš libbi* incantation No. A.1 "May the wind blow! May the grove quake!" ll. 33–39

Summary of the sections of manuscripts not included in the transliteration:

– Ms. A = STT 280
  i 8–17 = D Sympt.; prescr. 4 (Ms. E)
  i 18–21 = F prescr. 10–11 (Ms. E)
  i 52–55 = A prescr. 9–12, 16 (Ms. D)
  ii 10–21 = No. M.1 (Ms. B)
  ii 62–iii 23 = No. E.2 (Ms. H)
  iv 37–41 = No. A.1 (Ms. D)

– Ms. B = K. 9451+11676+Sm. 818+961
  obv i 1′ = Fragmentary prescriptions (see commentary *nīš libbi* F: 31)
  obv. i 2′–6′ = F prescr. 10–11 (Ms. D)
  obv. ii = Fragmentary
  rev. iv 7′–9′ = Part of an Ashurbanipal colophon similar to Hunger 1968, No. 318

– Ms. C = BM 68033
  1′–5′ = Fragmentary therapeutic ritual against witchcraft with incantation
  15′–17′ = Therapeutic prescription for soothing the anger of Marduk
  18′ = Fragmentary

– Ms. D = LKA 144
  obv. 1′–rev. 18–20 = Farber 1977a, Hauptritual B (with duplicates)
  rev. 34–36 = Fragmentary

– Ms. E = BAM 207
  obv. 1′–9′ = Abusch and Schwemer 2011, No. 2.5

– Ms. F = BAM 320
  obv. 1′–4′, 7′–12′ = Farber 1977a, Hauptritual B; Abusch and Schwemer 2011, No. 2.5

obv. 17′–rev. 2′ = Fragmentary recipes
rev. 3′–4′ = Recipe for a salve
rev. 15′–20′ (// Ms. G BAM 205 rev. 15′–20′) = Instruction for a therapeutic ritual against sexual impurity
rev. 21′ = Fragmentary

– Ms. G = BAM 205
obv. 1′–8′ = Abusch and Schwemer 2011, No. 2.5
obv. 9′–18′ = F prescr. 5, 8–9
9′–14′ = Farber 1977a, Hauptritual B (with duplicates)
15′–20′ (// BAM 320 rev. 15′–20′) = Instruction for a therapeutic ritual against sexual impurity

– Ms. H = BAM 318
obv. i–ii 37 = Rituals and recipes for curing different diseases
ii 38–rev. iii 10 = Fourteen prescriptions for curing impurity caused by witchcraft
iii 11–15 = Farber 1977a, Hauptritual B
iii 19–29 = Various medicinal drugs and their effects
iii 30–34 = Prescriptions for curing *ṣētu* of the head
iii 35–43 = Incantation rituals against the anger of the gods
iv 1–24 = Prescriptions and rituals against the anger of the gods
iv 25–36 = Instructions for a bath and an amulet against snakes portending evil

– Ms. I = BAM 319
obv. 1–rev. 2′ = Farber 1977a, Hauptritual B (with duplicates)

– Ms. J = BM 54650
obv. i–rev. iv 11′ = Farber 1977a, Hauptritual B (with duplicates)

– Ms. K = STT 95+295
see the edition and duplicates in Stadhouders forthcoming, see also Scurlock 2014b: 650–663; Abusch and Schwemer 2016: 39–44, No. 3.7 (with duplicates)

– Ms. L = BAM 311=KAR 186
obv. 1′–9′, 14′–22′ = A collection of protective *mêlu*-bags for a man who constantly suffers from the 'tenseness' and 'heartbreak'
obv. 23′–29′ = Prescriptions for leather bags against the "Lord of the roof"
obv. 30′–40′ = Against "Hand of the goddess"

obv. 41′–46′ = Against "anything evil"
obv. 47′–50′ = Against the evil *alû*-demon
rev. 51′–58′ = For the *bennu*-epilepsy
rev. 59′–76′ = For AN.TA.ŠUB.BA
rev. 77′–96′ = For several of the evils dealt with before

– Ms. M = K. 8907
see the edition and duplicates in Stadhouders forthcoming.
rev. 5′–11′ = Prescription against witchcraft that was given to eat, Abusch and Schwemer 2016: 444–445, No. A.2 (// BAM 161 iv 16′–29′ // AMT 29,5=BAM 436 rev. vi 12′–19′// BAM 282: 1′–8′, Abusch and Schwemer 2011, text 1.8.2: 16′–28′)

– Ms. N = KUB 4, 48
A i 1–iv 26 = N prescr. 1–23 (Ms. A) (note that ii 3–9 = K prescr. 31).
iv 27–31 = No. L.1 (Ms. B)
left e. 1–7 = N prescr. 24 (Ms. A)
lower e. 1–5 = N prescr. 25 (Ms. A)

– Ms. O = AMT 31, 4; 32, 1
see Abusch et al. 2020: 21–27, No. 3.15

Transliteration

I. A i 1–7 = Prescriptions 1–2

1.	A i 1	[DIŠ NA *ana* MUNUS-*šú* ŠÀ-*šú* N]U ÍL-*ma*
2.	A i 2	[. . . NA BI *kiš-pu* DAB-*s*]*u ana* TI-[*šú*]
3.	A i 3	[. . .] ^{na4}x x
4.	A i 4	[. . . *ina* G]Ú-*šú* GAR
5.	A i 5	[. . .] *ina* Ì $^{[gi]š}$ŠUR.MÌN
6.	A i 6	[. . . *ina* K]UŠ *ina* GÚ-*š*[*ú*] GAR-*an*
7.	A i 7	[DIŠ KI.MIN . . . NUM]UN$^{?}$ ^{giš}ḪAB [*ina* KAŠ] NAG

II. Symptoms (Ms. A i 8–17) = *nīš libbi* D Sympt.: 42–48

III. Two prescriptions (Ms. A i 18–21// Ms. B obv. i 2′–6′) = *nīš libbi* F prescr. 10–11: 31–35

IV. B obv. i 7′–16′ = Incantation and its ritual No. K.1

8.	B o. i 7′	[ÉN ^{d}]en-líl diĝir-maḫ nam-lú-ùlu-lu nam-še$_{23}$-a
9.	B o. i 8′	[á]$^{?}$-úr téš-a sè-ga bí-in-šu-du$_{7}$
10.	B o. i 9′	[x b]i mu-un-dab-ba šà-bi mu-un-dul$^{!}$-la$^{!}$
11.	B o. i 10′	x x x x ⸢šà⸣-zi-ga-b[i]$^{?}$ nam-e
12.	B o. i 11′	šà-bi mu-un-zi x [. . .] TU$_{6}$ ÉN
13.	B o. i 12′	DÙ.DÙ.BI PA SU.DIN^{mušen} ⸢*šá*⸣ x [x T]I-*qé*
14.	B o. i 13′	ḪÁD.A ⸢SÚD⸣ *ina* KAŠ ^{lú}KÚRUN.NA [NA]G-*šú*
15.	B o. i 14′	*ina* Ì+GIŠ [EŠ].MEŠ-*su ina* KUŠ *ina* G[Ú-*šú*] GAR-*an*
16.	B o. i 15′	ÉN GIM [dím-an-n]a GIM d[ím-an-na]
17.	B o. i 16′	x x [. . .] x dí[m$^{?}$ . . .]

Rest destroyed

V. B rev. iv 1′–6′ = Incantation fragment and its ritual No. K.2

18.	B r. iv 1′	traces
19.	B r. iv 2′	[. . . *k*]*a* [. . . TU$_{6}$ ÉN]
20.	B r. iv 3′	[KA.INIM.MA] ŠÀ.Z[I.GA]
21.	B r. iv 4′	[DÙ.DÙ.BI] *ina* IZI SAR-*šú-ma* ŠÀ.Z[I.GA]
22.	B r. iv 5′	[*ana* ŠÀ.ZI.GA T]UKU-*e* NUMUN ^{ú}ḪAŠḪUR$^{?}$ LÁ NUMUN Ú x [. . .]
23.	B r. iv 6′	[DIŠ NA *ana* MUNUS]-*šú* ŠÀ-*šú* ÍL-*šú-ma ana* MUNUS BAR-*ti* ŠÀ-*šú* N[U ÍL-*šú*]

VI. A i 22–51 // C rev.$^{?}$ 9′–11′ // D rev. 23–33 // E obv. 10′–rev. 3 // F obv. 13′–16′ = Prescriptions 3–18

24.	A i 22	[ ]*ka-šip-ma* UZU^{meš}-*šú* *tab-ku lu* *ina* DU-*šú lu* *ina* ⸢GUB⸣-[*šú*]
	C r.$^{?}$ 9′	DIŠ NA *ka-šip-ma* UZU^{meš}-*šú* *tab-ku lu-u ina* DU-*šú* *lu-u ina* KI.GUB-*šú*
	D r. 23	[*k*]*a-šip-ma* UZU^{meš}-*šú* *tab-ku lu* *ina* DU-*šu lu-u ina i-zu*(text ŠU)-⸢*zi-šu*⸣
25.	A i 23	[ ] KI.NÁ-*šú* *lu* *e-nu-ma* KÀŠ$^{!meš}$-*š*[*ú*] *i*-[*šat-t*]*i-nu*
	C r.$^{?}$ 9′–10′	*lu-u ina* K[I].N[Á-*šú*] / *lu* U$_{4}$-*ma*$^{!}$(text LU) KÀŠ *i-šá-tin-ni*
	D r. 24	[ ] *i-tul-i-šu* *lu-u i-nu*⟨*-ma*⟩ KÀŠ-[*š*]*u* *i-šá-ti-nu*
26.	A i 24	[*ri-ḫ*]*u-su* DU-*ak* GIM MUNUS *su-ʾu-su* *l*[*a e-l*]*il*
	C r.$^{?}$ 10′	*ri-ḫu-*⸢*us*⸣*-su* ⟨DU-*ak*⟩ GIM ⸢MUNUS *šu-ʾu-šú* ⸢*la e*⸣-[*el*]
	D r. 24–25	*ri-ḫu-su* ⸢DU-*ak*⸣ / [x] MUNUS *su-ʾu-su* NU *e-el*
27.	A i 25	[x B]I *ri-ḫu-su* KI ADDA *ina* KI ⸢*šu-nu-lat*⸣ *ana* TI-*šú*
	C	– – – – – – – – – – –
	D r. 25–26	NA BI *ri-ḫu-su* KI ÚŠ^{meš} *ina* KI ⸢*šu*⸣*-nu-lat*⸣ / [TI]-*šú*

28.	A i 26	[^{ú}LA]L ^{ú}DILI ^{ú}[SIK]IL ^{ú}N[A-*a*]-⸢*na* NUMUN⸣ ^{ú}IN$_{6}$.ÚŠ
	C r.$^{?}$ 11′	^{ú}LAL ^{ú}DILI ^{ú}SIKIL ^{ú}NA-*a-na* NUMUN ^{ú}IN$_{6}$.ÚŠ
	D r. 31	[^{ú}]LAL ^{ú}DILI ^{ú}SIKIL ^{ú}NA-*a-na* NUM[UN] ^{ú}IN.NU.ÚŠ
	E o. 10′–11′	⸢^{ú}⸣LAL ^{ú}DILI ⸢^{ú}⸣[x] / ⸢^{ú}⸣NA-*a-na* NU[MUN]
	F o. 13′–14′	^{ú}LAL ^{ú}DILI ^{ú}SIK[IL] / [NU]MUN ^{ú}IN$_{6}$.ÚŠ
29.	A i 27	[x ^{ú}]ŠAKIR SUḪUŠ ^{giš}[D]ÌḪ ⸢*ša*⸣ U[GU KI.MA]Ḫ [*ina*] KUŠ — *ina* GÚ-*šú* GAR —
	C r.$^{?}$ 11′	NUMUN ^{ú}⸢ŠAKIR SUḪUŠ⸣ ^{giš}DÌ[Ḫ — — —] — — — — —
	D r. 31–32	NUMUN ^{ú}ŠAKIR / [ ] *šá* UGU KI.MAḪ *ina* KUŠ DÙ.DÙ *ina* GÚ-*šú* GAR-*ma* TI
	E lo. e.	[ ] — — — /*ina*$^{?}$ [ ] — —
	F o. 14′–15′	NUMUN ^{ú}ŠA[KIR] / *šá* UGU KI.MA[Ḫ] — — — — —
30.	A i 28	[x UD]U.NÍTA$^{!}$ ŠIKA [x LÍMMU]-*tú* ⸢*šá*$^{!}$-*rat*$^{!}$⸣ UR.M[AḪ *ina*] KUŠ — *ina* GÚ-*šú* GAR-*an*—
	D r. 33	[x UD]U.NÍTA ŠIKA SILA LÍMMU SÍG UR.MAḪ *ina* KUŠ DÙ.DÙ *ina* GÚ-*šú* GAR-*an* TI
	E r. 2–3	*ina*$^{?}$ KUŠ$^{?}$ x [ ] / [ŠI]KA SILA LÍMMU SÍ[G]
	F o. 16′	ÉL[LAG] UDU.NÍTA ŠIKA SILA LÍMMU SÍ[G] — — — — —
31.	A i 29	DIŠ KI.MIN PEŠ$_{10}$^{d}⸢ÍD⸣ [BA.BA.Z]A$^{!}$. ^{d}ÍD *in*[*a* KUŠ *in*]*a* MÚRU-*šú* KEŠDA
32.	A i 30	DIŠ KI.MIN SU[ḪUŠ ^{ú}]EME.UR.GI$_{7}$ ⸢^{ú}⸣[LUM].ḪA SUḪUŠ ^{ú}SI.SÁ
33.	A i 31	SUḪUŠ ^{ú}[S]AG {SUḪUŠ ⸢^{ú}⸣SAG} [SUḪUŠ] ^{ú}[A].ZAL.LÁ
34.	A i 32	SUḪUŠ ^{ú}*kâm-ka-*⸢*du*⸣ S[UḪUŠ] ⸢^{ú}⸣AŠ.TÁL.[TÁL]
35.	A i 33	[SUḪUŠ] ^{ú}*ka-zal-*⸢*lá*⸣ [Ú ŠÀ.Z]I.GA *ina* [KUŠ *i*]*na* G[Ú-*šú* GAR-*an*]
36.	A i 34	[DIŠ NA *ana* MUNUS] TE-*ma* [. . .]
37–39.	A i 35–37	destroyed
40.	A i 38	DIŠ KI.MIN [. . .] ^{ú}*a-ra-ri-a-nu*
41.	A i 39	NUMUN ^{ú}[. . .] *ina* KUŠ
41.	A i 40	DIŠ KI.MIN ^{ú}x x [. . . ^{ú}]*a-ra-an-tú ina* KUŠ
43.	A i 41	DIŠ KI.MIN ^{ú}[Á]B.DUḪ [. . .] ^{ú}*kan-ka-du*
44.	A i 42	SUḪUŠ ^{ú}ŠAKIR ^{ú}x [. . .] *ina* Ì *ina* KUŠ
45.	A i 43	DIŠ KI.MIN ⸢^{ú}⸣NÍG.GIDRU ^{ú}*ka -*⸢*man*⸣$^{!}$*-du* [^{ú}*a*]*-la-mu-u ina* KUŠ
46.	A i 44	DIŠ ⸢KI.MIN⸣ ^{ú}SUMUN.DAR ^{ú}ÁB.DUḪ ^{ú}[A.ZAL-*ú*] ⸢^{ú}⸣DILI *ina* KUŠ
47.	A i 45	[DIŠ KI.MIN ^{giš}ḪA]ŠḪUR Ú.NAM.TI.LA [^{ú}]A.⸢ZAL⸣.[LÁ] ⸢Ú⸣.KU$_{6}$ ^{ú}ḪAB *ina* KUŠ

48.	A i 46	[DIŠ KI.MIN ^{n}]$^{\text{a4}}$NÍR $^{\text{na4}}$[DUR].MI.NA ^{n}[$^{\text{a4}}$Z]A.GÌN *ina* KUŠ
49.	A i 47	[DIŠ KI.MIN . . . ^{ú}IN$_6$].ÚŠ ^{ú}⌜DILI⌝ *ina* KUŠ
50.	A i 48	[DIŠ KI.MIN . . .] ^{ú}A.[ZAL.LÁ] *ina* KUŠ
51.	A i 49	[DIŠ KI].MIN ^{giš}ŠE.NÁ.A [. . .] *ina* KUŠ
52.	A i 50	DIŠ KI.MIN SUḪUŠ ^{giš}DÌḪ x KUR NAM ⌜*šá*⌝ x [. . .] *ina* KUŠ

VII. Five prescriptions (Ms. A i 52–55) = *nīš libbi* A prescr. 9–12, 16: 19–22, 27

VIII. A i 56–ii 9 // C rev.? 12′–13′ // D rev. 21–22 // F obv. 5′–6′, rev. 5′–11′ // G obv. 19′–rev. 2′, rev. 6′–8′// H rev. iii 16–18 // I rev. 3′–4′ // J rev. iv 12′–16′ // K obv. i 16–22 // L obv. 10′–13′ // M obv. 12′–16′ = Prescriptions 19–22

53.	A i 56	[DIŠ KI].MIN SUḪUŠ ^{ú}EME.UR.GI$_7$ SUḪUŠ Ú.KU$_6$ SUḪUŠ ^{ú}A[Š.TÁL.TÁL]
54.	A i 57	[SUḪ]UŠ ^{ú}NÍG.GIDRU! SUḪUŠ ^{ú}SI.SÁ *ina* KAŠ *ina* Ì [*ina* KUŠ]
55.	A i 58	[DIŠ KI.MIN ^{ú}]IGI-*lim* $^{\text{ḫe-p[í]}}$
56.	A i 59	[^{ú}SUMUN].DAR $^{\text{ḫe-pí}}$

57.	A ii 1	[ḪULUḪ.ḪULUḪ]-*ut* ŠÀ-*šú* *e-šú* — *ina* [ ]
	G o. 19′	[x] NA *ina* KI.NÁ-*sú* ḪULUḪ.ḪULUḪ-*ut* ŠÀ-*šú* *e*-(text GUR)-*šú* (text UT) — *ina* KI.NÁ-[*šú*]
	K o. i 16	DIŠ NA *ina* KI.NÁ-*sú* ḪULUḪ.ḪULUḪ-*ut* ŠÀ-*šú* *e-šu* *u* *ina* KI.NÁ-*šú*
	M o. 12′	[ḪULU]Ḫ.ḪULUḪ-*ut* ŠÀ-*šú* *e-šu* *u* *ina* KI.NÁ-*šú*
58.	A ii 2	[ ] BI *ki-mil-ti* ^{d}AMAR.UTU [ ]
	G o. 20′–21′	*ri-ḫu-su* GIN-*ak* NA BI DIB-*ti* ^{d}AMAR.[UTU] / *u* ^{d}*iš*$_8$-*tár*
	K o. i 16–18	A.RI.A-*su* GIN-*ak* / NA BI DIB-*tì* ^{d}AMAR.UTU / *u* ⌜^{d}*iš*$_8$-*tár*⌝
	M o. 12′–13′	[/ ^{d}AMAR.]⌜UTU *u* ^{d}⌝15
59.	A ii 3	⌜UGU-*šú*⌝ [ ] ^{ú}*tar-muš* ^{ú}ḪAR.ḪAR
	F rev. 5′	— — — — ^{ú}*tar-muš* ^{ú}DILI
	G o. 21′–22′	UGU-*šú* GÁL-*ši* *ana* TI-*šu* ^{ú}*tar*-⌜*muš*⌝ ^{ú}DILI
	K o. i 18	UGU-*šú* GÁL-*ši* *ana* TI.LA-*šú* ^{ú}*tar-muš* —
	L o. 10′	[DIŠ] KI.MIN ^{ú}*tar-muš* —
	M o. 13′	UGU-*šú* GÁL-*ši* *ana* TI.[LA-*šu*] —
60.	A ii 4	^{ú}ḪAR.[ḪUM.BA.ŠIR ^{ú}*ki*]-*ṣir* ^{giš}*bi-ni* ^{giš}DÌḪ SIG$_7$-*su*
	F rev. 5′–6′	[ ] / ZÚ.KEŠDA ^{giš}MA.NU ⟨^{giš}DÌḪ⟩ S[G$_7$-*su*]
	G o. 22′	^{ú}ḪAR.ḪUM.BA.ŠIR *ki-ṣir* ^{giš}*bi-ni* ^{giš}DÌḪ SIG$_7$-*su*

	K o. i 18–19	^{ú}ḪAR.ḪUM.BA.ŠIR / — ^{giš}MA.NU ILLU ^{giš}DÍḪ —
	L o. 10′	^{ú}ḪAR.ḪUM.BA.ŠIR^{sar} — ^{giš}*ḫe-[pí]*
	M o. 14′	[^{giš}MA.]⸢NU ILLU⸣ ^{giš}DÍḪ –
61.	A ii 5	^{na4}AD.BAR *ḫi-[ṣib* ^{na4}GU]G ^{ú}SIKIL ^{ú}IN$_6$.ÚŠ PA ^{giš}ŠE.NÁ.A
	F rev. 7′–8′	^{na4}AD.BAR *ḫi-ṣib*$^{!}$ ^{na4}[x] / ^{ú}SIKIL ^{ú}IN$_6$.ÚŠ [ ]
	G o. 23′–24′	⸢^{na4}⸣AD.BAR *ḫi-ṣib*$^{!}$ ^{na4}GUG ⸢^{ú}⸣SIKIL ^{ú}IN.NU.UŠ PA ^{giš}ŠE.NÁ.A
	K o. i 20–21	^{na4}AD.BAR *ḫi-ṣib*$^{!}$ ^{na4}GUG ^{ú}SIKIL – ⸢^{giš}ŠE.NÁ.A⸣ / ^{ú}IN.NU.UŠ
	L o. 11′–12′	⸢^{na4}⸣AD.BAR *ḫi-ṣib* ^{na4}GUG ^{ú}SIKIL – ^{giš}ŠE.NÁ.[A]/^{ú}IN.NU.UŠ *šá* KUR
	M o. 14′–15′	[/ ^{ú}IN.NU.U]Š
62.	A ii 6	^{na4}NÍG.BÙR.BÙR ^{ú}[A.ZAL.L]Á ^{ú}⸢EME⸣.UR.GI$_7$ ^{giš}KIŠI$_{16}$ *šá* É.G[AR$_8$]
	F rev. 9′–10′	⸢^{na4}⸣NÍG.BÙR.BÙR [ ] / ^{giš}KIŠI$_{16}$ *šá* UGU [ ]
	G o. 25′–r. 1′	⸢^{na4}⸣Ú.NÍG.BÙR.BÙR ^{ú}A.ZAL.LÁ ^{ú}EME.UR.GI$_7$ / [x x] x GÍR x ⸢KA⸣ x x
	K o. i 21	Ú.NÍG.BÙR.BÙR ^{ú}*a-zal-la* ^{ú}EME.UR.GI$_7$ ^{giš}DÌḪ *šá* IZ.ZI
	L o. 12′–13′	– ^{ú}A.ZAL.LÁ ^{ú}EME.UR.⸢GI$_7$⸣ ^{giš}DÌḪ – IZ.ZI
	M o. 15′–16′	Ú.NÍG.BÙR.B[ÙR] ⸢^{ú}*a*⸣-[*zal-la*]
63.	A ii 7	Ú.[ḪI].A ŠE[Š 1-*niš*] *ina* ÚŠ ^{giš}*eri$_4$-ni* ḪE.ḪE *ina* KUŠ
	F rev. 10′–11′	[ ] / *ina* ÚŠ ^{giš}EREN ⸢SUD⸣ [ ]
	G r. 1′–2′	[ ] x SÚD *ina* ÚŠ ^{giš}EREN SUD [ ]
	K o. i 22	1-*niš* $^{síg!}$AKÀ$^{!}$ NIGIN-*mi* — ÚŠ ^{giš}EREN SUD *ina* KUŠ
	L o. 13′	1-*niš* ^{síg}AKÀ ⟨NIGIN-*mi*⟩ — ÚŠ ^{giš}EREN SUD *ina* KUŠ
	M o. 16′	[ ]
64.	A ii 8	DIŠ KI.MIN ^{ú}⸢*si-ḫu*⸣ [ŠIM].^{d}MAŠ NITÁ *u* MUNUS SÍG UGU.DUL.BI
	C r.$^{?}$ 12′	— — ^{ú}*si-i-ḫu* ŠIM.^{d}MAŠ NITA *u* MUNUS SÍG UGU.DUL.BI
	D r. 21	— — [^{ú}*si-i-ḫ*]*u* Š[IM].^{d}MAŠ NITÁ *u* MUNUS SÍG UGU.DUL.BI
	F o. 5′	— — ⸢^{ú}⸣*si-ḫu* ŠIM.[^{d}MAŠ NI]TÁ *u* MUNUS SÍG [ ]
	G r. 6′	— — ^{giš}*si-ḫu* ŠIM.^{d}MAŠ NITA *u* MUNUS SÍG UGU.DUL.BI
	H r. iii 16	— — ^{giš}*si-ḫu* ŠIM.^{d}MAŠ NITÁ *u* MUNUS SÍG UGU.DUL.BI
	I r. 3′	— — [^{giš}*si*]-*ḫu* ŠIM.^{d}MAŠ NI[TÁ] UGU.DUL.BI
	J r. iv 12′–13′	— — ⸢^{ú}⸣*si-ḫ*[*u* Š]I[M].^{d}MAŠ NITA *u* MUNUS/[SÍG] UGU.DU[L.B]I
65.	A ii 9.	⸢KÙ.SI$_{22}$⸣ AN.BAR 1-*niš* *ina* [KU]Š — — DÙ.DÙ *ina* GÚ-*šú*(text ŠÁ) GAR-*an* — — — —
	C r.$^{?}$ 12′–13′	KÙ.SI$_{22}$ ⸢AN⸣.B[AR] / ^{munus}ÁŠ.GÀR GÌŠ.NU.ZU DÙ.DÙ *ina* GÚ-*šú* GAR-*an* LÚ N[U]
	D r. 21–22	KÙ.SI$_{22}$ AN.BAR TÈ[Š.BI] / [^{m}]^{unus}ÁŠ.GÀR GÌŠ.NU.ZU DÙ.DÙ *ina* GÚ-*šú* GAR LÚ NU KÙ *e-li-i*[*l*]

	F o. 6′	KÙ.SI$_{22}$ AN.BAR — *ina* KUŠ — — — — — — — — — —
	G r. 7′–8′	KÙ.SI$_{22}$ AN.BAR TÉŠ.BI *ina* KUŠ ^{munus}ÁŠ.GÀR GÌŠ.NU.ZU / DÙ.DÙ *ina* GÚ-*šú* GAR-*ma* LÚ NU KÙ *e-lil*
	H iii 17–18	KÙ.SI$_{22}$ AN.BAR TÉŠ.BI *ina* KUŠ ^{munus}ÁŠ.GÀR GÌŠ.NU.ZU / DÙ.DÙ *ina* GÚ-*šú* GAR-*an-ma* LÚ NU KÙ *el*
	I r. 3′–4′	KÙ.SI$_{22}$ [/ $^{m]unus}$ÁŠ.GÀR GÌŠ.NU.ZU [GÚ]-*šú* GAR-*an*-⸢*ma*⸣ [ ]
	J iv 13′–16′	KÙ.SI$_{22}$ AN.BAR TÉŠ.BI / *ina* KUŠ ^{munus}ÁŠ.GÀR GÌŠ.NU.ZU / DÙ.DÙ-*pí* *ina* GÚ-*šú* GAR-*an-ma* / LÚ *la el-lu$_4$ il-li-il*

IX. Ms. A ii 10–21 = *nīš libbi* incantation No. M.1 “Sexual desire! Sexual desire! The bed for the sexual desire” ll. 4–15

X. A ii 22–35 // F rev. 12′–14′ // G rev. 3′–5′ // O rev. iv 4–5 = Prescriptions 23–28

66.	A ii 22	DIŠ KI.MIN ŠÀ B[URU$_5$].ḪABRUD.DA^{mušen} NI[TÁ . . .]
67.	A ii 23	DIŠ KI.MIN ^{na4}[. . .] *tu*$^?$ *ina* Ì [*ina* KUŠ]
68.	A ii 24	DIŠ KI.MIN [SU]ḪUŠ [. . . *n*]*am šá mu sar* [. . .]
69.	A ii 25	DIŠ NA *ina* ⸢KI⸣.[NÁ-*šú* ḪULUḪ.ḪULUḪ-*ut*] ŠÀ-*šú e-*[*šu u ina*] KI.NÁ-[*šú*]
70	A ii 26	*ri-ḫu-s*[*u* GIN-*ak* NA B]I *ki-mil-ti* ^{d}AMAR.UTU *u* ^{d}*i*[*š-tar*]
71	A ii 27	UGU-*šú* G[ÁL-*ši ana* TI-*šú* ^{na4}GU]G.GAZI^{sar} ^{na4}ZA.GÌN ^{na4}GIŠ.N[U$_{11}$.GAL]
	F rev. 12′–13′	— — — — ^{na4}GUG.GAZI^{sar} ^{na4}[ ] / ^{na4}GIŠ.NU$_{11}$.GAL
	G rev. 3′	— — — — ^{na4}GUG.GAZI^{sar} ^{na4}ZA.GÌN ^{na4}[ ]
	O rev. iv 4	— — — — ⸢^{na4}GUG.GAZI^{sar}⸣ ^{na4}ZA.GÌN ^{na4}GIŠ.NU$_{11}$.GAL
72	A ii 28	^{na4}KUR-*n*[*u* DAB *i*]*na* DUR GADA È-*ak* *ina* G[Ú-*šú* —]
	F rev. 13′–14′	^{na4}KUR-*nu* DAB.[BA AN.]BAR / *ina* DUR GADA — *ina* GÚ-*šú* G[AR —]
	G rev. 4′–5′	^{na4}KA.GI.NA.DAB.BA AN.BAR *ina* ⸢DUR⸣ [x] / ⸢È-*ak*⸣ *ina* GÚ-*šú* GAR-*an-ma* [*e*]-⸢*lil*⸣
	O rev. iv 4–5	[/ DU]R GADA *ta-šá-kak ina* G[Ú-*šú*]
73	A ii 29	⸢*e*⸣ [. . . NA$_4$.^{d}]LAMMA ^{na4}ZA.GÌN ^{na4}PA [^{n}]^{a4}KUR-⟨*nu*⟩ DAB
74	A ii 30	[. . .] x x x x ^{na4}GIŠ.NU$_{11}$.GAL ^{na4}⸢NU.LUḪ⸣.ḪA
75.	A ii 31	[A$^?$] ^{šim}LI 10 Ú$^{!meš}$ ŠÀ.ZI$^!$.GA$^!$ *ina* SÍG SILA$_4$ [NI]TA *šá* ⟨*ana*⟩ U$_5$
76	A ii 32	ZI-*ú*$^!$ [*ina* GÚ-*šú*] GAR-*an*
77.	A ii 33	^{na4}ZA.GÌN ^{na4}GUG ^{na4}MUŠ.[GÍR ^{n}]^{a4}*aš-gì-gì* ^{na4}BABBAR.DIL
78.	A ii 34	⸢^{na4}⸣[SIK]IL$^?$ ^{na4}PA 7 NA$_4$$^{⸢meš⸣}$ ŠÀ.Z[I.G]A ÉN SU.Z[I MIN]
79.	A ii 35	[ÉN] ⸢ki⸣-in-da-⸢rab⸣ MIN 2 ÉN^{meš} [. . .] x [. . .]

XI. A ii 36–50 = Incantation and its ritual No. K.3

80.	A ii 36	[ÉN É.NU].RU ka ab ka [. . .] ma na
81.	A ii 37	[. . .] ⸢x x⸣ [. . .] x x [. . .] di di il din x
82.	A ii 38	lú bi ú [. . .] e te am
83.	A ii 39	lú bi ga x [. . . a]m na am
84.	A ii 40	ki ì bi ga ⸢lú⸣ [. . .] x ig ba
85.	A ii 41	la ba il kur te [. . .] bar ta
86.	A ii 42	la ba il ta [n]a [. . .] x si a
87.	A ii 43	la ba lu ka ma an [. . .] x x la ud
88.	A ii 44	la ba lu ka ma a[n x] x x x [TU$_6$] ÉN É.NU.RU
89.	A ii 45	Ú ⸢*tak-da*⸣-*na-nu* x [x] NUMUN$^?$ *lu-u* [. . .]
90.	A ii 46	*ana* ^{ú}EME.UR.GI$_7$ *ma* x ZI-*šú* x *šú* x [. . .]
91.	A ii 47	*ana* ZI-*šú* 7 ŠE KÙ.BABBAR [7 ŠE KÙ.S]I$_{22}$ *ana* IGI [. . .] x x [x]
92.	A ii 48	DU$_{11}$.GA 20 NÍG.BA x x [. . .] ⸢Ú⸣ ŠÀ.ZI.[GA]
93.	A ii 49	3-*šú* DU$_{11}$.GA Ú ŠÀ$^{!?}$.ZI.[GA^{m}]^{eš} *ana mi-n*[*a-ti*$^?$]-*šú*-⸢*nu*⸣ ŠUB
94.	A ii 50	Ú BI *ina* ^{síg}ÀKA UDU.[NIT]Á È-*ak ina*$^{!?}$ [MÚRU-*šú*] KEŠDA-*ma* $^{\text{ḫe-pí}}$[[ŠÀ.ZI.GA]]

XII. A ii 51–53 = Fragmentary incantation No. K.4

95.	A ii 51	ÉN É.NU.RU e ne $^{\text{ḫe-pí}}$
96.	A ii 52	za az zal ba al$^{!}$ $^{\text{ḫe-pí}}$
97.	A ii 53	KA.INIM.MA 7 $^{\text{[ḫe-pí]}}$

XIII. A ii 54–61 = Fragmentary incantation and its ritual No. K.5

98.	A ii 54	ÉN KI.NÁ $^{\text{[ḫe-pí]}}$
99.	A ii 55	KI.NÁ $^{\text{[ḫe-pí]}}$
100.	A ii 56	*ina* DAL.DAL SU $^{\text{[ḫe]-pí}}$
101.	A ii 57	DUMU $^{\text{ḫe-pí}}$
102.	A ii 58	KA.INIM.MA ŠÀ.ZI.GA
103.	A ii 59	DÙ.DÙ.BI NUMUN ^{ú}*pu-qut-tú* ⟨*ina*⟩ ÚŠ BURU$_5$.ḪABRUD.DA^{mušen} $^{\text{ḫe-pí}}$[[NITA ḪE.ḪE *ina* ^{síg}ÀKA NIGIN]]
104.	A ii 60	ÉN *an-ni-tú* 7-*šú ana muḫ-ḫi* ŠID-*nu* ^{giš}*bi-ni* $^{\text{ḫe-pí}}$
105.	A ii 61	[*m*]*u-ši* U$_4$ ŠÀ-*ka ul i-na-ḫa u ši-i* $^{\text{ḫe-pí}}$

XIV. Ms. A ii 62–iii 23 = *Nīš libbi* incantation No. E.2 "Light of the heavens, wise Ištar" ll. 18–45

XV. A iii 24–33 = Fragmentary incantation No. K.6

106.	A iii 24	NENNI [A NENNI . . .]
107.	A iii 25	x [x *r*]*a* x [. . .]
108.	A iii 26	x *ina* IGI x *ši šú* x [. . .]
109.	A iii 27	*šu-ú lim-gu-ug* GIM AN[ŠE . . .]
110.	A iii 28	GIM *ḫe-re-eb* ⟨*ana*⟩ MUŠEN *ḫur-ri an* [. . .]
111.	A iii 29	GIM KÙ.BABBAR *ana* MUN GIM KÙ.SI$_{22}$ [*ana* x x GIM]
112.	A iii 30	A.BÁR *ana* Ì+GIŠ [. . .]
113.	A iii 31	⸢LÚ⸣ SAR-[*šú-ma* . . .]
114.	A iii 32	KA.INIM.MA ŠÀ.[ZI.GA]
115.	A iii 33	x MU.MEŠ.NI ⸢72⸣ x [. . .]

XVI. A iii 34–42 = Incantation and its ritual No. K.7

116.	A iii 34	[ÉN] UR.MAḪ *lu-u* x x [. . .]
117.	A iii 35	*r*[*i-k*]*ib-ta-k*[*a*] *l*[*u* . . .]
118.	A iii 36	*i-*[*n*]*a q*[*í-b*]*it* AN.Z[ÍB ^{d}*iš-tar* ^{d}*na-na-a*]
119.	A iii 37	^{d}*ga*[*z-ba*]*-ba* ^{d}*k*[*a-ni-sur-ra*]
120.	A iii 38	ÉN *an-*⸢*ni*⸣*-ti* 3-*šú* ŠI[D-*nu*] x x
121.	A iii 39	DÙ.DÙ.B[I SÍ]K BABBAR SÍG SA$_5$ NU$^!$.NU$^!$ 7 [. . .]
122.	A iii 40	7-*ú* [. . .] *u* 7 ZÚ.KEŠDA KEŠ[DA . . .]
123.	A iii 41	ÉN *ina muḫ-ḫi* [ŠID-*nu* . . .]
124.	A iii 42	*ina* x x x KEŠDA-*ma* [ŠÀ.ZI.GA]

XVII. A iii 43–iv 7 // N ii 3–9 = Prescriptions 28–31

125.	A iii 43	1 SÌLA [. . .]
126.	A iii 44	SAḪAR KI.[MAḪ SA]ḪAR *šú-nu-*[*ti* . . .]
127.	A iii 45	*ana* Ì.GIŠ [ŠUB$^?$] NU NITA *u* MUNUS [DÙ-*uš* . . .]
128.	A iii 46	x x-*šú-nu-ti* NITA *u* MUNUS ZI-*ut*$^!$ Š[À]$^!$ x [. . .]
129.	A iii 47	*ana* ŠÀ.ZI.GA TUKU-*e a-bu-na-at* ⟨DÀRA.MAŠ⟩ SI DÀRA.[MAŠ]
130.	A iii 48	*ri-kib-ti*(text SI) DÀRA.MAŠ ^{ú}*a-ṣ*[*u-ṣu*]*-um-t*[*um*$^?$]
131.	A iii 49	*tam-ta-raq* GÌŠ *ana* UGU *ṭab-ti* ⸢*tal*⸣*-t*[*ap-pat*$^?$]
132.	A iii 50	*a-zap-pi* KUN NITA ZI-*aḫ* 1-*niš* ⸢DUR⸣ [^{túg}GADA . . .]
133.	A iii 51	[N]U.NU-*ma* {NA$_4$}$^?$ ^{na4}⸢AMAŠ⸣.PA.È *ina* ŠÀ ⸢È⸣-[*ak* . . .]
134.	A iii 52	*ina* MÚRU-*šú* KEŠDA-[*ma* ŠÀ.ZI.GA]
135.	A iv 1	[. . .]
	N ii 3	DIŠ KI.MIN *kap-pí* TE$_8$ NITÁ *kap-pí i-*[. . .]
136.	A iv 2	[ ] — *rit-ku*$^!$*-ba-*[*ti*]
	N ii 4–5	MUŠ.DÍM.KUR.RA EDIN *ri-it-ku-*[*ba-ti*] / ŠE$_{10}$ ARKAB^{mušen} ŠE$_{10}$

137. A iv 3 ⸢EME⸣.DIR.GÙN NUMUN ^{giš}x [. . .]
N ii 5–6 EME.DIR.GÙN.A NU[MUN . . .] / NUMUN ^{giš}MA.NU NUMUN ^{ú}IN.NU.UŠ

138. A iv 4 [ ] ^{ú}A.ZAL.LÁ NUMUN ^{giš}*mur-*[*du-di-i*]
N ii 7–8 NUMUN *a-zal-li* NUMUN ^{ú}MUL.DÙ.DÙ / NUMUN ^{ú}EME(text NAG).UR.GI$_7$

139. A iv 5 [NUMUN] ^{ú}ŠAKIR ^{ú}*ṣa-*⸢*ṣu-un*⸣-[*tu*]
N ii 8–9 — ^{ú}ŠAKIR.RA *ṣa-ṣ*[*u-un-tu*] / ^{ú}SUMUN.DAR^{sar}

140. A iv 6 [14] ⸢Ú. ḪI⸣.A ŠEŠ [. . .]
N ii 9 14 Ú.⟨ḪI⟩.A – TÉŠ.BI [. . .]

141. A iv 7 10 MU.MEŠ.NI [. . .]

XVIII A iv 8–23 = Fragmentary incantation and its ritual No. K.8

142. A iv 8 $^{ḫe-pí}$ *ú* [. . .] ŠÀ [. . .]
143. A iv 9 $^{ḫe-pí}$ *ul i-*[*d*]*e*$^{?}$*-e* [. . .]
144. A iv 10 $^{ḫe-pí}$ *ḫa-tum e-pu-*[*uš*] ⸢ÍD⸣
145. A iv 11 $^{ḫe-pí}$ *lìb-bi ta-ḫi-šá-ti ad* [. . .] ⸢*e*⸣
146. A iv 12 $^{ḫe-pí}$ ^{d}⸢15⸣$^{?}$ x [. . .] x *ia*
147. A iv 13 $^{ḫe-pí}$ *lip-šur-ka-ma* ⸢^{d}⸣[*iš-tar* ^{d}*na-na-a*] ^{d}*gaz-ba-ba*
148. A iv 14 ^{d}*ka-ni-sur-ra ana lìb-b*[*i* . . .] *ra pu-ḫa-lu*
149. A iv 15 $^{ḫe-pí}$ [[*ra-am* DÀRA$^{?}$]] 6-*šú* ⸢*ra*⸣*-am a-*[*a-li* 7-*šú ra-am*]
BURU$_5$.ḪABRUD. DA^{mušen}
150. A iv 16 $^{ḫe-pí}$ [[12-*šú* . . .]] *si-su-ú l*[*i* . . .] x *ra*
151. A iv 17 $^{ḫe-pí}$ *qab-la-at* [. . .] *šá* GIŠ.NÁ
152. A iv 18 $^{ḫe-pí}$ *la* 1 Ú 1 DÙ [. . . *u*]*b tar* ⸢*da*⸣
153. A iv 19 *la ta-šeb-ba-a* [*la-la*]-⸢*a*⸣$^{!}$-*šá*

154. A iv 20 KA.INIM.MA [ŠÀ.Z]I.G[A]

155. A iv 21 ⸢^{ḫe}⸣$^{-pí}$ [[DÙ.DÙ.BI]] [K]UN *pu-ḫa-l*[*i* ZI]-*i* TI-*qi* ⸢KUN⸣ GÍR.TAB
156. A iv 22 [. . .] ⸢*a*$^{!}$-*na*$^{!}$⸣ *ra*$^{!}$*-man-ni-k*[*a*] x ⸢*ga*⸣ *za*
157. A iv 23 [. . .] ⸢Ú⸣.ḪI.A [ŠE]Š
Rest destroyed

XIX. A iv 24–31 = Fragmentary incantation and its ritual No. K.9

158–162. A iv 24$^{?}$–28 Broken lines
163. A iv 29 KA.INIM.[MA ŠÀ.ZI.G]A

164. A iv 30 DÙ.DÙ.BI *ḫa-an-*[*du-ur bal-lu-ṣi-ti* . . .]
165. A iv 31 ÉN 7-*šú ana lìb-b*[*i* ŠID]-*nu* ^{giš}KUN^{meš}-*šú* [EŠ.M]EŠ-*ma* ŠÀ.ZI.G[A]

XX. A iv 32–36 = Prescriptions 32

166.	A iv 32	DIŠ KI.MIN *zi-qit* [. . .] x DÙ *zi-qit* NIM-LÀL *zi-qit* ^{na4}x [. . .]
167.	A iv 33	x [. . . *el*]-*lu-ti šik-kur-rat* SIKIL-*bu-ti* SUḪUŠ $^{ú!}$KUR.ZI
168.	A iv 34	*u* ⌜uzu⌝[. . .] x ÚŠ BURU$_5$.ḪABRUD.DA^{mušen} NITÁ EME.D[IR]$^?$ PA ^{giš}ḪAŠḪUR$^?$
169.	A iv 35	10 Ú.[ḪI].A ŠE[Š . . .] x *nu* SAR ÉN 7-*šú* [*ana m*]*uḫ-ḫi* ŠID-*nu*
170.	A iv 36	*ina* MÚRU-*šú* [KEŠ]DA-*ma* [Š]À.ZI.GA

XXI. Ms. A iv 37–41 = *nīš libbi* incantation No. A.1 "May the wind blow! May the grove quake!" ll. 33–39

Transcription

I. A i 1–7 = Prescriptions 1–2

1. [*šumma amēlu ana sinništīšu libbašu l*]*ā inaššī-ma*
2. [. . . *amēlu šū kišpu iṣbass*]*u ana bulluṭī*[*šu*]
3. [. . .] . . .
4. [. . . *ina kiš*]*ādīšu tašakkan*
5. [. . .] *ina šaman šurmēni*
6. [. . . *ina maš*]*ki ina kišādīš*[*u*] *tašakkan*

7. [*šumma* KI.MIN (= *amēlu ana sinništīšu libbašu lā inaššī-ma* [. . . *amēlu šū kišpu iṣbass*]*u*): . . . *zē*]*r ḫûrati* [*ina šikari*] *išatti*

II. Symptoms (Ms. A i 8–17) = *nīš libbi* D Sympt.: 42–48

III. Two prescriptions (Ms. A i 18–21// Ms. B obv. i 2′–6′) = *nīš libbi* F prescr. 10–11: 31–35

IV. B obv. i 7′–16′ = Incantation and its ritual No. K.1

8–12. Sumerian incantation

13. *dudubû*: *kappi šuttinni ša* . . . [. . . *teleq*]*qe*
14. *tubbal tasâk ina šikar sābî* [*išа*]*ttīšu*
15. *ina šamni* [*tapa*]*ššassu ina maški ina kiš*[*ādīšu*] *tašakkan*

16–17. Sumerian incantation

V. B rev. iv 1′–6′ = Incantation fragment and its ritual No. K.2

18. Traces
19. [. . .] . . . [. . . *tê šipti*]

20. [*šipat*] *n*[*īš*] *libbi*

21. [*dudubû*:] *ina išāti tuqattaršu-ma n*[*īš*] *libbi*

22. [*ana nīš libbi r*]*ašê*: *zēr ḫašḫūri maṭî zēr šammi* . . . [. . .]
23. [*šumma amēlu ana sinništī*]*šu libbašu inaššīšu-ma ana sinništi aḫīti libbašu l*[*ā inaššīšu*]

Translation

I. A i 1–7 = Prescriptions 1–2

1. [If a man does] not desire [his woman]
2. [. . . the witchcraft has seized that man], in order to cure [him]:
3. [. . .] . . .-stone
4. [. . .] you put (it) [around] his [neck].
5. [. . .] with cypress oil
6. [. . . in a leather ba]g, you put (it) around h[is] neck.
7. [If ditto (= a man does not desire his woman [. . . the witchcraft has seized that man])]: He drinks [. . . (and) se]eds of *ḫûratu* [with beer].

II. Symptoms (Ms. A i 8–17) = *nīš libbi* D Sympt.: 42–48

III. Two prescriptions (Ms. A i 18–21// Ms. B obv. i 2′–6′) = *nīš libbi* F prescr. 10–11: 31–35

IV. B obv. i 7′–16′ = Incantation and its ritual No. K.1

8. [Incantation]: (When) Enlil (and) Bēlet-ilī gave mankind a name,
9. they completely perfected (its) limbs,
10. they took? i[ts . . .], they covered its 'heart.'
11. (When) they commanded its sexual desire,
12. they made (the mankind) desires (lit. they raised its 'heart') . . . [. . .]. Incantation formula.
13. Its ritual: [You tak]e the wing of a bat, which . . . ..
14. you dry and pulverize (it), [he drink]s (it) in beer of the brewer,
15. you anoint it with oil and put (it) in a leather bag around [his] neck.

16–17. Incantation: Like crea[tion? of heav]en?! Like creatio[n of heaven]! . . . [. . .] . . . cre[ation . . .].

V. B rev. iv 1′–6′ = Incantation fragment and its ritual No. K.2

18–19. Traces

20. [Wording] of *n*[*īš*] *libbi* (incantation).

21. [Its ritual]: You fumigate him in fire and (he will get) sexual de[si]re.

22. [To g]et [sexual desire]: Seed of a small apple-tree, seed of . . .-plant [. . .].
23. [If a man] desires his [woman], but do[es not desire] another woman.

VI. A i 22–51 // D rev.$^{?}$ 9′–11′ // E rev. 23–33 // F obv. 10′–rev. 3 // G obv. 13′–15′ = Prescriptions 3–18

3. 24. [*šumma amēlu kašip-ma šīrūšu tabkū lū ina alākīšu lū ina uzuzzīšu*

25. *lū ina mayyālīšu lū enūma šīnātīšu išattinu*
26. *riḫûssu illak kīma sinništi su''ussu lā ēl* (var.: *illil*)
27. *amēlu šū riḫûssu itti mīti ina erṣeti šunullat ana bulluṭīšu*

28. *ašqulāla ēda sikilla amīlāna zēr maštakal*

29. *zēr šakirî šuruš balti ša eli kimaḫḫi ina maški tašappi ina kišādīšu tašakkan-ma iballuṭ*

---

4. 30. *ka*[*līt*] *immeri ḫaṣabti sūq erbetti šārāt nēši ina maški tašappi ina kišādīšu tašakkan-ma iballuṭ*

---

5. 31. *šumma* KI.MIN (= *amēlu kašip-ma šīrūšu tabkū lū ina alākīšu lū ina uzuzzīšu lū ina mayyālīšu lū enūma šīnātīšu išattinu riḫûssu illak kīma sinništi su''ussu lā amēlu šū riḫûssu itti mīti ina erṣeti šunullat*): *kibrīta* [*pappa*]*sīta in*[*a maški in*]*a qablīšu tarakkas*

---

6. 32. *šumma* KI.MIN (= *amēlu kašip-ma šīrūšu tabkū lū ina alākīšu lū ina uzuzzīšu lū ina mayyālīšu lū enūma šīnātīšu išattinu riḫûssu illak kīma sinništi su''ussu lā amēlu šū riḫûssu itti mīti ina erṣeti šunullat*): *šu*[*ruš*] *lišān-kalbi* [*barī*]*rāti šuruš šurdunî*

33. *šuruš* [S]AG [*šuruš*] [*a*]*zallî*
34. *šuruš kamkad*[*i*] *š*[*uruš*] *ardadi*[*lli*]
35. [*šuruš*] *kazalli* [*šammī n*]*īš* [*libbi*] *ina* [*maški i*]*na kišā*[*dīšu tašakkan*]

---

7. 36. [*šumma amēlu ana sinništi*] *iṭeḫḫe-ma* [. . .]
37–39. Destroyed

---

8. 40. *šumma* KI.MIN (= *amēlu ana sinništi iṭeḫḫe-ma* [. . .]): [. . .] *arariānu*
41. *zēr* [. . .] *ina maški*

---

9. 42. *šumma* KI.MIN (= *amēlu ana sinništi iṭeḫḫe-ma* [. . .]): . . . [. . .] *arantu ina maški*

---

10. 43. *šumma* KI.MIN (= *amēlu ana sinništi iṭeḫḫe-ma* [. . .]): [*kam*]*antu* [. . .] *kankadu*

44. *šuruš šakirî* . . . [. . .] *ina šamni ina maški*

---

11. 45. *šumma* KI.MIN (= *amēlu ana sinništi iṭeḫḫe-ma* [. . .]): *ḫaṭṭi-rē'i kamandu* [*a*] *lamû ina maški*

VI. A i 22–51 // D rev.? 9′–11′ // E rev. 23–33 // F obv. 10′–rev. 3 // G obv. 13′–15′ = Prescriptions 3–18

3. 24. If a man is bewitched and his flesh is weak, (and) neither when walking, nor standing
25. nor being on his bed, nor when urinating,
26. his sperm flows, like (that of) a woman (his) 'genital discharge' is impure,
27. the sperm of this man has been buried under the earth with a dead man. To cure him:
28. *ašqulālu*-plant, *ēdu*-plant, *sikillu*-plant, *amīlānu*-plant, seeds of *maš-takal*-plant,
29. seeds of *šakirû*-plant, root of *baltu*, which (grows) over a grave, you wrap up (them) in a leather bag (and) put (it) around his neck, then he will recover.

---

4. 30. You wrap up in a leather bag a ram's kid[ney], a sherd from a crossroads (and) lion hair, you put (it) around his neck, then he will recover.

---

5. 31. If ditto (= a man is bewitched and his flesh is weak, (and) neither when walking, nor standing nor being on his bed, nor when urinating, his sperm flows, like (that of) a woman (his) 'genital discharge' is impure, the sperm of this man has been buried under the earth with a dead man): Sulphur (and) *pappasītu*-white [gypsum] i[n a leather bag], you tie (it) around his waist.

---

6. 32. If ditto If ditto (= a man is bewitched and his flesh is weak, (and) neither when walking, nor standing nor being on his bed, nor when urinating, his sperm flows, like (that of) a woman (his) 'genital discharge' is impure, the sperm of this man has been buried under the earth with a dead man): Ro[ot of] "dog's-tongue"-plant, [*barī*]*rātu*-plant, root of *šurdunû*-plant,
33. root of [S]AG-plant, [root] of [*a*]*zallû*-plant,
34. root of *kamkad*[*u*]-plant, r[oot of] *ardadi*[*llu*]-plant,
35. [root] of *kazallu*-plant, you put [(these) plants for sexu]al [desire] in [a leather bag ar]ound [his] nec[k].

---

7. 36. [If a man] approaches [a woman: . . .]
37–39. Destroyed

---

8. 40. If ditto (= a man approaches [a woman . . .]): [. . .] *arariānu*-plant,
41. root of the [. . .]-plant [. . .] in a leather bag.

---

9. 42. If ditto (= a man approaches [a woman . . .]): . . .-plant, [. . .] *arantu*-plant in a leather bag.

---

10. 43. If ditto (= a man approaches [a woman . . .]): *kam*]*antu*-plant, [. . .] *kankadu*-plant,
44. root of *šakirû*-plant, . . .-plant [. . .] with oil in a leather bag.

---

11. 45. If ditto (= a man approaches [a woman . . .]): "Shepherd's-crook"-plant, *kamantu*-plant, [*a*]*lamû*-plant in a leather bag.

---

12. 46. *šumma* KI.MIN (= *amēlu ana sinništi iṭeḫḫe-ma* [. . .]): *šumuttu kamantu* [*azallû*] *ēdu ina maški*

13. 47. [*šumma* KI.MIN (= *amēlu ana sinništi iṭeḫḫe-ma* [. . .]): *ḫaš*]*ḫūru šammi balāṭi azallû šimru* (or *urânu*) *būšānu ina maški*

14. 48. [*šumma* KI.MIN (= *amēlu ana sinništi iṭeḫḫe-ma* [. . .]):] *ḫulālu* [*tur*]*minû* [*uq*]*nû ina maški*

15. 49. [*šumma* KI.MIN (= *amēlu ana sinništi iṭeḫḫe-ma* [. . .]): . . . *maš*]*takal ēdu ina maški*

16. 50. [*šumma* KI.MIN (= *amēlu ana sinništi iṭeḫḫe-ma* [. . .]): . . .] *a*[*zallû*] *ina maški*

17. 51. [*šumma* KI].MIN (= *amēlu ana sinništi iṭeḫḫe-ma* [. . .]): *šunû* [. . .] *ina maški*

18. 52. *šumma* KI.MIN (= *amēlu ana sinništi iṭeḫḫe-ma* [. . .]): *šuruš balti* . . . [. . .] *ina maški*

VII. Five Prescriptions (Ms. A i 52–55) = *nīš libbi* A prescr. 9–12, 16: 19–22, 27

VIII. A i 56–ii 9 // C rev.[?] 12′–13′ // D rev. 21–22 // F obv. 5′–6′, rev. 27′–33′ // G obv. 19′–rev. 2′, rev. 6′–8′// H rev. iii 16–18 // I rev. 3′–4′ // J rev. iv 12′–16′ // K obv. i 16–22 // L obv. 10′–13′ // M obv. 12′–16′ = Prescriptions 19–22

19. 53. [*šumma* KI].MIN (= *amēlu ana sinništi iṭeḫḫe-ma* [. . .]): *šuruš lišān-kalbi šuruš šimri* (or *urâni*) *šuruš ar*[*dadilli*]
54. [*šur*]*uš ḫaṭṭi-rē'i šuruš šurdunî ina šikari ina šamni* [*ina maški*]

20. 55. [*šumma* KI.MIN (= *amēlu ana sinništi iṭeḫḫe-ma* [. . .]):] *imḫur-līm* [ḫep[i]]
56. [. . . *šumu*]*ttu* [ḫepi]

21. 57. *šumma amēlu ina mayyālīšu igdanallut libbašu ešu u ina mayyālīšu*

58. *riḫûssu illak amēlu šū kimilti Marduk u Ištar*
59. *elīšu ibašši ana bulluṭīšu tarmuš ḫašû* (var.: *ēdu*)
60. *ḫarmunu kiṣir* (var.: omitted) *bīni* (var.: *ēri*) *baltu arqūssu* (var.: *ḫīl balti*)

61. *atbaru ḫiṣib sāmti sikillu maštakal* (var. adds: *ša šadî*) *ari šunî*

62. *pallišu azallû lišān-kalbi ašāgu* (var.: *baltu*) *ša* (var. adds: *muḫḫi*) *igāri*

63. *ištēniš* (var. adds: *tasâk*) *itqa talammi ina dām erēni tasallaḫ ina maški* (var.: 14 *ša*[*m*]*mī ann*[*ûti ištēniš*] *ina dām erēni taballal ina maški*)

12. 46. If ditto (= a man approaches [a woman . . .]): *šumuttu*-plant, *kamantu*-plant, [*azallû*]-plant, *ēdu*-plant in a leather bag.

13. 47. [If ditto (= a man approaches [a woman . . .]): Apple]-tree, “plant-of-life,” *azallû*[-plant], *šimru*-plant (or *urânu*-plant), *būšānu*-plant in a leather bag.

14. 48. [If ditto (= a man approaches [a woman . . .])]: *ḫulālu*-stone, [*tur*]*minû*-stone, [lapis] lazuli in a leather bag.

15. 49. [If ditto (= a man approaches [a woman . . .]): . . . *maš*]*takal*-plant, *ēdu*-plant in a leather bag.

16. 50. [If ditto (= a man approaches [a woman . . .]): . . .] *a*[*zallû*]-plant in a leather bag.

17. 51. [If dit]to (= a man approaches [a woman . . .]): Agnus castus [. . .] in a leather bag.

18. 52. If ditto (= a man approaches [a woman . . .]): Root of *baltu*-tree . . . [. . .] in a leather bag.

VII. Five Prescriptions (Ms. A i 52–55) = *nīš libbi* A prescr. 9–12, 16: 19–22, 27

VIII. A i 56–ii 9 // C rev.[?] 12′–13′ // D rev. 21–22 // F obv. 5′–6′, rev. 27′–33′ // G obv. 19′–rev. 2′, rev. 6′–8′// H rev. iii 16–18 // I rev. 3′–4′ // J rev. iv 12′–16′ // K obv. i 16–22 // L obv. 10′–13′ // M obv. 12′–16′ = Prescriptions 19–22

19. 53. [If dit]to (= a man approaches [a woman . . .]): Root of “dog’s-tongue”-plant, root of *šimru*-plant (or *urânu*-plant), root of *ar*[*dadillu*]plant,
54. [roo]t of “shepherd’s-crook”-plant, root of *šurdunû*-plant with beer, with oil [in a leather bag].

20. 55. [If ditto (= a man approaches [a woman . . .]): “Heals-a-thousand”-plant [bro[ken]]
56. [. . . *šumu*]*ttu*-plant [broken]

21. 57. If a man is repeatedly scared in his bed, his ‘heart’ (*libbu*) is confused and in his bed
58. his sperm comes out, over this man the wrath of Marduk and Ištar
59. has come. In order to cure him: *tarmuš*-plant, *ḫašû*-plant (var.: *ēdu*-plant),
60. *ḫarmunu*-plant, a knot (var.: omitted) of tamarisk (var.: *ēru*-tree), the green of the *baltu*-tree (var.: resin of *baltu*-tree),
61. basalt, cornelian chipping, *sikillu*-plant, *maštakal*-plant (var. adds: of the montain), a branch of agnus castus,
62. *pallišu*-stone, *azallû*-plant, “dog’s-tongue”-plant, *ašāgu*-acacia (var.: *baltu*-tree) from a wall,
63. (var. adds: you pulverize (them)), you wrap (them) together in fleece, you sprinkle (them) (var.: you mix [together] these 14 ingredients) with cedar ‘resin’ in a leather bag.

22. 64. *šumma* KI.MIN (= *amēlu ina mayyālīšu igdanallut libbašu ešu u ina mayyālīšu riḫûssu illak amēlu šū kimilti Marduk u Ištar elīšu ibašši*): *sīḫu nikipta zikar u sinniš šārāt pagî*

65. *ḫurāṣa parzilla ištēniš ina mašak unīqi lā petīti tašappi ina kišādīšu tašakkan amēlu lā ellu illil* (var.: *ēl*)

IX. Ms. A ii 10–21 = *nīš libbi* incantation No. M.1 "Sexual desire! Sexual desire! The bed for the sexual desire" ll. 4–15

X. A ii 22–35 // F rev. 12′–14′ // G rev. 3′–5′ // O rev. iv 4–5 = Prescriptions 23–27

23. 66. *šumma* KI.MIN (= *amēlu ina mayyālīšu igdanallut libbašu ešu u ina mayyālīšu riḫûssu illak amēlu šū kimilti Marduk u Ištar elīšu ibašši*): *libbi i*[*ṣṣur*] *ḫurri zik*[*ari* . . .]

24. 67. *šumma* KI.MIN (= *amēlu ina mayyālīšu igdanallut libbašu ešu u ina mayyālīšu riḫûssu illak amēlu šū kimilti Marduk u Ištar elīšu ibašši*): [. . .]. . . *ina šamni* [*ina maški*]

25. 68. *šumma* KI.MIN (= *amēlu ina mayyālīšu igdanallut libbašu ešu u ina mayyālīšu riḫûssu illak amēlu šū kimilti Marduk u Ištar elīšu ibašši*): [*šu*]*ruš* [. . .] . . . [. . .]

26. 69. *šumma amēlu ina mayyā*[*līšu igdanallut*] *libbašu e*[*šu u ina*] *mayyālī*[*šu*]

70. *riḫûss*[*u illak amēlu š*]*ū kimilti Marduk u I*[*štar*]
71. *elīšu ib*[*ašši ana bulluṭīšu aba*]*n kasî uqnû ašn*[*ugallu*]
72. *šadânu ṣābitu parzillu ina ṭurri kitê tašakkak ina kišādīšu tašakkan* (var. adds: *-ma* [*i*]*llil*)

27. 73. . . . [. . . *aban*] *lamassi uqnû ayyartu šadânu ṣābitu*
74. [. . .] . . . *ašnugalla nuḫurta*
75. [*mê*] *burāši* 10 *šammī nīš libbi ina šīpāt kalūmi* [*zik*]*ari ša ana rakābi*

76. *tebû* [*ina kišādīšu*] *tašakkan*

28. 77. *uqnû sāmtu muš*[*šaru*] *ašgigû pappardillû*
78. [*sik*]*illu ayyartu sebet abnī n*[*ī*]*š libbi šiptu*: *šalumma*[*tu šalummatu*]

79. [*šiptu*:] ki-in-da-rab MIN *šina šiptū* [. . .] . . . [. . .]

XI. A ii 36–50 = Incantation and its ritual No. K.3

80–88. Abracadabra
89. *takdanānu* . . . [. . .] *zēr* . . . [. . .]

22. 64. If ditto (= a man is repeatedly scared in his bed, his 'heart' is confused and in his bed his sperm comes out, over this man the wrath of Marduk and Ištar has come): (Resine of)[?] *sīḫu*-tree, "male" and "female" *nikiptu*-plant, hair of a monkey,

65. gold, silver, you wrap up (them) together in a leather bag from a female kid that has not yet mated (and) you put (it) around his neck, (then) the impure man will be pure.

IX. Ms. A ii 10–21 = *nīš libbi* incantation No. M.1 "Sexual desire! Sexual desire! The bed for the sexual desire" ll. 4–15

X. A ii 22–35 // F rev. 12′–14′ // G rev. 3′–5′ // O rev. iv 4–5 = Prescriptions 23–27

23. 66. If ditto (= a man is repeatedly scared in his bed, his 'heart' is confused and in his bed his sperm comes out, over this man the wrath of Marduk and Ištar has come): The innards of a ma[le] p[ar]tridge [. . .].

24. 67. If ditto (= a man is repeatedly scared in his bed, his 'heart' is confused and in his bed his sperm comes out, over this man the wrath of Marduk and Ištar has come): [. . .]-stone, . . . with oil [in a leather bag].

25. 68. If ditto (= a man is repeatedly scared in his bed, his 'heart' is confused and in his bed his sperm comes out, over this man the wrath of Marduk and Ištar has come): [Ro]ot of [. . .] . . . [. . .].

26. 69. If a man is repeatedly scare]d in [his] b[ed], his 'heart' is [confused and in his] bed

70. hi[s] sperm [comes out], over him the wrath of Marduk and I[štar]

71. has co[me. In order to cure him: *aba*]*n kasî*-stone, lapis lazuli, ala[baster],

72. magnetite, iron, you insert them in a linen string, you put (it) around his neck (var. adds: and [he] will pure).

27. 73. . . . [. . .] *lamassu*-stone, lapis lazuli, *ayyartu*-shell, magnetite,

74. [. . .] . . ., alabaster, *nuḫurtu*-plant,

75. juniper [water], the ten drugs for the sexual desire in wool of a [ma]le lamb which for mating

76. is reared-up, you put (it) [around his neck].

28. 77. Lapis lazuli, cornelian, *muš*[*šaru*-stone], *ašgigû*-stone, *pappardillû*-stone,

78. [*siki*]*llu*-stone, *ayyartu*-shell, these seven stones for the se[xu]al desire. Incantation: "Shi[ver! Shiver!]"

79. [Incantation:] ". . . . . .". The two incantations [. . .] . . . [. . .].

XI. A ii 36–50 = Incantation and its ritual No. K.3

80–88. Abracadabra

89. *takdanānu*-bush . . . [. . .] seeds of . . . [. . .].

90. *ana lišān-kalbi . . . tībīšu . . .* [. . .].
91. *ana tībīšu sebet uṭṭat kaspi* [*sebet uṭṭat ḫurāṣ*]*i ana maḫri* [. . .] . . . [. . .]
92. *iqabbi* 20 *qīštī . . .* [. . .] *šam nī*[*š*] *libbi*
93. *šalāšīšu iqabbi šammī nī*[*š*] *libbi ana min*[*âtī*]*šunu tanaddi*
94. *šamma šuāti ina itiq imm*[*er*]*i tašakkak ina* [*qablīšu*] *tarakkas-ma* ḫepi [[*nīš libbi*]]

XII. A ii 51–53 = Fragmentary incantation No. K.4

95. *šiptu* É.NU.RU e ne ḫepi
96. za az zal ba al! ḫepi
97. *šiptu: sebet*[ḫepi]

XIII. A ii 54–61 = Fragmentary incantation and its ritual No. K.5

98. *šiptu: mayyālu* [ḫepi]

---

99. *mayyālu* [ḫepi]
100. *ina napruši šīru* [ḫe]pi
101. *māru* ḫepi

---

102. *šipat nīš libbi*

---

103. *dudubû: zēr puqutti ina dām iṣṣūr ḫurri* ḫepi [[*zikari taballal ina itqi tal*[*ammi*]]

104. *šipta annītu sebîšu ana muḫḫi tamannu bīni* ḫepi
105. [*mū*]*ši urra libbaka ul inaḫḫa u šī* ḫepi

XIV. Ms. A ii 62–iii 23 = *nīš libbi* incantation No. E.2 “Light of the heavens, wise Ištar” ll. 18–45

XV. A iii 24–33 = Fragmentary incantation No. K.6

106. *annanna* [*mār annanna . . .*]
107. . . . [. . .] . . . [. . .]
108. . . . *ina pāni . . .* [. . .]
109. *šū limgug kīma imē*[*ri . . .*]
110. *kīma ḫerēb ana iṣṣūr ḫurri . . .* [. . .]
111. *kīma kaspi ana ṭābti kīma ḫurāṣi*[*ana . . . kīma*]
112. *abāru ana šamni* [. . .]
113. *amēlu tuqattar*[*šu-ma . . .*]

---

114. *šipat* [*nīš*] *libbi*

---

115. . . . *šumūšu* 72 . . . [. . .]

90. to? “dog’s-tongue”-plant for his ‘rising’ [. . .].
91. For his ‘rising’: Seven grains of silver, [seven grains of go]ld, in front of . . .] . . . [. . .]
92. he declares, 20 presents? . . . [. . .] the drug for the sexu[al] desire,
93. he declares three times, you apply the drugs for sexu[al] desire on their lim[bs],
94. you thread this plant on a r[a]m fleece, you tie (it) around [his waist] and broken [[(he will get his) sexual desire]].

XII. A ii 51–53 = Fragmentary incantation No. K.4

95. Enuru incantation . . . broken
96. Abracadabra
97. Incantation: Seven [broken]

XIII. A ii 54–61 = Fragmentary incantation and its ritual No. K.5

98. Incantation: Bed [broken]

---

99. Bed [broken]
100. By flying, the flesh [bro]ken
101. Son broken

---

102. Wording of *nīš libbi* (incantation).

---

103. Its ritual: [[You mix]] seed of the *puquttu*-thorn plant with blood of [[male]] partridge, broken [[you wrap it in a wad of wool]],
104. you recite this incantation over (it) seven times. Tamarisk broken
105. [nig]ht and day your desire (lit. *libbu*) will not abate and she broken

XIV. Ms. A ii 62–iii 23 = *nīš libbi* incantation E.2 “Light of the heavens, wise Ištar” ll. 18–45

XV. A iii 24–33 = Fragmentary incantation No. K.6

106. NN [son of NN . . .]
107. . . . [. . .]
108. . . . in front of . . . [. . .].
109. May he indeed mate like an as[s . . .],
110. like raven to partridge . . . [. . .],
111. like silver to salt, like gold [to . . . like]
112. lead to oil [. . .],
113. you fumigate the man [and . . .].

---

114. Wording of [*nīš*] *libbi* (incantation).
115. . . . its lines 72 . . . [. . .].

XVI. A iii 34–42 = Incantation and its ritual No. K.7

116. *[šiptu:] nēšu lû . . . [. . .]*
117. *r[ik]ibtāk[a] . . . [. . .]*
118. *i[n]a q[ib]īt telī[ti Ištar Nanāya]*
119. *Ga[zba]ba K[anisurra]*
120. *šipta annīta šalāšīšu tama[nnu] . . .*

---

121. *dudub[û: šīpā]ti peṣêti šīpāti sāmāti taṭammi sebet [. . .]*
122. *sebet . . . [. . .] u sebet riksī tarakk[as . . .]*
123. *šipta ina muḫḫi [tamannu . . .]*
124. *ina . . . tarakkas-ma [nīš libbi]*

XVII. A iii 43–iv 7 // N ii 3–9 = Prescriptions 29–31

29 125. 1 *qa* [. . .]
126. *eper ki[maḫi ep]erī šunū[ti . . .]*
127. *ana šamni [tanaddi] ṣalam zikari u sinništi [teppuš . . .]*

128. . . .-*šunūti zikaru u sinništu tibût*$^{?}$ *lib[bi*$^{?}$*] . . . [. . .]*

---

30 129. *ana nīš libbi rašê abunnat* ⟨*ayyali*⟩ *qaran ayya[li]*
130. *rikibti ayyali aṣ[uṣu]mt[a]*
131. *tamtarraq ušara ana muḫḫi ṭābta talt[appat*$^{?}$*]*

132. *azappi zibbati zikari tanassaḫ ištēniš ṭurri [kite . . .]*

133. *[taṭa]mmī-ma amašpâ ina libbi tašakk[ak . . .]*
134. *ina qablīšu tarakkas-[ma nīš libbi]*

---

31 135. *šumma* KI.MIN (= *amēlu ina mayyālīšu igdanallut libbašu ešu u ina mayyālīšu riḫûssu illak amēlu šū kimilti Marduk u Ištar elīšu ibašši*): *kappī našri zikari kappī . . . [. . .]*
136. *pizallurāt ṣēri ritkubā[ti] zê arkabi zê*

137. *ṣurārî barmi zēr [. . .] zēr ēri zēr maštakal*

138. *[zēr] azallî zēr mur[dudî zēr lišān-kalbi]*

139. *zēr šakirî ṣaṣun[ta] šumutta*
140. 14 *šammī annûti ištēniš [tubbal . . .]*
141. 10 *šumūšu [. . .]*

116. Incantation: Lion! Bull . . . [. . .].
117. Yo[ur] lo[ve]-making . . . [. . .].
118. At the command of wi[se Ištar, Nanāya],
119. Ga[zba]ba (and) K[anisurra]!
120. You recite this incantation three times.

---

121. It[s] ritual: You spin white [woo]l, red wool, seven [. . .],
122. seven . . . you ti[e] and seven bindings [. . .].
123. [You recite] the incantation over (this),
124. in . . . you tie and [sexual desire].

XVII. A iii 43–iv 7 // N ii 3–9 = Prescriptions 29–31

29 125. 1 litre [. . .]
126. dust of a gra[ve,] the[se] dus[ts . . .]
127. [you throw (it)] into oil, [you make] the figurines of the man and the woman [. . .]
128. . . . them, the "desire" (litt. "rising" of the 'hea[rt]') of the man and the woman . . . [. . .]

---

30 129. In order to get sexual desire: Navel of a stag, antler of a sta[g],
130. *rikibtu* of a stag, *aṣ[uṣu]mt[u]*-plant
131. you properly crush again and again, you sme[ar] the penis (of a stag$^{?}$) with salt,
132. you tear off hair of the tail of a male (stag$^{?}$) (and) together with a [linen] string [. . .]
133. [you s]pin and you insert *amašpû*-stone inside [. . .]
134. you tie (it) around his waist [and (he will get) sexual desire].

---

31 135. If ditto (= a man is repeatedly scared in his bed, his 'heart' is confused and in his bed his sperm comes out, over this man the wrath of Marduk and Ištar has come): Wings of a male eagle, wings of [. . .],
136. geckos of the steppe which copula[te], excrement of an *arkabu*-bat, excrement of
137. polychrome lizard, seeds of the . . .-tree, seeds of the *ēru*-tree, seeds of *maštakal*-plant,
138. [seeds] of *azallû*-plant, seeds of *mur[dudû*-plant, seeds of "dog's-tongue"-plant],
139. seeds of *šakirû*-plant, *ṣaṣun[tu*-plant], *šumuttu*-plant
140. [you dry] these 14 drugs together,
141. 10 its lines [. . .].

XVIII A iv 8–23 = Fragmentary incantation and its ritual No. K.8

142. ^{ḫepi} . . . [. . .] *libbu* [. . .]
143. ^{ḫepi} *ul ī*[*d*]*e* [. . .]
144. ^{ḫepi} *ḫātu ēp*[*uš* . . .] *nāru*
145. ^{ḫepi} *libbi* . . . [. . .] . . .
146. ^{ḫepi} *Ištar* . . . [. . .] . . .
147. ^{ḫepi} *lipšurka-ma* [*Ištar Nanāya*] *Gazbaba*
148. *Kanisurra ana libb*[*i* . . .] . . . *puḫālu*
149. ^{ḫepi} [[*râm turāḫi*]] *šeššīšu râm ayya*[*li sebîšu râm*] *iṣṣūr ḫurri*

150. ^{ḫepi} [[*šinšerīšu* . . .]] *sisû* . . . [. . .] . . .
151. ^{ḫepi} *qablat* [. . .] *ša erši*
152. *ḫepi lā* 1 . . . 1 . . . [. . .] . . .
153. *lā tašebbâ* [*lal*]*âša*
154. *šipat nī*[*š libbi*]
155. ^{ḫepi}[[*dudubû*]]: [*zibba*]*t puḫāl*[*i teb*]*î teleqqe zibbat zuqaqīpi*
156. [. . .] *ana ramānīk*[*a*] . . .
157. [. . .] *šammī* [*annû*]*ti*

Rest destroyed

XIX. A iv 24–31 = Fragmentary incantation and its ritual No. K.9

158–162. Broken lines
163. *šipa*[*t nī*]*š* [*libbi*]
164. *dudubû*: *ḫan*[*dur ballūṣīti* . . .]
165. *šipta sebîšu ana libb*[*i taman*]*nu rapāštīšu* [*taptanašša*]*š-ma nī*[*š*] *libbi*

XX. A iv 32–36 = Prescriptions 32

166. *šumma* KI.MIN (= *amēlu ina mayyālīšu igdanallut libbašu ešu u ina mayyālīšu riḫûssu illak amēlu šū kimilti Marduk u Ištar elīšu ibašši*): *ziqit* [. . .] . . . *ziqit nūbti ziqit* . . . [. . .]
167. . . . [. . . *el*]*lūti šikkurrat ebbūti šuruš samīdi*
168. . . . [. . .] . . . *dām iṣṣūr ḫurri zikari ṣurār*[*â*] *ari ḫašḫūri*
169. 10 *šamm*[*ī*] *annû*[*ti* . . .] . . . *tuqattar šipta sebîšu* [*ana m*]*uḫḫi tamannu*

170. *ina qablīšu* [*tara*]*kkas-ma nīš* [*lib*]*bi*

XXI. Ms. A iv 37–41 = *nīš libbi* incantation No. A.1 "May the wind blow! May the grove quake!" ll. 33–39

XVIII A iv 8–23 = Fragmentary incantation and its ritual No. K.8

142. [broken] . . . 'heart' [. . .]
143. [broken] I do not [kn]ow? [. . .]
144. [broken] I caus[ed] fear? [. . .] canal
145. [broken] the 'heart' . . . [. . .] . . .
146. [broken] Ištar . . . [. . .] . . .
147. [broken] May she release you and [Ištar, Nanāya], Gazbaba,
148. Kanisurra to the 'heart' [. . .] . . . a ram
149. [broken] [[the mating of the wild goat]] six times, the mating of the st[ag seven times, mating of the] partridge,
150. [broken] [[twelve times . . .]]. May the horse [. . .] . . .
151. [broken] waist? [. . .] of the bed
152. [broken] not 1 . . . 1 . . . [. . .] . . .
153. You will not be satisfied by her [lust]!

---

154. Wording of *nī[š libbi]* (incantation).

---

155. [broken] [[Its ritual]]: You take [the ta]il of a [reared] up ra[m], the tail of a scorpion,
156. [. . .] . . .
157. [. . .] [tho]se plants

Rest destroyed

XIX. A iv 24–31 = Fragmentary incantation and its ritual No. K.9

158–162. Broken lines
163. Word[ing] of *n[īš libbi]* (incantation).

---

164. Its ritual: [. . .] the spu[r of a *ballūṣītu*-bird],
165. [you reci]te the incantation over (it) seven times, you [repeatedly anoin]t his loins, and (he will get) sexual des[i]re.

XX. A iv 32–36 = Prescriptions 32

166. If ditto (= a man is repeatedly scared in his bed, his 'heart' is confused and in his bed his sperm comes out, over this man the wrath of Marduk and Ištar has come): Stinger of [. . .], . . ., stinger of bee, stinger of . . . [. . .]
167. . . . [. . . p]ure, pure *šikkuratu*-reeds, root of *samīdu*-plant
168. and (body part) [. . .] . . . blood of a male partridge, lizar[d], branch? of apple-tree;
169. you fumigate the[se] 10 drug[s . . .] . . ., you recite the incantation seven times [over i]t,
170. [you ti]e around his waist and (he will get) sexual [desi]re.

XXI. Ms. A iv 37–41 = *nīš libbi* incantation A.1 "May the wind blow! May the grove quake!" ll. 33–39

Commentary

8. See the catologue LKA 94 i 13: ÉN ^{d}BE ^{d}NIN.MAḪ nam-lú-ùlu-lu "Incantation: 'Enlil and Bēlet-ilī mankind.'" Here instead of DINGIR.MAḪ, we find ^{d}NIN.MAḪ, both to read Bēlet-ilī (see Biggs 1967: 45). Cf. *nīš libbi* M prescr. 5: 27: [. . . e]gir nam-tar-ra-zu al-gub ". . . a]fter$^{?}$ it your destiny is assigned."

10. It might be possible to restore with ŠÀ.

11. Biggs suggests the restoration la-la-bi on the basis of the Old Babylonian hymn to Ištar (*Inana C*): la-la šà-zi-ga níĝ-šu ĝál é$^{?}$ níĝ-gún ĝá-ĝá ^{d}Inana za-a-kam : *la-lu-*⸢*ú*⸣ *ni-iš li-bi-im* x [. . .] x x [. . .] *bi-ši-im ra-še-e ku-*[*ma*] IŠ$_{8}$.DAR "Attractiveness, sexual desire, to have goods and property are yours, Inana/Ištar" (ETCSL c.4.07.3: 121, see Sjöberg 1975: 190).

13. "Wing of bat" can be a 'secret name' for a plant. See URU.AN.NA: *maštakal* series: Ú *ka-mi-u-nu* : DILI : Á *šu-ti-ni* "*kamūnu*-fungus: secret name : "bat's wing" (l. 84, Rumor 2017: 16). Following Köcher's (1995: 204) suggestion, the Aš sign refers to DILI/*pirištu* 'secret,' thus indicating a secret name.

23. This is possibly a catchline (see Biggs 1967: 5 fn. 30), in fact Ms. B fragment Sm. 818 rev. 7′–9′ includes part of an Ashurbanipal colophon similar to Hunger 1968, No. 318. For this reason, we can restore, as suggested by Steinert (2018b: 264): [DUB X.KAM (. . .) DIŠ NA *ana* MUNUS]-*šú* ŠÀ-*šú* ÍL-*šú-ma ana* MUNUS BAR-*ti* ŠÀ-*šú* N[U ÍL-*šú*] "[Xth tablet of (. . .) 'If a man] desires his [woman], but do[es not desire] another woman." See also Q prescr. 1: 1–2: DIŠ NA *ana* MUNUS-*šú* DU-*ma a-*[*na* MUNUS-*šú* ŠÀ-*šú* ÍL-*šú-ma*] / *ana* MUNUS BAR-*ti* DU-*ma a-*[*na* MUNUS-*šú* ŠÀ-*šú* ÍL-*šú*] "If a man goes to his woman and [desires his woman, but] / he goes to another, but does not desire another woman." See the incipit of Tablet XXXIV of *Diagnostic Handbook* SA.GIG/*Sakikkû* in the catalogue l. 41, which can be restored from the fragmentary catchline of Tablet XXXIII: DIŠ ⸢NA⸣ *ana* ⸢MUNUS⸣-(*šú*) ŠÀ-*šú* ⸢ÍL⸣-*šú-ma* [*ana* MUNUS BAR$^{?}$]-*ti* ŠÀ-*šú* NU ÍL-*šú* MUNUS BI ŠÀ-[*šú* . . .] "If a man desires (his/a) woman, but does not desire [another$^{?}$] woman: this woman [. . .] his 'heart'/ desire$^{?}$ [. . .]" (see Schmidtchen 2018: 141).

24–27. This symptoms description, to which the prescriptions 3–21 are related, is not explicitly related to the absence of sexual desire in any manuscript except for Ms. A (STT 280). The text STT 280 is followed, in fact, by five prescriptions corresponding to the text *nīš libbi* A: 19–22, 27 (prescr. 9–12, 16): LKA 95 (Ms. A) obv. 17–22, 27–28 // LKA 96 (Ms. B) obv. 16 // STT 280 (Ms. D) i 51–55. In the text *nīš libbi* A (prescr. 1: 1), Ms. B describes as a symptom "if a man's (desire) to go to a woman is reduced," one of the typical symptoms of a loss of sexual desire. It is evident that this symptoms description, as well as the following one in lines

57–59 (prescr. 21–22), can be associated with the loss of sexual desire – although not exclusively as demonstrated by the other manuscripts concerning different diagnostic categories. It should be borne in mind, in fact, that therapies for the loss of sexual desire have not undergone a serialization process and consequently a tablet may contain remedies for multiple diagnostic categories.

26. GIM MUNUS *su-u'-su l[a e-l]il* (var.: *e-el*): it is not possible to define with certainty the meaning of the term *su''usu/suḫsu*: CAD S 249 *s.v. suḫsu*, 'bed'; Lambert 1975: 104, (BM 41005 ii 17) 'bed'; George 2013: 119 and 121, No. 19: 28; 235 and 245, No. 33: 40 'crotch, genitals.' See Civil 2006: 55 and fn. 3. Abusch and Schwemer (2011: 112) translate the phrase "his private parts are impure like those of a (menstruating) woman" (there is no proof the menstruating woman was considered impure, see Couto-Ferreira and Garcia-Ventura 2013: 517–519, *Pace* Abusch and Schwemer 2011: 112; Sallaberger 2011: 23). The text refers to an abnormal discharge (the patient's sperm continues to flow). The term should be understood as seminal fluids or genital secretions present in both women and men and here considered impure. See BAM 205 rev. 15′–16′: *lu* NITA *lu* MUNUS *su-u'-us re-ḫu-su-nu* / [*m*]*a-a'-da*[?!](text UŠ) DU-*ak ina* K[ÀŠ][meš]*-šu-nu* / [B]IR.[BIR . . .] "(If) either a man's or a woman's *seminal fluids* flow copiously (and) they spill it when urinatig [. . .]" (Biggs 1967: 68, i 24; Scurlock and Andersen 2005: 89–90; 4.3 var.: 81-2-4, 466: 3′). The text BAM 205 mentions the pathological and involuntary seminal secretions of both man and woman. Reynolds (2010: 301–302) considers the term a synonym of *riḫûtu* 'semen' (see also Stol 2000: 8; Steinert 2017b: 310–311 and 2018b: 275–276). See also in this corpus *nīš libbi* S prescr. 1: 2′. For the female vaginal discharges looking similar to male semen (*rihûtu*) see the *Aššur Medical Catalogue*, section "birth": "If a penis' semen-like (*kīma rihût ušari*) discharge [flows] from a woman's [vagina]. [. . . the] belly of a woman [. . .]" (l. 119, Steinert 2018b: 218). For a more detailed discussion see Steinert 2017b: 310–314. The aspect of purity is also pointed out in the following prescription, which provides an amulet with vegetable ingredients, including the *sikillu* and *maštakal* plants which have a purifying function. On the relationship between ejaculation and impurities see *Šumma ālu*: *šumma amēlu ginâ igtanallut amēlu šū lā ellu ḫīṭa magal irašši* "If a man constantly ejaculates, this man is not pure, it has a great sin" (tab. XLV 27, CT 39). On the relation between impurity and 'abnormal' sexuality see Sallaberger 2011: 24–25, 27–28. Regarding the connection between genital discharge, designated as 'fluid, water' (*mû*), and witchcraft see BAM 237: "If a woman has been given herbs of hate-magic to eat (and, because of this,) fluid flows heavily from her vagina (*mû ina libbi ūrīša magal illakū*)" (iv 29, Abusch et al. 2020: 90, No. 5.10).

29. KI.MAḪ/*kimaḫḫu*: source for magical-medical ingredients, particularly related to thorny bushes and in general to acacia varieties (i.e., *baltu* and *ašāgu*) (see CAD K 371).

30. In Ms. F a few fragmentary lines follow upon a second ruling.

31. [BA.BA.Z]A$^{!}$.^{d}ÍD: Abusch and Schwemer 2011 read [Ú]Ḫ$^{!?}$(x.'a').^{d}ÍD.

33. ^{ú}SAG is a not identified plant. In ḪAR-ra = *ḫubullu* XVII 25 (Landsberger et al. 1970 = MSL 10: 108) it is identified with *la-a-ar-tu*, perhaps corresponding to *lardu*, a high-alkaline plant used as soap (see CAD L 103; Landsberger 1937 = MSL 1: 224; Thompson 1949: 17).

35. See for other mentions of the expression Ú ŠÀ.ZI.GA "drug for the sexual desire": N prescr. 21 iv 13; K prescr. 27: 75: 10 Ú$^{!meš}$ ŠÀ.ZI$^{!}$.GA$^{!}$ "10 drugs for the sexual desire"; No. K.3: 92; A prescr. 17: 30 7 Ú^{meš} ŠÀ.ZI.GA *ina* KAŠ NAG "he drinks in beer the seven drugs (for) the sexual desire." See also in the *Aššur Medical Catalogue*, in the section XIX devoted to the loss of sexual desire: [. . .] x SAG MUŠEN DIŠ ⸢Ú⸣ ⸢ŠÀ⸣.ZI.G[A] ⸢*ana*⸣ [GÚ-š]*ú* GAR "[. . .] . . . the head of a bird. (Instructions) to place a drug for sexual desir[e] around his [neck]" (l. 100, Steinert 2018b: 217). See for specific plants for sexual desire's problem BAM 380 r. 42–44 (dupl. BAM 381 iii 37–40) (see Chapter III § "Plant ingredients").

46. K prescr. 12 = A prescr. 7: 7.

52. See K prescr. 25: 68.

56. Ms. A i 60 only traces.

75. See l. 35.

77. See the catalogue LKA 94 i 19, 26; A prescr. 19: 32; See also K prescr. 28: 77–79. See also *kunuk ḫalti* series ("*ḫaltu*-seal") K. 3010+ v 24–36′ (Schuster-Brandis 2008: 365, text 16 A, on kindarab text 4, p. 238 l. 88 a′-d′).

79. See the catalogue LKA 94 i 26–27

89. Note that the ritual begins without *dudubû* 'its ritual.'

92–93. See l. 35.

100. Biggs reads *aš ri ri su*. The interpretation DAL.DAL for the verb *naprušu* refers to the removed sexual desire. See for example: *simtī ippariš tarānu išḫiṭ* "My dignity has flown away, my protection has fled" (*Ludlul* II 48, Lambert 1960a: 32). The context is too fragmentary to find a solution.

103–105 See the same prescription in *nīš libbi* L prescr. 2: 6–7. See also the similarities between the Sumerian incantation L.1 and K.5 since both incantations have similar prescriptions. However, these similarities are very difficult to sustain due to the fragmentary nature of the incantations (see Abusch et al. 2020: 47).

103. The *puquttu*-thorn (CAD P: 515–516 'thorn, barb'; AHw. II 880 'Dornpflanze') is mentioned in the corpus also in *nīš libbi* L prescr. 2: 6, if we read the log-

ogram KIŠI$_{16}$.ḪAB as *puquttu* and not as *dadānu*, as attested in the medical commentary from Nippur 11N-T4 (l. 21, Civil 1974: 337).

105. For the verb *nâḫu* with *libbu* as its subject see CAD N/I 145–147 mng. 2a2´; AHw. II 716 mng. 4c; for the meaning of 'to diminish' referring to the excitement see CAD N/I 145–146, "let his desire not abate night or day." See for a similar expression No. M.1: 9.

109. The translation "may he indeed" is given by the use of *šū* before the verb with emphasis value.

113. LÚ as *amēlu* is uncertain here since the Ms. A uses the logogram NA.

128. Biggs proposes the following restoration: G[ÁL-*šú-ma ir-ta-nak-kab*$^{?}$].

129–130. *a-bu-na-at* ⟨DÀRA.MAŠ⟩ SI DÀRA.[MAŠ] /*ri-kib-ti*(text SI) DÀRA.MAŠ: The *emendatio a-bu-na-at* ⟨DÀRA.MAŠ⟩ is uncertain, since the navel of a stag never appears in this corpus; see however BAM 252 obv. 1–4: *a-bu-na-at* / DÀRA.MAŠ / *ap-pi* GÌŠ-*šú* / *ina* MURUB$_4$-*šú* KEŠDA-*ma* "The navel of stag, its glans penis you bind around his waist." The *emendatio ri-kib-ti*(text SI) DÀRA.MAŠ is based on the presence of the *rikibtu* of the stag in many *nīš libbi* texts. According to Chalendar (2018: 39–41 and fn. 143), the text should not be amended, since *abunnat qaran ayyali* (*a-bu-na-at* SI DÀRA.MAŠ) and *rikib qaran ayyali* (*ri-kib* SI DÀRA.MAŠ) would indicate two specific ingredients. The first one could indicate «la meule qui constitue l'extrémité basse du bois et fait office d'interface entre le bourrelet osseux (le pivot) et le merrain et qui porte les andouillers» (ibid. 40 fn. 143), while the second one the initial phase of the annual regrowth of the deer antlers, which is accompanied by new ramifications (her translation "'cornichon' de cerf"). She notes that the use of a specific part of animal horns is not extraneous to Mesopotamian pharmacopeia, since some prescriptions mention the use of animal horn tops (called *sapparti qaran*), from bulls (BAM 503 i 16′), deer and goats (BAM 237 iii 6). Although suggestive, Chalendar's hypothesis is not accepted here because in the *nīš libbi* corpus, as in other Mesopotamian medical documentation, both ingredients, to my knowledge, never appear. If my first *emendatio* remains uncertain, the second one seems more appropriate.For the restoration ^{ú}*a-ṣ*[*u-ṣu*]-*um-t*[*um*$^{?}$] see ibid. 45 fn. 181.

131. *tam-ta-raq* is Gtn-stem from the verb *marāqu* 'to crush.' The problem is that the Gtn-stem is not attested in the dictionaries. It seems that navel, antler and *rikibtu* of a stag together with the *aṣuṣumtu*-plant must be properly crushed several times before putting the powder into a bag(?) for the preparation of an amulet. Biggs (1967) reads *tab ti* ˹U$_5$˺ [. . .], but the suggestion by Chalendar (2018) is more appropriate: *ṭab-ti* ˹*ta*l˺-*t*[*ap-pat*$^{?}$]. The spelling *ṭab-ti* for salt is surprising since we usually find *ṭá-ab-ti* or MUN. It seems to refer to a method

of preservation by salting the stag(?)'s penis for the realization of an amulet. Salting an animal-based ingredient is well attested in this corpus.

132. Before NITA Biggs (1967) adds ⟨UDU(?)⟩. However, since in the prescriptions many ingredients come from the stag, Chalendar (2018) suggests that the hairs are from deer. If the text refers to the ram, we expect to find the term *šārtu*/SÍG; indeed, *azappu* is more suitable for a pig or a deer, whose both bristles and hairs are documented in this corpus.

134. Rest of the column is destroyed.

135–141 Ms. N (= *nīš libbi* N prescr. 8 ii 10–16) continues adding more details of the prescription: perhaps the fourteen drugs are pulverized and mixed with the *isqūqu*-flour. Then with this mixture, some small balls are made. Perhaps – the text is fragmentary and unique in the corpus – some (3?) are eaten on an empty stomach; others are tied to the patient's waist.

135. Biggs (1967) restores *i-[gi-ri-i]* 'heron,' which never appears in the corpus.

136. Similar passages have MUŠ.DÍM.GURUN.NA. MUŠ.DÍM.KUR.RA appears only in this text, maybe it is an incorrect writing for MUŠ.DA.GUR$_4$.RA (see Landsberger 1934: 115–116).

137. The word EME.DIR.GÙN.A/*ṣurāru barmu* indicates lizards in general, without any sex qualification, or indication of the species. Ebeling (1925) uses the word *ṣurīrītu* to read the Sumerogram EME.DIR. In OB texts we find the Sumerogram EME.ŠID; in the *Practical Vocabulary from Aššur* (401–402) EME.ŠID corresponds to *ṣurīrītu* and EME.DIR to *iṣṣû*. See Landsberger 1934: 114–115.

138. Note that in Ms. N (KUB 4, 48) ii 7 we find [ú]MUL.DÙ.DÙ instead of [ú]MUR.DÙ.DÙ. See in the same tablet (KUB 4 48), in *nīš libbi* N Ms. A i 22, the mistake BÁR.KA for MAŠ.SÌL ("BAR.KA$_4$," Akk. *naglabu*).

153. Biggs translates "do not satisfy her [lust]!". The literal translation is "you will not be satisfied of her [lust]!". It indicates the desire of the insatiability of woman's sexual pleasure and charm. On *šebû* see CAD Š/II 251–255 'to became satisfied'; AHw. III 1207 'sich sättigen an.' See, for example, in *Nergal and Ereškigal*: *Erra ḫāmeru lalêya/ ul ašbâ lalâšu ittalkanni* "Erra, husband, my lust, I was not satisfied with his charm, (when) he left me" (ll. 291–292, Ponchia and Luukko 2013: 19). See also Nougayrol 1968: 94; Hirsch 1973–1974: 68.

166. NA$_4$ is unclear here since after *ziqtu* 'sting' one expects an insect name, not a stone.

IV Texts from Uruk (with duplicates from other sites)

Nīš libbi L

List of manuscripts

Manuscript	Museum number	Publication	Tablet	Script	Date	Provenience	Incantations and prescriptions
A	W. 22307/9	SpTU 1, 10 1′–18′	Frg. of a single-col. tablet	LB	5th–4th cent.	Uruk Library of Anu-ikṣur	No. L.1; Prescr. 1–3, 5; No. L.2
B	Bo 4894	KUB 4, 48 iv 27–31	Two-col. tablet	'Not Hittite'[571]	13th cent.	Ḫattuša	No. L.1
C	VAT 13915+ 13933	LKA 98: 2–16	Frg. of a two-col.? tablet	NA	8th–7th cent.	Aššur Library N 4	Prescr. 3–6

Edition

Ebeling 1925: 55 (Ms. B)
Biggs 1967: 56, 63 (Ms. B, C)
Hunger 1976: 27–28 (Ms. A)
Abusch et al. 2020: 45–49, No. 4.3 (Ms. A)

Structure of the text

Text *nīš libbi* L provides a Sumerian incantation No. L.1 attested in Ms. A and B. It continues following Ms. A with six prescriptions with duplicate Ms. C and the incantation No. L.2 "Rearing onager, preeminent stallion, who roams the forests, who has dampened your ardour?".

A – B	No. L.1
A – C	Prescr. 1–6
A	No. L.2

571 Note the use of *bá*, see Schwemer 2013: 154.

https://doi.org/10.1515/9783110757262-009

I. Sumerian incantation No. L.1, ll. 1–4 (Ms. A 1′–4′ // B iv 27–31)

II. Prescriptions
   1. 5 (Ms. A 5′)
   2. 6–7 (Ms. A 6′–7′)
   3. 8–9 (Ms. A 8′–9′ // C 2–6)
   4. 10–13 (Ms. C 7–10)
   5. 14 (Ms. A 10′ // C 11–12)
   6. 15–18 (Ms. C 13–16)

III. Incantation No. L.2 “Rearing onager, preeminent stallion, who roams the forests, who has dampened your ardour?”, ll. 20–26 (Ms. A 11′–18′)

Summary of the sections of manuscripts not included in the transliteration:
- Ms B = KUB 4, 48
  i 1–iv 26 = N prescr. 1–23 (Ms. A) (note that ii 3–9 = K prescr. 31)
  left e. 1–7 = N prescr. 24 (Ms. A)
  lower e. 1–5 = N prescr. 25 (Ms. A)

Transliteration

I. A 1′–4′ // B iv 27–31 = Sumerian incantation L.1

1.	A 1′	[ÉN . . .] x [k]a [. . .]
2.	A 2′	[ÉN . . .] x te-en-te-en ki-ná še-ga k[i]-ná [. . .]
	B iv 27–29	[. . . te-en-t]e-en / [ ] x a še-g a(text KA) / [. . .]
3.	A 3′	[. . .]-li še-ga ki-in-gi uriki še-ga še-ga [. . .]
	B iv 29–30	[. . .] x x a še-ga še-ga / [. . .] x zi-zi-en-zi-en
4.	A 4′	[a-ri-a$^{?}$] ⸢ki⸣-in-gi a-ri-a [uriki$^{?}$ . . .]
	B iv 31	[TU$_6$ ÉN] ⸢É⸣.NU.RU

II. A 5′–10′ // C 2–16 = Prescriptions 1–6

5.	A 5′	[DÙ.DÙ.BI ŠÀ$^{?}$] ⸢UGA⸣^{mušen} GE$_6$ SÚD *ina* Ì ^{giš}ŠUR.MÌN UR.MÌN ḪE.ḪE ⸢ÉN⸣ [7-*šú* *ana* UGU ŠID-*ma* GÌŠ-*šú* ŠÉŠ-*ma* ŠÀ.ZI.GA]
6.	A 6′	[DIŠ KI.MIN NUMUN]⸢KIŠI$_{16}$⸣.ḪAB *ina* MÚD ⸢BURU$_5$⸣.ḪABRUD^{ru}.DA^{mušen} NITA ḪE.ḪE *ina* ^{síg}ÀKA NI[GIN ÉN 7-*šú* *ana* UGU ŠID-*nu* ^{giš}*bi-ni* . . .]
7.	A 7′	[*mu-ši ur-ra*] ŠÀ-*ka ul i-na-ḫi ù ši-i* [. . .]

8. A 8′ [DIŠ KI.MIN ${}^{\check{s}i}$]${}^{m?}$⸢GÍR⸣${}^{?}$ ⸢ú⸣*ṣa-ṣu-un-du* ^{ú}A.ZAL.LÁ Ú.KUR.RA ^{ú}ḪAR.SAG LAG.A.ŠÀ.G[A ...]
C 2–5 — — NUMUN ^{giš}KIŠI${}_{16}$.[ḪAB] / ^{ú}*ṣa-ṣu-um-t*[*ú*] / — Ú.KUR.RA ^{ú}Ḫ[AR.SAG / ^{ú}LAG.GÁ ḪÁD.A [...]

9. A 9′ [*ina* KAŠ] — — NAG-*ma* ŠÀ.[ZI.GA]
C 6 *ina* KAŠ *la pa-tan* ⸢NAG⸣-*m*[*a*]

---

10. C 7 ŠÀ ⸢BURU${}_{5}$.ḪABRUD.DA⸣[^{mušen} NITÁ ...]
11. C 8 *ina* MUN *te-te-*[*mir* ... *ina* MUL]
12. C 9 *tuš-bat* ⸢ÉN⸣ 3-*šú* [ŠID-*nu* ...]
13. C 10 [...]

---

14. A 10′ [DIŠ KI.MIN *ḫi-in-d*]*ur bal-lu-ṣi-ti ina* Ì.GIŠ SÚD GÌŠ-*šú* ŠÉŠ-*ma* [ ]
C 11–12 — — *ḫa-an-dur bal-lu-ṣi-*[*ti*] / *ina* Ì SÚD GÌŠ-*šú* EŠ-*aš-ma* ŠÀ.[ZI.GA]

---

15. C 13 KU.KU ⟨^{na4}⟩KUR-*nu* DAB KU.KU AN.[BAR ...]
16. C 14 *ina* Ì BUR ḪE.ḪE ÉN ^{d}IŠKUR [GÚ.GAL AN-*e* DUMU ^{d}*a-nim*]
17. C 15 7-*šú ana* ŠÀ ŠID-*nu* NIT[A GÌŠ-*šú*]
18. C 16 MUNUS GAL${}_{4}$.LA-*šá* EŠ-*aš-*[*ma ir-ta-na-ka-ab*]

Rest destroyed

III. A 11′–18′ = Incantation No. L.2

19. A 11′ [ÉN ANŠE.EDIN${}^{?}$].NA *pár-ḫu* ANŠE.KUR.RA *e-te-lu mut-tal-lu-ú* ^{giš}TIR^{meš} *man-nu* [*ú-ni-iḫ-ka*]
20. A 12′ [... *kaš-šá-pa*${}^{?}$]*-a-ti ki-i qé-e ra-mu-ti it-ba-la ni-iš lìb-b*[*i-ia*]
21. A 13′ [... *ina ka*]*l u*${}_{4}$*-me i-zer-ra-an-ni ina kal mu-ši i-pu-šá-an-ni ik-sa-an-ni ki-i k*[*a-mi-i*${}^{?}$]
22. A 14′ [*ki-i qé-e ra*]*-mu-ti it-ba-la ni-iš lìb-bi-ia at-ta lil-li-lu-ka u li-i*[*b*${}^{?}$*-bi-bu-ka*${}^{?}$]
23. A 15′ [^{ú}GÚR.GÚR${}^{?me}$]${}^{š?}$ *qu-ud-du-šú-tu* DUMU^{meš} KUR^{meš} *ḫur-sa-*[*a-ni*]
24. A 16′ [*ana da-ra-a*${}^{?}$]*-ti-im-ma ti-ku-ka lim-taq-qut-ka* ⸢TU${}_{6}$⸣ [ÉN]

---

25. A 17′ [KA.INIM.M]A ⸢ŠÀ⸣.ZI.GA : DÙ.DÙ.BI ^{ú}GÚR.GÚR *ina* NE SAR-*šú-ma* ŠÀ.Z[I.GA]

---

26. A 18′ [ÉN *a-kan-nu šá ana*] ⸢U${}_{5}$-*bi*⸣ ZI-*ú man-nu ú-ni-i*[*ḫ-k*]*a* x [...]

Rest destroyed

Transcription

I. A 1′–4′ // B iv 27–31 = Sumerian incantation L.1: 1–4

II. A 5′–10′ // C 2–16 = Prescriptions 1–6

1 5. [*dudubû: libbi*] *aribi ṣalmi tasâk ina šaman šurmēni taballal šipta* [*sebîšu ana muḫḫi tamannū-ma ušaršu tapaššaš-ma nīš libbi*]

---

2 6. [*šumma* KI.MIN *zēr*] *puqutti ina dām iṣṣūr ḫurri zikari taballal ina itqi tal*[*ammi šipta sebîšu ana muḫḫi tamannu bīni* . . .]

7. [*mūši urra*] *libbaka ul inâḫ u šī* [. . .]

---

3 8. [*šumma* KI.MIN . . . *a*]*su* (var.: *zēr puqutti*) *ṣaṣundu azallû nīnû azupīru kirbān eqli tubbal* [. . .]

9. [*ina šikari*] (var.: *lā patān*) *tašaqqīšu-ma* [*nīš*] *libbi*

---

4 10. *libbi iṣṣūr ḫurri* [*zikari* . . .]

11. *ina ṭābti tete*[*mir* . . . *ina kakkabī*]

12. *tušbât šipta šalāšīšu* [*tamannu* . . .]

13. [. . .]

---

5 14. [*šumma* KI.MIN] *ḫindūr ballūṣīti ina šamni tasâk ušaršu tapaššaš-ma* [*nīš*] *libbi*

---

6 15. *sīkti šadâni ṣābiti sīkti par*[*zilli* . . .]

16. *ina šaman pūri taballal šipta Adad* [*gugal šamê mār Ani*]

17. *sebîšu ana libbi tamannu zika*[*ru ušaršu*]

18. *sinništu biṣṣūraša iptaššaš*[*-ma irtanakkab*]

II. A 11′–18′ = Incantation No. L.2

19. [*šiptu: serrē*]*mu parḫu sisû etellu muttallû qišāti mannu* [*unīḫka*]

20. [. . . *kaššāp*]*āti kī qê ramûti itbala nīš lib*[*bīya*]

21. [. . . *ina ka*]*l ūme izerranni ina kal mūši īpušanni iksânni kī k*[*amî*]

22. [*kī qê ra*]*mûti itbala nīš libbīya attā lillilūka u li*[*bbibūka*]

Translation

I. A 1′–4′ // B iv 27–31 = Sumerian incantations L.1

1. [Incantation: "...] ... [...]."
2. [Incantation: ...] ... Cool down! In the benevolent bed! [In] the [benevolent] b[e]d [...],
3. [...] ... benevolent! (Let) Sumer and Akkad (be) benevolent! Benevolent! [...] (you two) be excited!
4. [Semen of] Sumer, semen [of Akkad ...]. // Enuru [incantation].

II. A 5′–10′ // C 2–16 = Prescriptions 1–6

1 5. [Its ritual]: You pound [the heart of] a black raven (and) mix it with cypress oil, [you recite] the incantation [seven times over (it). You rub his penis (with it) and (he will get) sexual desire].

2 6. [If ditto]: You mix [seeds] of *puquttu*-thorn with the blood of a male partridge, you w[rap] it in fleece. [You recite the incantation seven times over (it). Tamarisk wood ...].
7. [Night and day] desire (lit. *libbu*) will not abate and she [...].

3 8. [If ditto: ...] you dry [myrt]le, (var.: seeds of *puquttu*-thorn), *ṣaṣumtu*-plant, *azallû*-plant, *nīnû*-plant, *azupīru*-plant, *kirbān-eqli*-plant [...].
9. You make (him) drink (it) [in beer] (var.: on empty stomach) and (he will get) sexual [de]sire.

4 10. The innards of a [male] partridge [...]
11. you co[ver] with salt, [... under the stars]
12. you let it spend the night, [you recite] the incantation three times [...]
13. [...].

5 14. [If ditto]: You pound the spur of a *ballūṣītu*-bird in oil, you anoint his penis (with it) and [(he will get) sexual] desire.

6 15. Magnetite powder, iro[n] powder [...]
16. you mix with oil from the alabastron, the incantation "O Adad, [oversee of the canals of heaven, son of Anu"]
17. you recite seven times over it, the ma[n] anoints (with it) [his penis],
18. the woman her vulva [and he will have repeatedly sexual intercourses].

II. A 11′–18′ = Incantation No. L.2

19. [Incantation:] Rearing [onag]er, preeminent stallion, who roams the forests, who [has dampened your ardour]?
20. [The ... witche]s have taken away [my] sexual desire like slackened strings!
21. [... she (= witch)] practiced hate-magic against me [al]l day, bewitched me all night, bounds me like a pr[isoner].

23. [*kukr*]*ū quddušūtu mārū šadî ḫursā*[*nī*]
24. [*ana darâ*]*tim-ma tīkūka limtaqqutka tê* [*šipti*]
25. [*šipa*]*t nīš libbi*: *dudubû*: *kukra ina pēnti tuqattaršu-ma n*[*īš*] *libbi*

26. [*šiptu*: *akkannu ša ana*] *rakābi tebû mannu unī*[*ḫk*]*a* . . . [. . .]
Rest destroyed

22. [like sl]ackened [strings] they have taken away my sexual desire! You, may they purify you, may [they] c[leanse you],
23. the holy [*kukru*-plant]s, the inhabitants of the heights (and) mountain ran[ges].
24. May your rain keep falling for you(?) [forev]er! [Incantation] formula.

---

25. [Wording] of *nīš libbi* (incantation). Its ritual: You fumigate him with *kukru*-plant on charcoal and (he will get) sexual de[si]re.

---

26. [Incantation: "*Akkannu*-wild ass who] is reared-up for mating, who has dampe[ned your desi]re?" . . . [. . .]

Rest destroyed

Commentary

2–4. The reconstruction of this Sumerian incantation is very difficult due to the fragmentary state of the two manuscripts. Here I present an attempt without certainty. See the catalogue LKA 94 i 30: [ÉN ...] te-en-te "[Incantation: '...] cool down!'". This fragmentary Sumerian incantation should be compared, as suggested by Abusch et al. 2020: 47, to the similar fragmentary incantation K.5, which has the same ritual instruction (ll. 103–105) as this found here in prescr. 2: 6–7.

5. For the restoration of *libbu* 'heart/innards' see *nīš libbi* N prescr. 25. l. e. 1. Alternatively, one can restore UGU (*muḫḫu*) as in *nīš libbi* P presc. 7: 22. Another possible restoration is that SAG: *qaqqad aribi ṣalmi* "head of black raven" is a 'secret name' for the *ṣaṣumtu*-plant, see URU.AN.NA: *maštakal*: *ṣaṣumtu*-plant : 'secret name' head of a black raven" (III 103, Rumor 2017: 19). See the fragmentary incantation and its ritual No. K.5 (ll. 103–105) for the same prescription. For this reason, the logogram KIŠI_{16}.ḪAB has been read as *puquttu* and not as *dadānu* (as in l. 8 Ms. C).

8. Ms. C starts with the end of a prescription in line 1: [*i*]*na* IGI *i* [...].

14. CAD Ḫ 194 interprets the line of Ms. A as follows: *ḫi-in-du-ur* PA *al-lu-zi* "a spur? of a twig of *allazu*-plant," which is however without the determinative Ú. In CT 17, 44: 89 the word *ḫindūru* is associated with a bird: *ḫi-in-dur* MUŠEN *šakin* KUN UR *šakin* "(From his waist to his feet he (= the demon represented) is a dog), he has the spur? of a bird, he has the tail of dog" (CAD Ḫ 194). Hunger (1976: 28) translates "Kralle eines *b.*-Vogels." *Balluṣītu* (AHw. I 100, lit. "bei der etwas heraustritt" from *balāṣu*), may indicate a type of bird. It is in this way understood only from the *nīš libbi* texts by CAD B 65 and Biggs 1967. AHw. I 100 defines it as "'Wasservogel,' 'mit heraustretendem Steiß'" (with a protruding uropygium); Salonen (1973: 134–135) however as *Motacilla cinerea*. The term may also refer to a kind of pop-eyed lizard/gecko (see URU.AN.NA: *maštakal*, Landsberger 1934: 118). See *nīš libbi* P prescr. 10 rev. 5.

15–16. See for a similar prescription No. B.2: 38–40; No. D.2: 21–23; No.E.1: 15–17.

17. See No. B.2 rev. 6–16 "O Adad, locker keeper of the canals of heaven."

19. One could also restore *a-kan-na* "onager, wild ass," instead of ANŠE.EDIN.NA. The interpretation of BAR ḪU as *parḫu* 'rearing' (verbal adjective of *parāḫu* 'to sprout, to shoot up,' CAD P 145), understood as adjectival attribute in parallelism to *etellu* in the phrase *sisû etellu* is suggested by Abusch et al. 2020: 48, although not supported by comparable passages in the corpus.

20. The restoration *kaššāpāti* (plural), based only on the context, is suggested by Abusch et al. 2020: 48. The break offers space for the restoration of an additional sign, possibly a number.
The verb *tabālu* 'to remove, take away' is used to describe the loss of virility (*dūtu*), often because of a witch's attack: see *Maqlû* iii 9, 12; KAR 177 r. ii 7, r. iii 3f. (see CAD T 19 mng. 2 2´d; AHw. III 1297). For similar passages see No. A.3: 49–50; No. A.4: 76; No. D.2: 10–12.

21. *i-pu-šá-an-ni*: for the meaning of 'to bewitch' see CAD E 228 mng. 2f 1′.

23. The provenance of *kukru*-plant from the mountains is mentioned also in *Maqlû* (V 49, VI 22, 24–25, 34–35, 67, VIII 47′), where in two passages the mountains are designated as holy (*quddušūti*, VI 25, 35). The use of a plant name in the plural is not common but seems admissible in the present context where the *kukru*-plant is personified. See Abusch et al. 2020: 48.

24. The rain plays an important role in the incantations of this corpus, in fact, the sexual desire is metaphorically expressed with it. See No. A.1: 33–34; No. E.1: 1–2. See also No. D.2: 19: "Do not release the dew of the Daughters of Anu!". For atmospheric metaphors see Chapter II § "Third group: sexuality and nature.

26. See incantation No. A.2 "Incantation: *Akkannu*-wild ass who is reared-up for mating, who has dampened your desire?". Cf. catalogue LKA 94 i 17.

19–26. The incantation forms a dialogue. In line 19 the patient, addressed metaphorically as a stallion, that has lost its sexual desire, is asked who has caused his suffering. The patient answers in lines 20–22 attributing it to witchcraft. From the second half of line 22 (*attā*) to the end of the incantation the patient is addressed and assured of his imminent purification and the restoration of his sexual desire.

Nīš libbi M

List of manuscripts

Manuscript	Museum number	Publication	Tablet	Script	Date	Provenience	Incantations and prescriptions
A	W. 22656/5	SpTU 4, 135 obv. ii 1–rev. iii 12	Frg. of a two-col. tablet	NB/LB	4th–3rd cent.	Uruk U 18	Prescr. 1; No. M.1; Prescr. 2–6
B	SU 52/139+ 161+ 170+ 250+250A+ 323	STT 280 ii 10–21	Two-col. tablet	NA	8th–7th cent.	Sultantepe	No. M.1
C	BM 41279 rev. iv 20′–25′		Frg. of a two-col. tablet	NA	8th–7th cent.	Unknown provience	No. M.1

Edition

Biggs 1967: 44–45, No. 25 (Ms. B, mention of C[572])
von Weiher 1993: 45–47, No. 135 (Ms. A)
Prechel 1996: 154–156 (trans. of No. M.1)

Structure of the text

Text *nīš libbi* M is based on Ms. A. It contains a prescription, the incantation No. M.1 “Sexual desire! Sexual desire! The bed for the sexual desire” present also in Ms. B and C, and other five prescriptions.

A	Prescr. 1
A – B – C	No. M.1
A	Prescr. 2–5

572 See also Schwemer 2007a: 116; CAD P: 225.

I. Prescription 1 ll. 1–3 (A obv. ii 1–3)

II. Incantation and its ritual No. M.1 "Sexual desire! Sexual desire! The bed for the sexual desire" (Biggs 1967 No. 25) ll. 4–15 (Ms. A obv. ii 4–14// B ii 10–21 // C rev. iv 20′–25′)

III. Prescriptions
   2. 16–18 (Ms. A rev. iii 1–3)
   3. 19 (Ms. A rev. iii 4)
   4. 20 (Ms. A rev. iii 5)
   5. 21–22 (Ms. A rev. iii 6–7)
   6. 23–27 (Ms. A rev. iii 8–12)

Summary of the sections of manuscripts not included in the transliteration:

- Ms. A = SpTU 4, 135
  i = only traces of the signs of the first four lines are recognizable

- Ms B = STT 280i 1–7 = K prescr. 1–2 (Ms. A)
  i 8–17 = D Sympt.; prescr. 4 (Ms. E)
  i 18–21 = F prescr. 10–11 (Ms. E)
  i 22–51 = K prescr. 3–18 (Ms. A) (note i 44 = A prescr. 7 (Ms. D))
  i 52–55 = A prescr. 9–12, 16 (Ms. D)
  i 56–ii 9 = K prescr. 19–22 (Ms. A)
  ii 22–35 = K prescr. 23–28 (Ms. A)
  ii 36–50 = No. K.3 (Ms. A)
  ii 51–53 = No. K.4 (Ms. A)
  ii 54–61 = No. K.5 (Ms. A)
  ii 62–iii 23 = No. E.2 (Ms. H)
  iii 24–33 = No. K.6 (Ms. A)
  iii 34–42 = No. K.7 (Ms. A)
  iii 43–iv 7 = K prescr. 29–31 (Ms. A)
  iv 8–23 = No. K.8 (Ms. A)
  iv 24–31 = No. K.9 (Ms. A)
  iv 32–36 = K prescr. 32 (Ms. A)
  iv 37–41 = No. A.1 (Ms. D)

Transliteration

I. A obv. ii 1–3 = Prescription 1

1.	A ii 1	DÙ.DÙ.BI KU.KU AN.BAR IGI-*lim* PEŠ$_{10}$.^{d}ÍD ÚḪ.^{d}ÍD
2.	A ii 2	1-*niš* SÚD *ana lìb-bi* Ì ŠUB ÉN *an-nit 7-šú ana* ŠÀ ŠID
3.	A ii 3	*šá* NITA GÌŠ-*šú šá* MUNUS GAL$_{4}$.LA-*šá*(text ŠÚ) ŠÉŠ-*su*

II. A obv. ii 4–14// B ii 10–21 // C rev. iv 20′–25′ = Incantation and its ritual No. M.1

4.	A ii 4	ÉN ŠÀ.ZI.GA MIN KI.NÁ ŠÀ.ZI.GA —
	B ii 10	ÉN ŠÀ.ZI.GA M[IN K]I.NÁ ŠÀ.ZI.GA DÙ-*u*[*š*]
	C r. iv 20′	ÉN ŠÀ.ZI.GA ŠÀ.ZI.GA KI.NÁ ŠÀ.ZI.GA —
5.	A ii 5	— *i-p*[*u-u*]*š* ^{d}*iš-tar ana* ^{d}*dumu-zi*
	B ii 11	*ša* ^{d}*iš-tar* ⸢*a*⸣-[x] ^{d}*dumu-zi* DÙ-*u*[*š*]
	C r. iv 20′–21′	— ^{d}15 x / [x] ^{d}*dumu-zi* —
6.	A ii 6	— [*i-pu-u*]*š* ^{d}*na-na-a ana ḫa-i-ri-šá*(text ŠÚ)
	B ii 12	*ša* ^{d}*na-na-a* ⸢*a-na*⸣ *ḫa-ʾi -ri-šá* D[Ù-*uš*]
	C r. iv 21′	— ^{d}*na-na-a ana ḫa-mi-ri-šá*(text ŠÚ) —
7.	A ii 7–8	— [*i-pu-u*]*š* ^{d}*iš-ḫa-ra ana* ^{d}*al-ma-ni-šá*(text ŠÚ) / [*e-pu-u*]*š ana-ku*
	B ii 13	⸢*ša*⸣ ^{d}*iš-ḫa-*⸢*ra a-na*⸣ *al-ma-ni-šá* [—]
	C r. iv 21′–22′	— ^{d}*iš-ḫa-ra ana mu-ti-š*[*a*] / *a*[*na*]-⸢*ku*⸣ DÙ-⸢*uš*⸣
8.	A ii 8–9	*a-na* NENNI A NENNI / [*li-iḫ-m*]*u-ú* UZU^{meš}-*šú li-zaq-qí-pu ú-šar-šú*
	B ii 14	— ⸢NENNI⸣ A NENNI *li-*[*i*]*ḫ-mu-ú* UZU^{meš}-*šú* ⸢*li*⸣-[*zaq-qí-pu*]
	C r. iv 22′–23′	*ana* [*ḫ*]*a-mi-ri-ia lu-ú ḫa-mu-ú* UZU^{meš}-*š*[*ú*] / ⸢*lu-ú*! *za-qip*⸣ GÌŠ-*šú*
9.	A ii 10	[ ] *mu-ši u ur-ri ina qí-bit iq-bu-ú*
	B ii 15	[*a*]-*a*!? *i-na-a*[*ḫ* ŠÀ]-*šú mu-šá* [x *u*]*r-ra i-n*[*a*]
	C r. iv 23′–24′	*a-a i-nu-uḫ lìb-ba-šu* GE$_{6}$ *u im-mu ina q*[*í-bit*] / DU$_{11}$.GA-*ú*
10.	A ii 10–11	AN.ZÍB / [ ] ^{d}*na-na-a* ^{d}*gaz-ba-ba*
	B ii 16	[AN.]ZÍB ^{d}[*iš*]-*tar* ^{d}*na-na-*⸢*a*⸣ ^{d}*g*[*az-b*]*a-*⸢*ba*⸣
	C r. iv 24′	*te-lit* ^{d}15 ^{d}*na-na-a* ^{d}*gaz-ba-ba*
11.	A ii 11	— ^{d}*ka-ni-sur-r*[*a* — — x] ÉN
	B ii 17	— [^{d}*ka-ni-su*]*r-ra* — — T[U$_{6}$] ÉN
	C r. iv 24′–25′	⸢*ù*⸣ [ ] / *be-let* ^{mun}[^{us}U]Š$_{12}$.ZU^{meš}-*te* [x x]

12. B ii 18 K[A.INIM.M]A [ŠÀ.Z]I.GA

Ritual A

13. A ii 12 [DÙ.DÙ.B]I UZU.DIR.KUR.RA IGI-*lim ṣa-ṣu-un-tú* 3 Ú.[ḪI.A]
14. A ii 13 [1-*niš*] SÚD *ana* ŠÀ Ì BUR ŠUB-*di* NÍG.NA ^{šim}L[I GAR-*an* . . .]
15. A ii 14 [. . .] x ÉN *an-nit* 3-*š*[*ú* ŠID-*nu*]

Ritual B

13. B ii 19 DÙ.DÙ.[BI] ⸢^{ú}IGI⸣-*eš-ra* ^{ú}⸢*ni*⸣-[. . .] ⸢*e*⸣-*ti*
14. B ii 20 3 Ú.ḪI.[A SÚD] x [. . .] *iš*
15. B ii 21 *ina* IGI ^{d}⸢15⸣$^{?}$ [. . . *ina* K]UŠ

III. A rev. iii 1–10 = Prescriptions 1–5

16. A iii 1 [. . .] x
17. A iii 2 [. . .] *là* TE-*a*
18. A iii 3 [. . .] *ina kur-kur-ra-tu ina* KUŠ BAR-*ši*

---

19. A iii 4 [. . . *im-ḫu*]*r-lim kib-rit ina* KUŠ

---

20. A iii 5 [. . .] x ^{ú}*tar-muš* ^{ú}IGI-*lim* ^{ú}IGI.NIŠ

---

21. A iii 6 [. . . *n*]*u*$^{?}$ *tar-muš* KA *tam-tú* AN.BAR NI MAN$^{?}$
22. A iii 7 [. . .] *ina* GU GADA È *ina* GÚ-*šú* GAR

---

23. A iii 8 [. . . NÍ]TA$^{?}$ $^{\text{na4}}$URUDU NÍTA AN.BAR NÍTA *ina* KUŠ
24. A iii 9 [. . .] x MEŠ DIŠ NA *ina* MÁŠ.GI$_{6}$-*šú* ÚŠ^{meš} IGI.IGI
25. A iii 10 [. . . D]U$_{11}$.DU$_{11}$-*ub* ŠÍG.BABBAR ŠÍG.SA$_{5}$ NI.NU È
26. A iii 11 [. . . K]UD UD.DA DU$_{8}$ UD.DA BAD
27. A iii 12 [ÉN$^{?}$ . . . e]gir nam-tar-ra-zu al-gub

Transcription

I. A obv. ii 1–3 = Presciption 1

1. *dudubû: sīkti parzalli imḫur-līm kibrīta ru'tīta*
2. *ištēniš tasâk ana libbi šamni tanaddi šipta annita sebîšu ana libbi tamannu*
3. *ša zikari ušaršu ša sinništi ūrša tapaššassu*

II. A obv. ii 4–14// B ii 10–21 // C rev. iv 20′–25′ = Incantation and its ritual No. M.1

4. *šiptu: nīš libbi nīš libbi mayyāl nīš libbi*
5. *īpuš Ištar ana Dumuzi*
6. *īpuš Nanāya ana ḫā'irīša*
7. *[īpu]š Išḫara ana almānīša ēpuš anaku*
8. *annanna mār annanna* (var.: *ḫāmirīya*) *liḫmû šīrūšu lizaqqipu ušaršu*

9. *ay īnaḫ libbašu mūši u urri ina qibīt iqbû*
10. *telīti Ištar Nanāya Gazbaba*
11. *Kanisurra* (var.: *bēlet [kaš]šāpāti*) *t[ê] šipti*
12. *š[ipa]t nī[š libbi]*

Ritual A
13. *[dudub]û: kamūn šadî imḫur-līm ṣaṣuntu šalaš šammī*

14. *[ištēniš] tasâk ana libbi šaman pūri tanaddi nignak burā[ši tašakkan . . .]*

15. [. . .] . . . *šipta annīta šalāšīš[u tamannu* . . .]

Ritual B
13. *dudu[bû]: imḫur-ešrā* . . . [. . .] . . .
14. *šalaš šamm[ī tasâk]* . . . [. . .] . . .
15. *ina maḫar Ištar* [. . . *ina ma]ški*

III. A rev. iii 1–10 = Prescriptions 1–5

2 16. [. . .] . . .
17. [. . .] *lā ṭeḫê*
18. [. . .] *ina kurkurratu ina maški tumaššarši*

3 19. [. . . *imḫu]r-līm kibrīt ina maški*

4 20. [. . .] x *tarmuš imḫur-līm imḫur-ešrā*

Translation

I. A obv. ii 1–3 = Presciption 1

1. Its ritual: Iron powder, "heals-a-thousand"-plant, *ru'tītu*-sulphur
2. you pulverize together, you put (it) in oil, you recite the incantation seven times over it,
3. you anoint (with it) the penis of the man and the pelvic area of the woman.

II. A obv. ii 4–14// B ii 10–21 // C rev. iv 20′–25′ = Incantation and its ritual No. M.1

4. Incantation: Sexual desire! Sexual desire! The bed for the sexual desire,
5. (like the one that) Ištar did for Dumuzi,
6. (the one that) Nanāya did for her husband,
7. (the one that) Išḫara [di]d for her lover (var.: husband), I did!
8. May the flesh of NN, son of NN (var.: of my husband), be static, may (instead)$^{?}$ his penis be erect!
9. May his desire (lit. 'heart') not abate night and day! At the command of
10. the wise Ištar, Nanāya, Gazbaba
11. (and) Kanisurra (var.: lady of the [sor]ceresses). Incantation formula.
12. Word[in]g of *nī*[*š libbi*] (incantation).

Ritual A
13. I[ts ritual]: *kamūnu*-fungus from mountain, "heals-a-thousand"-plant, *ṣaṣuntu*-plant, the three drugs
14. you pulverize [together], you put into oil from the alabastron, [you set up] a censer with juniper[r . . .]
15. [. . .] . . . [you recite] this incantation three tim[es . . .].

Ritual B
13. [Its] ritual: "Heals-twenty"-plant, . . . -plant, [. . .] . . .
14. [you pulverize] the three drug[s] . . . [. . .] . . .
15. in front of Ištar [. . . in a leather b]ag.

III. A rev. iii 1–10 = Prescriptions 1–5

2 16. [. . .] . . .
17. [. . .] who does not approach,
18. [. . .] in the *kurkurratu*-pot, you leave (it)$^{?}$ in a leather bag.

3 19. [. . . "heal]s-a-thousand"-plant, sulphur in a leather bag.

4 20. [. . .] . . . *tarmuš*-plant, "heals-a-thousand"-plant, "heals-twenty"-plant.

5 21. [. . .] . . . *tarmuš imbû tâmti parzillu parzillu . . .*

22. [. . .] *ina qê kitê tašakkak ina kišādīšu tašakkan*

6 23. [. . . *zik*]*aru erû zikaru parzillu zikaru ina maški*

24. [. . .] . . . *šumma amēlu ina šuttīšu mītūti ītanammar*

25. [. . . *t*]*adabbub šīpāti peṣêti šīpāti sāmāti taṭammi tašakkak*

26. [. . . *tap*]*arras ṣētu lippaṭer ṣētu lissi*

27. [*šiptu*?: . . . e]gir nam-tar-ra-zu al-gub

5 21. [. . .] . . . *tarmuš*-plant, coral limestone, iron, . . .
22. [. . .] you align (them) on linen strings, you put (it) around his waist.

---

6 23. [Ma]le [. . .], male copper, male iron in a leather bag.
24. [. . .] . . . if a man repeatedly sees dead people in a dream,
25. [. . . you s]ay, you spin white wool and red wool (and) align
26. [. . . you c]ut, may the sun-heat be removed, the may sun-heat regress!
27. [Incantation: ". . . a]fter$^{?}$ it your destiny is assigned."

Commentary

4–12. For the analysis of the incantation and its ritual No. M.1 see Chapter II § "Forth group: *historiola* and its function." Incantation No. M.1 features two speakers: the female partner (ll. 4–7) and the therapeutic operator (ll. 8–11).

4. See the catalogue LKA 94 ii 15: ÉN ŠÀ.ZI.[G]A MIN KI.⌜NÁ⌝ MIN.

5–6. Sumerian literature is full of references to the bed and the bridal chamber in relation to Inana and Dumuzi. As for Nanāya, in an Old Babylonian love poem bequeathed to us, she features with her husband, Muati (see Lambert 1996: 41–51). According to *Nanāya Hymn of Sargon II* (Livingstone 1989: 14, No. 4 ii 3′) and the poem about the love between the goddess and Muati (Lambert 1996: 41–51), the husband of the goddess is the latter, associated to the god Nabû. Since the end of the second millennium, the goddess is in the group of Nabû (see Stol 1998–2001: 148).

7. The lover of the goddess Išḫara is Almānu (see the lexical list ḪAR-gud B vi 52, Reiner 1974b = MSL 11: 41, see also Prechel 1996: 149–150, 187). Ms. A and B differ as to the presence of the divine determinative in front of the name. Its presence in Ms. A urges us to interpret the word as a proper name, despite the presence of the possessive suffix (*-ša*) (see Ebeling 1932b: 71). CAD A/1 362, arguing that *almānu* is a proper name, rejects the translation as "widower," proposed by the AHw. I 38. According to Biggs (1967: 44), *almānu* is to be understood as a noun, meaning "man without family obligation," although in this passage it seems to be a synonym of "lover." Note that the Ms. C variant contains *mu-ti-š[a]* "her husband," Matouš (1970: 75) translates "husband" and identifies it with Dumuzi. The identification is uncertain, however, Išḫara may not be Ištar (as opposed to *Atram-ḫasīs* I 299–304, Lambert and Millard 1969: 649) here, but a different goddess, who shares with the latter the dominion of the sphere of loving and sexuality (in the same way Nanāya).

The association between the bed and the goddess Išḫara is confirmed by a passage of an Old Babylonian tablet of the *Epic of Gilgameš*: "For Išḫara the bed / has been prepared, / Gilgameš the young women / meets the night" (Pennsylvania v 196–199, George 2003: 178). Here, the bed clearly indicates the act of love, the place where sexual intercourse takes place. It should be noted that already in the Old Babylonian period, the goddess represents Ištar as a patron of the wedding period (note the mention of the spouses in the incantation) as confirmed by the following passage of the *Atram-ḫasīs*: "In the brid[al room ma]y the bed be placed, / may the wife and her husband lie together. / When the marriage (lit. wifehood and husbandhood) / pay attention to Ištar in the [wedding] house / may for nine days there will be joy, / may they call Ištar Išḫara" (I 299–304, Lambert and Millard 1969: 649).[573]

9. See for a similar expression No. K.5: 105 and commentary.

11. In Ms. B Biggs (1967) restores with Išḫara, however, the duplicates have: Ms. A ^{d}*ka-ni-sur-r*[*a*], Ms. C [. . .] *be-let* ^{mun}[^{us}U]Š$_{12}$.ZU^{meš}-*te*, Kanisurra's epithet. However, the space is too small to contain the four missing signs.

12. The fragmentary text Sp II 976 = BM 35394 (CCMAwR 3, pls. 2–3) obv. ii 1′–4′ is perhaps a duplicate of the final part of incantation No. M.1. It could be, alternatively, a duplicate of the incantation No. E.3, in fact in line 4′ we read [^{d}]*na-na-*[*a*] *be-let* Ḫ[I.LI].

13–15. The ritual prescriptions in the two manuscripts are very different, for this reason they are presented separately.

13. Ritual A: see the secret name of *kamūnu*-fungus: Ú *ka-mi-u-nu* : DILI : Á *šu-ti-ni* "*kamūnu*-fungus : secret name : "bat's wing" (URU.AN.NA: *maštakal* series l. 84, Rumor 2017: 16).

21. NI MAN may be a mistake for NÍTA, see von Weiher 1993: 47.

21–22. The ingredient group "*tarmuš*-plant, iron and coral limestone" is present also in F prescr. 10: 31.

573 For a discussion of the two passages see Prechel 1996: 58–60 and previous literature.

V Texts from Boghazköy

Nīš libbi N

List of manuscripts

Manuscript	Museum number	Publication	Tablet	Script	Date	Provenience	Incantations and prescriptions
A	Bo 4894	KUB 4, 48 i 1–ii 2; ii 10–iv 26; le. e; lo. e.	Two-col. tablet	'Not Hittite'[574]	13th cent.	Ḫattuša	Prescr. 1–25
B	Bo 5817	KUB 37, 80 1′–13′, 16′	Frg. of a single-col. tablet	NH	13th cent.	Ḫattuša	Prescr. 1–5, 7
C	(Bo 5885)+ AAA 3, No. 5	(KUB 37, 81)+ AAA 3, pl. 27 No. 5 2–4	Frg. part of the tablet containing KUB 37, 81	Ass.-Mitt.[?575]	13th cent.	Ḫattuša	Prescr. 8

Edition

Pinches 1910: 104–105 (Ms. C)[576]
Ebeling 1925: 46–55 (Ms. A)
Biggs 1967: 54–61 (Ms. A, B, C)
Chalendar 2018: 47 (Ms. A)

Structure of the text

Text *nīš libbi* N follows Ms. A's numeration. Only the prescriptions 1–5 and 7–8 have as duplicates Ms. B and C.

574 Note the use of *bá*, see Schwemer 2013: 154.
575 Note the use of *qè* in KUB 37, 81, see Schwemer 2013: 154.
576 For Pinches (1910: 105) Ms. C was a spell against a sickness. He interprets the expression ŠÀ.ZI.GA with *ina libbi itebbî* "from the midst he will depart."

https://doi.org/10.1515/9783110757262-010

A – B	Prescr. 1–5
A	Prescr. 6
A – B	Prescr. 7
A – C	Prescr. 8
A	Prescr. 9–23
A (+ dupl.)	= No. L.1
A	Prescr. 24–25

I. Prescriptions

1. i 1–7 (Ms. A i 1–7 // B 1′–2′)
2. i 8–11 (Ms. A i 8–11 // B 3′–4′)
3. i 12–16 (Ms. A i 12–16 // B 5′–7′)
4. i 17–22 (Ms. A i 17–22 // B 8′–10′)
5. i 23–27 (Ms. A i 23–27 // B 11′–13′)
6. i 28–32 (Ms. A i 28–32)
7. ii 1–2 (Ms. A ii 1–2 // B 16′)
8. ii 3–16 (Ms. A ii 3–16 // C 2–4) (note ii 3–9 = K prescr. 31)
9. ii 17–26 (Ms. A ii 17–26)
10. ii 27–31 (Ms. A ii 27–31)
11. ii 32 (Ms. A ii 32)
12. iii 1–6 (Ms. A iii 1–6)
13. iii 7–10 (Ms. A iii 7–10)
14. iii 11–23 (Ms. A iii 11–23)
15. iii 24–26 (Ms. A iii 24–26)
16. iii 27–31 (Ms. A iii 27–31
17. iii. 32–iv 2 (Ms. A iii 32–iv 2)
18. iv 3–4 (Ms. A iv 3–4)
19. iv 5–8 (Ms. A iv 5–6)
20. iv 9–10 (Ms. A iv 9–10)
21. iv 11–20 (Ms. A iv 11–20)
22. iv 21–23 (Ms. A iv 21–23)
23. iv 24–26 (Ms. A iv 24–26)

II. Ms. A iv 27–31 = Sumerian incantation No. L.1

III. Prescriptions

24. left e. 1–7 (Ms. A le. e 1–7)
25. lower e. 1–5 (Ms. A Ms. A lo. e. 1–5)

Summary of the sections of manuscripts not included in the transliteration:

- Ms A = KUB 4, 48
  ii 3–9 = K prescr. 31
  iv 27–31 = No. L.1 (Ms. B)
- Ms B = KUB 37, 80
  14′–15′ = Fragmentary prescription (see commentary ll. i 28–32)
  17′–18′ = Fragmentary prescription (see commentary ll. ii 1–2)
- Ms. C =
  AAA 3, pl. 27 No. 5 5–9 = V prescr. 2
  AAA 3, pl. 27 No. 5 10–12 = V prescr. 3
  KUB 37, 81 1′–8′ = W prescr. 1–4

Transliteration

A i 1–iv 26 // B 1′–16′ // C 2–4 = Prescriptions 1–23

Column i

1.	A i 1	DIŠ LÚ ŠÀ.ZI.GA *ina* ^{iti}BÁR.ZAG.G[AR]
2.	A i 2	TIL BURU$_5$.ḪABRUD.TA NITÁ *ta-ṣa-bat*
3.	A i 3	*kap-pa-šú ta-bá-qa-an-šú ta-ḫa-na-aq-šú-ma*
4.	A i 4	*tu-ra-qa-aq* MUN *ta-za-ru*
	B 1′	[ ]
5.	A i 5	*tu-bal* NUMUN ^{giš}KIŠI$_{16}$.ḪAB.KUR.RA
	B 1′	*tu-*[*bal*]
6.	A i 6	TÉŠ.BI *ta-sàk ina* KAŠ NAG-*šú-ma*
	B 2′	[ ]
7.	A i 7	LÚ BI ŠÀ.ZI.GA TUKU-*ši*
	B 2′	ZA BI [ ]
8.	A i 8	DIŠ KI.MIN BURU$_5$.ḪABRUD.DA NITÁ *ša a-na* U$_5$ ZI.GA
	B 3′	— KI.MIN [BURU$_5$.ḪABRU]D.DA NITÁ [ ]
9.	A i 9	*tu-bal ta-sàk a-na* ŠÀ A^{meš}
	B 3′–4′	[ ] / *ana* ŠÀ ⌜A⌝^{meš}
10.	A i 10	*ša ú-ri bu-ut-tu$_4$* ŠUB-*ma* NAG-*šú-ma*
	B 4′	*šá ú-ri b*[*u-ut-tu$_4$*]
11.	A i 11	LÚ BI ŠÀ.ZI.GA TUKU-*ši*
	B 4′	[ ]
12.	A i 12	[DIŠ] KI.MIN BURU$_5$.ḪABRUD.DA NITÁ SAG.DU-*sú* KUD-*is*
	B 5′	— KI.MIN BURU$_5$.ḪABRUD$^{!}$.DA NITÁ SAG.D[U-*sú*]
13.	A i 13	ÚŠ^{meš}-*šu a-na* ŠÀ *me-e* ŠUB-*ma*
	B 5′	[ ]
14.	A i 14	ŠÀ-*šú ta-a-al-lu-ut-ma* A^{meš} *šu-nu-ti*
	B 6′	ŠÀ(text 3)-*šú ta-al-lu-ut-ma* A$^{me[š}$]

15.	A i 15	*ina* MUL *tuš-bat ki-ma* ^{d}UTU È
	B 6′	[ ]
16.	A i 16	NAG-*šú-ma* ŠA.ZI.GA TUKU-*ši*
	B 7′	NAG-*šú-ma* TI-*uṭ*
17.	A i 17	DIŠ KI.MIN *mu-ša-ar* BURU$_5$.ḪABRUD.DA NITÁ
	B 8′	— KI.MIN ŠIR MUŠEN *ḫur-ri* NITÁ
18.	A i 18	*ru-pu-uš-ti* GU$_4$ ZI.GA
	B 8′	*ru-pu-u*[*š-ti*]
19.	A i 19	*ru-pu-uš-ti* UDU ZI.GA [*rupušti* MÁŠ.NITÁ ZI.GA]
	B 8′	[ ]
20.	A i 20	*ina* A^{meš} NAG-*šú-ma* *ina ša*$_{10}$-*aḫ-r*[*a-at*]
	B 9′	— — — *ina šaḫ-ra-at* [KU]N *ù*
21.	A i 21	SÍG *šab-ri-šu ša* UDU *ta-lam-m*[*e-ma*]
	B 9′	SÍG *š*[*ab-ri-šu*]
22.	A i 22	*ina* MAŠ.SÌL(text BÁR.KA) GAR-*an*-⟨*ma*⟩ ŠÀ.ZI.GA TUKU-*ši*
	B 10′	*ina* MÚRU-*šú* *t*[*a-ša-kan-ma*] TI-*uṭ*
23.	A i 23	DIŠ KI.MIN AL.DI.RÍ.GA^{mušen} *ta-ba-qà-an*
	B 11′	— KI.MIN AL.DI(text KI).{IŠ}RÍ.GA^{mušen} *ta*-[*ba-qa-an*]
24.	A i 24	*ta-zar-ra-ak-ma* MUN Ú.KUR.RA
	B 11′–12′	[ ] / MUN Ú.KU[R.R]A
25.	A i 25	TAG.GA-*šú* *tu-bal ta-sàk*
	B 12′	*tâ-lap-pa-at*-[*šú*]
26.	A i 26	*ina* $^{zi!}$NÍG.ŠE.SA.A ḪE.ḪE-*ma*
	B 12′	[ ]
27.	A i 27	NAG-*šú-ma* — — ŠÀ.ZI.GA TUKU-*ši*
	B 12′–13′	[ ] / ZA BI [ŠÀ.ZI.G]A TUKU-*ši*
28.	A i 28	DIŠ KI.MIN NAM.GEŠTIN^{mušen} *ta-ba-qà-an ta-za-ra-ak*
29.	A i 29	MUN *a-ma-ni* Ú.KUR.RA *ta-la-pat*
30.	A i 30	*ta-sàk* ŠE $^{zi!}$NÍG.ŠE.SA.A

31.	A i 31	$^{[gi]š}$KIŠI$_{16}$.ḪAB.KUR.RA *bá-lu$_4$ pa-tan*
32.	A i 32	[N]AG-*šú-ma* ŠÀ.ZI.GA TUKU-*ši*

Column ii

1.	A ii 1	DIŠ KI.MIN ^{ú}EME.UR.GI$_7$ ⸢*tu*⸣-[*bal*$^?$ *ina* KAŠ/A$^{meš?}$]
	B 16′	— KI.MIN 〈^{ú}〉⸢EME⸣.UR.G[I$_7$]
2.	A ii 2	NAG-*šú-ma* ŠÀ.ZI.GA TUKU-[*ši*]
	B 16′	[ ]

3–9 = *nīš libbi* K prescr. 31: 135–140.

10.	A ii 10	*ta-pa-a-aṣ ina* ZÌ.KUM ḪE.ḪE *tu-ka*[*p-pa-at*]
11.	A ii 11	3 *ku-up-pa-ti-in-ni* DÙ-*u*[*š* . . .]
12.	A ii 12	*ina* ŠÀ *ku-up-pa-ti-in-ni* [. . .]
13.	A ii 13	DÙ *bá-lu$_4$ pa-ta-a-an ta-*[. . .]
14.	A ii 14	*a-na ku-up-pí-ta-an-ni* [. . .]
	C 2–3	[*ku*]-*up-p*[*í-ta-an-ni* . . . / . . .]
15.	A ii 15	*tàra-kas$_4$ ina* ⸢MÚRU-*ka*⸣ [. . .]
	C 3–4	[*tàra*]-*kas$_4$ ina* MÚ[RU-*ka* . . . / . . .]
16.	A ii 16	ŠÀ.ZI.G[A TUKU-*ši*]
	C 4	ŠÀ.ZI.GA TUKU-[*ši*]

17.	A ii 17	DIŠ KI.MIN *e-n*[*u-ma* . . .]
18.	A ii 18	traces
19–24.	A ii 19–24	destroyed
25.	A ii 25	*te-le*[*q-qé* . . .]
26.	A ii 26	NAG-*šú-ma* [ŠÀ.ZI.GA TUKU-*ši*]

27.	A ii 27	DIŠ KI.MIN *e-nu-ma* UDU NITÁ *ina*! [*muḫ-ḫi* U$_8$ *iš-ḫi-ṭu*]
28.	A ii 28	SÍG KUN-*šú ni ba*$^?$ [. . .]
29.	A ii 29	GU-*šú* ⸢*ri*⸣-*ta-t*[*u$_4$*$^?$ . . .]
30.	A ii 30	*ina* MÚR[U-*šú* GAR-*an-ma*]
31.	A ii 31	[ŠÀ.ZI.GA TUKU-*ši*]

32.	A ii 32	DIŠ KI.MIN *e-*[*nu-ma* . . .]

Rest destroyed

Ms. A column iii

1. ŠÀ BURU$_5$.ḪABRUD.DA^{mu}[^{šen} NITÁ *ta-al-lu-ut*]
2. *ḫa*$^!$(text KI)-*aḫ-ḫu* GU$_4$ ZI.GA [*te-leq-qé a-na* A^{meš} ŠUD-*di*]
3. *ina ú-ri ana* MUL [*tuš-bat*]
4. *ki-ma* ^{d}UTU *it-*[*tap-ḫu ina* UGU PA^{meš} ^{giš}KIŠI$_{16}$]
5. *iz-zi-zu a-n*[*a* IGI ^{d}UTU . . .]
6. *ina* [A^{meš} NAG-*šú* ŠÀ.ZI.GA TUKI-*ši*]

---

7. B[E KI.MIN] x ⸢UR.GI$_7$⸣ *ra-*[*ki-bi*$^?$ . . .]
8. *tu-bal ta-sàk ta-*x [. . .]
9. *tu-bal-lal-ma ina* A NAG-*šú-*[*ma*]
10. ŠÀ.ZI.GA TUKU-[*ši*]

---

11. BE KI.MIN *ri-kib-ti* ARKAB^{mušen} ^{ú}*e-li-*[*ku-la*$^?$]
12. *te-er-te-en-na* ^{giš}ŠINIG [. . .]
13. ^{ú}IN.NU.UŠ GIŠIMMAR ^{giš}GAN.[U$_5$]
14. *u* ⸢*mím-ma*⸣$^?$ NUMUN^{meš} *ina* A^{meš} ŠUB-*ma*
15. *ana* MUL *tuš-bat ina ú-ri ta-ša-*[*kan*]
16. LÚ *ša-a-šu* TUŠ.A *ina* A^{meš} G[AZI^{sar}]
17. *ka-la* SU-*šú tu-šáḫ-*[*ḫa-aṭ* . . .]
18. *ri-kib-ti* ARKAB^{mušen} [. . .]
19. *miš-la* NAG-*ma miš-*[*la* ŠÉŠ.ŠÉŠ-*sú*$^?$]
20. Ì.SAG$^?$(text EZEN) *u ter-te-e*[*n-na* ^{giš}ŠINIG]
21. *ina* GÌR$^?$ GÙB-*šú a* x x [. . .]
22. *ta-sàk-ma ina* Ì.GIŠ *p*[*u-ri* SU-*šú*]
23. ŠÉŠ.ŠÉŠ-*sú-ma* L[Ú BI ŠÀ.ZI.GA TUKU-*ši*]

---

24. BE KI.MIN UGU BURU$_5$.ḪABRUD.DA^{mu}[^{šen} NITÁ . . .]
25. *ina* Ì.GIŠ *pu-ri* SU-*šú*
26. ŠÉŠ.ŠÉŠ-*sú-ma* L[Ú BI ŠÀ.ZI.GA TUKU-*ši*]

---

27. BE KI.MIN *ni-ši lìb-bi i-ṭi*$_4$*-ir* [. . .]
28. *mu-ša-ar* BURU$_5$.ḪABRUD.[DA^{mušen} NITÁ]
29. ^{uzu}*nap-šat ša* UDU.NITÁ [*ina* ^{síg}ḪÉ.ME.DA]
30. NIGIN-*ma ina* MÚRU GAR-[*an-ma*]
31. LÚ BI ŠÀ.ZI.GA [TUKU-*ši*]

---

32. BE KI.MIN MUŠ.DÍM.KUR.RA E[DIN *ri-it-ku-ba-ti*]
33. *ina* IGI DÙ x [. . .]

Ms. A column iv

1. [. . . *bá*]*-lu*$_4$ *pa-ta-a-an*
2. [NAG-*ma* Š]À.ZI.GA TUKU

---

3. [BE KI.MIN *ru-pu-uš-ti*$^?$ UDU$^?$ ZI.G]A SI DÀRA.MAŠ
4. [. . . ŠÀ.ZI].GA

---

5. [BE KI.MIN . . .]
6. destroyed
7. [. . . *ina* MÚRU-*šú*] *tàra-kas*$_4$
8. destroyed

---

9. DIŠ LÚ *ni-š*[*i lìb-bi e-ṭi*$_4$*-ir* . . .]
10. *e-nu-*[*ma* . . .]

---

11. BE LÚ *ša n*[*i-iš lìb-bi-šú eṭ-ru* . . .]
12. *it-ti* x x [. . . *i-l*]*e-ʾe*$^?$*-e*
13. Ú ŠÀ.ZI.GA [. . .] x
14. *ni-iš lìb-b*[*i* . . .]
15. *šum-ma it-ti* x [. . .] x
16. *ana ni-iš* ŠÀ-*šú* [TUKU-*e* . . .]
17. ^{ú}KASKAL.MUNUS$^?$ x [. . .]
18. *ta-maḫ-ḫar* [. . .]
19. *ra*-x [. . .]
20. Š[À.ZI.GA TUKU-*ši*]

21–23. traces

---

24. [BE LÚ ŠE$_{10}$$^?$ EME.DIR.]GÙN.KUR.RA
25. [. . . *ina* MÚRU-*šú t*]*àr-kas*$_4$*-ma*
26. [. . . ŠÀ.ZI].GA$^!$

II. A iv 27–31 = Sumerian incantation No. L.1

III. A left e. 1–7, lower e. 1–5 = Prescriptions 24–25

Ms. A left edge
1. [*e-n*]*u-ma* MUŠEN *ḫur-ri ir-ta-na-kab-*[*bu* . . .]
2. [MUŠEN^{meš} *ḫur*]*-ri* NITÁ *ina* ^{iti}GU$_4$.SI.SÁ *ta-*[*ṣa-bat ta-ba-qa-an*]
3. ⌜ÚŠ^{meš}⌝ *la tu-maš-šar ir-ri-šu-nu la* [. . . *l*]*a a tu*$^!$-x x
4. *te-ʾè-il-ma lu-ú* 2 ITI *lu-ú* [3 ITI . . . *ṣ*]*u-up-ri ap-pí kar-ši*
5. *še-er-a-ni u ir-ri ina* NA$_4$.NA$_4$ *ta-sàk* [. . .] *ni ta-*⟨*ra*⟩*-bá-ak-ma*
6. 3 ⌜ŠU⌝ NUMUN ^{ú}*a-lu-zi-in-ni* 2 ŠU *ḫu-*[. . .] x ḪE.ḪE *ina* KAŠ *ki-ma ka-ia-n*[*am-ma*]
7. *ta-maḫ-ḫaṣ bá-lu*$_4$ *pa-tan* NAG-*šú-ma* L[Ú BI ŠÀ-*šú i*]*-na-aš-ši*

Ms. A lower edge
1. [DIŠ KI.MIN *lìb*]*-bi* UGA(text Ú.NAGA.⟨GA⟩)^{mušen} NITA ÚŠ MÁŠ.NITÁ ZI.GA
2. [ÚŠ M]UŠEN *ḫur-ri* NITÁ ⟨*ri*⟩*-kib*(text KAL)*-ti* ARKAB^{mušen}
3. NUMUN ^{giš}KIŠI$_{16}$.ḪAB TÉŠ.BI ḪE.ḪE ⟨GIŠ⟩.KUN LÚ$^!$ *šu*(text ZU)*-lu-uš-šu*
4. ŠÉŠ-⟨*ma*⟩ ŠÀ.ZI.GA TUKU-*ši* KA.INIM.MA ŠÀ.ZI.GA
5. DUB 1.KAM DIŠ LÚ ŠÀ.ZI.GA

Transcription

I. A i 1–iv 26 // B 1′–16′ // C 2–4 = Prescriptions 1–23

Column i

1 1. *šumma amēlu nīš libbi ina Nisan[ni]*
2. *iqti iṣṣūr ḫurri zikara taṣabbat*
3. *kappašu tabaqqanšu taḫannaqšu-ma*
4. *turaqqaq ṭabta tazarru*
5. *tubbal zēr dadān šadî*
6. *ištēniš tasâk ina šikari išattīšu-ma*
7. *amēlu šū nīš libbi irašši*

2 8. *šumma* KI.MIN (= *amēlu nīš libbi ina Nisanni iqti*) *iṣṣūr ḫurri zikara ša ana rakābi tebû*
9. *tubbal tasâk ana libbi mê*
10. *ša ūri buttu tanaddī-ma išattīšu-ma*
11. *amēlu šū nīš libbi irašši*

3 12. [*šumma*] KI.MIN (= *amēlu nīš libbi ina Nisanni iqti*) *iṣṣūr ḫurri zikara qaqqassu tanakis*
13. *dāmīšu ana libbi mê tanaddī-ma*
14. *libbašu taʾallut-ma mê šunūti*
15. *ina kakkabī tušbāt kīma Šamšu uṣṣû*
16. *išattīšu-ma nīš libbi irašši* (var.: *iballuṭ*)

4 17. *šumma* KI.MIN (= *amēlu nīš libbi ina Nisanni iqti*) *mušar* (var.: *išik*) *iṣṣūr ḫurri zikara*
18. *rupušti alpi tebî*
19. *rupušti immeri tebî* [*rupušti dašši tebî*]
20. *ina mê išattīšu-ma* (var.: omitted) *ina šaḫrat* [*zib*]*bati u*
21. *šīpāt šabrīšu ša immeri talamm*[*ī-ma*]
22. *ina naglabi* (var.: *qablīšu*) *tašakkan-ma nīš libbi irašši* (var.: *iballuṭ*)

5 23. *šumma* KI.MIN (= *amēlu nīš libbi ina Nisanni iqti*) *diqdiqqa tabaqqan*
24. *tazarrak-ma ṭabta nīnâ*
25. *talappassu tubbal tasâk*
26. *ina qalīti taballal-ma*
27. *išattīšu-ma* (var.: *amēlu šū*) *nīš libbi irašši*

Translation

I. A i 1–iv 26 // B 1′–16′ // C 2–4 = Prescriptions 1–23

Column i

1 1. If a man's sexual desire has finished in the month Nisan[nu]:
2. You take a male partridge,
3. you pluck its wings, you strangle and
4. flatten (it), you spread (over it) the salt,
5. you dry (it), together with seeds of mountain *dadānu*-acacia
6. you pulverize (it), he drinks (it) in beer and
7. this man will get sexual desire.

2 8. If ditto (= a man's sexual desire has finished in the month Nisannu): A male partridge who is reared-up for the mating
9. you dry, you pulverize (it), in the water,
10. which on the roof spent several nights, you put (it) and he drinks this and
11. this man will get sexual desire.

3 12. [If] ditto (= a man's sexual desire has finished in the month Nisannu): You behead a male partridge,
13. you put its blood in the water,
14. you ingest its innards and this water
15. you leave out during the night, when the sun rises
16. he drinks this and will get sexual desire (var.: he will recover).

4 17. If ditto (= a man's sexual desire has finished in the month Nisannu): The penis (var.: testicles) of a male partridge,
18. the saliva of a reared-up bull,
19. the saliva of a reared-up ram, [the saliva of a reared-up buck],
20. he drinks (them) in water (var.: omitted) and with the hair of the [ta]il and
21. the wool of the perineum of a sheep you wrap u[p (it) and]
22. you put (it) on the hip (var.: around his waist) and he will get sexual desire (var.: he will recover).

5 23. If ditto (= a man's sexual desire has finished in the month Nisannu): You pluck a wren,
24. you eviscerate (it) and you apply *amānu*-salt (and) *nīnû*-plant (on it),
25. you dry, you pulverize,
26. you mix (it) with toasted grain and
27. he drinks this and he (var.: that man) will get sexual desire.

6 28. *šumma* KI.MIN (= *amēlu nīš libbi ina Nisanni iqti*) NAM.GEŠTIN^{mušen} *tabaqqan tazarrak*
29. *ṭabti amāni nīnâ talappat*
30. *tasâk qalīta*
31. *dadān šadî balu patān*
32. [*iš*]*attīšu-ma nīš libbi irašši*

Column ii

7 1. *šumma* KI.MIN (= *amēlu nīš libbi ina Nisanni iqti*) *lišān-kalbi tu*[*bbal ina šikari/mê*]
2. *išattīšu-ma nīš libbi iraš*[*ši*]

8 3–9 = *nīš libbi* K prescr. 31: 135–140.

10. *tapâṣ ina isqūqi taballal tuka*[*ppat*]
11. *šalaš kuppatinnī teppu*[*š* . . .]
12. *ina libbi kuppatinnī* [. . .]
13. *kalâ balu patān* . . . [. . .]
14. *ana kuppitannī* [. . .]
15. *tarrakas ina qablīka* [. . .]
16. *nī*[*š*] *libbi iraš*[*ši*]

9 17. *šumma* KI.MIN (= *amēlu nīš libbi ina Nisanni iqti*) *en*[*ūma* . . .]
18. Traces
19–24. Destroyed
25. *tele*[*qqe* . . .]
26. *išattīšu-ma* [*nīš libbi irašši*]

10 27. *šumma* KI.MIN (= *amēlu nīš libbi ina Nisanni iqti*) *enūma immeru ina* [*muḫḫi laḫri išḫiṭu*]
28. *šīpāt zibbatīšu* . . . [. . .]
29. *qêšu* . . . [. . .]
30. *ina qabl*[*īšu tašakkan-ma*]
31. [*nīš libbi irašši*]

11 32. *šumma* KI.MIN (= *amēlu nīš libbi ina Nisanni iqti*) *e*[*nūma* . . .]
Rest destroyed

6 28. If ditto (= a man's sexual desire has finished in the month Nisannu): You pluck and eviscerate a NAM.GEŠTIN-bird,
29. you apply *amānu*-salt (and) *nīnû*-plant,
30. you pulverize (it) with barley, roasted flour,
31. mountain *dadānu*-[plan]t, on an empty stomach
32. [he d]rinks (it) and he will get sexual desire.

Column ii

7 1. If ditto (= a man's sexual desire has finished in the month Nisannu): You [dry] "dog's-tongue"-plant, [in beer/water]
2. he drinks it and he will ge[t] sexual desire.

8 3–9 = *nīš libbi* K prescr. 31: 135–140 = If ditto (a man's sexual desire in Nisannu month): the wings of a male eagle, the wings of [. . .] /geckos of the steppe which copula[te], / excrement of an *arkabu*-bat, excrement of polychrome lizard, seeds of the . . .-tree, / seeds of the *ēru*-tree, seeds of *maštakal*-plant, / [seeds] of *azallû*-plant, seeds of *murdudû*-plant, / seeds of "dog's-tongue"-plant, seeds of *šakirû*-plant, *ṣaṣun*[*tu*-plant], /*šumuttu*-plant, / [you dry] these 14 drugs together,
10. you pulverize (it), mix with *isqūqu*-flour and roll into [balls],
11. you mak[e] three *kupatinnu*-pellets, [. . .]
12. in the middle of the *kupatinnu*-pellets [. . .]
13. all on an empty stomach you [. . .]
14. towards *kupatinnu*-pellets [. . .]
15. you tie, around your waist [. . .]
16. [he will get] sexual de[si]re.

9 17. If ditto (= a man's sexual desire has finished in the month Nisannu): whe[n . . .]
18. Traces
19–24. Destroyed
25. you tak[e . . .]
26. he drinks it [and he will get sexual desire]

10 27. If ditto (= a man's sexual desire in Nisannu month): when a ram [mounts the sheep]
28. wool of the tail . . . [. . .]
29. its yarn . . . [. . .]
30. [you put] around [his] waist [and]
31. [he will get sexual desire].

11 32. If ditto (= a man's sexual desire has finished in the month Nisannu): w[hen . . .]
Rest destroyed

Column iii

12 1. *libbi iṣṣūr ḫurri* [*zikari ta'allut*]
2. *ḫaḫḫi alpi tebî* [*teleqqe ana mê tanaddi*]
3. *ina ūri ana kakkabī* [*tušbāt*]
4. *kīma Šamšu it*[*tapḫu ina muḫḫi arī ašāgi*]
5. *izzizu an*[*a pān Šamši* . . .]
6. *ina* [*mê išattīšu-ma nīš libbi irašši*]

13 7. *šum*[*ma* KI.MIN (= *amēlu nīš libbi ina Nisanni iqti*)] . . . *kalbi rā*[*kibi* . . .]

8. *tubbal tasâk* . . . [. . .]
9. *tuballal-ma ina mê išattīšu-*[*ma*]
10. *nīš libbi iraš*[*ši*]

14 11. *šumma* KI.MIN (= *amēlu nīš libbi ina Nisanni iqti*) *rikibti arkabi eli*[*kula*]

12. *tertenna bīni* [. . .]
13. *maštakal gišimmara bukā*[*na*]
14. *u mimma zērī ina mê tanaddī-ma*
15. *ana kakkabi tušbāt ina ūri tašak*[*kan*]
16. *amēlu šâšu tušaššab ina mê k*[*asî*]
17. *kala zumuršu tušaḫ*[*ḫaṭ* . . .]
18. *rikibti arkabi* [. . .]
19. *mišla išattī-ma miš*[*la tapaššassu*]
20. *šamna rūšta u terte*[*nna bīni*]
21. *ina šēp šumēlīšu* . . . [. . .]
22. *tasâk-ma ina šamni p*[*ūri zumuršu*]
23. *tapaššassu-ma amē*[*lu šū nīš libbi irašši*]

15 24. *šumma* KI.MIN (= *amēlu nīš libbi ina Nisanni iqti*) *muḫḫi iṣṣūr ḫurri* [*zikari* . . .]

25. *ina šamni pūri zumuršu*
26. *tapaššassu-ma amē*[*lu šū nīš libbi irašši*]

16 27. *šumma* KI.MIN (= *amēlu nīš libbi ina Nisanni iqti*) *nīš libbi eṭir* [. . .]

28. *mušar iṣṣūr ḫ*[*urri zikari*]
29. *napišta ša immeri* [*ina nabāsi*]
30. *talammī-ma ina qabli tašakka*[*n-ma*]
31. *amēlu šū nīš libbi* [*irašši*]

17 32. *šumma* KI.MIN (= *amēlu nīš libbi ina Nisanni iqti*) *pizallurāt ṣ*[*ēri ritkubāti*]

33. *ina pān* . . . [. . .]

Column iii

[If ditto (= a man's sexual desire has finished in the month Nisannu): . . .]

12 1. [you ingest] the innards of a [male] partridge,
2. [you take] the phlegm of a reared-up bull, [you put (it) in the water],
3. [you let] it [spend the night] under the stars on the roof,
4. when the sun ri[ses, on the branches of *ašāgu*-acacia]
5. he stands, in front [of the sun . . .]
6. with [the water he drinks this and will get sexual desire].

13 7. I[f ditto (= a man's sexual desire has finished in the month Nisannu)]: . . . of a dog which co[pulates . . .]
8. you dry (it), you pulverize (it), you . . . [. . .]
9. you mix and he drinks it with water [and]
10. will ge[t] sexual desire.

14 11. If ditto (= a man's sexual desire has finished in the month Nisannu): *rikibtu* of an *arkabu*-bat, *eli*[*kulla*]-plant,
12. . . . of tamarisk [. . .]
13. *maštakal*-plant, date palm, *bukā*[*nu*]-wood
14. and you put in water all seeds and
15. let them spend the night under the stars, you pu[t] (it) on the roof,
16. you let that man sit down, with *k*[*asû*]-water
17. you [wash] all his body [. . .]
18. the *rikibtu* of an *arkabu*-bat [. . .]
19. he drinks half and [you anoint$^{?}$ him with the other] ha[lf]
20. premium oil and . . . [of tamarisk . . .]
21. on his left foot . . . [. . .]
22. you pulverize and with [oil] from the a[labastron his body],
23. you anoint (it) and [that] m[an will get sexual desire].

15 24. If ditto (= a man's sexual desire has finished in the month Nisannu): The skull of a [male] partridge [. . .]
25. with oil from the alabastron his body
26. you anoint and [that] m[an will get sexual desire].

16 27. If ditto (= a man's sexual desire has finished in the month Nisannu) (and his) sexual desire is taken away [. . .]
28. the penis of a [male] partri[dge]
29. the throat of a ram [with red wool],
30. you wrap up and put (them) around his waist [and]
31. that man [will get] sexual desire].

17 32. If ditto (= a man's sexual desire has finished in the month Nisannu): A gecko of the step[pe which copulates]
33. in front of . . . [. . .],

Column iv

1. [. . . *ba*]*lu patān*
2. [*išattī-ma*] *nīš libbi irašši*

18 3. [*šumma* KI.MIN (= *amēlu nīš libbi ina Nisanni iqti*) *rupušti immeri teb*]*î qaran ayyali*
4. [. . . *nī*]*š* [*libbi*]

19 5. [*šumma* KI.MIN (= *amēlu nīš libbi ina Nisanni iqti*) . . .]
6. Destroyed
7. [. . . *ina qablīšu*] *tarakkas*
8. Destroyed

20 9. *ana amēlu nīš* [*libbi eṭir* . . .]
10. *enū*[*ma* . . .]

21 11. *šumma amēlu ša n*[*īš libbīšu eṭru*]
12. *itti* . . . [. . . *il*]e’’*i*
13. *šammi nīš libbi* [. . .] . . .
14. *nīš libb*[*i* . . .]
15. *šumma itti* . . . [. . .]
16. *ana nīš libbīšu* [*rašê* . . .]
17. $^{\acute{u}}$KASKAL.MUNUS . . . [. . .]
18. *tamaḫḫar* [. . .]
19. . . . [. . .]
20. [*nīš*] *lib*[*bi irašši*]

22 21–23. Traces

23 24. [*šumma amēlu* (*ša nīš libbīšu eṭru*) *zê ṣurāri*] *barmi šadê*

25. [. . . *ina qablīšu t*]*arakkas-ma*
26. [. . . *nī*]*š* [*libbi*]

II. A iv 27–31 = Sumerian incantation No. L.1

III. A left e. 1–7, lower e. 1–5 = Prescriptions 24–25

Column iv

1. [. . . on an] empty [sto]mach
2. [he drinks and] he will get sexual desire.

18 3. [If ditto (= a man's sexual desire has finished in the month Nisannu): saliva of a rear]ed up [ram], antler of a stag,
4. [. . . sexu]al [desire]

19 5. [If ditto (= a man's sexual desire has finished in the month Nisannu): . . .]
6. [. . .]
7. [. . . around his waist] you bind,
8. [. . .]

20 9. For a man whose sexu[al desire is taken away]
10. wh[en . . .].

21 11. If a man, se[xual desire is taken away]
12. with . . . [. . . he ca]n/cannot$^{?}$
13. pharmaceutical for sexual desire [. . .] . . .
14. the sexual desi[re . . .]
15. if with . . . [. . .]
16. in order to [get] his sexual desire [. . .]
17. KASKAL.MUNUS-plant . . . [. . .]
18. you catch [. . .]
19. . . . [. . .]
20. [he will get [sexual] desi[re].

22 21–23. Traces

23 24. [If a man('s sexual desire is taken away): Excrement of a] polychrome [lizard] of the mountain,
25. [. . . around his waist yo]u bind and
26. [. . . sexu]al [desire].

II. A iv 27–31 = Sumerian incantation No. L.1

III. A left e. 1–7, lower e. 1–5 = Prescriptions 24–25

Left edge

24 1. [*en*]*ūma iṣṣūr ḫurri irtanakkab*[*u* . . .]
2. [*iṣṣūr ḫur*]*ri zikarūti ina ayyāri ta*[*ṣabbat tabaqqan*]
3. *dāmī lā tumaššar irrīšunu lā* [. . . *l*]*ā* . . .
4. *te’’il-ma lū šina arḫū lū* [*šalaš arḫū* . . . *ṣ*]*uprī appī karšī*

5. *šer’ānī u irrī ina abni tasâk* [. . .] . . . *tarabbak-ma*
6. *šalaš qāt zēr aluzinni šina qāt* . . . [. . .] . . . *taballal ina šikari kīma kayyān*[*amma*]

7. *tamaḫḫaṣ balu patān išattīšu-ma amē*[*la šuāti libbašu i*]*našši*

Lower edge

25 1. [*šumma* KI.MIN (= *amēlu ša nīš libbīšu eṭru*) *lib*]*bi aribi zikari dām dašši tebî*

2. [*dām i*]*ṣṣūr ḫurri zikari rikibti arkabi*
3. *zēr dadāni ištēniš taballal rapašti amēli šullušû*

4. *tapaššaš-*⟨*ma*⟩ *nīš libbi irašši šipat nīš libbi*
5. *tuppu naḫru šumma amēlu nīš libbi*

Left edge

24 1. [Whe]n the partridge mates [. . .]

2. you [take . . . of] the male [partrid]ges in the Ayyāru [and pluck (it)],
3. you do not make the blood flow, you do not [. . .] their entrails, you do not . . .
4. you tie, either two months or [three months . . .] you crush [the ta]lons, the beaks, the stomachs,
5. the muscles/tendons and the intestines on the stone [. . .], you soak in . . .
6. three handfuls of seeds of *aluzinnu*-plant, two handfuls. . . [. . .] . . . you mix, in beer in the usual wa[y]
7. you dilute, he drinks it on an empty stomach, and [that] m[an will desi]re.

Lower edge

25 1. [If ditto (= a man's sexual desire is taken away): The hear]t of a male raven, the blood of a reared-up buck,

2. [the blood of a] male [partri]dge, the *rikibtu* of an *arkabu*-bat,
3. seed of the *dadānu*-acacia, you mix (them) together, the pelvis of that man for three times
4. you rub with it ⟨and⟩ he will get sexual desire. Wording of *nīš libbi* (incantation).
5. Tablet 1st of "If a man's sexual desire."

Commentary

i 1–2. See the tablet inventory BM 103690 rev. iii 5: 1 ŠÀ.ZI.GA *ina* ^itiBÁRA.ZAG.GAR "One (tablet for) 'Sexual desire in the month of Nisannu'" (Finkel 2018: 31). The inventory quotes the incipit in abbreviated form, for this reason we could read the first sign DIŠ *šumma* 'if' instead of 'one' (tablet). See however in the same text KUB 4, 48 prescr. 25 lo. e. 5: DUB 1.KAM DIŠ LÚ ŠÀ.ZI.GA.

i 10. *ša ú-ri bu-ut-tu*$_4$: Krecher (1970: 355) suggests emending *ša* (*ina*) *ūri* (*šu*) *buttu* "das die Nacht über auf dem Dach gestanden hatte." Note however that in Ms. B we also find *b*[*u-ut-tu*$_4$], thus stative D-stem from *biātu*.

i 14. "You ingest": While at first it appears that the exorcist ingests the heart, this is a mistake for 3rd person.

i 20. In Ms. A *ina* A^meš NAG-*šú-ma* should probably be omitted since it does not make sense here and it is absent in Ms. B.

i 22. In Ms. A the signs BÁR.KA could be a mistake for MAŠ.SÌL (BAR.KA$_4$, Akk. *naglabu*), since, as suggested by Biggs (1967: 59), perhaps the scribes often wrote logograms as they pronounced them. Note that in the same tablet KUB 4, 48 in *nīš libbi* K prescr. 31: 138 (Ms. M ii 7) (= *nīš libbi* N prescr. 8 ii 7) we find ^úMUL.DÙ.DÙ instead of ^úMUR.DÙ.DÙ. Biggs (1967: 59) suggests to emend on the basis of the duplicate with MÚRU-*ka* (see in the same text prescr. 8: 15. *ina* ⌜MÚRU-*ka*⌝), but it seems to me improbable.

i 24. On the verb *zarāku* (or *zarāqu*) (also in line i 28), Biggs (1967: 59) argues that it should be kept separate from *sarāqu* 'to scatter' and *zarāqu* 'to sprinkle (liquids)' since both verbs require the object to be expressed. The verb *zarāku* (or *zarāqu*) obviously describes a part of the treatment of the bird before it is salted and dried and probably means 'to eviscerate' or 'to dress (a fowl).'

i 28–32. This prescription in Ms. B is too fragmentary: 14′–15′. KI.MIN [. . .] *ta* [. . .] / *ina* A^me[š . . .].

ii 1–2. In Ms. B another fragmentary prescription follows in lines 17′–18′: KI.MIN *ṣe-e-e*[*t* . . .] / *ina* IZI x x [. . .]. Ms. B's rest is destroyed.

ii 3–9. = *nīš libbi* K prescr. 31: 135–140.

ii 11–14. *kupatinnu*-pellets (of clay) appear also in P prescr. 11 rev. 7.

ii 14–16. Ms. C has in line 1: [. . .] *ni* [. . .]. Pinches (1910) reads the sign *ta*, Biggs (1967) *bi*.

ii 15. "Around your waist" is probably a mistake. It should refer to the patient's waist.

ii 17–26. This prescription is too fragmentary, it may be a duplicate of *nīš libbi* W prescr. 2.

iii 1–6. Restorations from F prescr. 12: 26–41; E prescr. 6: 73–78.

iii 5. The parallel F prescr. 12: 41 has GUB-*su-ma*. It is not possible to restore *a-n*[*a* IGI ^{d}UTU NAG-*ma*] because the following line 6 starts with *ina*.

iii 9. *tu-bal-lal-ma*: According to von Soden (1968: 457), *tu-bal-lal-ma* must be emended to *ta-bal-lal-ma* since the D-stem is not used to express the logogram ḪE.ḪE (*contra* CAD B 42–43), which should be read instead *taballal*. However, see Abusch and Schwemer 2011: 29, No. 1.1: 16′ where the traces of KUB 37, 44(+) obv. i 16′ suggest *t*[*u-bal-lal*] (see Abusch and Schwemer 2016: 442). Biggs (1967: 55) does not emend *tu-bal-lal-ma* (see also CAD B 42 mng. 3b.1′). In this volume the logogram ḪE.ḪE is read with the G-stem *taballal*. In this passage instead I prefer to keep *tuballal* because I think that the choice to write the verb syllabically is meant to make the D-stem unequivocally understandable. It is plausible, in fact, that both *taballal* and *tuballal* were used in the context of mixing ingredients in the technical language of the prescriptions, perhaps with nuances of meaning that we are unable to understand.

iii 11. Possible restoration [*la*], [*na*/*i*], or [*kul-la*] (see BM 44204). Ebeling (1925) restores *e-li*-[*ku-nu*].

iii 12. It is difficult to understand the meaning of the word *te-er-te-en-na* in this context. This is probably a variant of the term *tardennu* (*terdennu*) (see CAD T 225–228; AHw. III 1329). The adjective means "second" (of social class, size, age) or "second quality." and designates human beings (also as noun: "successor" or a "type of official"), animals, or objects (such as textiles). It may also refer to the second course of a meal. In our text the word appears two times (iii 12, 20) and is associated with tamarisk (^{giš}ŠINIG). Ebeling (1925) translates "Sproß" ("sprout").

iii 27–31. *nīš libbi* F prescr. 16: 42–43 presents the same prescription but giving more details.

iii 33–iv 2. Maybe at the beginning of column iv the prescription of the end of column iii continues.

iv 13. See for other mentions of the expression Ú ŠÀ.ZI.GA "drug for the sexual desire": K prescr. 6: 35; K prescr. 27: 75: 10 Ú$^{!meš}$ ŠÀ.ZI$^{!}$.GA$^{!}$ "10 drugs for the sexual desire"; No. K.3: 92; A prescr. 17: 30 7 Ú^{meš} ŠÀ.ZI.GA *ina* KAŠ NAG "he drinks in beer the seven drugs (for) the sexual desire." See also in the *Aššur Medical Catalogue*, in the section XIX devoted to the loss of sexual desire: [. . .] x SAG MUŠEN DIŠ ⸢Ú⸣ ⸢ŠÀ⸣.ZI.G[A] ⸢*ana*⸣ [GÚ-š]*ú* GAR "[. . .] . . . the head of a bird. (Instructions) to place a drug for sexual desir[e] around his [neck]" (l. 100, Steinert 2018b: 217). See for specific plants for sexual desire's problem BAM 380 r. 42–44 (dupl. BAM 381 iii 37–40) (see Chapter III § "Plant ingredients").

iv 17. Biggs (1967) restores *ina* ^{dug}BUR.ZI, but it is uncertain. The term appears only in No. E.2: 23.

iv 24. The restoration is uncertain, cf. in the same text *nīš libbi* N prescr. 8 ii 5 (= K prescr. 31: 137–138. Here the polychrome lizard seems to come from the mountain.

Left e.

1. "When the partridge mates" (*enūma iṣṣūr ḫurri irtanakkabu*) is a Gtn-stem in an iterative sense. Here I refer to the coupling period. *Iṣṣūr ḫurri* should be understood as a term to indicate the species. Similarly, Ebeling translates (1925): "Sobald sich Höhlenvögel begotten." Biggs (1967) translates in singular: "[Wh]en a partridge(?) is copulating."

3. *irrīšunu*: the possessive suffix (3 m. pl.) indicates that the entrails used in the ritual belong to more animals. Ebeling (1925) and Biggs (1967) emphasize the strangeness of this plural possessive. I do not consider the plural possessive inconsistent in the text for the following reasons: 1. It is possible to integrate the gap at the beginning of line 2 with MUŠEN^{meš} and therefore read the logogram NITÁ as *zikarūti*; 2. This would also explain the plural ÚŠ^{meš}; 3. The terms *ṣuprī*, *appī*, *karšī*, *šer'ānī*, and *irrī* are all plural and belong to more animals.

4. 3 ITI : Biggs's suggestion.

5. *ina* NA_4.NA_4/*abnī*: the doubling of the Sumerogram is not clear to me. Ebeling (1925) translates "mit Steinen"; Biggs (1967) "in a mortar"; CAD A/I 187 "with a pestle."

6. Perhaps the *aluzinnu*-plant, attested only in this text, is related to *alluzzu*-plant (see Thompson 1949: 185–186).

7. For the meaning 'to dilute' of the verb *maḫāṣu* see CAD M/I 78–70 n. 3-e.

Lower e.

2. ⟨*ri*⟩-*kib*(text KAL)-*ti*: Otherwise, it is possible to read KALAG-*ti*/*dannati* 'strong' in reference to the male partridge (*iṣṣūr ḫurri*). Note, however, that it is unlikely that, on the contrary of the other texts, no part of *arkabu*-bat is mentioned as an ingredient.

5. The reading should be DUB 1.KAM DIŠ LÚ ŠÀ.ZI.GA (*contra* Biggs 1967: DUB 2.KAM), according to tablet inventory BM 103690 rev. iii 5 1 ŠÀ.ZI.GA *ina* ^{iti}BARA.ZAG.GAR "One (tablet for) 'Sexual desire in the month of Nisannu'" (Finkel 2018: 31), where the first sign DIŠ could be read *šumma* 'if' instead of 'one' (tablet) (see the same text KUB 4, 48 prescr. 1 i 1–2).

Nīš libbi O

List of manuscripts

Manuscript	Museum number	Publication	Tablet	Script	Date	Provenience
A	Bo 61/r	KBo. 36, 27 1–26	A single-col. tablet	'Not Hittite'[577]	13th cen.	Ḫattuša Geb. E

Edition

Schwemer 2004: 59–79
Wasserman 2016: 239–241, No. 27

Structure of the text

I. Prescriptions
   1. 1
   2. 2
   3. 3
   4. 4–6
   5. 7–8
   6. 9–13
   7. 14

II. Incantation and its ritual No. O.1 "I have sex with you, Nanāya!" ll. 15–26

Transliteration

I. A 1–14 = Prescriptions 1–7

1. [. . .] x x [. . .]
2. [. . .] x x x x ⸢KI GA⸣ [. . .]
3. [^{ú}]EME.⸢UR.GI$_7$ KI GA⸣ [. . .]
4. [NUMUN] ⸢^{ú}⸣EME.UR.GI$_7$ NUMUN ^{ú}⸢*la*⸣-*ap-t*[*i*]
5. [NUMUN] ^{ú}*sa-mì-di* NUMUN ^{ú}*a-nu-zi-ni*

577 'Non-Hittite' ductus: North Syria (scribe whose mother tongue is neither Akkadian nor Hittite).

6. [KI G]EŠTIN *u šum-ma* KAŠ

7. ⸢KIŠI$_{16}$⸣.ḪAB ^{ú}GAMUN(⟨DIN⟩.⸢TIR⸣) *ka-bu-ul-ta*$^{?}$

8. ⸢NUMUN⸣ ^{ú}SULLIM$^{?}$ KI GEŠTIN *u šum-ma* KAŠ

9. KUN GÍR.TAB KUN *a-du-um-mì*

10. KUN NIM.LÀL SAG$^{?}$.DU *kúl-bi-bi* SA$_{5}$

11. *ina* SÍG(text ZI) $^{⸢udu⸣(text\ 3)}$*pu-ḫa-li ta-*⸢*la*⸣*-pa-a*[*s-su*]

12. *ina* DUR BABBAR 7 *ki-iṣ-ri ta-*⸢*kaṣ*⸣-[*ṣar*]

13. *ina* MÚRU$^{?}$-*šú ta-ra-*⸢*kas*$_{4}$⸣-[*su*]

14. ⸢Ì⸣ *a-sí* MÚRU-*šú* ŠÉŠ$^{?}$.MEŠ

II. A 15–16 = Incantation and its ritual No. O.1

15. ⸢TU$_{6}$⸣$^{?}$ ÉN É.NU.RU *ag-ra-aḫ-ki na-na-a*

16. *ag-ra-aḫ-ki na-na-a* ⸢*ki*⸣*-ma im-me-ri*

17. *a-la-*⸢*la*⸣*-ma* ⸢*ki*⸣*-ma a-ri-ti ta-nu-qa-tum*(text ⸢AM⸣)-*ma*

18. *ak*$^{!}$*-kum bá-*⸢*ki*⸣*-ti-ia pí-ta-a-*⸢*ki*⸣

19. *ak*$^{!}$*-kum bá-*⸢*ki*⸣*-ti-ia ma-ia-a-al-ki*

20. *a-ra-am* ⸢*ki-li-li*⸣ *a-na-ak ki-li-li*

21. gìš te gìš ⸢te da⸣ ga an ni

22. da zi da an ni {. . .}

23. gá an ga da ⸢ga⸣ an ni ⸢maḫ⸣ ḫa

MAN MAN

24. 1 ⸢GÍN$^{?}$(text ZU) Ú⸣.KUR.R[A$^{?}$ $^{ú?}$. . .] x-*tu* ^{ú}*kur-ka-na-a*

25. ⸢^{ú}*tar-muš*⸣ [. . .] x-*ra-na* TÉŠ.BI *ta* (text MAR/IŠ)-*s*[*àk*]

26. ⸢*ù*⸣ [*ina* A^{meš} ŠUB-*ma*$^{?}$ *ina* M]UL ⸢*tuš*⸣*-bat*

27-rev 56. Fragmentary

Transcription

I. A 1–14 = Prescriptions 1–7

1 1. [. . .] . . . [. . .]

2 2. [. . .] . . . *itti šizbi* [. . .]

3 3. *lišān-kalbi itti šizbi* [. . .]

4 4. [*zēr*] *lišān-kalbi zēr lapt*[*i*]
5. [*zēr*] *samīdi zēr anuzinni*
6. [*itti k*]*arāni u šumma šikari*

5 7. *dadāna kamūna kabulta*
8. *zēr šambalilti itti karāni u šumma šikari*

6 9. *zibbat zuqaqīpi zibbat adummi*
10. *zibbat nūbti qaqqad kulbībi sāmi*
11. *ina šīpāt puḫāli talappa*[*ssu*]
12. *ina ṭurri peṣî sebet kiṣrī takaṣ*[*ṣar*]
13. *ina qablīšu tarakkas*[*su*]

7 14. *šaman asi qabalšu taptanaššaš*

II. A 15–16 = Incantation and its ritual No. 0.1

15. *tê šipti* É.NU.RU: *agraḫki Nanāya*
16. *agraḫki Nanāya kīma immeri*
17. *alāla-ma kīma arīti tanūqātum-ma*
18. *akkūm bakkitīyia pītāki*
19. *akkūm bakkitīyia mayyālki*
20. *arâm Kilili anâk Kilili*
21–23. Abracadabra

---

24. 1 *šiqil nin*[*â . . .*] . . . *kurkanâ*
25. *tarmuš* [. . .] . . . *ištēniš tas*[*âk*]
26. *u* [*ina mê tanaddī-ma ina k*]*abbabī tušbāt*
27-rev 56. Fragmentary

Translation

I. A 1–14 = Prescriptions 1–7

1 1. [. . .] . . . [. . .]

2 2. [. . .] . . . with milk [. . .].

3 3. "dog's-tongue"-plant with milk [. . .].

4 4. [Seeds] of "dog's-tongue"-plant, seeds of *lapt*[*u*]-plant,
5. [seeds] of *samīdu*-plant, seeds of *anuzinnu*-plant
6. [in w]ine or in beer.

5 7. *dadānu*-acacia *kamūnu*-plant, *kabultu*-plant
8. seeds of fenugreek, in wine or in beer.

6 9. Tail of a scorpion, stinger of a wasp,
10. stinger a bee, head of a red ant,
11. you apply it (= group of ingredients) in wool of a ram,
12. you ma[ke] seven knots in a white string,
13. you tie [it] around his waist.

7 14. You anoint his waist repeatedly with myrtle oil.

II. A 15–16 = Incantation and its ritual No. O.1

15. Enuru incantation formula: I have sex with you, oh Nanāya!
16. I have sex with you, oh Nanāya! Like (that) of a ram
17. it is a joyful song and like (that) of a pregnant (one) is a battle cry!
18. Instead of my 'wailing woman' – your two openings!
19. Instead of my 'wailing woman' – your bed!
20. I love (you), oh Kilili, I make love with (you), oh Kilili
21–23. Abracadabra

---

24. 1 shekel of *nīn*[*û*-plant, . . .] . . . *kurkanû*-plant,
25. *tarmuš*-plant, [. . .] . . . you cru[sh] together
26. and [put (it) in water, under the s]tars you let it spend the night.
27–rev 56. Fragmentary

Commentary

It is difficult to state with certainty if this incantation belongs to the corpus of therapies for the recovery of sexual desire, for Wasserman (2016, No. 27) and Schwemer (2004) it is a love incantation. The goddess Nanāya certainly plays a central role in the *nīš libbi* corpus, while the name Kilili never appears. It must be noted that in this incantation the patient addresses the woman, with whom sexual intercourse is invoked, as if she were the goddess Nanāya or Kilili. It is also possible, as suggested by Wasserman (2016: 240), that "Nanāya!" (ll. 15–16) and "Kilili!" (l. 20) are not direct object but invocations or exclamation of excitement.

Regarding the ingredients, although some of them are present in other prescriptions (such as the "dog's-tongue"-plant, ll. 3–4), others are exceptions (such as milk, sting of wasp, and head of red ant, ll. 2–3, 9–10). In any case, there is no doubt that the text serves to charge the man's sexual desire.

2. Is it possible to restore with a form of the verb *šatû*(NAG) 'to drink.' Usually, it has the preposition *ina*, but cases with *itti* exist as well (see Schwemer 2004: 65).

5. *anuzinnu*-plant is a byform for *amuzinnu* (see Schwemer 2004: 65).

7. TIR is a defective writing of logogram ^{ú}GAMUN(DIN.TIR) for *kamūnu* 'cumin' (see Schwemer 2004: 65). It could also be a qualification of the previous plant, like "of the forest," and therefore to be read *qištu*. Another possibility is that it is a defective writing for $NENNI_5$(|TIR.TIR|) to be read *ašlu*, a medical ingredient (see CAD A/II 449), which, however, as *kamūnu*, does not appear in the corpus. It remains unclear.
kabultu: it could be a byform for the *kabullu*-plant or a form of the word *kabaltu*, a secondary product of flour production (see Stol 1987–1990: 328; Schwemer 2004: 65). The first possibility is more probable because the *kabullu*-plant appears several times in this corpus.

9. *adummu* may be a variant for *adammūmu* 'wasp.' Here *zibbatu* indicates both the abdomen and the stinger of the wasp. Note that K prescr. 32: 166 we find as ingredient the stinger (*ziqtu*) of a bee (NIM.LÀL/*nūbtu*). In general, these are not specific ingredients of the corpus.

11. TA ZI : It should be emended *ina* SÍG/*šīpāt* "in wool (of a ram)." A verbal form *ta-ṣí* from verb *naṣû* 'to scratch' or *eṣû* 'to cut' is to be excluded in this context (see Schwemer 2004: 65–66, also fn. 12).

14. *agraḫki* is a form of the rare verb *garāḫu*, synonym to *niāku* 'to have (illicit) intercourse' (see Schwemer 2004: 66; Wasserman 2016: 240).

17. *alālu* 'work song, work cry' is mentioned in love context also in *The Moussaieff Love Song* obv. 9 (Wasserman 2016: 133) and in KAR 158 viii 20′ (Wasserman 2016: 213).
tanūqātu 'a battle cry' (CAD T 176 mng. b) is derived from the verb *nâqu* A 'to cry, to groan' (CAD N/1, 341, A mng. a), which as suggested by Wasserman (2016: 240) could testify to a semantic play with the quasi-homonymous *niāku* 'to have (illicit) intercourse.' In *nīš libbi* corpus the sexual sphere is often expressed using battle metaphors. This passage testifies to this metaphorical correlation between the two domains. In fact, the terms metaphorically refer to the love moans (see Schwemer 2004: 66). In the second millennium love literature the term *ernettu* 'cry of joy after military triumph' refers to the shout of excitement at the moment of the orgasm (see Wasserman 2016: 38).

18–19. Schwemer (2004) reads ⸢IM⸣-KUM PA-⸢KI⸣-*ti-ia* BI[!?]-*ta-a*-⸢*ki*⸣ / IM-KUM PA-⸢KI⸣-*ti-ia ma-ia-a-al-ki*. I follow Wasserman's reading (2016: 240): the first sign in this couplet is not IM, but probably a malformed AK (*akkū*(*m*) < *an*(*a*) *kûm* 'instead of').
pītāki: "Your two openings," perahaps they metaphorically indicate the two erogenous zones of the woman (see that "gate" refers to female genitalia, see Wasserman 2016: 240). See the verb *petû* in reference to the vulva CAD P 346, mng. 1d4′. For other possible readings see Schwemer 2004: 67.
According to Wasserman (2016: 240) *bakkītu* 'wailing woman, mourner' could be a derogatory designation of the man's wife, as opposed to the man's lover. However, if we consider the text an incantation for the recovery of male sexual desire, it is possible that the two lines metaphorically describe the passage from the patient's suffering condition (and consequently that of the female lover) to a condition of a restored well-being and sexual activity, expressed by the "two openings" and the bed. In other words, the incantation, recited by the patient, refers to the healing by emphasizing the passage from the woman's lament (for the absence of desire) to the sexual intercourse. Moreover, we underlined above how the absence of male sexual desire involves the woman as well, who is an important actor in the therapeutic process.

20. Since the verb *niāku* requires an object, which Kilili cannot be, Schwemer (2004, 68) suggests that a haplography occurred: the scribe had in mind *arâmki* and *anâkki* but does not write the pronoun *-ki* to avoid the clash with the initial syllable in Kilili. The direct object of the verbs may be the female lover, whose "two openings" and "bed" are mentioned. For Kilili see Lambert 1976–1980: 591; Schwemer 2004: 72–75.

21–23. It is a (Peudo-)Sumerian abracadabra with some elements with meaning: the sign UŠ can be interpreted as gìš 'penis' or nita 'man'; te as *ṭeḫe* 'to approach (sexually)'; da-ga-an-ni as a phonetic spelling of da-gán 'her sleeping chamber, dwelling'; da-zi-da-an-ni "his right side"; gá-an-ga infinitive of the verb ĝá-ĝá 'to put' or phonetic spelling for ga-na "come on"; maḫ-ḫa infinitive form from maḫ+a 'to be superior, exalted.' For the interpretations of the abracadabra see Schwemer 2004: 68; Wasserman 2016: 240. Wasserman translates: "Penis! Approach! Penis! Approach her bedroom! / (Approach) her right side! / Come on! (Approach) her exalted bedroom!".

24. Ú.KUR.RA: Schwemer (2004) reads *samīmu*.

VI Ritual fragments from Nineveh and Aššur

Nīš libbi P

List of manuscripts

Manuscript	Museum number	Publication	Tablet	Script	Date	Provenience
A	BM 098571 + = K 3350 + 1905–04–09, 77	AMT 62, 3 obv. 1–rev. 17	Frg. of a single-col. tablet	NA	8th–7th cent.	Nineveh

Edition

Thompson 1934: 91–93, 148–149 No. 181
Biggs 1967: 51

Structure of the text

I. Enuru incantation end (obv. 1–2) and prescriptions
   1. obv. 3–4
   2. 5–7
   3. 8–10
   4. 11–15
   5. 16–17
   6. 18–21
   7. 22–23
   8. rev. 1–2
   9. 3–4
   10. 5
   11. 6–10
   12. 11–13
   13. 14–17

https://doi.org/10.1515/9783110757262-011

Transliteration

I. A obv. 1–rev. 17 = Enuru incantation end and prescriptions 1–13

obverse

1. traces
2. [TU$_6$ ÉN] É.NU.RU
3. [. . .]
4. [. . . *k*]*i*$^?$ *te el ti im m*[*a* . . .]
5. [DÙ.DÙ.B]I NUMUN ^{giš}KIŠI$_{16}$.ḪAB [. . .]
6. ⸢*i-na*⸣ *qa-an-ni-ka tàra-kás* [. . .]
7. MUNUS *ši-i* [. . .]
8. [NUMUN] ^{giš}KIŠI$_{16}$.ḪAB ^{ú}*sà-as-sà-ta* ^{ú}⸢A⸣[.ZAL.LÁ . . .]
9. [. . .] x ^{ú}ḪUR.SAG^{sar} [. . .]
10. [Ú.ḪI.A] *an-nu-tì tu-ḫal* : *ta-sàk ana* ŠÀ [A^{meš} ŠUB$^?$]
11. [*a-na*] ŠÀ NITA *ù* MUNUS *šu-up-šu-ri* [. . .]
12. [NUMUN$^?$] ^{ú}SIKIL *ù* Ì : ^{giš}MA.NU SIG$_7$-⟨*su*⟩ TI-[*qé* . . .]
13. [*t*]*a-qa-at-ta-ap ma-la-ma-*⸢*liš*⸣$^?$ [. . .]
14. [T]I-*qé-ma ta-šà-ak* [. . .]
15. [NI]TA *ù* MUNUS NAG-*šú-nu-ti-ma* [. . .]
16. DIŠ KI.MIN *i-na bi-rit* NITA *u* MUNUS [. . .]
17. [Š]À-*šú-nu* ZI *i-na* UGU GAR-*an an-ni-*[*ta* . . .]
18. DIŠ KI.MIN *a-na lìb-bi* NITA ZI-*bi* ⟨*u*⟩ *a-na*(text BE) MUN[US GIN-*šu*]
19. ^{uzu}*nap-šá-at* UDU.NITÁ *te-bi-i-im* [*ina* ^{síg}ḪÉ.ME.DA]
20. NIGIN-*ma i-na* MÚRU-*šú tàr-kas*$_4$-*ma* [. . .]
21. *ina ši-bu-ra-ti* [. . .]
22. UGU *e-ri-bi* ÚŠ MUŠEN *ḫur*$^?$-*ri* [NITÁ . . .]
23. U$_5$ ARKAB^{mušen} *ta-sàk* x [. . .]

reverse

1. *an-nu-tu*$_4$ [. . .]
2. [*š*]*a*$^?$ *ana ku* [. . .]
3. ŠÀ MUŠEN *ḫu-ur-ri* NITÁ
4. *ina* MUL *tuš-bat* ÉN 3-*šú ana* ŠÀ Š[ID-*nu* . . .]
5. DIŠ KI.MIN *ḫi-in-du-ur ba-al-lu-ṣí-t*[*i* . . .]
6. DIŠ KI.MIN *šu um du šak-ka-di-ir-ru* [. . .]
7. *ka-pa-ti-in-ni ša* IM x [. . .]
8. ÉN *an-ni-ta* 3-*šú ana* UGU ŠID(text ŠUB)-*ma ana* [. . .]
9. x x x *im-ma i*[*p* . . .]
10. ŠÀ.ZI.GA KUŠ-*šú* UZU^{meš}-*šú ù* SA^{meš}-[*šú* . . .]
11. *an-nu-tì* KÌD.KÌD.BI-*šu-nu-tì ša* x [. . .]
12. *na tur qu* SIG$_5$^{meš} *an-nu-tì ar* [. . .]

13. *ul-ta*(text TE)-⸢*ti*⸣?-*ik-šu-*[*nu-ti* . . .]

---

14. [. . .] LÁ-*ṭu ana/*DIŠ NA GIG [. . .]
15. [. . .] x SAG SÚD IGI GIG [. . .]
16. [. . .] x *an su ma* x [. . .]
17. [. . .] KÁR [. . .]

Rest destroyed

Transcription

I. A obv. 1–rev. 17 = Enuru incantation end and prescriptions 1–13

obverse

1. Traces

2. [*tê šipti*] É.NU.RU

---

1 3. [. . .]
4. [. . .] . . . [. . .]

---

2 5. [*dudub*]*û*: *zēr dadāni* [. . .]
6. *ina qannīka tarakkas* [. . .]
7. *sinništu* . . . [. . .]

---

3 8. [*zēr*] *dadāni sassata a*[*zallâ* . . .]
9. [. . .] . . . *azupīra* [. . .]
10. [*šammī*] *annūti tuḫāl* : *tasâk ana libbi* [*mê tanaddi*]

---

4 11. [*ana*] *libbi zikari u sinništi šupšuri* [. . .]
12. [*zēr*] *sikilla u šamna* : *ēra arqūssu tele*[*qqe* . . .]
13. [*t*]*aqattap malamališ* [. . .]
14. [*teleq*]*qē-ma tašâk* [. . .]
15. [*zika*]*ru u sinništu išattûšunūti-ma* [. . .]

---

5 16. *šumma* KI.MIN *ina birīt zikari u sinništi* [. . .]
17. [*libba*]*šunu itebbû ina muḫḫi tašakkan annī*[*ta* . . .]

---

6 18. *šumma* KI.MIN *ana libbi zikari tebî u ana sinniš*[*ti alākīšu*]

19. *napišta ša immeri tebîm* [*ina nabāsi*]
20. *tulammī-ma ina qablīšu tarakkas-ma* [. . .]
21. *ina šibburrati* [. . .]

---

7 22. *muḫḫi aribi dām iṣṣūr ḫurri* [*zikari* . . .]
23. *rikib arkabi tasâk* . . . [. . .]

Reverse

8 1. *annûtu* [. . .]
2. [*š*]*a anāku* [. . .]

---

9 3. *libbi iṣṣūr ḫurri zikari*
4. *ina kakkabī tušbāt šipta šalāšīšu ana libbi tam*[*annu* . . .]

---

10 5. *šumma* KI.MIN *ḫindūr ballūṣīt*[*i* . . .]

---

11 6. *šumma* KI.MIN . . . *šakkadirru* [. . .] *šakkadirru* [. . .]
7. *kappatinnī ša ṭīdi* . . . [. . .]
8. *šipta annīta šalāšīšu ana muḫḫi tamannū-ma ana* [. . .]
9. . . . [. . .]

Translation

I. A obv. 1–rev. 17 = Enuru incantation end and prescriptions 1–13

obverse
1. Traces
2. Enuru [incantation formula]

1 3–4. Traces

2 5. [Its ritua]l: . . . seeds of *dadānu*-acacia [. . .],
6. around your edge (of the dress)? you tie (them) [. . .]
7. the woman . . . [. . .].

3 8. [Seeds] of *dadānu*-acacia, grass, *a*[*zallû*]-plant,
9. [. . .] . . . *azupīru*-plant [. . .]
10. you melt these [plants], you pulverize (and) [put] it in [water].

4 11. [In order to] release the 'heart' of the man and of the woman: [. . .]
12. [seeds] of *sikillu*-plant and oil : you ta[ke] the green of *ēru*-tree [. . .],
13. you [g]ather, equally [. . .]
14. [you ta]ke and you pulverize [. . .],
15. [the ma]n and the woman drink them and [. . .].

5 16. If ditto: Between the man and the woman [. . .]
17. they [desi]re, you put over it, thes[e . . .].

6 18. If ditto: For the 'rising' of 'heart' of the man and for [his going to] a woma[n]:
19. The throat of reared-up ram [with red wool]
20. you encircle and enclose his waist and [. . .]
21. with the rue [. . .].

7 22. Skull of a raven, blood of a [male] partridge [. . .]
23. guano? of an *arkabu*-bat you pulverize . . . [. . .].

Reverse

8 1. these [. . .]
2. [. . .] . . . [. . .].

9 3. The innards of a male partridge
4. you place it under the stars overnight, you re[cite] the incantation three times over it [. . .].

10 5. If ditto: Spur of the *ballūṣīt*[*u*]-bird [. . .].

11 6. If ditto: . . . *šakkadirru*-lizard [. . .]
7. *kupatinnu*-pellets of clay . . . [. . .],
8. you re[cite] the incantation three times over it and over [. . .]
9. . . . [. . .]

10. *nīš libbi mašakšu šīrīšu u šer'ānī[šu . . .]*

12 11. *annûti kikiṭṭîšunū* . . . [. . .]
12. . . . *damqūti annûti* . . . [. . .]
13. *ultattikšu[nūti* . . .]

13 14. [. . .] *imaṭṭi ana amēli marṣi* [. . .]
15. [. . .] . . . *rēšu zâku īnu marṣatu* [. . .]
16. [. . .] . . . [. . .]
17. [. . .] *riksu* [. . .]
Rest destroyed

10. sexual desire, his skin, his limbs and [his] tendons [. . .].

12 11. These ritual procedures . . . [. . .]
12. . . . these good (results) . . . [. . .]
13. he has tested th[em . . .].

13 14. [. . .] it decreases, towards the sick man [. . .]
15. [. . .] . . . [. . .]
16. [. . .] . . . free head, sick eye [. . .]
17. [. . .] binding [. . .]
Rest destroyed

Commentary

Obv.

6. Thompson reads: [. . .] *ka-an-ni-ka . . . bi*. However, I read, following Biggs, *i-na*, at the beginning of the line, although it is broken. Biggs reads the last sign KÁS, but in Thompson's copy it is BI.

8. For the restoration [NUMUN] see l. 5.

17. Biggs (1967) restores DÙ-*uš*, but the restoration is uncertain.

18. *a-na*(text BE) MUN[US GIN-*šu*]: Biggs (1967) reads the sign BE in the copy as NU written over erased ZI and argues that the next sign is the beginning of *ḫu*. The latter could be also MUNUS.

19. For the restoration see F prescr. 13: 42, see also N prescr. 16: 29.

22. The head (*qaqqadu*) or the skull (*muḫḫu*) of the raven (*ēribu*) (see CAD A/II 265–267, *s.v. āribu*; AHw. I 68; Salonen 1973: 124–31) is a widely used ingredient in many medical texts. It is often pulverized along with other ingredients (AMT 5, 1: 14; AMT 69: 4; BAM 237 iv 34; see CAD A mng. 1c; AHw. I 68). The reason for the presence of the raven is unclear. In omens the raven has a negative value since it can damage the harvest of the fields (see De Zorzi 2009: 118–119 and previous bibliography). In *Šumma ibzu*, the raven indicates plague and greed (see De Zorzi 2014: 166, in particular 12: 11; 18: 79′). It cannot be excluded that it is a 'secret name' for a plant, in this regard see URU.AN.NA: *maštakal* : *ṣaṣumtu*-plant : 'secret name' head of a black raven" (III 103, Rumor 2017: 19) (note that another 'secret name' of the *ṣaṣumtu*-plant is "wool of a virgin ewe," ibid. III 123).

23. U_5 ARKAB^{mušen} has been read *rikibti arkabi* by the previous literature. As Chalendar (2018) pointed out, there is no correlation between U_5 and *rikibtu* in the lexical lists. On the contrary U_5 can be read *rikbu*. According to her, *rikib arakabi* could refer to bat guano.

Rev.

5. See *nīš libbi* L prescr. 5: 14.
7. Biggs (1967) restores *te-pu-uš*.
   kupatinnu-pellets appear also in N prescr. 8: 11–12, 14.
8. *ana* could be the beginning of another sign.
9. Biggs (1967) reads the first sign *li*.
12. *na tur qu* : Thompson (1934) *na-at*$^{?}$*-qu*. Biggs (1967) Ú *tur-qu*.

Nīš libbi Q

List of Manuscripts

Manuscript	Museum number	Publication	Tablet	Script	Date	Provenience
A	BM – Sm 1514	AMT 66, 1: 1–5, 11	Small frg. of a single-col. tablet	NA	8th–7th cent.	Nineveh

Edition

Thompson 1934: 83–84, 132–133, No. 173
Biggs 1967: 52

Structure of the text

I. Prescriptions
   1. ll. 1–5
      2–3 6–8, 9–10 = A prescr. 22–23: 63–64, 65–66
   4. 11

Transliteration

I. A 1–11 = Prescriptions 1–4

1. DIŠ NA *ana* MUNUS-*šú* DU-*ma a*-[*na* MUNUS-*šú* ŠÀ-*šú* ÍL-*šú-ma*]
2. *ana* MUNUS BAR-*ti* DU-*ma a*-[*na* MUNUS-*šú* ŠÀ-*šú* ÍL-*šú*]
3. ^{ú}*ka-bul-lu* ^{ú}E[ME.UR.GI$_{7}$ . . .]
4. KI KAŠ ḪE.ḪE-*ma* [NAG-*šú*]
5. EGIR-*šú* GEŠTIN *dan-nu* [NAG-*ma* SILIM-*im*]

6–10 = A prescr. 22–23: 63–64, 65–66.

11. ⸢^{ú}⸣IGI-*lim* ^{ú}x [. . .]

Rest destroyed

Transcription

I. A 1–11 = Prescriptions 1–4

1	1. *šumma amēlu ana sinništīšu illak-ma a[na sinništīšu libbašu lā inaššīšu]* 2. *ana sinništi aḫīti illak-ma a[na sinništi aḫīti libbašu lā inaššīšu]* 3. *kabulla li[šān-kalbi . . .]* 4. *itti šikari taballal-ma [išattīšu]* 5. *arkīšu karāna dannu [išattī-ma išallim]*
2	6–8 = A prescr. 22: 63–64
3	9–10 = A prescr. 23: 65–66
4	11. *imḫur-līm . . . [. . .]* Rest destroyed

Translation

I. A 1–11 = Prescriptions 1–4

1 1. If a man goes to his woman, and [desires his woman, but]
2. he goes to another woman, but does not desire another woman.
3. *kabullu*-plant, ["dog's]-to[ngue"]-plant,
4. you mix (them) with beer and [he drinks it],
5. after that [he drinks] strong wine [and will be healthy].

2 6–8 = A prescr. 22: 63–64

3 9–10 = A prescr. 23: 65–66

4 11. "Heals-a-thousand"-plant, [. . .]-plant [. . .]
Rest destroyed

Commentary

1–2. The restoration is suggested by Steinert 2018b: 264. See also *nīš libbi* J.2: 23: [DUB X.KAM (. . .) DIŠ NA *ana* MUNUS]-*šú* ŠÀ-*šú* ÍL-*šú-ma ana* MUNUS BAR-*ti* ŠÀ-*šú* N[U ÍL-*šú*] "[Xth tablet of (. . .) 'If a man] desires his [woman], but do[es not desire] another woman.'" See the incipit of Tablet XXXIV of *Diagnostic Handbook* SA.GIG/*Sakikkû* in the catalogue l. 41, which can be restored from the fragmentary catchline of Tablet XXXIII: DIŠ ⌜NA⌝ *ana* ⌜MUNUS⌝-(*šú*) ŠÀ-*šú* ⌜ÍL⌝-*šú-ma* [*ana* MUNUS BAR?]-*ti* ŠÀ-*šú* NU ÍL-*šú* MUNUS BI ŠÀ-[*šú* . . .] "If a man desires (his/a) woman, but does not desire [another?] woman: this woman [. . .] his 'heart'/desire? [. . .]" (see Schmidtchen 2018: 141).

Nīš libbi R

List of Manuscripts

Manuscript	Museum number	Publication	Tablet	Script	Date	Provenience
A	K. 9036	TCS 2, pl. 1 1′–13′	Fragment	NA	8th–7th cent.	Aššur

Edition

Biggs 1967: 65

Structure of the text

I. Prescription 1: 1′–5′
Bow ritual: 6′–13′

Transliteration

I. A 1′–13′ = Prescriptions 1 and bow ritual

1′. traces
2′. [. . .] x *ana* IGI ^{d}UTU *pa-a*[*n* . . .]
3′. [. . .] x *ta-na-suk ki-a-a*[*m* . . .]
4′. [. . .] x *ú ša-mu-ú* DU_{11}.GA *a*[*n* . . .]
5′. [*ana* Í]L ŠÀ TUKU-*e* LÚ BI NAG [. . .]

6′. [*ana* ŠÀ].ZI.GA TUKU-*e ana* IGI ^{d}15 [. . .]
7′. [A^{meš} ^{š}]^{im}LI KÙ SUD NÍG.NA ^{šim}L[I GAR-*an*]
8′. [*mi*]-*iḫ-ḫa* BAL-*qí* ^{giš}BAN *š*[*á* . . .]
9′. [. . .] SA.MUD MAŠ.DÀ *šá* GÙB *ma-ta-an-*[*šá* . . .]
10′. [. . .] x TAG.GA *ur-ba-te ḫ*[*u*$^{?}$ . . .]
11′. [. . . *ra-ma*]*n-ni a-a ip-pa-šìr* [. . .]
12′. [. . . M]UL *šá* ^{d}⌜15⌝ [. . .]
13′. traces
Rest destroyed

Transcription

I. A 1′–13′ = Prescriptions 1 ll. 1′–5′ and bow ritual ll. 6′–13′

1	1′. Traces
	2′. [. . .] . . . *ana pān Šamši pā*[*n* . . .]
	3′. [. . .] . . . *tanassuk kīa*[*m* . . .]
	4′. [. . .] . . . *u šamû taqabbi* . . . [. . .]
	5′. [*ana n*]*īš libbi rašê amēlu šū išatti* [. . .]
Bow ritual	6′. [*ana*] *nīš* [*libbi*] *rašê ana pān Ištar* [. . .]
	7′. [*mê bu*]*rāši ellūti tasallaḫ nignak burāš*[*i tašakkan*]
	8′. [*mi*]*ḫḫa tanaqqi qašta* . . . [. . .]
	9′. [. . .] *šer'ān eqbi ṣabīti ša šumēli matan*[*ša* . . .]
	10′. [. . .] . . . *šiltāḫ urbate* . . . [. . .]
	11′. [. . . *rama*]*nni ay ippašir* [. . .]
	12′. [. . . *kak*]*kabu ša Ištar* [. . .]
	13′. Traces
	Rest destroyed

Translation

I. A 1′–13′ = Prescriptions 1 ll. 1′–5′ and bow ritual ll. 6′–13′

1	1′. Traces
	2′. [. . .] . . . in front of the sun, in fro[nt of . . .]
	3′. [. . .] . . . you throw in this wa[y . . .]
	4′. [. . .] . . . and the heaven, you say . . . [. . .]
	5′. [in order to] get [sex]ual desire, that man drinks [. . .].
Bow ritual	6′. [In order to] get sexual [desire]: In front of Ištar [. . .]
	7′. you sprinkle pure [ju]niper [water], [you put] junipe[r] in a censer,
	8′. you libate [*mi*]*ḫḫu*-beer, the bow o[f . . .]
	9′. [. . .] a tendon of the left hock of a gazelle is [his] string [. . .]
	10′. [. . .] . . . a reed arrow . . . [. . .]
	11′. [. . .] may [himse]lf be released [. . .]
	12′. [. . . the sta]r of Ištar [. . .]
	13′. Traces
	Rest destroyed

Nīš libbi S

List of Manuscripts

Manuscript	Museum number	Publication	Tablet	Script	Date	Provenience
A	K. 5901	TCS 2, pl. 3 1′–11′	Fragment	NA	8th–7th cent.	Aššur

Edition

Biggs 1967: 65

Structure of the text

I. 1 prescription ll. 1′–11′

Transliteration

I. A 1′–11′ = Prescription 1

1′. traces
2′. [. . .] x *su-uḫ*-[*su* . . .]
3′. [. . .] *ši nu* x [. . .]
4′. [. . .] IZI *d*[*i*? . . .]
5′. [. . .] ⌜*i*⌝ NUNUZ *di* [. . .]
6′. [. . . *l*]*i* GAB ÉN [. . .]
7′. [. . .] TAR-*su* [. . .]
8′. [. . .] DÁLA(text MI)^{meš} *u* NA$_4$.{BE}^{meš} [. . .]
9′. [. . .] x ÉN ki-in-da-⌜rab⌝ [. . .]
10′. [. . . Š]À.ZI.GA
11′. [. . .] x A *ana* UGU NA$_4$ x x [. . .]
Rest destroyed

Transcription

I. A 1′–11′ = Prescription 1

1′. Traces
2′. [. . .] . . . *suḫ*[*su* . . .]
3′. [. . .] . . . [. . .]
4′. [. . .] *išātu* . . . [. . .]
5′. [. . .] . . . *per*ʾ*u* . . . [. . .]
6′. [. . .] . . . *irtu šiptu* [. . .]
7′. [. . .] *ta*/*iprusu*/*ū* [. . .]
8′. [. . .] *ṣillū u abnū* [. . .]
9′: [. . .] . . . *šiptu*: ki-in-da-rab [. . .]
10′. [. . .] *nīš* [*lib*]*bi*
11′. [. . .] . . . *mû ana muḫḫi* . . . [. . .]
Rest destroyed

Translation

I. A 1′–11′ = Prescription 1

1′. Traces
2′. [. . .] . . . 'genital [discharge' . . .]
3′. [. . .] . . . [. . .]
4′. [. . .] fire . . . [. . .]
5′. [. . .] . . . sprout . . . [. . .]
6′. [. . .] . . . chest, incantation [. . .]
7′. [. . .] you/he/they cut/s [. . .]
8′. [. . .] thorns and stones. . . [. . .]
9′: [. . .] incantation: "kindarab" [. . .]
10′. [. . .] sexual [desi]re
11′. [. . .] water over the . . .-stone [. . .]
Rest destroyed

Commentary

2′. For the term *suḫsu* see discussion in commentary *nīš libbi* J pescr. 3: 26.

9′. See for the mention of the incantation the catalogue LKA 94 i 27 and K prescr. 28: 79.

VII Ritual fragments from Sippar and Uruk

Nīš libbi T

List of manuscripts

Manuscript	Museum number	Publication	Tablet	Script	Date	Provenience
A	81–7-1, 270(BM 42510) + F 224 (81–1-1 unnumbered) obv. 1′–rev. 11′	Finkel 2000	Frg. of a single-col. tablet	LB	Achaemenid (before Xerxes)	Sippar

Edition

Finkel 2000: 160–161
Chalendar 2018: 46

Structure of the text

I. Prescriptions
   1. obv. 1′–3′
   2. 4′–rev. 11′

Transliteration

I. A obv. 1′–rev. 11′ = Prescription 1–2

obverse
1′. [x] x x x ⸢KÙ$^{?}$⸣.G[A$^{?}$ . . .]
2′. Ì.UDU ŠAḪ $^{⟨ú⟩}$ḪAB [. . .]
3′. NAGAR-⟨*nu*⟩ *šá* ŠÀ.Z[I.GA$^{?}$]

4′. SUḪUŠ ^{ú}KUR.ZI *šá* ⸢*ina*$^{!}$ *mu-sa*⸣-*r*[*i*]
5′. ḪÁD.DU GAZ $^{⟨ú⟩}$SIKIL *ina* KAŠ x [. . .]
6′. ^{na4}*mu-ṣa* ^{na4}*šada-nu* DAB ^{na4}x
7′. ⸢*lìb*⸣$^{?}$-[*bi*] BURU$_5$.ḪABRUD.DA NITA *ina* KAŠ

https://doi.org/10.1515/9783110757262-012

edge
8′. *lu-ú ina* GEŠTIN NAG ^{giš}KIŠI$_{16}$.Ḫ[AB]

reverse
9′. NUMUN ^{ú}NÍG.GÁN.GÁN NUMUN ^{ú}DILI [. . .]
10′. *ri-kib-tú* DÀRA.MAŠ x [. . .]
11′. GAZ SIM *ina* GEŠTIN.KALAG.GA [NAG . . .]

Transcription

I. A obv. 1′–rev. 11′ = Prescription 3

obverse	
1	1′. [. . .] . . . *ell*[*u* . . .]
	2′. *lipi šaḫî būšānu* [. . .]
	3′. *allānu ša n*[*īš*] *libbi*
3	4′. *šuruš samīdi ša ina musar*[*î*]
	5′. *tubbal taḫaššal sikilla ina šikari* . . . [. . .]
	6′. *mūṣa šadâna ṣābita* . . .
	7′. *lib*[*bi*] *iṣṣūr ḫurri zikari ina šikari*
edge	8′. *lū ina karāni išatti dadā*[*na*]
reverse	9′. *zēr egemgiri zēr ēdi* [. . .]
	10′. *rikibti ayyali* . . . [. . .]
	11′. *taḫaššal tanappi ina karāni danni* [*išatti* . . .]

Translation

I. A obv. 1′–rev. 11′ = Prescription 3

obverse

1 1′. [. . .] pur[e] . . . [. . .]
2′. fat of pig, *būšānu*-plant [. . .].
2 3′. Suppository for the se[xual] desire.
3 4′. Root of *samīdu*-plant, which is in a garden-pl[ot],
5′. you dry and crush, *sikillu*-plant in beer . . . [. . .]
6′. *mūṣu*-stone, magnetite, . . .-stone,
7′. [innards/blood] of a male partridge in beer,
edge 8′. or he drinks (them) in wine, *dadā*[*nu*-acacia],
reverse 9′. seeds of *egemgiru*-plant, seeds of *ēdu*-plant [. . .]
10′. *rikibtu* of a stag . . . [. . .]
11′. you crush (and) sift (them), [he drinks it] in strong wine [. . .].

Commentary

Finkel (2000) considered this text as prescriptions for a "sick interior." At line 3′ one could restore ŠÀ.Z[I.GA]. It may be in fact a *nīš libbi* text, as already proposed by Biggs (2003–2005: 604–605), because of the use of ingredients such as the *rikibtu* of the stag and the blood or the heart (*libbu*) of a male partridge.

3′. The *nīš libbi* texts do not mention the use of a suppository. This is a unique case. The logogram NAGAR needs a phonetic addition (–*nu*).

7′. It is possible to read at the beginning of the lines Ú[š^{meš}] instead of ⌜*lìb*⌝$^{?}$-[*bi*] (see Chalendar 2018: 46).

NĪŠ LIBBI U

List of manuscripts

Manuscript	Museum number	Publication	Tablet	Script	Date	Provenience
A	W. 22277a	SpTU 1, 20 obv. 1–rev. 9	Fragment	LB	4th–3rd cent.	Uruk

Edition

Hunger 1976: 33, No. 20

Structure of the text

I. 1$^?$ Prescription obv. 1–rev. 9

Transliteration

I. A obv. 1–rev. 9 = 1$^?$ prescription

obverse
1. [. . .] x [. . .]
2. [. . .] GU$_4^?$ [. . .]
3. [. . .] x *ṣa at* DU [. . .]
4. [. . . ŠÀ$^?$].ZI.GA [. . .]
5–10. [. . .] SU [. . .]

reverse
1. [. . .] x [. . .]
2. [. . . ŠÀ.Z]I$^?$.GA [. . .]
3. [. . .] x É x [. . .]
4. [. . .] *šú* ZI ZI$^?$ [. . .]
5. [. . .] *šú* KUR-*i* EN ZI ZI [. . .]
6. [. . . KUR]-*i* EN ZI ZI [. . .]
7. [. . .] x *te* ÉN [. . .]
8. [. . . ŠÀ.Z]I$^?$.GA [. . .]
9. [. . .] x x *ú*$^?$ [. . .]

VIII Ritual fragments from Boghazköy

Nīš libbi V

List of manuscripts

Manuscript	Museum number	Publication	Tablet	Script	Date	Provenience
A	(Bo 5885)+AAA 3, No. 5	(KUB 37, 81)+ AAA 3, pl. 27 No. 5 5–12	Frg. part of the tablet containing KUB 37, 81	Ass.-Mitt.?	13th cent.	Ḫattuša

Edition

Pinches 1910: 104–105
Biggs 1967: 60–61

Structure of the text

I. Prescriptions
   1. 1–4
   2. 5–9
   3. 10–12

Summary of the sections of manuscripts not included in the transliteration:

– Ms. A =
  AAA 3, pl. 27 No. 5 1–4 = N prescr. 3: 14–16
  KUB 37, 81 1′–8′ = W prescr. 1–4

Transliteration

I. A 1–12 = Prescriptions 1–3

1–4 = *nīš libbi* N prescr. 3: 14–16

5. [DIŠ KI.MIN *e-nu-ma* UD]U.NITÁ *ina muḫ-ḫi* U₈ [*iš-ḫi-ṭu*]
6. [SÍG KUN *ša*] *ki-la-li-šu-nu* [*ta-na-saḫ*?]
7. [. . . *k*]*i-ip-la ta-ṭ*[*á-mi* . . .]

https://doi.org/10.1515/9783110757262-013

8. [*ina* MÚRU-*šú*] *tàra-kas*$_4$*-ma* [. . .]
9. [. . . Š]À.ZI.GA [TUKU-*ši*]

---

10. [. . .] x GIG *lìb-bi* [. . .]
11. [. . .] ⸢*a*⸣-*na* GIG *ni* x [. . .]
12. [. . .] ⸢ŠÀ⸣.ZI.GA
Rest destroyed

Transcription

I. A 1–12 = Prescriptions 1–3

1 1–4 = *nīš libbi* N prescr. 3: 14–16

2 5. [*šumma* KI.MIN *enūma im*]*meru ina muḫḫi immerti* [*išḫiṭu*]
6. [*šīpāt zibbati ša*] *kilallīšunu* [*tanassaḫ*]
7. [. . . *k*]*ipla taṭ*[*ammi* . . .]
8. [*ina qablīšu*] *tarakkas-ma* [. . .]
9. [. . . *nī*]*š libbi* [*iraššī*]

3 10. [. . .] . . . *muruṣ libbi* [. . .]
11. [. . .] *ana murṣi* . . . [. . .]
12. [. . .] *nīš libbi*
Rest destroyed

Translation

I. A 1–12 = Prescriptions 1–3

1 1–4 = *nīš libbi* N prescr. 3: 14–16

2 5. [If ditto: When the ra]m [has mounted] the female sheep,
6. [you tear off wool of] both their [tails],
7. [. . . a s]ting, you s[pin. . .]
8. you tie [around his waist] and [. . .]
9. [. . . he will get sexual] desire.

3 10. [. . .] . . . 'heart' (*libbu*)'s sickness [. . .]
11. [. . .] for the . . .-sickness [. . .]
12. [. . .] (and he will get) sexual desire.
Rest destroyed

NĪŠ LIBBI W

List of manuscripts

Manuscript	Museum number	Publication	Tablet	Script	Date	Provenience
A	Bo 5885+(AAA 3, No. 5)	KUB 37, 81+(AAA 3, pl. 27 No. 5) 1′–8′	Frg. part of the tablet containing AAA 3, No. 5	Ass.-Mitt.?[578]	13th cent.	Ḫattuša

Edition

Biggs 1967: 60

Structure of the text

I. Prescriptions
   1. 1′
   2. 2′–3′
   3. 4′–7′

4. 8′

Summary of the sections of manuscripts not included in the transliteration:

– Ms. A = AAA 3, pl. 27 No. 5
  1–4 = N prescr. 3: 14–16
  5–9 = V prescr. 2
  10–12 = V prescr. 3

578 Note the use of *qè*, see Schwemer 2013: 154.

Transliteration

I. A 1′–8′ = Prescriptions 1–4

1′. [LÚ BI ŠÀ.Z]I.GA ⸢TUKU⸣-[*ši*]

2′. [DIŠ KI.MIN MÚR]U-*šú te zi ni u*[*r* . . .]

3′. [. . .] *nam* LÚ BI ŠÀ.ZI.[GA TUKU-*ši*]

4′. [DIŠ KI.MIN UZ]U$^{?}$.DIR *ša* EGIR-*šú* x [. . .]

5′. [. . .] x *šu te-leq-qè ur* [. . .]

6′. [. . .] x NAG-*šú-ma* [. . .]

7′. [LÚ BI Š]À.ZI.GA TUKU-[*ši*]

8′. Traces

Transcription

I. A 1′–8′ = Prescriptions 1–4

1 1′. [*amēlu šū nīš*] *libbi iraš*[*ši*]

2 2′. [*šumma* KI.MIN *qabl*]*īšu* . . . [. . .]
3′. [. . .] . . . *amēlu šū nīš lib*[*bi iraššī*]

3 4′. [*šumma* KI.MIN *ka*]*mūnu ša arkīšu* . . . [. . .]
5′. [. . .] . . . *teleqqe* . . . [. . .]
6′. [. . .] . . . *išattīšu-ma* [. . .]
7′. [*amēlu šū n*]*īš libbi iraš*[*ši*]

4 8′. Traces

Translation

I. A 1′–8′ = Prescriptions 1–4

1 1′. [that man] will ge[t sexual] desire.

---

2 2′. [If ditto]: his [waist] . . . [. . .]
3′. [. . .] . . . that man [will get] sexual desir[e].

---

3 4′. [If ditto: *ka*]*mūnu*-fungus, which after this[?] . . . [. . .]
5′. [. . .] . . . you take . . . [. . .]
6′. [. . .] . . . he drinks it and [. . .]
7′. [that man] will get [sexual] desire.

---

4 8′. Traces

Commentary

4′. UZU.DIR *ša* EGIR-*šú*/*kamūnu ša arkīšu*, a kind of *kamūnu*-fungus (see EGIR-*šú* UZU.DIR in CT 30, 50 K.957: 15).

NĪŠ LIBBI X

List of manuscripts

Manuscript	Museum number	Publication	Tablet	Script	Date	Provenience
A	621/b	KUB 37, 82: 1′–11′	Fragment	MA[579]	14th cent.	Ḫattuša Room 4

Edition

Biggs 1967: 61

Structure of the text

I. Prescriptions
   1. 1′–3′
   2. 4′–11′

Transliteration

I. A 1′–11′ = Prescriptions 1–2

1′. traces
2′. [. . . U$_4$.X.K]AM UD.N[Á.A$^?$ . . .]
3′. [. . .] x ⸢7⸣-*šú ù* 7-[*šú* . . .]

4′. [DIŠ LÚ ŠÀ-*šú* NU Í]L-*ma* ŠU ^{d}*iš*$_8$-*tár a-na* [TI-*šú*]
5′. [. . . *zap*]-*pi ša i-na* UGU *ú*-[*ru-ul-li*$^?$ . . .]
6′. [. . . *i*]-*na* UGU *ú-ru-ul-li* [. . .]
7′. [. . . TÉŠ.B]I *ta-pát-taḫ* A^{meš}-*šu* x [. . .]
8′. [. . .] x *šu-a-ti šìr* ^{d}NIN.[KILIM . . .]
9′. [. . .] x *sú tà-sap-pi* [. . .]
10′. [. . . PÉ]Š.ÙR.RA *tu-u*[*r-ar* . . .]
11′. traces
Rest destroyed

579 Note use of *tà* in *tà-šap-pi* in 1. 9′ (cf. KUB 28, 58 rev. VI 6, see Schwemer 2013: 154). See also Fincke 2010: 48.

Transcription

I. A 1′–11′ = Prescriptions 1–2

1 1′. Traces

2′. [. . . U$_4$.X.K]AM *bibbu*[*lu* . . .]

3′. [. . .] . . . *sebîšu u sebî*[*šu* . . .]

---

2 4′. [*šumma amēlu libbašu lā ina*]*ššī-ma qāt Ištar ana* [*bulluṭīšu*]

5′. [. . . *zap*]*pī ša ina muḫḫi u*[*rulli* . . .]

6′. [. . . *i*]*na muḫḫi urulli* [. . .]

7′. [. . . *ištē*]*niš tapattaḫ mêšu* . . . [. . .]

8′. [. . .] . . . *šuāti šīr šik*[*kî* . . .]

9′. [. . .] . . . *tasappi* [. . .]

10′. [. . . *ar*]*raba tu*[*rrar* . . .]

11′. Traces

Rest destroyed

Translation

I. A 1′–11′ = Prescriptions 1–2

1 1′. Traces
2′. [. . .] day of new [moon . . .]
3′. [. . .] . . . seven times and seven [time . . .].

---

2 4′. [If a man does not desi]re and "Hand of Ištar," in order to [cure him]:
5′. [. . . hai]rs which over the fo[reskin . . .]
6′. [. . . o]ver the foreskin [. . .]
7′. [. . .] you puncture [toge]ther, his fluids . . . [. . .]
8′. [. . .] . . . him, meat of a mongo[ose . . .]
9′. [. . .] you pluck . . . [. . .]
10′. [. . .] you [parch an *arrabu*]-mouse [. . .]
11′. Traces
Rest destroyed

Commentary

5′. Another possible restoration is *u-ri* (see Biggs 1967: 61).

8′. *šīr* ^{d}NIN.KILIM/*šikkî* "meat of a mongoose" is perhaps a coded name for liquorice root (BAM 574 I 8, Böck 2011: 694).

NĪŠ LIBBI Y

List of manuscripts

Manuscript	Museum number	Publication	Tablet	Script	Date	Provenience
A	178/b	KUB 37, 201: 1′–13′	Frg. of a single-col. tablet	MA[580]	14th cent.	Ḫattuša Room 4

Edition

Fincke 2010: 48–49 No. 41

Structure of the text

I. 1 Prescription ll. 1′–13′

Transliteration

I. A 1′–13′ = 1 prescription

1′. [NA BI *kiš-pu* D]AB.⸢DAB⸣ x [. . .]
2′. [. . .] x *zé* UR.GI$_7$ [. . .]
3′. [. . . *i-na* KUŠ] TAG *šap-pi* [. . .]
4′. [. . .] x TI-*uṭ* [. . .]
5′. [. . .] x-*sú* ŠIR LÚ *ša* x [. . .]
6′. [. . .] *lu-u* SUḪUŠ$^{!?}$ *ú-ra-ni*[*m* . . .]
7′. [. . .] x NÍG.⸢BÚN$^{?}$⸣.NA [. . .]
8′. [. . .] x *i-na* ⸢KUŠ TAG⸣ *ša*[*p-pi* . . .]
9′. [. . .] TI-*u*[*ṭ* . . .]
10′. [. . .] x ⸢AḪ$^{?}$⸣ [*t*]*u*$^{?}$ [*š*]*a*$^{?}$-*a*-⸢*šu*⸣ [. . .]
11′. [. . .] SA[G.D]U x [. . .] x [. . .]
12′. [. . .] *i-na* x [x x] x x [. . .]
13′. [. . . T]I-*uṭ* [. . .]
Rest destroyed

580 See Fincke 2010: 48.

Transcription

I. A 1′–13′ = 1 prescription

1′. [*amēlu šū kišpu iṣ*]*bassu* [. . .]
2′. [. . .] . . . *zê kalbi* [. . .]
3′. [. . . *ina maški*] *talappat šappi* [. . .]
4′. [. . .] . . . *iballuṭ* [. . .]
5′. [. . .] . . . *iški amēli ša* . . . [. . .]
6′. [. . .] . . . *šuruš urâni*[*m* . . .]
7′. [. . .] . . . *šeleppû* [. . .]
8′. [. . .] . . . *ina maški talappat ša*[*ppi* . . .]
9′. [. . .] *iballu*[*ṭ* . . .]
10′. [. . .] . . . [*š*]*âšu* [. . .]
11′. [. . .] *qa*[*qqa*]*du* . . . [. . .] . . . [. . .]
12′. [. . .] *ina* . . . [. . .] . . . [. . .]
13′. [. . . *ib*]*alluṭ* [. . .]
Rest destroyed

Translation

I. A 1′–13′ = 1 prescription

1′. [The witchcraft has] continually seized [that man . . .]
2′. [. . .] . . . excrement of a dog [. . .]
3′. [. . .] you touch [with the leather], *šappu*-container [. . .]
4′. [. . .] . . . he will recover [. . .]
5′. [. . .] . . . testiscles of the man who . . . [. . .]
6′. [. . .] . . . root of *urân*[*u* . . .]
7′. [. . .] . . . turtle [. . .]
8′. [. . .] . . . you touch with the leather, *ša*[*ppu*-container . . .]
9′. [. . .] he will hea[l . . .]
10′. [. . .] . . . [to/of h]im [. . .]
11′. [. . .] he[a]d . . . [. . .] . . . [. . .]
12′. [. . .] in . . . [. . .] . . . [. . .]
13′. [. . . he will he]al [. . .].
Rest destroyed

Commentary

5′. None of texts of the corpus mention the patient's testicles.

7′. None of the *nīš libbi* texts mentions ingredients from a turtle (NÍG.BÚN.NA/*šeleppû*). For therapeutic uses of the animal see Fincke 2010: 49.

List of mineral and botanical ingredients

aban kasî (^{na4}GUG.GAZI^{sar}): precious red stone, 'red carnelian,' see Schuster-Brandis 2008: 414.
aban lamassi (NA$_4$.^{d}LAMMA): pink/rose-colored stone, 'pink chalcedony,' see Schuster-Brandis 2008: 427–428.
abāru (A.BÁR): lead, see CAD A/I 36 *s.v. abāru* A; AHw.I 4 *s.v. abāru* I.
abašmû (^{na4}*a-ba-aš-mu*): a greenish stone, see Schuster-Brandis 2008: 392–393.
aktam (^{ú}*ak-tam*): an unidentified medicinal plant; Thompson 1949: 130–133 suggests an identification with the castor oil plant, but it remains uncertain, see CAD A/I 282–283; AHw. I 30.
alamû (^{ú}*a-la-mu*): an unidentified plant, perhaps an aquatic one. It is rarely used as medicinal plant, see CAD A/I 333; AHw. I 35 'eine Wasserpflanze'; Thompson 1949: 243 'Anchusa.'
algamišu (^{na4}ALGAMEŠ): a soft stone, see Schuster-Brandis 2008: 393–394.
aluzinnu (^{ú}*a-lu-zi-in-ni*): an unidentified plant, it appears only in the *nīš libbi* text N (KUB 4, 48) prescr. 24 left. e. 6; see CAD A/I 392, mng. 2; AHw. I 39–49 mng. 2.
amašpû (or *ašpû*) (^{na4}AMAŠ.PA.È; ^{na4}*aš-pu-u*): a precious stone, perhaps a kind of chalcedony from light blue to gray used in a magical context against specific diseases, see Thompson 1936: 167; Schuster-Brandis 2008: 401–402.
amḫara (^{ú}*am-ḫa-ra*): an unidentified medical plant, see CAD A/II 45–46; AHw. I 43.
amīlānu (^{ú}LÚ.U$_{19}$.LU): lit. "like-man-(plant)," perhaps mandrake for its resemblance of the human form; see CAD A/II 46; von Soden 1957–1958: 394; Köcher BAM 3: xxiv fn. 56.
amuzinnu: (^{ú}*a-nu-zi-ni*): an unidentified plant. In *Šammu šikinšu* text 2, § 28′ (Stadhouders 2011: 22) it is stated that it is good for the sexual desire and to remove feebleness. In BAM 1 iii 32 (Attia and Buisson 2012: 29) it is considered a therapeutic ingredient for the weakness of the flesh.
ankinūtu (^{ú}*an-ki-nu-tu*): the Akkadian name is a loan from the Sumerian an-ki-nu-di "reaching neither heaven nor earth." According to CAD A/II 123–124 it would, based on the name, be an epiphyte, or climbing plant. According to Thompson (1949: 234) it would rather be identified with *Nelumbo nucifera*. See *Šumma šikinšu* text 1 § 4: 1–3; see Stadhouders 2011: 6; ibid. 19: 1; text STT 93, also text 1 § 3: 1–3 (referring to the *ašqulālu*-plant). For the association with the aquatic *ašqulālu*-plant see CAD A/II 124; Stol 2000: 54 and fn. 35. The logogram AN.KI.DU.TI is also used to indicate the *ašqulālu*-plant, mostly written ^{ú}LAL. See also URU.AN.NA: *maštakal* CT 14, 21 Sm 1328:11ff.; KADP 2 ii 24–28. It is a plant used in prescriptions within many diagnostic categories for the production of ointments, amulets, potions and for fumigation.
arantu (^{ú}*a-ra-an-tu*): a kind of grass. It does not appear in the anti-witchcraft texts, but still has a vast therapeutic use (see BAM 380 r. 38 and dupl. ibid. 381 iii 30 where it is used to counter problems with pregnancy and childbirth; see Stol 2000: 53). See BAM 379 iv 3–5, Stadhouders 2011: 36; ibid. 19: 17. See CAD A/II 231; AHw. I 64; Thompson 1949: 16–17 and 148–149. The grass also appears in omens in the *Šumma ālu* series (CT 39, 6: 7, Langdon 1916: 31), although the mentioning of the grass in the omens does not seem to be related to its therapeutic use.
arariānu (^{ú}*a-ra-ri-a-nu*): its identification is not certain. Its cognate in Syriac *'â'ărînâ* is identified with *Onobrychis*, cockshead. According to Thompson (1949: 125–126) it is lupine (see also CAD A/II 232–233; AHw. I 65). See BAM 379 iii 34′–35′ (Stadhouders 2011: 36 and ibid. 19: 17). The plant is common in the therapeutic context, often crushed and drunk in potion.
ardadillu (^{ú}AŠ.TÁL.TÁL; ^{ú}*ar-da-dil-lum*): an unidentified plant, growing in reed thickets and in the 'steppe' according to Sumerian literary texts. In URU.AN.NA: *maštakal* I 679 the plant is similar to the *lišān-kalbi*-plant. See *Šumma šikinšu* text I § 38′ (Stadhouders 2011: 13, see ibid. text II § 5 and § 16). It does not have a wide therapeutic use, but it is used for rectal problems and flatulence in AMT 47, 1 ii 8 (Geller 2005: 150). It is frequently used in anti-witchcraft texts, where it always appears together with other plants. In a *nīš libbi* incantation (No. E.2) it is considered a remedy against witchcraft. See CAD A/II 241; AHw. I 67; Thompson 1949: 257.

https://doi.org/10.1515/9783110757262-014

ašāgu (^{giš}KIŠI$_{16}$): a thorny shrub, probably a variety of acacia, see CAD A/II 408–410 'a kind of acacia'; AHw. I 77 'ein Dornstrauch, Kameldorn'(?); Landsberger 1937–1939: 139; Thompson 1949: 182–184 *Lycium*; Adams 1965; Civil 1987: 41–42. According to CAD, it can be identified with the modern Arabic *šok* (*Prosopis farcta* or *stephaniana*), a type of acacia, widely used as fuel. We find some pharmacological information in the text BAM 1 i 7, 38, 42. The plant is used as an ingredient, in the anti-witchcraft corpus, against ghost attacks (see Scurlock 2006a: 776), in *namburbû*-rituals (Maul 1994: 510), and in many medical texts for various sicknesses (for its medical use see Thompson 1949: 182–183). Note its 'secret name' in URU.AN.NA: *maštakal*: "Idem (= *galgaltu*) (or black spot?) from the crotch of a donkey" (III 17, Rumor 2017: 7). See *Šammu šikinšu*, text II § 24′, it is a plant for purification.

ašgigû (^{na4}*aš-gì-gì*): a green stone, perhaps turquoise, see Schuster-Brandis 2008: 400–401.

ašnugallu (^{na4}GIŠ.NU$_{11}$.GAL): alabaster, see Schuster-Brandis 2008: 412–413.

ašqulālu (^{ú}LAL, also ^{ú}AN.KI.NU.DI): a marine plant, whose medical use is uncommon. According to *Šammu šikinšu*, the plant is not found underwater and is visible on the surface (see STT 93 r. 79′–81′ and dupl. BRAM 4 32: 18, CAD A/II 453). An anti-witchcraft incantation underlines the resistance of the plant to natural forces (see Abusch and Schwemer 2011: 192, No. 7.8.6: 14′–24′). See Thompson 1949: 239–240; Kinnier Wilson 1957: 47, according to whom the plant was the "Dead Sea Apple"; Oppenheim 1956: 286 fn. 134; Perdibon 2019: 167–168.

ašpû see *amašpû*

asu (^{šim}GÍR): myrtle, see see CAD A/II 342–344; AHw. I 76.

atā'išu (Ú.KUR.KUR): an unidentified plant. It has a wide spectrum of use in the therapeutic context: for the sources and possible identification see CAD A/II 8081; Thompson 1949: 151–154 "hellebore 'Sneeze-plant.'" In *Maqlû* one who recites the incantation is identified with the plant: "Incantation: Pure sulphur (and) *atā'išu*, the holy (*quddušu*) plant, I am" (VI 85, Abusch 2016: 159, see also Abusch 2002: 207–211).

atbaru (^{na4}AD.BAR): basalt, see Schuster-Brandis 2008: 393.

ayyartu (^{na4}PA): a kind of shell, maybe *Cypraeidae*, see Oppenheim 1963; Schuster-Brandis 2008: 438.

azallû (^{ú}A.ZAL-*ú*; ^{ú}A.ZAL.LÁ): an unidentified medicinal plant, see CAD A/II 524–525; AHw. I 92: 'Haschisch'(?); Thompson 1949: 220–222 *Cannabis Indica*). Thompson's theory that the plant is to be identified with cannabis is based on a passage of the *Šammu šikinšu*: "The *azallû*-plant, (whose appearance) is like (that of) *kanašû* and that is red – (its name) *azallû;* it is against melancholy" (BRM iv, 32, 19, Stadhouders 2011: 17 fn. 81, see URU.AN.NA: *maštakal* II 1ff.). He interprets *kanašû* as opium and translates *nissatu* with "depression of the spirit." The traditional identification with cannabis, however, is questionable (see Farber 1981: 271). In BAM 316 (iii 23–25) the plant is used against depression (see Couto-Ferreira 2000c: 32). In the same text it used again against *hīp libbi* ("breaking of his insides") if crushed and drunk in beer (BAM 316 iii 19, see also *Šammu šikinšu* text II, § 8 (1–2), Stadhouders 2011; 17). In BAM 1 iii 35 the plant is associated with the sexual desire: *azallâ* : KI.MIN (= *nīš libbi*) : *nissata lā īši* "*azallû*-plant : for ditto (= sexual desire) : he will have no pains" (see Attia and Buisson 2012: 29, see BAM i 59; ibid. 318 iii 29). Except for these passages, as CAD A/II 525 reminds us, there is more evidence for considering the plant a narcotic. In fact, against the interpretation of Thompson, Landsberger (1967a: 51–52 fn. 183) points out the correspondence between this plant and *ḫaṭṭi-rē'î* (lit. "shepherd's-crook," *Equisetum*, CAD Ḫ 156) proposing its identification with *Polygonum*, a plant characterized, as the *Equisetum*, by nodes. Böck (2014: 142) argues that both *azallû* and *ḫaṭṭi-rē'î* are names for the plant *lišān-kalbi*. Seeds, leaves and roots of the plant are also used in medical texts.

azupīru (^{ú}ḪAR.SAG): traditionally identified with saffron or with safflower (*Carthamus tinctorius*) for etymological reasons (see CAD A/II 530–531; AHw. I 93; Thompson 1949: 159; Landsberger 1964: 260 fn. 56, Bottéro 1957–1971: 341a, 344a; Vincente 1991: 351), but as pointed out by CAD A/II 531 the identification is not supported by its use (see the mentioning of seeds). It is used both as a spice and in the therapeutic context. It is frequently associated with the Ú.KUR.RA/*nīnû*-plant. Interestingly, it is mentioned in *Maqlû* for its use against witchcraft: *kīma azupīri liṣappiruši kispūša* "Like *azupīru*-plant may her witchcraft cut her down" (V 28, Abusch 2016: 136). Note the 'secret names' of the plant are "dust around$^{?}$ the tracks of a male goat / dust from the tracks of a wronged/oppressed woman$^{?}$" (URU.AN.NA: *maštakal* III 21–21, Rumor 2017: 7) and in fact the male goat is an animal which often appears in this corpus.

baltu (^{giš}DÌḪ): thorny bush, perhaps a variety of acacia, maybe *Acacia erioloba* (*Acaia giraffae*: camelthorn), see CAD B 65–66; AHw. I 100; Civil 1987: 41.

barīrātu ($^{\text{giš/ú}}$LUM.ḪA): an unidentified aromatic plant, maybe *Ferula persicana* and its gummy resin (sagapenum), see CAD B 111; AHw. I 107; Thompson 1949: 361–364; Kinnier Wilson 2005: 20–21. See *Šammu šikinšu*, text II § 24′ it is plant for purification.

billi ($^{\text{na4}}$*bil-lí*): an identified stone used for amulet necklaces, see CAD B 228–229; AHw. I 126.

bīnu (^{giš}*bi-nu*, also ^{giš}ŠINIG): tamarisk, see CAD B 239–242; AHw. I 127; Perdibon 2019: 146–149 with previous bibliography.

bukānu (^{giš}GAN.U$_5$): it is difficult to determine the reading of the Sumerogram; according to Stol (2009b: 167) it would be the pseudo-logogram for *kam'atu* (**kam'u*) 'truffle,' whereas according to Scurlock (2006a: 216) it stands for *kiškanû*, an unidentified tree (see CAD K 453). In the opinion of Abusch and Schwemer (2011: 469), which I follow here, it stands for a variant of $^{\text{giš/ú}}$GAN.NA/*bukānu* (lit. "pestle"-plant), an unidentified medicinal plant (see CAD B 308; AHw. I 136; Schwemer 2007b: 114; Schwemer 2009: 66). The *bukānu*-plant is used twice in *nīš libbi* prescriptions and in an incantation in which the plant has the power to undo the fascination performed by the witch, No. E.2. This role against witchcraft is underscored by its use in the anti-witchcraft ritual: "(I have equipped myself against you (= witches)) with the *bukānu*-wood which that undoes witchcraft" (KAL 2, 36, rev. v 40′, Abusch and Schwemer 2011: 166, No. 7.8.1: 23′).

buluḫḫu (^{šim}BULUḪ): an aromatic plant, see CAD B 74–75 'a tree and its resin, galbanum'; AHw. I 101 'Galbanum-Kraut'; Thompson 1949: 342–344 *Ferula gummosa*.

burāšu (^{šim}LI): juniper, see CAD B 326–328; AHw. I 139; Perdibon 2019: 159–160 with previous bibliography.

būšānu (^{ú}ḪAB): an unidentified medicinal plant, see CAD B 351; AHw. I 143; Thompson 1949: 150–151 *Helleborus orientalis* or *foetidu*; Kinnier Wilson 1966: 52–54 'wild grapes,' but Lambert 1969: 37 does not agree. See BAM 1 iii 20–21 where the plant is called "dog of Ninigizibara," whereas in URU.AN.NA: *maštakal* II 109–110 it is "dog of Gula." See also AMT 19, 7: 4: "The *būšānu*-plant whose [name is] 'name of Gu[la]" (for other possible names of the plant, see URU.AN.NA: *maštakal* II 108: 120, Böck 2014: 132–133, among these names we are reminded of DÙ.A.BI SIG$_5$/*kal damiq* "good for anything"). Böck (2014: 157–158) argues that *būšānu* is another name of *lišān-kalbi*. The term comes from *ba'āšu* 'to stink, smell badly' perhaps a reference to the particular smell of the plant, lit. "smelly-plant." The plant has a vast medical use (for the text references in which the plant is used alone see Thompson 1949: 272–273; Böck 2014: 131–140; CAD B 351). It has precisely the same name as *būšānu*-sickness (see Köcher BAM 6: xvi; Kämmerer 1995: 157; Volk, 1999: 27–30; Scurluck and Andersen 2005: 40–42; CAD B 351) and is used to eradicate it. The plant does not often appear in the anti-witchcraft texts.

dadānu (^{giš}KIŠI$_{16}$.ḪAB): it is possible that this is a subspecies of *ašāgu*, maybe acacia, based on the use in its logogram of the sign KIŠI$_{16}$. See its mention in URU.AN.NA: *maštakal* I 178 in the *ašāgu* group. See CAD D 17 'a subspecies of the false carob'; AHw. I 148 'Sterndistel'(?); Thompson 1949: 184–185 'centurea'; Civil 1987: 47 *Prosopis farcta*). The plant occurs in medical texts for both external and internal use, in the latter case often to cure "restriction" (*ḫiniqtu*, see Geller 2005).

ēdu (^{ú}DILI; ^{ú}*e-di*): lit. "solitary-(plant)," an unidentified medical plant. According to URU.AN.NA: *maštakal* I 236, it is to be identified with *ḫaṭṭi-rē'i* "shepherd's-crook," associated with the *azallû*-plant (ḪAR-ra = *ḫubullu* XVII 106–107, see Landsberger et al. 1970 = MSL 10: 186). See Thompson 1949: 353–354; Abusch and Schwemer 2011: 469. See *Šammu šikinsu* text 1 § 2–3, text II § 25′ and 39′.

egemgiru (^{ú}NÍG.GÁN.GÁN): the plant has been identified as rocket (*Eruca sativa*) by the Syriac and Arabic cognates (see CAD E 43 'rocket'; AHw. I 189 'Rauke'; Holma 1913: 67; Thompson 1949: 211–212). The plant is used for anus problems in BAM 1 iii 6, 14–16. The use of the plant in therapeutic contexts is not frequent, although in BAM 1 the plant is mentioned for anus problems. In the corpus published by Geller (2005), however, there is no mention of this ingredient.

elikulla (^{ú}*e-li-ku-la*) (also *elkulla, elikulla, ilikulla, erkulla*): an unidentified medical plant, see CAD E 100; AHw. I 203. In medical texts sometimes treated as separate drugs, see Abusch et al. 2020: 471.

ennēnu (^{še}IN.NU.ḪA) (also *inninnu*): a kind of barley, see CAD I-J 151; AHw. I 219.

erēnu (^{giš}EREN, ^{giš}*eri*$_4$*-ni*): cedar, see CAD E 274, 279; AHw. I 237; Thompson 1949: 282–285; Perdibon 2019: 153–156 with previous bibliography.

ēru (^{giš}MA.NU): it is a tree native to Mesopotamia, whose wood is used mainly to produce canes. CAD E 320 suggests that it is a variety of cornelian cherry (*Cornus mas*), since a variety of *ēru* is called *murrānu*, which appears as a foreign loan in Aramaic, *murrānā* 'cornelian cherry' and in Syriac *mūrrāniṯa* meaning *baculus, corno facta, hasta*. Thompson (1949: 298–300) identifies the plant with laurel, for both the similarity to Arabic *ghār*, *Laurus nobilis*, and its medical use (see also Salonen 1939: 99 and 152; Birot 1993: 50d). Another interpretation, as the ash tree, is offered by CAD M/I 221; CAD M/II 220; CAD S 202; AHw. I 676. Kinnier Wilson (1988: 81 and fn. 24) considers it 'wild pomegranate.' According to Steinkeller (1987: 91–92), none of the above identifications can be accepted, so instead he proposes the willow (*Salix acmophylla*). See also Postgate 1992: 185–186; Powell 1992: 102–103; Jiménez 2017: 217–223; Perdibon 2019: 160–162. The tree is used for therapeutic purposes as an ingredient. The bud, leaf, root, seeds, and, as in our case, the green "part" (probably referring to the still green branches) are used. Figurines of the seven sages (*ūmu-apkallū*) are made from its wood (KAR 298: 2, CAD E 320; see Wiggermann 1992: 65). The tree is called "weapon of the gods": "Oh *ēru*-tree, strong weapon of the gods, created for the sake of your sweet shade" (KAR 252 ii 51–52, CAD E 319); "He placed on his head the *ēru*-wood, the august symbol of Anu" (Geller 1980: 29, ll. 75′–76′). For more textual sources where the tree has a magical significance see Wiggermann 1992: 79–85. Generally speaking, the tree seems endowed with intrinsic magic-prophylactic properties, which is why it is used in the magic-therapeutic field, in order to chase away evil. The use of this tree, along with the *sikillu*-plant (and other ingredients), in the *nīš libbi* context demonstrates the purifying and prophylactic function of such a therapeutic prescription.

erû (^{na4}URUDU): copper, see CAD E 321–323 *s.v. erû* A.

^{ú}GAB.LAM: an unidentified medical plant.

gabû (IM.SAḪAR.NA$_4$.KUR.RA): alum, see CAD G 7, 279; AHw. I 272.

gišimmaru (GIŠIMMAR): date palm, see CAD G 102–104; AHw. I 292; Landsberger 1967a; Perdibon 2019: 150–153 with previous bibliography.

ḫarmunu (^{ú}ḪAR.ḪUM.BA.ŠIR): an unidentified plant, see CAD Ḫ 104–105; AHw. I 326.

ḫašḫūr api (^{ú}GIŠ.ḪAŠḪUR.GIŠ.GI): lit. "marsh apple," see CAD Ḫ 139; AHw. I 534. Thompson (1949: 255) suggests an identification with the 'gall-apple,' but it remains uncertain. In BAM 380 r. 42–44 (dupl. BAM 381 iii 37–40) the plant is considered suitable treatment for the loss of sexual desire. The plant is used frequently in the anti-witchcraft corpus in large medical prescriptions. Table LXIII of the *Ušburruda* series, from the anti-witchcraft corpus, mentions the plant along with other ingredients, many of which appear in *nīš libbi* prescriptions. The symptoms are varied, but some of them are closely related to the *nīš libbi* ones. In addition to a general state of psycho-physical discomfort, the patient's "(desire) to go to a woman is reduced, does not desire a woman" (tablet LXV of the canonical *Ušburruda* series, Abusch and Schwemer 2011: 118, No. 7.2). The symptom "(his desire) to go to a woman is reduced," can also be found, always amongst others, in other anti-witchcraft rituals. One of the evil actions consists of manufacturing male figurines and giving them to a dead man. This action of the witch is typical of *nīš libbi* texts (BAM 460 obv. 16, Abusch and Schwemer 2011: 417, No. 10.4.1: 16). In an *Ušburruda* incantation the plant is taken and given to the patient to eat by the god Asalluḫi (Abusch and Schwemer 2011: 183, No. 7.8.9: 31″). In a medical prescription (*Ušburruda*) it is used, along with the *aprušu*-plant in form of an oil-based ointment to cure a bewitched man, suffering from abdominal as well as epigastric problems (BAM 434 obv. iii 13 and ibid. 90 r. 11′, Abusch and Schwemer 2011: 215, No. 7.10.1: 43″). The plant is described as an ingredient used for stomach problems in BAM 1: ii 48. Another indication can be found in CT 14, 36 79-7-8,22 r. 3 for stopping the blood. It is even mentioned in the *Šummu šikinšu* series but unfortunately the text is fragmentary (Stadhouders 2011: 4, § 30).

ḫašḫūru (GIŠ.ḪAŠḪUR): apple-tree, see CAD Ḫ 139–140; AHw. I 333–334.

ḫašû (ᵘḪAR.ḪAR): the plant does not have a broad strictly medical use; it is used particularly in texts against ghost attacks (Scurlock 2006a: 778) and against witchcraft (Abusch and Schwemer 2011) as well as as a spice (see CAD Ḫ 144–145 B; AHw. I 335 III 'Thymian'; Thompson 1949: 74 'thyme'). In BAM 1 i 54 it is used against inflammation; ii 10 against the *šibiṭ šāri*-sickness. It is also a remedy for a bleeding anus if pulverized and drunk in beer (BAM 99 r. 38, Geller 2005: 214). The plant is used for problems related to the anus (BAM 99 r. 38, Geller 2005: 214) and inflammation caused by the sun's heat, diagnostic categories related to the absence of sexual desire (for other sources see Thompson 1949: 74). The ingredient also has a magical value, as it is expressly employed against witchcraft. The poetic language is interesting, expressed by a paronomasia, in *Maqlû*: "Like *ḫašû*-plant may her witchcraft chop her up (*liḫaššûši*)" (V 32, Abusch 2016: 136). The power of the plant against witchcraft is stressed in two *Maqlû* incantations: "I am sending against you (= witch) *ḫašû*-plant and sesame (*šamaššammū*)" (V 4, Abusch 2016: 134); "I am lifting up against you *kukru*, the offspring of the mountain, (and) *ḫašû*-plant, the nourishment of the land (V 49, Abusch 2016: 139). Note the 'secret names' of the plant in URU.AN.NA: *maštakal*: "Black (spot/hair?) from the leg of a donkey / claw of black dog" (III 46–47, Rumor 2017: 10) and the fact both animals, donkey and dog, often appear in the corpus in question and that the animal substances often come from the thigh of animals which mount their females.

ḫaṭṭi-rē'i (ᵘNÍG.GIDRU): litt. "shepherd's-crook," an unidentified plant. Böck (2014: 141–142, 147–148, 157–158, 167) argues that *ḫaṭṭi-rē'î* is a name for the plant *būšānu* (as *lišān-kalbi*, *azallû*, and *ṣaṣuntu*). See also CAD Ḫ 156.

ḫulālu (ⁿᵃ⁴NÍR): semi-precious stone with black and white stripes, a type of agate, see CAD Ḫ 226–227; AHw. I 353; Frahm 1997: 147–148; Schuster-Brandis 2008: 436.

ḫurāṣu ((ⁿᵃ⁴)KÙ.SI$_{22}$): gold, see CAD Ḫ 245–247; AHw. I 358. When with the determinative NA$_4$, like in No. C.1: 11, it designates a gold bead or perhaps a stone bead whose appearance is similar to gold.

ḫûratu (ᵍⁱˢḪAB): an unidentified plant, see CAD Ḫ 247; AHw. I 358 'Gerber-Sumach'; CDA 121 'madder (*Rubia tinctorum*).'

imbû tâmti (ⁿᵃ⁴KA A.AB.BA; KA *tam-tim*): it is present in ḪAR-ra = *ḫubullu* XI after words referring to sediments in rivers, probably indicating some mineral of the sea coast. In *nīš libbi* texts it appears along with the determinative na$_4$ (stone); while in URU.AN.NA: *maštakal* (I 664ff.) with the determinative ú (*imbû* indicates the fiber of the date palm). There are reasons to think that it refers to the coral or coral limestone. According to other scholars, it is a type of seaweed or seagrass (AHw. I 375 'Alge, Algenschlamm'; Schuster-Brandis 2008: 421–422 'alga'; cf. CAD I-J 108–109 'a mineral'). In URU.AN.NA: *maštakal* it appears along with ᵘKU.SA A.AB.BA : ᵘMUL *tam-tim* 'starfish' (I 667–668).

imḫur-ešrā (ᵘIGI.NIŠ): an unidentified medical plant, lit. "it heals (or counteracts) twenty (sickness)," see CAD I-J 117–118; AHw. I 376; Thompson 1949 122–125. See its mentioned in URU.AN.NA: *maštakal*, along with the plants *azallû*, *ḫassu* and *imḫur-līm* (Kinnier Wilson 2005: 49). See *Šumma šikinšu* text II, § 15 (1–2) (Stadhouders 2011: 17, see also text I § 20 (1–3)). See BAM 379: "The *imḫur-ešrā*-plant – it is good as a drug against feeble-mindedness, [Deputy Pow]er of Gula. (You apply) likewise" (ii 56, Stadhouders 2011: 35, ibid. 19: 16). In BAM 1 it is an ingredient for sickness of the stomach (TÙN/*tākaltu*) (ii 51, Attia and Buisson 2012: 28). The plant has a wide therapeutic function, it is used in prescriptions against the "ghost hand" always together with the *imḫur-līm* and *tarmuš* plants (as in many other prescriptions), and often in groups of plants containing *urnû* (see Scurlock 2006a: 70–71 fn. 1222; Herrero 1984: 46–47; KADP 36 iii 3–5; *Practical Vocabulary* ll. 95–97). The same three plants are mentioned in recipes to relieve kidney and rectal problems, particularly "restriction" of the bladder (see Geller 2005: No. 2 i 33, 35, ii 4) and "flatulence" of the kidneys (No. 9 iv 18′). It is a remedy against witchcraft, like the *imḫur-līm* and *tarmuš* plants (KADP 1 rev. V 19, 24, 27–28). See KAL 2 36: "(I arm myself against you witches) with the *imḫur-ešrā*-plant that does not allow the magic to approach (the body)" (r. v 41′ and dupl., Abusch and Schwemer 2011: 166, No. 7.8.1: 24′). These three plants appear frequently in the texts of anti-witchcraft. See in particular Abusch and Schwemer 2011: 126–127, No. 7.5: 1′–16′: the plant is held in the left hand (while he keeps *tarmuš*-plant in his mouth) of the one who recites the incantation dedicated to the god Šamaš. This ritualistic procedure is also reflected in a *Bīt rimki* text, whose ritual protagonist is the king, and which is followed by the incantation *Bēl bēlī šar šarrī Šamaš* (W. 22730/6 r. iii 28–41, von Weiher 1983, No. 12). Another incantation against witchcraft emphasizes the properties of the plant to act against it: "*imḫur-ešrā*-plant that does not allow magic to come near (var. adds: the body)" (Abusch and Schwemer 2011, No. 7.8.1: 24′). The identity of the plant is uncertain, according to Thompson it would be *Chrysanthemum segetum* (1949: 125), but there is not enough evidence to be sure.

imḫur-līm (ᵘIGI-*lim*): an unidentified medical plant, lit. "it heals (or counteracts) thousand (sickness)," see CAD I-J 118; AHw. I 376; Thompson 1949: 122–125; Perdibon 2019: 116. It is a widely used ingredient in *nīš libbi* rituals, and anti-witchcraft texts. In *Maqlû* it is also used against witchcraft and defined as "the plant which releases" (VI 102, Abusch 2016: 161). See also Abusch and Schwemer 2011 No. 7.8.3: 17′–30′; No. 8.7: 110‴, 113‴. It reoccurs in a group of seven plants in a *nīš libbi* incantation which states that the cause of suffering is witchcraft, No. E.2. Note its 'secret name': "dust from the tracks of a wolf" (URU.AN.NA: *maštakal* III 23, Rumor 2017: 7), the wolf is mentioned several times in the corpus. The identification of the plant is still unknown, despite Thompson's idea that it could be a type of heliotrope (1949 124–125). See *Šammu šikinšu* text 1, § 24′: 62′, § 23′: 58′–61′ (Stadhouders 2011: 10–11, see also ibid. text II § 14 and BAM 379, ii 55′). In *nīš libbi* texts, the plant is never used alone, in contrast to other medical texts in which it is employed for external use against the sting of the scorpion (AMT 91, 1 r. 12), as well as against itching. While its internal use, drunk in beer or wine, is good against snake bites (AMT 92, 7: 8); urinary problems; "restriction" of the bladder (instead of flour of the "male" mandrake and of the *dadānu*-acacia in beer) (AMT 59, 1: 30, see Geller 2005: 45, No. 2); birth with uterine problems (KAR 195 r. 18); menstruation (KAR 194: 37). The emphasis in the latter two examples is connected to the female world and in particular to birth and pregnancy: "Plant for a woman who does not bear" (BAM 380 r. 27; dupl. ibid. 381 iii 19, see Stol 2000: 53). The plant is a remedy against the "restriction" of the bladder, crushed and drunk in wine (BAM i 25); against the inflammation by sun-heat, a possible cause of the absence of sexual desire (ibid. I 58); for the stomach (TÙN/*tākaltu*) (ibid. ii 50).

isqūqu (ZÌ.KUM): a fine quality of flour, see CAD I-J 202–203; AHw. I 389.

iṣ pišri (GIŠ BÚR): an unidentified medicinal plant, lit. "wood-of-release." The plant names *bukānu* and *iṣ pišri* are interchangeable. See Abusch and Schwemer 2011 No. 7.10: 97′;

7.8.1: 23′; 7.8.4: 69′–73′; Abusch and Schwemer 2016, No. 3.4: 17 (cf. also No. 8.18: 17′). GIŠ BÚR/*iṣ pišri*/*gišburru* is as a designation of the exorcist's ceremonial mace (*gamlu*). See its variant name ^{giš}ŠÍTA 'weapon' in Sumerian. For *bukānu* = ^{ú}GAN.NA = ^{ú}GAN.U$_{5}$ = *iṣ pišri* = ^{giš}ŠÍTA/ ^{giš}ŠITA see Schwemer 2007b: 114; 2009: 66, commentary on BM 40568 rev. 7. See Perdibon 2019: 167.

kabullu (^{ú}*ka-bul-lu*): an unidentified medicinal plant, see CAD K 131; AHw. I 434. Thompson (1949 14: 132) suggests the identification with the camphor, in Arabic *kāfur*, plant with aphrodisiac properties. Interestingly, the plant only appears in the *nīš libbi* texts, all of which makes identification impossible. The suggestion by Thompson that it is an aphrodisiac is supported by the fact that it appears in this corpus. In any case, whether or not an aphrodisiac, the use of the plant can be associated with sexual desire.

kammantu (^{ú}ÁB.DUḪ; ^{ú}*ka-man-du*): an unidentified medicinal plant, see CAD K 109–110; AHw. I 432 'eine Gemüsepflanze'; Thompson 1949 163–164 '*Rhus coriaria*, sumach'; Donbaz and Stolper 1993; Scurlock 2007 *Lawsonia inermis L.*/'henna.'

kamkadu (^{ú}*kám-ka-du*, ^{ú}*kan-ka-du*): an unidentified medicinal plant, see CAD K 123–124; AHw. I 432 'eine in Felsspalten wachsende Pflanze'; Thompson 1949: 166–168 '*Colchicum*.'

kammu ša aškāpi (*kam-ma šá* ^{lú}AŠGAB): fungus of tanner (see KADP 11 iii 4–5), see CAD K 125 *s.v.* *kammu* A mng. b; AHw. I 433.

kamūnu (^{ú}GAMUN): cumin, see CAD K 131–132 *s.v. kamūnu* A; AHw. I 434 *s.v. kamūnu* I.

kamūnu (UZU.DIR): a fungus, see CAD K 132 *s.v. kamūnu* B; AHw. I 434 *s.v. kamūnu* II.

^{ú}KASKAL.MUNUS: an unidentified medical plant.

kaspu (($^{\text{na4}}$)KÙ.BABBAR): silver, see CAD K 245–247; AHw. I 454. When with the determinative NA$_{4}$, like in No. C.1: 11, it designates a silver bead or perhaps a stone bead whose appearance is similar to silver.

kasû ($^{\text{(ú)}}$GAZI^{sar}): an unidentified plant used as a condiment and for medicinal purposes. Several identifications have been proposed, for example, Landsberger and Gurney (1957–1958: 337–338) identify it with 'mustard' (see also CAD K 250; Thompson 1949: 194–197; Landsberger 1967b: 151–152, No. 70). Choukassizian Eypper (2019) identifies it with tamarind.

kazallu (^{ú}*ka-zal-lu*): an unidentified medicinal plant, see CAD K 309; AHw. I 467; Thompson 1949: 15; Postgate 1973: 174.

kibrītu (PIŠ$_{10}$.^{d}ÍD; *kib-rit*): sulfur, see Schuster-Brandis 2008: 423. It is not an ingredient much used in these kinds of rituals. In the magical context sulfur is often mentioned as related to purification as in KAR 43: "May you be as clean as sulphur" (l. 28, Ebeling 1931: 16, 26); or in *Maqlû*: "I am the holy Sulphur, daughter of great heaven" (VI 69, Abusch 2016: 158), capable of delivering a man from witchcraft: "Sulphur, Sulphur, Sulphur, daughter of River, Sulphur, daughter-in-law of Rive [. . .]. May Sulphur rele[ase] the sorcery that the seven and seven have performed against me" (VI 78, 83). In *Maqlû*, the one who recites the incantation is always identified with sulfur: "Sulphur my physique" (VI 98); "Like pure Sulphur my ha[ir] (is pure)" (VI 102, also VI 85). In medical texts, in addition to its use for fumigation which often has the purpose of removing the evil spirits by means of its unpleasant smell, sulfur powder is also used, often along with other ingredients, for the preparation of balms with which the body is rubbed, or used on parts of the patient's body, such as the head and the skin (see AMT 1, 2: 8; BAM 159 vi 48; ibid. 156: 40; ibid. 156:45; ibid. 199: 4, see Thompson 1936: 40–41).

kibrītu ru'tītu (PEŠ$_{10}$.^{d}ÍD ÚḪ.^{d}ÍD): a kind of sulfur with a green-yellowish color, see Thompson 1936: 38; Faber 1975: 190; Schuster-Brandis 2008: 451.

kirbān-eqli (^{ú}LAG.A.ŠÀ.(GA)/GÁN/GÁ): unidentified medicinal plant, lit. "field clod"-plant, see CAD K 401–404 *s.v. kirbānu* mng. 3; AHw. I 483–484 *s.v. kirbānu* mng. 8.

kubtu (*ku-ub-tú*, also IM.DUGUD): lump of metal, see CAD K 487; AHw. I 495.

kukru (^{šim}GÚR.GÚR): an aromatic plant, often used for fumigation, see Thompson 1949: 262–265. The provenance of *kukru*-plant as being from the mountains is mentioned also in *Maqlû* (V 49, VI 22, 24–25, 34–35, 67, VIII 47′), where in two passages the mountains are designated as holy (*quddušūti*, VI 25, 35) and in No. K.2: 23. See No. K.2: 22–23 and 25: "You, may they purify you may [they] c[leanse you], / the holy [*kukru*-plant]s, the inhabitants of the heights (and) mountain ran[ges]." The plant is used for purification practices. Note that "green frog" (*muṣa''irānu arqu*) is maybe a 'secret name' for the *kukru*-plant and for the drug *baqqu*-gnat (URU.AN.NA: *maštakal* III 43–43a, Rumor 2017: 10).

kurkanû (^{ú}*kur-ka-na-a*): unidentified medicinal and aromatic plant, maybe turmeric, see CAD K 560–561; AHw. I 510; Thompson 1949: 157–161; Scurlock 2020.

kušru (^{ú}KU.GAG): *kušru:* unidentified plant, see CAD K 600; AHw. I 517; Thompson 1949: 203.

laptu (^{ú}*la-ap-ti*, also LU.ÚB^{sar}): maybe turpin, it is a rare ingredient used in the corpus, see CAD K 96 *s.v. laptu* A 'turpin'; AHw. I 537 *s.v. laptu* II 'Speiserübe.'

lipāru (^{ú}MI.PÀR): a fruit tree or shrub, see CAD L 198 'a fruit tree'; AHw. I 554 'ein Baum oder Strauch'; Gelb 1982: 79–82 'Apfel'(?); Postgate 1987: 119–120 'Maulbeere.'

lišān-kalbi (^{ú}EME.UR.GI$_{7}$): lit. "dog tongue"-plant, see CAD L 209; AHw. I 556. KADP 2: 40–42 (Stadhouders 2011: 38) states that the *lišān-kalbi*-plant, upon which the gecko lies (MUŠ.DÍM.GURUN.NA/*pizallurtu*), and whose other name is ^{ú}NÍG.GIDIR/*ḫaṭṭi-rē'î* "shepherd's-crook," is good for sexual desire if pulverized and rubbed with oil upon the patient. In KADP 4 we find a similar passage, but in association with quite a different diagnostic problem – a woman "who does not give birth": "The plant upon which the gecko usually lies: *ḫaṭṭi-rē'î* ('shepherd's-crook') / its name is *lišān-kalbi*, for a woman who cannot give birth" (r. 36–37, Böck 2014: 141). In BAM IV 380 (and dupl. ibid. 381 iii), the section preceding the one on the sexual desire, concerns pregnancy and childbirth (see Stol 2000: 52–53–54) and mentions the *lišān-kalbi*-plant and its seeds, used for a woman with a difficult labor (BAM IV 380 r. 31–32; ibid. 381 iii 23–24). Böck, in her study on the goddess Gula (2014: 129–158), analyzes the plants *būšānu*, *lišān-kalbi* and *ṣaṣuntu* because, according to her, they are linked to the goddess. She argues that *būšānu* and *lišān-kalbi* are two names for the same plant (on the criticism of this interpretation see infra Chapter III fn. 369). The *lišān-kalbi*-plant has a wide medical use for a variety of problems and diagnostic categories. The plant is often used alone (see Böck 2014: 142–156) for internal use for stomach and urinary troubles, jaundice, childbirth, menstruation, and cough; externally for swellings, and as a poultice for the eyes. As an amulet it is useful for dog and snake bites. According to Thompson (1949: 26–27), the plant can be identified as the Cinoglossa (lit. "dog tongue," in Arabic this plant is also called *lisān al-kalb*, also Böck 2014: 157–158). According to Böck, both *lišān-kalbi*-plant and the absence of sexual desire are related to the goddess Gula (*contra* Böck's interpretation of the relationship between the goddess and the loss of desire see Chapter III fn. 369).

maštakal (^{ú}IN$_{6}$.ÚŠ; ^{ú}IN.NU.UŠ; also ^{ú}URU.AN.NA): probably a kind of soapwort, see CAD M/I 391–320; AHw. II 630; Thompson 1949: 39–43 'Struthium, Saponaria'; Maul 1994: 65; Panayotov 2018: 208–209, on the contrary, *Salsola*; Perdibon 2019: 163–166. It is considered a purifying plant (see Maul 1994: 65) and is used against witchcraft (KADP 1 rev. v 19). This becomes evident since in the lexical lists it is associated with plants that have the same function, such as tamarisk (*bīnu*) and the *sikillu*-plant (ḪAR-ra = *ḫubullu* XVII 131ff.; URU.AN.NA: *maštakal* III 419). The medical use of the seeds of the plant is also confirmed by a letter of Urad-Nanâ to King Esarhaddon on Prince Eṭlu-šamê-erṣeti-muballissu's health issues (Parpola 1970: 194, No. 251 r. 10). Its purifying function is emphasized in many incantations and rituals: "May the tamarisk purify him, may the *maštakal*-plant absolve him" (*Lipšur Litanies*, Reiner 1956 136: 74). In *Maqlû*: instructions for tablet VIII 157′–159′, Abusch 2016: 223; VIII 47′–49′, ibid. 197–198; I 21–23, ibid. 29; I 46–49, ibid. 33–34; III 173, ibid. 109. See also *Bīt rimki*, Borger 1967: 10, 6–9; Abusch and Schwemer 2011: 188–189 No. 7.8.4: 3′–4′; ibid. 331 No. 8.6: 95′–96′; ibid.

347 No. 8.7: 105‴–112‴; CUSAS 32, II.A.4 No. 5i v 7′–36′, George 2016: 60. See also the ritual commentary in Livingstone 1986: 176–177 l. 5 where the plant is equated to the god Ea.

mekku: a mineral, frit or raw glass used in the production of glass and as an amulet stone, see Schuster-Brandis 2008: 430.

mikû (*mi-ki-í*): an unidentified medical plant, see CAD M/II 8 *s.v. mekû*; AHw. II 642 *s.v. mekû* II 'eine bittere Pflanze.'

murdudû (^{ú}MUL.DÙ.DÙ in Bogh./ normal writing ^{giš}*mur-du-du*): an unidentified plant used against witchcraft, see CAD M/II 219; AHw. II 675–676.

nurmû (^{giš}NU.ÚR.MA): pomegranate-(tree), see CAD N/II 345–347; AHw. II 804–805.

murru (^{šim}SES): the term has been identified with 'myrrh' but it is not certain. The mention of its seeds also suggests a "native 'bitter' plant" (CAD M 222). In many texts it is difficult to understand, whether myrrh or a plant with astringent qualities is meant. See BAM 1 i 22 where use of myrrh is made for bladder problems; *Šammu šikinšu* text II § 1 for purging the bowels; text IIIa § 4–5 good for the rectum. See also AHw. II 676; Thompson 1949: 309–310; Van Beek 1958; Farber 1993–1997: 536–537; Jursa 2009: 163.

muššaru ($^{\text{na4}}$MUŠ.GÍR): semiprecious stone with red and white stripes, see Schuster-Brandis 2008: 433).

mūṣu ($^{\text{na4}}$*mu-ṣa*) the dictionaries (CAD M/II 246–247; AHw. II 670) give two meanings: the first one indicates a sickness, the second one a mineral. The *mūṣu*-sickness, characterized by incontinence, has been identified as urinary schistosomiasis (see Kinner Wilson 1968: 245–246) or a urinary infection caused by stones or pyelonephritis (see Geller and Cohen 1995: 1812–1813). The sickness also affects women (see URU.AN.NA: *maštakal* III 161). The *mūṣu*-stone, however, is employed for therapeutic and magical use in a variety of texts. According to CAD M/II 247, it may refer to kidney stone (in relation to the sickness) and consequently to the same mineral. Schuster-Brandis (2008: 432) thinks that the two should be separated. See also Herrero 1975: 49–50.

nikiptu (ŠIM.^{d}MAŠ; also ^{d}NIN.URTA): an unidentified aromatic, gum-yielding plant, possibly a *Euphorbia* shrub (spurge), see Thompson 1949: 364.

nīnû (Ú.KUR.RA): an unidentified plant used in medical contexts and as a spice, see CAD N/II 241; AHw. II 791 'Ammi, Zahnstocherdolde'; Thompson 1949: 67–69 '*ammi*'; Kinnier Wilson 2005: 50–51 'mint.' In BAM 1 I 46 is a medicament for an attack(?) of *ašru*-sickness (see CAD A/II 460, for other readings see Attia and Buisson 2012: 36); iii 9–10 for colorectal sickness in a suppository to put into the rectum. Its use for rectal problems is confirmed by certain prescriptions, for example STT 100: 3′–5′and 9′–10′, Geller 2005: 232–233, No. 43. It must be said that the plant has a broad therapeutic use; that is, it is considered an important ingredient in many diagnostic categories (see Thompson 1949: 67–68). In *Maqlû* it is mentioned with a metaphor: "Like *nīnû*-plant may her witchcraft give way" (V 27, Abusch 2016: 136). Its anti-witchcraft function is emphasized in a *nīš libbi* incantation to be recited by a man hit by a curse, No. E.2. In the *nīš libbi* prescriptions it is an important ingredient. For example, it is applied to the flesh of the wren (*diqdiqqu*) and the NAM.GEŠTIN-bird in *nīš libbi* N prescr. 5–6. Jiménez (2017: 342) emphasized the relationship between the wren and the plant. In fact, the name of the bird can be written NI.NI-*qu*, thus recalling the name of the plant. In the lexical commentary ḪAR.gud (C l. 296, Landsberger 1962 = MSL 8/2: 179) the bird is called *iṣṣūr samēdi* "bird of the *samēdu*-plant." This last plant has a logographical writing similar to that of the *nīnû*-plant (^{ú}KUR.RA): ^{ú}KUR.ZI.

nuḫurtu (^{ú}NU.LUḪ.ḪA): an unidentified plant used mainly in a magic-therapeutic context, see CAD N/II 322; AHw. II 802; Thompson 1949: 354–358 *Asafoetida*. It is considered an important remedy against witchcraft. Its healing power against witchcraft is underlined in *Maqlû*: "I have scattered *nuḫurtu*-plant (upon a censer) at the head of my bed – *nuḫurtu*-plant is especially strong, it will house all your witchcraft to wither" (VI 133′–134′, Abusch 2016: 163) and in the instructions: "Incantation: 'Ha! My witch, my inseminatrix'. / He recites (it) over *nuḫurtu*-plant, you then place (it) upon the censer that is at the head of the bed. / You surround the bed with (colored) twine" (V 115′–117′, Abusch 2016: 219). In this series the plant is also the subject of an analogy through appearance: "Like *nuḫurtu*-plant may her lips be made to shrivel" (V 35, Abusch 2016: 137, var.: "like root of *nuḫurtu*-plant may [his witchcraft] cause him to shrivel"). In this regard Worthington writes: «The same idea of inherence assumes a more articulate form when *nuḫurtu* is called upon it to *nuḫḫuru* sorcery 'make it shrivel' – the name and function are obviously linked, and one must have been assigned on the strength of other» (2003: 6–7). In an Old Babylonian prescription, we read: "If a man is bewitched: He drinks root of *nuḫurtu*-plant in sesame oil and will recover" (BAM 393 r. 13–14, Abusch and Schwemer 2011: 65, No. 2.1.2). In BAM 1 its medical properties are emphasized: "The resin ingredient of the *nuḫurtu*-plant – medicament for the restriction of the bladder : he drinks it in the beer (or) you anoint it with oil (or) insert it into his penis by means of a small tube" (i 21); "The resin ingredient of the *nuḫurtu*-plant – medicament for the man who suffers from the *saḫḫu*-sickness : he drinks beer with honey, oil and beer" (ii 41, for *saḫḫu*-sickness see Scurlock and Andersen 2005: 704, No. 109, text 6.101). In medical texts that compile treatments of kidney and rectal problems it appears as a frequently-used ingredient (see Geller 2005). It is also present in the texts against ghost attacks (see Scurlock 2006a: 779). See, on the use of the medical plant, also Thompson 1949: 354–355. The fact that it appears in the *nīš libbi* texts is due to its ability to defeat witch's attacks, and probably for its medicinal properties against bladder problems, as well as urino-genital apparatus.

nuṣābu (^{ú}*nu-ṣa-bu*): an unidentified medical plant, see CAD N/II 353–354; AHw. II 806 'eine Anemone'(?); Thompson 1949: 139–140.

pallišu (^{na4}NÍG.BÙR.BÙR): an unidentified stone, see Schuster-Brandis 2008: 435.

pappardillû (^{na4}BABBAR.DIL): black semi-precious stone with white streaks, see Schuster-Brandis 2008: 403–404.

pappasītu (BA.BA.ZA.^{d}ÍD): a kind of sulphur, see CAD P 111.

parzillu (AN.BAR): iron, see CAD P 212–213; AHw. II 837–838.

puquttu (^{ú}*pu-qut-tú*, ^{giš}KIŠI$_{16}$.ḪAB): a thorn, see CAD P 515–516; AHw. II 880.

riḫu (^{na4}*ri-ḫu*): an unidentified stone, note that it is not attested. The reading is uncertain.

^{ú}SAG: an unidentified plant. The plant is equated with *la-a-ar-tu* in ḪAR-ra = *ḫubullu* XVII (Landsberger et al. 1970 = MSL 10: 108, l. 25), see also CAD L 103 *s.v. lardu*.

saḫḫû (^{na4}*saḫ-ḫu-u*): an unidentified mineral or stone, see Schuster-Brandis 2008: 441.

samīdu (^{ú}KUR.ZI; ^{ú}*sa-mì-dì*): an unidentified vegetable, spice and medicinal plant, maybe a soapwort. It is a rare ingredient used in the corpus, see Tsukimoto 1985: 436.

sāmtu (^{na4}GUG): cornelian, Schuster-Brandis 2008: 413–414.

sassatu (^{ú}*sà-as-sà-ta*; also ^{ú}KI.KAL): a salt grass commonly growing along Babylonian waterways, see CAD S 194; AHw. II 1032 'Bodenbedeckung, Gras'; Civil 1987: 48 *Poa sinaica* ("meadow grass" or "bulbous bluegrass"); Landsberger (1934: 65) thinks that the meaning of the name of the plant is "Pflanze des unbebauten Landes." For other references see Landsberger 1933: 227; 1949: 275 fn. 83. Grass is not often used in a therapeutic context; however, it has the power to deliver a patient from evil: "Grass that receives every evil receive (mine) from me and, grass, carry off my evil" (KAR 165: 14, Scurlock and Anderson 2005: 117; see also KAR 165: 21). The plant is also connected to purity: "May he make me as pure as *sassatu*-grass" (BMS 11: 25, cit. in CAD S 194); "In your presence (= Gods of the Night) I have (now) become pure like *sassatu*-grass" (*Maqlû* I 25, Abusch 2016: 29, see Landsberger 1934: 66). In *Maqlû* the "raising" of the patient, which contrasts the evil action of the magician, is likened to the grass growing on the bank of the canal (VI 93, Abusch 2016: 160). Lamaštu finds its refuge in the *sassatu*-grass: "Her abode is (in) the marshes, her lair in the grass" (*Lamaštu* II 121, Farber 2014: 174–175). In *Šurpu* the god Girru is invoked to placate the *sassatu*-grass, in this context negative, growing in the grass: "Ditto (= may placate) also the weed, sprung up in the grass" (V-VI 192, Reiner 1958: 35). In *Šurpu*, among the many oaths, that of "the tearing up the grass" is invoked (VIII 50, Reiner 1958: 42). In *Šumma ālu* it appears quite frequently (see Guinan 1996). In CT 39, 6: 7 grass is associated with *arantu*-grass (see Langdon 1916: 31). The problem is that often grass is written by means of the logogram KI.KAL, whose reading can be both, *sassatu* as well as *arantu*.

sīḫu ([ú]*si-ḫu*): unidentified medicinal plant, see CAD S 241–242 *s.v.* *sīḫu* B; AHw. II 1040 *s.v.* *sīḫu* II 'Wermut (Artemisia)'(?); Thompson 1949: 261–262 'pin blanc, *Pinus Halepensis*.'

sikillu ([ú]SIKIL): unidentified medicinal plant, lit. "the pure (herb)," see CAD S 243–244; AHw. II 1041 'eine Pflanze (Zizyphus-Art?),' In URU.AN.NA: *maštakal* I 6 it is associated with the *maštakal*-plant; ibid. I 19 it is defined as *maltakal šadî* "mountain *maštakal*-plant" (see CAD S 244–245), while at ibid. III 4 its 'secret name' is "fat of 'nest' snake" (Rumor 2017: 4). The *Šammu šikinšu* remarks that the leaves of the plant are similar to those of the *ḫašḫūr api*-plant and that it is a plant for purification and to dispel witchcraft (text I, § 19 (1), Stadhouders 2011: 7). The connection with purity is present even in the lexical lists: [ú]SIKIL = KI.MIN (= *ma-al-ta-kal*), *ú-si-ki-lu*, *šam-mu el-lu*, [ú]SIKIL.E.DÈ = *šam-me te-lil-te*, Ú.SIKIL = *šam-mu el-lu* (ḪAR-ra = *ḫubullu* XVII 134ff.). Note the association with other plants such as tamarisk and *maštakal*, whose purifying function against witchcraft is known. In a *nīš libbi* incantation, No. E.2, the plant's role for the purification against witchcraft is mentioned.

sû ([na4]*su-u*; also [na4]ŠU.SAL.LA): probably a red sandstone, see Schuster-Brandis 2008: 442–443.

supālu: the term (CAD S 390–391; AHw. II 1059–1060) refers to two different plants: a variety of juniper ([giš]ZA.BA.LUM; [ú]NIGIN[sar]) and the *supālu*-plant ([ú]MUNZER). For the reading [ú]MUNZER of the signs [ú]KI.[d]NANNA see Civil 1966: 122–123. On the Sumerian term [ú]munzer to be identified with liquorice see Civil 1987: 46; with *Lecanora esculenta* see Kinnier Wilson 2005: 5. The *supālu*-plant is mentioned in the *Šumma ālu* series (CT 39, 9: 18, see Guinan 1996). Note the "secret name" of the plant in URU.AN.NA: *maštakal*: "dust of crossroad from burnt ashes" (Rumor III 19, 2017: 7) (see also URU.AN.NA: *maštakal* CT 14, 46 r. 1–7). The plant is associated with the *ardadillu*/*aštatillu*-plant in ḪAR.ra = *ḫubullu* XVII 44–47 (Landsberger et al. 1970 = MSL 10: 84).

ṣaṣuntu ([ú]*ṣa-ṣu-un-tú*; [ú]*ṣa-ṣu-ut-tú*; [ú]*ṣa-ṣu-um-tú*; [ú]*a-ṣu-ṣu-um-tú*/*tum*): an unidentified medicinal plant, see CAD: Ṣ 116; AHw. III 1087a; Thompson 1949: 277–278; Böck 2014: 158–163. In her study of the goddess Gula, Böck (2014) analyzes the plant's relationship with the goddess of medicine. In fact, in URU.AN.NA: *maštakal* (l. 42a) the plant is called "Gula's plant." The same text states that the plant is used against the *sāmānu*-sickness (l. 42), as BAM 1 ii 19 (Attia and Buisson 2012: 27, see KADP 1 v: 15; CT 14, 41 Rm. 362 l. 6′). Even in URU.AN.NA: *maštakal* (l. 286) the plant is mentioned for its ability to cure the *bibirru*-sickness. *Ṣaṣuntu* is also a medical ingredient against inflammation by sun-heat as stressed by BAM 1 i 49. The plant is also used against the *asû*-disease (BAM 1 ii 3, Attia and Buisson 2012: 27). Another interesting use of the plant is against wild animals that come close to a house (KADP 1 15–17). In this regard Böck writes: «Two explanations are possible, one referring to the smell of the plant, which was so unpleasant that it repelled the animals and another one, related to the ideational level. The use of Gula's plant *ṣaṣuntu* could have been motivated by the association with her dog, the plant would turn into an actual animal which as guard dog or sheep dog would keep (wild) animals away» (2014: 163). See its 'secret names': "head of a black raven" (URU.AN.NA: *maštakal* III 103); "wool of a virgin ewe" (ibid. III 123).

ṣurru ([na4]ZÚ): black stone, maybe obsidian, see Schuster-Brandis 2008: 457. In the catalogue KAR 158 it designates the male sexual organ: *râmka lū ṣurru* / *ṣīḫātuka lū ḫurāṣu* "Your love is *ṣurru*-black stone, your lovemaking is golden" (vii 43′–44′, Wasserman 2016: 212).

šadânu ṣābitu ([na4]KA.GI.NA.DAB.(BA)): magnetite, see CAD Š/I 36b 'magnetite, lodestone'; AHw. III 1123a 'Magneteisenstein'(?); George 1979: 134 fn. 47; Postgate 1997: 212 and fn. 45–46;

Salvini 1999: 380 and fn. 29; Schuster-Brandis 2008: 424–425. In *Abnu šikinšu* (BAM 194 vii 15′–18′) is defined "stone of truth" (Schuster-Brandis 2008: 33 and 39, text E). The connection between, justice, represented by the god Šamaš, and the magnetite can be found in *Lugal-e* 497–512 (Seminara 2001: 176–177). The magnetite is used in many prescriptions, in particular for the making of amulets and necklaces (see Schuster-Brandis 2008: 425). An incantation against witchcraft attacks underlines the magical power of magnetite: "Incantation: I have equipped myself with magnetite, [...] ... [...]/ which does not allow evil magic to come near. You in heaven, pay attention to [me], you of the earth, listen to me!" (Abusch and Schwemer 2011: 189, No. 7.8.4: 25′–26′). The ritual that follows involves making figurines of the male and the female magician whose mouths are sealed by a seal of magnetite, sent, according to the incantation, by the very Šamaš himself. The stone is also used for amulets to acquire sexual desire: BAM 419 ii′ 13′–iii′ 5 (also BM 56148+V 41–42; CTMMA 2, 32 iii 12–14; K. 3010+v 22′–36′, see Schuster-Brandis 2008: 137, Kette 127). These stones must be wrapped into a woolen band of a reared-up ram. The ritual must be accompanied by reciting the "SU.ZI" and "KI.IN.DA.RAB" incantations (K. 3010+v 25′–26′). See that most of the ointments in *nīš libbi* corpus are composed of iron and magnetite powder to rub on the male and female genital organs. The use of both ingredients is clearly to stimulate sexual attraction.

šakirû (^{ú}ŠAKIR): an unidentified plant, see CAD Š/I 167–168; AHw. III 1140 'Bilsenpflanze'(?); Thompson 1949: 230 *Hyoscyamus niger*; contrary to this identification see Civil 1987: 42.

šammi balāṭi (Ú.NAM.TI.LA): lit. "plant-of-life." The name "plant of life" is mentioned in Neo-Assyrian royal inscriptions and in a letter addressed to the king in the form of similitude (see KWatanabe 1994: 588–589; Kübel 2007: 190). The plant is mentioned in an incantation dedicated to the goddess Gula (KAR 73:30–31: "Gula 4" see Mayer 1976: 387; BAM 404): "I took this 'plant of life' of my lady (Gula) [...] and I came alive" (see KWatanabe 1994: 590). The only explicit reference to the plant in literature is in the mythological Sumerian composition *Inana's Descent into the Netherworld*. In this text, Enki creates *kurgarra* and *galaturra* and assigns them the "plant of life" and the "water of life" to revive Inana, which allows her to escape the underworld (ll. 224–225, 252–253). According to Watanabe, however, there is no relationship between the use of the plant on the one hand in the inscriptions, the letter and the mythological epic and on the other hand in incantations and therapeutic texts in general. Her opinion is that in the first case it is a similitude, an allegorical function, in the second a real substance. Clearly, then, one must distinguish between the real vegetal substance on one hand and the periphrastic or allegorical use of the expression "plant of life" on the other. Also, in mythological and epic literature, one can trace a therapeutic function for these substances, as also is the case with the "plant of life," and one cannot draw a clear dividing line regarding its mention in texts of different nature (see KWatanabe 1994: 595). The plant is certainly used as a drug in therapeutic prescriptions (see *Šumma šakinšu* text IIIa, § 15′, Stadhouders 2011: 27 and ibid. 2012: 13). In this regard, Thompson (1949: 227–228) identifies the plant ingredient as opium. In BAM 248: 34–35 (dupl. AMT 67, 1 iv 27–28) it is used to treat a woman who is giving birth when her labor is difficult: "You have received my present for you; give me the 'plant-of-life,' so that NN, daughter of NN, may have an easy delivery" (see for the ritual Veldhuis 1991: 253; Reiner 1996: 37–38; Stol 2000: 71).

šambaliltu (^{ú}SULLIM): fenugreek, see CAD Š/I 310–311; AHw. III 1156; Thompson 1949: 64–65.

šarmadu (^{ú}GUR$_5$.UŠ): an unidentified plant, used in therapeutics, see CAD Š/II 64–65; AHw. III 1187; Thompson 1949: 228–230.

šibburratu (^{ú}LUḪ.MAR.TU/TÚ): an aromatic plant, not identified with certainty, see CAD Š/II 376; AHw. III 1226. According to Thompson (1949: 75–76), it is probably rue, possibly a syriac variety, *šabbârâ, Paganum harmala*. The plant is of therapeutic use, against the "restriction" of the bladder along with at least twenty other ingredients to drink in wine or strong beer on an empty stomach (AMT 59, 1: 39, see Geller 2005: 47, No. 2); for an enema (along with other ingredients, BAM 409 r. 22); against urinary problems (along with 27 other ingredients, AMT 82, 1 + duplicates ii 35′, Geller 2005: 97, No. 9); against witchcraft (among many others: KUB 37 43 i 9; ibid. 44: 23; BAM 434 iv 16). For an extensive study on rue in a general Semitic context see Stiehler-Alegria 2006.

šikkurratu (*šik-kur-rat*): a reed or rush, see CAD Š/II 435; AHw. III 1234 'ein Teil des Schilfrohrs(?).'

šimru (Ú.KU$_6$): lit. "fish-plant," maybe fennel. Another possible reading is *urânu* 'anise' or 'fennel' (also written ^{ú}TÁL.TÁL). See URU.AN.NA: *maštakal* I 316–317. See CAD Š/III 8–9, AHw. III 1238, 1430. It is thus not a widely used plant in this corpus.

šumuttu (^{ú}SUMUN.DAR): maybe beetroot, see CAD Š/III 301–302 'beetroot'(?); AHw. III 1276 'eine Gemüsepflanze Bete'(?); Thompson 1949: 49–51 '*Beta vulgaris*, beetroot'; Kinnier Wilson 2005: 47–48.

šunû (^{giš}ŠE.NÁ.A): agnus castus, see CAD Š/III 309–310; AHw. III 1277; Thompson 1949: 296–298.

šurdunû (^{ú}SI.SÁ): an unidentified medicinal plant, possibly rocket, see CAD Š/III 343–344.; AHw. III 1283 'Rauke'(?); Thompson 1949: 210–212 '*Eruca sativa*' (cf. *egingīru*-plant).

šurmēnu (ŠUR.MÌN): cypress, see Thompson 1949: 286–287; Perdibon 2019: 160.

takdanānu (^{ú}*tak-da-na-nu*): an unidentified plant, see CAD T 70; AHw. III 1306 'ein Strauch'; Thompson 1949: 118.

tarmuš (^{ú}*tar-muš*; ^{ú}*tar-muš*$_8$): a plant identified as lupine by the comparison with Arabic and Aramaic, see CAD T 238–239; AHw. III 1331 'Lupine'; Thompson 1949 121–126 'lupine'. See *Šummu šikinšu* BAM 379, ii 57 (Stadhouders 2011: 35 and ibid. 19: 16). In BAM 1 ii 52 it is drug for the stomach (Attia and Buisson 2012: 28). Note the 'secret name' of the plant in URU.AN.NA: *maštakal*: "Fat of a male pig mottled with red / fat from the kidneys of a white pig mottled with red" (III 51–51a, Rumor 2017: 11). In the anti-witchcraft texts, it is frequently used together with the *imḫur-līm* and *imḫur-ešrā* plants (see Herrero 1984: 46–47, also Thompson 1949: 121–126). In fact, all three are considered remedies against witchcraft (KADP 1 rev. V 19, 24, 27–28). Interestingly it also occurs in a ritual in which the bewitched patient in front of Šamaš has in his mouth the *tarmuš*-plant, in his left hand the *imḫur-ešrā*-plant and in his right hand beer (K 3661 r. iv 1′–10′, Abusch and Schwemer 2011: 126–127, No. 7.5).

tīyatu (^{ú}*ti-ia-tú*; also ^{ú}KU.NU.LUḪ(.ḪA)): an unidentified plant, used in medical contexts and as a spice, regularly combined and associated with *nuḫurtu*, see CAD T 400; AHw. III 1357; Thompson 1936: 354.

tullal (^{ú}*tu-lal*): an unidentified medicinal plant, see CAD T 464; AHw. III 1369; Thompson 1949 42–43. It is maybe a soapweed plant (like *maštakal*). In BAM 1 i 34 is medicine for gall-bladder (*martu*) (for a discussion on reading and interpreting the signs see Attia and Buisson 2012: 34; I fellow here CAD T 464). For other uses of the medical plants see Thompson 1949: 42. It is also employed in the anti-witchcraft texts in potions or amulets along with many other ingredients. The association between the *tullal* and *maštakal* plants is found in URU.AN.NA: *maštakal* 4–5. The plants' functions are similar and both are associated with washing and purification (see Maul 1994: 63 fn. 42, also Thompson 1949: 43; Böck 2011: 693). In BBR No. 11 iii 6 the *bārû*-diviner purifies (*ūtallal*, Dt-stem from *elēlu*) himself with tamarisk and the *tullal*-plant. In the *Epic of Gilgameš* the goddess Ninsun purifies herself before her prayer to Šamaš: "She went seven times into the bathhouse, / [she cleansed] herself in water (perfumed with) tamarisk and *tullal*-plant" (III 37–38, George 2003: 576–577). The plant has in fact been associated, through popular etymology, with the verb *elēlu* 'to purify' (CAD T 464; Thompson 1949 42–43), hence its meaning literal meaning: "you make pure."

turminû ($^{\text{na4}}$DUR.MI.NA): detritic rock or marble, see CAD T 487; AHw. III 1373; Thompson 1936: 193; Landsberger 1970: 20; Schuster-Brandis 2008: 406.

ṭabti amāni (MUN *a-ma-ni*(*m*)): a reddish salt, see CAD A/II 2; Thomposn 1936: 5–6.

uḫūlu (^{ú}NAGA): a plant ("salt-plant") and its product (potassium), see Thompson 1949: 31–35.

uqnû ($^{\text{na4}}$ZA.GÌN): blue stone conventionally translated as `lapis lazuli,' see CAD U 195–202; AHw. III 1426; Oppenheim 1970: 9–14; Polvani 1980; Rölling 1987–1990b: 488–489; Schuster-Brandis 2008: 453–454. In incantation No. E.2 reference to its purifying quality is made.

urânu (*ú-ra-nim*, also Ú.KU$_{6}$/ ^{ú}TÁL.TÁL): anise, see CAD U 206–207. Another possible reading of the logogram Ú.KU$_{6}$ is *šimru*, maybe 'fennel.' It is not a widely used in this corpus.

urnû (^{ú}BÚR): unidentified medicinal plant, see CAD U 234–235; AHw. III 1432; Thompson 1949: 77–79. The text BAM 1 mentions the plant twice: iii 1 for the sickness of the rectum; iii 17–18 for an aching rectum (see Attia and Buisson 2012: 28). The plant is in fact used in many prescriptions, along with other ingredients, for kidney and rectal problems. In *Šumma šikinšu* it is good for expelling *urbatu*-worms (BAM 379 ii 11′–12′, dupl. CTN IV, 195 + 196 iii 8′, Stadhouders 2011: 22, text II § 27′: 1–2). The same function can be found in BAM 380 r. 60–61 (dupl. ibid. 381 iv 15–16). The plant is also used in anti-witchcraft rituals for the creation of potions, along with other common ingredients. According to Thompson (1949: 77–79), the plant would be mint (*Mentha pulegium* or *Mentha piperita*), whereas for Abusch and Schwemer (2011: 473) it suggests an identification with *Ammi*.

urṭû (^{ú}*ur-ṭú*): an unidentified medicinal plant, see CAD U 256; AHw. III 1434; Thompson: 1949: 350–351 *Raetam Retama*. It is described in BRM 4 32:10: "The *urṭû*-plant is like tamarisk, but red." The plant is used against parasites that attack humans in BAM 1 iii 33 (Attia and Buisson 2012: 28, see also STT 92 iii 8′; AMT 1, 2: 7). The association with the *arariānu*-plant can be found in URU.AN.NA: *maštakal*: ^{ú}*a-ra-ri-a-nu* = ^{ú}*ur-tu-u* (KADP 2 i 24). The plant, as the *arariānu*-plant, usually is not mentioned in anti-witchcraft texts.

ušû (^{giš}ESI): ebony, see CAD U 326; AHw. III 1442 'Dunkelholzbaum'; Thompson 1949: 289–290.

zumbi-ḫurāṣi (NIM.KÙ.SI$_{22}$): an unidentified medicinal plant, lit. "gold-fly," see CAD Z 155; AHw. III 1535. It only appears in a *nīš libbi* prescription as a medical ingredient. The determinative Ú, in the expression ^{ú}NIM.KÙ.GI A DIR "'gold-fly' to dissolve in water" (KADP 12 iv 68′), suggests that it is the name of a plant.

Index of ingredients

aban kasî (^{na4}GUG.GAZI^{sar}): K prescr. 26: 71.
aban lamassi (NA$_4$.^{d}LAMMA): K prescr. 27: 73.
abāru (A.BÁR, lead): A prescr. 15: 25.
abašmû (^{na4}*a-ba-aš-mu*): F prescr. 17: 55.
adammūmu (wasp): (a-*du-um-mi*): stinger (KUN/*zibbatu*): O prescr. 6: 9.
akalu (NINDA, bread): B prescr. 1: 24; *ennēnu*-barley bread (NINDA ^{še}IN.NU.ḪA/*akal ennēni*): D prescr. 3: 41.
aktam (^{ú}*ak-tam*): No. D.4: 46.
alamû (^{ú}*a-la-mu*): K prescr. 11: 45.
algamišu (^{na4}ALGAMEŠ): F prescr. 17: 56 (restored).
alpu (GU$_4$): hair from the leg (*šarta ša purīdi*) of a breeding bull (GU$_4$/*alpu puḫālu*): A prescr. 2: 6–7; thigh hair of a black bull (*šārat rapalte ša* GU$_4$.NITÁ GE$_6$/*alpi ṣalmi*): E prescr. 2: 63; saliva (*rupuštu*): F prescr. 12: 38–39; saliva of a reared-up bull (*rupušti* GU$_4$ ZI.GA/*alpi tebi*): N prescr. 4 i 18; phlegm of a reared-up bull (*ḫaḫḫu* GU$_4$ ZI.GA/*alpi tebi*): N prescr. 12 iii 2.
aluzinnu (^{ú}*a-lu-zi-in-ni*): seeds (NUMUN/*zēru*): N prescr. 24 le. e. 6 (3 handfulls).
amašpû (or *ašpû*) (^{na4}AMAŠ.PA.È; ^{na4}*aš-pu-u*): catalogue LKA 94 i 23; F prescr. 17: 54; K prescr. 30: 133.
amḫara (^{ú}*am-ḫa-ra*): C prescr. 2: 18.
amīlānu (^{ú}LÚ.U$_{19}$.LU): B prescr. 4 left e. ii 2; K prescr. 3: 28.
amuzinnu: (^{ú}*a-nu-zi-ni*): seeds (NUMUN/*zēru*): O prescr. 4: 5.
ankinūtu (^{ú}*an-ki-nu-tu*): A prescr. 5: 15; root (SUḪUŠ/*šuršu*) A prescr. 17: 28; A prescr. 23: 65 = Q prescr. 3: 9.
arantu (^{ú}*a-ra-an-tu*): H prescr. 3: 5 (restored); K prescr. 9: 42.
arariānu (^{ú}*a-ra-ri-a-nu*): H prescr. 2: 4; K prescr. 8: 40.
ardadillu (^{ú}AŠ.TÁL.TÁL; ^{ú}*ar-da-dil-lum*): A prescr. 4: 14; A prescr. 22: 64 = Q prescr. 2: 6; A prescr. 24: 70; No. E.2: 39; F prescr. 5: 22; F prescr. 11: 34; *aštatillu* (^{ú}*aš-ta-til-la*): F prescr. 3: 14; root (SUḪUŠ/*šuruš*): K prescr. 6: 34; K prescr. 19: 53.
arkabu (ARKAB^{mušen}, a bat): guano (ŠE$_{10}$/*zû*): K prescr. 31: 136 = N prescr. 8 ii 5; guano$^{?}$ (*rikbu*): P prescr. 7 obv. 23; *rikibtu*: N prescr. 14 iii 11, 18; N prescr. 25 lo. e. 2.
arrabu (PÉŠ.ÙR.RA, a mouse): X prescr. 1: 10′; tendon (SA/*šer'ānu*): E Bow ritual: 58.
ašāgu (^{giš}KIŠI$_{16}$, acacia): branch (PA/*aru*): E prescr. 6: 73–75; F prescr. 12: 37, 40–41; N prescr. 12 iii 2, 6 (restored); from a wall (*šá* É.GAR$_8$/*ša igāri*): K prescr. 21: 62 (Ms A).
ašgigû (^{na4}*aš-gì-gì*): K prescr. 28: 77.
ašnugallu (^{na4}GIŠ.NU$_{11}$.GAL, alabaster): No. E.2: 38; K prescr. 26: 71; K prescr. 27: 74.
ašqulālu (^{ú}LAL): K prescr. 3: 28.
ašpû see *amašpû*
asu (^{šim}GÍR, myrtle): L prescr. 3: 8 (Ms A); myrtle oil (Ì/*šaman asi*): O prescr. 6: 14.
asu (AZ, bear): feat of a dancing$^{?}$ bear (*šaman asi*/Ì AZ *muttalliki*): D prescr. 2: 29.
atā'išu (Ú.KUR.KUR): No. D.4: 46; F prescr. 8: 26; F prescr. 9: 30.
atbaru (^{na4}AD.BAR, basalt): A prescr. 11: 21; K prescr. 21: 61.
ayyalu (DÀRA.MAŠ): navel (*abunnatu*): K prescr. 30: 129; penis (GÌŠ/*ušaru*): A prescr. 16: 27; K prescr. 30: 131(?); antler (SI/*qarnu*): A prescr. 16: 27; K prescr. 30: 129; N prescr. 18 iv 3; *rikibtu* (*ri-kib-ti*): A prescr. 16: 27; No. B.2 variant B r. 34; No. C.1: 8, 11; No. F.3: 73(?); No. H.2: 18; K prescr. 30: 130; T prescr. 2 rev. 10′; hair of the tail of a male (stag?) (*azappi zibbati zikari*/KUN NITA): K prescr. 30: 132.
ayyartu (^{na4}PA): F prescr. 14: 48; F prescr. 17: 55 (restored); K prescr. 27: 73; K prescr. 28: 78.
azallû (^{ú}A.ZAL-*ú*; ^{ú}A.ZAL.LÁ): A prescr. 7: 17; A prescr. 8: 18; A prescr. 9: 19; A prescr. 14: 24; F prescr. 4: 17; K prescr. 12: 46; K prescr. 13: 47; K prescr. 16: 50; K prescr. 21: 62; L prescr. 3: 8; P prescr. 3 obv. 8; root (SUḪUŠ/*šuršu*): A prescr. 17: 28; K prescr. 6: 33; seeds (NUMUN/*zēru*): B prescr. 1: 25; K prescr. 31: 138 = N prescr. 8 ii 7.
azupīru (^{ú}ḪAR.SAG): L prescr. 3: 8; P prescr. 3 obv. 8.

ballūṣītu (*bal-lu-ṣi-ti*, a bird): spur (*ḫanduru*/*ḫinduru*): No. K.9: 164; L prescr. 5: 14; P prescr. 10 rev. 5.

https://doi.org/10.1515/9783110757262-015

baltu (^{giš}DÌḪ): root (SUḪUŠ/*šuruš*): K prescr. 18: 52; root of *baltu* (growing) over a grave (SUḪUŠ ^{giš}DÌḪ *ša* UGU KI.MAḪ/*šuruš balti ša eli kimaḫḫi*): K prescr. 3: 29; from a wall (*šá* IZ.ZI/*ša igāri*): K prescr. 21: 62 (Ms. K, L); green (part) (SIG$_7$/*arqūtu*): K prescr. 21: 60 (Ms. A, F, G); resine (ILLU/*ḫīlu*): K prescr. 21: 60 (Ms. K, M).

barīrātu (^{ú}LUM.ḪA): K prescr. 6: 32.

billi (^{na4}*bil-lí*): F prescr. 17: 56.

bīnu (^{giš}*bi-nu*, tamarisk): B prescr. 1: 27; No. K.5: 104; *tertennu* (?): N prescr. 14 iii 12, 20; tamarisk water (A ^{giš}*bi-ni*/*mê bīni*): No. D.1: 8; L prescr. 2: 6 (restored); knot (*kiṣru*): K prescr. 21: 60 (Ms. A, G).

bukānu (^{giš}GAN.U$_5$): No. E.2: 40 (Ms. A, D); F prescr. 7: 25; N prescr. 14 iii 13.

buluḫḫu (^{šim}BULUḪ): A prescr. 12: 22.

burāšu (^{šim}LI, juniper): A prescr. 24: 71; No. E.2: 20; F prescr. 9: 29; No. F.4: 92; No. M.1 ritual A: 14; R Bow ritual: 7′; juniper water (A^{meš} ^{šim}LI/*mê burāši*): No. H.1: 3; K prescr. 27: 75; pure juniper water (A^{meš} ^{šim}LI KÙ/*mê burāši ellūti*): R Bow ritual: 7′.

burzigallu (^{dug}BUR.ZI.GAL, a bowl): unfired (NU AL.ŠEG$_6$.GÁ/*lā ṣariptu*): No. E.2: 23 (Ms. G).

būšānu (^{ú}ḪAB): K prescr. 13: 47; T prescr. 1 obv. l 2′; seeds (NUMUN/*zēru*): A prescr. 22: 63 = Q prescr. 2: 6.

dadānu (^{giš}KIŠI$_{16}$.ḪAB): O prescr. 5: 7; T prescr. 2 e. 8′; fruit (GURUN/*inbu*): E Bow ritual: 55; from mountain (KUR/*šadû*): N prescr. 6 i 31; seeds (NUMUN/*zēru*): L prescr. 2: 6; N prescr. 25 lo. e. 3; P prescr. 2 obv. 5; P prescr. 3 obv. 8; seeds from mountain *d.* (KUR/*šadû*): N prescr. 1 i 5.

daššu (MÁŠ.NÍTA, buck): hair of a reared-up buck (SÍG MÁŠ.NÍTA ZI-*i*/*šārat dašši tebî*): No. D.4: 61 (restored); "something little" of penis (NÍG.TUR *šá* GÌŠ/*ṣeḫertu ša ušari*): No. D.4: 61; saliva of a reared-up buck (*rupušti* MÁŠ.NITÁ ZI.GA/*dašši tebî*): N prescr. 4 i 19 (restored); blood of a reared-up buck (ÚŠ MÁŠ.NITÁ ZI.GA/*dām dašši tebî*): N prescr. 25 lo. e. 1.

diqdiqqu (AL.TI.RÍ.GA^{mušen}, wren): E prescr. 4: 68–69; N prescr. 5 i 23–27.

ēdu (^{ú}DILI; ^{ú}*e-di*): A prescr. 7: 17; K prescr. 3: 28; K prescr. 12: 46; K prescr. 21: 59 (Ms. G); root (SUḪUŠ/*šuršu*): A prescr. 17: 29; K prescr. 15: 49; seeds (NUMUN/*zēru*): T prescr. 2 rev. 9′.

egemgiru (^{ú}NÍG.GÁN.GÁN): A prescr. 24: 69; seeds (NUMUN/*zēru*): T prescr. 2 rev. 9′.

elikulla (^{ú}*e-li-ku-la*): N prescr. 14 iii 11.

eper askuppi (SAḪAR I.DIB): of house(?) (É/*bītu*): D prescr. 2: 29.

eper kimaḫi (SAḪAR KI.MAḪ, dust of a grave): K prescr. 29: 126.

aribu (UGA^{mušen}): innards (or head) of a black raven (ŠÀ (or SAG) UGA^{mušen} GE$_6$/*libbi*(or *qaqqad*) *aribi ṣalmi*): L prescr. 1: 5; innards of a male raven (*lìbbi* UGA^{mušen} NITA/*aribi zikari*): N prescr. 25 lo. e. 1; skull (UGU/*muḫḫu*): P prescr. 7 obv. 22.

erēnu (^{giš}EREN; ^{giš}*eri$_4$-ni*, cedar): oil (Ì.GIŠ/*šamnu*): F prescr. 8: 27; 'resine' (ÚŠ/*dāmu*): K prescr. 21: 63.

eršu (^{giš}NÁ, bed): No. E.2: 41–45.

ēru (^{giš}MA.NU): knot (ZÚ.KEŠDA/*kiṣru*): K prescr. 21: 60 (Ms. F; omitted in K; restored in M); seeds (NUMUN/*zēru*): K prescr. 31: 137 = N prescr. 8 ii 6; green (branches) (SIG$_7$/*arqūtu*): P prescr. 4 obv. 12.

erû (^{na4}URUDU, copper): male (NÍTA/*zikaru*): M prescr. 6: 23.

^{ú}GAB.LAM: F prescr. 15: 50.

gabû (IM.SAḪAR.NA$_4$.KUR.RA, alum): A prescr. 6: 16.

gišimmaru (GIŠIMMAR, date palm): N prescr. 14 iii 13.

ḫarmunu (^{ú}ḪAR.ḪUM.BA.ŠIR): K prescr. 21: 60.

ḫašḫūr api (^{ú}GIŠ.ḪAŠḪUR.GIŠ.GI): A prescr. 9: 19; E Bow ritual: 55; F prescr. 8: 27.

ḫašḫūru (GIŠ.ḪAŠḪUR, apple-tree): K prescr. 13: 47; leaves (Ú): A prescr. 8: 18(?); seed of a small apple tree (NUMUN ^{ú}ḪAŠḪUR LÁ/*zēr ḫašḫūri maṭî*): No. K.2: 22; branch (PA/*aru*): K prescr. 32: 168.

ḫašû (^{ú}ḪAR.ḪAR): B prescr. 3 left e. i 1; No. D.4: 46; K prescr. 21: 59 (Ms A); root (SUḪUŠ/*šuršu*): A prescr. 17: 28; F prescr. 5: 21.

ḫaṭṭi-rēʾî (ᵘNÍG.GIDRU): K prescr. 11: 45; root (SUḪUŠ/*šuršu*): A prescr. 17: 29; K prescr. 19: 54.
ḫulālu (ⁿᵃ⁴NÍR): K prescr. 14: 48.
ḫurāṣu ((ⁿᵃ⁴)KÙ.SI$_{22}$, gold): A prescr. 15: 25; No. C.1: 11; No. E.2: 38; K prescr. 22: 65; seven grains (7 ŠE/*uṭṭatu*): No. K.3: 91.
ḫûratu (ᵍⁱˢḪAB): seeds (NUMUN/*zēru*): A prescr. 4: 14; K prescr. 2: 7.

imbû tâmti (ⁿᵃ⁴KA A.AB.BA; KA *tam-tim*, coral limestone): F prescr. 6: 24; F prescr. 10: 31; M prescr. 5: 21.
imēru (ANŠE, donkey): hair(?): A prescr. 2: 8.
imḫur-ešrā (ᵘIGI.NIŠ): A prescr. 6: 16; A prescr. 13: 23; No. E.2: 39; E Bow ritual: 54; No. M.1 ritual B: 13; M prescr. 4: 20.
imḫur-līm (ᵘIGI-*lim*): A prescr. 24: 69; No. E.1 var. B rev. 23; No. E.2: 39; E Bow ritual: 54; F prescr. 5: 21; F prescr. 10: 31; F prescr. 14: 48; H prescr. 1: 1; K prescr. 20: 55; M prescr. 1: 1; No. M.1 ritual A: 13; M prescr. 3: 19; M prescr. 4: 20; Q prescr. 4: 11.
immeru (UDU, sheep): wool from forehead ((SÍG) *ina* SAG.KI UDU KU$_5$-*si*/(*šīpāt*) *ina pūt immeri parsī*): No. E.2: 42–44; hair of the tail (*šaḫrat* KUN/*zibbatī*): N prescr. 4 i 20; wool of the perineum (SÍG/*šīpāt šabrīšu*): N prescr. 4 i 21.
immeru (UDU.(NITÁ), ram): hair(?): A prescr. 2: 8; wool of a reared-up ram (SÍG UDU.NÍTA ZI-*i*/*šīpāt immeri tebî*): No. D.4: 62; F prescr. 17: 57; wool of a ram (SÍG/*šīpāt puḫālī*): O prescr. 6: 11; wool from forehead ((SÍG) *ina* SAG.KI/(*šīpāt*) *ina pūt puḫāla*): No. E.2: 42–44; fleece (ˢⁱᵍÀKA/*itqu*): No. K.3: 94; tail wool (SÍG KUN, *šīpāt zibbatī*): N prescr. 10 ii 27–28; U prescr. 2 5–6 (restored); tail of a reared-up ram (KUN *pu-ḫa-li* ZI-*i*/*zibbat puḫāli tebî*): No. K.8: 155; blood (ÚŠ /*dāmu*): E prescr. 3: 65; throat (*napštu*): F prescr. 13: 42; N prescr. 16 iii 29; P prescr. 6 obv. 19; kidney (ÉLLAG/*kalītu*): K prescr. 4: 30; saliva of a reared-up ram (*rupušti* UDU ZI.GA/*immeri tebî*): N prescr. 4 i 19; N prescr. 18 iv 3.
isqūqu (ZÌ.KUM, a flour): N prescr. 8 ii 10.
iṣ pišri (GIŠ BÚR): No. E.2: 40 (Ms. G).
iṣṣūr ḫurri (BURU$_5$.ḪABRUD⁽ʳᵘ⁾.(DA)ᵐᵘˢᵉⁿ, MUŠEN *ḫur-ri*, partridge): general (often male, (NITÁ/*zikaru*): A prescr. 1: 3; N prescr. 1 i 2–6; N prescr. 3 i 12–16; N prescr. 24 le. e. 1–7; reared-up for the mating (*šá a-na* U$_5$ ZI.(GA)/*ša ana rakābi tebû*): E prescr. 5: 70–72; N prescr. 2 i 8–10; innards (ŠÀ/*libbu*): A prescr. 1: 4; A prescr. 10: 20; A prescr. 18: 31; E Bow ritual: 56; E prescr. 6: 76; F prescr. 12: 38; K prescr. 23: 66; L prescr. 4: 10; N prescr. 12 iii 1; P prescr. 9 rev. 9; S prescr. 2 obv. 7′; blood (ÚŠ/*dāmu*): A prescr. 21: 62; B prescr. 1: 14; E Bow ritual: 55; E prescr. 6: 74; F prescr. 10: 32; F prescr. 12: 37; No. F.2: 63; No. K.5: 103; K prescr. 32: 168; L prescr. 2: 6; N prescr. 25 lo. e. 2; P prescr. 7 obv. 22; head (SAG.DU/*qaqqadu*): No. C.1: 10; skull (UGU/*muḫḫu*): N prescr. 15 iii 24; penis (GÌŠ/*ušaru, mušaru*): F prescr. 13: 42 (restored); N prescr. 4 i 17 (Ms A); N prescr. 16 iii 28; testicles (ŠIR/*išku*): N prescr. 4 i 17 (Ms B).
išātu (IZI, fire): A prescr. 3: 13; No. K.2: 21; S prescr. 1: 4′.
itqu (ˢⁱᵍÀKA, fleece): A prescr. 3: 10; K prescr. 21: 63; L prescr. 2: 6.

kabullu (ᵘ*ka-bul-lu*): A prescr. 24: 70; F prescr. 3: 14; Q prescr. 1: 3; (*ka-bu-ul-ta*ˀ): O prescr. 5: 7.
kalbu (UR.GI$_7$, dog): N prescr. 13 iii 7; hair(?): A prescr. 2: 8; excrement (*zû*): Y presc. 1: 2′.
kalūmu (SILA$_4$): wool of a male lamb which is reared-up for mating (SÍG SILA$_4$ NITA *šá ana* U$_5$ ZI-*ú*/*šīpāt kalūmi zikari ša ana rakābi tebû*): K prescr. 27: 75–76.
kamantu (ᵘÁB.DUḪ; ᵘ*ka -man-du*): A prescr. 7: 17; A prescr. 14: 24; K prescr. 10: 43; K prescr. 11: 45; K prescr. 12: 46.
kamkadu (ᵘ*kám-ka-du*; ᵘ*kan-ka-du*): K prescr. 10: 43; root (SUḪUŠ/*šuruš*): K prescr. 6: 34.
kammu ša aškāpi (*kam-ma šá* ˡᵘ́AŠGAB, fungus of tanner): D prescr. 2: 29.
kamūnu (ᵘGAMUN, cumin): O prescr. 5: 7.
kamūnu (UZU.DIR, a fungus): W prescr. 3: 4′; from mountain (UZU.DIR.KUR.RA/*kamūn šadî*): No. M.1 ritual A: 13.
karānu (ᵍⁱˢGEŠTIN, wine): A prescr. 20: 61; A prescr. 21: 62; A prescr. 22: 64 = Q

prescr. 2: 8; A prescr. 24: 72; O prescr. 4: 6; N prescr. 5: 8; T prescr. 2 e. 8′; stong wine (^{giš}GEŠTIN/*karānu dannu*): Q prescr. 1: 5; T prescr. 2 rev. 11′.

karānu ṣaḫtu (^{giš}GEŠTIN.ŠUR.RA, grape juice): D prescr. 4: 47; E prescr. 2: 64 (not/*lū*); F prescr. 5: 23; precious (KAL/*aqru*): B prescr. 1: 15.

^{ú}KASKAL.MUNUS: N prescr. 21 iv 17.

kaspu ((^{na4})KÙ.BABBAR, silver): A prescr. 15: 25; No. C.1: 11; No. E.2: 38; K prescr. 22: 65; seven grains (7 ŠE/*uṭṭatu*): No. K.3: 91.

kāsu (*ka-si*; cup): unfired (NU AL.ŠEG$_{6}$.GÁ/*lā ṣariptu*): No. E.2: 23 (Ms. F).

kasû (GAZI^{sar}, mustard): mustar water (A GAZI^{sar}/*mê kasî*): No. D.3: 37; N prescr. 14 iii 16.

kazallu (^{ú}*ka-zal-lu*): A prescr. 17: 28; root (SUḪUŠ/*šuruš*): K prescr. 6: 35.

kibrītu (PIŠ$_{10}$.^{d}ÍD; *kib*-rit, sulfur): A prescr. 1: 2; A prescr. 3: 13; No. E.1 var. B rev. 23; H prescr. 1: 1; K prescr. 5: 31; M prescr. 3: 19.

kibrītu ru'tītu (PEŠ$_{10}$.^{d}ÍD ÚḪ.^{d}ÍD, a sulphur): M prescr. 1: 1.

kirbān-eqli (LAG.A.ŠÀ.GA): L prescr. 3: 8.

kitû (^{túg}GADA, flax): No. E.2: 40; K prescr. 26: 72; string (DUR/*ṭurru*): K prescr. 26: 72; K prescr. 30: 132 (restored); string (GU/*qû*): M prescr. 5: 22.

kubtu (*ku-ub-tú*, lump of metal): B prescr. 2: 44.

kukru (^{šim}GÚR.GÚR): F prescr. 9: 29; No. L.2: 25.

kulbību (NIM.LÀL, ant): head of a red ant (SAG.DU/*qaqqad kulbībi* SA$_{5}$/*sāmi*): O prescr. 6: 10.

kupatinnu (*ku-up-pa-ti-in-ni*; *ku-up-pí-ta-an-ni*, *ka-pa-ti-in-ni*, a kind of pellet): N prescr. 8: 11–12, 14; of clay (*ša* IM/*ṭīdi*): P prescr. 11 rev. 7.

kurkanû (^{ú}*kur-ka-na-a*): No. O.1: 24.

kurkurratu (a pot): M prescr. 2: 18.

kušru (^{ú}KU.GAG): A prescr. 10: 20.

laptu (^{ú}*la-ap-ti*): O prescr. 4: 4.

lipāru (^{ú}MI.PÀR): A prescr. 10: 20.

līš kunāši (NÍG.SILA$_{11}$.GÁ ZÍZ.AN.NA, emmer dough): D Diagn: 24.

lišān-kalbi (^{ú}EME.UR.GI$_{7}$): A prescr. 5: 15; A prescr. 22: 63 = Q prescr. 2: 7; A prescr. 23: 65 = Q prescr. 3: 9; No. D.4: 46; D prescr. 5: 49; F prescr. 11: 34; F prescr. 14: 48; K prescr. 21: 62; No. K.3: 90; N prescr. 7 ii 1; P prescr. 3: 3; Q prescr. 1: 3 (restored); root (SUḪUŠ/*šuršu*): A prescr. 20: 60; A prescr. 24: 69; K prescr. 6: 32; K prescr. 19: 53; seeds (NUMUN/*zēru*): K prescr. 31: 138 (restored) = N prescr. 8 ii 8; O prescr. 4: 4 (restored).

mašku (KUŠ, leather bag): A prescr. 3: 13; 4: 14; 5: 15; 6: 16, 7: 17, 8: 18; 9: 19; 10: 20; 11: 21; 12: 22; 13: 23; 14: 24; 15: 26; 16: 27; 23: 66; No. C.1: 11; F prescr. 4: 18; F prescr. 7: 25; F prescr. 10: 33; F prescr. 11: 35; F prescr. 14: 49 (restored); F prescr. 15: 51; F prescr. 15: 51; F prescr. 16: 53 (restored); K prescr. 1: 6 (restored); No. K.1: 15; K prescr. 3: 29; K prescr. 4: 30; K prescr. 5: 31; K prescr. 6: 35 (restored); K prescr. 8: 41; K prescr. 9: 42; K prescr. 10: 44; K prescr. 11: 45; K prescr. 12: 46; K prescr. 13: 47; K prescr. 13: 47; K prescr. 15: 49; K prescr. 16: 50; K prescr. 19: 54 (restored); K prescr. 21: 63; K prescr. 24: 67 (restored); No. M.1 ritual B: 15; M prescr. 2: 18; M prescr. 3: 19; M prescr. 6: 23; Y presc. 1: 3′ (restored), 8′; from a female kid that has not yet mated (^{munus}ÁŠ.GÀR GÌŠ.NU.ZU/*unīqi lā petīti*): K prescr. 22: 65.

maštakal (^{ú}IN$_{6}$.ÚŠ; ^{ú}IN.NU.UŠ): D prescr. 3: 41; F prescr. 11: 34; F prescr. 15: 50; K prescr. 15: 49; K prescr. 21: 61 (from the mointain Ms. L); N prescr. 14 iii 13; seeds (NUMUN/*zēru*): B prescr. 1: 25; K prescr. 3: 28; K prescr. 31: 137= N prescr. 8 ii 6.

mekku (^{na4}*me-ek-ki*): No. F.2: 62.

miḫḫu (*mi-iḫ-ḫa*, a beer): R Bow ritual: 8′.

mikû (*mi-ki-i*): B prescr. 1: 24.

mû (A(^{meš}), water): B prescr. 1: 26, 27; E prescr. 3: 67; E prescr. 6: 73; F prescr. 12: 37, 39; No. I.3: 37 (restored); N prescr. 2 i 9; N prescr. 3 i 13–14; N prescr. 4 i 20; N prescr. 12 iii 2, 6 (restored); N prescr. 14 iii 15; No. O.1: 26 (restored); P prescr. 3 obv. 10 (restored); S prescr. 1: 11′; pure (KÙ/*ellu*): No. D.4: 63; river water (A ÍD/*mê nāri*): E prescr. 6: 74.

murdudû (^{ú}MUL.DÙ.DÙ): seeds (NUMUN/*zēru*): K prescr. 31: 138 = N prescr. 8 ii 7.

nurmû (^{giš}NU.ÚR.MA, pomegranate): pomegranate sweet juice (A ^{giš}NU.ÚR.MA KU$_7$.KU$_7$/*mê nurmî matqī*): D prescr. 2: 30.
murru (^{šim}SES): A prescr. 12: 22.
muṣa''irānu arqu (BIL.ZA.ZA SIG$_7$, green frog): No. D.1: 6; tadpole (*atmu*): F prescr. 8: 26.
muššaru (^{na4}MUŠ.GÍR): K prescr. 28: 77.
mūṣu (^{na4}*mu-ṣa*): F prescr. 6: 24; T prescr. 2 obv. 6′.
nabāsu (SÍG.ḪÉ.ME.DA, red(-dyed) wool): No. D.4: 62; F prescr. 13: 42; N prescr. 16 iii 29 (restored).
NAM.GEŠTIN^{mušen} (a bird): N prescr. 6 i 28–32.
napšašti taskarinni (^{giš}DÍLIM.Ì.ŠÉŠ ^{giš}TAŠKARIN, boxwood container for unguents): No. F.4: 91.
našru (TE$_8$, eagle): wings of a male eagle (*kappī* TE$_8$/*našri* NITÁ/*zikari*): K prescr. 31: 135 = N prescr. 8 ii 3.
nēšu (UR.MAḪ, lion): hair (*šārtu*): K prescr. 4: 30.
nignakku (NÍG.NA, censer): A presscr. 24: 71; No. E.2: 20; No. M.1 ritual A: 14; R Bow ritual: 7′.
nikiptu (ŠIM.^{d}MAŠ): “male” and “female” (NITÁ *u* MUNUS): K prescr. 22: 64.
nīnû (Ú.KUR.RA): N.1: 24 (1 shekel); root (SUḪUŠ/*šuršu*): A prescr. 20: 60; No. E.2: 40; F prescr. 5: 21; L prescr. 3: 8; N prescr. 5 i 24; N prescr. 6 i 29; green (SIG$_7$/*arqūtu*): B prescr. 2: 42.
nīqu (^{udu}SISKUR, sheep): sacrificed sheep (shoulder, caul fat, the roast meat ^{uzu}ZAG/*imittu*, ^{uzu}ME.ḪÉ/*ḫimṣu*, ^{uzu}KA.NE/*šumû*): No. E.2: 21.
nūbtu (NIM.LÀL, bee): stinger (*ziqtu*): K prescr. 32: 166; stinger (KUN/*zibbatu*): O prescr. 6: 10.
nuḫurtu (^{ú}NU.LUḪ.ḪA): F prescr. 5: 22; K prescr. 27: 74.
nuṣābu (^{ú}*nu-ṣa-bu*): A prescr. 13: 23.

pallišu (^{na4}NÍG.BÙR.BÙR): K prescr. 21: 62.
pappardillû (^{na4}BABBAR.DIL): K prescr. 28: 77.
pappasītu (BA.BA.ZA.^{d}ÍD): K prescr. 5: 31.
parzillu (AN.BAR, iron): A prescr. 15: 25; No. B.1: 16; No. D.2: 21; F prescr. 6: 24; F prescr. 10: 31; K prescr. 26: 72 (restored); L prescr. 4: 21; powder (KU.KU/*sīktu*): No. B.2: 38; No. E.1: 15 and var. B rev. 22; L prescr. 6: 15; M prescr. 1: 1; M prescr. 5: 21; male (NÍTA/*zikaru*): M prescr. 6: 23.
patar siparri (GÍR ZABAR, bronze knife): D prescr. 5: 50.
pāṭiru (^{gi}DU$_8$, a portable altar): No. E.2: 19.
pēntu (NE, charcoal): No. L.2: 25.
per'u (NUNUZ, sprout): S prescr. 1: 5′.
pizallurtu (MUŠ.DÍM.GURUN.NA, gecko): copulating (*ritkubātu*): D prescr. 6: 51–52; from the steppe (EDIN/*ṣēru*): K prescr. 31: 136 = N prescr. 8 ii 4; N prescr. 17 iii 32.
puquttu (^{ú}*pu-qut-tú*, a thorn): seeds (NUMUN/*zēru*): No. K.5: 103; L prescr. 3: 8 (Ms. C).
pursītu (^{dug}BUR.ZI, a bowl): unfired (NU AL.ŠEG$_6$.GÁ/*lā ṣariptu*): No. E.2: 23 (Ms. A); E prescr. 3: 65.

qalītu (^{zì}NÍG.ŠE.SA.A, toasted grain): N prescr. 5 i 26; N prescr. 6 i 30; toasted flour (ŠE ^{zì}NÍG.ŠE.SA.A/*še'a qalīti*): N prescr. 6 i 30.
qulipti nūni (BAR KU$_6$, scaly skin of a fish): B prescr. 1: 24.

riḫu (^{na4}*ri-ḫu*): F prescr. 17: 56 (?).

^{ú}SAG (a plant): root (SUḪUŠ/*šuruš*): K prescr. 6: 33.
saḫḫû (^{na4}*saḫ-ḫu-u*): F prescr. 17: 56.
samīdu (^{ú}KUR.ZI, ^{ú}*sa-mì-dí*): root (SUḪUŠ/*šuršu*): K prescr. 32: 167; (root) in a garden-plot (*ina musarî*): T prescr. 2 obv. 4′; seeds (NUMUN/*zēru*): O prescr. 4: 5 (restored).
ṣabītu (MAŠ.DÀ): tendon of the left hock (SA.MUD MAŠ.DÀ ŠA GÙB/*šer'ān eqbi ṣabīti ša šumēlī*): R Bow ritual: 9′.
sāmtu (^{na4}GUG, cornelian): K prescr. 28: 77; chipping (*ḫiṣbu*): K prescr. 21: 61.
sassatu (^{ú}*sà-as-sà-ta*, a grass): P prescr. 3 obv. 9.
sīḫu (^{ú}*si-ḫu*): K prescr. 22: 64.
sikillu (^{ú}SIKIL): No. E.2: 40; F prescr. 3: 14; K prescr. 3: 28; K prescr. 21: 61; K prescr. 28: 78; T prescr. 2 obv. 5′; seeds (NUMUN/*zēru*): P prescr. 4 obv. 12.
sisû (ANŠE.KUR.RA, horse): edge of the remains of urine (*siḫir middu'ri* KÀŠ^{meš}-*šú*/*šīnātišu*): E prescr. 1: 60–62.
sû (^{na4}*su-u*): A prescr. 6: 16.
sukannīnu (TU.KUR$_4$^{mušen}, dove): blood (ÚŠ/*dāmu*): A prescr. 21: 62.

supālu (^{ú}MUNZER): A prescr. 22: 63 = Q prescr. 2: 6.

supālu (^{ú}NIGIN^{sar}, a juniper): F prescr. 5: 22.

sūq erbetti (SILA LÍMMU.BA/*-tú*, a crossroads): dust (SAḪAR/*eperu*): H prescr. 1: 1; sherd (ŠIKA/*ḫaṣabtu*): K prescr. 4: 30.

ṣalmu (NU/figurine): tallow (Ì.UDU/*lipû*), wax (DUḪ.LÀL/*iškūru*), bitumen (*kupru*), gypsum (IM.BABBAR/*gaṣṣu*), clay (IM/*ṭīdu*), dough (NÍG.ŠILA$_{11}$.GÁ/*līšu*), cedar wood (^{giš}ERIN/*erēnu*): No. E.2: 22–23.

ṣaṣuntu (^{ú}*ṣa-ṣu-un-tú*; ^{ú}*ṣa-ṣu-ut-tú*; ^{ú}*ṣa-ṣu-um-tú*; ^{ú}*a-ṣu-ṣu-um-tú/tum*): A prescr. 4: 14; A prescr. 9: 19; A prescr. 18: 31; E Bow ritual: 54; F prescr. 3: 15; K prescr. 30: 130; K prescr. 31: 139 = N prescr. 8 ii 8; L prescr. 3: 8; No. M.1 ritual A: 13.

ṣillu (thorn): S prescr. 1: 8′ (emended); bow of thorn (^{giš}BAN *šá* ^{giš}DÁLA/*qaštu ša ṣillî*): E Bow ritual: 57; No. I.2: 13; R Bow ritual: 9′ (restored).

ṣurārû (EME.DIR, lizard): K prescr. 32: 168; excrement of polychrome lizard (ŠE$_{10}$ EME.DIR.GÙN.A/*zê ṣurāri barmi*): K prescr. 31: 136–137 = N prescr. 8 ii 5; from the mountain (KUR/*šadû*): N prescr. 23 iv 24 (restored).

ṣurru (^{na4}ZÚ): F prescr. 17: 54.

šadânu ṣābitu (^{na4}KA.GI.NA.DAB.(BA); ^{na4}KUR-*nu* DIB; ^{na4}*šada-nu* DAB, magnetite): catalogue LKA 94 i 24; A prescr. 23: 65 = Q prescr. 3: 9; No. A.2.: 47; No. B.1: 14; No. D.2: 21; F prescr. 17: 55; K prescr. 26: 72; K prescr. 27: 73; T prescr. 2 obv. 6′; powder (KU.KU/*sīktu*): A prescr. 12: 22; No. B.2: 38; No. E.1: 15 and var. B rev. 22; L prescr. 6: 15.

šaḫû (ŠAḪ, pig): bristles(?): A prescr. 2: 8; bristles of a pig reared up for mating (*zappī* ŠAḪ *šá ana* U$_5$ ZI-*ú*/*šaḫî ša ana rakābi tebû*): F prescr. 10: 32; fat (Ì.UDU/*lipu*): T prescr. 1 obv. l 2′;

šakirû (^{ú}ŠAKIR): root (SUḪUŠ/*šuršu*): F prescr. 15: 50; K prescr. 10: 44; seeds (NUMUN/*zēru*): K prescr. 3: 29; J prescr. 31: 139 = N prescr. 8 ii 8.

šakkadirru (*šak-ka-di-ir-ru*, a lizard): P prescr. 11 rev. 6.

šammi balāṭi (Ú.NAM.TI.LA): A prescr. 8: 18; K prescr. 13: 47; root (SUḪUŠ/*šuršu*): H prescr. 3: 5.

šambaliltu (^{ú}SULLIM, fenugreek): seeds (NUMUN/*zēru*): O prescr. 5: 8.

šamnu (Ì; Ì.GIŠ, oil): A prescr. 4: 14; A prescr. 5: 15; A prescr. 6: 16; A prescr. 7: 17; A prescr. 8: 18 (restored); A prescr. 10: 20; A prescr. 12: 22; A prescr. 14: 24 (restored); No. A.2: 47; No. A.3: 58; A prescr. 23: 66; No. D.1: 8; No. E.1 var. B rev. 23; E prescr. 3: 66; F prescr. 13: 43; F prescr. 14: 49; F prescr. 15: 51 (restored); F prescr. 15: 51 (restored); F prescr. 16: 52; No. K.1: 15; K prescr. 10: 44; K prescr. 19: 54; K prescr. 24: 67; L prescr. 5: 14; M prescr. 1: 2; N prescr. 4 obv. 12; oil from the alabastron (Ì BUR/*šamni pūri*): No. B.1: 14, 16; No. B.2: 39; No. D.1: 7; No. D.2: 22; No. E.1: 16; No. F.4: 91; L prescr. 6: 16; No. M.1 ritual A: 14; N presc. 14: 22; N prescr. 15 iii 25; premium oil (Ì.SAG/*šamnu rūštu*): N prescr. 14 iii 20.

šappu (*šap-pi*, a container): Y presc. 1: 3′, 8′.

šārāt pagî (SÍG UGU.DUL.BI, monkey's hair): K prescr. 22: 64.

šarmadu (^{ú}GUR$_5$.UŠ): A prescr. 18: 31.

šeleppû (NÍG.BÚN.NA, turtle): Y presc. 1: 7′.

šer'ān sammî (SA ZÀ.MÍ, harp string): No. A.1: 38.

šibburratu (*ši-bu-ra-ti*, rue): P prescr. 6 obv. 21.

šikaru (KAŠ, beer): A prescr. 17: 30; A prescr. 18: 31; A prescr. 20: 60; A prescr. 22: 64 = Q prescr. 2: 7; A prescr. 24: 72; D prescr. 4: 47; No. E.2: 20 (var. SAG/*rēštû* 'premium'); E prescr. 2: 64 (not/ *lū*); F prescr. 3: 16; F prescr. 4: 17; F prescr. 14: 49; F prescr. 15: 51 (restored); F prescr. 15: 51 (restored); F prescr. 16: 52; K prescr. 2: 7 (restored); K prescr. 19: 54; L prescr. 3: 8 (restored); N prescr. 1 i 6; N prescr. 24 le. e. 6; O prescr. 4: 6; O prescr. 5: 8; Q prescr. 1: 4; T prescr. 2 obv. 5′; T prescr. 2 obv. 7′; good beer (KAŠ.DU$_{10}$(G)/*šikaru ṭābu*): B prescr. 2: 44; premium beer (KAŠ.SAG/*šikaru rēštû*): E Bow ritual: 57; E prescr. 5: 72; beer of the brewer (KAŠ ^{lú}KÚRUN.NA/*šikar sābî*): No. K.1: 14.

šikkurratu (*šik-kur-rat*, a reed): pure (SIKIL/*ebbu*): K prescr. 32: 167.

šiltāḫ urbate (TAG.GA *ur-ba-te*, reed arrow): R Bow ritual: 10′.

šimru or *urānu* (Ú.KU$_6$): A prescr. 8: 18; F prescr. 11: 34; K prescr. 13: 47; root (SUḪUŠ/*šuruš*): K prescr. 19: 53.
šīpātu (SÍG, wool): white (BABBAR/*peṣû*): No. K.7: 103; M prescr. 6: 25; red (SA5/*sāmu*): No. K.7: 103; M prescr. 6: 25.
šīr šikkî (*šīr* ^{d}NIN.KILIM, mongoose meat): X prescr. 1: 8′.
šizbu (GA, milk): O prescr. 2: 2; O prescr. 3: 3.
šumuttu (^{ú}SUMUN.DAR): A prescr. 7: 17; A prescr. 10: 20; A prescr. 14: 24; K prescr. 12: 46; K prescr. 20: 56; K prescr. 31: 139 = N prescr. 8 ii 9.
šunû (^{giš}ŠE.NÁ.A, agnus castus): K prescr. 17: 51; branch (PA/*aru*): K prescr. 21: 61.
šurdunû (^{ú}SI.SÁ): root (SUḪUŠ/*šuruš*): K prescr. 6: 32; K prescr. 19: 54.
šurmēnu (ŠUR.MÌN, cypress): oil (Ì.GIŠ/*šamnu*): F prescr. 9: 30; K prescr. 1: 5, L prescr. 1: 5.
šuttinnu (SU.TIN^{mušen}, bat): excrement (*pītu*): E Bow ritual: 54; wing (PA/*kappu*): No. K.1: 13.

takdanānu (^{ú}*tak-da-na-nu*): A prescr. 16: 27; No. K.3: 89.
tarmuš (^{ú}*tar-muš*; ^{ú}*tar-muš*$_8$): A prescr. 11: 21; A prescr. 24: 69; No. D.4: 46; F prescr. 7: 25; F prescr. 10: 31; K prescr. 21: 59; M prescr. 4: 20; M prescr. 5: 21; No. O.1: 25.
tinūru (NINDU, oven): B prescr. 1: 27.
tīyatu (^{ú}*ti-ia-tú*): A prescr. 13: 23.
tullal (^{ú}*tu-lal*): F prescr. 3: 14; F prescr. 5: 22.
turminû (^{na4}DUR.MI.NA): K prescr. 14: 48.
ṭabti amāni (MUN *a-ma-ni*(*m*), a salt): F prescr. 9: 29; N prescr. 6 i 29.
ṭabtu (MUN, salt): A prescr. 18: 31; K prescr. 30: 131; L prescr. 4: 11; N prescr. 1 i 4; N prescr. 5 i 24.
ṭīd kullati (IM KI.GAR, potter's clay): D Diagn: 24.
ṭurru peṣû (DUR BABBAR, white string): O prescr. 6: 12.

uḫūlu (^{ú}NAGA): B prescr. 1: 26.
uqnû (^{na4}ZA.GÌN, lapis lazuli): catalogue LKA 94 i 25; No. E.2: 38; F prescr. 17: 54; K prescr. 13: 47; K prescr. 26: 71; K prescr. 27: 73; K prescr. 28: 77.
urânu (*ú-ra-nim*): root (SUḪUŠ/*šuruš*): Y presc. 1: 6′.
urânu or *šimru* (Ú.KU$_6$): A prescr. 8: 18; F prescr. 11: 34; K prescr. 13: 47; root (SUḪUŠ/*šuruš*): K prescr. 19: 53.
urnû (^{ú}BÚR): A prescr. 11: 21; from montains (*urnâ ša* KUR-*e*/*šadê*): E Bow ritual: 54.
urṭû (^{ú}*ur-ṭú*): H prescr. 2: 4.
ušû (^{giš}ESI, ebony): F prescr. 10: 31.

zappu (*zap-pi*, hair): over the foreskin of an animal (*ina* UGU/*muḫḫi urullī*): X prescr. 1: 5′.
zisurrû (ZÍD.SUR.RA, magical circle of flour): A prescr. 3: 12.
zumbi-ḫurāṣi (NIM.KÙ.SI$_{22}$): A prescr. 24: 70.
zuqaqīpu (GÍR.TAB, scorpion): tail (KUN/*zibbatu*): No. K.8: 155; O prescr. 6: 9.

Concordances

TEXT NUMBER	MUSEUM NUMBER	PUBLICATION
Catalogue	Ass 13955 kb	LKA 94
Text A Ms. A	Ass 13955	LKA 95
Text A Ms. B	VAT 13758	LKA 96
Text A Ms. C	Ass 13955	LKA 101
Text A Ms. D	SU 52/139+161+170+250+250A+323	STT 280
Text A Ms. E	W. 22307/4+68	SpTU 1, 9
Text A Ms. F	K. 5991	AMT 88, 3
Text A Ms. G	VAT 13731	LKA 100
Text A Ms. H	Sp II 976 = BM 35394	CCMAwR 3, pls. 2–3
Text A Ms. I	BM –	AMT 66, 1
Text B Ms. A	VAT 13610	LKA 102
Text B Ms. B	VAT 8916	KAR 70
Text B Ms. C	A 483	BAM 369 = CCMAwR 3, pl. 1
Text C	Ass 1395	LKA 103
Text D Ms. A	Ass 13955	LKA 101
Text D Ms. B	K. 2499	TCS 2, pl. 1
Text D Ms. C	W. 22307/4+68	SpTU 1, 9
Text D Ms. D	VAT 13721	LKA 97
Text D Ms. E	SU 52/139+161+170+250+250A+323	STT 280
Text D Ms. F	VAT 8916	KAR 70
Text E Ms. A	VAT 8233	KAR 236
Text E Ms. B	VAT 8916	KAR 70
Text E Ms. C	A 483	BAM 369 = CCMAwR 3, pl. 1
Text E Ms. D	Sp II 976 = BM 35394	CCMAwR 3, pls. 2–3
Text E Ms. E	VAT 8265	KAR 243
Text E Ms. F	VAT 13643bis	LKA 99b
Text E Ms. G	VAT 10697+10830	LKA 99d = CCMAwR 3, pls. 4–5
Text E Ms. H	SU 52/139+161+170+250+250A+323	STT 280
Text E Ms. I	K. 11076	TCS 2, pl. 3
Text E Ms. J	VAT 10090	BAM 272
Text E Ms. K	K 916	AMT 73, 2
Text F Ms. A	VAT 8916	KAR 70
Text F Ms. B	81-7-27, 73	TCS 2, pl. 2
Text F Ms. C	Ass 13955hl	BAM 205
Text F Ms. D	K. 9451+ 11676+Sm. 818 + 961	TCS 2, pl. 1
Text F Ms. E	SU 52/139+161+170+250+250A+323	STT 280
Text F Ms. F	BM 46911 (= 81-8-30, 377)	TCS 2, pl. 3
Text G	Ass 13955hl	LKA 99c
Text H Ms. A	K. 8790	AMT 65, 7

https://doi.org/10.1515/9783110757262-016

Text H Ms. B	VAT 13703	WVDOG 147 (= KAL 7) No. 22
Text I	K. 8698	TCS 2, pl. 3
Text J Ms. A	K.10002	TCS 2, pl. 2
Text J Ms. B	K. 9415+10791	TCS 2, pl. 2+CT 13, 31 and Thompson 1930 pl. 17
Text K Ms. A	SU 52/139+161+170+250+250A+323	STT 280
Text K Ms. B	K. 9451+11676+Sm. 818 + 961	TCS 2, pl. 1
Text K Ms. C	BM 68033	CCMAwR 1, pl. 18
Text K Ms. D	VAT 13616	LKA 144
Text K Ms. E	VAT 13917	BAM 207, BID, pl. 24
Text K Ms. F	VAT 13893+13982	BAM 320
Text K Ms. G	A 2715	BAM 205
Text K Ms. H	A 522	BAM 318 = CCMAwR 2, pls. 53–60
Text K Ms. I	VAT 14111	BAM 319
Text K Ms. J	BM 54650	BID, pls. 19–21
Text K Ms. K	SU 1951, 93+SU unnumbered	STT 95+295
Text K Ms. L	VAT 08914	BAM 311 = KAR 186
Text K Ms. M	K. 8907	Stadhouders forthcoming
Text K Ms. N	Bo 4894	KUB 4, 48
Text K Ms. O	K. 2417	AMT 31, 4; 32, 1
Text L Ms. A	W. 22307/9	SpTU 1, 10
Text L Ms. B	Bo 4894	KUB 4, 48
Text L Ms. C	VAT 13915+13933	LKA 98
Text M Ms. A	W. 22656/5	SpTU 4, 135
Text M Ms. B	SU 52/139+161+170+250+250A+323	STT 280
Text M Ms. C	BM 41279	–
Text N Ms. A	Bo 4894	KUB 4. 48
Text N Ms. B	Bo 5817	KUB 37, 80
Text N Ms. C	(Bo 5885)+AAA 3 No. 5	AAA 3, pl. 27 No. 5
Text O	Bo 61/r	KBo 36, 27
Text P	BM 98571+ = K 3350+ 1905-04-09, 77	AMT 62, 3
Text Q	BM –	AMT 66, 1
Text R	K.9036	TCS 2, pl. 1
Text S	K. 5901	TCS 2, pl. 3
Text T	81-7-1, 270 (BM 42510)+ F 224 (81-1-1 unnumbered)	Finkel 2000
Text U	W. 22277a	SpTU 1, 20
Text V	(Bo 5885)+AAA 3 No. 5	AAA 3, pl. 27 No. 5
Text W	Bo 5885+(AAA 3 No. 5)	KUB 37, 81
Text X	621/b	KUB 37, 82
Text Y	178/b	KUB 37, 201

MUSEUM NUMBER	PUBLICATION	TEXT NUMBER
178/b	KUB 37, 201	Text Y
621/b	KUB 37, 82	Text X
81-7-1, 270 (BM 42510)+ F 224 (81-1-1 unnumbered)	Finkel 2000	Text T
81-7-27, 73	TCS 2, pl. 2	Text F Ms. B
A 483	BAM 369 = CCMAwR 3, pl. 2	Text B Ms. C; Text E Ms. C
A 522	BAM 318 = CCMAwR 2, pls. 53–60	Text K Ms. H
Ass 1395	LKA 103	Text C
Ass 13955	LKA 101	Text A Ms. C; Text D Ms. A
Ass 13955 kb	LKA 94	Catalogue
Ass 13955	LKA 95	Text A Ms. A
Ass 13955hl	BAM 205	Text F Ms. C; Text K Ms. G
Ass 13955hl	LKA 99c	Text G
BM –	AMT 66, 1	Text A Ms. I; Text Q
BM 46911 (= 81-8-30, 377)	TCS 2, pl. 3	Text F Ms. F
BM 54650	BID, pls. 19–21	Text K Ms. J
BM 68033	CCMAwR 1, pl. 18	Text K Ms. C
BM 98571+ = K 3350+ 1905-04-09, 77	AMT 62, 3	Text P
BM 41279	–	Text M Ms. C
Bo 61/r	KBo 36, 27	Text O
Bo 4894	KUB 4, 48	Text K Ms. N; Text L Ms. B; Text N Ms. A
Bo 5817	KUB 37, 80	Text N Ms. B
(Bo 5885)+AAA 3 No. 5	AAA 3, pl. 27 No. 5	Text N Ms. C; Text U
Bo 5885+(AAA 3 No. 5)	KUB 37, 81	Text W
K. 916	AMT 73, 2	Text E Ms. K
K. 2417	AMT 31, 4; 32, 1	Text K Ms. O
K. 2499	TCS 2, pl. 1	Text D Ms. B
K. 5901	TCS 2, pl. 3	Text S
K. 5991	AMT 88, 3	Text A Ms. F
K. 8698	TCS 2, pl. 3	Text I
K. 8790	AMT 65, 7	Text H Ms. A
K. 8907	Stadhouders forthcoming	Text K Ms. M
K. 9036	TCS 2, pl. 1	Text R
K. 9415+10791	TCS 2, pl. 2+CT 13, 31 and Thompson 1930 pl. 17	Text J Ms. B
K. 9451+11676+ Sm. 818+961	TCS 2, pl. 1	Text F Ms. D; Text K Ms. B
K. 10002	TCS 2, pl. 2	Text J Ms. A
K. 11076	TCS 2, pl. 3	Text E Ms. I
Sp II 976 = BM 35394	CCMAwR 3, pls. 2–3	Text A Ms. H; Text D Ms. G; Text E Ms. D

SU 1951, 93+SU unnumbered	STT 95+295	Text K Ms. K
SU 52/139+161+170+ 250+250A+323	STT 280	Text A Ms. D; Text D Ms. E; Text E Ms. H; Text F Ms. E; Text K Ms. A; Text M Ms. B
VAT 8233	KAR 236	Text E Ms. A
VAT 8265	KAR 243	Text E Ms. E
VAT 8914	BAM 311 = KAR 186	Text K Ms. L
VAT 8916	KAR 70	Text B Ms. B; Text D Ms. F; Text E Ms. B; Text F Ms. A
VAT 10090	BAM 272	Text E Ms. J
VAT 10697+10830	LKA 99d = CCMAwR 3, pls. 4–5	Text E Ms. G
VAT 13610	LKA 102	Text B Ms. A
VAT 13616	LKA 144	Text K Ms. D
VAT 13643bis	LKA 99b	Text E Ms. F
VAT 13703	WVDOG 147 (= KAL 7) No. 22	Text H Ms. B
VAT 13721	LKA 97	Text D Ms. D
VAT 13731	LKA 100	Text A Ms. G
VAT 13758	LKA 96	Text A Ms. B
VAT 13893+13982	BAM 320	Text K Ms. F
VAT 13915+13933	LKA 98	Text L Ms. C
VAT 13917	BAM 207, BID, pl. 24	Text K Ms. E
VAT 14111	BAM 319	Text K Ms. I
W. 22277a	SpTU 1, 20	Text U
W. 22307/4+68	SpTU 1, 9	Text A Ms. E; Text D Ms. C
W. 22307/9	SpTU 1, 10	Text L Ms. A
W. 22656/5	SpTU 4, 135	Text M Ms. A

PUBLICATION	MUSEUM NUMBER	TEXT NUMBER
—	BM 41279	Text M Ms. C
AAA 3, pl. 27 No. 5	(Bo 5885)+AAA 3 No. 5	Text N Ms. C; Text V
AMT 31, 4; 32, 1	K. 2417	Text K Ms. O
AMT 62, 3	BM 98571+ = K 3350 + 1905-04-09, 77	Text P
AMT 65, 7	K. 8790	Text H Ms. A
AMT 66, 1	BM –	Text A Ms. I; Text Q
AMT 73, 2	K. 916	Text E Ms. K
AMT 88, 3	K. 5991	Text A Ms. F
BAM 205	Ass 13955hl	Text F Ms. C; Text K Ms. G
BAM 207, BID, pl. 24	VAT 13917	Text K Ms. E
BAM 272	VAT 10090	Text E Ms. J
BAM 311 = KAR 186	VAT 8914	Text K Ms. L
BAM 318 = CCMAwR 2, pls. 53–60	A 522	Text K Ms. H
BAM 319	VAT 14111	Text K Ms. I
BAM 320	VAT 13893+13982	Text K Ms. F

BAM 369 = CCMAwR 3, pl. 1	A 483	Text B Ms. C; Text E Ms. C
BID, pls. 19–21	BM 54650	Text K Ms. J
CCMAwR 1, pl. 18	BM 68033	Text K Ms. C
CCMAwR 3, pls. 2–3	Sp II 976 = BM 35394	Text A Ms. H; Text D Ms. G; Text E Ms. D
Finkel 2000	81-7-1, 270 (BM 42510)+ F 224 (81-1-1 unnumbered)	Text T
KAR 70	VAT 8916	Text B Ms. B; Text D Ms. F; Text E Ms. B; Text F Ms. A
KAR 186 = BAM 311	VAT 8914	Text K Ms. L
KAR 236	VAT 8233	Text E Ms. A
KAR 243	VAT 8265	Text E Ms. E
KBo 36, 27	Bo 61/r	Text O
KUB 4, 48	Bo 4894	Text K Ms. N; Text L Ms. B; Text N Ms. A
KUB 37, 80	Bo 5817	Text N Ms. B
KUB 37, 81	Bo 5885+(AAA 3 No. 5)	Text W
KUB 37, 82	621/b	Text X
KUB 37, 201	178/b	Text Y
LKA 94	Ass 13955 kb	Catalogue
LKA 95	Ass 13955	Text A Ms. A
LKA 96	VAT 13758	Text A Ms. B
LKA 97	VAT 13721	Text D Ms. D
LKA 98	VAT 13915+13933	Text L Ms. C
LKA 99b	VAT 13643bis	Text E Ms. F
LKA 99c	Ass 13955hl	Text G
LKA 99d = CCMAwR 3, pls. 4–5	VAT 10697+10830	Text E Ms. G
LKA 100	VAT 13731	Text A Ms. G
LKA 101	Ass 13955	Text A Ms. C; Text D Ms. A
LKA 102	VAT 13610	Text B Ms. A
LKA 103	Ass 1395	Text C
LKA 144	VAT 13616	Text K Ms. D
SpTU 1, 9	W. 22307/4+68	Text A Ms. E; Text D Ms. C
SpTU 1, 10	W. 22307/9	Text L Ms. A
SpTU 1, 20	W. 22277a	Text U
SpTU 4, 135	W. 22656/5	Text M Ms. A
Stadhouders forthcoming	K. 8907	Text K Ms. M
STT 95+295	SU 1951, 93 +SU unnumbered	Text K Ms. K
STT 280	SU 52/139+161+170+250+ 250A+323	Text A Ms. D; Text D Ms. E; Text E Ms. H; Text F Ms. E; Text K Ms. A; Text M Ms. B
TCS 2, pl. 1	K. 2499	Text D Ms. B
TCS 2, pl. 1	K. 9036	Text R
TCS 2. pl. 1	K. 9451+ K. 11676+ Sm. 818+961	Text F Ms. D; Text K Ms. B
TCS 2, pl. 2	81-7-27, 73	Text F Ms. B

TCS 2, pl. 2	K. 10002	Text J Ms. A
TCS 2, pl. 2+CT 13, 31 and Thompson 1930 pl. 17	K. 9415+10791	Text J Ms. B
TCS 2, pl. 3	K. 5901	Text S
TCS 2, pl. 3	K. 8698	Text I
TCS 2, pl. 3	BM 46911 (= 81-8-30, 377)	Text F Ms. F
TCS 2, pl. 3	K. 11076	Text E Ms. I
WVDOG 147 (= KAL 7) No. 22	VAT 13703	Text H Ms. B

Bibliography

Abrahami 2003: Philippe Abrahami, "A propos des fonctions de l'*asû* et de l'*āšipu*: La conception de l'auteur de l'hymne sumérien dédié à Ninisina", *Journal des Médecines Cunéiformes* 2 (2003), pp. 19–20.

Abrahami 2006: Philippe Abrahami, "Le cochon dans les collections d'oracles de la Mésopotamie", in Lion and Michel 2006: 267–282.

Abusch 1985: Tzvi Abusch, "Dismissal by Authorities: *Šuškunu* and Related Matters", *Journal of Cuneiform Studies* 37, 1 (1985), pp. 91–100.

Abusch 1998: Tzvi Abusch, "Ghost and God: Some Observations on a Babylonian Understanding of Human Nature", in Albert I. Baumgarten, Jan Assmann, and Gedaliahu G. Stroumsa (eds.), *Self, Soul and Body in Religious Experience*. Leiden, Boston, Köln: Brill, 1998, pp. 363–383.

Abusch 1999: Tzvi Abusch, "Witchcraft and the Anger of the Personal God", in Tzvi Abusch and Karel van der Toorn (eds.), *Mesopotamian Magic. Textual, Historical, and Interpretative Perspectives* (Ancient Magic and Divination 1). Groningen: Styx, 1999, pp. 81–121.

Abusch 2002: Tzvi Abusch, *Mesopotamian Witchcraft. Toward a History and Understanding of Babylonian Witchcraft Beliefs and Literature* (Ancient Magic and Divination 5). Leiden, Boston, Köln: Brill, 2002.

Abusch 2016: Tzvi Abusch, *The Magical Ceremony Maqlû. A Critical Edition* (Ancient Magic and Divination 10). Leiden, Boston: Brill, 2016.

Abusch and Schwemer 2011: Tzvi Abusch and Daniel Schwemer, *Corpus of Mesopotamian Anti-witchcraft Rituals*, vol. 1 (Ancient Magic and Divination 8.1). Leiden, Boston: Brill, 2011.

Abusch and Schwemer 2016: Tzvi Abusch and Daniel Schwemer, *Corpus of Mesopotamian Anti-witchcraft Rituals*, vol. 2 (Ancient Magic and Divination 8.2). Leiden, Boston: Brill, 2016.

Abusch et al. 2020: Tzvi Abusch, Daniel Schwemer, Mikko Luukko, and Greta Van Buylaere, *Corpus of Mesopotamian Anti-witchcraft Rituals*, vol. 3 (Ancient Magic and Divination 8.3). Leiden, Boston: Brill, 2020.

Acherknecht 1942: Erwin H. Ackerknecht, "Problems of Primitive Medicine", *Bulletin of the History of Medicine* 11 (1942), pp. 503–521.

Acherknecht 1946: Erwin H. Ackerknecht, "Natural Diseases and Rational Treatment in Primitive Medicine", *Bulletin of the History of Medicine* 19 (1946), pp. 467–497.

Acherknecht 1971: Erwin H. Ackerknecht, *Medicine and Ethnology. Selected essays*. Baltimore: Johns Hopkins University Press, 1971.

Adams 1965: Robert McCormick Adams, *Land behind Baghdad. A History of Settlement on the Diyala Plains*. Chicago, London: University of Chicago Press, 1965.

Ahern 1979: Emily M. Ahern, "The Problem of Efficacy: Strong and Weak Illocutionary Acts", *Man* N.S. 14, 1 (1979), pp. 1–17

Akrich 1995: Madeleine Akrich, "Petite anthropologie du médicament", *Techniques & Culture* 25–26 (1995), pp. 129–157.

Alaura and Bonechi 2012: Silvia Alaura and Marco Bonechi, "Il carro del dio del sole nei testi cuneiformi dell'Età del Bronzo", *Studi micenei ed egeo-anatolici* 54 (2012), pp. 5–115.

Alberti 2006: Benjamin Alberti, "Archaeology, Men, and Masculinities", in Sarah Milledge Nelson (ed.), *Handbook of Gender in Archaeology*. Lanham: Altamira Press, 2006, pp. 401–434.

Alcoff and Potter 1993: Linda Alcoff and Elizabeth Potter (eds.), *Feminist Epistemologies*. New York; Routledge, 1993.

Alster 1975: Bendt Alster, *Studies in Sumerian Proverbs* (Mesopotamia 3). Copenhagen: Akademisk Forlag, 1975.

Alster 1983: Bendt Alster, "The Mythology of Mourning", *Acta Sumerologica* 5 (1983), pp. 1–16.

https://doi.org/10.1515/9783110757262-017

Alster 1993: Bendt Alster, "Marriage and Love in the Sumerian Love Songs", in Mark E. Cohen, Daniel C. Snell, and David B. Weisberg (eds.), *The Tablet and the Scroll. Near Eastern Studies in Honour of William W. Hallo*. Bethesda: CDL Press, 1993.

Alster 1997: Bendt Alster, *Proverbs of Ancient Sumer. The World's Earliest Proverb Collections*, 2 vols. Bethesda: CDL Press.

Alster and Vanstiphout 1987: Bendt Alster and Herman L.J. Vanstiphout, "Lahar and Ashnan. Presentation and Analysis of the Sumerian Disputation", *Acta Sumerologica* 8 (1987), pp. 1–43.

Althusser 1970: Louis Althusser, "Idéologie et appareils idéologiques d'Etat", *La Pensée* 151 (1970).

Ambos 2003: Claus Ambos, "Nanaja – eine ikonographische Studie zur Darstellung einer altorientalischen Göttin in hellenistisch-parthischer Zeit", *Zeitschrift für Assyriologie und vorderasiatische Archäologie* 93, 2 (2003), pp. 231–272.

Ambos 2007: Claus Ambos, "Types of Ritual Failure and Mistakes in Ritual in Cuneiform Sources", in Ute Hüsken (ed.), *When Ritual Goes Wrong. Mistakes, Failure, and the Dynamics of Ritual*. Leiden, Boston: Brill, 2007, pp. 99–47.

Anzaldúa 1987: Gloria E. Anzaldúa, *Borderlands/La Frontera. The New Mestiza*. San Francisco: Aunt Lute Books, 1987.

Appadurai 1986: Arjun Appadurai (ed.), *The Social Life of Things. Commodities in Cultural Perspective*, Cambridge: Cambridge University Press, 1986.

Arnaud 1987: Daniel Arnaud, *Recherches au pays d'Aštata* (Emar VI/4). Paris: Éditions Recherche sur les civilisations, 1987.

Arnaud 1989: Daniel Arnaud, *Altbabylonische Rechts- und Verwaltungsurkunden aus dem Musée du Louvre* (Berliner Beiträge zum Vorderen Orient, Texte 1). Berlin: Dietrich Reimer, 1989.

Arnaud 2012: Daniel Arnaud, "Une amulette néo-assyrienne contre l'impuissance", *Aula orientalis: revista de estudios del Próximo Oriente Antiguo* 30, 1 (2012), pp. 185–187.

Asher-Greve 1998: Julia M. Asher-Greve, "The Essential Body: Mesopotamian Conceptions of the Gendered Body", in Maria Wyke (ed.), *Gender and the Body in the Ancient Mediterranean*. Oxford: Blackwell Publishers, 1998, pp. 8–37.

Asher-Greve 2000: Julia M. Asher-Greve, "Stepping into the Maelstrom: Gender, Woman and Ancient Near Eastern Scholarship", *NIN: Journal of Gender Studies in Antiquity* 1 (2000), pp. 1–22.

Asher-Greve 2002: Julia M. Asher-Greve (with the collaboration of Mary Frances Wogec), "Women and Gender in Ancient Near Eastern Cultures: Bibliography 1885 to 2001 AD", *NIN: Journal of Gender Studies in Antiquity* 3 (2002), pp. 33–114.

Asher-Greve 2008: Julia M. Asher-Greve, "Images of Men, Gender Regimes, and Social Stratification in the Late Uruk Period", in Diane R. Bolger (ed.), *Gender through Time in the Ancient Near East*. Lanham, New York, Toronto, Plymouth: Altamira Press, 2008, pp. 119–171.

Assante 2002: Julia A. Assante, "Sex, Magic and the Liminal Body in the Erotic Art and Texts of the Old Babylonian Period", in Simo Parpola and Robert M. Whiting (eds.), *Sex and Gender in the Ancient Near East. Proceedings of the 47th Rencontre Assyriologique Internationale, Helsinki, July 2–6, 2001*. Helsinki: Neo-Assyrian Text Corpus Project, 2002, pp. 27–52.

Assante 2007: Julia A. Assante, "The Lead Inlays of Tukulti-Ninurta I: Pornography as Imperial Strategy", in Jack Cheng and Marian H. Feldman (eds.), *Ancient Near Eastern Art in Context. Studies in Honor of Irene J. Winter by Her Students*. Leiden, Boston: Brill, 2007, pp. 369–407.

Assante 2017: Julia A. Assante, "Men Looking at Men: the Homoerotics of Power in the State of Arts of Assyria", in Zsolnay 2017: 42–82.

Attia 2018: Annie Attia, "The *libbu* our Second Brain? (part 1)", *Journal des Médecines Cunéiformes* 31 (2018), pp. 67–88.

Attia 2019: Annie Attia, "The *libbu* our Second Brain? Appendix part 2", *Journal des Médecines Cunéiformes* 33 (2019), pp. 50–92.

Attia and Buisson 2004: Annie Attia and Gilles Buisson, “Du bon usage des médecins en assyriologie”, *Journal des Médecines Cunéiformes* 4 (2004), pp. 9–15.
Attia and Buisson 2012: Annie Attia and Gilles Buisson, “BAM 1 et consorts en transcription”, *Journal des Médecines Cunéiformes* 19 (2012), pp. 22–50.
Attinger 1984: Pascal Attinger, “Enki et Ninḫursag̃a”, *Zeitschrift für Assyriologie und vorderasiatische Archäologie* 74, 1 (1984), pp. 1–52.
Augé 1980: Marc Augé, “Persona”, *Enciclopedia Einaudi*, vol. 10. Torino: Einaudi, 1980, pp. 651–672.
Augé 1982: Marc Augé, *Génie du paganisme*. Paris: Gallimard, 1982.
Augé 1994: Marc Augé, “Ordre biologique, ordre social : la maladie forme élémentaire de l’événement”, in Marc Augé and Claudine Herzlich (eds.), *Le sens du mal. Anthropologie, histoire, sociologie de la maladie*. Paris: Éditions des archives contemporaines, 1994, pp. 35–91.
Austin 1962: John L. Austin, *How to Do Things with Words. The William James Lectures delivered at Harvard University in 1955*. Oxford: Oxford University Press, 1962.
Avalos 1995: Hector Avalos, *Illness and Health Care in The Ancient Near East. The Role of the Temple in Greece, Mesopotamia, and Israel* (Harvard Semitic Monographs 54). Atlanta: Scholars Press, 1995.
Bácskay 2018: András Bácskay, “‘Seize a Frog!’ The Use of the Frog in Medical and Magical Texts”, *Journal des Médecines Cunéiformes* 32 (2018), pp. 1–16.
Bahrani 2000: Zainab Bahrani, “The Whore of Babylon: Truly All Woman and of Infinite Variety”, *NIN: Journal of Gender Studies in Antiquity* 1 (2000), pp. 95–106.
Bahrani 2001: Zainab Bahrani, *Women of Babylon. Gender and Representation in Mesopotamia*. London: Routledge, 2001.
DBaker 1976: David W. Baker, *Idiomatic Expression in Hebrew and Akkadian Relating to the Head*. Ph.D. dissertation, University of London, 1976.
SBaker 1993: Steve Baker, *Picturing the Beast. Animals, Identity and Representation*. Urbana: University of Illinois Press, 1993.
Ball 1987: Charles J. Ball, “Inscriptions of Nebuchadrezzar II”, *Proceedings of the Society of Biblical Archaeology* 10 (1987), pp. 87–129.
Baragli 2020: Beatrice Baragli, “Abracadabra Incantations: Nonsense or Healing Therapies?”, *Kaskal. Rivista di storia, ambienti e culture del Vicino Oriente antico* 16 (2019), pp. 293–321.
Barjamovic 2015: Gojko Barjamovic, “Contextualizing Tradition: Magic, Literacy and Domestic Life in Old Assyrian Kanesh”, in Paul Delnero and Jacob Lauinger (eds.), *Texts and Contexts. The Circulation and Transmission of Cuneiform Texts in Social Space* (Studies in Ancient Near Eastern Records 9). Boston, Berlin: De Gruyter, pp. 48–86.
Barré 2001: Michael L. Barré, “‘Wandering about’ as a Topos of Depression in Ancient Near Eastern Literature and in the Bible”, *Journal of Near Eastern Studies* 60, 3 (2001), pp. 177–187.
Barthes 1984: Roland Barthes, *Le bruissement de la langue*. Paris: Seuil, 1984.
Battini 2009: Laura Battini, “La conception des animaux domestiques et des animaux de compagnie dans la Mésopotamie d’époque historique”, *Res Antiquae* 6 (2009), pp. 7–37.
Beaulieu 2003: Paul-Alain Beaulieu, *The Pantheon of Uruk During the Neo-Babylonian Period* (Cuneiform Monographs 23). Leiden, Boston: Brill, 2003.
Beckman 1983: Gary M. Beckman, *Hittite Birth Rituals (Second Revised Edition)* (Studien zu dem Boğazköy-Texten 29). Wiesbaden: Harrassowitz, 1983.
Beckman and Foster 1988: Gary M. Beckman and Benjamin R. Foster, “Assyrian Scholarly Texts in the Yale Babylonian Collection”, in Erle Leichty, Maria deJ. Ellis, and Pamela Gerardi (eds.), *A Scientific Humanist. Studies in Memory of Abraham Sachs* (Occasional Publications of the Samuel Noah Kramer Fund 9). Philadelphia: University of Pennsylvania Museum, 1988, pp. 1–26.

Behrens 1978: Hermann Behrens, *Enlil und Ninlil. Ein sumerischer Mythos aus Nippur* (Studia Pohl: Series Maior 8). Rome: Biblical Institute Press, 1978.

Behrens 1998: Hermann Behrens, *Die Ninegalla-Hymne. Die Wohnungnahme Inannas in Nippur in altbabylonischer Zeit* (Freiburger Altorientalische Studien 21). Stuttgart: Franz Steiner, 1998.

bell hooks 1996: bell hooks, "Choosing the Margin as a Space of Radical Openness", in Ann Garry and Marilyn Pearsall (eds.), *Woman, Knowledge and Reality. Explorations in Feminist Philosophy*, New York, London: Routledge, 1996, pp. 48–55.

Beneduce 2005: Roberto Beneduce, "Dall'efficacia simbolica alle politiche del sé", in Roberto Beneduce and Élisabeth Roudinesco (eds.), *Antropologia della cura*. Torino: Bollati Boringhieri, 2005, pp. 7–27.

Beneduce 2007: Roberto Beneduce, *Etnopsichiatria. Sofferenza mentale e alterità fra Storia, dominio e cultura*. Roma: Carocci, 2007.

Benito 1969: Carlos Alfredo Benito, *'Enki and Ninmah' and 'Enki and the World Order'*. Ph.D. dissertation, University of Pennsylvania, 1969.

Bennett 2020: Eleanor Bennett, "'I am a Man': Masculinities in the Titulary of the Neo-Assyrian Kings in the Royal Inscriptions", *Kaskal. Rivista di storia, ambienti e culture del Vicino Oriente antico* 16 (2019), pp. 373–392.

Besnier 2002: Marie-Françoise Besnier, "Temptation's Garden: The Gardener, a Mediator Who Plays an Ambiguous Part", in Simo Parpola and Robert M. Whiting (eds.), *Sex and Gender in the Ancient Near East. Proceedings of the 47th Rencontre Assyriologique Internationale, Helsinki, July 2–6, 2001*. Helsinki: Neo-Assyrian Text Corpus Project, 2002. pp. 59–70.

Bettini 1998: Maurizio Bettini, *Nascere. Storie di donne, donnole, madri ed eroi*. Torino: Einaudi, 1998.

Bibeau 1982: Gilles Bibeau, "A Systems Approach to Ngbandi Medicine", in P. Stanley Yoder (ed.), *African Health and Healing Systems. Proceedings of a Symposium*. Los Angeles: African Studies Center, 1982, pp. 43–84.

Bibeau 1983: Gilles Bibeau, "L'activation des mécanismes endogènes d'auto-guérison dans les traitements rituels des Angbandi", *Culture* 3, 1 (1983), pp. 33–49 (Italian translation: "L'attivazione dei meccanismi endogeni di autoguarigione nei trattamenti rituali degli Angbandi", in Vittorio Lanternari and Maria Luisa Ciminelli (eds.), *Medicina, Magia, Religione, Valori. II. Dall'antropologia medica all'etnopsichiatria*. Napoli: Liguori, 1998, pp. 131–158).

Biggs 1967: Robert Biggs, *ŠÀ.ZI.GA. Ancient Mesopotamian Potency Incantations* (Texts from Cuneiform Sources 2). Locust Valley, New York: J.J. Augustin, 1967.

Biggs 1995: Robert Biggs, "Medicine, Surgery and Public Health in Ancient Mesopotamia", in Jack M. Sasson (eds.), *Civilizations of the Ancient Near East*. New York: Charles Scribner's Sons, 1995, pp. 1911–1924.

Biggs 2002: Robert Biggs, "The Babylonian Sexual Potency Texts", in Simo Parpola and Robert M. Whiting (eds.), *Sex and Gender in the Ancient Near East. Proceedings of the 47th Rencontre Assyriologique Internationale, Helsinki, July 2–6, 2001*. Helsinki: Neo-Assyrian Text Corpus Project, 2002, pp. 71–78.

Biggs 2003–2005: Robert Biggs, "Potenzerhöhung", in *Reallexikon der Assyriologie und vorderasiatischen Archäologie*, vol. 10. Berlin: De Gruyter, 2003–2005, pp. 604–605.

Biggs 2006: Robert Biggs, "The Human Body and Sexuality in the Babylonian Medical Texts", in Laura Battini and Pierre Villard (eds.), *Médecine et médecins au Proche-Orient ancien. Actes du colloque international organisé à Lyon les 8 et 9 novembre 2002, Maison de l'Orient et de la Méditerranée*. Oxford: Archaeopress, 2006, pp. 39–52.

Birot 1993: Maurice Birot, *Correspondance des gouverneurs de Qaṭṭunân* (Archives Royales de Mari 27). Paris: Éditions Recherche sur les civilisations, 1993.

Black 1996: Jeremy Black, "The Imagery of Birds in Sumerian Poetry", in Marianna E. Vogelzang and Herman L.J. Vanstiphout (eds.), *Mesopotamian Poetic Language: Sumerian and Akkadian* (Cuneiform Monographs 6). Groningen: Styx, 1996, pp. 23–46.

Böck 2000: Barbara Böck, *Die babylonisch-assyrische Morphoskopie* (Archiv für Orientforschung Beiheft 27). Wien: Institut für Orientalistik der Universität Wien, 2000.

Böck 2001–2002: Barbara Böck, review of Schwemer 1998, *Archiv für Orientforschung* 48/49 (2001–2002), pp. 228–232.

Böck 2007: Barbara Böck, *Das Handbuch Muššu'u "Einreibung". Eine Serie sumerischer und akkadischer Beschwörungen aus dem 1. Jt. vor Chr.* (Biblioteca del Próximo Oriente Antiguo 3). Madrid: Consejo Superior de Investigaciones Científicas, 2007.

Böck 2008: Barbara Böck, "Babylonisch-assyrische Medizin in Texten und Untersuchungen: Erkrankungen des uro-genitalen Traktes, des Enddarmes und des Anus", *Wiener Zeitschrift für die Kunde des Morgenlandes* 98 (2008), pp. 295–346.

Böck 2009: Barbara Böck, "On Medical Technology in Ancient Mesopotamia", in Annie Attia and Gilles Buisson (eds.), *Advances in Mesopotamian Medicine from Hammurabi to Hippocrates. Proceedings of the International Conference "Oeil malade et mauvais œil"; Collège de France, Paris, 23rd June 2006* (Cuneiform Monographs 37). Leiden, Boston: Brill, 2009, pp. 105–128.

Böck 2011: Barbara Böck, "Sourcing, Organizing and Administering Medicinal Ingredients", in Karen Radner and Eleanor Robson (eds.), *The Oxford Handbook of Cuneiform Culture*. Oxford: Oxford University Press, pp. 690–705.

Böck 2014: Barbara Böck, *The Healing Goddess Gula. Towards an Understanding of Ancient Babylonian Medicine* (Culture and History of the Ancient Near East 67). Leiden, Boston: Brill, 2014.

Boer 2015: Roland Boer, "From Horse Kissing to Beastly Emissions: Paraphilias in the Ancient Near East", in Mark Masterson, Nancy Sorkin Rabinowitz, and James Robson (eds.), *Sex in Antiquity. Exploring Gender and Sexuality in the Ancient World*. London: Routledge, 2015, pp. 67–79.

Bolger 2008: Diane R. Bolger, "Introduction", in Eadem (ed.), *Gender through Time in the Ancient Near East*. Lanham, New York, Toronto, Plymouth: Altamira Press, 2008, pp. 1–20.

Borger 1956: Rykle Borger, *Die Inschriften Asarhaddons Königs von Assyrien* (Archiv für Orientforschung Beiheft 9). Graz: Im Selbstverlage des Herausgebers, 1956.

Borger 1964: Rykle Borger, review of Lambert 1960a, *Journal of Cuneiform Studies* 18 (1964), pp. 49–56.

Borger 1967: Rykle Borger, "Das Dritte 'Haus' der Serie *bīt rimki*", *Journal of Cuneiform Studies* 21 (1967), pp. 1–17.

Borger 1971: Rykle Borger, "Gott Marduk und Gott-König Šulgi als Propheten: Zwei prophetische Texte", *Bibliotheca Orientalis* 28 (1971), pp. 3–24.

Borger 2004: Rykle Borger, "Eine altorientalische und antike Fabel Zaunkönig/Mücke und Elephant/ Stier", *Bibliotheca Orientalis* 61 (2004), pp. 461–474.

Bottéro 1957–1971: Jean Bottéro, "Gewürze", in *Reallexikon der Assyriologie und vorderasiatischen Archäologie*, vol. 3. Berlin: De Gruyter, 1957–1971, pp. 340–344.

Bottéro 1987–1990: Jean Bottéro, "Magie. A. In Mesopotamien", in *Reallexikon der Assyriologie und vorderasiatischen Archäologie*, vol. 7. Berlin: De Gruyter, 1987–1990, pp. 200–234.

Bourdieu 1972: Pierre Bourdieu, *Esquisse d'une théorie de la pratique. Précédé de Trois études d'ethnologie kabyle*. Genève: Librairie Droz, 1972.

Bourdieu 1980: Pierre Bourdieu, *Le Sens pratique*. Paris: Éditions de Minuit, 1980.

Bourdieu 1998: Pierre Bourdieu, *La domination masculine*. Paris: Seuil, 1998.

Bourdieu 2001: Pierre Bourdieu, "Objectiver le sujet de l'objectivation", in Idem *Science de la science et réflexivité*. Paris: Raisons d'Agir, 2001, pp. 173–184.

Braidotti 2002: Rosi Braidotti, *Nuovi Soggetti Nomadi*. Roma: L. Sossella, 2002.

Brandes 1981: Stanley H. Brandes, "Like Wounded Stags: Male Sexual Ideology in an Andalusian Town", in Sherry B. Ortner and Harriet Whitehead (eds.), *Sexual Meanings: The Cultural Construction of Gender and Sexuality*. Cambridge: Cambridge University Press, 1981, pp. 216–239.

Brandes 1984: Stanley H. Brandes, "Animal Metaphors and Social Control in Tzintzuntzan", *Ethnology* 23, 3 (1984), pp. 207–215.

Brown 1973: Mary K. Brown, *Symbolic Lions: A Study in Ancient Mesopotamian Art and Literature*. Ph.D. dissertation, Harvard University, 1973.

Buccellati 1976: Giorgio Buccellati: "Towards a Formal Typology of Akkadian Similes", in Barry L. Eichler, Jane W. Heimerdinger, and Åke W. Sjöberg (eds.), *Kramer Anniversary Volume. Cuneiform Studies in Honor of Samuel Noah Kramer* (Alter Orient und Altes Testament 25). Kevelaer: Butzon & Bercker, 1976, pp. 59–60.

Budin 2015: Stephanie Lynn Budin, "Sexuality: Ancient Near East (except Egypt)", in Patricia Whelehan and Anne Bolin (eds.), *The International Encyclopedia of Human Sexuality*. Chichester: Wiley-Blackwell, 2015, pp. 1–5.

Budin and Webb 2016: Stephanie Lynn Budin and Jennifer Webb (eds.), *Near Eastern Archaeology* (Special Issue: Gender Archaeology) 79, 3 (2016).

Buisson 2016: Gilles Buisson, "À la recherche de la mélancolie en Mésopotamie ancienne", *Journal des Médecines Cunéiformes* 28 (2916), pp. 1–54.

Busoni 2016: Mila Busoni, *Genere, sesso, cultura. Uno sguardo antropologico*, Roma: Carocci, 2016.

Butler 1986: Judith Butler, "Sex and Gender in Simone de Beauvoir's *Second Sex*", *Yale French Studies* 72 (1986), pp. 35–50.

Butler 1990: Judith Butler, *Gender Trouble. Feminism and Subversion of Identity*. London, New York: Routledge, 1990.

Butler 1999: Judith Butler, "Performativity's Social Magic", in Richard Shusterman (ed.), *Bourdieu: A Critical Reader*. Oxford: Blackwell, pp. 113–128.

Butler 2004: Judith Butler, *Undoing Gender*. London, New York: Routledge, 2004.

Bühler 1934: Karl Bühler, *Sprachtheorie. Die Darstellungsfunktion der Sprache*. Jena: Gustav Fischer, 1934.

Cagni 1969: Luigi Cagni, *L'Epopea di Erra* (Studi Semitici 34). Roma: Istituto di Studi del Vicino Oriente, Università di Roma, 1969.

Cammarosano 2018: Michele Cammarosano, *Hittite Local Cults* (Writings from the Ancient World 40). Atlanta: Society of Biblical Literature Press, 2018.

Canguilhem 1966: Georges Canguilhem, *Le normal et le pathologique*. Paris: Presses universitaires de France, 1966.

Capomacchia 2009: Anna M.G. Capomacchia (ed.), *Animali tra mito e simbolo*. Roma: Carocci, 2009.

Capraro 1998: Martina Capraro, "Aššur-šākin-šumi", in Karen Radner (ed.), *The Prosopography of the Neo-Assyrian Empire*, Vol. 1, Part I. Helsinki: The Neo-Assyrian Text Corpus Project, 1998, p. 216.

Cardona 2006: Giorgio R. Cardona, *I sei lati del mondo. Linguaggio ed esperienza*. Roma, Bari: Laterza, 2006.

Casaburi 2002–2005: Maria C. Casaburi, "Early Evidence of Astrological Aspects in a Neo-Assyrian Medical Hemerology", *State Archives of Assyria Bulletin* 14 (2002–2005), pp. 63–88.

Cassin 1981: Elena Cassin, "Le roi et le lion", *Revue de l'histoire des religions* 198, 4 (1981), pp. 353–401.

Cavigneaux 1981: Antoine Cavigneaux, *Textes Scolaires du Temple de Nabû ša Ḫarê*. Baghdad: State Organization of Antiquites & Heritage, 1981.

Cavigneaux 1999: Antoine Cavigneaux "A Scholars Library in Meturan?", in Tzvi Abusch and Karel van der Toorn (eds.), *Mesopotamian Magic. Textual, Historical, and Interpretative Perspectives* (Ancient Magic and Divination 1). Groningen: Styx, 1999, pp. 258–273.

Cavigneaux and al-Rawi 1993 : Antoine Cavigneaux and Farouk N.H. al-Rawi, "Gilgameš et le Taureau de Ciel (šul.mè.kam). Textes de Tell Haddad IV", *Revue d'Assyriologie et d'Archéologie orientale* 87 (1993), pp. 97–129

Cavigneaux and al-Rawi 2002: Antoine Cavigneaux and Farouk N.H. al-Rawi, "Liturgies exorcistiques agraires (Textes de Tell Haddad IX)", *Zeitschrift für Assyriologie und vorderasiatische Archäologie* 92, 1 (2002), pp. 1–59.

Cavigneaux at al. 1985: Antoine Cavigneaux, Hans G. Güterbock, Martha T. Roth, and Gertrud Farber, *The Series Erim-huš = anantu and An-ta-gál = šaqû* (Materialien zum sumerischen Lexikon 17). Rome: Biblical Institute Press, 1985.

Ceccarelli 2009: Manuel Ceccarelli, "Einige Bemerkungen zum Synkretismus BaU/Ninisina", in Paola Negri Scafa and Salvatore Viaggio (eds.), *Dallo Stirone al Tigri, dal Tevere all'Eufarte. Studi in onore di Claudio Saporetti*, Roma: Aracne, 2009, pp. 31–54.

Ceravolo 2020: Marinella Ceravolo, *Il mito e il rito alla luce dell'historiola: il caso dell'antica Mesopotamia*. Ph.D. dissertation, Sapienza Università di Roma, 2020.

Chalendar 2013: Vérène Chalendar, "'Un aperçu de la neuropsychiatrie assyrienne': Une édition du texte BAM III-202", *Journal des Médecines Cunéiformes* 21 (2013), pp. 1–60.

Chalendar 2016: Vérène Chalendar, "What reality for animals in the Mesopotamian medical texts? Plant vs animal", *Anthropozoologica* 51, 2 (2016), pp. 97–103.

Chalendar 2018: Vérène Chalendar, "Éléments de pharmacopée mésopotamienne : retour sur l'ingrédient *rikibtu*", *Journal des Médecines Cunéiformes* 32 (2018), pp. 24–55.

Chapman 2004: Cynthia R. Chapman, *The Gendered Language of Warfare in the Israelite-Assyrian Encounter* (Harvard Semitic Monographs 62). Winoka Lake: Eisenbrauns, 2004.

Cholidis 1992: Nadja Cholidis, *Möbel in Ton. Untersuchungen zur archäologischen und religionsgeschichlichen Bedeutung der Terrakottamodelle von Tischen, Stühlen und Betten aus dem Alten Orient* (Altertumskunde des Vorderen Orients 1). Münster: Ugarit-Verlag, 1992.

Choukassizian Eypper 2019: Sona Choukassizian Eypper, "*kasû* (Ú.GAZI.SAR) Revisited", *Journal des Médecines Cunéiformes* 33 (2019), pp. 35–49.

Cifarelli 1998: Megan Cifarelli, "Gesture and Alterity in the Art of Ashurnasirpal II of Assyria", *The Art Bulletin* 80, 2 (1998), pp. 210–228.

Ciminelli 1997: Maria Luisa Ciminelli, "*Culture-bound syndromes*: un concetto vago e di dubbia utilità", *AM. Rivista della Società italiana di antropologia medica* 3–4 (1997), pp. 247–280.

Ciminelli 1998: Maria Luisa Ciminelli, "La decostruzione del concetto di *culture-bound syndrome*", in Vittorio Lanternari and Maria Luisa Ciminelli (eds.), *Medicina, Magia, Religione, Valori. II. Dall'antropologia medica all'etnopsichiatria*. Napoli: Liguori, 1998, pp. 85–108.

Civil 1960: Miquel Civil, "Prescriptions médicales sumériennes", *Revue d'Assyriologie et Archéologie orientale* 54 (1960), pp. 57–72.

Civil 1966: Miquel Civil, "Notes on Sumerian Lexicography, I", *Journal of Cuneiform Studies* 20 (1966), pp. 119–124.

Civil 1969: Miquel Civil, *The Series lú = ša and Related Texts* (Materialien zum sumerischen Lexikon 12). Rome: Biblical Institute Press, 1969.

Civil 1974: Miquel Civil, "Medical Commentaries from Nippur", *Journal of Near Eastern Studies* 33 (1974), pp. 329–338.

Civil 1984: Miquel Civil, "On Some Terms for 'Bat' in Mesopotamia", *Aula orientalis: revista de estudios del Próximo Oriente Antiguo* 2 (1984), pp. 5–9.

Civil 1987: Miquel Civil, "Feeding Dumuzi's Sheep: The Lexicon as a Source of Literary Inspiration", in Francesca Rocheberg-Halton (ed.), *Language, Literature, and History: Philological and Historical Studies Presented to Erica Reiner* (American Oriental Series 67). New Haven: American Oriental Society, 1987, pp. 37–55.

Civil 2006: Miquel Civil, "be$_5$/pe-en-zé-er = *biṣṣūru*", in Ann K. Guinan, Maria deJ. Ellis, A.J. Ferrara, Sally M. Freedman, Matthew T. Rutz, Leonhard Sassmannshausen, Steve Tinney, and Matthew W. Waters (eds.), *If a Man Builds a Joyful House. Assyriological Studies in Honor of Erle Verdun Leichty* (Cuneiform Monographs 31). Leiden, Boston: Brill, 2006, pp. 55–61.

Civil et al. 1979: Miquel Civil, Margaret W. Green, and Wilfred G. Lambert, *Ea A = nâqu, Aa A = nâqu, with their Forerunners and Related Texts* (Materialien zum sumerischen Lexikon 14). Rome: Biblical Institute Press, 1979.

Clifford 1994: Richard J. Clifford, *Creation Accounts in the Ancient Near East and in the Bible*. Washington: Catholic University of America, 1994.

Clifford and Marcus 1986: James Clifford and George E. Marcus (eds.) *Writing Culture. The Poetics and Politics of Ethnography*. Berkley, London: University of California Press, 1986.

Cohen 1994: Anthony P. Cohen, *Self Consciousness. An Alternative Anthropology of Identity*. London: Routledge, 1994.

Coleman 2005: Mary Coleman, "Lettre aux éditeurs : 'Reply to Nils P. Heeßel'", *Journal des Médecines Cunéiformes* 6 (2005), pp. 43–48.

BCollins 2002: Billie J. Collins (ed.), *A History of the Animal World in the Ancient Near East*. Leiden, Boston, Köln: Brill, 2002.

BCollins 2003: Billie J. Collins, "Hittite Canonical Compositions: Rituals", in William W. Hallo (ed.), *The Context of Scripture. Volume I: Canonical Compositions from the Biblical World*. Leiden, Boston: Brill, 2003, pp. 160–168.

BCollins 2014: Billie J. Collins, "Woman in Hittite Ritual", in Mark W. Chavalas (ed.), *Women in the Ancient Near East. A Sourcebook*. London, New York: Routledge, 2014, pp. 246–271.

TCollins 1999: Timothy Joseph Collins, *Natural Illness in Babylonian Medical Incantations*. Ph.D. dissertation, University of Chicago, 1999.

Connell 1995: Raewyn Connell, *Masculinities*. Berkeley: University of California Press, 1995.

Cooper 1972: Jerrold S. Cooper, "Bilinguals from Boghazköi II", *Zeitschrift für Assyriologie und vorderasiatische Archäologie* 62, 1 (1972), pp. 62–81.

Cooper 1977: Jerrold S. Cooper, "Gilgamesh Dreams of Enkidu: The Evolution and Dilution of Narrative", in Maria deJ. Ellis (ed.), *Essays on the Ancient Near East in Memory of Jacob Joel Finkelstein*. Hamden: Published for the Academy by Archon Books, 1977: pp. 39–44.

Cooper 1989: Jerrold S. Cooper, "Enki's Member: Eros and Irrigation in Sumerian Literature", in Hermann Behrens, Darlene Loding, and Martha T. Roth (eds.), DUMU-E$_2$-DUB-BA-BA-A. *Studies in Honor of Åke W. Sjöberg*. Philadelphia: University of Pennsylvania Museum, 1989, pp. 87–89.

Cooper 1996: Jerrold S. Cooper, "Magic and M(is)use: Poetic Promiscuity in Mesopotamian Ritual", in Marianna E. Vogelzang and Herman L.J. Vanstiphout (eds.), *Mesopotamian Poetic Language: Sumerian and Akkadian* (Cuneiform Monographs 6). Groningen: Styx, 1996, pp. 47–57.

Cooper 1997: Jerrold S. Cooper, "Gendered Sexuality in Sumerian Love Poetry", in Irving L. Finkel and Markham J. Geller (eds.), *Sumerian Gods and Their Representations* (Cuneiform Monographs 7). Groningen: Styx, 1997, pp. 85–97.

Cooper 2002: Jerrold S. Cooper, "Virginity in Ancient Mesopotamia", in Simo Parpola and Robert M. Whiting (eds.), *Sex and Gender in the Ancient Near East. Proceedings of the 47th Rencontre Assyriologique Internationale, Helsinki, July 2–6, 2001*. Helsinki: Neo-Assyrian Text Corpus Project, 2002, pp. 91–112

Cooper 2017: Jerrold S. Cooper, "Female Trouble and Troubled Males: Roiled Seas, Decadent Royals, and Mesopotamian Masculinities in Myth and Practice", in Zsolnay 2017: 112–124.

Corò 2005: Paola Corò, "Il 'bestiario' di Mari. I. Le valenze simboliche", in Ettore Cingano, Antonella Ghersetti, and Lucio Milano (eds.), *Animali tra zoologia, mito e letteratura nella cultura classica e orientale*. Padova: Sargon, 2005: pp. 33–45.

Couto-Ferreira 2007: M. Érica Couto-Ferreira, "Conceptos de transmisión de la enfermedad en Mesopotamia: algunas reflexiones", *Historiae* 4 (2007), pp. 1–24.

Couto-Ferreira 2009: M. Érica Couto-Ferreira, *Etnoanatomía y partonomía del cuerpo humano en sumerio y acadio. El léxico Ugu-mu*. Ph.D. dissertation, Universitat Pompeu Fabra, Barcelona, 2009.

Couto-Ferreira 2010a: M. Érica Couto-Ferreira, review of Capomacchia 2009, *Rivista di Studi Orientali* 83 (2010), pp. 455–460.

Couto-Ferreira 2010b: M. Érica Couto-Ferreira, "It is the Same for a Man and a Woman: Melancholy and Lovesickness in Ancient Mesopotamia", *Quaderni di Studi Indo-Mediterranei* 3 (2010), pp. 21–39.

Couto-Ferreira 2013: M. Érica Couto-Ferreira, "The River, the Oven, the Garden: Female Body and Fertility in a Late Babylonian Ritual Text", in Claus Ambos and Lorenzo Verderame (eds.), *Approaching rituals in ancient cultures. Questioni di rito: rituali come fonte di conoscenza delle religioni e delle concezioni del mondo nelle culture antiche. Proceedings of the Conference, November 28–30 2011, Roma* (Rivista di Studi Orientali Nuova Serie 86, Supplemento 2). Pisa, Roma: Fabrizio Serra editore, 2013, pp. 97–116.

Couto-Ferreira 2015: M. Érica Couto-Ferreira, "Agency, Performance and Recitation as Textual Tradition in Mesopotamia: An Akkadian Text of the Late Babylonian Period to Make a Woman Conceive", in Magali de Haro Sanchez (ed.), *Écrire la magie dans l'Antiquité. Actes du colloque international (Liège, 13–15 octobre 2011)*. Liège: Presses Universitaires de Liège, 2015, pp. 187–199.

Couto-Ferreira 2015–2016: M. Érica Couto-Ferreira, "In the Womb: Embryological Ideas in Mesopotamian Cuneiform Texts", *Korot* 23 (2015–2016), pp. 47–71.

Couto-Ferreira 2017: M. Érica Couto-Ferreira, "'Let Me Be Your Canal': Some Thoughts on Agricultural Landscape and Female Bodies in Sumero-Akkadian Sources", in Lluis Feliu, Fumi Karahashi, and Gonzalo Rubio (eds.), *The First 90 Years. A Sumerian Celebration in Honor of Miquel Civil* (Studies in Ancient Near Eastern Records 12). Berlin: De Gruyter, 2017, pp. 54–69.

Couto-Ferreira 2018a: M. Érica Couto-Ferreira, "Cuerpos mansos: sobre la domesticación sexual y reproductiva de la mujer en las fuentes sumerias", *Claroscuro* 17 (2018), pp. 1–16.

Couto-Ferreira 2018b: M. Érica Couto-Ferreira, "Politics of the Body Productive: Agriculture, Royal Power, and the Female Body in Sumerian Sources", in Stephanie Lynn Budin, Megan Cifarelli, Agnès Garcia-Ventura, and Adelina Millet Albà (eds.), *Gender and Methodology in the Ancient Near East. Approaches from Assyriology and beyond*. Barcelona: Edicions de la Universitat de Barcelona, 2018, pp. 1–9.

Couto-Ferreira 2020: M. Érica Couto-Ferreira, "Disturbing Disorders: Reconsidering the Problems of 'Mental Diseases' in Ancient Mesopotamia", in Ulrike Steinert (ed.), *Cultural Systems of Classification. Sickness, Health and Local Epistemologies*. London, New York: Routledge, 2020, pp. 261–278.

Couto-Ferreira and Garcia-Ventura 2013: M. Érica Couto-Ferreira and Agnès Garcia-Ventura, "Engendering Purity and Impurity in Assyriological Studies: A Historiographical Overview", *Gender & History* 25, 3 (2013), pp. 513–528.

Crapanzano 1986: Vincent Crapanzano, "Hermes' Dilemma: The Masking of Subversion in Ethnographic Description", in Clifford and Marcus 1986: 51–76.

Csordas 1990: Thomas J. Csordas, "Embodiment as a Paradigm for Anthropology", *Ethos: Journal of the Society for Psychological Anthropology* 18 (1990), pp. 5–47.

Csordas 1997: Thomas J. Csordas, "Prophecy and the Performance of Metaphor", *American Anthropologist* N.S. 99, 2 (1997), pp. 321–332.

Csordas 1999: Thomas J. Csordas, "Embodiment and Cultural Phenomenology", in Gail Weiss and Honi Fern Haber (eds.), *Perspectives on Embodiment. The Intersection of Nature and Culture*. New York, London: Routledge, 1999, 143–162.

Csordas 2011: Thomas J. Csordas, "Cultural Phenomenology. Embodiment: Agency, Sexual Difference, and Illness", in Frances E. Mascia-Lees (ed.), *A Companion to the Anthropology of the Body and Embodiment*. Chichester: Wiley-Blackwell, 2011, pp. 137–156.

Csordas and Kleinman 1990: "The Therapeutic Process", in Thomas M. Johnson and Carolyn F. Sargent (eds.), *Medical Anthropology. Contemporary Theory and Method*. New York: Praeger, 1990, pp. 11–25 (Italian translation: "Il processo terapeutico", in Vittorio Lanternari and Maria Luisa Ciminelli (eds.), *Medicina, Magia, Religione, Valori. II. Dall'antropologia medica all'etnopsichiatria*. Napoli: Liguori, 1998, pp. 109–129).

Cunningham 1997: Graham Cunningham, *'Deliver Me from Evil' Mesopotamian Incantations 2500–1500 BC* (Studia Pohl: Series Maior 17). Rome: Biblical Institute Press, 1997.

Çağirgan 1976: Galip Çağirgan, *Babylonian Festivals*. Ph.D. dissertation, University of Birmingham, 1976.

Daxelmüller and Thomsen 1982: Christoph Daxelmüller and Marie-Louise Thomsen, "Bildzauber im alten Mesopotamien", *Anthropos* 77, 1/2 (1982), pp. 27–64.

Dein 2002: Simon Dein, "The Power of Words: Healing Narratives among Lubavitcher Hasidim", *Medical Anthropology Quarterly* N.S. 16, 1 (2002), pp. 41–63.

de Lauretis 1990: Teresa de Lauretis, "Eccentric Subjects: Feminist Theory and Historical Consciousness", *Feminist Studies* 16, 1 (1990), pp. 115–150.

Delphy 1991: Christine Delphy, "Penser le genre : quels problèmes?", in Marie-Claude Hurtig, Michèle Kail, and Hélène Rouch (eds.), *Sexe et genre. De la hiérarchie entre les sexes*. Paris: CNRS Éditions, 1991, pp. 89–101.

Delphy 1996: Christine Delphy, "Rethinking Sex and Gender", in Diana Leonard and Lisa Adkins (eds.), *Sex in Question. French Materialist Feminism*. London: Taylor & Francis, 1996, pp. 30–41.

DelVecchio Good et al. 1992: Mary-Jo Delvecchio Good, Paul E. Brodwin, Byron J. Good, and Arthur Kleinman (eds.), *Pain as Human Experience. An Anthropological Perspective*. Berkeley: University of California Press, 1992.

De Martino 1948: Ernesto De Martino, *Il mondo magico. Prolegomeni a una storia del magismo*. Torino: Einaudi, 1948.

De Martino 1953–1954: Ernesto De Martino, "Fenomenologia religiosa e storicismo assoluto", *Studi e materiali di storia delle religioni* 24–25 (1953–1954), pp. 1–25.

De Martino 1959: Ernesto De Martino, *Sud e magia*. Milano: Feltrinelli, 1959 (English translation: *Magic. A Theory from the South*. Chicago: Hau Books, 2015).

De Martino 1962: Ernesto De Martino, *Furore Simbolo Valore*. Milano: Il Saggiatore, 1962.

De Martino 1977: Ernesto De Martino, *La fine del mondo. Contributo all'analisi delle apocalissi culturali*. Torino: Einaudi, 1977.

Derrida 1972: Jacques Derrida, "Signature événement contexte", in Idem, *Marges de la philosophie*. Paris: Éditions de Minuit, 1972, pp. 365–393.

Descola 2005: Philippe Descola, *Par-delà nature et culture*. Paris: Gallimard, 2005.

Devereux 1967: George Devereux, *From Anxiety to Method in the Behavioral Sciences*. Paris: Mouton & Co and École Pratique des Hautes Études, 1967.

De Zorzi 2014: Nicla De Zorzi, *La serie teratomatica Šumma izbu. Testo, tradizione, orizzonti culturali*, 2 vols (History of the ancient Near East. Monographs 15). Padova: Sargon, 2014.

De Zorzi 2017: Nicla De Zorzi, "Teratomancy at Tigunānum: Structure, Hermeneutics, and *Weltanschauung* of a Northern Mesopotamian Omen Corpus", *Journal of Cuneiform Studies* 69 (2017), pp. 125–150.

Dierauer 1977: Urs Dierauer, *Tier und Mensch im Denken der Antike. Studien zur Tierpsychologie, Anthropologie und Ethik* (Studien zur Antiken Philosophic 6). Amsterdam: B.R. Grüner, 1977.

Dieterlen 1973 : Germaine Dieterlen (ed.), *La notion de personne en Afrique Noire*. Paris: Éditions du Centre national de la recherche scientifique, 1973.

Dietrich 2010: Manfried Dietrich, “Die Dichotomie ‘Leib’ und ‘Seele’ in der mesopotamischen Literatur”, *Mitteilungen für Anthropologie und Religionsgeschichte* 20 (2010), pp. 19–36.

Di Nola 1983 : Alfonso M. Di Nola, *L’arco di rovo. Impotenza e aggressività in due rituali del Sud.* Torino: Bollati Boringhieri, 1983.

Donbaz and Stolper 1993: Veysel Donbaz and Matthew W. Stolper, “Gleanings from Muraßû Texts in the Collections of the Istanbul Archaeological Museums”, *Nouvelles Assyriologiques Brèves et Utilitaires* 1993, 4, no. 102, pp. 85–96.

Dossin 1978 : Georges Dossin, *Correspondance féminine* (Archives Royales de Mari 10). Paris: Geuthner, 1978.

Douglas 1973: Mary Douglas, *Natural Symbols. Explorations in Cosmology.* Harmondsworth: Penguin, 1973

Dozon 1987: Jean-Pierre Dozon, “Ce que valoriser la médecine traditionnelle veut dire”, *Politique Africaine* 28 (1987), pp. 9–20.

Durand 1988: Jean-Marie Durand, *Archives épistolaires de Mari I/1* (Archives Royales de Mari 26/1). Paris: Éditions Recherche sur les civilisations, 1988.

Duranti 1993: Alessandro Duranti, “Intentions, Self, and Responsibility: An Essay in Samoan Ethnopragmatics”, in Jane H. Hill (ed.), *Responsibility and Evidence in Oral Discourse.* Cambridge: Cambridge University Press, 1993, pp. 24–47.

Ebeling 1915–1923: Erich Ebeling, *Keilschrifttexte aus Assur religiösen Inhalts*, Ausgrabungen der Deutschen Orient-Gesellschaft in Assur. Leipzig: Hinrichs, 1915–1923.

Ebeling 1918: Erich Ebeling, *Quellen zur Kenntnis der babylonischen Religion* (Mitteilungen der vorderasiatischen Gesellschaft 23). Leipzig: Hinrichs, 1918.

Ebeling 1925: Erich Ebeling, *Liebeszauber im Alten Orient* (Mitteilungen der altorientalischen Gesellschaft 1/1). Leipzig: Eduard Pfeiffer, 1925.

Ebeling 1931: Erich Ebeling, *Aus dem Tagewerk eines assyrischen Zauberpriesters* (Mitteilungen der altorientalischen Gesellschaft 5/3). Osnabrück: Otto Zeller, 1931.

Ebeling 1932a: Erich Ebeling, “Ba’u”, in *Reallexikon der Assyriologie und vorderasiatischen Archäologie*, vol. 1. Berlin: De Gruyter, 1932, pp. 432–433.

Ebeling 1932b: Erich Ebeling, “Almanu”, in *Reallexikon der Assyriologie und vorderasiatischen Archäologie*, vol. 1. Berlin: De Gruyter, 1932, p. 71.

Ebeling 1953a: Erich Ebeling, *Die akkadische Gebetserie ‘Handerhebung’* (Deutsche Akademie der Wissenschaften, Institut für Orientforschung, Veröffentlichung 20). Berlin: Akademie-Verlag, 1953.

Ebeling 1953b: Erich Ebeling, *Literarische Keilschrifttexte aus Assur.* Berlin: Akademie-Verlag, 1953.

Ebeling 1954: Erich Ebeling, “Ein Hymnus auf die Suprematie des Sonnengottes in Exemplaren aus Asur und Boghazköi”, *Orientalia* 23 (1954), pp. 209–216.

Edzard 1976–1980a: Dietz O. Edzard, “Kanisurra”, in *Reallexikon der Assyriologie und vorderasiatischen Archäologie*, vol. 5. Berlin: De Gruyter, 1976–1980, p. 389.

Edzard 1976–1980b: Dietz O. Edzard, “Keule”, in *Reallexikon der Assyriologie und vorderasiatischen Archäologie*, vol. 5. Berlin: De Gruyter, 1976–1980, pp. 578–579.

Edzard 1998–2001: Dietz O. Edzard,“Nin-isina”, in *Reallexikon der Assyriologie und vorderasiatischen Archäologie*, vol. 9. Berlin: De Gruyter, 1998–2001, pp. 387–388.

Edzard and Kammenhuber 1976–1980: Dietz O. Edzard and Annelies Kammenhuber, “Hurriter”, in *Reallexikon der Assyriologie und vorderasiatischen Archäologie*, vol. 5. Berlin: De Gruyter, 1976–1980, pp. 509–510.

Eichmann 1997: Ricardo Eichmann, “Ein Hund, ein Schwein, ein Musikant”, in Beate Pongratz-Leisten, Hartmut Kühne, and Paolo Xella (eds.), Ana šadî Labnāni lū allik. *Beiträge zu altorientalischen und mittelmeerischen Kulturen, Festschrift für Wolfgang Röllig* (Alter Orient und Altes Testament 247). Kevelaer: Butzon & Bercker, 1997, pp. 97–108.

Eisenberg 1977: Leon Eisenberg, "Disease and Illness: Distinction between Professional and Popular Ideas of Sickness", *Culture, Medicine, and Psychiatry* 1 (1977), pp. 19–23.
Eisenberg and Kleinman 1981: Leon Eisenberg and Arthur Kleinman (eds.), *The Relevance of Social Sciences for Medicine*. Dordrecht: Springer, 1981.
Etkin 1988: Nina L. Etkin, "Cultural Construction of Efficacy", in Sjaak van der Geest and Susan Reynolds Whyte (eds.), *The Context of Medicines in Developing Countries. Studies in Pharmaceutical Anthropology*. Dordrecht, Boston, London: Kluwer Academic Publishers, 1988, pp. 299–326.
Elat 1982: Moshe Elat, "Mesopotamische Kriegsrituale", *Bibliotheca Orientalis* 39 (1982), pp. 5–25.
Enfield 2002: Nick J. Enfield, "Avoiding the Exoticisms of 'Obstinate Monosemy' and 'Onlineextension'", *Pragmatics & Cognition* 10, 1–2 (2002), pp. 85–106.
Enfield and Wierzbicka 2002: Nick J. Enfield and Anna Wierzbicka, "The Body in Description of Emotion", *Pragmatics & Cognition* 10, 1–2 (2002), pp. 1–25.
Engel 1987: Burkhard J. Engel, *Darstellung von Dämonen und Tieren in assyrischen Palästen und Tempeln nach den schriftlichen Quellen*. Mönchengladbach: Hackbarth, 1987.
Evans-Pritchard 1937: Edward E. Evans-Pritchard, *Witchcraft, Oracles and Magic among the Azande*. Oxford: Oxford University Press (reprint: Oxford: Clarendon Press, 1976).
Evans-Pritchard 1940: Edward E. Evans-Pritchard, *The Nuer. A Description of the Modes of Livelihood and Political Institutions of a Nilotic People*. Oxford: Clarendon Press, 1940.
Fabrega 1972: Horacio Fabrega, "Concepts of Disease: Logical Features and Social Implications", *Perspectives in Biology and Medicine* 22 (1972), pp. 583–616.
Fales 2015: Frederick M. Fales, "Anatomy and Surgery in Ancient Mesopotamia: a Bird's-eye View", in Hélène Perdicoyianni-Paléologou (eds), *Anatomy and Surgery from Antiquity to the Renaissance*. Amsterdam: Adolf M. Hakkert, 2015, pp. 3–71.
Falkenstein 1931: Adam Falkestein, *Die Haupttypen der sumerischen Beschwörung* (Leipziger semitistische Studien 1). Leipzig: Hinrichs, 1931.
Falkenstein 1966: Adam Falkestein, *Die Inschriften Gudeas von Lagaš* (Analecta Orientalia 30). Rome: Biblical Institute Press, 1966.
Fanon 1952: Frantz Fanon, *Peau noire, masques blancs*. Paris: Seuil, 1952.
Fanon 1961: Frantz Fanon, *Les damnés de la terre*. Paris: Éditions Maspero, 1961.
Faraone 1991: Christopher A. Faraone, "Binding and Burying the Forces of Evil: The Defensive Use of 'Voodoo Dolls' in Ancient Greece", *Classical Antiquity* 10, 2 (1991), pp. 165–205, 207–220.
Farber 1973: Walter Farber, "*ina* KUŠ.DÙ.DÙ(.BI) = *ina maški tašappi*", *Zeitschrift für Assyriologie und vorderasiatische Archäologie* 63, 1 (1973), pp. 59–68.
Farber 1977a: Walter Farber, *Beschwörungsrituale an Ištar und Dumuzi*. Attī Ištar ša harmaša Dumuzi (Akademie der Wissenschaften und der Literatur. Veröffentlichungen der Orientalischen Kommission 30). Wiesbaden: Franz Steiner, 1977.
Farber 1977b: Walter Farber,"Drogerien in Babylonien und Mesopotamien", *Iraq* 39 (1977), pp. 223–228.
Farber 1979: Walter Farber, review of Hunger 1976, *Zeitschrift für Assyriologie und vorderasiatische Archäologie* 69, 2: 300–304.
Farber 1981: Walter Farber, "Drogen im alten Mesopotamien – Sumerer und Akkader", in Gisela Völger, Karin von Welck, and Aldo Legnaro (eds.), *Rausch und Realität. Drogen im Kulturvergleich*. Köln: Rautenstrauch-Joest-Museum für Völkerkunde, 1981, pp. 270–280.
Farber 1990: Walter Farber,"*Mannam lušpur ana Enkidu*. Some New Thoughts about an Old Motif", *Journal of Near Eastern Studies* 49: 299–321.
Farber 1993–1997: Walter Farber, "Myrrhe", in *Reallexikon der Assyriologie und vorderasiatischen Archäologie*, vol. 8. Berlin: De Gruyter, 1993–1997, pp. 534–537.
Farber 2010: Walter Farber, "Ištar und due Ehekrise. Bemerkungen zu STT 257, Ra 18, 21ff. ("Tisserant 17"), und ST 249", in Dahlia Shehata, Frauke Weiershäuser, and Kamran V. Zand (eds.), *Von Göttern und Menschen. Beiträge zu Literatur und Geschichte des Alten Orients*.

Festschrift für Brigitte Groneberg (Cuneiform Monographs 41). Leiden, Boston: Brill, 2010, pp. 73–85.

Farber 2014: Walter Farber, *Lamaštu. An Edition of the Canonical Series of Lamaštu Incantations and Rituals and Related Texts from the Second and First Millennia B.C.* (Mesopotamian Civilizations 17). Winoka Lake: Eisenbrauns, 2014.

Fechner 2017: Josephine Fechner, "Zaunkönig", in *Reallexikon der Assyriologie und vorderasiatischen Archäologie*, vol. 15. Berlin: De Gruyter, 2017, p. 234.

Feldman 2015: Marian Feldman, "In Pursuit of Luxury in Mesopotamia", in Richard Jasnow and Kathlyn M. Cooney (eds.), *Joyful in Thebes: Egyptological Studies in Honor of Betsy M. Bryan*. Bristol: Lockwood Press, 2015, pp. 121–126.

Feldt 2007: Laura Feldt, "On Divine-referent Bull Metaphors in the ECTSL Corpus", in Jarle Ebeling and Graham Cunningham (eds.), *Analysing Literary Sumerian Corpus-based Approaches*, London: Equinox, 2007, pp. 184–214.

Fernandez 1974: James W. Fernandez, "The Mission of Metaphor in Expressive Culture", *Current Anthropology* 15, 2 (1974), pp. 119–145.

Fernandez 1986: James W. Fernandez, *Persuasions and Performances. The Play of Tropes in Culture*. Bloomington: Indiana University Press, 1986.

Fincke 2010: Jeanette C. Fincke, "KUB 37, 201: Ein weiteres ŠÀ-ZI.GA-Fragment aus Ḫattuša", *Nouvelles Assyriologiques Brèves et Utilitaires* 2010, 2, no. 41, pp. 48–49.

Finet 1993: André Finet, "Le comportement du chien, facteur de son ambivalence en Mésopotamie", in Liliane Bodson (ed.), *L'histoire de la connaissance du comportement animal. Actes du colloque international, Université de Liège, 11–14 mars 1992*. Liège: Université de Liège, 1993, pp. 133–142.

Finkel 1976: Irving L. Finkel, *ḪUL.BA.ZI.ZI. Ancient Mesopotamian Exorcistic Incantations*. Ph.D. dissertation, University of Birmingham.

Finkel 1980: Irving L. Finkel, "The Crescent Fertile", *Archiv für Orientforschung* 27 (1980), pp. 35–52.

Finkel 2000: Irving L. Finkel, "On a Late Babylonian Medical Training", in Andrew R. George and Irving L. Finkel (eds.), *Wisdom, Gods and Literature. Studies in Honour of W.G. Lambert*. Winona Lake: Eisenbrauns, 2000, pp. 137–223.

Finkel 2018: Irving L. Finkel, "On Three Tablet Inventories", in Ulrike Steinert (ed.), *Assyrian and Babylonian Scholarly Text Catalogues* (Die babylonisch-assyrische Medizin in Texten und Untersuchungen 9). Boston, Berlin: De Gruyter, 2018, pp. 25–41.

Finkelstein 1966: Jacob Joel Finkelstein, "Sex Offenses in Sumerian Laws", *Journal of the American Oriental Society* 86 (1966), pp. 355–372.

BFoster 2002: Benjamin R. Foster, "Animals in Mesopotamian Literature", in BCollins 2002: 271–288.

BFoster 2005: Benjamin R. Foster, *Before the Muses. An Anthology of Akkadian Literature*. Bethesda: CDL Press, 2005.

BFoster 2007: Benjamin R. Foster, *Akkadian Literature of Late Period* (Guides to the Mesopotamian Textual Record 2). Münster: Ugarit-Verlag, 2007.

GFoster 1976: George M. Foster, "Disease Etiologies in Non-Western Medical Systems", *American Anthropologist* 78, 4 (1976), pp. 773–782.

Foster and Salgues 2006: Benjamin R. Foster and Emmanuelle Salgues, "'Everything except the Squeal' Pigs in Early Mesopotamia", in Lion and Michel 2006: 131–134.

Foucault 1963: Michel Foucault, *Naissance de la clinique. Une archéologie du regard médical*. Paris: Presses universitaires de France, 1963.

Foucault 1969: Michel Foucault, *L'Archéologie du savoir*. Paris: Gallimard, 1969.

Foucault 1971: Michel Foucault, *L'Ordre du discours*. Paris: Gallimard, 1971.

Foucault 1972: Michel Foucault, *Histoire de la folie à l'âge classique*. Paris: Gallimard, 1972.

Foucault 1976: Michel Foucault, *Histoire de la sexualité*, vol. 1: *La volonté de savoir*. Paris: Gallimard, 1976.

Frahm 1997: Eckart Frahm, *Einleitung in die Sanherib-Inschriften* (Archiv für Orientforschung Beiheft 27). Wien: Institut für Orientalistik der Universität Wien, 1997.

Frahm 2011: Eckart Frahm, *Babylonian and Assyrian Text Commentaries, Origins of Interpretation* (Guides to the Mesopotamian Textual Record 5). Münster: Ugarit-Verlag, 2011.

Franco 2003a: Cristriana Franco, "Animali e analisi culturale", in Fabio Gasti and Elisa Romano (eds.), *'Buoni per pensare'. Gli animali nel pensiero e nella letteratura dell'antichità. Atti della II Giornata ghisleriana di Filologia classica (Pavia, 18–19 aprile 2002)*. Pavia, Como: Ibis, pp. 63–81.

Franco 2003b: Cristriana Franco, *Senza ritegno. Il cane e la donna nell'immaginario della Grecia antica*. Bologna: Il Mulino, 2003 (English translation: *Shameless. The Canine and the Feminine in Ancient Greece*. Berkeley: University of California Press, 2014).

Franco 2006: Cristriana Franco, "Il verro e il cinghiale. Immagini di caccia e virilità nel mondo greco", *Studi italiani di Filologia classica* 4,1 (2006), pp. 5–31.

Franco 2008a: Cristriana Franco, "Questioni di genere e metafore animali nella letteratura greca", *Annali online dell'Università di Ferrara. Sezione Lettere* 1 (2008), pp. 73–94.

Franco 2008b: Cristriana Franco, "Cani e porci. Temi etnozoologici dal mondo antico", in Eadem (ed.), *Gli animali e i loro uomini*. Siena: Protagon Editori Toscani, pp. 45–51.

Franco 2015: Cristriana Franco, "L'argomento convincente. Gli animali come allievi di *physis* e 'vox naturae' nella tradizione greca e romana", *Studi italiani di Filologia classica* 13, 2 (2015), pp. 185–212.

Frankfurter 1995: David Frankfurter, "Narrating Power: The Theory and Practice of the Magical Historiola in Ritual Spells", in Marvin Meyer and Paul Mirecki (eds.), *Ancient Magic and Ritual Power* (Religions in the Graeco-Roman World 129). Leiden: Brill, 1995, pp. 457–476.

Frankfurter 2016: David Frankfurter, "Narratives That Do Things", in Sarah Iles Johnston (ed.), *Religion: Narrating Religion*. Farmington Hills: Macmillan Reference, 2016, pp. 95–106.

Frymer-Kensky 2000: Tikva Frymer-Kensky, "Lolita-Inanna", *NIN: Journal of Gender Studies in Antiquity* 1 (2000), pp. 91–94.

Frayne 1990: Douglas Frayne, *The Royal Inscriptions of Mesopotamia. Old Babylonian Period (2003–1595 BC)* (The Royal Inscriptions of Mesopotamia. Early Periods 4). Toronto: University of Toronto Press, 1990.

Freedman 2017: Sally M. Freedman, *If a City Is Set on a Height. The Akkadian Omen Series* Šumma Alu ina Mēlê Šakin. *Volume 3, Tablets 41–63*. Winona Lake: Eisenbrauns, 2017.

Fuhr 1977: Ilse Fuhr, "Der Hund als Begleiter der Göttin Gula und anderer Heilgottheiten", in Barthel Hrouda (ed.) *Isin-Išān Bahrīyāt I: Die Ergebnisse der Ausgrabungen 1973–1974*. München: Bayerische Akademie der Wissenschaften, 1977, pp. 135–145.

Gadotti 2014: Alhena Gadotti, *'Gilgamesh, Enkidu and the Netherworld' and the Sumerian Gilgamesh Cycle* (Untersuchungen zur Assyriologie und vorderasiatischen Archäologie 10). Boston, Berlin: De Gruyter, 2014.

Gardner 1983: Donald S. Gardner, "Performativity in Ritual: The Mianmin Case", *Man* N.S. 18, 2 (1983), pp. 346–360.

García-Selgas 2004: Fernando J. García-Selgas, "Feminist Epistemologies for Critical Social Theory: From Standpoint Theory to Situated Knowledge", in Sandra Harding (ed.), *The Feminist Standpoint Theory Reader. Intellectual and Political Controversies*. London, New York: Routledge, 2004, pp. 293–308.

Garcia-Ventura 2014: Agnès Garcia-Ventura, "Constructing Masculinities through Textile Production in the Ancient Near East", in Mary Harlow, Cécile Michel, and Marie-Louise Nosch (eds.), *Prehistoric, Ancient Near Eastern and Aegean Textiles and Dress. An Interdisciplinary Anthology*. Oxford: Oxbow Books, 2014, pp. 167–183.

Garcia-Ventura and Svärd 2018: Agnès Garcia-Ventura and Saana Svärd, "Theoretical Approaches, Gender, and the Ancient Near East: An Introduction", in Eaedem (eds.), *Studying Gender in the Ancient Near East*. Winona Lake: Eisenbrauns, 2018, pp. 1–13.

Garcia-Ventura and Zisa 2017: Agnès Garcia-Ventura and Gioele Zisa, "Gender and Women in Ancient Near Eastern Studies: Bibliography 2002–2016", *Akkadica* 138, 1 (2017), pp. 37–67.

Geertz 1973: Clifford Geertz, *The Interpretation of Culture*. New York: Basic Books, 1973.

Geertz 1988: Clifford Geertz, *Works and Lives. The Anthropologist as Author*. Stanford: Stanford University Press, 1988.

Gehlken 2012: Erlend Gehlken, *Weather Omens of Enūma Anu Enlil: Thunderstorms, Wind and Rain (Tablets 44–49)* (Cuneiform Monographs 43). Leiden, Boston: Brill, 2012.

Gelb 1982: Ignace J. Gelb, "Sumerian and Akkadian Words for 'String of Fruit'", in Govert Van Driel, Theo J.H. Krispijn, Marten Stol, Klaas R. Veenhof (eds.), *Zikir šumin. Assyriological Studies Presented to F.R. Kraus on the Occasion of his Seventieth Birthday*, Leiden: Brill, 1982, pp. 67–82.

Geller 1980: Markham J. Geller, "A Middle Assyrian Tablet of *Utukkū Lemnūti*, Tablet 12", *Iraq* 42 (1980), pp. 23–51.

Geller 1985: Markham J. Geller, *Forerunners to Udug-hul. Sumerian Exorcistic Incantations* (Freiburger altorientalische Studien 12): Stuttgart: Franz Steiner, 1985.

Geller 1999: Markham J. Geller, "Freud and Mesopotamian Magic", in Tzvi Abusch and Karel van der Toorn (eds.), *Mesopotamian Magic. Textual, Historical, and Interpretative Perspectives* (Ancient Magic and Divination 1). Groningen: Styx, 1999, pp. 49–55.

Geller 2002: Markham J. Geller, "Mesopotamian Love Magic: Discourse or Intercourse?", in Simo Parpola and Robert M. Whiting (eds.), *Sex and Gender in the Ancient Near East. Proceedings of the 47th Rencontre Assyriologique Internationale, Helsinki, July 2–6, 2001*. Helsinki: Neo-Assyrian Text Corpus Project, 2002, pp. 129–139.

Geller 2005: Markham J. Geller, *Renal and Rectal Disease Texts* (Die babylonisch-assyrische Medizin in Texten und Untersuchungen 7). Berlin, Boston: De Gruyter, 2005.

Geller 2007a: Markham J. Geller, "Médecine et magie : l'*asû*, l'*āšipu* et le *mašmâšu*", *Journal des Médecines Cunéiformes* 9 (2007), pp. 1–8.

Geller 2007b: Markham J. Geller, "Comment et de quelle façon les praticiens gagnaient-ils leur vie?", *Journal des Médecines Cunéiformes* 10 (2007), pp. 34–41.

Geller 2014: Markham J. Geller, *Melothesia in Babylonia. Medicine, Magic, and Astrology in the Ancient Near East* (Science, Technology, and Medicine in Ancient Cultures 2). Berlin, Boston: De Gruyer, 2014.

Geller 2015: Markham J. Geller, *Ancient Babylonian Medicine. Theory and Practice*. Chichester: Wiley-Blackwell, 2015.

Geller 2016: Markham J. Geller, *Healing Magic and Evil Demons. Canonical Udug-hul Incantations* (Die babylonisch-assyrische Medizin in Texten und Untersuchungen 8). Berlin, Boston: De Gruyter, 2016

Geller 2018: Markham J. Geller, "The Exorcist's Manual (KAR 44)", in Ulrike Steinert (ed.), *Assyrian and Babylonian Scholarly Text Catalogues* (Die babylonisch-assyrische Medizin in Texten und Untersuchungen 9). Boston, Berlin: De Gruyter, 2018, pp. 292–312.

Geller and Cohem 1995: Markham J. Geller and Simon L. Cohen, "Kidney and Urinary Tract Disease in Ancient Babylonia, with Translations of the Cuneiform Sources", *Kidney International* 47 (1995), pp. 1811–1815.

George 1979: Andrew R. George, "Cuneiform Texts in the Birmingham City Museum", *Iraq* 41 (1979), pp. 121–140.

George 1991: Andrew R. George, "Babylonian Texts from the Folios of Sidney Smith. Part Two: Prognostic and Diagnostic Omens, Tablet I", *Revue d'Assyriologie et d'Archéologie orientale* 85 (1991), pp. 137–167.

George 1999: Andrew R. George, "The Dogs of Ninkilim: Magic Against Field Pests in Ancient Mesopotamia", in Horst Klengel and Johannes Renger (eds.), *Landwirtschaft im Alten Orient: ausgewählte Vorträge der XLI. Rencontre Assyriologique Internationale Berlin, 4.-8.7.1994*. Berlin: D. Reimer, 1999, pp. 341–354.

George 2000: Andrew R. George, "Four Temple Rituals from Babylon", in Andrew R. George and Irving L. Finkel (eds.), *Wisdom, Gods and Literature. Studies in Honour of W.G. Lambert*. Winona Lake: Eisenbrauns, 2000, pp. 259–300.

George 2003: Andrew R. George, *The Babylonian Gilgamesh Epic. Introduction, Critical Edition and Cuneiform Texts*, 2 vols. Oxford: Oxford University Press, 2003.

George 2009: Andrew R. George, *Babylonian Literary Texts in the Schøyen Collection* (Cornell University Studies in Assyriology and Sumerology 10). Bethesda: CDL Press, 2009.

George 2013: Andrew R. George, *Babylonian Divinatory Texts Chiefly in the Schøyen Collection* (Cornell University Studies in Assyriology and Sumerology 18). Bethesda: CDL Press, 2013.

George 2016: Andrew R. George, *Mesopotamian Incantations and Related Texts in the Schøyen Collection* (Cornell University Studies in Assyriology and Sumerology 32). Bethesda: CDL Press, 2016.

Gluckman 1956: Max Gluckman, *Custom and Conflict in Africa*. Oxford: Basil Blackwell, 1956 (reprint: 1970).

Goetze 1955: Albrecht Goetze, "An Incantation Against Diseases", *Journal of Cuneiform Studies* 9 (1955), pp. 8–18.

Goltz 1974: Dietlinde Goltz, *Studien zur altorientalischen und griechischen Heilkunde. Therapie, Arzneibereitung, Rezeptstruktur* (Sudhoffs Archiv, Zeitschrift für Wissenschaftsgeschichte 16). Wiesbaden: Franz Steiner, 1974.

Good 1977: Byron J. Good, "The Heart of What's the Matter: The Semantics of Illness in Iran", *Culture, Medicine and Psychiatry* 1, 1 (1977), pp. 25–58.

Good 1994: Byron J. Good, *Medicine, Rationality, and Experience. An Anthropological Perspective*. Cambridge: Cambridge University Press, 1994 (reprint: 2003).

Good and DelVecchio Good 1981: Byron J. Good and Mary-Jo DelVecchio Good, "The Semantics of Medical Discourse", in Everett MendelsohnYehuda Elkana (eds.), *Sciences and Culture. Anthropological and Historical Studies of the Sciences* (Sociology of the Sciences 5). Dordrecht, Boston, London: D. Reidel, 1981, pp. 177–212.

Good et al. 1985: Byron J. Good, Mary-Jo DelVecchio Good and Robert Moradi, "The Interpretation of Iranian Depressive Illness and Dysphoric Affect", in Arthur Kleinman and Byron J. Good (eds.) *Culture and Depression. Studies in the Anthropology and Cross-cultural Psychiatry of Affect and Disorder*, Berkeley: University of California Press, 1985, pp. 369–490.

Göhde 1998: Hildegard Göhde, *Vom Hirtenhund zum Göttersymbol. Die Bedeutung des Hundes im Alten Mesopotamien von Beginn bis zum Untergang*. Ph.D. dissertation, Universität Münster, 1998.

Gramsci 1975: Antonio Gramsci, *Quaderni del carcere*, edited by Valentino Gerratana, 4 vols. Torino: Einaudi, 1975.

Grayson 2000: Albert Kirk Grayson, "Murmuring in Mesopotamia", in Andrew R. George and Irving L. Finkel (eds.), *Wisdom, Gods and Literature. Studies in Honour of W.G. Lambert*. Winona Lake: Eisenbrauns, 2000, pp. 301–308.

Groneberg 1986: Brigitte Groneberg, "Notes brèves : *ḫabābu-ṣabāru*", *Revue d'Assyriologie et d'Archéologie orientale* 80 (1986), pp. 188–190.

Groneberg 1997: Brigitte Groneberg, *Lob der Ištar. Gebet und Ritual an die altbabylonische Venusgöttin* (Cuneiform Monographs 8). Groningen: Styx, 1997.

Groneberg 1999: Brigitte Groneberg, "'Brust'(*irtum*)-Gesänge", in Barbara Böck, Eva Christiane Cancik-Kirschbaum, and Thomas Richter (eds.), *Munuscula Mesopotamica. Festschrift für Johannes Renger* (Alter Orient und Altes Testament 267). Münster: Ugarit-Verlag, 1999, pp. 169–196.

Groneberg 2007: Brigitte Groneberg, "Liebes- und Hundbeschwörungen im Kontext", in Martha T. Roth, Walter Farber, Matthew W. Stolper, and Paula von Bechtolsheim (eds.), *Studies Presented to Robert D. Biggs, June 4, 2004* (Assyriological Studies 27). Chicago: University of Chicago Press, 2007, pp. 91–107.

Gruber 1980: Mayer I. Gruber, *Aspects of Nonverbal Communication in the Ancient Near East* (Studia Pohl 12). Rome: Biblical Institute Press, 1980.

Gruber 1990: Mayer I. Gruber, "Fear, Anxiety and Reverence in Akkadian, Biblical Hebrew and Other North-West Semitic Languages", *Vetus Testamentum* 40, 4 (1990), pp. 411–422.

Guggino 2006: Elsa Guggino, *Fate, sibille e altre strane donne*. Palermo: Sellerio, 2006.

Guichard and Marti 2013: Michaël Guichard and Lionel Marti, "Purity in Ancient Mesopotamia: The Paleo-Babylonian and Neo-Assyrian Periods", in Christian Frevel and Christophe Nihan (eds.), *Purity and the Forming of Religious Traditions in the Ancient Mediterranean World and Ancient Judaism* (Dynamics in the History of Religions 3). Leiden, Boston: Brill, 2013, pp. 47–113.

Guinan 1996: Ann K. Guinan, "Social Constructions and Private Designs: The House Omens of *Šumma ālu*", in Klaas R. Veenhof (ed.), *Houses and Households in Ancient Mesopotamia. Papers read at the 40e Rencontre Assyriologique Internationale, Leiden, July 5–8, 1993*. Istanbul: Nederlands Historisch-Archaeologisch Instituut, 1996, pp. 61–68.

Guinan 1997: Ann K. Guinan, "Auguries of Hegemony: The Sex Omens of Mesopotamia", *Gender & History* 9 (1997), pp. 462–476.

Guinan 2002: Ann K. Guinan, "Erotomancy: Scripting the Erotic", in Simo Parpola and Robert M. Whiting (eds.), *Sex and Gender in the Ancient Near East. Proceedings of the 47th Rencontre Assyriologique Internationale, Helsinki, July 2–6, 2001*. Helsinki: Neo-Assyrian Text Corpus Project, pp. 185–201.

Gurney 1960: Oliver Robert Gurney, "A Tablet of Incantation Against Slander", *Iraq* 22 (1960), pp. 221–227.

Gutmann 1997: Matthew C. Gutmann, "Trafficking in Men: The Anthropology of Masculinity", *Annual Review of Anthropology* 26 (1997), pp. 385–409.

Haas 1994: Volkert Haas, *Geschichte der hethitischen Religion* (Handbuch der Orientalistik 15). Leiden: Brill, 1994.

Haas 1999: Volkert Haas, *Babylonischer Liebesgarten. Erotik und Sexualität im Alten Orient*. München: C.H. Beck, 1999.

Haas 2003: Volkert Haas, *Materia Magica et Medica Hethitica*, Berlin: De Gruyter, 2003.

Hahn 1985: Robert A. Hahn, "Culture-bound Syndromes unbound", *Social Science & Medicine* 21, 2 (1985), pp. 165–171.

Halberstam 1998: Jack Halberstam, *Female Masculinity*. Durham: Duke University Press, 1998.

Hallo 1966: William W. Hallo, "New Hymns to the Kings of Isin", *Bibliotheca Orientalis* 23 (1966), pp. 239–247.

Haraway 1988: Donna Haraway, "Situated Knowledges: The Science Question in Feminism and the Privilege of Partial Perspective", *Feminist Studies* 14 (1988), pp. 575–599.

Haraway 1991: Donna Haraway, "A Cyborg Manifesto: Science, Technology, and Socialist-Feminism in the Late Twentieth Century", in Eadem, *Simians, Cyborgs, and Women. The Reinvention of Nature*. New York: Routledge, 1991, pp. 149–181.

Harding 1986: Sandra Harding, *The Science Question in Feminism*. Ithaca: Cornell University Press, 1986.

Harris 2006: Rivkah Harris, *Gender and Aging in Mesopotamia. The Gilgamesh Epic and Other Ancient Literature*. Norman: University of Oklahoma Press, 2006.

Hartsock 1983: Nancy C.M. Hartsock, "The Feminist Standpoint: Developing the Ground for a Specifically Feminist Historical Materialism", in Sandra Harding and Merrill B. Hintikka (eds.), *Discovering Reality: Feminist Perspectives on Epistemology, Metaphysics, Methodology, and Philosophy of Science*. Dordrecht: Kluwer Academic Publishers, 1983, pp. 283–310.

Haussperger 1997: Martha Haussperger, "Die mesopotamische Medizin und ihre Ärzte aus heutiger Sicht", *Zeitschrift für Assyriologie und vorderasiatische Archäologie* 87, 2 (1997), pp. 196–218.

Heeßel 2000: Nils P. Heessel, *Babylonisch-assyrische Diagnostik* (Alter Orient und Altes Testament 43). Münster: Ugarit-Verlag, 2000.

Heeßel 2004a: Nils P. Heessel, "Diagnosis, Divination and Disease: Towards an Understanding of the Rationale Behind the Babylonian Diagnostic Handbook", in Manfred Horstmanshoff and Marten Stol (eds.), *Magic and Rationality in Ancient Near Eastern and Graeco-Roman Medicine* (Studies in Ancient Medicine 27). Leiden: Brill, 2004, pp. 97–116.

Heeßel 2004b: Nils P. Heessel, "Reading and Interpreting Medical Cuneiform Texts, Methods and Problems", *Journal des Médecines Cunéiformes* 3 (2004), pp. 2–9.

Heeßel 2007: Nils P. Heessel, "The Hands of the Gods: Disease Names, and Divine Anger", in Markham J. Geller and Irving L. Finkel (eds.), *Disease in Babylon* (Cuneiform Monographs 36). Leiden, Boston: Brill, 2007, pp. 120–130.

Heeßel 2010a: Nils P. Heessel, "Neues von Esagil-kīn-apli. Die ältere Version der physiognomischen Omenserie *aladimmû*", in Stefan M. Maul (ed.), *Assur-Forschungen. Arbeiten aus der Forschungsstelle "Edition literarischer Keilschrifttexte aus Assur" der Heidelberger Akademie der Wissenschaften*. Wiesbaden: Harrassowitz, 2010, pp. 139–187.

Heeßel 2010b: Nils P. Heessel, "Rechts oder links, wörtlich oder dem Sinn nach? Zum Problem der kulturellen Gebundenheit bei der Übersetzung von medizinischen Keilschrifttexten", in Annette Imhausen and Tanja Pommerening (eds.), *Writings of Early Scholars in the Ancient Near East, Egypt, Rome, and Greece* (Beiträge zur Altertumskunde 286). Berlin, New York: De Gruyter, 2010, pp. 175–188.

Heeßel 2016: Nils P. Heessel, "Medizinische Texte aus dem Alten Mesopotamien", in Annette Imhausen and Tanja Pommerening (eds.), *Translating Writings of Early Scholars in the Ancient Near East, Egypt, Greece and Rome. Methodological Aspects with Examples* (Beiträge zur Altertumskunde 344). Berlin, New York: De Gruyter, 2010, pp. 17–73.

Heeßel 2018: Nils P. Heessel, "Identifying Divine Agency: The Hands of the Gods in Context", in Greta Van Buylaere, Mikko Luukko, Daniel Schwemer, and Avigail Mertens-Wagschal (eds.), *Sources of Evil. Studies in Mesopotamian Exorcistic Lore* (Ancient Magic and Divination 15). Leiden, Boston: Brill, 2018, pp. 134–139.

Heimpel 1968: Wolfgang Heimpel, *Tierbilder in der sumerischen Literatur* (Studia Pohl. 2). Rome: Biblical Institute Press, 1968.

Heimpel 1970: Wolfgang Heimpel, review of Biggs 1967, *Zeitschrift der Deutschen Morgenländischen Gesellschaft* 120 (1970), pp. 189–191.

Heimpel 1972–1975a: Wolfgang Heimpel, "Hirsch", in *Reallexikon der Assyriologie und vorderasiatischen Archäologie*, vol. 4. Berlin: De Gruyter, 1972–1975, pp. 418–421.

Heimpel 1972–1975b: Wolfgang Heimpel, "Hund", in *Reallexikon der Assyriologie und vorderasiatischen Archäologie*, vol. 4. Berlin: De Gruyter, 1972–1975, pp. 494–497.

Heimpel 1987–1990a: Wolfgang Heimpel, "Löwe. A.I. Mesopotamien", in *Reallexikon der Assyriologie und vorderasiatischen Archäologie*, vol. 7. Berlin: De Gruyter, 1987–1990, pp. 80–85.

Heimpel 1987–1990b: Wolfgang Heimpel, "Maus", in *Reallexikon der Assyriologie und vorderasiatischen Archäologie*, vol. 7. Berlin: De Gruyter, 1987–1990, pp. 605–609.

Heimpel 1998–2001: Wolfgang Heimpel, "Ninigizibara I und II", in *Reallexikon der Assyriologie und vorderasiatischen Archäologie*, vol. 9. Berlin: De Gruyter, 1998–2001, pp. 382–384.

Held 1961: Moshe Held, "A Faithful Lover in an Old Babylonian Dialogue", *Journal of Cuneiform Studies* 15 (1961), pp 1–26.

Held 1962: Moshe Held, "A Faithful Lover in an Old Babylonian Dialogue (JCS 15 1–26) Addenda et Corrigenda", *Journal of Cuneiform Studies* 16 (1962), pp. 37–39.

Helle 2018: Sophus Helle, "'Only in Dress?' Methodological Concerns Regarding Non-Binary Gender", in Stephanie Lynn Budin, Megan Cifarelli, Agnès Garcia-Ventura, and Adelina Millet Albà (eds.), *Gender and Methodology in the Ancient Near East. Approaches from Assyriology and beyond*. Barcelona: Edicions de la Universitat de Barcelona, 2018, pp. 41–54.

Helle 2019: Sophus Helle, "Weapons and Weaving Instruments as Symbols of Gender in the Ancient Near East", in Megan Cifarelli (ed.), *Fashioned Selves. Dress and Identity in Antiquity*, Oxford, Havertown: Oxbow Books, 2019, pp. 105–115.

Helle 2020: Sophus Helle, "Marduk's Penis: Queering *Enūma Eliš*", *Distant Worlds Journals* 4 (2020), pp. 53–77.

Herrero 1975: Pablo Herrero, "Tablette médicale assyrienne inédite", *Revue d'Assyriologie et d'Archéologie orientale* 69 (1975), pp. 41–53.

Herrero 1984: Pablo Herrero, *La thérapeutique mésopotamienne*. Paris: Éditions Recherche sur les civilisations, 1984.

Hertz 1909 : Robert Hertz, "La prééminence de la main droite : Étude sur la polarité religieuse", *Revue Philosophique de la France Et de l'Etranger* 68 (1909), pp. 553–580.

Hill Collins 2000: Patricia Hill Collins, *Black Feminist Thought. Knowledge, Consciousness, and the Politics of Empowerment*. New York, London: Routledge, 2000.

Hillers 1973: Delbert R. Hillers, "The Bow of Aqhat: The Meaning of a Mythological Theme", in Harry A. Hoffner (ed.), *Orient and Occident. Essays Presented to Cyrus H. Gordon on the Occasion of His Sixty-fifth Birthday* (Alter Orient und Altes Testament 22). Kevelaer: Butzon & Bercker, 1973 pp. 71–80.

Hirvonen 2019: Joonas Hirvonen, "Animals and Demons. Faunal Appearances, Metaphors, and Similes in Lamaštu Incantations", in Mattila et al. 2019: 314–343.

Hoffner 1966: Harry A. Hoffner, "Symbols for Masculinity and Feminity: Their Use in Ancient Near Eastern Sympathetic Magic Rituals", *Journal of Biblical Literature* 85 (1966), pp. 326–334.

Hoffner 1973: Harry A. Hoffner, "Incest, Sodomy and Bestiality in the Ancient Near East", in Idem (ed.), *Orient and Occident. Essays Presented to Cyrus H. Gordon on the Occasion of His Sixty-fifth Birthday* (Alter Orient und Altes Testament 22). Kevelaer: Butzon & Bercker, 1973 pp. 81–90.

Hoffner 1987: Harry A. Hoffner, "Paskuwatti's Ritual against Sexual Impotence (CTH 406)", *Aula orientalis: revista de estudios del Próximo Oriente Antiguo* 5 (1987), pp. 271–287.

Holma 1911: Harri Holma, *Die Namen der Körperteile im Assyrisch-Babylonischen. Eine lexikalisch-etymologische Studie*. Helsinki: Suomalainen Tiedeakatemia, 1911.

Holma 1913: Harri Holma, *Kleine Beiträge zum Assyrischen Lexikon*. Helsinki: Suomalainen Tiedeakatemia, 1913.

Hood-Williams 1996: John Hood-Williams, "Goodbye to Sex and Gender", *The Sociological Review* 44, 1 (1996), pp. 1–16.

Horowitz 1998: Wayne Horowitz, *Mesopotamian Cosmic Geography* (Mesopotamian Civilizations 8). Winoka Lake: Eisenbrauns, 2014.

Hsu 2010: Elisabeth Hsu, "Introduction. Plants in Medical Practice and Common Sense: on the Interface of Ethnobotany and Medical Anthropology", in Elisabeth Hsu and Stephen Harris (eds.). *Plants, Health and Healing. On the Interface of Ethnobotany and Medical Anthropology*. New York: Berghahn Books, 2010, pp. 1–48.

Hsu and Llop Raudà 2020: Shih-Wei Hsu and Jaume Llop Raduà, *The Expression of Emotions in Ancient Egypt and Mesopotamia* (Culture and History of the Ancient Near East 116). Leiden, Boston: Brill, 2020.

Hunger 1968: Hermann Hunger, *Babylonische und assyrische Kolophone* (Alter Orient und Altes Testament 2). Kevelaer: Butzon & Bercker, 1968.
Hunger 1976: Hermann Hunger, *Spätbabylonische Texte aus Uruk, I* (Ausgrabungen der Deutschen Forschungsgemeinschaft in Uruk-Warka 9). Berlin: Gebr. Mann Verlag, 1976.
Inglese 2012: Salvatore Inglese, "Sindromi culturalmente caratterizzate (*Culture-Bound Sindromes*)", in Donatella Cozzi (ed.), *Le parole dell'antropologia medica. Piccolo dizionario*. Perugia: Morlacchi, 2012, pp. 253–267.
Ingold 1994: Tim Ingold (ed.), *What is an Animal?*. London, New York: Routledge, 1994.
Jacobs 2010: John Jacobs, "Traces of the Omen Serie *Šumma izbu* in Cicero, *De Divinatione*", in Amar Annus (ed.), *Divination and Interpretation of Signs in the Ancient World*. Chicago: Oriental Institute of the University of Chicago, 2010, pp. 317–339.
Jacobsen 1930: Thorkild Jacobsen, "How did Gilgameš oppress Uruk?", *Acta Orientalia* 8 (1930), pp. 62–74.
Jacobsen 1976: Thorkild Jacobsen, *The Treasures of Darkness. A History of Mesopotamian Religion*. New Haven: Yale University Press.
Jakobson 1960: Roman Jakobson, "Closing Statements: Linguistics and Poetics", in Thomas A. Sebeok (ed.), *Style in Language*. Cambridge: MIT Press, 1960, pp. 350–377.
Janković 2004: Bojana Janković, *Vogelzucht und Vogelfang in Sippar im 1. Jahrtausend v. Chr.* (Alter Orient und Altes Testament 315). Münster: Ugarit-Verlag, 2004.
Jaques 2006: Margaret Jacques, *Le vocabulaire des sentiments dans les textes sumériens: Recherche sur le lexique sumérien et akkadien* (Alter Orient und Altes Testament 332). Münster: Ugarit-Verlag, 2006.
CJean 2006: Cynthia Jean, *La magie néo-assyrienne en contexte. Recherches sur le métier d'exorciste et le concept d'*āšipūtu (State Archives of Assyria Studies 17). Helsinki : Neo-Assyrian Text Corpus Project, 2006.
C-FJean 1926: Charles-François Jean, *Contrats de Larsa* (Textes Cunéiformes 11). Paris: Geuthner, 1926.
C-FJean 1950: Charles-François Jean, *Archives Royales de Mari II, Lettres Diverses* (Archives Royales de Mari 2). Paris: Imprimerie Nationale, 1950.
Jiménez 2017: Enrique Jiménez, *The Babylonian Disputation Poems. With Editions of the Series of the Poplar, Palm and Vine, the Series of the Spider, and the Story of the Poor, Forlorn Wren* (Culture and History of the Ancient Near East 87). Leiden, Boston: Brill, 2017.
Joannès 2006: Francis Joannès, "Les porcs dans la documentation néo-babylonienne", in Lion and Michel 2006: 131–134.
Jursa 2009: Michael Jursa, "Die Kralle des Meeres und andere Aromata", in Werner Arnold, Michael Jursa, Walter W. Muller, and Stephan Prochazk (eds.), *Philologisches und Historisches zwischen Anatolien und Sokotra. Analecta Semitica in Memoriam Alexander Sima*. Wiesbaden: Harrassowitz, 2009, pp. 147–180.
Karahashi 2000: Fumi Karahashi, *Sumerian Compound Verbs with Body-Part Terms*. Ph.D. dissertation, University of Chicago, 2000.
Kessler 1990: Suzanne J. Kessler, "The Medical Construction of Gender: Case Management of Intersexed Infants", *Signs* 16, 1 (1990), pp. 3–26.
Kämmerer 1995: Thomas R. Kämmerer, "Die erste Pockendiagnose stammt aus Babylonien", *Ugarit-Forschungen* 27 (1995), pp. 129–168.
Kimmel 1996: Michael Kimmel, *Manhood in America. A Cultural History*. New York: Free Press, 1996.
King 1896: Leonard W. King, *Babylonian Magic and Sorcery. The Prayers of the Lifting of the Hand*. London: Luzac and Co., 1896 (reprint: Hildesheim: Olms, 1975).
Kinnier Wilson 1957: James V. Kinnier Wilson, "Two Medical Texts from Nimrud (continued)", *Iraq* 19, 1 (1957), pp. 40–49.

Kinnier Wilson 1962a: James V. Kinnier Wilson, "The Kurba'il Statue of Shalmaneser III", *Iraq* 24 (1962), pp. 90–115.

Kinnier Wilson 1962b: James V. Kinnier Wilson, "Hebrew and Akkadian Philological Notes", *Journal of Semitic Studies* 7 (1962), pp. 173–183.

Kinnier Wilson 1965: James V. Kinnier Wilson, "An Introduction to Babylonian Psychiatry", in Hans G. Güterbock and Thorkild Jacobsen (eds.), *Studies in Honor of Benno Landsberger on His Seventy-Fifth Birthday, April 21, 1965* (Assyriological Studies 16). Chicago: University of Chicago Press, 1965, pp. 289–298.

Kinnier Wilson 1966: James V. Kinnier Wilson, "Leprosy in Ancient Mesopotamia", *Revue d'Assyriologie et d'Archéologie orientale* 60 (1966), pp. 47–58.

Kinnier Wilson 1967: James V. Kinnier Wilson, "Mental Diseases of Ancient Mesopotamia", in Don R. Brothwell and Andrew T. Sandison (eds.), *Diseases in Antiquity. A Survey of the Diseases, Injuries and Surgery of Early Populations*. Springfield: Thomas, 1967, pp. 723–732.

Kinnier Wilson 1982: James V. Kinnier Wilson, "Medicine in the Land and Times of the Old Testament", in Tomoo Ishida (ed.), *Studies in the Period of David and Solomon and Other Essays*. Winona Lake: Eisenbrauns, 1982, pp. 337–365.

Kinnier Wilson 1988: James V. Kinnier Wilson, "Lines 40–52 of the Banquet Stele of Aššurnaṣirpal II", *Iraq* 50 (1988), pp. 79–82.

Kinnier Wilson 2005: James V. Kinnier Wilson, "The Assyrian Pharmaceutical Series URU.AN.NA: *maštakal*", *Journal of Near Eastern Studies* 64 (2005), pp. 45–51.

Kinnier Wilson and Reynolds 2007: James V. Kinnier Wilson and Edward H. Reynolds, "On Stroke and Facial Palsy in Babylonian Texts", in Markham J. Geller and Irving L. Finkel (eds.), *Disease in Babylon* (Cuneiform Monographs 36). Leiden, Boston: Brill, 2007, pp. 67–99.

Kipfer 2017: Sara Kipfer (ed.), *Visualizing Emotions in the Ancient Near East* (Orbis Biblicus et Orientalis 285). Göttingen: Vandenhoeck & Ruprecht, 2017.

Kirmayer 1992: Laurence J. Kirmayer, "The Body's Insistence on Meaning: Metaphor as Presentation and Representation in Illness Experience", *Medical Anthropology Quarterly* N.S. 6, 4 (1992), pp. 323–346.

Kirmayer 1993a: Laurence J. Kirmayer, "Healing and the Invention of Metaphor: The Effectiveness of Symbols Revisited", *Culture, Medicine and Psychiatry* 17, 2 (1993), pp. 161–195.

Kirmayer 1993b: Laurence J. Kirmayer, "La folie de la métaphore", *Anthropologie et Sociétés* 17, 1–2 (1993), pp. 43–55.

Kirmayer 2008: Laurence J. Kirmayer, "Culture and the Metaphoric Mediation of Pain", *Transcultural Psychiatry* 45, 2 (2008), pp. 318–338.

Klein 1982: Jacob Klein, "'Personal God' and Individual Prayer in Sumerian Religion", *Archiv für Orientforschung* 19 (1982), pp. 295–306.

Kleinman 1973: Arthur Kleinman, "Medicine's Symbolic Reality. On a Central Problem in the Philosophy of Medicine", *Inquirity* 16 (1973), pp. 206–213.

Kleinman 1978: Arthur Kleinman, "Concepts and a Model for the Comparison of Medical Systems as Cultural Systems", *Social Science and Medicine* 12 (1978), pp. 85–93

Kogan and Militarev 2002: Leonid Kogan and Alexander Militarev, "Akkadian Terms for Genitalia: New Etymologies, New Textual Interpretations", in Simo Parpola and Robert M. Whiting (eds.), *Sex and Gender in the Ancient Near East. Proceedings of the 47th Rencontre Assyriologique Internationale, Helsinki, July 2–6, 2001*. Helsinki: Neo-Assyrian Text Corpus Project, 2002, pp. 311–319.

Konstantopoulos 2017: Gina Konstantopoulos, "Shifting Alignments: The Dichotomy of Benevolent and Malevolent Demons in Mesopotamia", in Siam Bhayro and Catherine Rider (eds.), *Demons and Illness from Antiquity to the Early-Modern Period*. Leiden, Boston: Brill, pp. 17–38.

Köcher 1955: Franz Köcher, *Keilschrifttexte zur assyrisch-babylonischen Drogen- und Pflanzenkunde. Texte der Serien* uru.an.na : maltakal, ḪAR.ra : ḫubullu *und* Ú GAR-šú (Veröffentlichung/

Deutsche Akademie der Wissenschaften, Institut für Orientforschung 28). Berlin: Akademie-Verlag, 1995.

Köcher 1963–1980: Franz Köcher, *Die babylonisch-assyrische Medizin in Texten und Untersuchungen*, 6 vols. Berlin, Boston: De Gruyter, 1963–1980.

Köcher 1978: Franz Köcher, "Spätbabylonische medizinische Texte aus Uruk", in Christa Habrich, Frank Marguth, Jörn Henning Wolf, and Renate Wittern (eds.), *Medizinische Diagnostik in Geschichte und Gegenwart. Festschrift für Heinz Goerke*. München: Fritsch, 1978, pp. 17–39.

Köcher 1995: Franz Köcher, "Ein Text medizinischen Inhalts aus dem neubabylonischen Grab 405", in Rainer Michael Boehmer, Friedhelm Pedde, and Beate Salje (eds.), *Uruk. Die Gräber* (Ausgrabungen in Uruk-Warka 10). Mainz am Rhein: P. von Zabern, 1995, pp. 203–217.

Kramer 1973: Samuel Noah Kramer, "The Jolly Brother: A Sumerian Dumuzi Tale", *Journal of Ancient Near Eastern Society of Columbia University* 5 (1973), pp. 243–253.

Kraus 1951: Fritz Rudolf Kraus, "Nippur und Isin nach altbabylonischen Rechtsurkunden", *Journal of Cuneiform Studies* 3 (1951), pp. 1–228.

Krebernik 1984: Manfred Krebernik, *Die Beschwörungen aus Fara und Ebla: Untersuchungen zur ältesten keilschriftlichen Beschwörungsliteratur*. Hildesheim: Olms, 1984.

Krebernik 1993–1997: Manfred Krebernik,"Muttergöttin A. I. In Mesopotamien", in *Reallexikon der Assyriologie und vorderasiatischen Archäologie*, vol. 8. Berlin: De Gruyter, 1993–1997, pp. 502–516.

Krebernik 2009–2011: Manfred Krebernik,"Šazu", in *Reallexikon der Assyriologie und vorderasiatischen Archäologie*, vol. 12. Berlin: De Gruyter, 2009–2011, p. 110.

Krebernik 2018: Manfred Krebernik, "Eine neue elamische Beschwörung aus der Hilprecht-Sammlung (HS 2338) im Kontext alloglotter Texte der altbabylonischen Zeit", in Behzad Mofidi Nasrabadi, Doris Prechel, and Alexander Pruss (eds.), *Elam and its Neighbors. Recent Research and New Perspectives. Proceedings of the International Congress Held at Johannes Gutenberg University Mainz, September 21–23, 2016* (Elamica 8). Hildesheim: Verlag Franzbecker, 2018, pp. 13–48.

Krecher 1970: Joachim Krecher, review of Biggs 1967, *Orientalische Literaturzeitung* 65 (1970), pp. 350–355.

Kselman 2002: John S. Kselman, "'*Wandering about' and Depression. More Examples*", *Journal of Near Eastern Studies* 61, 4 (2002), pp. 275–277.

Küchler 1904: Friedrich Küchler, *Beiträge zur Kenntnis der assyrisch-babylonischen Medizin Texte mit Umschrift, Übersetzung und Kommentar* (Assyriologische Bibliothek 18). Leipzig: Hinrichs, 1904.

Kübel 2007: Paul Kübel, *Metamorphosen der Paradieserzählung* (Orbis Biblicus et Orientalis 231). Göttingen: Vandenhoeck & Ruprecht, 2007.

Labat 1951: René Labat, *Traité akkadien de diagnostics et pronostics médicaux*. Paris: Académie internationale d'histoire des science, 1951.

Labat 1957–1971: René Labat, "Geschlechtskrankheiten", in *Reallexikon der Assyriologie und vorderasiatischen Archäologie*, vol. 3. Berlin: De Gruyter, 1957–1971, pp. 221–223.

Labat 1959: René Labat, "Le premier chapitre d'un précis médical assyrien", *Revue d'Assyriologie et d'Archéologie orientale* 53 (1959), pp. 1–18.

Labat 1968: René Labat, review of Biggs 1967, *Bibliotheca Orientalis* 25, 5–6 (1968), pp. 356–358.

Laderman 1987: Carol Laderman, "The Ambiguity of Symbols in the Structure of Healing", *Social Science and Medicine* 24 (1987), pp. 293–301.

Lakoff and Johnson 1980: George Lakoff and Mark Johnson, *Metaphors we live by*. Chicago: University of Chicago Press, 1980.

Lambert 1959: Wilfred G. Lambert, "Divine Love Lyrics from Babylon", *Journal of Semitic Studies* 4 (1959), pp. 1–15.

Lambert 1959–1960: Wilfred G. Lambert, "Three Literary Prayers of the Babylonians", *Archiv für Orientforschung* 19 (1959–1960), pp. 38–51.
Lambert 1960a: Wilfred G. Lambert, *Babylonian Wisdom Literature*. Oxford: Clarendon Press, 1960.
Lambert 1960b: Wilfred G. Lambert, "A Catalogue of Texts and Authors", *Journal of Cuneiform Studies* 16 (1960), pp. 59–109.
Lambert 1966: Wilfred G. Lambert, "Divine Love Lyrics from the Reign of Abi-ešuḫ", *Mitteilungen des Instituts für Orientforschung* 12 (1966), pp. 41–56.
Lambert 1967: Wilfred G. Lambert, "The Gula Hymn of Bulluṭsa-rabi", *Orientalia* N.S. 36 (1967), pp. 105–132.
Lambert 1969: Wilfred G. Lambert, "A Middle Assyrian Medical Text", *Iraq* 31 (1969), pp. 28–39.
Lambert 1970: Wilfred G. Lambert, "Fire Incantations", *Archiv für Orientforschung* 23 (1970), pp. 39–45.
Lambert 1975: Wilfred G. Lambert, "The Problem of the Love Lyrics", in Hans Goedicke and Jimmy Jack MacBee Roberts (eds.), *Unity and Diversity: Essays in the History, Literature, and Religion of the Ancient Near East*. Baltimore: Johns Hopkins University Press, 1975, pp. 98–135.
Lambert 1976–1980: Wilfred G. Lambert, "Išḫara", in *Reallexikon der Assyriologie und vorderasiatischen Archäologie*, vol. 5 Berlin: De Gruyter, 1976–1980, pp. 176–177.
Lambert 1982: Wilfred G. Lambert, "Sum. ninda = Akk. *ittû 'father'*", *Revue d'Assyriologie et d'Archéologie orientale* 76 (1982), p. 94.
Lambert 1983: Wilfred G. Lambert, "Exorcistic Mumbo Jumbo", *Revue d'Assyriologie et d'Archéologie orientale* 77 (1983), pp. 94–95.
Lambert 1987a: Wilfred G. Lambert, "Devotion: The Languages of Religion and Love", in Murray Mindlin, Markham J. Geller, and John E. Wansbrough (eds.), *Figurative Language in the Ancient Near East*. London: Taylor and Francis, 1987, pp. 25–39.
Lambert 1987b: Wilfred G. Lambert, "A Further Attempt at the Babylonian 'Man and His God'", in Francesca Rochberg-Halton (ed.), *Language, Literature and History. Philological and Historical Studies Presented to Erica Reiner* (American Oriental Series 67). New Haven: American Oriental Society, 1987, pp. 187–202.
Lambert 2007: Wilfred G. Lambert, *Babylonian Oracle Questions* (Mesopotamian Civilizations 13). Winoka Lake: Eisenbrauns, 2007.
Lambert 2013: Wilfred G. Lambert, *Babylonian Creation Myths* (Mesopotamian Civilizations 16). Winoka Lake: Eisenbrauns, 2014.
Lambert and Millard 1969: Wilfred G. Lambert and Alan Ralph Millard, *Atra-Ḫasīs. The Babylonian Story of the Flood*. Oxford: Clarendon Press, 1969.
Landsberger 1933: Benno Landsberger, "Lexikalisches Archiv. 5. *ṣarātu, teṣ/zû, naš/sāḫu*", *Zeitschrift für Assyriologie und vorderasiatische Archäologie* 41, 1–4 (1933), pp. 222–233.
Landsberger 1934: Benno Landsberger, *Die Fauna des alten Mesopotamien nach der 14. Tafel der Serie* ḪAR-ra = ḫubullu. Leipzig: Hirzel, 1934.
Landsberger 1935–1936: Benno Landsberger, "Studien zu den Urkunden aus der Zeit des Ninurta-tukul-Aššur", *Archiv für Orientforschung* 10 (1935–1936), pp. 140–159.
Landsberger 1937: Benno Landsberger, *Die Serie ana ittišu* (Materialien zum sumerischen Lexikon 1). Rome: Biblical Institute Press, 1937.
Landsberger 1937–1939: Benno Landsberger, "Keilschrifttexte nach Kopien von T. G. Pinches. 9 Texte zur Serie ḪAR.ra = *ḫubullu* (mit 2 Tafeln)", *Archiv für Orientforschung* 12 (1937–1939), pp. 135–144.
Landsberger 1951: Benno Landsberger, *Die Serie* Ur-e-a = nâqu (Materialien zum sumerischen Lexikon 2). Rome: Biblical Institute Press, 1951.
Landsberger 1954: Benno Landsberger, "Assyrische Königsliste und 'Dunkles Zeitalter'", *Journal of Cuneiform Studies* 8 (1954), pp. 106–133.

Landsberger 1958: Benno Landsberger, "Corrections to the Article, 'An Old Babylonian Charm against Merhu'", *Journal of Near Eastern Studies* 17, 1 (1958), pp. 56–58.

Landsberger 1959: Benno Landsberger, *The Series* ḪAR-ra = ḫubullu. *Tablets VIII-XII* (Materialien zum sumerischen Lexikon 7). Rome: Biblical Institute Press, 1959.

Landsberger 1960: Benno Landsberger, *The Fauna of Ancient Mesopotamia. First Part.* ḪAR-ra = ḫubullu *Tablet XIII* (Materialien zum sumerischen Lexikon 8/1). Rome: Biblical Institute Press, 1960.

Landsberger 1962: Benno Landsberger, *The Fauna of Ancient Mesopotamia. Second Part.* ḪAR-ra = ḫubullu *Tablet XIV and XVIII* (Materialien zum sumerischen Lexikon 8/2). Rome: Biblical Institute Press, 1962.

Landsberger 1966: Benno Landsberger, "Einige unerkannt gebliebene oder verkannte Nomina des Akkadischen", *Die Welt des Orients* 3, 3 (1966), pp. 246–268.

Landsberger 1967a: Benno Landsberger, *The Date Palm and Its By-products according to the Cuneiform Sources* (Archiv für Orientforschung Beiheft 17). Graz: Im Selbstverlage des Herausgebers, 1967.

Landsberger 1967b: Benno Landsberger, "Über Farber im Sumerisch-akkadischen", *Journal of Cuneiform Studies* 21 (1967), pp. 139–173.

Landsberger 1968: Benno Landsberger, "Jungfräulichkeit: Ein Beitrag zum Thema 'Beilager und Eheschließung'", in Johan A. Ankum, Robert Feenstra, and Wilhelmus François Leemans (eds.), *Symbolae iuridicae et historicae Martino David dedicatae*, vol. 1. Leiden: Brill, pp. 41–105.

Landsberger and Gurney 1957–1958: Benno Landsberger and Oliver Robert Gurney, "Practical Vocabulary of Assur", *Archiv für Orientforschung* 18 (1957–1858), pp. 328–341.

Landsberger and Jacobsen 1955: Benno Landsberger and Thorkild Jacobsen, "An Old Babylonian Charm against Merhu", *Journal of Near Eastern Studies* 14, 1 (1955), pp. 14–21.

Landsberger et al. 1970: Benno Landsberger, Miguel Civil, and Erica Reiner, *The Series* ḪAR-ra = ḫubullu *Tablets XVI, XVII, XIX and Related Texts* (Materialien zum sumerischen Lexikon 10). Rome: Biblical Institute Press, 1970.

Landsberger et al. 1971: Benno Landsberger, Miguel Civil, and Erica Reiner, Izi = išātu, Ká-gal = abullu *and* Níg-ga = makkuru (Materialien zum sumerischen Lexikon 13). Rome: Biblical Institute Press, 1971.

Lapinkivi 2004: Pirjo Lapinkivi, *The Sumerian Sacred Marriage in the Light of Comparative Evidence* (State Archives of Assyria Studies 15). Helsinki: Neo-Assyrian Text Corpus Project, 2004.

Lapinkivi 2010: Pirjo Lapinkivi, *The Neo-Assyrian Myth of Ištar's Descent and Resurrection* (State Archives of Assyria Cuneiform Texts 6). Helsinki: Neo-Assyrian Text Corpus Project, 2010.

Laqueur 1990: Thomas Laqueur, *Making Sex. Body and Gender from the Greeks to Freud.* Cambridge, London: Harvard University Press, 1990.

Laessøe 1953: Jørgen Laessøe, "Reflexions on Modern and Ancient Oriental Water Works", *Journal of Cuneiform Studies* 7 (1953), pp. 5–26.

Leach 1964: Edmund R. Leach, "Anthropological Aspects of Language: Animal Categories and Verbal Abuse", in Eric Heinz Lenneberg (ed.), *New Directions in the Study of Language.* Cambridge: The Massachusets Institute of Technology Press, 1964, pp. 23–63.

Leach 1966: Edmund R. Leach, "Ritualization in Man in Relation to Conceptual and Social Development", *Philosophical Transactions of the Royal Society B: Biological Sciences* 251, 772 (1966), pp. 403–408.

Le Breton 1991 : David Le Breton, "Corps et anthropologie : de l'efficacité symbolique", *Diogène* 153 (1991), pp. 92–107.

Lecompte 2016: Camille Lecompte, "Representation of Women in Mesopotamian Lexical Lists", in Brigitte Lion and Cécile Michel (eds.) *The Role of Women in Work and Society in the Ancient Near East* (Studies in Ancient Near Eastern Records 13). Boston, Berlin: De Gruter, pp. 29–56.

Leenhardt 1939: Maurice Leenhardt, “La personne mélanésienne”, *École pratique des hautes études, Section des sciences religieuses*, Annuaire 1940–1941 (1939), pp. 5–36.

Legrain 1930: Leon Legrain, *Terracottas from Nippur*. Philadelphia: The University Museum Babylonian Section 16, Pennsylvania, 1930.

Leichty 1988: Erle Leichty, “Guaranteed to Cure”, in Erle Leichty, Maria deJ. Ellis, and Pamela Gerardi (eds.), *A Scientific Humanist. Studies in Memory of Abraham Sachs* (Occasional Publications of the Samuel Noah Kramer Fund 9). Philadelphia: University of Pennsylvania Museum, 1988, pp. 261–264.

Leick 1994: Gwendolyn Leick, *Sex and Eroticism in Mesopotamian Literature*. London, New York: Routledge, 1994.

Leick 2015: Gwendolyn Leick, “Too Young, too Old? Sex and Age in Mesopotamian Literature”, in Mark Masterson, Nancy Sorkin Rabinowitz, and James Robson (eds.), *Sex in Antiquity. Exploring Gender and Sexuality in the Ancient World*. London: Routledge, 2015, pp. 80–96.

Lenzi 2010: Alan Lenzi, “*Šiptu ul yuttum*. Some Reflections on a Closing Formula in Akkadian Incantations”, in Jeffrey Stackert, Barbara N. Porter, and David P. Wright (eds), *Gazing on the Deep. Ancient Near Eastern and Other Studies in Honor of Tzvi Abusch*. Bethesda: CDL Press, 2020, pp. 131–166.

Lenzi 2011: Alan Lenzi, *Reading Akkadian Prayers and Hymns. An Introduction*. Atlanta: Society of Biblical Literature, 2011.

Lévi-Strauss 1949a: Claude Lévi-Strauss, “Le sorcier et sa magie ”, *Les Temps Modernes*, 4e année, 41 (1949), pp. 385–406 (reprint: Idem, *Anthropologie structurale*, Paris: Plon, 1958, pp. 183–203).

Lévi-Strauss 1949b: Claude Lévi-Strauss, “L’efficacité symbolique”, *Revue de l’histoire des religions* 135, 1 (1949), pp. 5–27 (reprint: Idem, *Anthropologie structurale*, Paris: Plon, 1958, pp. 205–226).

Lévi-Strauss 1962a: Claude Lévi-Strauss, *Le totémisme aujourd’hui*. Paris: Presses universitaires de France, 1962.

Lévi-Strauss 1962b: Claude Lévi-Strauss, *La pensée sauvage*. Paris: Plon, 1962.

Levy at al. 2006: Jacques Levy, Annie Attia, and Gilles Buisson, “L’usage médical des cochons ”, in Lion and Michel 2006: 195–203.

Li Causi 2005: Pietro Li Causi, “Corpi, spazi, luoghi, animali. Dall’animale come spazio visivo localizzato alle funzioni dell’anima”, *Athenaeum* 96, 1 (2005), pp. 55–75.

Limet 1968: Henri Limet, *L’Anthroponymie sumérienne dans les documents de la 3e Dynastie d’Ur*. Paris: Les Belles Lettres, 1968.

Lion 2007: Brigitte Lion, “La notion de genre en assyriologie”, in Violaine Sebillotte Cuchet and Nathalie Ernoult (eds.), *Problèmes du genre en Grèce ancienne*. Paris: Éditions de la Sorbonne, 2007, pp. 51–64.

Liverani 2011: Mario Liverani, “Portrait du héros comme un jeune chien”, in Jean-Marie Durand, Thomas Römer, and Michel Langlois (eds.), *Le jeune héros. Recherches sur la formation et la diffusion d’un thème littéraire au Proche-Orient ancien* (Orbis Biblicus et Orientalis 250). Göttingen: Vandenhoeck & Ruprecht, 2011, pp. 11–26.

Livingstone 1986: Alasdair Livingstone, *Mystical and Mythological Explanatory Works of Assyrian and Babylonian Scholars*. Oxford: Clarendon Press, 1986.

Livingstone 1989: Alasdair Livingstone, *Court Poetry and Literary Miscellanea* (State Archives Of Assyria 3). Helsinki: Helsinki University Press, 1989.

Livingstone 1991: Alasdair Livingstone, “An Enigmatic Line in a Mystical / Mythological Explanatory Work as Agriculture Myth”, *Nouvelles Assyriologiques Brèves et Utilitaires* 1991, 1, no. 6, pp. 5–6.

Livingstone 1999: Alasdair Livingstone, “The Magic of Time”, in Tzvi Abusch and Karel van der Toorn (eds.), *Mesopotamian Magic. Textual, Historical, and Interpretative Perspectives* (Ancient Magic and Divination 1). Groningen: Styx, 1999, pp. 131–137.

Lion and Michel 2006: Brigitte Lion and Cécile Michel (eds.), *De la domestication au tabou. Le cas des suidés au Proche-Orient ancien*. Paris: De Boccard, 2006.

Lloyd 1991: Geoffrey E.R. Lloyd, "The Invention of Nature", in Idem, *Methods and Problems in Greek Science*. Cambridge: Cambridge University Press, 1991, pp. 417–434.

Lock 1987: Margaret Lock, "DSM-III as a Culture-Bound Construct: Commentary on Culture-Bound Syndromes and International Disease Classifications", *Culture, Medicine, and Psychiatry* 11, 1 (1987), pp. 35–42.

Lock and Gordon 1988: Margaret Lock and Deborah Gordon (eds.), *Biomedicine examined*. Dordrecht, Boston, London: Springer, 1988.

Lotman 1972: Yury Lotman, *Analiz poeticheskogo teksta. Struktura stikha*. Leningrad: Prosveshchenie, 1972 (Enghlish translation: *Analysis of the Poetic Text*. Ann Arbor: Ardis, 1976).

Lupo 1993: Alessandro Lupo, "The Importance of Prayers in the Study of Cosmologies and Religious Systems of Native Oral Cultures", in Jon Davies and Isabel Wollaston (eds.), *The Sociology of Sacred Texts*. Sheffield: Sheffield Academic Press, 1993, pp. 83–93.

Lupo 1995: Alessandro Lupo, "La oración: estructura, forma y uso. Entre tradición escrita y oral", in Carmelo Lisón Tolosana (ed.), *Antropología y Literatura*. Zaragoza: Diputación General de Aragón. Departamento de Educación y Cultura, 1995, pp. 49–66.

Lupo 2009: Alessandro Lupo, *Il mais nella croce. Pratiche e dinamiche religiose nel Messico indigeno*. Roma: CISU, 2009.

Lupo 2012: Alessandro Lupo, "Malattia ed efficacia terapeutica", in Donatella Cozzi (ed.), *Le parole dell'antropologia medica. Piccolo dizionario*. Perugia: Morlacchi, 2012, pp. 127–155.

Lutz 1988: Catherine A. Lutz, *Unnatural Emotions. Everyday Sentiments on a Micronesian Atoll and Their Challenge to Western Theory*. Chicago: University of Chicago Press, 1988.

Macherey 2009: Pierre Macherey, *Da Canguilhem à Foucault, la force des normes*. Paris: Fabrique, 2009.

Maier 2009: John R. Maier, "A Mesopotamian Hero for a Melancholy Age", in Stephen Bertman and Lois Parker (eds.), *The Healing Power of Ancient Literature*. Newcastle: Cambridge Scholars, 2009, pp. 23–44.

Malinowski 1965: Bronisław Malinowski, *Coral Gardens and Their Magic*, 2 vols. Bloomington: Indiana University Press, 1965.

Marchesi 2002: Gianni Marchesi, "On the Divine Name ^{d}BA.Ú", *Orientalia* N.S. 71 (2002), pp. 161–172.

Marcus 1977: David Marcus, "Animal Similes in Assyrian Royal Inscriptions", *Orientalia* 46 (1977), pp. 86–106.

Margalit 1989: Baruch Margalit, *The Ugaritic Poem of AQHT. Text, Translation, Commentary*. Berlin, New York: De Gruyter, 1989.

Martin 2000: Emily Martin, "Mind-Body Problems", *American Ethnologist* 27, 3 (2000), pp. 569–590.

Mathieu 1991: Nicole-Claude Mathieu, *L'anatomie politique. Catégorisations et idéologies du sexe*. Paris: Indigo & Côté-femmes, 1991.

Mattila et al. 2019: Raija Mattila, Sanae Ito, and Sebastian Fink (eds.), *Animals and their Relation to Gods, Humans and Things in the Ancient World*. Wiesbaden: Springer, 2019.

Mattingly and Garro 2000: Cheryl Mattingly and Linda C. Garro (eds.), *Narrative and Cultural Construction of Illness and Healing*. Berkeley: University of California Press, 2000.

Matsushima 1988: Eiko Matsushima,"Les rituels du mariage divin dans les documents accadiens", *Acta Sumerologica* 10 (1988), pp. 95–128.

Maul 1994: Stefan M. Maul, *Zukunftsbewältigung. Eine Untersuchung altorientalischen Denkens anhand der babylonisch-assyrischen Löserituale (Namburbi)* (Baghdader Forschungen 18). Mainz am Rhein: P. von Zabern, 1994.

Maul 1996: Stefan M. Maul, "Die babylonische Heilkunst: Medizinische Keilschrifttexte auf Tontafeln", in Heinz Schott (ed.), *Meilensteine der Medizin*. Dortmund: Harenberg, 1996, pp. 32–39.

Maul 2004: Stefan M. Maul, “Die ‘Lösung vom Bann’: Überlegungen zu altorientalischen Konzeptionen von Krankheit und Heilkunst”, in Manfred Horstmanshoff and Marten Stol (eds.), *Magic and Rationality in Ancient Near Eastern and Graeco-Roman Medicine* (Studies in Ancient Medicine 27). Leiden: Brill, 2004, pp. 79–95.

Maul 2009: Stefan M. Maul, “Die Lesung der Rubra DÙ.DÙ.BI und KÌD.KÌD.BI”, *Orientalia* N.S. 78 (2009), pp. 69–80.

Maul 2010: Stefan M. Maul, “Die Tontafelbibliothek aus dem sogenannten ‘Haus des Beschwörungspriesters’”, in Stefan M. Maul and Nils P. Heeßel (eds.), *Assur-Forschungen. Arbeiten aus der Forschungsstelle “Edition literarischer Keilschrifttexte aus Assur” der Heidelberger Akademie der Wissenschaften*. Wiesbaden: Harrassowitz, 2020, pp. 189–228.

Maul and Strauß 2011: Stefan M. Maul and Rita Strauß, *Ritualbeschreibungen und Gebete I. Mit Beiträgen von Daniel Schwemer* (Keilschrifttexte aus Assur literarischen Inhalts 4, Wissenschaftliche Veröffentlichungen der Deutschen Orient-Gesellschaft 133). Wiesbaden: Harrassowitz, 2011.

Mauss 1934: Marcel Mauss, “Les techniques du corps”, *Journal de Psychologie* 32, 3–4 (1934), pp. 271–293 [the pages refer to http://classiques.uqac.ca/classiques/mauss_marcel/socio_et_anthropo/6_Techniques_corps/techniques_corps.pdf].

Mauss 1938: Marcel Mauss, “Une catégorie de l’esprit humain : la notion de personne celle de ‘moi’”, *Journal of the Royal Anthropological Institute* 68 (1938), pp. 263–281.

Mauss and Hubert 1902–1903: Marcel Mauss and Henri Hubert, “Esquisse d’une théorie générale de la magie”, *L’Année sociologique* 7 (1902–1903), pp. 1–146 [the pages refer to http://classiques.uqac.ca/classiques/mauss_marcel/socio_et_anthropo/1_esquisse_magie/esquisse_magie.pdf].

May 2018a: Natalie N. May, “Female Scholars in Mesopotamia?”, in Stephanie Lynn Budin, Megan Cifarelli, Agnès Garcia-Ventura, and Adelina Millet Albà (eds.), *Gender and Methodology in the Ancient Near East. Approaches from Assyriology and beyond*. Barcelona: Edicions de la Universitat de Barcelona, 2018, pp. 149–162.

May 2018b: Natalie N. May, “Exorcists and Physicians at Assur: More on their Education and Interfamily and Court Connections”, *Zeitschrift für Assyriologie und vorderasiatische Archäologie* 108, 1 (2018), pp. 63–80.

Mayer 1976: Werner R. Mayer, *Untersuchungen zur Formensprache der babylonischen “Gebetsbeschwörungen”* (Studia Pohl: Series Maior 5). Rome: Biblical Institute Press, 1976.

Mayer 1999: Werner R. Mayer, “Das Ritual *KAR 76* mit dem Gebet ‘Marduk 24’”, *Orientalia* 68 (1999), pp. 145–163.

McCown et al. 1967: Donald E. McCown, Richard C. Haines, and Donald P. Hansen (eds.), *Nippur I, Temple of Enlil, Scribal Quarter, and Soundings* (Oriental Institute Publications 78). Chicago: The University of Chicago Press, 1967.

McLaren 2007: Angus McLaren, *Impotence: a Cultural History*. Chicago: The University of Chicago Press, 2007.

McNay 2004: Lois McNay, “Agency and Experience: Gender as a Lived Relation”, in Lisa Adkins and Beverley Skeggs (eds.), *Feminism After Bourdieu*. Oxford: Blackwell Publishers, 2004, pp. 175–190.

Meier 1939: Gerhard Meier, “Ein akkadisches Heilungsritual aus Boğazköy”, *Zeitschrift für Assyriologie und vorderasiatische Archäologie* 45, 2–3 (1939), pp. 195–215.

Meinhold 2017: Wiebke Meinhold, *Ritualbeschreibungen und Gebet II* (Keilschrifttexte aus Assur literarischen Inhalts 7, Wissenschaftliche Veröffentlichungen der Deutschen Orient-Gesellschaft 147). Wiesbaden: Harrassowitz, 2017.

Meissner 1925: Bruno Meissner, *Babilonien und Assyrien*, vol. 2. Heidelberg: Carl Winters Universitätsbuchhandlung, 1925.

Melville 2004: Sarah C. Melville, "Neo-Assyrian Royal Women and Male Identity: Status as a Social Tool", *Journal of the American Oriental Society* 124 (2004), pp. 37–57.

Merleau-Ponty 1945: Maurice Merleau-Ponty, *Phénoménologie de la perception*. Paris: Gallimard, 1945.

Mertens-Wagshal 2018: Avigail Mertens-Wagschal, "The Lion, the Witch, and the Wolf: Aggressive Magic and Witchcraft in the Old Babylonian Period", in Greta Van Buylaere, Mikko Luukko, Daniel Schwemer, and Avigail Mertens-Wagschal (eds.), *Sources of Evil. Studies in Mesopotamian Exorcistic Lore* (Ancient Magic and Divination 15). Leiden, Boston: Brill, 2018, pp. 158–169.

Michalowski 1981: Piotr Michalowski, "Carminative Magic: Towards an Understanding of Sumerian Poetics", *Zeitschrift für Assyriologie und vorderasiatische Archäologie* 71, 1 (1981), pp. 1–18.

Michalowski 1998: Piotr Michalowski, "Literature as a Source of Lexical Inspiration: Some Notes on a Hymn to the Goddess Inana", in Jan Braun, Krystyna Lyczkowska, Maciej Popko, and Piotr Steinkeller (eds.), *Written on Clay and Stone. Ancient Near Eastern Studies Presented to Krystyna Szarzyńska on the Occasion of her 80th Birthday*. Warsaw: Agade, 1998, pp. 65–73.

Michel 2004: Cécile Michel, "Deux incantations paléo-assyriennes. Une nouvelle incantation pour accompagner la naissance", in Jan Gerrit Dercksen (ed.), *Assyria and Beyond. Studies Presented to Morgens Trolle Larsen*. Leiden: Nederlands Instituut voor het Nabije Oosten, 2004, pp. 395–420.

Michel 2016: Cécile Michel, "Women Work, Men are Professionals in the Old Assyrian Archives", in Brigitte Lion and Cécile Michel (eds.) *The Role of Women in Work and Society in the Ancient Near East* (Studies in Ancient Near Eastern Records 13). Boston, Berlin: De Gruter, pp. 193–208.

Milano 2005: Lucio Milano, "Il nemico bestiale. Su alcune caratteristiche animalesche del nemico nella letteratura sumero-accadica", in Ettore Cingano, Antonella Ghersetti, and Lucio Milano (eds.), *Animali tra zoologia, mito e letteratura nella cultura classica e orientale*. Padova: Sargon, 2005: pp. 47–67.

Miller 2010: Jared L. Miller, "Paskuwatti's Ritual: Remedy for Impotence or Antidote to Homosexuality?", *Journal of Ancient Near Eastern Religions* 10 (2010), pp. 83–89.

Minh-ha 1989: Trinh T. Minh-ha, Woman, *Native Other. Writing Postcoloniality and Feminism*. Bloomington: Indiana University Press, 1989.

Minunno 2013: Giuseppe Minunno, *Ritual Employs of Birds in Ancient Syria-Palestine* (Alter Orient und Altes Testament 402). Münster: Ugarit-Verlag, 2013.

Moerman 1979: Daniel E. Moerman, "Anthropology of Symbolic Healing", *Current Anthropology* 20, 1 (1979), pp. 59–66.

Moerman 2002: Daniel E. Moerman, *Meaning, Medicine, and the "Placebo-effect"*. Cambridge: Cambridge University Press, 2002.

Mouton 2007: Alice Mouton, *Rêves hittites. Contribution à une histoire et une anthropologie du rêve en Anatolie ancienne* (Culture and History of the Ancient Near East 28). Leiden, Boston: Brill, 2007.

Nathan 1995: Tobie Nathan, "Manifeste pour une psychopathologie scientifique", in Tobie Nathan and Isabelle Stengers, *Médecins et sorciers*. Paris: Les Empêcheurs de penser en rond, 1995.

Nathan 2001: Tobie Nathan, *Nous ne sommes pas seuls au monde*. Paris: Seuil, 2001.

Nissinen 2001: Martti Nissinen, "Akkadian Rituals and Poetry of Divine Love", in Robert M. Whiting (ed.), *Mythology and Mythologies.Methodological Approaches to Intercultural Influences* (Melammu Symposia 2). Helsinki: Neo-Assyrian Text Corpus Project, 2001, pp. 93–136

Nougayrol 1945–1946 : Jean Nougayrol, "Textes hépatoscopiques d'époque ancienne", *Revue d'Assyriologie et d'Archéologie orientale* 40 (1945–1946), pp. 56–97.

Nougayrol 1968: Jean Nougayrol, review of Biggs 1967, *Revue d'Assyriologie et d'Archéologie orientale* 62 (1968), pp. 92–94.

N'Shea 2016: Omar N'Shea, "Royal Eunuchs and Elite Masculinity in the Neo-Assyrian Empire", *Near Eastern Archaeology* 79, 3 (2016), pp. 214–221.

N'Shea 2019: Omar N'Shea, "Dressed to Dazzle, Dressed to Kill: Staging Assurbanipal in the Royal Lion Hunt Reliefs from Nineveh", in Megan Cifarelli (ed.), *Fashioned Selves. Dress and Identity in Antiquity. Oxford*, Havertown: Oxbow Books, 2019, pp. 175–183.

Oberhuber 1972: Karl Oberhuber, *Die Kultur des alten Orients*. Frankfurt am Main: Akademische Verlagsgesellschaft Athenaion, 1972.

Obeyesekere 1985: Gananath Obeyesekere, "Depression, Buddhism and the Work of Culture in Sri Lanka", in Arthur Kleinman and Byron Good (eds.), *Culture and Depression. Studies in the Anthropology and Cross-cultural Psychiatry of Affect and Disorder*. Berkley: University of California Press, 1985, pp. 134–152.

Obeyesekere 1990: Gananath Obeyesekere, *The Work of Culture. Symbolic Transformation in Psychoanalysis and Anthropology*. Chicago: The University of Chicago Press, 1990.

Oettinger 1976: Norbert Oettinger, *Die militärischen Eide der Hethiter* (Studien zu den Boğazköy-Texten 22). Wiesbaden: Harrassowitz, 1976.

Ong 1982: Walter J. Ong, *Orality and Literacy. The Technologizing of the World*. London: Methuen & Co. Ltd, 1982 (reprint: London, New York: Routledge, 2002).

Opificius 1961: Ruth Opificius, *Das Altbabylonische Terrakottarelief* (Untersuchungen zur Assyriologie und vorderasiatische Archäologie 2). Berlin: De Gruyter, 1961.

Oppenheim 1941: Adolf L. Oppenheim, "Idiomatic Accadian (Lexicographical Researches)", *Journal of the American Oriental Society* 61, 4 (1941), pp. 251–271.

Oppenheim 1956: Adolf L. Oppenheim, *The Interpretation of Dreams in the Ancient Near East. With a Translation of an Assyrian Dream-Book, Transactions of the American Philosophical Society NS* 46, 3 (1956), pp. 179–373.

Oppenheim 1962: Adolf L. Oppenheim, "On the Observation of the Pulse in Mesopotamian Medicine", *Orientalia* 31 (1962), pp. 27–33.

Oppenheim 1963: Adolf L. Oppenheim, "Mesopotamian Conchology", *Orientalia* N.S. 32 (1963), pp. 407–412.

Oppenheim 1964: Adolf L. Oppenheim, *Ancient Mesopotamia. Portrait of a Dead Civilization*. Chicago: University of Chicago Press, 1964.

Oppenheim 1970: Adolf L. Oppenheim, "Glasses in Mesopotamian Sources", in Adolf L. Oppenheim, Robert Howard Brill, Dan Barag, and Axel von Saldern (eds.), *Glass and Glassmaking in Ancient Mesopotamia. An Edition of the Cuneiform Texts which contain Tnstructions for Glassmakers with a Catalogue of Surviving Objects*. New York: The Corning Museum of Glass, pp. 9–21.

Ortner and Whitehead 1981: Sherry B. Ortner and Harriet Whitehead, "Introduction: Accounting for Sexual Meanings", in Eaedem (eds.), *Sexual Meaning. The Cultural Construction of Gender and Sexuality*. Cambridge: Cambridge University Press, 1981, pp. 1–27.

Oshima 2014: Takayoshi M. Oshima, *Babylonian Poems of Pious Suffers. Ludlul Bēl Nēeqi and the Babylonian Theodicy* (Orientalische Religionen in der Antike 14). Tübingen: Mohr Siebeck, 2014.

Panayotov 2013: Strahil Panayotov, "A Ritual for A Flourishing Bordello", *Bibliotheca Orientalis* 70 (2013), pp. 285–309.

Panayotov 2018: Strahil Panayotov, "Magico-medical Plants and Incantations on Assyrian House Amulets", in Greta Van Buylaere, Mikko Luukko, Daniel Schwemer, and Avigail Mertens-Wagschal (eds.), *Sources of Evil. Studies in Mesopotamian Exorcistic Lore* (Ancient Magic and Divination 15). Leiden, Boston: Brill, 2018, pp. 193–222.

Pangas 1988: Julio C. Pangas, "Aspectos de la sexualidad en la antiqua Mesopotamia", *Aula orientalis: revista de estudios del Próximo Oriente Antiguo* 6 (1988), pp. 211–226.

Parayre 2000a: Dominique Parayre (ed.), *Les animaux et les hommes dans le monde syro-mésopotamien aux époques historiques* (Topoi. Orient-Occident, Supplement 2). Lyon: De Boccard, 2000.

Parayre 2000b: Dominique Parayre, "Les suidés dans le monde syro-mésopotamien aux époques historiques", in Parayre 2000a: 141–206.

Parpola 1970: Simo Parpola, *Letters from Assyrian scholars to the kings Esarhaddon and Assurbanipal. Volume 1: Texts* (Alter Orient und Altes Testament 5/1). Kevelaer: Butzon und Bercker; Neukirchen-Vluyn: Neukirchener Verlag des Erziehungsvereins, 1970.

Parpola 1971: Simo Parpola, *Letters from Assyrian scholars to the kings Esarhaddon and Assurbanipal. Part II A: Introduction and Appendixes* (Alter Orient und Altes Testament 5/2). Kevelaer: Butzon und Bercker; Neukirchen-Vluyn: Neukirchener Verlag des Erziehungsvereins, 1971.

Parpola and Watanabe 1988: Simo Parpola and Kazuko Watanabe, *Neo-Assyrian Treaties and Loyalty Oaths* (State Archives of Assyria 2). Helsinki: Helsinki University Press, 1988.

Parys 2017: Magalie Parys, "Introduction aux symptômes mentaux en Mésopotamie", in Olga Drewnowska and Małgorzata Sandowicz (eds.), *Fortune and Misfortune in the Ancient Near East. Proceedings of the 60th Rencontre Assyriologique Internationale at Warsaw 21–25 July 2014*. Winona Lake: Eisenbrauns, 2017, pp. 105–117.

Paul 1997: Shalom M. Paul, "A Lover's Garden of Verse: Literal and Metaphorical Imagery in Ancient Near Eastern Love Poetry", in Mordechai Cogan, Barry L. Eichler, and Jeffrey H. Tigay (eds.), *Tehillah le-Mosche. Biblical and Judaic Studies in Honor of Mosche Greenberg*. Winoka Lake: Eisenbrauns, 1997, pp. 99–110.

Paul 2002: Shalom M. Paul, "The Shared Legacy of Sexual Metaphors and Euphemisms in Mesopotamian and Biblical Literature", in Simo Parpola and Robert M. Whiting (eds.), *Sex and Gender in the Ancient Near East. Proceedings of the 47th Rencontre Assyriologique Internationale, Helsinki, July 2–6, 2001*. Helsinki: Neo-Assyrian Text Corpus Project, 2002, pp. 489–498.

Peled 2016: Ilan Peled, *Masculinities and Third Gender. The Origins and Nature of an Institutionalized Gender Otherness in the Ancient Near East* (Alter Orient und Altes Testament 435). Münster: Ugarit-Verlag, 2016.

Peled 2018: Ilan Peled, "Identifying Gender Ambiguity in Texts and Artifacts", in Stephanie Lynn Budin, Megan Cifarelli, Agnès Garcia-Ventura, and Adelina Millet Albà (eds.), *Gender and Methodology in the Ancient Near East. Approaches from Assyriology and beyond*. Barcelona: Edicions de la Universitat de Barcelona, 2018, pp. 55–64.

Perdibon 2019: Anna Perdibon, *Mountains and Trees, Rivers and Springs. Animist Beliefs and Practices in Mesopotamian Religion* (Leipziger altorientalistische Studien 11). Wiesbaden: Harrassowitz, 2019.

Peterson 2008: Jeremiah Peterson, "An Early ša$_3$-zi-ga Prescription from Nippur", *Zeitschrift für Assyriologie und vorderasiatische Archäologie* 98, 2 (2008), pp. 195–200.

Pettinato 1971: Giovanni Pettinato, *Das altorientalische Menschenbild und die sumerischen und akkadischen Schöpfungsmythen*. Heidelberg: Carl Winter Universitätsverlag, 1971.

Pettinato 1981: Giovanni Pettinato, *Testi lessicali monolingui della biblioteca L. 2769* (Materiali Epigrafici di Ebla 3): Napoli: Istituto universitario orientale, 1981.

Pfitzner 2019: Judith Pfitzner, "Holy Cow! On Cattle Metaphors in Sumerian Literary Texts", in Mattila et al. 2019: 137–173.

Pinches 1910: Theophilus G. Pinches, "Notes upon the Fragments of Hittite Cuneiform Tablets from Yuzgat, Boghaz Keui", *Annals of Archaeology and Anthropology* 3 (1910), pp. 99–106.

Pizza 2005: Giovanni Pizza, *Antropologia medica. Saperi, pratiche e politiche del corpo*. Roma: Carocci, 2005.

Pizza 2012: Giovanni Pizza, "La medicina popolare, una riflessione", in Donatella Cozzi (ed.), *Le parole dell'antropologia medica. Piccolo dizionario*. Perugia: Morlacchi, 2012, pp. 181–204.

Poebel 1933–1934: Arno Poebel, "Eine sumerische Inschrift Samsuilunas über die Erbauung der Festung Dur-Samsuiluna", *Archiv für Orientforschung* 9 (1933–1934), pp. 241–292.

Pollock 2008: Susan Pollock, "Wer hat Angst vorm bösen Wolf? Gendesr und Feminismus in der vorderasiatischen Archäologie", in Rainer Brunner, Jens P. Laut, and Maurus Reinkowski (eds.), *XXX. Deutscher Orientalistentag. Orientalistik im 21. Jahrhundert. Welche Vergangenheit, Welche Zukunft*. Freiburg: Albert-Ludwigs-Universität, 2008, pp. 1–16.

Polvani 1980: Anna M. Polvani, "La pietra ZA.GÌN nei testi di Hattusa", *Mesopotamia* 15 (1980), pp. 73–91.

Ponchia 2019: Simonetta Ponchia, "Gilgameš and Enkidu. The Two-thirds-god and the Two-thirds-animal", in Mattila et al. 2019: 187–210.

Ponchia and Luukko 2013: Simonetta Ponchia and Mikko Luukko, *The Standard Babylonian Myth of Nergal and Ereškigal* (State Archives of Assyria Cuneiform Texts 8). Helsinki: The Neo-Assyrian Text Corpus Project, 2013.

Pool and Geissler 2005: Robert Pool and Wenzel Geissler, *Medical Anthropology*. Maidenhead: Open University Press, 2005.

Porter 2002: Barbara N. Porter, "Beds, Sex, and Politics: The Return of Marduk's Bed to Babylon", in Simo Parpola and Robert M. Whiting (eds.), *Sex and Gender in the Ancient Near East. Proceedings of the 47th Rencontre Assyriologique Internationale, Helsinki, July 2–6, 2001*. Helsinki: Neo-Assyrian Text Corpus Project, 2002, pp. 523–535.

Postgate 1973: John N. Postgate, "Assyrian texts and Fragments", *Iraq* 35 (1973), pp. 13–36.

Postgate 1987: John N. Postgate, "Notes on Fruit in the Cuneiform Sources", *Bulletin on Sumerian Agriculture* 3 (1987), pp. 115–144.

Postgate 1992: John N. Postgate, "Trees and Timber in the Assyrian Texts", *Bulletin on Sumerian Agriculture* 6 (1992), pp. 177–192.

Postgate 1997: John N. Postgate, "Mesopotamian Petrology: Stages in the Classification of the Material World", *Cambridge Archaeological Journal* 7, 2 (1997), pp. 205–224.

Powell 1992: Marvin G. Powell, "Timber Production in Presargonic Lagaš", *Bulletin on Sumerian Agriculture* 6 (1992), pp. 99–122.

Powell 1993: Marvin G. Powell, "Drugs and Pharmaceuticals in Ancient Mesopotamia", in Irene Jacob and Walter Jacob (eds.), *The Healing Past. Pharmaceuticals in the Biblical and Rabbinic World* (Studies in Ancient Medicine 7). Leiden, Boston: Brill, 1993, pp. 47–67.

Prechel 1996: Doris Prechel, *Die Göttin Išḫara. Ein Beitrag zur altorientalischen Religionsgeschichte* (Abhandlungen zur Literatur Alt-Syrien-Palästinas und Mesopotamiens 11). Münster: Ugarit-Verlag, 1996.

Prechel and Richter 2011: Doris Prechel and Thomas Richter, "Abrakadabra oder Althurritisch: Betrachtungen zu einigen altbabylonishen Beschwörungstexten", in Thomas Richter, Doris Prechel, and Jörg Klinger (eds.), *Kulturgeschichten. Altorientalische Studien für Volkert Haas zum 65 Geburtstag*. Saarbrücken: Saarbrücker Druckerei und Verlag, 2011, pp. 333–372.

Preciado 2001: Paul B. Preciado, *Manifeste contra-sexuel*. Paris: Balland, 2001.

Prince 1982: Raymond Prince, "Shamans and Endorphins: Hypotheses for a Synthesis", *Ethos* 10, 4 (1982), pp. 409–423.

Radner 1997: Karen Radner, *Die Neuassyrischen Privatrechtsurkunden als Quelle für Mensch und Umwelt* (State Archives of Assyria Studies 6). Helsinki: The Neo-Assyrian Text Corpus Project, 1997.

Reiner 1956: Erica Reiner, "*Lipšur* Litanies", *Journal of Near Eastern Studies* 15 (1956), pp. 129–149.

Reiner 1958: Erica Reiner, *Šurpu. A Collation of Sumerian and Akkadian Incantations* (Archiv für Orientforschung Beiheft 11). Graz: Im Selbstverlage des Herausgebers, 1958.

Reiner 1960: Erica Reiner, "Fortune Telling in Mesopotamia", *Journal of Near Eastern Studies* 19 (1960), pp. 23–35.

Reiner 1974a: Erica Reiner, "A Sumero-Akkadian Hymn of Nanâ", *Journal of Near Eastern Studies* 33 (1974), pp. 221–236.

Reiner 1974b: Erica Reiner, *The Series* ḪAR-ra = ḫubullu. *Tablets XX-XXIV* (Materialien zum sumerischen Lexikon 11). Rome: Biblical Institute Press, 1974.

Reiner 1985: Erica Reiner, "The Heart Grass", in Eadem, *Your Thwarts in Pieces, Your Mooring Rope Cut. Poetry from Babylonia and Assyria*. Ann Arbor: University of Michigan, 1985, pp. 94–100.

Reiner 1995: Erica Reiner, *Astral Magic in Babylonia* (Transactions of the American Philosophical Society Philadelphia, New Series 85, 4). Philadelphia: American Philosophical Society, 1995.

Reisner 1896: George A. Reisner, *Sumerisch-babylonische Hymnen nach Tontafeln griechischer Zeit*. Berlin: W. Spemann, 1896.

Rendu Loisel 2013: Anne-Caroline Rendu Loisel, "Heurs et malheurs du jardinier dans la littérature sumérienne", in Daniel Barbu, Philippe Borgeaud, and Youri Volokhine (eds.), *Mondes Clos. Cultures et Jardins* (Supplément à Asdiwal Revue Genevoise d'Anthropologie et d'Histoire des Religions 1). Gollion: Infolio, pp. 67–84.

Rendu Loisel 2016: Anne-Caroline Rendu Loisel, *Les chants du monde. Le paysage sonore de l'ancienne Mésopotamie*. Paris: Presses universitaires du Midi, 2016.

Reynolds 2010: Frances Reynolds, "A Divine Body: New Joins from the Sippar Collection", in Heather D. Baker, Eleanor Robson, and Gábor Zólyomi (eds.), *Your Praise Is Sweet. A Memorial Volume for Jeremy Black from Students, Colleagues and Friends*. London: British Institute for the Study of Iraq, 2010, pp. 291–302.

Reynolds and Kinnier Wilson 2004: Edward H. Reynolds and James V. Kinnier Wilson, "Stroke in Babylonia", *Arch Neurol* 61 (2004), pp. 597–601.

Reynolds and Kinnier Wilson 2008: Edward H. Reynolds and James V. Kinnier Wilson, "Psychoses of Epilepsy in Babylon. The Oldest Account of the Disorder", *Epilepsia* 49 (2008), pp. 1488–1490.

Reynolds and Kinnier Wilson 2012: Edward H. Reynolds and James V. Kinnier Wilson, "Obsessive Compulsive Disorder and Psychopathic Behavior in Babylon", *J Neurol Neurosurg Psychiatry* 83 (2012) 199–201.

Reynolds and Kinnier Wilson 2014: Edward H. Reynolds and James V. Kinnier Wilson, "Neurology and Psychiatry in Babylon", *Brain. A Journal of Neurology* 137 (2014), pp. 2611–2619.

Rhodes 1990: Lorna A. Rhodes "Studying Biomedicine as a Cultural System", in Carolyn F. Sargent and Thomas M. Johnson (eds.), *Medical Anthropology. Contemporary Theory and Method*. Westport, London: Praeger, 1990, pp. 165–180.

Rich 1980: Adrienne Rich, "Compulsory Heterosexuality and Lesbian Existence", *Signs: Journal of Women in Culture and Society* 5, 4 (1980), pp. 631–660.

Rich 1987: Adrienne Rich, "Notes Toward a Politics of Location", in Eadem, *Blood, Bread and Poetry. Selected Prose 1979–1985*. London: Virago Press, 1987, pp. 210–231.

Richardson 2019: Seth Richardson, "Nature Engaged and Disengaged: The Case of Animals in Mesopotamian Literatures", in Tristan Schmidt and Johannes Pahlitzsch (eds.), *Impious Dogs, Ridiculous Monkeys and Exquisite Fish. Evaluative Perception and Interpretation of Animals in Ancient and Medieval Thought*. Berlin, Boston: De Gruyter, pp. 11–40.

MRichter 1973: Maurice N. Richter, *Science as a Cultural Process*. London: F. Muller, 1973.

TRichter 2004: Thomas Richter, *Untersuchungen zu den lokalen Panthea Süd- und Mittelbabyloniens* (Alter Orient und Altes Testament 257). Münster: Ugarit-Verlag.

TRichter 2014–2016: Thomas Richter, "Tutu", in *Reallexikon der Assyriologie und vorderasiatischen Archäologie*, vol. 14. Berlin: De Gruyter, 2014–2016, pp. 241–242.

Riemschneider 2004: Kaspar K. Riemschneider, *Die akkadischen und hethitischen Omentexte aus Boğazköy* (Dresdner Beiträge zur Hethitologie 12). Dresden: Technische Universität Dresden, 2004.

Ritter 1965: Edith K. Rittter, "Magical Expert (*āšipu*) and Physician (*asû*): Notes on Two Complementary Professions in Babylonian Medicine", in Hans G. Güterbock and Thorkild Jacobsen (eds.), *Studies in Honor of Benno Landsberger on His Seventy-Fifth Birthday, April 21, 1965* (Assyriological Studies 16). Chicago: University of Chicago Press, 1965, pp. 299–321.

Ritter and Kinnier Wilson 1980: Edith K. Rittter and James V. Kinnier Wilson, "Prescription for an Anxiety State: A Study of BAM 234", *Anatolian Studies* 30 (Special Number in Honour of the Seventieth Birthday of Professor O.R. Gurney, 1980), pp. 23–30.

Rivers 1924: William H. Rivers, *Medicine, Magic and Religion*. New York: Harcourt, Brace and Co., 1924.

Robson 2008: Eleanor Robson, "Mesopotamian Medicine and Religion: Current Debates, New Perspectives", *Religion Compass* 2, 4 (2008), pp. 455–483.

Rochberg 2015: Francesca Rochberg, "The Babylonians and the Rational: Reasoning in Cuneiform Scribal Scholarship", in J. Cale Johnson (ed.), *In the Wake of the Compendia. Infrastructural Contexts and the Licensing of Empiricism in Ancient and Medieval Mesopotamia* (Science, Technology, and Medicine in Ancient Cultures 3). Berlin, Boston: De Gruyter, pp. 209–246.

Rölling 1987–1990b: Wolfgang Rölling, "Lapislazuli A. Philologisch", in *Reallexikon der Assyriologie und vorderasiatischen Archäologie*, vol. 6. Berlin: De Gruyter, 1987–1990, pp. 488–489.

Römer 1965: Willem H.Ph. Römer, *Sumerische 'Königshymnen' der Isin-Zeit*. Leiden: Brill, 1965.

Römer 1969: Willem H.Ph. Römer, "Einige Beobachtungen zur Göttin Nini(n)sina auf Grund von Quellen der Ur III-Zeit und der altbabylonischen Periode", in Wolfgang Rölling (ed.), Lišān mitḫurti. *Festschrift Wolfram Freiherr von Soden zum 19.VI.1968 gewidmet von Schülern und Mitarbeiten* (Alter Orient und Altes Testament 1). Kevelaer: Butzon & Bercker, 1969, pp. 279–305.

Römer 1988: Willem H.Ph. Römer, "Sumerische Hymnen II", *Bibliotheca Orientalis* 45 (1988), pp. 24–60.

Römer 2001: Willem H.Ph. Römer, *Hymnen und Klagelieder in sumerischer Sprache* (Alter Orient und Altes Testament 276). Münster: Ugarit-Verlag, 2001.

Rosaldo 1982: Michelle Z. Rosaldo, "The Things We Do with Words: Ilongot Speech Acts and Speech Act Theory in Philosophy", *Language in Society* 11 (1982), pp. 203–237.

Rosaldo 1984: Michelle Z. Rosaldo, "Toward an Anthropology of Self and Feeling", in Richard A. Shweder and Robert A. LeVine (eds.), *Culture Theory. Essay on Mind, Self and Emotion*. Cambridge: Cambridge University Press, 1984, pp. 137–157.

Roth 1987: Martha T. Roth, "Age at Marriage and the Household: A Study of Neo-Babylonian and Neo-Assyrian Forms", *Comparative Studies in Society and History* 29 (1987), pp. 715–747.

Rubin 1975: Gayle Rubin, "The Traffic in Women: Notes on the 'Political Economy' of Sex", in Rayna R. Reiter (ed.), *Toward an Anthropology of Women*. New York: Monthly Review, 1975, pp. 157–210.

Rumor 2015: Maddalena Rumor, *Babylonian Pharmacology in Graeco-Roman Dreckapotheke. With an Edition of Uruanna III 1–143 (138)*. Ph.D. dissertation, Freie Universität Berlin, 2015.

Rumor 2017: Maddalena Rumor, "The 'AŠ section' of Uruanna III in Partitur", *Journal des Médecines Cunéiformes* 29 (2017), pp. 1–34.

Salin 2020: Silvia Salin, *Le espressioni della sofferenza individuale nei testi assiro-babilonesi*. Verona: Alteritas, 2020.

Salin forthcoming: Silvia Salin, "The Akkadian *libbu* and Concepts Related to the Centre of the Body", in Antonio Panaino, Claudia Fabrizio, Hans Christian Luschützky, Céline Redard, and Velizar Sadovski (eds.), *Linguistic Studies of Iranian and Indo-European Languages. Proceedings of the Symposium in memoriam Xavier Tremblay (1971–2011)*. Wien: Österreichische Akademie der Wissenschaften Wien.

Sallaberger 1993: Walther Sallaberger, *Der kultische Kalender der Ur III-Zeit*, 2 vols (Untersuchungen zur Assyriologie und vorderasiatische Archäologie 7). Berlin, Boston: De Gruyter.

Sallaberger 2000: Walther Sallaberger, "Das Erscheinen Marduks als Vorzeichen: Kultstatue und Neujahrfest in der Omenserie *Šumma ālu*", *Zeitschrift für Assyriologie und vorderasiatische Archäologie* 90, 2 (2000), pp. 227–262.

Sallaberger 2007: Walther Sallaberger, "Reinheit. A. Mesopotamien", in *Reallexikon der Assyriologie und vorderasiatischen Archäologie*, vol. 11. Berlin: De Gruyter, 2007, pp. 295–299.

Sallaberger 2011: Walther Sallaberger, "Körperliche Reinheit und soziale Grenzen in Mesopotamien", in Peter Burschel and Christoph Marx (eds.), *Reinheit*. Wien, Köln, Weimer: Böhlau Verlag Ges.m.b.H. und Co.KG, 2011, pp. 17–45.

Salonen 1939: Armas Salonen, *Die Wasserfahrzeuge in Babylonien nach sumerisch-akkadischen Quellen (mit besonderer Berücksichtigung der 4. Tafel der Serie* ḪAR-ra = ḫubullu): *Eine lexikalische und kulturgeschichtliche Untersuchung* (Studia Orientalia 8, 4). Hensilki: Societas Orientalis Fennica, 1939.

Salonen 1973: Armas Salonen, *Vögel und Vogelfang im Alten Mesopotamien*. Helsinki: Academia Scientiarum Fennica, 1973.

Sanders 2001: Seth L. Sanders, "A Historiography of Demons: Preterit-Thema, Para-Myth, and Historiola in the Morphology of Genres", in Tzvi Abusch, Paul-Alain Beaulieu, John Huehnergard, Peter Machinist, and Piotr Steinkeller (eds.), *Proceedings of the XLV*[e] *Rencontre Assyriologique Internationale*. Part I: *Historiography in the Cuneiform World*. Bethesda: CDL Press, 2001, pp. 429–440.

Saunders 1993: George R. Saunders "'Critical Ethnocentrism' and the Ethnology of Ernesto De Martino", *American Anthropologist* 95, 4 (1993), pp. 875–893.

Saxl 1957: Fritz Saxl, "Macrocosm and Microcosm in Medieval Pictures", in Idem, *Lectures*, vol. 1. London: The Warburg Institute – University of London, 1957, pp. 58–72.

Scarry 1985: Elaine Scarry, *The Body in Pain. The Making and Unmaking of the World*. Oxford: Oxford University Press, 1985.

Scheil 1921: Vincent Scheil, "Catalogue de la Collection Eugène Tisserant", *Revue d'Assyriologie et d'Archéologie orientale* 18 (1921), pp. 1–48.

Scheper-Hughes and Lock 1987: Nancy Scheper-Hughes and Margaret Lock, "The Mindful Body: A Prolegomenon to Future Work in Medical Anthropology", *Medical Anthropology Quarterly* N.S. 1, 1 (1987), pp. 6–41.

Schirripa 1996: Pino Schirripa, "Promesse e minacce della medicina tradizionale africana", *AM. Rivista della società italiana di antropologia medica* 1–2 (1996), pp. 155–178.

Schirripa 2015: Pino Schirripa, *La vita sociale dei farmaci. Produzione, circolazione, consumo degli oggetti materiali della cura*. Lecce: Argo, 2015.

Schmidtchen 2018: Eric Schmidtchen, "Esagil-kīn-apli's Catalogue of *Sakikkû* and *Alamdimmû*", in Ulrike Steinert (ed.), *Assyrian and Babylonian Scholarly Text Catalogues* (Die babylonisch-assyrische Medizin in Texten und Untersuchungen 9). Boston, Berlin: De Gruyter, 2018, pp. 137–157.

Schuster-Brandis 2008: Anais Schuster-Brandis, *Steine als Schutz- und Heilmittel. Untersuchung zu ihrer Verwendung in der Beschwörungskunst Mesopotamiens im 1. Jt. vor Chr.* (Alter Orient und Altes Testament 46). Münster: Ugarit-Verlag, 2008.

Schwemer 1998: Daniel Schwemer, *Akkadische Rituale aus Ḫattuša. Die Sammeltafel K Bo XXXVI 29 und verwandte Fragmente* (Texte des Hethiter 23). Heidelberg: Carl Winter Universitätsverlag, 1988.

Schwemer 2001: Daniel Schwemer, *Die Wettergottgestalten Mesopotamiens und Nordsyriens im Zeitalter der Keilschriftfulturen*. Wiesbaden: Harrassowitz, 2001.

Schwemer 2004: Daniel Schwemer, "Ein akkadischer Liebeszauber aus Ḫattuša", *Zeitschrift für Assyriologie und vorderasiatische Archäologie* 94, 1 (2004), pp. 59–79.
Schwemer 2006–2008: Daniel Schwemer, "Šāla. A. Philologisch", in *Reallexikon der Assyriologie und vorderasiatischen Archäologie*, vol. 11. Berlin: De Gruyter, 2006–2008, pp. 565–567.
Schwemer 2007a: Daniel Schwemer, *Abwehrzauber und Behexung. Studien Zum Schadenzauberglauben im alten Mesopotamien*. Wiesbaden: Harrassowitz.
Schwemer 2007b: Daniel Schwemer, *Rituale und Beschwörungen gegen Schadenzauber* (Keilschrifttexte aus Assur literarischen Inhalts 2, Wissenschaftliche Veröffentlichungen der Deutschen Orient-Gesellschaft 117). Wiesbaden: Harrassowitz, 2007.
Schwemer 2009: Daniel Schwemer, "Washing, Defiling and Burning: Two Bilingual Anti-Witchcraft Incantations", *Orientalia* N.S. 78 (2009), pp. 44–68.
Schwemer 2010: Daniel Schwemer, "Therapien gegen Impotenz", in Bernd Janowski and Daniel Schwemer (Eds.), *Texte zur Heilkunde* (Texte aus der Umwelt des Alten Testaments Neue Folge 5). München: Gütersloher Verlagshaus, 2010, pp. 115–122.
Schwemer 2013: Daniel Schwemer, "Gauging the Influence of Babylonian Magic: the Reception of Mesopotamian Traditions in Hittite Ritual Practice", in Eva Cancik-Kirschbaum, and Gerfrid G.H. Müller (eds.), *Diversity and Standardization. Perspectives on Ancient Near Eastern Cultural History*. Berlin, Boston: De Gruyter, 2013, pp. 145–171.
Schwemer 2014: Daniel Schwemer, "'From Follows Function'? Rhetoric and Poetic Language in First Millennium Akkadian Incantations", *Die Welt des Orients* 44 (2014), pp. 263–288.
Scott 1986: Joan W. Scott, "Gender: A Useful Category of Historical Analysis", *The American Historical Review* 91, 5 (1986), pp. 1053–1105.
Scurlock 1989–1990: JoAnn Scurlock, "Was There a "Love-Hungry" Ēntu-priestess Named Eṭirtum?", *Archiv für Orientforschung* 36 (1989–1990), pp. 107–112.
Scurlock 1999: JoAnn Scurlock, "Physician, Exorcist, Conjurer, Magician: A Tale of Two Healing Professionals", in Tzvi Abusch and Karel van der Toorn (eds.), *Mesopotamian Magic. Textual, Historical, and Interpretative Perspectives* (Ancient Magic and Divination 1). Groningen: Styx, 1999, pp. 69–79.
Scurlock 2002a: JoAnn Scurlock, "Animals in Ancient Mesopotamian Religion", in BCollins 2002: 361–388.
Scurlock 2002b: JoAnn Scurlock, "Animal Sacrifice in Ancient Mesopotamian Religion", in BCollins 2002: 389–404.
Scurlock 2005–2006: JoAnn Scurlock, "Sorcery in the Stars: STT 300, BRM 4, 19–20 and the Mandaic Book of the Zodiac", *Archiv für Orientforschung* 51 (2005–2006), pp. 125–146.
Scurlock 2006a: JoAnn Scurlock, *Magico-medical Means of Treating Ghost-induced Illnesses in Ancient Mesopotamia* (Ancient Magic and Divination 3). Leiden, Boston: Brill, 2006.
Scurlock 2006b: JoAnn Scurlock, "Whatever Possessed Them? Progress and Regress in the History of Medicine", *Journal des Médecines Cunéiformes* 7 (2006), pp. 11–6.
Scurlock 2007: JoAnn Scurlock, "A Proposal for Identification of a Missing Plant: *Kamantu*/ Ú ÁB.DUḪ = *Lawsonia inermis* L./'henna'", *Wiener Zeitschrift für die Kunde des Morgenlandes* 97 (2007), pp. 491–520.
Scurlock 2011: JoAnn Scurlock, "Bestiality I: Ancient Near East", *Encyclopaedia of the Bible and Its Reception* 3. Berlin, New York: De Gruyter, pp. 935–938.
Scurlock 2014a: JoAnn Scurlock, "Medicine and Healing Magic", in Mark W. Chavalas (ed.), *Women in the Ancient Near East. A Sourcebook*. London, New York: Routledge, 2014, pp. 101–143.
Scurlock 2014b: JoAnn Scurlock, *Sourcebook for Ancient Mesopotamian Medicine* (Writings from the Ancient World 36). Atlanta: Society of Biblical Literature Press, 2014.
Scurlock 2020: JoAnn Scurlock, "Blind Mice and Despairing Rats: The Uses of *kurkanû*-Turmeric in Ancient and Modern Medicine", *Journal des Médecines Cunéiformes* 35 (2020), pp. 34–68.

Scurlock and Andersen 2005: JoAnn Scurlock and Burton R. Andersen, *Diagnoses in Assyrian and Babylonian Medicine. Ancient Sources, Translations, and Modern Medical Analyses*. Urbana, Chicago: University of Illinois Press, 2005.

Searle 1969: John R. Searle, *Speech Acts. An Essay in the Philosophy of Language*. Cambridge: Cambridge University Press, 1969.

Searle 1975: John R. Searle, "A Taxonomy of Illocutionary Acts", *Language in Society* 5, 1 (1975), pp. 1–23.

Sefati 1998: Yitzhak Sefati, *Love Songs in Sumerian Literature. Critical Edition of the Dumuzi-Inanna Songs*. Rāmat-Gan: Bar-Ilan University Press, 1988.

Sefati and Klein 2002: Yitzhak Sefati and Jacob Klein, "The Role of Women in Mesopotamian Witchcraft", in Simo Parpola and Robert M. Whiting (eds.), *Sex and Gender in the Ancient Near East. Proceedings of the 47th Rencontre Assyriologique Internationale, Helsinki, July 2–6, 2001*. Helsinki: Neo-Assyrian Text Corpus Project, 2002, pp. 569–587.

Selz 2019: Gebhard J. Selz, "Reflections on the Pivotal Role of Animals in Early Mesopotamia", in Mattila et al. 2019: 23–56.

Seminara 2001: Stefano Seminara, *La versione accadica del Lugal-e. La tecnica babilonese della traduzione dal sumerico e le sue 'regole'* (Materiali per il vocabolario sumerico 8). Roma: Università degli studi di Roma "La Sapienza", Dipartimento di studi orientali, 2001.

ASeppilli 1962: Anira Seppilli, *Poesia e magia*. Torino: Einaudi, 1962 (reprint: 1971).

TSeppilli 1996: Tullio Seppilli, "Antropologia medica: i fondamenti per una strategia", *AM. Rivista della Società italiana di antropologia medica* 1–2 (1996), pp. 7–22.

Seux 1976 : Marie-Joseph Seux, *Hymnes et prières aux dieux de Babylonie et d'Assyrie* (Littératures anciennes du Proche-Orient 8). Paris: Éditions du Cerf, 1976.

Severi 2004: Carlo Severi, *Il percorso e la voce. Un'antropologia della memoria*. Torino: Einaudi, 2004.

Sharifian et al. 2008: Farzad Sharifian, René Dirven, Ning Yu, and Susanne Niemeier (eds.), *Culture, Body, and Language Conceptualizations of Internal Body Organs across Cultures and Languages* (Applications of Cognitive Linguistics 7). Berlin: De Gruyter, 2008.

Sibbing Plantholt 2017: Irene Sibbing-Plantholt, "Black Dogs in Mesopotamia and Beyond", in David Kertai and Olivier Nieuwenhuyse (eds.), *From the Four Corners of the Earth. Studies in the Iconography of the Ancient Eastern Mediterranean and Near East in Honour of F.A.M. Wiggermann* (Alter Orient und Altes Testament 441). Münster: Ugarit-Verlag, 2017, pp. 165–180.

Simon 2017: Zsolt Simon, "Why Did Paškuwatti's Patient Fail in the Matrimonial Bed?", in Olga Drewnowska and Małgorzata Sandowicz (eds.), *Fortune and Misfortune in the Ancient Near East. Proceedings of the 60th Rencontre Assyriologique Internationale at Warsaw 21–25 July 2014*. Winona Lake: Eisenbrauns, 2017, pp. 97–103.

Sindzingre and Zempléni 1981: Nicole Sindzingre and Andras Zempléni, "Modèles et pragmatique, activation et répétition : réflexions sur la causalité de la maladie chez les Senoufo de Côte d'Ivoire", *Social Science & Medicine* 15, 3 (1981), pp. 279–293.

Singer 1990: Merrill Singer, "Reinventing Medical Anthropology: Toward a Critical Realignment", *Social Science and Medicine* 30, 2 (1990, pp. 179–187.

Sjöberg 1975: Åke W. Sjöberg, "in-nin šà-gur$_4$-ra. A Hymn to the Goddess Inanna by the En-Priestess Enḫeduanna", *Zeitschrift für Assyriologie und vorderasiatische Archäologie* 65, 2 (1975), pp. 161–253.

Sjöberg 1977: Åke W. Sjöberg, "Miscellaneous Sumerian Texts II", *Journal of Cuneiform Studies* 29 (1977), pp. 3–45.

Sjöberg and Bergmann 1969: Åke W. Sjöberg and Eugen Bergmann, *The Collection of the Sumerian Temple Hymns* (Texts from Cuneiform Sources 3), Locust Valley, New York: J.J. Augustin, 1969.

Sommer and Ehelolf 1924: Ferdinand Sommer and Hans Ehelolf, *Das hethitische Ritual des Pāpanikri von Komana (KBo V 1 = Bo 2001). Text, Übersetzungsversuch, Erläuterungen* (Boghazköi-Studien 10). Leipzig: Hinrichs, 1924.

Sommerfeld 1982: Walter Sommerfeld, *Der Aufstieg Marduks. Die Stellung Marduks in der babylonischen Religion des zweiten Jahrtausends v. Chr.* (Alter Orient und Altes Testament 213). Kevelaer: Butzon und Bercker, 1982.

Sontag 1978: Susan Sontag, *Illness as Metaphor*. New York: Farrar, Straus & Giroux, 1978.

Sontag 1989: Susan Sontag, *AIDS and Its Metaphors*. New York: Farrar, Straus & Giroux, 1989.

Spivak 1988: Gayatri Ch. Spivak, "Can the Subaltern Speak?", in Cary Nelson and Lawrence Grossberg (eds.), *Marxism and the Interpretation of Culture*. Urbana: University of Illinois Press, 1988, pp. 271–313.

Stadhouders 2011: Henry Stadhouders, "The Pharmacopoeia Handbook *Šammu šikinšu* – An Edition", *Journal des Médecines Cunéiformes* 18 (2011), pp. 1–55.

Stadhouders 2012: Henry Stadhouders, "The Pharmacopoeia Handbook *Šammu šikinšu* – A Translation", *Journal des Médecines Cunéiformes* 19 (2012), pp. 3–51.

Stadhouders forthcoming: Henry Stadhouders, "Edition of STT 95+295", *Journal des Médecines Cunéiformes*.

Stamm 1939: Johann J. Stamm, *Die akkadische Namengebung* (Mitteilingen der vorderasiatisch-aegytischen Gesellschaft 44). Leipzig: Hinrichs, 1939.

Steinert 2012a: Ulrike Steinert, *Aspekte des Menschseins im Alten Mesopotamien. Eine Studie zu Person und Indentität im 2. und 1. Jt. v. Chr.* (Cuneiform Monographs 44). Leiden, Boston: Brill, 2012.

Steinert 2012b: Ulrike Steinert, "K. 263+10934, A Tablet with Recipes Against the Abnormal Flow of a Woman's Blood", *Sudhoffs Archiv. Zeitschrift für Wissenschaftsgeschichte* 96, 1 (2012), pp. 64–94.

Steinert 2012c: Ulrike Steinert, "'Zwei Drittel Gott, ein Drittel Mensch': Überlegungen zum altmesopotamischen Menschenbild", in Bernd Janowski (ed.), *Der ganze Mensch. Zur Anthropologie der Antike und ihrer europäischen Nachgeschichte*. Berlin: Akademie-Verlag, 2012, pp. 59–81.

Steinert 2013: Ulrike Steinert, "Fluids, Rivers, and Vessels: Metaphors and Body Concepts in Mesopotamian Gynaecological Texts", *Jounal des Médecines Cunéiformes* 22 (2013), pp. 1–23.

Steinert 2014: Ulrike Steinert, review of Böck 2014, *Journal of Near Eastern Studies* 73 (2014), pp. 357–364.

Steinert 2015: Ulrike Steinert, "'Tested' Remedies in Mesopotamian Medical Texts: A Label for Efficacy Based on Empirical Observation?", in J. Cale Johnson (ed.), *In the Wake of the Compendia. Infrastructural Contexts and the Licensing of Empiricism in Ancient and Medieval Mesopotamia* (Science, Technology, and Medicine in Ancient Cultures 3). Berlin, Boston: De Gruyter, pp. 103–145.

Steinert 2016: Ulrike Steinert, "Körperwissen, Tradition und Innovation in der babylonischen Medizin", in Almut-Barbara Renger and Christoph Wulf (eds.), *Körperwissen. Transfer und Innovation, Paragrana. Internationale Zeitschrift für Historische Anthropologie* 25/1, Berlin: De Gruyter, 2016, pp. 195–254.

Steinert 2017a: Ulrike Steinert, "Cows, Women and Wombs: Interrelations between Texts and Images from the Ancient Near East", in David Kertai and Olivier Nieuwenhuyse (eds.), *From the Four Corners of the Earth. Studies in the Iconography of the Ancient Eastern Mediterranean and Near East in Honour of F.A.M. Wiggermann* (Alter Orient und Altes Testament 441). Münster: Ugarit-Verlag, 2017, pp. 205–258.

Steinert 2017b: Ulrike Steinert, "Concepts of the Female Body in Mesopotamian Gynaecological Texts", in John Z. Wee (ed.), *The Comparable Body. Analogy and Metaphor in Ancient*

Mesopotamian, Egyptian, and Greco-Roman Medicine (Studies in Ancient Medicine 49). Leiden, Boston: Brill,2017, pp. 275–357.

Steinert 2018a: Ulrike Steinert, "Catalogues, Texts and Specialists: Some Thoughts on the Assur Medical Catalogue, Mesopotamian Medical Texts and Healing Professions", in Greta Van Buylaere, Mikko Luukko, Daniel Schwemer, and Avigail Mertens-Wagschal (eds.), *Sources of Evil. Studies in Mesopotamian Exorcistic Lore* (Ancient Magic and Divination 15). Leiden, Boston: Brill, 2018, pp. 48–132 = in Ulrike Steinert. (ed.), *Assyrian and Babylonian Scholarly Text Catalogues* (Die babylonisch-assyrische Medizin in Texten und Untersuchungen 9). Boston, Berlin: De Gruyter, 2018, pp. 158–200.

Steinert 2018b: Ulrike Steinert, "The Assur Medical Catalogue (AMC)", in Eadem (ed.), *Assyrian and Babylonian Scholarly Text Catalogues* (Die babylonisch-assyrische Medizin in Texten und Untersuchungen 9). Boston, Berlin: De Gruyter, 2018, pp. 203–291.

Steinert 2020: Ulrike Steinert, "Pounding Hearts and Burning Livers: The 'Sentimental Body' in Mesopotamian Medicine and Literature", in Shih-Wei Hsu and Jaume Llop Raduà (eds.), *The Expression of Emotions in Ancient Egypt and Mesopotamia* (Culture and History of the Ancient Near East 116). Leiden, Boston: Brill, 2020, pp. 410–469.

Steinkeller 1987: Piotr Steinkeller, "The Foresters of Umma: Toward a Definition of Ur III Labor", in Marvin G. Powell (ed.), *Labor in Ancient Near East* (American Oriental Series 68). New Haven: American Oriental Society, 1987, pp. 73–115.

Stengers 1995: Isabelle Stengers, "Le médecin et le charlatan", in Tobie Nathan and Isabelle Stengers, *Médecins et sorciers*. Paris: Les Empêcheurs de penser en rond, 1995.

Stiehler-Alegria 2006: Gisela Stiehler-Alegría, "Weinraute oder Harmelraute, welche Spezies verbirgt sich hinter den Termini *šibburratu*, *š'mbr'* und *sauma*?", *Altorientalische Forschungen* 33 (2006), pp. 125–143.

Stol 1971: Martin Stol, "Notes brèves: Le 'roitelet' et l'éléphant", *Revue d'Assyriologue et d'Archéologie orientale* 65 (1971), p. 180.

Stol 1987–1990: Martin Stol, "Malz", in *Reallexikon der Assyriologie und vorderasiatischen Archäologie*, vol. 7. Berlin: De Gruyter, 1987–1990, pp. 322–229.

Stol 1991–1992: Martin, Stol, "Diagnosis and Therapy in Babylonian Medicine", *Jaarbericht van het Voor-Aziatisch-Egyptisch-Genootschap Ex Oriente Lux* 32 (1991–1992), pp. 42–65.

Stol 1993: Martin Stol, *Epilepsy in Babylonia* (Cuneiform Monographs 2). Groningen: Styx, 1993.

Stol 1995: Martin Stol, "Woman in Mesopotamia", *Journal of the Economic and Social History of the Orient* 38 (1995), pp. 123–144.

Stol 1998–2001: Martin Stol, "Nanaja", in *Reallexikon der Assyriologie und vorderasiatischen Archäologie*, vol. 9. Berlin: De Gruyter, 1998–2001, pp. 146–151.

Stol 1999: Martin Stol, "Psychosomatic Suffering in Ancient Mesopotamia", in Tzvi Abusch and Karel van der Toorn (eds.), *Mesopotamian Magic. Textual, Historical, and Interpretative Perspectives* (Ancient Magic and Divination 1). Groningen: Styx, 1999, pp. 57–68.

Stol 2000: Martin Stol, *Birth in Babylonia and the Bible. Its Mediterranean Setting* (Cuneiform Monographs 14). Groningen: Styx, 2000.

Stol 2003–2005a: Martin Stol, "Pflanzenkunde A. Nach schriftlichen", in *Reallexikon der Assyriologie und vorderasiatischen Archäologie*, vol. 10. Berlin: De Gruyter, 2003–2005, pp. 503–506.

Stol 2003–2005b: Martin Stol, "Pharmakologie", in *Reallexikon der Assyriologie und vorderasiatischen Archäologie*, vol. 10. Berlin: De Gruyter, 2003–2005, pp. 524–525.

Stol 2007: Martin Stol, "Fevers in Babylonia", in Markham J. Geller and Irving L. Finkel (eds.), *Disease in Babylon* (Cuneiform Monographs 36). Leiden, Boston: Brill, 2007, pp. 1–39.

Stol 2009a: Martin Stol, "Insanity in Babylonian Sources", *Journal des Médecines Cunéiformes* 13 (2009), pp. 1–12.

Stol 2009b: Martin Stol, review of Schwemer 2007b, *Bibliotheca Orientalis* 66, 1–2 (2009), pp. 167–168.
Stol 2015: Martin Stol, "Die Waffen in alten Mesopotamien. A. Die Waffen", *Bibliotheca Orientalis* 72, 5–6 (2015), pp. 613–622.
Stol 2016: Martin Stol, *Women in the Ancient Near East*. Berlin, Boston: De Gruyter, 2016.
Strawn 2005: Brent A. Strawn, *What is Stronger than a Lion? Leonine Image and Metaphor in the Hebrew Bible and the Ancient Near East* (Orbis Biblicus et Orientalis 212). Freiburg: Vandenhoeck & Ruprecht, 2005.
MStreck 1916: Maximilians Streck, *Assurbanipal und die letzten Assyrischen Könige bis zum Untergange Niniveh's*, II Teil (Vorderasiatische Bibliothek 7). Leipzig: Hinrichs, 1916.
MPStreck 1998–2001: Michael P. Streck, "Ninurta/Ninĝirsu. A. in Mesopotamien", in *Reallexikon der Assyriologie und vorderasiatischen Archäologie*, vol. 9. Berlin: De Gruyter, 1998–2001, pp. 512–522.
MPStreck 1999: Michael P. Streck, *Die Bildersprache der akkadischen Epik* (Alter Orient und Altes Testament 264). Münster: Ugarit-Verlag, 1999.
MPStreck 2006–2008: Michael P. Streck, "Salz, Versalzung A. Nach Schriftquellen", in *Reallexikon der Assyriologie und vorderasiatischen Archäologie*, vol. 11. Berlin: De Gruyter, 2006–2008, pp. 592–599.
Streck and Wasserman 2012: Michael P. Streck, and Nathan Wasserman, "More Light on Nanâya", *Zeitschrift für Assyriologie und vorderasiatische Archäologie* 102, 2 (2012), pp. 183–201.
Suter 2012: Claudia E. Suter, "The Royal Body and Masculinity in Early Mesopotamia", in Angelika Berlejung, Jan Dietrich, and Joachim Friedrich Quack (eds.), *Menschenbilder und Körperkonzepte im Alten Israel, in Ägypten und im Alten Orient*. Tübingen: Mohr Siebeck, 2012, pp. 433–458.
Svärd 2010: Saana Svärd, "'Maid of the king' (GÉME *ša šarrī*) in the Neo-Assyrian texts", in Şevket Dönmez (ed.), *DUB.SAR É.DUB.BA. Studies Presented in Honour of Veysel Donbaz*. Istanbul: Ege Yayınları, 2010, pp. 251–260.
Svärd 2015: Saana Svärd, *Women and Power in Neo-Assyrian Palaces* (State Archives of Assyria Studies 23). Helsinki: The Neo-Assyrian Text Corpus Project, 2015.
Tambiah 1968: Stanley J. Tambiah, "The Magical Power of Words", *Man* N.S. 3, 2 (1968), pp. 175–208.
Tambiah: 1969: Stanley J. Tambiah, "Animals are Good to Think and Good to Prohibit", *Ethnology* 8 (1969), pp. 424–459.
Tambiah 1981: Stanley J. Tambiah, "A Performative Approach to Ritual", *Proceedings of British Academy 1979*, 65 (1981), pp. 113–169.
Tambiah 1985: Stanley J. Tambiah, "Form and Meaning of Magical Acts", in Idem, *Culture, Thought, and Social Action. An Anthropological Perspective*. Cambridge: Harvard University Press, 1985, pp. 60–86.
Tapper 1988: Richard Tapper, "Animality, Humanity, Morality, Society", in Ingold 1994: 47–62.
Taussig 1980: Michael T. Taussig, "Reification and the Consciousness of the Patient", *Social Science and Medicine* 14b (1980), pp. 3–13.
Tavernier 2008: Jan Tavernier, "KADP 36: Inventory, Plant List, or Lexical Exercise", in Robert D. Biggs, Jennie Myers, and Martha T. Roth (eds.), *Proceedings of the 51st Rencontre Assyriologique Internationale held at the Oriental Institute of the University of Chicago, July 18–22, 2005* (Studies in Ancient Oriental Civilization 62). Chicago: The Oriental Institute of the University of Chicago, 2008, pp. 191–202.
Thompson 1923: Reginald Campbell Thompson, *Assyrian Medical Texts from the Originals of the British Museum*. London: Milford, 1923.
Thompson 1930: Reginald Campbell Thompson, *Epic of Gilgamish*. Oxford: Clarendon Press, 1930.

Thompson 1930–1931: Reginald Campbell Thompson, "Assyrian Prescriptions for Treating Bruises or Swellings", *The American Journal of Semitic Languages and Literatures* 47 (1930–1931), pp. 1–25.

Thompson 1934: Reginald Campbell Thompson, "Assyrian Prescriptions for Diseases of the Urine, etc", *Babyloniaca. Études de philologie assyro-babylonienne* 14 (1934), pp. 57–151.

Thompson 1936: Reginald Campbell Thompson, *A Dictionary of Assyrian Chemistry and Geology*. Oxford: Clarendon Press, 1936.

Thompson 1936–1937: Reginald Campbell Thompson, "Assyrian Prescriptions for Stone in the Kidneys, for the "Middle", and for Pneumonia", *Archiv für Orientforschung* 11 (1936–1937), pp. 336–340.

Thompson 1949: Reginald Campbell Thompson, *A Dictionary of Assyrian Botany*. London: British Academy, 1949.

Thomsen 1987: Marie-Louise Thomsen, *Zauberdiagnose und schwarze Magie in Mesopotamien* (Carsten Niebuhr Institute Publications 2). Copenhagen: Museum Tusculanum Press, 1987.

Thureau-Dangin 1925: Francois Thureau-Dangin, "Un hymne à Ištar de la haute époque babylonienne", *Revue d'Assyriologue et d'Archéologie orientale* 22 (1925), pp. 169–177.

Todorov 1973: Tzvetan Todorov, "Le Discours de la magie", *L'Homme* 13, 4 (1973), pp. 38–65.

Torri 2003: Giulioa Torri, *La similitudine nella magia analogica ittita* (Studia asiana 2). Roma: Herder, 2003.

Tsukimoto 1985: Akio Tsukimoto, *Untersuchungen zur Totenpflege (kispum) in alten Mesopotamien* (Alter Orient und Altes Testament 216). Kevelaer: Butzon und Bercker, 1985.

Tsouparopoulou 2020: Christina Tsouparopoulou, "The Healing Goddess, Her Dogs and Physicians in Late Third Millennium BC Mesopotamia", *Zeitschrift für Assyriologie und vorderasiatische Archäologie* 110, 1 (2020), pp. 14–24.

Turner 1967: Victor W. Turner, *The Forest of Symbols. Aspects of Ndembu Ritual*. Ithaca: Cornell University Press, 1967.

Turner 1975: Victor W. Turner, *Revelation and Divination in Ndembu Ritual*. Ithaca: Cornell University Press, 1975.

Tutrone 2016: Fabio Tutrone, "*Vox Naturae*: The Myth of Animal Nature in the Late Roman Republic", in Patricia A. Johnston, Attilio Mastrocinque, and Sophia Papaioannou (eds.), *Animals in Greek and Roman Religion and Myth*. Newcastle: Cambridge Scholars, 2016, pp. 51–84.

Unger 1957–1971: Eckhard Unger, "Farben (Symbolik)", in *Reallexikon der Assyriologie und vorderasiatischen Archäologie*, vol. 3. Berlin: De Gruyter, 1957–1971, pp. 24–26.

van Beek 1958: Gus W. van Beek, "Frankincense and Myrrh in Ancient South Arabia", *Journal of the American Oriental Society* 78 (1958), pp. 141–152.

van Binsbergen and Wiggermann 1999: Wim van Binsbergen and Frans A.M. Wiggermann, "Magic in History. A Theoretical Perspective, and Its Application to Ancient Mesopotamia", in Tzvi Abusch and Karel van der Toorn (eds.), *Mesopotamian Magic. Textual, Historical, and Interpretative Perspectives* (Ancient Magic and Divination 1). Groningen: Styx, 1999, pp. 1–34.

Van Buylaere 2020: Greta Van Buylaere, "Depression at the Royal Courts of Esarhaddon and Assurbanipal", in Shih-Wei Hsu and Jaume Llop Radua (eds.), *The Expression of Emotions in Ancient Egypt and Mesopotamia* (Culture and History of the Ancient Near East 116). Leiden, Boston: Brill, 2020, pp. 201–219.

Van de Mieroop 2013: Marc Van de Mieroop, "Recent Trends in the Study of Ancient Near Eastern History: Some Reflections", *Journal of Ancient History* 1, 1 (2013), pp. 83–98.

van der Geest and Whyte 1989: Sjaak van der Geest and Susan Reynolds Whyte, "The Charm of Medicines: Metaphors and Metonyms", *Medical Anthropology Quarterly* 3 (1989), pp. 345–367.

van der Geest et al. 1996: Sjaak van der Geest, Susan Reynolds Whyte, and Anita Hardon, "The Anthropology of Pharmaceuticals: a Biographical Approach", *Annual Review of Anthropology* 25 (1996), pp. 153–178.
van der Toorn 1985: Karel van der Toorn, *Sin and Sanction in Israel and Mesopotamia*. Assen: Van Gorcum Ltd, 1985.
van der Toorn 1996: Karel van der Toorn, *Family Religion in Babylonia, Syria and Israel. Continuity and Change in Forms of Religious Life* (Studies in the History and Culture of the Ancient Near East 7). Leiden, New York, Köln: Brill, 1996.
van Dijk 1975: Johannes J.A. van Dijk, "Incantations accompagnant la naissance de l'homme", *Orientalia* 44 (1975), pp. 52–79.
van Dijk et al. 1985: Johannes J.A. van Dijk, Albrecht Götze, and Mary I. Hussey, *Early Mesopotamian Incantations and Rituals* (Yale Oriental Series 11). New Haven: Yale University Press, 1985.
van Driel 1993: Govert van Driel, "Neo-Babylonian Sheep and Goats", *Bulletin on Sumerian Agriculture* 7 (1993), pp. 219–258.
van Laere 1980: Ralf van Laere, "Techniques hydrauliques en Mésopotamie ancienne", *Orientalia Lovaniensia Periodica* 11 (1980), pp. 11–53.
Vanstiphout 1992: Herman L.J. Vanstiphout, "Repetition and Structure in the Aratta Cycle: Their Relevance for the Orality Debate", in Marianna E. Vogelzang and Herman L.J. Vanstiphout (eds.), *Mesopotamian Epic Literature: Oral or Aural?*. Lewiston: Edwin Mellen Press, 1992, pp. 247–264.
Veldhuis 1990: Niek Veldhuis, "The Heart Grass and Related Matters", *Orientalia Lovaniensia Periodica* 21 (1990), pp. 27–44.
Veldhuis 1991: Niek Veldhuis, *A Cow of Sîn*. Groningen: Styx, 1991.
Veldhuis 1993: Niek Veldhuis, "The Fly, the Worm and the Chain", *Orientalia Lovaniensia Periodica* 24 (1993), pp. 41–64.
Veldhuis 1999: Niek Veldhuis, "Poetry of Magic", in Tzvi Abusch and Karel van der Toorn (eds.), *Mesopotamian Magic. Textual, Historical, and Interpretative Perspectives* (Ancient Magic and Divination 1). Groningen: Styx, 1999, pp. 35–48.
Veldhuis 2004: Niek Veldhuis, *Religion, Literature and Scholarship. The Sumerian Composition Nanše and the Birds* (Cuneiform Monographs 22). Leiden: Brill, Styx, 2004.
Verderame 2004a: Lorenzo Verderame, *Il ruolo degli 'esperti' (ummânu) nel periodo neo-assiro*. Ph.D. dissertation, Sapienza Università di Roma, 2004.
Verderame 2004b: Lorenzo Verderame, "I colori nell'astrologia mesopotamica", in Hartmut Waetzoldt (ed.), *Von Sumer nach Ebla und zurück. Festschrift Giovanni Pettinato zum 27 September 1999 gewidmet von Freunden, Kollegen und Schülern* (Heidelberger Studien zum Alten Orient 9). Heidelberg: Heidelberger Orientverlag, 2004, pp. 327–332.
Verderame 2011: Lorenzo Verderame (ed.) *Demoni Mesopotamici* (Studi e materiali di storia delle religioni 77, 2). Brescia: Morcelliana, 2011.
Verderame 2013a: Lorenzo Verderame, "Osservazioni a margine dei concetti di "ibrido" e "mostro" in Mesopotamia", in Igor Baglioni (ed.), *Monstra. Costruzione e percezione delle entità ibride e mostruose nel Mediterraneo antico*. Roma: Edizioni Quasar, 2013, pp. 160–172.
Verderame 2013b: Lorenzo Verderame, "'Their Divinity is Different, Their Nature is Distinct!' Nature, Origin, and Features of Demons in Akkadian Literature", *Archiv fur Religionsgeschichte* 14, 1 (2013), pp. 117–127.
Verderame 2013c: Lorenzo Verderame, "Means of substitution: The Use of Figurines, Animals, and Human Beings as Substitutes in Assyrian Rituals", in Claus Ambos and Lorenzo Verderame (eds.), *Approaching rituals in ancient cultures. Questioni di rito: rituali come fonte di conoscenza delle religioni e delle concezioni del mondo nelle culture antiche. Proceedings of the Conference, November 28–30 2011, Roma* (Rivista di Studi Orientali Nuova Serie 86, Supplemento 2). Pisa, Roma: Fabrizio Serra editore, 2013, pp. 301–323.

Verderame 2017a: Lorenzo Verderame, "The Seven Attendants of Hendursaĝa: A Study of Animal Symbolism in Mesopotamian Cultures", in Lluís Feliu, Fumi Karahashi, and Gonzalo Rubio (eds.) *The First Ninety Years. A Sumerian Celebration in Honor of Miguel Civil* (Studies in Ancient Near East Eastern Records 12). Berlin, Boston: De Gruyter, 2017, pp. 389–408.

Verderame 2017b: Lorenzo Verderame, "Demons at Work in Ancient Mesopotamia", in Siam Bhayro and Catherine Rider (eds.), *Demons and Illness from Antiquity to the Early-Modern Period*. Leiden, Boston: Brill, 2017, pp. 61–78.

Verderame 2017c: Lorenzo Verderame, "On the Early History of the Seven Demons (Sebettu)", in David Kertai and Olivier Nieuwenhuyse (eds.), *From the Four Corners of the Earth. Studies in the Iconography of the Ancient Eastern Mediterranean and Near East in Honour of F.A.M. Wiggermann* (Alter Orient und Altes Testament 441). Münster: Ugarit-Verlag, 2017, pp. 283–296.

Verderame 2017d: Lorenzo Verderame, "Le symbolisme de la porte dans les rituels assyro-babyloniens", in Patrick M. Michel (ed.), *Rites aux portes* (Etudes genevoises sur l'Antiquité 4). Bern: Peter Lang, 2017, pp. 83–92.

Verderame 2018: Lorenzo Verderame, "Ninmaḫ and her Imperfect Creatures: The bed Wetting Man and Remedies to Cure Enuresis (STT 238)", in Strahil V. Panayotov, Gene Trabich, and Luděk Vacín (eds.), *Mesopotamian Medicine and Magic. Studies in Honor of Markham J. Geller* (Ancient Magic and Divination 14). Leiden, Boston: Brill, 2018, pp. 779–800.

Vigo 2016: Matteo Vigo, "Sources for the Study of the Role of Women in the Hittite Administration", in Brigitte Lion and Cécile Michel (eds.), *The Role of Women in Work and Society in the Ancient Near East* (Studies in Ancient Near Eastern Records 13). Boston, Berlin: De Gruyter, pp. 328–353.

Villard 2000: Pierre Villard, "Le chien dans la documentation néo-assyrienne", in Parayre 2000a: 235–249.

Vincente 1991: Claudine A. Vincente, *Tell Leilan tablets dated by the limmu of Habil-kinu*. Ph.D. dissertation, Yale University, New Haven.

Vogel 2012a: Helga Vogel, "Frauen- und Genderforschung in der Altorientalistik und vorderasiatischen Archäologie", in Hans J. Nissen (ed.), *Geschichte Alt-Vorderasiens*. Münster: Oldenbourg Wissenschaftsverlag, 2012, pp. 208–211, 242–245.

Vogel 2012b: Helga Vogel, "Das Konzept 'Gender' in der vorderasiatischen Archäologie", in Marita Günther-Saeed and Esther Hornung (eds.), *Zwischenbestimmungen. Identität und Geschlecht jenseits der Fixierbarkeit?*. Würzburg: Königshausen u. Neumann, 2012, pp. 121–137.

Volk 1995: Konrad Volk, *Inanna und Šukaletuda. Zur historisch-politischen Deutung eines sumerischen Literaturwerkes* (Santag 3). Wiesbaden: Harrassowitz, 1995.

Volk 1999: Konrad Volk, "Kinderkrankheiten nach der Darstellung babylonisch-assyrischer Keilschrifttexte", *Orientalia* 68 (1999), pp. 1–30.

von Hirsch 1973–1974: Hans von Hirsch, "Einiges Nebenbei zu akkadischen, 'Liebesbeschwörungen'", *Wiener Zeitschrift für die Kunde des Morgenlandes* 65–66 (1973–1974), pp. 59–68.

von Soden 1950: Wolfram von Soden, "Ein Zwiegespräch Hammurabis mit einer Frau (Altbabylonische Dialektdichtungen 2)", *Zeitschrift für Assyriologie und vorderasiatische Archäologie* 49 (1950), pp. 151–194.

von Soden 1957–1958: Wolfram von Soden, "Die 'Schwertpflanze'", *Archiv für Orientforschung* 18 (1958–1959), p. 394.

von Soden 1965–1981: Wolfram von Soden, *Akkadisches Handwörterbuch*, 3 vols. Wiesbaden: Harrassowitz, 1965–1981.

von Soden 1968: Wolfram von Soden, "Review: *The Assyrian Dictionary, Vol. 2: B*", *Orientalistische Literaturzeitung* 63: 457–459.

von Soden 1974: Wolfram von Soden,"Duplikate aus Ninive", *Journal of Near Eastern Studies* 33 (1974), pp. 339–344.

von Soden 1995: Wolfram von Soden, *Grundriss der akkadischen Grammatik* (Analecta Orientalia 33). Rome: Biblical Institute Press, 1995.

von Weiher 1983: Egbert von Weiher, *Spätbabylonische Texte aus Uruk*. Teil II (Ausgrabungen der Deutschen Forschungsgemeinschaft in Uruk-Warka 10). Berlin: Gebr. Mann, 1983.

von Weiher 1988: Egbert von Weiher, *Spätbabylonische Texte aus Uruk*. Teil III (Ausgrabungen der Deutschen Forschungsgemeinschaft in Uruk-Warka 10). Berlin: Gebr. Mann, 1988.

von Weiher 1993: Egbert von Weiher, *Uruk, spätbabylonische Texte aus dem Planquadrat U 18/4*, (Ausgrabungen in Uruk-Warka 12). Mainz am Rhein: P. Von Zabern, 1993.

Waetzoldt 2006–2008: Hartmut Waetzoldt, "Rind. A. In mesopotamischen Quellen des 3. Jahrtausends", in *Reallexikon der Assyriologie und vorderasiatischen Archäologie*, vol. 11. Berlin: De Gruyter, 2006–2008, pp. 375–388.

Waldram 2000: James B. Waldram, "The Efficacy of Traditional Medicine: Current Theoretical and Methodological Issues", *Medical Anthropology Quarterly* N.S. 14, 4 (2000), pp. 603–625.

Wasserman 2003: Nathan Wasserman, *Style and Form in Old-Babylonian Literary Texts* (Cuneiform Monographs 27). Leiden, Boston: Brill, Styx, 2003.

Wasserman 2016: Nathan Wasserman, *Akkadian Love Literature of the Third and Second Millennium BCE* (Leipziger altorientalistische Studien 4). Wiesbaden: Harrassowitz, 2016.

ChWatanabe 2000: Chikako E. Watanabe, "The Lion Metaphor in the Mesopotamian Royal Context", in Parayre 2000a: 399–409.

ChWatanabe 2002: Chikako E. Watanabe, *Animal Symbolism in Mesopotamia. A Contextual Approach*. Wien: Institut für Orientalistik der Universität Wien, 2002.

KWatanabe 1994: Kazuko Watanabe, "Lebenspendende und todbringende Substanzen in Altmesopotamien", *Baghdader Mitteilungen* 25 (1994), pp. 579–596.

Wee 2015: John Z. Wee, "Phenomena in Writing: Creating and Interpreting Variants of the Diagnostic Series Sa-gig", in J. Cale Johnson (ed.), *In the Wake of the Compendia. Infrastructural Contexts and the Licensing of Empiricism in Ancient and Medieval Mesopotamia* (Science, Technology, and Medicine in Ancient Cultures 3). Berlin, Boston: De Gruyter, pp. 247–287.

Weidner 1924–1925: Ernst F. Weidner, "Altbabylonische Götterlisten", *Archiv für Keilschriftforschung* 2 (1924–1925), pp. 71–123.

Weidner 1932–1933: Ernst F. Weidner, "Der Staatvertrag Aššurnirâris VI. von Assyrien mit Mati'ilu von Bît-Agusi", *Archiv für Orientforschung* 8 (1932–1933), pp. 17–34.

Weidner 1957–1971: Ernst F. Weidner,"Gazbaba", in *Reallexikon der Assyriologie und vorderasiatischen Archäologie*, vol. 3. Berlin: De Gruyter, 1957–1971, p. 153.

Westenholz 1992: Joan G. Westenholz, "Metaphorical Language in the Poetry of Love in Ancient Near East", in Dominique Charpin and Francis Joannès (eds.), *La circulation des biens, des personnes et des idées dans le Proche-Orient ancien. Actes de la XXXVIIIe Rencontre Assyriologique Internationale (Paris, 8–10 juillet 1991)*. Paris: Éditions Recherche sur les civilisations 1992, pp. 381–387.

Westenholz 1995: Joan G. Westenholz, "Love Lyrics from the Ancient Near East", in Jack M Sasson (eds.), *Civilizations of the Ancient Near East*. New York: Charles Scribner's Sons, 1995, pp. 2471–2484.

Westenholz 1996: Joan G. Westenholz, "Symbolic Language in Akkadian Narrative Poetry: The Metaphorical Relationship between Poetical Images and Real World", in Marianna E. Vogelzang and Herman L.J. Vanstiphout (eds.), *Mesopotamian Poetic Language: Sumerian and Akkadian* (Cuneiform Monographs 6). Groningen: Styx, 1996, pp. 183–206.

Westenholz 1997: Joan G. Westenholz, "Nanaya: Lady of Mistery", in Irving L. Finkel and Markham J. Geller (eds.), *Sumerian Gods and Their Representations* (Cuneiform Monographs 7). Groningen: Styx, 1997, pp. 57–84.

Westenholz 2000: Joan G. Westenholz, "King by Love of Inanna, an Image of Female Empowerment?", *NIN: Journal of Gender Studies in Antiquity* 1 (2000), pp. 75–89.

Westenholz 2009: Joan G. Westenholz, "Construction of Masculine and Feminine Ritual Roles in Mesopotamia", in Bernhard Heininger (ed.), *Ehrenmord und Emanzipation. Die Geschlechterfrage in Ritualen von Parellelgesellschaften*. Berlin: Lit, 2009, pp. 73–98.

Westenholz and Westenholz 1977: Joan G. Westenholz and "Help for Rejected Suitors. The Old Akkadian Love Incantation MAD V 8", *Orientalia* N.S. 46 (1977), pp. 198–219.

Westenholz and Zsolnay 2017: Joan G. Westenholz and Ilona Zsolnay, "Categorizing Men and Masculinities in Sumer", in Zsolnay 2017: 12–41.

Weszeli 2003–2005: Michaela Weszeli, "Pferd. A. In Mesopotamien", in *Reallexikon der Assyriologie und vorderasiatischen Archäologie*, vol. 10. Berlin: De Gruyter, 2003–2005, pp. 469–481.

Weszeli 2006–2008: Michaela Weszeli, "Rind. B. In mesopotamischen Quellen des 2. und 1. Jahrtausends", in *Reallexikon der Assyriologie und vorderasiatischen Archäologie*, vol. 11. Berlin: De Gruyter, 2006–2008, pp. 388–406.

Weszeli 2009–2011: Michaela Weszeli, "Schwein. A. In Mesopotamien", in *Reallexikon der Assyriologie und vorderasiatischen Archäologie*, vol. 12. Berlin: De Gruyter, 2009–2011, pp. 319–328.

Whyte and van der Geest 1994: Susan Reynolds Whyte and Sjaak van der Geest, "Injections: Issues and Methods for Anthropological Research", in Michael L. Tan and Nina L. Etkin (eds.), *Medicines. Meanings and Contexts*. Quezon City: Health Action Information Network, 1994, pp. 137–161.

Whyte et al. 2002: Susan Reynolds Whyte, Sjaak van der Geest, and Anita Hardon, *The Social Life of Medicines*. Cambridge: Cambridge University Press, 2002.

Wierzbicka 1999: Anna Wierzbicka, *Emotions Across Languages and Cultures. Diversity and Universals*. Cambridge: Cambridge University Press, 1999.

Wiggermann 1992: Frans A.M. Wiggermann, *Mesopotamian Protective Spirits. The Ritual Texts* (Cuneiform Monographs 1). Groningen: Styx, 1992.

Wiggermann 1999–2001: Frans A.M. Wiggermann, "Nackte Göttin", in *Reallexikon der Assyriologie und vorderasiatischen Archäologie*, vol. 9. Berlin: De Gruyter, 1998–2001, pp. 47–53.

Wiggermann 2009–2011: Frans A.M. Wiggermann, "Sexualiltät A. In Mesopotamien", *Reallexikon der Assyriologie und vorderasiatischen Archäologie*, vol. 13. Berlin: De Gruyter, 2009–2011, pp. 410–426.

Wijeiewardene 1968: Gehan Wijeyewardene, "Address, Abuse, and Animal Categories in Northern Thailand", *Man* 3 (1968), pp. 76–93.

Wilcke 1976–1980: Claus Wilcke, "Inanna/Ištar, A Philologisch", in *Reallexikon der Assyriologie und vorderasiatischen Archäologie*, vol. 5. Berlin: De Gruyter, 1976–1980, pp. 74–87.

Wilcke 1985: Claus Wilcke, "Liebesbeschwörungen aus Isin", *Zeitschrift für Assyriologie und vorderasiatische Archäologie* 75, 2 (1985), pp. 188–209.

Wilcke 1987: Claus Wilcke, "A Riding Tooth: Metaphor, Metonymy and Synecdoche, Quick and Frozen in Everyday Language", in Murray Mindlin, Markham J. Geller, and John E. Wansbrough (eds.), *Figurative Language in the Ancient Near East*. London: Taylor and Francis, 1987, pp. 7–102.

Willis 1994: Roy Willis (ed.), *Signifying Animals. Human Meaning in the Natural World*. London, New York: Routledge, 1994.

Winter 1996: Irene Winter, "Sex, Rhetoric, and the Public Monument: The Alluring Body of Naram-Sîn of Agade", in Natalie Kampen (ed.), *Sexuality in Ancient Art: Near East, Egypt, Greece, and Italy*, Cambridge: Cambridge University Press, 1996, pp. 11–26.

Wittig 1980: Monique Wittig, "On ne naît pas femme", *Questions Féministes* 8, 8 (1980), pp. 75–84 (English translation: "One Is Not Born a Woman", in Carole R. McCann, Seung-Kyung Kim, and Emek Ergun (eds.), *Feminist Theory Reader. Local and Global Perspectives*. New York: Routledge, Taylor & Francis, 2013, pp. 246–251.

Worthington 2003: Martin Worthington, "A Discussion of Aspects of the UGU Series", *Journal des Médecines Cunéiformes* 2 (2003), pp. 2–13.

Yuhong 2001: Wu Yuhong, "Rabies and Rabid Dogs in Sumerian and Akkadian Literature", *Journal of the American Oriental Society* 121, 1 (2001), pp. 32–43.

Xella 1979: Paolo Xella, *Problemi del mito nel Vicino Oriente antico*. Napoli: Istituto orientale di Napoli, 1979.

Yanagisako and Collier 1987: Sylvia J. Yanagisako and Jane F. Collier, "Toward a Unified Analysis of Gender and Kinship", in Eaedem (eds.), *Gender and Kinship. Essays toward a Unified Analysis*. Stanford: Stanford University Press, 1987, pp. 14–50.

Yap 1969: Pow Ming Yap, "The Culture Bound Syndromes", in William A. Caudill and Tsung-Yi Lin (eds.). *Mental Health Research in Asia and The Pacific*. Honolulu: East-West Center Press, 1969 pp. 33–53.

Young 1982: Allan Young, "The Anthropologies of Illness and Sickness", *Annual Reviews of Anthropology* 11 (1982), pp. 257–285.

Young 1997: Allan Young, "Modi del ragionare e antropologia interpretativa", *AM. Rivista della Società italiana di antropologia medica* 3–4 (1997), pp. 11–27.

Yuste and Garrido 2010: Piedad Yuste and Ángel Garrido, "Brain Diseases in Mesopotamian Societies", *Broad Research in Artificial Intelligence and Neuroscience* 1, 3 (2010), pp. 75–82.

Zempléni 1969: Andras Zempléni, "La thérapie traditionnelle des troubles mentaux chez les Wolof et les Lebou (Sénégal)", *Social Science and Medicine* 3 (1969), pp. 191–205.

Zempléni 1985: Andras Zempléni, "La 'maladie' et ses 'causes': Introduction", in Idem (ed.), *Causes, origines et agents de la maladie chez les peuples sans écriture* (Ethnographie 81). Paris: Société d'ethnographie de Paris, 1985, pp. 13–44.

Zimmern 1915–1916: Heinrich Zimmern, "Zu den 'Keilschrifttexten aus Assur religiösen Inhalts'", *Zeitschrift für Assyriologie und vorderasiatische Archäologie* 30, 2 (1915–1916), pp. 184–229.

Zimmern 1927: Heinrich Zimmern, "Simat, Sima, Tyche, Manat", *Islamica* 2 (1927), pp. 574–584.

Zisa 2012: Gioele Zisa, "Sofferenza, malessere e disgrazia. Metafore del dolore e senso del male nell'opera paleo-babilonese. 'Un uomo e il suo dio': un approccio interdisciplinare", *Historiae* 9 (2012), pp. 1–30.

Zisa 2020: Gioele Zisa, "Going, Returning, Rising: The Movement of the Organs in the Mesopotamian Anatomy", *Kaskal. Rivista di storia, ambienti e culture del Vicino Oriente antico* 16 (2019), pp. 453–476.

Zisa 2021: Gioele Zisa, "Il corpo sessuato delle dee. Agricoltura, pastorizia e mondo vegetale nella Mesopotamia antica", in Daniela Bonanno and Ignazio E. Buttitta (eds.), *Narrazioni e rappresentazioni del sacro femminile. Atti del convegno internazionale di studi in memoria di Giuseppe Martorana* (Nanaya. Studi e materiali di antropologia religiosa 2). Palermo: Edizioni Museo Pasqualino, 2021, pp. 29–60.

Zomer 2018: Elyze Zomer, *Corpus of Middle Babylonian and Middle Assyrian Incantations* (Leipziger altorientalistische Studien 9). Wiesbaden: Harrassowitz, 2018.

Zsolnay 2017: Ilona Zsolnay (ed.), *Being a Man. Negotiating Ancient Constructs of Masculinity*. London, New York: Routledge.

Index of Akkadian terms and expressions discussed

adirtu 65–66, *69–70*, 323
āšipu 4, 6, 33, 36, *82–86*, 112, *145–146*, 155, 158, 167, 192, 194, 201, 252
āšipūtu see *āšipu*
asû 6, 33
asûtu see *asû*
ašuštu 63, 65, 67, *69–70*, 72, 224

bunzerru 205, *329–331*

diliptu 65, 67, *69–70*, 246
dilûtu 109, 122–123, 149, 221, 292, 296
dūtu 8, 40, 121, 132, 425

ekēmu 40
eṭēru 40

ḫabābu *127–129*, 222
ḫašāḫu 48

ištaru *250–251*

lalû 46, 55, 416
libbašu sinništa ḫašiḫ 48, 60, 64
libbu VIII, XIX, 9, 24, 28, *42–43*, 45, 48, 51, *52–53*, 55–56, 60–63, 66–67, 69, 75–79, 85, 140, 151, 164, 179, 184, 191, 245–246, 255, 293, 300, 415, 424, 490

manû *82–84*
muṭṭu 10, *46–47*, 60, 64, 66, 73

nâḫu 245, *267–268*, *293–294*, 415
našû *37–38*, 42
– *libbu (lā) inaššīšu* 10, 37–39, *42–43*, 46–48, 51, *59–60*, 64, 75
– *libbašu itanašši* 42
nissatu *70–71*, 515
nīšu *37–38*

paqādu 350

qabû *82–84*, 124
qatû 40

ramû *117–119*
rašû 40
rihûtu 413
rikbu 168, 472
rikibtu 5, 80, 91, 107–108, 130, 146, 166, *168–173*, 181–183, 185, 205, 415, 472, 490
ruʾāmu 252–253

suʾʾusu/suḫsu 413, 484

ṣabātu 40, 268, 328

šadādu *118–119*
šīmtu *250–251*

tabālu 40, 425
tebû *48–51*, 104, 123, 128, 174, 194, 248, 268, 326
tibût libbi *51–52*

https://doi.org/10.1515/9783110757262-018

Index

Abracadabra 87–88, *153–156*, *206–207*, 212, 334, 342–343, 373, 376–377, 385, 404–405, 407–408, 460–461, 464
Adad 38, 138, 149, 221, 229, 257–258, 264–265, 299, 322, 334–335, 420–421, 424
Agent of suffering 54–57, 87, 108–123, 203
akkannu-wild ass 49, 94, *100–101*, 108, 207, 217–218, 222, 224, 228–229, 239, 241, 247–249, 254, 268, 279, 287, 295, 423, 425
Amulet IX, 4, 50, 82–83, *90–92*, 124, 157, 159, 163–168, 170, 175, 179, *184–186*, 191, 194, 196, 203–204, 221, 223, 351, 413, 415–416, 509, 511, 516–517, 520–521
Analogical thinking see Analogy
Analogical magic see Analogy
Analogy 13, 32, 56, 66, *89–93*, *94–95*, 98, 100, *118–121*, 137, 190–191, 202, *204–206*, 211, 328, 518
Animal metaphors 80, 87, *93–108*, 110, *123–129*, 147, 207, 211, 248, 267, 271, 295, *326–331*, *353–354*
Animal metonymies 87, *93–108*, 296, 327, 329, *353–354*
Anxiety 17, 24, 29, 53, 58, *67–69*, 75–77, 82, 153, 201–203, 210
Apple 293, 351, 510
arkabu-bat 5, 167, *168–173*, 181–183, 185, *205–206*, 409, 447, 449, 453, 457, 469
arrabu-mouse 79, 119–120, 168, *195–196*, 319, 503
Asalluḫi 63, 135, 137–138, *143–144*, 238–239, *248–252*, 288–289, 290–291, 348–349, 513
Astral influence IX, 164, *176–180*, 183
Authority 82, *84–86*, 136, 155–156, 207, 212

balluṣītu-bird 44, 166, 181, 411, 421, 424
Battle 23, 93, 101, 104–105, 137, 141, *195–200*, 203, 205, 208, 211, 330, 353, 461, 463
Bau *141–143*, 144, 161
Bed 41, 49, 51, 55, 62–63, 65, 67, 74, 80, 83, 89, 91, 105, 123, *124–127*, *150–151*, 164, 181, 207, 217, 219, 221, 225, 267, 302, 315, 317, 325–326, 349, 353, 377, 401, 403–405, 407, 409, 411, 413, 421, 426–427, 431, *434–435*, 461, 463, 517–518
Being-acted-upon 23, 117, 189
Bēlet-ilī 41, 43, 78, 137–138, *150–151*, 181, 207, 218, 222, 251, 322, 383–384, 399, 412
Belief IX, 22, *31–32*, 92, 155, 156, 207, 212
Bestiality *127–129*
Binding 23, 61, 91, 110, *113–121*, 125, 157, 166, 197, 205, 222, 237, 239, 252, 409, 415, 471
Biomedicine VIII, XI, *13–17*, *19–21*, *25–27*, 33, 58, 112, 157–158, 201, 209–212
Bitch 44, 80, 100, 192, 218, 223, 268, 315, *326–328*
Body *vs.* mind XX, 13–14, *25–28*, *52–53*, 209
Bondage *114–115*, 116–117
Bow 59, 61, 79, 91, 101, 119–121, 157, 159, 168, 186, *194–200*, 203, 205, 208, 219, 225, 319, 377, 479
Bowstring *119–121*, 168, 195, 205
Buck 49–50, 125–128, 147, 166–167, 169, 181, 185, 207, 221, 225, 291, 315, 322, 326, 334–335, 347, 353–354, 445, 453

Canal 98, *121–123*, 132, *149–150*, 217, 245, 257–258, 265, 292, 334–335, 365, 367, 411, 421, 424, 518
Cause of the suffering 10, 22, 25, 27, *54–59*, 61–63, 66–67, 69, 75, 81, 108, 133, 163, 187, 202–203, 211, 249–250, 252, 255, 514
Citationality 24, 34, 96
Cold water 63, 66, 241, 246, 254, 255
Conative function 88, 91, 211
Couples therapy *78–82*
Critical ethnocentrism 16
Crossroad 365–366, 401, 519
Culture-bound syndromes/CBS *30–31*
Curing *vs.* healing XX, 20, 28, 209

Deer see Stag
Dehistoricization *150–153*, 207, 211
Depression 9, 30, *68–70*, 75–76, 510
Dew 147, 287, *298–300*, 425
Diagnostic and Statistical Manual of Mental Disorders/DSM 9, 27, 30–31
Disease *19–22*, 209
Distress 23, 30, 63, *65–73*, 82, 224, 263, 323

https://doi.org/10.1515/9783110757262-019

Divine anger/wrath 26–27, *54–55*, 57–58, 63, 69–70, 72–73, 187
Dog 44, 80, 96–98, 100, 126, 161, 166–167, 191–193, 207, 218, 221, 223, 237, 268, 315, *326–329*, 366, 424, 449, 507, 511, 513, 516, 519
Domestic bull 101–104, 132
Dove 98, 166, 241, 246–247
Dumuzi 80, 99, 101, 103, 105, 107, 131–132, 137, 140, 151, 323, 434

Ea 113, 135–138, 141, 143–144, 190, 288–289, 290–291, 344–345, 248–349, 517
Effectiveness X–XI, XX, 17, 20, 27, 30, 32, *85–86*, 93, 123, *136–138*, 144, 154–156, 158–159, 173, 177, 190, *201–202*, 206, *208–213*, 353
Efficacy see Effectiveness
Ejaculation 11, 67, 99, 147, 161, 328, 413
Embodiment/embodied subject 25, *28–29*, 53, 112
Enkidu 50, 70, 101, 129, 207, 222
Enlil 41, 78, 132, 138, 141, *150–151*, 180, 207, 218, 222, 251, 383–384, 399, 412
Erectile dysfunction VIII, XI, 10, 27, 95, 201
Erection 10–11, 27, 42, 44–47, 50–52, 58, 60–61, 63, 80, 95, 199, 246
Erotic animal similes 87, *123–129*
Ethno-epistemology 15–16, 26
Ethno-Psychiatry XIX–XX, 14–15, 17, 29, 31
Etiology 19, 21–22, 54–55, 69, 112, 115, 126, 157, 163, *187–194*, 203, 207, 211
Exorcist 4, 6, 7, 56, 82–86, 88, 114, 116, 135–136, 138, 144–146, 155–156, 207, 212, 247, 295, 300, 323, 454, 515
Explanatory models of illness 20, *22–23*, 202

Fatigue 23, 61, 63, 72, 77, 93, 119, 203
Fear 17, 53, 63, *65–73*, 75–77, 89–90, 149, 202–203, 210, 241, 254, 294, 317, 323, 411
Feminism XX, *17–19*, 35
Femininity 148, 329
Feminization 97, 193–194, 197
Fertility 103–104, 106, 147, 192
Figurine 54, 56, 73–74, 76, 79, 84, 89, 109, 115, 124, 157, 182, 187, *188–191*, 194–195, 211, 287, 317, 409, 512–513, 520
Forschungsgeschichte *8–12*, 13, 27, 52, 198

Garden 6, *146–149*, 171, 217, 221, 245, 228–229, 239, 252, 368, 371–372, 386, 397, 410–411
Gazbaba 138–139, *141–142*, 410–411, 430–431
Gender identity XX, 17–19, 25, *34–36*, 94, *95–98*, *197–200*, 207
Gender Studies XX, *17–19*, 35, 93, *95–96*, 207
Genitalia IX, 35, 59, 62, 65, 150, 164–165, 171, 183, 201, 205, 223, 268, 323, 326, 330, 412, 463, 518, 520
Grief 9, 63, 70, 73, 160
Grove see Garden
Gula 6, 38, 133, *143–145*, 159, *161–162*, 204, 327, 511, 514, 516, 519, 520

Hand of DN/god 54, *79*, 187, 196, 249, 366,
– Hand of Ištar 54, 187, 224, 289, 503
– Hand of personal god/goddess 250, 387
Harp 45, 60, 91, 119–121, 157, 239
Heavenly Daughters of Anu 122, 147, 150, *295–199*, 287, 300, 425
Historiolae 87, *150–153*, *206–207*
Horse 94, 96–98, 100, *104–106*, 109, 113, 124, 149, 167, 207, 218, 222, 239, 245, 248–249, 319, 327, 411, 417–418, 421, 425
Hunting 94, 100–101, 108, 196, 198–199, 205, 218, 279, 287, 295, *329–331*

Ibex 106–107, 129–130, 146, 131, 275, 347
Illness X, *19–23*, 112, 159, 202, 209
Illocutionary X, XI, *91–92*, 210
Immobility 23, 61–63, 65, 77, 113, *117–119*
Inana 46, 55, 102–103, 105, 107–108, 131–132, 139, 293, 351, 412, 434, 520
Inflammation by sun–heat 46, *57–58*, 241, 433, 513–514, 519
Insomnia *65–73*, 75–76, 108, 246, 241, 254
Instigator of the aggression 56, 81, 248
Išḫara 89, 138–139, 141, 151, 207, 262–263, 271, 430–431, *434–435*
Ištar 38, 46, 54–55, 57–58, 61, 63, 67, 69, 80, 83–84, 89–90, 99–101, 103–106, 108, 124, 132, 135, 137–141, 143, 151, 157, 180, 182, 186–188, 196–198, 207, 218–219, 224, 238–241, 249, 250–253, 288–291, 297, 302, 314–317, 322, 348–349, 352–353, 376–379, 385, 394, 402–412, 430–431, 434–435, 478–479, 502–503

Kanisurra 137–140, *141–142*, 262–263, 271, 410–411, 430–431, 435
Kilili 460–463

Lack of appetite 12, 47, 63, 66, 72
Laughs 72, 80, 197, 205, 315, 322, 329
Left hand 78, 181, 239, 289, 514, 521
Legitimacy 85–86, *135–137*
Libations 55, 89, 124, 157, 159, 180, *186–187*, 224, 243, 246, 292, 315, 479
Limbs 45, 50–51, 61, 63, 68, 74–77, 94, 113, *115–117*, 182, 239, 248–249, 251, 253, 263, 270, 349, 353, 355, 359, 360, 399, 471
Lion 94, 100, *105–106*, 118, 122–123, 140, 166–167, 207, 263, 267, 270–271, 327, 385, 401, 409
Locutionary 91–93
Logical thinking 19, *31–33*, 172
Loosening/relaxation/slackening of tendons 93, 117–121, 239, 248, 249
Love-sickness *72–73*, 94

Magnetism 10, 183–184, 202, 223, 270
Manhood XX, 17, 48, 59, 97, 119, 121–122, 133, 193–194, *197–200*, 208, 251
Marduk 38, 55, 57–58, 63, 67, 69, 113, 115, 122, 127, 138, 144, 147, 220, 252, 292, 299, 328, 386, 402–405, 408–411
martû-wood 44, 60, 347, 349, 353
mašgašu-weapon 44, 60, 80, 187, 315, 329, 330
Medical Anthropology VII, IX, XIII, XIX–XX, *12–17*, *19–21*, 25, 31, 154
Melancholy 30, *69–73*, 74–75, 77, 510
Mental disorders/suffering 8–9, *29–31*, 68, 120
Merchant's leather money pouch X, 108, 118–119, 243, 256
Mindful body 25, 28
mungu-paralysis 47, 55, 61–62, 65, 191, 289, 343, 345

Nanāya 80, 98, 106, *137–142*, 143, 148, 151, 207, 240–241, 246, 252–254, 299, 316–317, 408–411, 430–431, 434, 458, 460–462
Navel IX, 44, 48, 63, 78–79, 107, 139, 166, 170, 180, 181–182, 184–185, 202, 263, 270, 294, 319, 325, 409 415
neḫēs narkabti-sickness 46, 57, 58, 241
Ninĝirsu 142–144, 219, 225, 334–335, 346–349

Ointment IX, 3– 4, 43–44, 76, 78, 82, 157, 159, 163, 165–168, *180–184*, 185, 188, 200, 202, 223, 248, 269–270, 292, 295, 353, 366, 509, 513, 520
Old age 23, 46, *57–58*, *133–134*, 241
Onager 94, 100, 101, 108–109, 167, 207, 279, 287, 296, 417–418, 424
Orchard see Garden
Ox see Domestic bull

Panic attacks 24, 65–70, 73, 75–77, 85, 182, 289, 294
Partridge 49, 102, 106–108, 129–130, 146, 165, 167, 170–171, *173–175*, 176–177, 180–181, 195, 204, 237, 239, 241, 275, 319, 321, 345, 347, 407, 411, 421, 445, 449, 453, 456–457, 469, 489–490
Pelvic area 43, 79, 182–183, 330–331, 431
Penetration 44–45, 50, 197–198, 265, 328
Penis X, 10–11, 27, 37, *43–46*, 47, 49, 51, 60, 63, 74, 80, 150, 161, 165–166, 170, 172, 174–175, 180–185, 190, 196–198, 202, 208, 237, 239, 246, 248, 263, 265, 267–271, 287, 289, 291–293, 295, 315, 319, 325–329, 345, 347, 349, 353, 355, 365–367, 380, 409, 413, 415–416, 421, 431, 445, 449, 464, 518
Performative VII, 32, 91–92, 96, 138
Perlocutionary 92–93, 210–211
Persuasive analogy 91, 93, 206, 211
Pharmaceutical IX–X, XX, 43–44, 78–79, 82, *157–159*, 163, 172, 177, 186, 203–204, 209, 211, 223, 243, 269, 451
Phenomenological approach 25, 28–29, 31, 111
Pig 5, 49, 54, 79, 98, 100, 102, 107, 126, 130, 157, 166, 185, 187, *191–194*, 207, 211, 237, 287, 289, 317, 327–329, 345, 416, 489, 521
Plowing 100, 103–104, 131–132, 191–192, 328
Poetry *87–93*, 112–113, 126, 131, 206, 210–211, 252, 354, 513
Pomegranate 181, 220, 225, 289, 293, 351, 512, 517
Post–Feminism see Feminism
Potion 78, 82, 157, 159, 163, 165–169, 171, *173–178*, 182, 185–186, 191, 195, 203–204, 221, 293, 509, 521–522
Psychosomatic medicine 12, 25, 27

Quarry 89, 197, 205, 315, 329

Queer Theories/Studies XIII, XX, *17–19*, 34–36, 93, 95–96, 207

Rain 146–149, 298–299, 325–326, 423, 425
Ram 44, 49–50, 83, 89, 90, 102, 107, 124–127, 130, 141, 166–167, 182, 185, 196, 204, 207, 217, 221, 302, 315, 317, 319, 326–327, 347, 353–354, 401, 411, 416, 445, 447, 449, 451, 461–462, 469, 520
Raven 166, 169, 180–181, 196, 407, 421, 424, 453, 469, 472, 519
Releaser 5, 114, *142–146*, 225, 252, 334–335, 347, 349, 352
Right hand 44, 78–79, 181–182, 289, 329, 521
River water 41, 45, 60, 149–150, 176–177, 218, 224, 239, 245, 321, 366
Roaring 49, 106, 123–124, 128, 224, 257, 263, 267, 270, 334–335, 347

Semantic illness network *23–24*, 59, 77
Sexual excitement XI, 42, *50–52*, 80, 97, 191, 196, 295, 415, 421, 462–463
Sexual potency XI, 8, 10–12, 28, 42, 45, 48, 52, 57, 59, 94–95, 100–102, 104–106, 108, 119–121, 125–126, 130, 132–135, 149, 166, 171, 173–174, 192, 197–199, 203–205, 207–208, 248, 329, 354
Sexually excited animals 10, 27, 95
Sheep 83, 89, 91, 96, 99, 102, 105, 107, 124–125, 130, 167, 185–187, 221, 223, 315, 317, 345, 445, 447, 495, 519
Sickness *19–22*, 23, 112
Sleeplessness see Insomnia
Snake 48, 98–99, 109, 123, 128–129, 145, 167, 175, 263, 267, 270–271, 294, 387, 514, 516, 519
Sperm 5, 44–45, 55–56, 61–63, 65, 67, 99, 102, 150, 155, 164, 169, 171–172, 190, 241, 263, 270–271, 323, 350, 359–360, 401, 403, 405, 409, 411, 413
Stag 8, 49, 80, 91, 94, 98, 100, 102, 104, *106–108*, 114, 123–124, 127–131, 146, 166–168, 170–173, 185, 198, 204–205, 207, 218–219, 222, 224, 237, 244, 263, 265, 270–271, 275, 347, 352–353, 356, 359, 409, 411, 415–416, 451, 489–490
Stallion see Horse
Statuette see Figurine
Stick 44, 46, 57–58, 60, 188, 297, 329, 347, 349, 353
Subjectification see Subjectivity
Subjectivity 17, 19, 29, 95, 96
Šamaš 38, 76, 110, 113, 115, 127, 135, 137–138, 143–144, 178, 185, 190, 252, 288–291, 297, 299, 348–349, 366, 514, 520–521

Tiredness 62–63, 65, 67, 120, 160, 263

Vagina see Vulva
Vigor see Sexual potency
Vulva 43–44, 78–80, 98–99, 103, 126, 132, 134, 167, 183, 202, 205, 265, 268–269, 287, 295, 315, 326–328, 330–331, 413, 421, 463

War see Battle
Way of reasoning 32, 89, 204
Weakness 24, 51, 61–62, 73–77, 93, 101, 117–120, 160, 203, 209, 509
Wild bull 49–50, 61, 94, 98, 100, *101–104*, 106–108, 116, 123–124, 130, 140, 167, 196, 207, 217, 219, 222, 224, 228–229, 241, 254, 263, 267, 271, 317, 347, 349, 352–354, 356, 359
Wild goat see Ibex
Wildness 96–97, 100, 106, 108, 207
Wind X, 6, 37, 146–149, 192, 217, 221, 224, 228–229, 239, 245, 300, 302, 315, 325–326, 334–335, 361–362, 365, 368, 371–372, 386, 397, 410–411
Wolf 123, 130, 140, 167, 263, 267, 270–271, 377, 514
Wren 165, 175, 206, 445, 517

Youth 28, 57, 118, *131–135*

www.ingramcontent.com/pod-product-compliance
Lightning Source LLC
Chambersburg PA
CBHW081943230426
43669CB00019B/2905
* 9 7 8 3 1 1 1 2 7 8 2 8 5 *